# Lecture Notes in Computer Science 1880

Edited by G. Goos, J. Hartmanis and J. van Leeuwen

W0055472

# Springer

*Berlin*
*Heidelberg*
*New York*
*Barcelona*
*Hong Kong*
*London*
*Milan*
*Paris*
*Singapore*
*Tokyo*

Mihir Bellare (Ed.)

# Advances in Cryptology – CRYPTO 2000

20th Annual International Cryptology Conference
Santa Barbara, California, USA, August 20-24, 2000
Proceedings

 Springer

Series Editors

Gerhard Goos, Karlsruhe University, Germany
Juris Hartmanis, Cornell University, NY, USA
Jan van Leeuwen, Utrecht University, The Netherlands

Volume Editor

Mihir Bellare
University of California, Department of Computer Science and Engineering, 0114
9500 Gilman Drive, La Jolla, CA 92093, USA
E-mail: mihir@cs.ucsd.edu

Cataloging-in-Publication Data applied for

Die Deutsche Bibliothek - CIP-Einheitsaufnahme

Advances in cryptology : proceedings / CRYPTO 2000, 20th Annual
International Cryptology Conference, Santa Barbara, California, USA,
August 20 - 24, 2000. Mihir Bellare (ed.). [IACR]. - Berlin ;
Heidelberg ; New York ; Barcelona ; Hong Kong ; London ; Milan ;
Paris ; Singapore ; Tokyo : Springer, 2000
    (Lecture notes in computer science ; Vol. 1880)
    ISBN 3-540-67907-3

CR Subject Classification (1998): E.3, G.2.1, D.4.6, K.6.5, F.2.1-2, C.2, J.1

ISSN 0302-9743
ISBN 3-540-67907-3 Springer-Verlag Berlin Heidelberg New York

This work is subject to copyright. All rights are reserved, whether the whole or part of the material is
concerned, specifically the rights of translation, reprinting, re-use of illustrations, recitation, broadcasting,
reproduction on microfilms or in any other way, and storage in data banks. Duplication of this publication
or parts thereof is permitted only under the provisions of the German Copyright Law of September 9, 1965,
in its current version, and permission for use must always be obtained from Springer-Verlag. Violations are
liable for prosecution under the German Copyright Law.

Springer-Verlag is a company in the BertelsmannSpringer publishing group.
© Springer-Verlag Berlin Heidelberg 2000
Printed in Germany

Typesetting: Camera-ready by author, data conversion by Steingräber Satztechnik GmbH, Heidelberg
Printed on acid-free paper      SPIN: 10722418      06/3142      5 4 3 2 1 0

# Preface

Crypto 2000 was the 20th Annual Crypto conference. It was sponsored by the International Association for Cryptologic Research (IACR) in cooperation with the IEEE Computer Society Technical Committee on Security and Privacy and the Computer Science Department of the University of California at Santa Barbara.

The conference received 120 submissions, and the program committee selected 32 of these for presentation. Extended abstracts of revised versions of these papers are in these proceedings. The authors bear full responsibility for the contents of their papers.

The conference program included two invited lectures. Don Coppersmith's presentation "The development of DES" recorded his involvement with one of the most important cryptographic developments ever, namely the Data Encryption Standard, and was particularly apt given the imminent selection of the Advanced Encryption Standard. Martín Abadi's presentation "Taming the Adversary" was about bridging the gap between useful but perhaps simplistic threat abstractions and rigorous adversarial models, or perhaps, even more generally, between viewpoints of the security and cryptography communities. An abstract corresponding to Martín's talk is included in these proceedings.

The conference program also included its traditional "rump session" of short, informal or impromptu presentations, chaired this time by Stuart Haber. These presentations are not reflected in these proceedings.

An electronic submission process was available and recommended, but for the first time used a web interface rather than email. (Perhaps as a result, there were no hardcopy submissions.) The submission review process had three phases. In the first phase, program committee members compiled reports (assisted at their discretion by sub-referees of their choice, but without interaction with other program committee members) and entered them, via web forms, into web-review software running at UCSD. In the second phase, committee members used the software to browse each other's reports, discuss, and update their own reports. Lastly there was a program committee meeting to discuss the difficult cases.

I am extremely grateful to the program committee members for their enormous investment of time, effort, and adrenaline in the difficult and delicate process of review and selection. (A list of program committee members and sub-referees they invoked can be found on succeeding pages of this volume.) I also thank the authors of submitted papers —in equal measure regardless of whether their papers were accepted or not— for their submissions. It is the work of this body of researchers that makes this conference possible.

I thank Rebecca Wright for hosting the program committee meeting at the AT&T building in New York City and managing the local arrangements, and Ran Canetti for organizing the post-PC-meeting dinner with his characteristic gastronomic and oenophilic flair.

The web-review software we used was written for Eurocrypt 2000 by Wim Moreau and Joris Claessens under the direction of Eurocrypt 2000 program chair Bart Preneel, and I thank them for allowing us to deploy their useful and colorful tool.

I am most grateful to Chanathip Namprempre (aka. Meaw) who provided systems, logistical, and moral support for the entire Crypto 2000 process. She wrote the software for the web-based submissions, adapted and ran the web-review software at UCSD, and compiled the final abstracts into the proceedings you see here. She types faster than I speak.

I am grateful to Hugo Krawczyk for his insight and advice, provided over a long period of time with his usual combination of honesty and charm, and to him and other past program committee chairs, most notably Michael Wiener and Bart Preneel, for replies to the host of questions I posed during the process. In addition I received useful advice from many members of our community including Silvio Micali, Tal Rabin, Ron Rivest, Phil Rogaway, and Adi Shamir. Finally thanks to Matt Franklin who as general chair was in charge of the local organization and finances, and, on the IACR side, to Christian Cachin, Kevin McCurley, and Paul Van Oorschot.

Chairing a Crypto program committee is a learning process. I have come to appreciate even more than before the quality and variety of work in our field, and I hope the papers in this volume contribute further to its development.

June 2000                                             MIHIR BELLARE
                                         Program Chair, Crypto 2000

# CRYPTO 2000

August 20–24, 2000, Santa Barbara, California, USA

Sponsored by the
*International Association for Cryptologic Research (IACR)*

in cooperation with
*IEEE Computer Society Technical Committee on Security and Privacy,
Computer Science Department, University of California, Santa Barbara*

### General Chair
Matthew Franklin, Xerox Palo Alto Research Center, USA

### Program Chair
Mihir Bellare, University of California, San Diego, USA

### Program Committee

Alex Biryukov ......................... Weizmann Institute of Science, Israel
Dan Boneh ........................................ Stanford University, USA
Christian Cachin ............................... IBM Research, Switzerland
Ran Canetti .......................................... IBM Research, USA
Ronald Cramer ............................... ETH Zurich, Switzerland
Yair Frankel ............................................... CertCo, USA
Shai Halevi .......................................... IBM Research, USA
Arjen Lenstra ............................................... Citibank, USA
Mitsuru Matsui .................... Mitsubishi Electric Corporation, Japan
Paul Van Oorschot ........................... Entrust Technologies, Canada
Bart Preneel ...................... Katholieke Universiteit Leuven, Belgium
Phillip Rogaway ....................... University of California, Davis, USA
Victor Shoup ..................................... IBM Zurich, Switzerland
Jessica Staddon ......................... Bell Labs Research, Palo Alto, USA
Jacques Stern ........................... Ecole Normale Supérieure, France
Doug Stinson ............................... University of Waterloo, Canada
Salil Vadhan .................. Massachusetts Institute of Technology, USA
David Wagner ....................... University of California, Berkeley, USA
Rebecca Wright ......................... AT&T Laboratories Research, USA

*Advisory members*

Michael Wiener (Crypto 1999 program chair) .. Entrust Technologies, Canada
Joe Kilian (Crypto 2001 program chair) ................. Intermemory, USA

## Sub-Referees

Bill Aiello, Jeehea An, Olivier Baudron, Don Beaver, Josh Benaloh, John Black, Simon Blackburn, Alexandra Boldyreva, Nikita Borisov, Victor Boyko, Jan Camenisch, Suresh Chari, Scott Contini, Don Coppersmith, Claude Crépeau, Ivan Damgård, Anand Desai , Giovanni Di Crescenzo, Yevgeniy Dodis, Matthias Fitzi, Matt Franklin, Rosario Gennaro, Guang Gong, Luis Granboulan, Nick Howgrave-Graham, Russell Impagliazzo, Yuval Ishai, Markus Jakobsson, Stas Jarecki, Thomas Johansson, Charanjit Jutla, Joe Kilian, Eyal Kushilevitz, Moses Liskov, Stefan Lucks, Anna Lysyanskaya, Philip MacKenzie, Subhamoy Maitra, Tal Malkin, Barbara Masucci, Alfred Menezes, Daniele Micciancio, Sara Miner, Ilia Mironov, Moni Naor , Phong Nguyen, Rafail Ostrovsky, Erez Petrank, Birgit Pfitzmann, Benny Pinkas, David Pointcheval, Guillaume Poupard, Tal Rabin, Charlie Rackoff, Zulfikar Ramzan, Omer Reingold, Leo Reyzin, Pankaj Rohatgi, Amit Sahai, Louis Salvail, Claus Schnorr, Mike Semanko, Bob Silverman, Joe Silverman, Dan Simon, Nigel Smart, Ben Smeets, Adam Smith, Martin Strauss, Ganesh Sundaram, Serge Vaudenay, Frederik Vercauteren, Bernhard von Stengel, Ruizhong Wei, Susanne Gudrun Wetzel, Colin Williams, Stefan Wolf, Felix Wu, Yiqun Lisa Yin, Amir Youssef, Robert Zuccherato

# Table of Contents

# The XTR Public Key System

Arjen K. Lenstra[1] and Eric R. Verheul[2]

[1] Citibank, N.A., 1 North Gate Road, Mendham, NJ 07945-3104, U.S.A.,
arjen.lenstra@citicorp.com
[2] PricewaterhouseCoopers, GRMS Crypto Group, Goudsbloemstraat 14, 5644 KE
Eindhoven, The Netherlands,
Eric.Verheul@[nl.pwcglobal.com, pobox.com]

**Abstract.** This paper introduces the XTR public key system. XTR is
based on a new method to represent elements of a subgroup of a multiplicative group of a finite field. Application of XTR in cryptographic
protocols leads to substantial savings both in communication and computational overhead without compromising security.

## 1 Introduction

The Diffie-Hellman (DH) key agreement protocol was the first published practical solution to the key distribution problem, allowing two parties that have
never met to establish a shared secret key by exchanging information over an
open channel. In the basic DH scheme the two parties agree upon a generator
$g$ of the multiplicative group $GF(p)^*$ of a prime field $GF(p)$ and they each send
a random power of $g$ to the other party. Assuming both parties know $p$ and $g$,
each party transmits about $\log_2(p)$ bits to the other party.

In [7] it was suggested that finite extension fields can be used instead of prime
fields, but no direct computational or communication advantages were implied.
In [22] a variant of the basic DH scheme was introduced where $g$ generates a
relatively small subgroup of $GF(p)^*$ of prime order $q$. This considerably reduces
the computational cost of the DH scheme, but has no effect on the number of
bits to be exchanged. In [3] it was shown for the first time how the use of finite
extension fields and subgroups can be combined in such a way that the number of
bits to be exchanged is reduced by a factor 3. More specifically, it was shown that
elements of an order $q$ subgroup of $GF(p^6)^*$ can be represented using $2\log_2(p)$
bits if $q$ divides $p^2 - p + 1$. Despite its communication efficiency, the method
of [3] is rather cumbersome and computationally not particularly efficient.

In this paper we present a greatly improved version of the method from [3]
that achieves the same communication advantage at a much lower computational
cost. We refer to our new method as XTR, for Efficient and Compact Subgroup
Trace Representation. XTR can be used in conjunction with any cryptographic
protocol that is based on the use of subgroups and leads to substantial savings in
communication and computational overhead. Furthermore, XTR key generation
is very simple. We prove that using XTR in cryptographic protocols does not
affect their security. The best attacks we are aware of are Pollard's rho method
in the order $q$ subgroup, or the Discrete Logarithm variant of the Number Field

M. Bellare (Ed.): CRYPTO 2000, LNCS 1880, pp. 1–19, 2000.
© Springer-Verlag Berlin Heidelberg 2000

Sieve in the full multiplicative group $GF(p^6)^*$. With primes $p$ and $q$ of about $1024/6 \approx 170$ bits the security of XTR is equivalent to traditional subgroup systems using 170-bit subgroups and 1024-bit finite fields. But with XTR subgroup elements can be represented using only about $2 * 170$ bits, which is substantially less than the 1024-bits required for their traditional representation.

Full exponentiation in XTR is faster than full scalar multiplication in an Elliptic Curve Cryptosystem (ECC) over a 170-bit prime field, and thus substantially faster than full exponentiation in either RSA or traditional subgroup discrete logarithm systems of equivalent security. XTR keys are much smaller than RSA keys of comparable security. ECC keys allow a smaller representation than XTR keys, but in many circumstances (e.g. storage) ECC and XTR key sizes are comparable. However, XTR is not affected by the uncertainty still marring ECC. Key selection for XTR is very fast compared to RSA, and orders of magnitude easier and faster than for ECC. As a result XTR may be regarded as the best of two worlds, RSA and ECC. It is an excellent alternative to either RSA or ECC in applications such as SSL/TLS (Secure Sockets Layer, Transport Layer Security), public key smartcards, WAP/WTLS (Wireless Application Protocol, Wireless Transport Layer Security), IPSEC/IKE (Internet Protocol Security, Internet Key Exchange), and SET (Secure Electronic Transaction).

In [14] it is argued that ECC is the only public key system that is suitable for a variety of environments, including low-end smart cards and over-burdened web servers communicating with powerful PC clients. XTR shares this advantage with ECC, with the distinct additional advantage that XTR key selection is very easy. This makes it easily feasible for all users of XTR to have public keys that are not shared with others, unlike ECC where a large part of the public key is often shared between all users of the system. Also, compared to ECC, the mathematics underlying XTR is straightforward, thus avoiding two common ECC-pitfalls: ascertaining that unfortunate parameter choices are avoided that happen to render the system less secure, and keeping abreast of, and incorporating additional checks published in, newly obtained results. The latest example of the latter is [8], where yet another condition affecting the security of ECC over finite fields of characteristic two is described. As a consequence the draft IKE protocol (part of IPSec) for ECC was revised. Note that Odlyzko in [16] advises to use ECC key sizes of at least 300 bits, even for moderate security needs.

XTR is the first method we are aware of that uses $GF(p^2)$ arithmetic to achieve $GF(p^6)$ security, without requiring explicit construction of $GF(p^6)$. Let $g$ be an element of order $q > 6$ dividing $p^2 - p + 1$. Because $p^2 - p + 1$ divides the order $p^6 - 1$ of $GF(p^6)^*$ this $g$ generates an order $q$ subgroup of $GF(p^6)^*$. Since $q$ does not divide any $p^s - 1$ for $s = 1, 2, 3$ (cf. [11]), the subgroup generated by $g$ cannot be embedded in the multiplicative group of any true subfield of $GF(p^6)$. We show, however, that arbitrary powers of $g$ can be represented using a single element of the subfield $GF(p^2)$, and that such powers can be computed efficiently using arithmetic operations in $GF(p^2)$ while avoiding arithmetic in $GF(p^6)$.

In Section 2 we describe XTR, and in Section 3 we explain how the XTR parameters can be found quickly. Applications and comparisons to RSA and

ECC are given in Section 4. In Section 5 we prove that using XTR does not have a negative impact on the security. Extensions are discussed in Section 6.

## 2    Subgroup Representation and Arithmetic

### 2.1    Preliminaries

Let $p \equiv 2 \bmod 3$ be a prime such that the sixth cyclotomic polynomial evaluated in $p$, i.e., $\phi_6(p) = p^2 - p + 1$, has a prime factor $q > 6$. In subsection 3.1 we give a fast method to select $p$ and $q$. By $g$ we denote an element of $\mathrm{GF}(p^6)^*$ of order $q$. Because of the choice of $q$, this $g$ is not contained in any proper subfield of $\mathrm{GF}(p^6)$ (cf. [11]). Many cryptographic applications (cf. Section 4) make use of the subgroup $\langle g \rangle$ generated by $g$. In this section we show that actual representation of the elements of $\langle g \rangle$ and of any other element of $\mathrm{GF}(p^6)$ can be avoided. Thus, there is no need to represent elements of $\mathrm{GF}(p^6)$, for instance by constructing a sixth or third degree irreducible polynomial over $\mathrm{GF}(p)$ or $\mathrm{GF}(p^2)$, respectively. A representation of $\mathrm{GF}(p^2)$ is needed, however. This is done as follows.

From $p \equiv 2 \bmod 3$ it follows that $p \bmod 3$ generates $\mathrm{GF}(3)^*$, so that the zeros $\alpha$ and $\alpha^p$ of the polynomial $(X^3 - 1)/(X - 1) = X^2 + X + 1$ form an optimal normal basis for $\mathrm{GF}(p^2)$ over $\mathrm{GF}(p)$. Because $\alpha^i = \alpha^{i \bmod 3}$, an element $x \in \mathrm{GF}(p^2)$ can be represented as $x_1\alpha + x_2\alpha^p = x_1\alpha + x_2\alpha^2$ for $x_1, x_2 \in \mathrm{GF}(p)$. In this representation of $\mathrm{GF}(p^2)$ an element $t$ of $\mathrm{GF}(p)$ is represented as $-t\alpha - t\alpha^2$, e.g. 3 is represented as $-3\alpha - 3\alpha^2$. Arithmetic operations in $\mathrm{GF}(p^2)$ are carried out as follows.

For any $x = x_1\alpha + x_2\alpha^2 \in \mathrm{GF}(p^2)$ we have that $x^p = x_1^p\alpha^p + x_2^p\alpha^{2p} = x_2\alpha + x_1\alpha^2$. It follows that $p^{\mathrm{th}}$ powering in $\mathrm{GF}(p^2)$ does not require arithmetic operations and can thus be considered to be for free. Squaring of $x_1\alpha + x_2\alpha^2 \in \mathrm{GF}(p^2)$ can be carried out at the cost of two squarings and a single multiplication in $\mathrm{GF}(p)$, where as customary we do not count the cost of additions in $\mathrm{GF}(p)$. Multiplication in $\mathrm{GF}(p^2)$ can be done using four multiplications in $\mathrm{GF}(p)$. These straightforward results can simply be improved to three squarings and three multiplications, respectively, by using a Karatsuba-like approach (cf. [10]): to compute $(x_1\alpha + x_2\alpha^2) * (y_1\alpha + y_2\alpha^2)$ one computes $x_1 * y_1$, $x_2 * y_2$, and $(x_1 + x_2) * (y_1 + y_2)$, after which $x_1 * y_2 + x_2 * y_1$ follows using two subtractions. Furthermore, from $(x_1\alpha + x_2\alpha^2)^2 = x_2(x_2 - 2x_1)\alpha + x_1(x_1 - 2x_2)\alpha^2$ it follows that squaring in $\mathrm{GF}(p^2)$ can be done at the cost of two multiplications in $\mathrm{GF}(p)$. Under the reasonable assumption that a squaring in $\mathrm{GF}(p)$ takes 80% of the time of a multiplication in $\mathrm{GF}(p)$ (cf. [4]), two multiplications is faster than three squarings. Finally, to compute $x * z - y * z^p \in \mathrm{GF}(p^2)$ for $x, y, z \in \mathrm{GF}(p^2)$ four multiplications in $\mathrm{GF}(p)$ suffice, because, with $x = x_1\alpha + x_2\alpha^2$, $y = y_1\alpha + y_2\alpha^2$, and $z = z_1\alpha + z_2\alpha^2$, it is easily verified that $x * z - y * z^p = (z_1(y_1 - x_2 - y_2) + z_2(x_2 - x_1 + y_2))\alpha + (z_1(x_1 - x_2 + y_1) + z_2(y_2 - x_1 - y_1))\alpha^2$. Thus we have the following.

**Lemma 2.1.1** *Let* $x, y, z \in \mathrm{GF}(p^2)$ *with* $p \equiv 2 \bmod 3$.

*i. Computing* $x^p$ *is for free.*

***ii.*** *Computing $x^2$ takes two multiplications in $GF(p)$.*
***iii.*** *Computing $x * y$ takes three multiplications in $GF(p)$.*
***iv.*** *Computing $x * z - y * z^p$ takes four multiplications in $GF(p)$.*

For comparison purposes we review the following well known results.

**Lemma 2.1.2** *Let $x, y, z \in GF(p^6)$ with $p \equiv 2 \bmod 3$, and let $a, b \in \mathbf{Z}$ with $0 < a, b < p$. Assume that a squaring in $GF(p)$ takes 80% of the time of a multiplication in $GF(p)$ (cf. [4]).*

***i.*** *Computing $x^2$ takes 14.4 multiplications in $GF(p)$.*
***ii.*** *Computing $x * y$ takes 18 multiplications in $GF(p)$.*
***iii.*** *Computing $x^a$ takes an expected $23.4 \log_2(a)$ multiplications in $GF(p)$.*
***iv.*** *Computing $x^a * y^b$ takes an expected $27.9 \log_2(\max(a, b))$ multiplications in $GF(p)$.*

**Proof.** Since $p \equiv 2 \bmod 3$, $GF(p^6)$ can be represented using an optimal normal basis over $GF(p)$ so that the 'reduction' modulo the minimal polynomial does not require any multiplications in $GF(p)$. Squaring and multiplication in $GF(p^6)$ can then be done in 18 squarings and multiplications in $GF(p)$, respectively, from which $i$ and $ii$ follow. For $iii$ we use the ordinary square and multiply method, so we get $\log_2(a)$ squarings and an expected $0.5 \log_2(a)$ multiplications in $GF(p^6)$. For $iv$ we use standard multi-exponentiation, which leads to $\log_2(\max(a, b))$ squarings and $0.75 \log_2(\max(a, b))$ multiplications in $GF(p^6)$.

## 2.2    Traces

The *conjugates* over $GF(p^2)$ of $h \in GF(p^6)$ are $h$, $h^{p^2}$, and $h^{p^4}$. The *trace* $Tr(h)$ over $GF(p^2)$ of $h \in GF(p^6)$ is the sum of the conjugates over $GF(p^2)$ of $h$, i.e., $Tr(h) = h + h^{p^2} + h^{p^4}$. Because the order of $h \in GF(p^6)^*$ divides $p^6 - 1$, i.e., $p^6 \equiv 1$ modulo the order of $h$, we have that $Tr(h)^{p^2} = Tr(h)$, so that $Tr(h) \in GF(p^2)$. For $h_1, h_2 \in GF(p^6)$ and $c \in GF(p^2)$ we have that $Tr(h_1 + h_2) = Tr(h_1) + Tr(h_2)$ and $Tr(c * h_1) = c * Tr(h_1)$. That is, the trace over $GF(p^2)$ is $GF(p^2)$-linear. Unless specified otherwise, conjugates and traces in this paper are over $GF(p^2)$.

The conjugates of $g$ of order dividing $p^2 - p + 1$ are $g$, $g^{p-1}$ and $g^{-p}$ because $p^2 \equiv p - 1 \bmod p^2 - p + 1$ and $p^4 \equiv -p \bmod p^2 - p + 1$.

**Lemma 2.2.1** *The roots of $X^3 - Tr(g)X^2 + Tr(g)^p X - 1$ are the conjugates of $g$.*

**Proof.** We compare the coefficients of $X^3 - Tr(g)X^2 + Tr(g)^p X - 1$ with the coefficients of the polynomial $(X - g)(X - g^{p-1})(X - g^{-p})$. The coefficient of $X^2$ follows from $g + g^{p-1} + g^{-p} = Tr(g)$, and the constant coefficient from $g^{1+p-1-p} = 1$. The coefficient of $X$ equals $g * g^{p-1} + g * g^{-p} + g^{p-1} * g^{-p} = g^p + g^{1-p} + g^{-1}$. Because $1 - p \equiv -p^2 \bmod p^2 - p + 1$ and $-1 \equiv p^2 - p \bmod p^2 - p + 1$, we find that $g^p + g^{1-p} + g^{-1} = g^p + g^{-p^2} + g^{p^2-p} = (g + g^{-p} + g^{p-1})^p = Tr(g)^p$, which completes the proof.

Similarly (and as proved below in Lemma 2.3.4.$ii$), the roots of $X^3 - Tr(g^n)X^2 + Tr(g^n)^p X - 1$ are the conjugates of $g^n$. Thus, the conjugates of $g^n$ are fully determined by $X^3 - Tr(g^n)X^2 + Tr(g^n)^p X - 1$ and thus by $Tr(g^n)$. Since $Tr(g^n) \in \mathrm{GF}(p^2)$ this leads to a compact representation of the conjugates of $g^n$. To be able to use this representation in an efficient manner in cryptographic protocols, we need an efficient way to compute $Tr(g^n)$ given $Tr(g)$. Such a method can be derived from properties of $g$ and the trace function. However, since we need a similar method in a more general context in Section 3, we consider the properties of the polynomial $X^3 - cX^2 + c^p X - 1$ for general $c \in \mathrm{GF}(p^2)$ (as opposed to $c$'s that are traces of powers of $g$).

## 2.3 The Polynomial $F(c, X)$

**Definition 2.3.1** For $c \in \mathrm{GF}(p^2)$ let $F(c, X)$ be the polynomial $X^3 - cX^2 + c^p X - 1 \in \mathrm{GF}(p^2)[X]$ with (not necessarily distinct) roots $h_0$, $h_1$, $h_2$ in $\mathrm{GF}(p^6)$, and let $\tau(c, n) = h_0^n + h_1^n + h_2^n$ for $n \in \mathbf{Z}$. We use the shorthand $c_n = \tau(c, n)$.

In this subsection we derive some properties of $F(c, X)$ and its roots.

**Lemma 2.3.2**
*i.* $c = c_1$.
*ii.* $h_0 * h_1 * h_2 = 1$.
*iii.* $h_0^n * h_1^n + h_0^n * h_2^n + h_1^n * h_2^n = c_{-n}$ for $n \in \mathbf{Z}$.
*iv.* $F(c, h_j^{-p}) = 0$ for $j = 0, 1, 2$.
*v.* $c_{-n} = c_{np} = c_n^p$ for $n \in \mathbf{Z}$.
*vi.* Either all $h_j$ have order dividing $p^2 - p + 1$ and $> 3$ or all $h_j \in \mathrm{GF}(p^2)$.
*vii.* $c_n \in \mathrm{GF}(p^2)$ for $n \in \mathbf{Z}$.

**Proof.** The proofs of $i$ and $ii$ are immediate and $iii$ follows from $ii$. From $F(c, h_j) = h_j^3 - ch_j^2 + c^p h_j - 1 = 0$ it follows that $h_j \neq 0$ and that $F(c, h_j)^p = h_j^{3p} - c^p h_j^{2p} + c^{p^2} h_j^p - 1 = 0$. With $c^{p^2} = c$ and $h_j \neq 0$ it follows that $-h_j^{3p}(h_j^{-3p} - ch_j^{-2p} + c^p h_j^{-p} - 1) = -h_j^{3p} * F(c, h_j^{-p}) = 0$, which proves $iv$.

From $iv$ it follows, without loss of generality, that either $h_j = h_j^{-p}$ for $j = 0, 1, 2$, or $h_0 = h_0^{-p}$, $h_1 = h_2^{-p}$, and $h_2 = h_1^{-p}$, or that $h_j = h_{j+1 \bmod 3}^{-p}$ for $j = 0, 1, 2$. In either case $v$ follows. Furthermore, in the first case all $h_j$ have order dividing $p + 1$ and are thus in $\mathrm{GF}(p^2)$. In the second case, $h_0$ has order dividing $p + 1$, $h_1 = h_2^{-p} = h_1^{p^2}$ and $h_2 = h_1^{-p} = h_2^{p^2}$ so that $h_1$ and $h_2$ both have order dividing $p^2 - 1$. It follows that they are all again in $\mathrm{GF}(p^2)$. In the last case it follows from $1 = h_0 * h_1 * h_2$ that $1 = h_0 * h_2^{-p} * h_0^{-p} = h_0 * h_0^{p^2} * h_0^{-p} = h_0^{p^2 - p + 1}$ so that $h_0$ and similarly $h_1$ and $h_2$ have order dividing $p^2 - p + 1$. If either one, say $h_0$, has order at most 3, then $h_0$ has order 1 or 3 since $p^2 - p + 1$ is odd. It follows that the order of $h_0$ divides $p^2 - 1$ so that $h_0 \in \mathrm{GF}(p^2)$. But then $h_1$ and $h_2$ are in $\mathrm{GF}(p^2)$ as well, because $h_j = h_{j+1 \bmod 3}^{-p}$. It follows that in the last case either all $h_j$ have order dividing $p^2 - p + 1$ and $> 3$, or all $h_j$ are in $\mathrm{GF}(p^2)$, which concludes the proof of $vi$.

If all $h_j \in \mathrm{GF}(p^2)$, then *vii* is immediate. Otherwise $F(c, X)$ is irreducible and its roots are the conjugates of $h_0$. Thus $c_n = Tr(h_0^n) \in \mathrm{GF}(p^2)$ (cf. 2.2). This concludes the proof of *vii* and Lemma 2.3.2.

**Remark 2.3.3** It follows from Lemma 2.3.2.*vi* that $F(c, X) \in \mathrm{GF}(p^2)[X]$ is irreducible if and only if its roots have order dividing $p^2 - p + 1$ and $> 3$.

**Lemma 2.3.4**
*i.* $c_{u+v} = c_u * c_v - c_v^p * c_{u-v} + c_{u-2v}$ for $u, v \in \mathbf{Z}$.
*ii.* $F(c_n, h_j^n) = 0$ for $j = 0, 1, 2$ and $n \in \mathbf{Z}$.
*iii.* $F(c, X)$ is reducible over $\mathrm{GF}(p^2)$ if and only if $c_{p+1} \in \mathrm{GF}(p)$.

**Proof.** With the definition of $c_n$, $c_n^p = c_{-n}$ (cf. Lemma 2.3.2.*v*), and Lemma 2.3.2.*ii*, the proof of *i* follows from a straightforward computation.

For the proof of *ii* we compute the coefficients of $(X - h_0^n)(X - h_1^n)(X - h_2^n)$. We find that the coefficient of $X^2$ equals $-c_n$ and that the constant coefficient equals $-h_0^n * h_1^n * h_2^n = -(h_0 * h_1 * h_2)^n = -1$ (cf. Lemma 2.3.2.*ii*). The coefficient of $X$ equals $h_0^n * h_1^n + h_0^n * h_2^n + h_1^n * h_2^n = c_{-n} = c_n^p$ (cf. Lemma 2.3.2.*iii* and *v*). It follows that $(X - h_0^n)(X - h_1^n)(X - h_2^n) = F(c_n, X)$ from which *ii* follows.

If $F(c, X)$ is reducible then all $h_j$ are in $\mathrm{GF}(p^2)$ (cf. Remark 2.3.3 and Lemma 2.3.2.*vi*). It follows that $h_j^{(p+1)p} = h_j^{p+1}$ so that $h_j^{p+1} \in \mathrm{GF}(p)$ for $j = 0, 1, 2$ and $c_{p+1} \in \mathrm{GF}(p)$. Conversely, if $c_{p+1} \in \mathrm{GF}(p)$, then $c_{p+1}^p = c_{p+1}$ and $F(c_{p+1}, X) = X^3 - c_{p+1}X^2 + c_{p+1}X - 1$. Thus, $F(c_{p+1}, 1) = 0$. Because the roots of $F(c_{p+1}, X)$ are the $(p+1)^{\text{st}}$ powers of the roots of $F(c, X)$ (cf. *iv*), it follows that $F(c, X)$ has a root of order dividing $p + 1$, i.e., an element of $\mathrm{GF}(p^2)$, so that $F(c, X)$ is reducible over $\mathrm{GF}(p^2)$. This proves *iii*.

Lemma 2.3.2.*v* and Lemma 2.3.4.*i* lead to a fast algorithm to compute $c_n$ for any $n \in \mathbf{Z}$.

**Corollary 2.3.5** Let $c$, $c_{n-1}$, $c_n$, and $c_{n+1}$ be given.
*i.* Computing $c_{2n} = c_n^2 - 2c_n^p$ takes two multiplications in $\mathrm{GF}(p)$.
*ii.* Computing $c_{n+2} = c * c_{n+1} - c^p * c_n + c_{n-1}$ takes four multiplications in $\mathrm{GF}(p)$.
*iii.* Computing $c_{2n-1} = c_{n-1} * c_n - c^p * c_n^p + c_{n+1}^p$ takes four multiplications in $\mathrm{GF}(p)$.
*iv.* Computing $c_{2n+1} = c_{n+1} * c_n - c * c_n^p + c_{n-1}^p$ takes four multiplications in $\mathrm{GF}(p)$.

**Proof.** The identities follow from Lemma 2.3.2.*v* and Lemma 2.3.4.*i*: with $u = v = n$ and $c_0 = 3$ for *i*, with $u = n + 1$ and $v = 1$ for *ii*, $u = n - 1$, $v = n$ for *iii*, and $u = n + 1$, $v = n$ for *iv*. The cost analysis follows from Lemma 2.1.1.

**Definition 2.3.6** Let $S_n(c) = (c_{n-1}, c_n, c_{n+1}) \in \mathrm{GF}(p^2)^3$.

**Algorithm 2.3.7 (Computation of $S_n(c)$ given $c$)** If $n < 0$, apply this algorithm to $-n$ and use Lemma 2.3.2.$v$. If $n = 0$, then $S_0(c) = (c^p, 3, c)$ (cf. Lemma 2.3.2.$v$). If $n = 1$, then $S_1(c) = (3, c, c^2 - 2c^p)$ (cf. Corollary 2.3.5.$i$). If $n = 2$, use Corollary 2.3.5.$ii$ and $S_1(c)$ to compute $c_3$ and thereby $S_2(n)$. Otherwise, to compute $S_n(c)$ for $n > 2$ let $m = n$. If $m$ is even, then replace $m$ by $m - 1$. Let $\bar{S}_t(c) = S_{2t+1}(c)$ for $t \in \mathbf{Z}$, $k = 1$, and compute $\bar{S}_k(c) = S_3(c)$ using Corollary 2.3.5.$ii$ and $S(2)$. Let $(m - 1)/2 = \sum_{j=0}^{r} m_j 2^j$ with $m_j \in \{0, 1\}$ and $m_r = 1$. For $j = r - 1, r - 2, \ldots, 0$ in succession do the following:

- If $m_j = 0$ then use $\bar{S}_k(c) = (c_{2k}, c_{2k+1}, c_{2k+2})$ to compute $\bar{S}_{2k}(c) = (c_{4k}, c_{4k+1}, c_{4k+2})$ (using Corollary 2.3.5.$i$ for $c_{4k}$ and $c_{4k+2}$ and Corollary 2.3.5.$iii$ for $c_{4k+1}$) and replace $k$ by $2k$.
- If $m_j = 1$ then use $\bar{S}_k(c) = (c_{2k}, c_{2k+1}, c_{2k+2})$ to compute $\bar{S}_{2k+1}(c) = (c_{4k+2}, c_{4k+3}, c_{4k+4})$ (using Corollary 2.3.5.$i$ for $c_{4k+2}$ and $c_{4k+4}$ and Corollary 2.3.5.$iv$ for $c_{4k+3}$) and replace $k$ by $2k + 1$,

After this iteration we have that $2k + 1 = m$ so that $S_m(c) = \bar{S}_k(c)$. If $n$ is even use $S_m(c) = (c_{m-1}, c_m, c_{m+1})$ to compute $S_{m+1}(c) = (c_m, c_{m+1}, c_{m+2})$ (using Corollary 2.3.5.$ii$) and replace $m$ by $m + 1$. As a result we have $S_n(c) = S_m(c)$.

**Theorem 2.3.8** *Given the sum $c$ of the roots of $F(c, X)$, computing the sum $c_n$ of the $n^{th}$ powers of the roots takes $8 \log_2(n)$ multiplications in $GF(p)$.*

**Proof.** Immediate from Algorithm 2.3.7 and Corollary 2.3.5.

**Remark 2.3.9** The only difference between the two different cases in Algorithm 2.3.7 (i.e., if the bit is off or on) is the application of Corollary 2.3.5.$iii$ if the bit is off and of Corollary 2.3.5.$iv$ if the bit is on. The two computations involved, however, are very similar and take the same number of instructions. Thus, the instructions carried out in Algorithm 2.3.7 for the two different cases are very much alike. This is a rather unusual property for an exponentiation routine and makes Algorithm 2.3.7 much less susceptible than usual exponentiation routines to environmental attacks such as timing attacks and Differential Power Analysis.

## 2.4   Computing with Traces

It follows from Lemma 2.2.1 and Lemma 2.3.4.$ii$ that

$$S_n(Tr(g)) = (Tr(g^{n-1}), Tr(g^n), Tr(g^{n+1}))$$

(cf. Definition 2.3.6). Furthermore, given $Tr(g)$ Algorithm 2.3.7 can be used to compute $S_n(Tr(g))$ for any $n$. Since the order of $g$ equals $q$ this takes $8 \log_2(n \bmod q)$ multiplications in $GF(p)$ (cf. Theorem 2.3.8). According to Lemma 2.1.2.$iii$ computing $g^n$ given $g$ can be expected to take $23.4 \log_2(q)$ multiplications in $GF(p)$. Thus, computing $Tr(g^n)$ given $Tr(g)$ is almost three times faster than computing $g^n$ given $g$. Furthermore, $Tr(g^n) \in GF(p^2)$ whereas $g^n \in GF(p^6)$. So representing, storing, or transmitting $Tr(g^n)$ is three times cheaper than it is for $g^n$. Unlike the methods from for instance [2], we do not assume that $p$

has a special form. Using such primes leads to additional savings by making the arithmetic in $GF(p)$ faster (cf. Algorithm 3.1.1).

Thus, we replace the traditional representation of powers of $g$ by their traces. The ability to quickly compute $Tr(g^n)$ based on $Tr(g)$ suffices for the implementation of many cryptographic protocols (cf. Section 4). In some protocols, however, the product of two powers of $g$ must be computed. For the standard representation this is straightforward, but if traces are used, then computing products is relatively complicated. We describe how this problem may be solved in the cryptographic applications that we are aware of. Let $Tr(g) \in GF(p^2)$ and $S_k(Tr(g)) \in GF(p^2)^3$ (cf. Definition 2.3.6) be given for some secret integer $k$ (the private key) with $0 < k < q$. We show that $Tr(g^a * g^{bk})$ can be computed efficiently for any $a, b \in \mathbf{Z}$.

**Definition 2.4.1** Let $A(c) = \begin{pmatrix} 0 & 0 & 1 \\ 1 & 0 & -c^p \\ 0 & 1 & c \end{pmatrix}$ and $M_n(c) = \begin{pmatrix} c_{n-2} & c_{n-1} & c_n \\ c_{n-1} & c_n & c_{n+1} \\ c_n & c_{n+1} & c_{n+2} \end{pmatrix}$ be $3 \times 3$-matrices over $GF(p^2)$ with $c$ and $c_n$ as in Definition 2.3.1, and let $C(V)$ denote the center column of a $3 \times 3$ matrix $V$.

**Lemma 2.4.2** $S_n(c) = S_m(c) * A(c)^{n-m}$ and $M_n(c) = M_m(c) * A(c)^{n-m}$ for $n, m \in \mathbf{Z}$.

**Proof.** For $n - m = 1$ the first statement is equivalent with Corollary 2.3.5.$ii$. The proof follows by induction to $n - m$.

**Corollary 2.4.3** $c_n = S_m(c) * C(A(c)^{n-m})$.

**Lemma 2.4.4** The determinant of $M_0(c)$ equals $D = c^{2p+2} + 18c^{p+1} - 4(c^{3p} + c^3) - 27 \in GF(p)$. If $D \neq 0$ then

$$M_0(c)^{-1} = \frac{1}{D} * \begin{pmatrix} 2c^2 - 6c^p & 2c^{2p} + 3c - c^{p+2} & c^{p+1} - 9 \\ 2c^{2p} + 3c - c^{p+2} & (c^2 - 2c^p)^{p+1} - 9 & (2c^{2p} + 3c - c^{p+2})^p \\ c^{p+1} - 9 & (2c^{2p} + 3c - c^{p+2})^p & (2c^2 - 6c^p)^p \end{pmatrix}.$$

**Proof.** This follows from a simple computation using Lemma 2.3.2.$v$ and Corollary 2.3.5 combined with the fact that $x \in GF(p)$ if $x^p = x$.

**Lemma 2.4.5** $\det(M_0(Tr(g))) = (Tr(g^{p+1})^p - Tr(g^{p+1}))^2 \neq 0$.

**Proof.** This follows by observing that $M_0(Tr(g))$ is the product of the Vandermonde matrix $\begin{pmatrix} g^{-1} & g^{-p^2} & g^{-p^4} \\ 1 & 1 & 1 \\ g & g^{p^2} & g^{p^4} \end{pmatrix}$ and its inverse, and therefore invertible. The determinant of the Vandermonde matrix equals $Tr(g^{p+1})^p - Tr(g^{p+1})$.

**Lemma 2.4.6** $A(Tr(g))^n = M_0(Tr(g))^{-1} * M_n(Tr(g))$ can be computed in a small constant number of operations in $GF(p^2)$ given $Tr(g)$ and $S_n(Tr(g))$.

**Proof.** $Tr(g^{n\pm2})$ and thus $M_n(Tr(g))$ can be computed from $S_n(Tr(g))$ using Corollary 2.3.5.$ii$. The proof follows from Lemmas 2.4.2, 2.4.4, 2.4.5, and 2.1.1.$i$.

**Corollary 2.4.7** $C(A(Tr(g))^n) = M_0(Tr(g))^{-1} * (S_n(Tr(g)))^T$.

**Algorithm 2.4.8 (Computation of $Tr(g^a * g^{bk})$)** Let $Tr(g)$, $S_k(Tr(g))$ (for unknown $k$), and $a, b \in \mathbf{Z}$ with $0 < a, b < q$ be given.

1. Compute $e = a/b \bmod q$.
2. Compute $S_e(Tr(g))$ (cf. Algorithm 2.3.7).
3. Compute $C(A(Tr(g))^e)$ based on $Tr(g)$ and $S_e(Tr(g))$ using Corollary 2.4.7.
4. Compute $Tr(g^{e+k}) = S_k(Tr(g)) * C(A(Tr(g))^e)$ (cf. Corollary 2.4.3).
5. Compute $S_b(Tr(g^{e+k}))$ (cf. Algorithm 2.3.7), and return $Tr(g^{(e+k)b})$ $= Tr(g^a * g^{bk})$.

**Theorem 2.4.9** *Given* $M_0(Tr(g))^{-1}$, $Tr(g)$, *and* $S_k(Tr(g)) = (Tr(g^{k-1})$, $Tr(g^k), Tr(g^{k+1}))$ *the trace* $Tr(g^a * g^{bk})$ *of* $g^a * g^{bk}$ *can be computed at a cost of* $8\log_2(a/b \bmod q) + 8\log_2(b) + 34$ *multiplications in* $\mathrm{GF}(p)$.

**Proof.** The proof follows from a straightforward analysis of the cost of the required matrix vector operations and Theorem 2.3.8.

Assuming that $M_0(Tr(g))^{-1}$ is computed once and for all (at the cost of a small constant number of operations in $\mathrm{GF}(p^2)$), we find that $Tr(g^a * g^{bk})$ can be computed at a cost of $16\log_2(q)$ multiplications in $\mathrm{GF}(p)$. According to Lemma 2.1.2.$iv$ this computation would cost about $27.9\log_2(q)$ multiplications in $\mathrm{GF}(p)$ using the traditional representation. Thus, in this case the trace representation achieves a speed-up of a factor 1.75 over the traditional one. We conclude that both single and double exponentiations can be done substantially faster using traces than using previously published techniques.

## 3   Parameter Selection

### 3.1   Finite Field and Subgroup Size Selection

We describe fast and practical methods to select the field characteristic $p$ and subgroup size $q$ such that $q$ divides $p^2 - p + 1$. Denote by $P$ and $Q$ the sizes of the primes $p$ and $q$ to be generated, respectively. To achieve security at least equivalent to 1024-bit RSA, $6P$ should be set to about 1024, i.e., $P \approx 170$, and $Q$ can for instance be set at 160. Given current cryptanalytic methods we do not recommend choosing $P$ much smaller than $Q$.

**Algorithm 3.1.1 (Selection of $q$ and 'nice' $p$)** Find $r \in \mathbf{Z}$ such that $q = r^2 - r + 1$ is a $Q$-bit prime, and next find $k \in \mathbf{Z}$ such that $p = r + k * q$ is a $P$-bit prime that is 2 mod 3.

Algorithm 3.1.1 is quite fast and it can be used to find primes $p$ that satisfy a degree two polynomial with small coefficients. Such $p$ lead to fast arithmetic operations in $\mathrm{GF}(p)$. In particular if the search for $k$ is restricted to $k = 1$ (i.e., search for an $r$ such that both $r^2 - r + 1$ and $r^2 + 1$ are prime and such that $r^2 + 1 \equiv 2 \bmod 3$) the primes $p$ have a very nice form; note that in this case $r$ must be even and $p \equiv 1 \bmod 4$. On the other hand, such 'nice' $p$ may be undesirable from a security point of view because they may make application of the Discrete Logarithm variant of the Number Field Sieve easier. Another method to generate $p$ and $q$ that does not have this disadvantage (and thus neither the advantage of fast arithmetic modulo $p$) is the following.

**Algorithm 3.1.2 (Selection of $q$ and $p$)** First, select a $Q$-bit prime $q \equiv 7 \bmod 12$. Next, find the roots $r_1$ and $r_2$ of $X^2 - X + 1 \bmod q$. It follows from $q \equiv 1 \bmod 3$ and quadratic reciprocity that $r_1$ and $r_2$ exist. Since $q \equiv 3 \bmod 4$ they can be found using a single $((q + 1)/4)^{\text{th}}$ powering modulo $q$. Finally, find a $k \in \mathbf{Z}$ such that $p = r_i + k * q$ is a $P$-bit prime that is $2 \bmod 3$ for $i = 1$ or $2$.

The run time of Algorithms 3.1.1 and 3.1.2 is dominated by the time to find the primes $q$ and $p$. A precise analysis is straightforward and left to the reader.

## 3.2    Subgroup Selection

We consider the problem of finding a proper $Tr(g)$ for an element $g \in \mathrm{GF}(p^6)$ of order $q$ dividing $p^2 - p + 1$ and $> 3$. Note that there is no need to find $g$ itself, finding $Tr(g)$ suffices. Given $Tr(g)$ for an unspecified $g$, a subgroup generator can be computed by finding a root in $\mathrm{GF}(p^6)$ of $F(Tr(g), X)$. We refer to this generator as $g$ and to the order $q$ subgroup $\langle g \rangle$ as the *XTR group*. Note that all roots of $F(Tr(g), X)$ lead to the same XTR group.

A straightforward approach to find $Tr(g)$ would be to find a third degree irreducible polynomial over $\mathrm{GF}(p^2)$, use it to represent $\mathrm{GF}(p^6)$, to pick an element $h \in \mathrm{GF}(p^6)$ until $h^{(p^6-1)/q} \neq 1$, to take $g = h^{(p^6-1)/q}$, and to compute $Tr(g)$. Although conceptually easy, this method is less attractive from an implementation point of view. A faster method that is also easier to implement is based on the following lemma.

**Lemma 3.2.1** *For a randomly selected $c \in \mathrm{GF}(p^2)$ the probability that $F(c, X) \in \mathrm{GF}(p^2)[X]$ is irreducible is about one third.*

**Proof.** This follows from a straightforward counting argument. About $p^2 - p$ elements of the subgroup of order $p^2 - p + 1$ of $\mathrm{GF}(p^6)^*$ are roots of monic irreducible polynomials of the form $F(c, X)$ (cf. Lemma 2.2.1 and Lemma 2.3.4.*ii*). Since each of these polynomials has three distinct roots, there must be about $(p^2-p)/3$ different values for $c$ in $\mathrm{GF}(p^2)\backslash\mathrm{GF}(p)$ such that $F(c, X)$ is irreducible.

With Remark 2.3.3 it follows that it suffices to pick a $c \in \mathrm{GF}(p^2)$ until $F(c, X)$ is irreducible and until $c_{(p^2-p+1)/q} \neq 3$ (cf. Definition 2.3.1), and to take $Tr(g) = c_{(p^2-p+1)/q}$. The resulting $Tr(g)$ is the trace of some $g$ of order $q$, but explicit computation of $g$ is avoided. As shown in [13] the irreducibility test for $F(c, X) \in$

$GF(p^2)[X]$ can be done very fast, but, obviously, it requires additional code. We now present a method that requires hardly any additional code on top of Algorithm 2.3.7.

**Algorithm 3.2.2 (Computation of $Tr(g)$)**
1. Pick $c \in GF(p^2) \setminus GF(p)$ at random and compute $c_{p+1}$ using Algorithm 2.3.7.
2. If $c_{p+1} \in GF(p)$ then return to Step 1.
3. Compute $c_{(p^2-p+1)/q}$ using Algorithm 2.3.7.
4. If $c_{(p^2-p+1)/q} = 3$, then return to Step 1.
5. Let $Tr(g) = c_{(p^2-p+1)/q}$.

**Theorem 3.2.3** *Algorithm 3.2.2 computes an element of $GF(p^2)$ that equals $Tr(g)$ for some $g \in GF(p^6)$ of order $q$. It can be expected to require $3q/(q-1)$ applications of Algorithm 2.3.7 with $n = p + 1$ and $q/(q-1)$ applications with $n = (p^2 - p + 1)/q$.*

**Proof.** The correctness of Algorithm 3.2.2 follows from the fact that $F(c, X)$ is irreducible if $c_{p+1} \notin GF(p)$ (cf. Lemma 2.3.4.*iii*). The run time estimate follows from Lemma 3.2.1 and the fact that $c_{p+1} \notin GF(p)$ if $F(c, X)$ is irreducible (cf. Lemma 2.3.4.*iii*).

In [13] we present an even faster method to compute $Tr(g)$ if $p \not\equiv 8 \mod 9$.

## 3.3 Key Size

The XTR public key data contain two primes $p$ and $q$ as in 3.1 and the trace $Tr(g)$ of a generator of the XTR group (cf. 3.2). In principle the XTR public key data $p$, $q$, and $Tr(g)$ can be shared among any number of participants, just as in DSA (and EC-DSA) finite field (and curve), subgroup order, and subgroup generator may be shared. Apart from the part that may be shared, someone's XTR public key may also contain a public point $Tr(g^k)$ for an integer $k$ that is kept secret (the private key). Furthermore, for some applications the values $Tr(g^{k-1})$ and $Tr(g^{k+1})$ are required as well (cf. Section 4). In this section we discuss how much overhead is required for the representation of the XTR public key in a certificate, i.e., on top of the user ID and other certification related bits.

The part $(p, q, Tr(g))$ that may be shared causes overhead only if it is not shared. In that case, $(p, q, Tr(g))$ may be assumed to belong to a particular user or group of users in which case it is straightforward to determine $(p, q, Tr(g))$, during initialization, as a function of the user (or user group) ID and a small number of additional bits. For any reasonable choice of $P$ and $Q$ (cf. 3.1) the number of additional bits on top of the user ID, i.e., the overhead, can easily be limited to 48 (6 bytes) (cf. [13]), at the cost of a one time application of Algorithm 2.3.7 with $n = (p^2 - p + 1)/q$ by the recipient of the public key data.

We are not aware of a method to reduce the overhead caused by a user's public point $Tr(g^k) \in GF(p^2)$. Thus, representing $Tr(g^k)$ in a certificate requires representation of $2P$ bits. The two additional values $Tr(g^{k-1}), Tr(g^{k+1}) \in GF(p^2)$, however, can be represented using far fewer than $4P$ bits, at the cost of a very reasonable one time computation by the recipient of the public key.

This can be seen as follows. Since $\det(A(c)^k) = 1$, the equation from Lemma 2.4.6 leads to a third degree equation in $Tr(g^{k-1})$, given $Tr(g)$, $Tr(g^k)$, and $Tr(g^{k+1})$, by taking the determinants of the matrices involved. Thus, at the cost of a small number of $p^{\text{th}}$ powerings in $GF(p^2)$, $Tr(g^{k-1})$ can be determined based on $Tr(g)$, $Tr(g^k)$, and $Tr(g^{k+1})$ and two bits to indicate which of the roots equals $Tr(g^{k-1})$. In [13] we present, among others, a conceptually more complicated method to determine $Tr(g^{k-1})$ based on $Tr(g)$, $Tr(g^k)$, and $Tr(g^{k+1})$ that requires only a small constant number of operations in $GF(p)$, and a method to quickly determine $Tr(g^{k+1})$ given $Tr(g)$ and $Tr(g^k)$ that works if $p \not\equiv 8 \bmod 9$. Because this condition is not unduly restrictive we may assume that the two additional values $Tr(g^{k-1}), Tr(g^{k+1}) \in GF(p^2)$ do not have to be included in the XTR public key data, assuming the public key recipient is able and willing to carry out a fast one time computation given the XTR public key data $(p, q, Tr(g), Tr(g^k))$. If this computation if infeasible for the recipient, then $Tr(g^{k+1})$ must be included in the XTR public key data; computation of $Tr(g^{k-1})$ then takes only a small constant number of operations in $GF(p)$.

# 4   Cryptographic Applications

XTR can be used in any cryptosystem that relies on the (subgroup) discrete logarithm problem. In this section we describe some applications of XTR in more detail: Diffie-Hellman key agreement in 4.1, ElGamal encryption in 4.2, and Nyberg-Rueppel message recovery digital signatures in 4.3, and we compare XTR to RSA and ECC (cf. [15]).

## 4.1   XTR-DH

Suppose that Alice and Bob who both have access to the XTR public key data $p$, $q$, $Tr(g)$ want to agree on a shared secret key $K$. This can be done using the following XTR version of the Diffie-Hellman protocol:

1. Alice selects at random $a \in \mathbf{Z}$, $1 < a < q - 2$, uses Algorithm 2.3.7 to compute $S_a(Tr(g)) = (Tr(g^{a-1}), Tr(g^a), Tr(g^{a+1})) \in GF(p^2)^3$, and sends $Tr(g^a) \in GF(p^2)$ to Bob.
2. Bob receives $Tr(g^a)$ from Alice, selects at random $b \in \mathbf{Z}$, $1 < b < q - 2$, uses Algorithm 2.3.7 to compute $S_b(Tr(g)) = (Tr(g^{b-1}), Tr(g^b), Tr(g^{b+1})) \in GF(p^2)^3$, and sends $Tr(g^b) \in GF(p^2)$ to Alice.
3. Alice receives $Tr(g^b)$ from Bob, uses Algorithm 2.3.7 to compute $S_a(Tr(g^b)) = (Tr(g^{(a-1)b}), Tr(g^{ab}), Tr(g^{(a+1)b})) \in GF(p^2)^3$, and determines $K$ based on $Tr(g^{ab}) \in GF(p^2)$.
4. Bob uses Algorithm 2.3.7 to compute $S_b(Tr(g^a)) = (Tr(g^{a(b-1)}), Tr(g^{ab}), Tr(g^{a(b+1)})) \in GF(p^2)^3$, and determines $K$ based on $Tr(g^{ab}) \in GF(p^2)$.

The communication and computational overhead of XTR-DH are both about one third of traditional implementations of the Diffie-Hellman protocol that are based on subgroups of multiplicative groups of finite fields, and that achieve the same level of security (cf. Subsection 2.4).

## 4.2 XTR-ElGamal Encryption

Suppose that Alice is the owner of the XTR public key data $p$, $q$, $Tr(g)$, and that Alice has selected a secret integer $k$, computed $S_k(Tr(g))$, and made public the resulting value $Tr(g^k)$. Given Alice's XTR public key data $(p, q, Tr(g), Tr(g^k))$, Bob can encrypt a message $M$ intended for Alice using the following XTR version of the ElGamal encryption protocol:

1. Bob selects at random $b \in \mathbf{Z}$, $1 < b < q - 2$, and uses Algorithm 2.3.7 to compute $S_b(Tr(g)) = (Tr(g^{b-1}), Tr(g^b), Tr(g^{b+1})) \in \mathrm{GF}(p^2)^3$.
2. Bob uses Algorithm 2.3.7 to compute $S_b(Tr(g^k)) = (Tr(g^{(b-1)k}), Tr(g^{bk}), Tr(g^{(b+1)k})) \in \mathrm{GF}(p^2)^3$.
3. Bob determines a symmetric encryption key $K$ based on $Tr(g^{bk}) \in \mathrm{GF}(p^2)$.
4. Bob uses an agreed upon symmetric encryption method with key $K$ to encrypt $M$, resulting in the encryption $E$.
5. Bob sends $(Tr(g^b), E)$ to Alice.

Upon receipt of $(Tr(g^b), E)$, Alice decrypts the message in the following way:

1. Alice uses Algorithm 2.3.7 to compute $S_k(Tr(g^b)) = (Tr(g^{b(k-1)}), Tr(g^{bk}), Tr(g^{b(k+1)})) \in \mathrm{GF}(p^2)^3$.
2. Alice determines the symmetric encryption key $K$ based on $Tr(g^{bk}) \in \mathrm{GF}(p^2)$.
3. Alice uses the agreed upon symmetric encryption method with key $K$ to decrypt $E$, resulting in the encryption $M$.

The message $(Tr(g^b), E)$ sent by Bob consists of the actual encryption $E$, whose length strongly depends on the length of $M$, and the overhead $Tr(g^b) \in \mathrm{GF}(p^2)$, whose length is independent of the length of $M$. The communication and computational overhead of XTR-ElGamal encryption are both about one third of traditional implementations of the ElGamal encryption protocol that are based on subgroups of multiplicative groups of finite fields, and that achieve the same level of security (cf. Subsection 2.4).

**Remark 4.2.1** XTR-ElGamal encryption as described above is based on the common hybrid version of ElGamal's method, i.e., where the key $K$ is used in conjunction with an agreed upon symmetric key encryption method. In more traditional ElGamal encryption the message is restricted to the key space and 'encrypted' using, for instance, multiplication by the key, an invertible operation that takes place in the key space. In our description this would amount to requiring that $M \in \mathrm{GF}(p^2)$, and by computing $E$ as $K * M \in \mathrm{GF}(p^2)$. Compared to non-hybrid ElGamal encryption, XTR saves a factor three on the length of both parts of the encrypted message, for messages that fit in the key space (of one third of the 'traditional' size).

**Remark 4.2.2** As in other descriptions of ElGamal encryption it is implicitly assumed that the first component of an ElGamal encrypted message represents $Tr(g^b)$, i.e., the conjugates of a power of $g$. This should be explicitly verified in some situations, by checking that $Tr(g^b) \in \mathrm{GF}(p^2) \setminus \mathrm{GF}(p)$, that $Tr(g^b) \neq 3$, and by using Algorithm 2.3.7 to compute $S_q(Tr(g^b)) = (Tr(g^{b(q-1)}), Tr(g^{bq}), Tr(g^{b(q+1)}))$ and to verify that $Tr(g^{bq}) = 3$. This follows using methods similar to the ones presented in Section 3.

### 4.3   XTR-Nyberg-Rueppel Signatures

Let, as in 4.2, Alice's XTR public key data consist of $p$, $q$, $Tr(g)$, and $Tr(g^k)$. Furthermore, assume that $Tr(g^{k-1})$ and $Tr(g^{k+1})$ (and thus $S_k(Tr(g))$) are available to the verifier, either because they are part of the public key, or because they were reconstructed by the verifier (either from $(p, q, Tr(g), Tr(g^k), Tr(g^{k+1}))$ or from $(p, q, Tr(g), Tr(g^k))$). We describe the XTR version of the Nyberg-Rueppel (NR) message recovery signature scheme, but XTR can also be used in other 'ElGamal-like' signature schemes. To sign a message $M$ containing an agreed upon type of redundancy, Alice does the following:

1. Alice selects at random $a \in \mathbf{Z}$, $1 < a < q - 2$, and uses Algorithm 2.3.7 to compute $S_a(Tr(g)) = (Tr(g^{a-1}), Tr(g^a), Tr(g^{a+1})) \in \mathrm{GF}(p^2)^3$.
2. Alice determines a symmetric encryption key $K$ based on $Tr(g^a) \in \mathrm{GF}(p^2)$.
3. Alice uses an agreed upon symmetric encryption method with key $K$ to encrypt $M$, resulting in the encryption $E$.
4. Alice computes the (integer valued) hash $h$ of $E$.
5. Alice computes $s = (k * h + a) \bmod q \in \{0, 1, \ldots, q - 1\}$.
6. Alice's resulting signature on $M$ is $(E, s)$.

To verify Alice's signature $(E, s)$ and to recover the signed message $M$, the verifier Bob does the following.

1. Bob checks that $0 \le s < q$; if not failure.
2. Bob computes the hash $h$ of $E$.
3. Bob replaces $h$ by $-h \bmod q \in \{0, 1, \ldots, q - 1\}$.
4. Bob uses Algorithm 2.4.8 to compute $Tr(g^s * g^{hk})$ based on $Tr(g)$ and $S_k(Tr(g))$.
5. Bob uses $Tr(g^s * g^{hk})$ (which equals $Tr(g^a)$) to decrypt $E$ resulting in $M$.
6. The signature is accepted $\iff$ $M$ contains the agreed upon redundancy.

XTR-NR is considerably faster than traditional implementations of the NR scheme that are based on subgroups of multiplicative groups of finite fields of the same security level. The length of the signature is identical to other variants of the hybrid version of the NR scheme (cf. Remark 4.2.1): an overhead part of length depending on the desired security (i.e., the subgroup size) and a message part of length depending on the message itself and the agreed upon redundancy. Similar statements hold for other digital signature schemes, such as DSA.

### 4.4   Comparison to RSA and ECC

We compare XTR to RSA and ECC. For the RSA comparison we give the run times of 1020-bit RSA and 170-bit XTR obtained using generic software. For ECC we assume random curves over prime fields of about 170-bits with a curve subgroup of 170-bit order, and we compare the number of multiplications in $\mathrm{GF}(p)$ required for 170-bit ECC and 170-bit XTR applications. This 'theoretical' comparison is used because we do not have access to ECC software.

If part of the public key is shared (ECC or XTR only), XTR and ECC public keys consist of just the public point. For ECC its $y$-coordinate can be derived from the $x$-coordinate and a single bit. In the non-shared case, public keys may be ID-based or non-ID-based[1]. For ECC, the finite field, random curve, and group order take $\approx 595$ bits, plus a small number of bits for a point of high order. Using methods similar to the one alluded to in Subsection 3.3 this can be reduced to an overhead of, say, 48 bits (to generate curve and field based on the ID and 48 bits) plus 85 bits for the group order information. For XTR the sizes given in Table 1 follow from Subsection 3.3.   For both RSA and XTR 100 ran-

**Table 1.** RSA, XTR, ECC key sizes and RSA, XTR run times.

|  | shared keysize | ID-based keysize | non-ID-based keysize | key selection | encrypting (verifying) | decrypting (signing) |
|---|---|---|---|---|---|---|
| 1020-bit RSA | n/a | 510 bits | 1050 bits | 1224 ms | 5 ms | 40 (no CRT: 123) ms |
| 170-bit XTR | 340 | 388 bits | 680 bits | 73 ms | 23 ms | 11 ms |
| 170-bit ECC | 171 | 304 bits | 766 bits |  |  |  |

**Table 2.** 170-bit ECC, XTR comparison of number of multiplications in $GF(p)$.

|  | encrypting | decrypting | encryption overhead | signing | verifying | signature overhead | DH speed | DH size |
|---|---|---|---|---|---|---|---|---|
| ECC | 3400 | 1921 (1700) | 171 (340) bits | 1700 | 2575 | 170 bits | 3842 (3400) | 171 (340) bits |
| XTR | 2720 | 1360 | 340 bits | 1360 | 2754 | 170 bits | 2720 | 340 bits |

dom keys were generated. (ECC parameter generation is much slower and more complicated than for either RSA or XTR and not included in Table 1.) For RSA we used random 32-bit odd public exponents and 1020-bit moduli picked by randomly selecting 510-bit odd numbers and adding 2 until they are prime. For XTR we used Algorithm 3.1.2 with $Q = 170$ and $P \geq 170$ and the fast $Tr(g)$ initialization method mentioned at the end of Subsection 3.2. For each RSA key 10 encryptions and decryptions of random 1020-bit messages were carried out, the latter with Chinese remaindering (CRT) and without (in parentheses in Table 1). For each XTR key 10 single and double exponentiations (i.e., applications of Algorithms 2.3.7 and 2.4.8, respectively) were carried out for random exponents $< q$. For RSA encryption and decryption correspond to signature verification and generation, respectively. For XTR single exponentiation corresponds to decryption and signature generation, and double exponentiation corresponds to signature verification and, approximately, encryption. The average run times are in milliseconds on a 450 MHz Pentium II NT workstation. The ECC figures in Table 2 are based on the results from [4]; speed-ups that may be obtained at the cost of specifying the full $y$-coordinates are given between parentheses. The time or number of operations to reconstruct the full public keys from their compressed versions (for either system) is not included.

---

[1] ID based key generation for RSA affects the way the secret factors are determined. The ID based approach for RSA is therefore viewed with suspicion and not generally used. A method from [23], for instance, has been broken, but no attack against the methods from [12] is known. For discrete logarithm based methods (such as ECC and XTR) ID-based key generation affects only the part of the public key that is not related to the secret information, and is therefore not uncommon for such systems.

## 5   Security

### 5.1   Discrete Logarithms in GF($p^t$)

Let $\langle\gamma\rangle$ be a multiplicative group of order $\omega$. The security of the Diffie-Hellman protocol in $\langle\gamma\rangle$ relies on the *Diffie-Hellman* (DH) problem of computing $\gamma^{xy}$ given $\gamma^x$ and $\gamma^y$. We write $DH(\gamma^x, \gamma^y) = \gamma^{xy}$. Two other problems are related to the DH problem. The first one is the *Diffie-Hellman Decision* (DHD) problem: given $a, b, c \in \langle\gamma\rangle$ determine whether $c = DH(a, b)$. The DH problem is at least as difficult as the DHD problem. The second one is the *Discrete Logarithm* (DL) problem: given $a = \gamma^x \in \langle\gamma\rangle$ with $0 \le x < \omega$, find $x = DL(a)$. The DL problem is at least as difficult as the DH problem. It is widely assumed that if the DL problem in $\langle\gamma\rangle$ is intractable, then so are the other two. Given the factorization of $\omega$, the DL problem in $\langle\gamma\rangle$ can be reduced to the DL problem in all prime order subgroups of $\langle\gamma\rangle$, due to the Pohlig-Hellman algorithm [17]. Thus, for the DL problem we may assume that $\omega$ is prime.

Let $p$, $q$, $Tr(g)$ be (part of) an XTR public key. Below we prove that the security of the XTR versions of the DL, DHD, and DH problem is equivalent to the DL, DHD, and DH problem, respectively, in the XTR group (cf. Subsection 3.2). First, however, we focus on the DL problem in a subgroup $\langle\gamma\rangle$ of prime order $\omega$ of the multiplicative group GF($p^t$)* of an extension field GF($p^t$) of GF($p$) for a fixed $t$. There are two approaches to this problem (cf. [1], [5], [9], [11], [16], [19], [21]): one can either attack the multiplicative group or one can attack the subgroup. For the first attack the best known method is the Discrete Logarithm variant of the Number Field Sieve. If $s$ is the smallest divisor of $t$ such that $\langle\gamma\rangle$ can be embedded in the subgroup GF($p^s$)* of GF($p^t$)*, then the heuristic expected asymptotic run time for this attack is $L[p^s, 1/3, 1.923]$, where $L[n, v, u] = \exp((u + o(1))(\ln(n))^v(\ln(\ln(n)))^{1-v})$. If $p$ is small, e.g. $p = 2$, then the constant 1.923 can be replaced by 1.53. Alternatively, one can use one of several methods that take $O(\sqrt{\omega})$ operations in $\langle\gamma\rangle$, such as Pollard's Birthday Paradox based rho method (cf. [18]).

This implies that the difficulty of the DL problem in $\langle\gamma\rangle$ depends on the size of the minimal surrounding subfield of $\langle\gamma\rangle$ and on the size of its prime order $\omega$. If GF($p^t$) itself is the minimal surrounding subfield of $\langle\gamma\rangle$ and $\omega$ is sufficiently large, then the DL problem in $\langle\gamma\rangle$ is as hard as the general DL problem in GF($p^t$). If $p$ is not small the latter problem is believed to be as hard as the DL problem with respect to a generator of prime order $\approx \omega$ in the multiplicative group of a prime field of cardinality $\approx p^t$ (cf. [6], [20]). The DL problem in that setting is generally considered to be harder than factoring $t * \log_2(p)$-bit RSA moduli.

The XTR parameters are chosen in such away that the minimal surrounding field of the XTR group is equal to GF($p^6$) (cf. Section 1), such that $p$ is not small, and such that $q$ is sufficiently large. It follows that, if the complexity of the DL problem in the XTR group is less than the complexity of the DL problem in GF($p^6$), then the latter problem is at most as hard as the DL problem in GF($p^3$), GF($p^2$), or GF($p$), i.e., the DL problem in GF($p^6$) collapses to its true subfields. This contradicts the above mentioned assumption about the complexity of computing discrete logarithms in GF($p^t$). It follows that the DL problem in the XTR group may be assumed to be as hard as the DL problem in $GF(p^6)$, i.e., of complexity $L[p^6, 1/3, 1.923]$. Thus, with respect to known attacks, the DL problem in the XTR group is generally considered to be more difficult than factoring a $6 * \log_2(p)$-bit RSA modulus, provided the prime order $q$ is sufficiently large. By comparing the computational effort required for both algorithms mentioned above, it turns out that if $p$ and $q$ each are about 170 bits long, then the DL problem in the XTR group is harder than factoring an RSA modulus of $6 * 170 = 1020$ bits.

## 5.2  Security of XTR

Discrete logarithm based cryptographic protocols can use many different types of subgroups, such as multiplicative groups of finite fields, subgroups thereof (such as the XTR group), or groups of points of elliptic curves over finite fields. As shown in Section 4 the XTR versions of these protocols follow by replacing elements of the XTR group by their traces. This implies that the security of those XTR versions is no longer based on the original DH, DHD, or DL problems but on the XTR versions of those problems. We define the *XTR-DH* problem as the problem of computing $Tr(g^{xy})$ given $Tr(g^x)$ and $Tr(g^y)$, and we write $XDH(g^x, g^y) = g^{xy}$. The *XTR-DHD* problem is the problem of determining whether $XDH(a, b) = c$ for $a, b, c \in Tr(\langle g \rangle)$. Given $a \in Tr(\langle g \rangle)$, the *XTR-DL* problem is to find $x = XDL(a)$, i.e., $0 \le x < q$ such that $a = Tr(g^x)$. Note that if $x = DL(a)$, then so are $x * p^2 \bmod q$ and $x * p^4 \bmod q$.

We say that problem $\mathcal{A}$ is $(a, b)$-*equivalent* to problem $\mathcal{B}$, if any instance of problem $\mathcal{A}$ (or $\mathcal{B}$) can be solved by at most $a$ (or $b$) calls to an algorithm solving problem $\mathcal{B}$ (or $\mathcal{A}$).

**Theorem 5.2.1** *The following equivalences hold:*
*i. The XTR-DL problem is $(1, 1)$-equivalent to the DL problem in $\langle g \rangle$.*
*ii. The XTR-DH problem is $(1, 2)$ equivalent to the DH problem in $\langle g \rangle$.*
*iii. The XTR-DHD problem is $(3, 2)$-equivalent to the DHD problem in $\langle g \rangle$.*

**Proof.** For $a \in GF(p^2)$ let $r(a)$ denote a root of $F(a, X)$.

To compute $DL(y)$, let $x = XDL(Tr(y))$, then $DL(y) = x * p^{2j} \bmod q$ for either $j = 0$, $j = 1$, or $j = 2$. Conversely, $XDL(a) = DL(r(a))$. This proves *i*.

To compute $DH(x, y)$, compute $d_i = XDH(Tr(x * g^i), Tr(y))$ for $i = 0, 1$, then $r(d_i) \in \{(DH(x, y) * y^i)^{p^{2j}} : j = 0, 1, 2\}$, from which $DH(x, y)$ follows. Conversely, $XDH(a, b) = Tr(DH(r(a), r(b)))$. This proves *ii*.

To prove *iii*, it easily follows that $DH(x, y) = z$ if and only if $XDH(Tr(x), Tr(y)) = Tr(z)$ and $XDH(Tr(x*g), Tr(y)) = Tr(z*y)$. Conversely, $XDH(a, b) = c$ if and only if $DH(r(a), r(b)) = r(c)^{p^{2j}}$ for either $j = 0$, $j = 1$, or $j = 2$. This proves *iii* and completes the proof of Theorem 5.2.1.

**Remark 5.2.2** It follows from the arguments in the proof of Theorem 5.2.1 that an algorithm solving either DL, DH, or DHD with non-negligible probability can be transformed in an algorithm solving the corresponding XTR problem with non-negligible probability, and vice versa.

It follows from the arguments in the proof of Theorem 5.2.1.*ii* that in many practical situations a single call to an XTR-DH solving algorithm would suffice to solve a DL problem. As an example we mention DH key agreement where the resulting key is actually used after it has been established.

**Remark 5.2.3** Theorem 5.2.1.*ii* states that determining the (small) XTR-DH key is as hard as determining the whole DH key in the representation group $\langle g \rangle$. From the results in [24] it actually follows that determining the image of the XTR-DH key under any non-trivial GF($p$)-linear function is also as hard as the whole DH key. This means that, for example, finding the $\alpha$ or the $\alpha^2$ coefficient of the XTR-DH key is as hard as finding the whole DH key, implying that cryptographic applications may be based on just one of the coefficients.

# 6    Extensions

The methods and techniques described in this paper can be extended in various straightforward ways to the situation where the underlying field GF($p$) is itself an extension field, say of the form GF($p^e$) for some integer $e$. The resulting field will then be of the form GF($p^{6e}$) instead of GF($p^6$). The parameters $p$, $q$, and $e$ should be generated so that

- $q$ is a prime dividing the $6e^{\text{th}}$ cyclotomic polynomial $\phi_{6e}(X)$ evaluated in $p$ (cf. [11]).
- $\log_2(q)$ and $6e * \log_2(p)$ are sufficiently large, e.g. $\log_2(q) \geq 160$ and $6e * \log_2(p) > 1000$.

By doing so, the parameter $p$ can be chosen smaller to achieve the same security. Note that for large choices of $e$ fewer suitable primes are available, while the savings obtained, if any, depend strongly on the choice that is made. In particular the choice $p = 2$ is an option, which has the property (cf. [24]) that bits of the XTR-DH exchanged key are as hard as the whole key. However, for such very small $p$ one should take into account that they make computation of discrete logarithms easier (cf. [5]), and that $6e * \log_2(p)$ should be at least 1740 to get security equivalent to 1024-bit RSA moduli. As an example, $\phi_{6*299}(2)$ is divisible by a 91-digit prime.

Because $\phi_{6e}(X)$ divides $X^{2e} - X^e + 1$, one may replace $p$ by $p^e$ in many expressions above, since conditions that hold modulo $p^2 - p + 1$ still hold if $p$ and $p^2 - p + 1$ are replaced by $p^e$ and $p^{2e} - p^e + 1$ respectively. The (mostly straightforward) details of these and other generalizations are left to the reader.

**Acknowledgment**

We are greatly indebted to Mike Wiener for his permission to include his improvements of our earlier versions of Algorithms 2.3.7 and 2.4.8.

# References

1. L.M. Adleman, J. DeMarrais, *A subexponential algorithm for discrete logarithms over all finite fields*, Proceedings Crypto'93, LNCS 773, Springer-Verlag 1994, 147-158.
2. D.V. Bailey, C. Paar, *Optimal extension fields for fast arithmetic in public-key algorithms*, Proceedings Crypto'98, LNCS 1462, Springer-Verlag 1998, 472-485.
3. A.E. Brouwer, R. Pellikaan, E.R. Verheul, *Doing more with fewer bits*, Proceedings Asiacrypt99, LNCS 1716, Springer-Verlag 1999, 321-332.
4. H. Cohen, A. Miyaji, T. Ono, *Efficient elliptic curve exponentiation using mixed coordinates*, Proceedings Asiacrypt'98, LNCS 1514, Springer-Verlag 1998, 51-65.
5. D. Coppersmith, *Fast evaluation of logarithms in fields of characteristic two*, IEEE Trans. Inform. Theory 30 (1984), 587-594.
6. D. Coppersmith, personal communication, March 2000.
7. T. ElGamal, *A Public Key Cryptosystem and a Signature scheme Based on Discrete Logarithms*, IEEE Transactions on Information Theory 31(4), 1985, 469-472.
8. P. Gaudry, F. Hess, N.P. Smart, *Constructive and destructive facets of Weil descent on elliptic curves*, manuscript, January, 2000, submitted to Journal of Cryptology.
9. D. Gordon, *Discrete logarithms in GF(p) using the number field sieve*, SIAM J. Discrete Math. 6 (1993), 312-323.
10. D.E. Knuth, *The art of computer programming, Volume 2, Seminumerical Algorithms*, second edition, Addison-Wesley, 1981.
11. A.K. Lenstra, *Using cyclotomic polynomials to construct efficient discrete logarithm cryptosystems over finite fields*, Proceedings ACISP97, LNCS 1270, Springer-Verlag 1997, 127-138.
12. A.K. Lenstra, *Generating RSA moduli with a predetermined portion*, Proceedings Asiacrypt '98, LNCS 1514, Springer-Verlag 1998, 1-10.
13. A.K. Lenstra, E.R. Verheul, *Key improvements to XTR*, in preparation.
14. A.J. Menezes, *Comparing the security of ECC and RSA*, manuscript, January, 2000, available as www.cacr.math.uwaterloo.ca/ ajmeneze/misc/cryptogram-article.html.
15. A.J. Menezes, P.C. van Oorschot, S.A. Vanstone, *Handbook of applied cryptography*, CRC Press, 1997.
16. A.M. Odlyzko, *Discrete Logarithms: The past and the future*, Designs, Codes and Cryptography, 19 (2000), 129-145.
17. S.C. Pohlig, M.E. Hellman, *An improved algorithm for computing logarithms over GF(p) and its cryptographic significance*, IEEE Trans. on IT, 24 (1978), 106-110.
18. J.M. Pollard, *Monte Carlo methods for index computation (mod p)*, Math. Comp., 32 (1978), 918-924.
19. O. Schirokauer, *Discrete logarithms and local units*, Phil. Trans. R. Soc. Lond. A 345, 1993, 409-423.
20. O. Schirokauer, personal communication, March 2000.
21. O. Schirokauer, D. Weber, Th.F. Denny, *Discrete logarithms: the effectiveness of the index calculus method*, Proceedings ANTS II, LNCS 1122 Springer-Verlag 1996.
22. C.P. Schnorr, *Efficient signature generation by smart cards*, Journal of Cryptology, 4 (1991), 161-174.
23. S.A. Vanstone, R.J. Zuccherato, *Short RSA keys and their generation*, Journal of Cryptology, 8 (1995), 101-114.
24. E. Verheul, *Certificates of recoverability with scalable recovery agent security*, Proceedings of PKC 2000, LNCS 1751, Springer-Verlag 2000, 258-275.

# A Chosen-Ciphertext Attack against NTRU

Éliane Jaulmes[1] and Antoine Joux[2]

[1] SCSSI, 18 rue du Docteur Zamenhof
F-92131 Issy-les-Moulineaux cedex, France
eliane.jaulmes@wanadoo.fr
[2] SCSSI, 18 rue du Docteur Zamenhof
F-92131 Issy-les-Moulineaux cedex, France
Antoine.Joux@ens.fr

**Abstract.** We present a chosen-ciphertext attack against the public key cryptosystem called NTRU. This cryptosystem is based on polynomial algebra. Its security comes from the interaction of the polynomial mixing system with the independence of reduction modulo two relatively prime integers $p$ and $q$. In this paper, we examine the effect of feeding special polynomials built from the public key to the decryption algorithm. We are then able to conduct a chosen-ciphertext attack that recovers the secret key from a few ciphertexts/cleartexts pairs with good probability. Finally, we show that the OAEP-like padding proposed for use with NTRU does not protect against this attack.

## 1 Overview

In [7], Hoffstein, Pipher and Silverman have presented a public key cryptosystem based on polynomial algebra called NTRU. The security of NTRU comes from the interaction of the polynomial mixing system with the independence of reduction modulo $p$ and $q$. In [7], the authors have studied different possible attacks on their cryptosystem.

First the brute force attack, which can be eased by the meet-in-the-middle principle, may be used against the private key or against a single message. However, for a suitable choice of parameters this attack will not succeed in a reasonable time.

Then there is a multiple transmission attack, which will provide the content of a message that has been transmitted several time. Thus multiple transmissions are not advised. It is also one of the reasons why NTRU recommends a preprocessing scheme.

Finally, several attacks make use of the LLL algorithm of Lenstra-Lenstra-Lovász [10] which produces a reduced basis for a given lattice. They can either recover the secret key from the public key or decipher one given message. However the authors of NTRU claim that the time required is exponential in the degree of the polynomials. For most lattices, it is indeed very difficult to find extremely short vectors. Thus for suitably large degrees, this attack is expected to fail and does fail in practice. Another idea, described by Coppersmith and

M. Bellare (Ed.): CRYPTO 2000, LNCS 1880, pp. 20–35, 2000.
© Springer-Verlag Berlin Heidelberg 2000

Shamir in [3] would be to use LLL to find some short vector in the lattice which could act as a decryption key, but the authors of NTRU claim that experimental evidence suggests that the existence of such spurious keys does not pose a security threat.

However, we show now that it is possible to break the system using a chosen-ciphertext attack. Such attacks have already been used for example in [9] and [5]. They work as follows: The attacker constructs invalid cipher messages. If he can know the plaintexts corresponding to his messages, he can recover some information about the decryption key or even retrieve the private key. In [5], the authors point out that finding the plaintext corresponding to a given ciphertext can reasonably be achieved. This possibility is even increased if decryption is done on a smart card. The standard defense against such attacks is to require redundancy in the message and this is why there exists a padded version of NTRU. The chosen-ciphertext attack we present here has a good probability of recovering the private key from one or two well chosen ciphertexts on the unpadded version of NTRU. It is also able to recover the key on the padded version from a reasonable number of chosen ciphertexts.

This paper is organized as follows: we first recall the main ideas of the cryptosystem without preprocessing, then we present our chosen-ciphertext attack on the unpadded version and give an example of this attack. Finally we study the case where the OAEP-like padding is used and explain how our attack can still recover the private key in this situation.

## 2  Description of the Cryptosystem

### 2.1  Notations

The NTRU cryptosystem depends on three integers parameters $(N, p, q)$ and four sets of polynomials of degree $(N - 1)$ with integer coefficients, called $\mathcal{L}_f$, $\mathcal{L}_g$, $\mathcal{L}_\phi$, $\mathcal{L}_m$.

The parameters $p$ and $q$ are chosen with $\gcd(p, q) = 1$ and $q$ is much larger than $p$. All polynomials are in the ring

$$R = \mathbb{Z}[X]/(X^N - 1).$$

We write $\circledast$ to denote multiplication in $R$. In the system, some multiplications will be performed modulo $q$ and some modulo $p$.

The sets $\mathcal{L}_f, \mathcal{L}_g, \mathcal{L}_\phi$ and $\mathcal{L}_m$ are chosen as follows. The space of messages $\mathcal{L}_m$ consists of all polynomials modulo $p$. Assuming $p$ is odd, it is most convenient to take

$$\mathcal{L}_m = \left\{ m \in R : \begin{array}{c} m \text{ has coefficients lying between} \\ -\tfrac{1}{2}(p-1) \text{ and } \tfrac{1}{2}(p-1) \end{array} \right\}.$$

To describe the other samples spaces, we will use sets of the form

$$\mathcal{L}(d_1, d_2) = \left\{ F \in R : \begin{array}{c} F \text{ has } d_1 \text{ coefficients equal to } 1 \\ d_2 \text{ coefficients equal to } -1, \text{ the rest } 0 \end{array} \right\}.$$

With this notation, we choose three positive integers $d_f$, $d_g$, $d$ and set

$$\mathcal{L}_f = \mathcal{L}(d_f, d_f - 1), \quad \mathcal{L}_g = \mathcal{L}(d_g, d_g), \quad \text{and} \quad \mathcal{L}_\phi = \mathcal{L}(d, d).$$

We take $\mathcal{L}_f = \mathcal{L}(d_f, d_f - 1)$ instead of $\mathcal{L}(d_f, d_f)$ because we want $f$ to be invertible and a polynomial satisfying $f(1) = 0$ can never be invertible.

## 2.2   The Key Generation

To create an NTRU key, one chooses two polynomials $f \in \mathcal{L}_f$ and $g \in \mathcal{L}_g$. The polynomial $f$ must have inverses modulo $p$ and $q$. We will denote these inverses by $F_p$ and $F_q$. So we have:

$$F_p \circledast f \equiv 1 \pmod{p} \quad \text{and} \quad F_q \circledast f \equiv 1 \pmod{q}.$$

The public key is then the polynomial:

$$h \equiv F_q \circledast g \pmod{q}.$$

Of course, the parameters $N$, $p$, $q$ are public too.

The private key is the polynomial $f$, together with $F_p$.

## 2.3   Encryption and Decryption Procedure

**Encryption.** The encryption works as follows. First, we select a message $m$ from the set of plaintexts $\mathcal{L}_m$. Next we choose randomly a polynomial $\phi \in \mathcal{L}_\phi$ and use the public key to compute:

$$e \equiv p\phi \circledast h + m \pmod{q}.$$

$e$ is our encrypted message.

**Decryption.** We have received an encrypted message $e$ and we want to decrypt it using our private key $f$. To do this, we should have precomputed the polynomial $F_p$ as described in 2.2. In order to decrypt $e$, we compute :

$$a \equiv f \circledast e \pmod{q},$$

where we choose the coefficients of $a$ in the interval from $-q/2$ to $q/2$. Now, treating $a$ as a polynomial with integer coefficients, we recover the message by computing:

$$F_p \circledast a \pmod{p}.$$

**How Decryption Works.** The polynomial $a$ verifies

$$a \equiv f \circledast e \equiv f \circledast p\phi \circledast h + f \circledast m \pmod{q}$$
$$= f \circledast p\phi \circledast F_q \circledast g + f \circledast m \pmod{q}$$
$$= p\phi \circledast g + f \circledast m \pmod{q}.$$

For appropriate parameter choices, we can ensure that all coefficients of the polynomial $p\phi \circledast g + f \circledast m$ lie between $-q/2$ and $q/2$. So the intermediate value $p\phi \circledast g + f \circledast m \bmod q$ is in fact the true (non modular) value of this polynomial. This means that when we compute $a$ and reduce its coefficients into this interval, we recover exactly the polynomial $p\phi \circledast g + f \circledast m$. Hence its reduction modulo $p$ give us $f \circledast m \bmod p$ and the multiplication by $F_p$ retrieves the message $m$.

The basic idea for the attack presented here will be to construct intermediate polynomials such that the modular values differ from the true values.

### 2.4   Sets of Parameters for NTRU

The authors of NTRU have defined different sets of parameters for NTRU providing various security levels. Theses parameters are given in [12].

| Name | N | p | q | $\mathcal{L}_f$ | $\mathcal{L}_g$ | $\mathcal{L}_\phi$ |
|------|-----|---|-----|----------------|-----------------|--------------------|
| Case A | 107 | 3 | 64 | $\mathcal{L}(15,14)$ | $\mathcal{L}(12,12)$ | $\mathcal{L}(5,5)$ |
| Case B | 167 | 3 | 128 | $\mathcal{L}(61,60)$ | $\mathcal{L}(20,20)$ | $\mathcal{L}(18,18)$ |
| Case C | 263 | 3 | 128 | $\mathcal{L}(50,49)$ | $\mathcal{L}(24,24)$ | $\mathcal{L}(16,16)$ |
| Case D | 503 | 3 | 256 | $\mathcal{L}(216,215)$ | $\mathcal{L}(72,72)$ | $\mathcal{L}(55,55)$ |

In the original formulation of the NTRU public key cryptosystem [7], it was suggested that one could use $N = 107$ to create a cryptosystem with moderate security. Such a system can be broken by lattice attacks in a few hours. Thus the use of case A is not recommended anymore but we will still use it to describe our attack in its simple version.

## 3   The Chosen-Ciphertext Attack

### 3.1   Principle

As stated in 2.3, we want to build cipher texts such that the intermediate values in the deciphering process will differ from the true values. We first consider the effect of deciphering a cipher text of the form $ch + c$, where $c$ is an integer and $h$ is the public key. The decryption algorithm first multiplies by $f$ modulo $q$:

$$a \equiv f \circledast ch + cf \pmod{q}$$
$$\equiv cg + cf \pmod{q},$$

where $g$ and $f$ both have coefficients equal to 0, 1 or $-1$. Hence the polynomial $cf + cg$ have coefficients equal to 0, $c$, $-c$, $2c$ or $-2c$. We then need to reduce the coefficients of $a$ between $-q/2$ and $q/2$. If $c$ has been chosen such that $c < q/2$ and $2c > q/2$, we will have to reduce only the coefficients equal to $2c$ or $-2c$.

If we now suppose that a single coefficient in $a$ is $\pm 2c$, say $a_i = +2c$, then the value of $a \bmod q$ is $cg + cf - qx^i$. The deciphering process outputs

$$cg \circledast F_p + c - qx^i \circledast F_p \quad (\bmod \ p)$$

If $c$ has been chosen as a multiple of $p$, then the output is

$$-qx^i \circledast F_p \quad (\bmod \ p).$$

Since $\gcd(p, q) = 1$, we can recover $x^i \circledast F_p \equiv x^i/f \bmod p$ and compute its inverse $f/x^i \bmod p$. Since all the coefficients of $f$ are 1 or $-1$, it is the true value of the polynomial. We can then compute

$$g/x^i = h \circledast f/x^i \quad (\bmod \ q),$$

which is also the true value of $g/x^i$. Going back to the key process described in section 2.2, we can see that $(f, g)$ and $(f/x^i, g/x^i)$ are equivalent keys.

Of course, in general, the polynomial $cf + cg$ may have none or several coefficients equal to $\pm 2c$, and then the above attack does not work anymore. In the next section, we will analyze the attack and generalize it to make it work for all the security parameters proposed for NTRU in [7].

### 3.2   Analysis of the Attack

We say that two polynomials $P_1$ and $P_2$ have a *collision* when they have the same non zero coefficient at the same degree.

We now define the *intersection* polynomial $k$ of $(P_1, P_2)$ by:

$$k = \sum k_i x^i,$$

where

$$k_i = \begin{cases} 1 \text{ if } P_1 \text{ and } P_2 \text{ both have their } i^{\text{th}} \text{ coefficient equal to 1} \\ -1 \text{ if } P_1 \text{ and } P_2 \text{ both have their } i^{\text{th}} \text{ coefficient equal to -1} \\ 0 \text{ otherwise} \end{cases}$$

Using this notation, we write again the result of the first decryption step of $c + ch$, as seen in section 3.1. $a \equiv cg + cf \bmod q = c + ch - qk$

The decrypted message obtained is then

$$m \equiv cF_p \circledast f + cF_p \circledast g - qF_p \circledast k \quad (\bmod \ p)$$
$$\equiv c + ch - qF_p \circledast k \quad (\bmod \ p)$$

Since $c$ has been chosen such that $c \equiv 0 \bmod p$,

$$m = -qF_p \circledast k \quad (\bmod \ p).$$

The private key $f$ can then be obtained from $f \equiv -qk \circledast m^{-1} \mod p$

When $f$ and $g$ have few common coefficients, the polynomial $k$ has only a few non zero coefficients. By testing different values for $k$, we can compute possible polynomials $f$. The private key is likely the one that satisfies the condition $f \in \mathcal{L}_f$. It is then a simple matter to verify our guess by trying to decrypt a message with $f$ or by computing $h \circledast f \mod q = g'$. Then if $g' = \pm x^i \circledast g$, we know we have a correct key.

Let us study the probability of success of our attack over the sets of parameters given in section 2.4.

The probability of $f$ and $g$ having one and only one collision is the following:

$$p = p_1 + p_{-1},$$

where $p_1$, the probability of collision of two 1, is:

$$\sum_{k=0}^{min(d_f-1,d_g)} \frac{\binom{d_g}{1}\binom{d_g}{k}\binom{N-2d_g}{d_f-1-k}\binom{N-d_f-d_g+k}{d_f-1}}{\binom{N}{d_f}\binom{N-d_f+1}{d_f-1}}$$

and $p_{-1}$, the probability of collision of two $-1$, is:

$$\sum_{k=0}^{min(d_f-2,d_g)} \frac{\binom{d_g}{1}\binom{d_g}{k}\binom{N-2d_g}{d_f-2-k}\binom{N-d_f+1-d_g+k}{d_f}}{\binom{N}{d_f}\binom{N-d_f+1}{d_f-1}}$$

There are similar formulas for more collisions. However, they are quickly cumbersome to compute.

Another approach is to evaluate the expected number of collisions between $f$ and $g$. An heuristic approximation of this number is

$$\frac{(2d_f-1)d_g}{N}.$$

In case A, we find an average number of collisions of 3.25. We can thus expect $k$ to have around three non zero coefficients.

The table below shows the different probabilities of collisions in the different proposed cases. It also gives the average expected number of collisions.

|                              | Case A | Case B          | Case C          | Case D           |
|------------------------------|--------|-----------------|-----------------|------------------|
| Average number of collisions | 3.25   | 14.5            | 9.03            | 61.7             |
| Probability of 0 collision   | 0.026  | $9.3 * 10^{-9}$ | $3.1 * 10^{-5}$ | $2 * 10^{-36}$   |
| Probability of 1 collision   | 0.13   | $5.8 * 10^{-7}$ | $5 * 10^{-4}$   | $1.1 * 10^{-33}$ |
| Probability of 2 collisions  | 0.25   | $9.5 * 10^{-6}$ | $3 * 10^{-3}$   | $1.5 * 10^{-31}$ |
| Probability of 3 collisions  | 0.28   | $8.6 * 10^{-5}$ | 0.011           | $1.2 * 10^{-29}$ |
| Probability of 4 collisions  | 0.22   | $5.1 * 10^{-4}$ | 0.028           | $7.3 * 10^{-28}$ |

For example, with the parameters of NTRU 107, which has a key security of $2^{50}$ against a meet-in-the-middle attack, we have a one-collision probability of $p = 0.13$. It means one over ten cipher messages will produce a polynomial $k$ with a single non zero coefficient and the simple case described in section 3.1 will apply. We can see that the attack, as it has currently been described, will fail in cases B, C and D. In section 3.3, we generalize our idea to make it work in those cases.

In general, $k$ may have more than one coefficient, and we need to enumerate the possible $k$ and compute $f' = k/m \bmod p$, where $m$ is our decrypted message. When $f' \in \mathcal{L}_f$, we have found a likely polynomial. We just need to verify that $f'$ is able to decrypt messages. If we now analyze the number of possible polynomials $k$ we need to test in order to recover the private key, we can first note that the polynomials of the form $x^i f \bmod x^N - 1$ have as many coefficients equal to 1 and $-1$ as $f$. As the multiplication by $x^i$ will not change the value of the coefficients of $a$ and as the decryption proceeding consists in multiplying and dividing by $f$, the rotated key $f' = x^i f \bmod x^N - 1$ can be used to decrypt any message encrypted with $f$. Hence we can assume $k(0) \neq 0$.

So if we assume that $k$ has $n$ non zero coefficients, we will have to try $2^n \binom{N-1}{n-1}$ different values for $k$.

We can see in the table below the approximate number of polynomials we need to test function of the expected number of collisions.

| Expected no of collisions | Case A   | Case B   | Case C   | Case D   |
|---------------------------|----------|----------|----------|----------|
| 1 collision               | 2        | 2        | 2        | 2        |
| 2 collisions              | $2^9$    | $2^{10}$ | $2^{10}$ | $2^{11}$ |
| 3 collisions              | $2^{16}$ | $2^{17}$ | $2^{18}$ | $2^{20}$ |
| 4 collisions              | $2^{22}$ | $2^{24}$ | $2^{26}$ | $2^{29}$ |

The message $c + ch$ can fail to produce the private key, if $f$ and $g$ have too many collisions. We can then try again with $cx + ch$ and more generally with polynomials of the form $cx^i + ch$. This means considering collisions between $g$ and $x^i f \bmod x^N - 1$. So there is a compromise between the number of possible collisions we will test and the number of cipher texts we will need. Many ciphertexts are likely to produce at least a polynomial whose number of non zero coefficient is below the average value. If we have only one ciphertext, it may take more time to test possible polynomials before finding the key.

### 3.3    Extending to Higher Security Parameters

As seen in section 3.2, the parameters proposed in [7] for higher security give us a very high number of collisions. This means that there will be an extremely low probability of having only a few collisions. Therefore, we can no longer use messages of the form $cx^i + ch$. Instead, we reduce the average number of collisions by testing messages of the form

$$chx^{i_1} + \cdots + chx^{i_n} + cx^{j_1} + cx^{j_2} + \cdots + cx^{j_m},$$

where $c$ is a multiple of $p$ that verifies

$$(n + m - 1)c < q/2 \quad \text{and} \quad (n + m)c > q/2.$$

We choose the numbers $n$ and $m$ in order to get a good probability of having only one or two collisions. As before, we do not explicitly compute these probabilities, but we estimate the average number of collisions. When this number is near 1, it means that the $n$ and $m$ are correctly chosen. An heuristic approximation of the number of collisions is given by:

$$\frac{2d_f^m d_g^n}{N^{n+m-1}}$$

## 4    Example

### 4.1    Detailed Example of Case D

In [7], it is claimed that the highest security level will be obtained with the set of parameters D.

We now give an example that shows, with this set of parameters, that our attack can recover the secret key.

Here is the private key $(f, g)$ we have used:

$$
\begin{aligned}
f = {}&-x^{502} + x^{501} + x^{500} - x^{499} - x^{498} + x^{497} - x^{496} - x^{495} - x^{494} - x^{493} - x^{492} \\
&-x^{491} + x^{490} - x^{488} + x^{487} - x^{486} - x^{485} - x^{482} + x^{481} - x^{480} - x^{479} + x^{477} \\
&+x^{475} + x^{474} + x^{472} - x^{471} + x^{470} + x^{468} - x^{467} + x^{466} + x^{464} - x^{463} + x^{462} \\
&+x^{461} - x^{460} - x^{459} + x^{458} + x^{457} - x^{455} - x^{454} - x^{453} - x^{451} - x^{450} + x^{449} \\
&+x^{448} + x^{447} + x^{446} + x^{445} - x^{444} - x^{443} + x^{442} + x^{441} + x^{440} - x^{439} + x^{438} \\
&-x^{437} - x^{436} + x^{435} + x^{434} + x^{433} - x^{430} - x^{429} + x^{428} - x^{425} + x^{424} + x^{423} \\
&-x^{422} - x^{421} - x^{420} - x^{418} - x^{417} - x^{416} + x^{415} - x^{414} + x^{412} - x^{411} - x^{409} \\
&-x^{408} + x^{407} + x^{406} - x^{405} + x^{404} - x^{402} - x^{401} - x^{400} + x^{399} - x^{398} + x^{397}
\end{aligned}
$$

$$+x^{396} - x^{394} + x^{393} - x^{391} + x^{390} + x^{389} + x^{388} - x^{387} - x^{386} + x^{385} + x^{384}$$
$$+x^{383} + x^{381} - x^{380} - x^{379} + x^{378} - x^{377} + x^{376} + x^{374} - x^{373} + x^{372} + x^{371}$$
$$+x^{370} - x^{369} + x^{368} - x^{367} + x^{366} - x^{365} + x^{364} - x^{363} - x^{362} + x^{361} - x^{360}$$
$$+x^{359} - x^{358} - x^{357} + x^{356} + x^{355} - x^{354} + x^{353} - x^{352} + x^{350} - x^{349} - x^{348}$$
$$-x^{346} - x^{345} + x^{344} - x^{343} - x^{342} + x^{341} - x^{340} - x^{339} - x^{338} + x^{337} + x^{336}$$
$$+x^{334} + x^{333} - x^{332} - x^{331} - x^{330} + x^{329} - x^{328} + x^{327} + x^{326} + x^{325} - x^{324}$$
$$+x^{323} - x^{322} + x^{321} - x^{320} - x^{319} + x^{318} + x^{317} + x^{316} - x^{315} - x^{313} - x^{311}$$
$$-x^{310} - x^{309} + x^{308} - x^{306} - x^{305} + x^{304} - x^{303} + x^{302} - x^{301} + x^{300} - x^{299}$$
$$-x^{298} - x^{297} + x^{294} - x^{293} - x^{292} - x^{291} - x^{290} - x^{288} + x^{287} - x^{286} - x^{285}$$
$$+x^{284} - x^{283} + x^{282} + x^{280} - x^{279} + x^{277} - x^{276} + x^{275} + x^{274} + x^{273} + x^{272}$$
$$-x^{271} + x^{270} - x^{269} + x^{268} - x^{267} - x^{266} + x^{264} + x^{263} - x^{262} + x^{261} - x^{260}$$
$$+x^{259} - x^{257} + x^{256} - x^{255} - x^{254} + x^{253} + x^{252} + x^{251} + x^{249} + x^{248} - x^{247}$$
$$+x^{246} - x^{245} + x^{243} - x^{242} + x^{240} + x^{238} - x^{237} - x^{236} + x^{234} - x^{233} - x^{232}$$
$$+x^{231} - x^{230} + x^{229} - x^{228} - x^{227} + x^{226} - x^{225} + x^{223} + x^{222} - x^{221} + x^{220}$$
$$+x^{219} + x^{218} - x^{217} - x^{215} - x^{214} + x^{213} - x^{212} + x^{210} - x^{209} + x^{208} + x^{207}$$
$$-x^{206} - x^{205} + x^{203} + x^{202} - x^{201} - x^{200} + x^{199} + x^{198} - x^{197} + x^{196} + x^{195}$$
$$-x^{194} + x^{193} + x^{192} + x^{191} + x^{190} + x^{188} + x^{187} - x^{186} + x^{185} - x^{184} + x^{183}$$
$$+x^{182} + x^{181} + x^{180} - x^{179} - x^{178} + x^{177} - x^{176} + x^{175} + x^{174} - x^{173} + x^{172}$$
$$-x^{170} + x^{169} + x^{168} + x^{167} + x^{166} - x^{165} - x^{164} + x^{161} + x^{160} - x^{159} + x^{158}$$
$$-x^{155} + x^{154} + x^{152} + x^{151} - x^{150} + x^{149} + x^{148} + x^{147} - x^{145} - x^{142} + x^{141}$$
$$-x^{140} - x^{139} + x^{138} + x^{137} - x^{136} - x^{135} + x^{133} - x^{132} + x^{131} + x^{130} + x^{128}$$
$$+x^{127} - x^{126} + x^{125} + x^{124} + x^{123} - x^{121} + x^{120} + x^{118} - x^{116} + x^{115} - x^{114}$$
$$-x^{113} - x^{112} + x^{110} + x^{109} + x^{108} + x^{107} - x^{106} - x^{105} - x^{103} + x^{102} + x^{100}$$
$$+x^{99} + x^{98} + x^{96} + x^{95} - x^{94} - x^{93} - x^{92} + x^{91} - x^{90} - x^{89} - x^{88} - x^{87}$$
$$+x^{86} - x^{85} + x^{84} + x^{83} - x^{82} + x^{81} - x^{80} + x^{79} + x^{78} + x^{77} + x^{75} + x^{74}$$
$$-x^{73} - x^{72} - x^{71} - x^{69} - x^{68} - x^{67} + x^{66} + x^{65} + x^{64} + x^{63} + x^{62} - x^{60}$$
$$-x^{59} + x^{58} - x^{57} + x^{56} + x^{55} + x^{54} + x^{53} - x^{51} - x^{50} + x^{49} + x^{48} - x^{47}$$
$$+x^{46} + x^{45} + x^{44} - x^{43} - x^{42} + x^{41} + x^{40} - x^{39} - x^{38} + x^{37} - x^{36} + x^{35}$$
$$-x^{34} - x^{32} - x^{31} + x^{30} - x^{29} - x^{28} + x^{27} - x^{25} - x^{24} - x^{23} - x^{21} + x^{20}$$
$$-x^{19} + x^{18} - x^{17} - x^{16} - x^{15} - x^{14} + x^{13} + x^{12} - x^{11} - x^{10} + x^9 - x^8$$
$$-x^7 - x^6 - x^5 - x^3 + x^2 - 1$$

$$g = -x^{499} + x^{496} + x^{495} - x^{487} + x^{486} + x^{484} - x^{480} + x^{478} + x^{470} - x^{466} + x^{465}$$
$$-x^{462} + x^{461} + x^{460} + x^{451} - x^{446} - x^{431} - x^{428} + x^{421} + x^{415} + x^{412} - x^{411}$$
$$-x^{406} - x^{403} - x^{402} - x^{398} - x^{397} - x^{395} + x^{392} + x^{373} - x^{371} - x^{370} + x^{367}$$
$$+x^{366} - x^{364} - x^{359} - x^{355} + x^{352} + x^{351} + x^{349} + x^{347} + x^{340} + x^{339} + x^{338}$$
$$+x^{335} + x^{328} + x^{326} + x^{323} + x^{317} - x^{314} - x^{309} - x^{308} + x^{307} + x^{306} + x^{304}$$
$$-x^{303} - x^{302} - x^{299} - x^{295} - x^{292} + x^{291} - x^{289} + x^{288} + x^{283} + x^{281} + x^{280}$$
$$-x^{277} + x^{266} + x^{264} - x^{262} - x^{260} - x^{257} + x^{256} - x^{255} - x^{251} - x^{250} - x^{249}$$
$$-x^{236} - x^{235} + x^{233} - x^{232} + x^{230} + x^{227} + x^{226} - x^{224} + x^{217} + x^{216} - x^{215}$$
$$-x^{212} + x^{206} - x^{205} + x^{203} + x^{196} - x^{194} + x^{193} + x^{190} + x^{185} - x^{183} - x^{177}$$
$$-x^{172} - x^{169} - x^{168} + x^{165} - x^{163} - x^{157} + x^{156} + x^{155} - x^{138} + x^{136} - x^{135}$$
$$+x^{134} + x^{132} - x^{131} - x^{123} + x^{119} - x^{117} - x^{111} - x^{102} - x^{99} + x^{97} - x^{95}$$
$$-x^{94} + x^{92} + x^{91} - x^{89} - x^{88} - x^{86} + x^{84} + x^{83} - x^{78} + x^{76} - x^{66} + x^{60}$$
$$-x^{52} + x^{51} - x^{47} + x^{46} - x^{36} - x^{35} - x^{34} + x^{30} + x^{28} + x^{16} + 1$$

We do not give here values of $F_p$, $F_q$ or of the public key $h$ since they are big and they can easily be computed from $f$ and $g$.

If we use messages of the form $c + chx^{i_1} + chx^{i_2} + chx^{i_3}$, our heuristic estimates the average number of collisions by 1.26.

We want $c$ to verify $c \bmod p = 0$, $3c < q/2$ and $4c > q/2$. We chose $c = 33$, which satisfies this conditions.

We use the chosen ciphertext $e = 33h + 33 + 33hx + 33hx^4$.

Let $m$ be the decoded message. We find then that

$$(1 + x^{67})/m \quad (\bmod \ p)$$

is a possible value $f'$.

That gives us the following value for $f'$

$$
\begin{aligned}
f' = & \ x^{501} - x^{500} - x^{499} + x^{498} - x^{497} + x^{496} - x^{495} - x^{494} + x^{493} - x^{492} + x^{490} \\
& + x^{489} - x^{488} + x^{487} + x^{486} + x^{485} - x^{484} - x^{482} - x^{481} + x^{480} - x^{479} + x^{477} \\
& - x^{476} + x^{475} + x^{474} - x^{473} - x^{472} + x^{470} + x^{469} - x^{468} - x^{467} + x^{466} + x^{465} \\
& - x^{464} + x^{463} + x^{462} - x^{461} + x^{460} + x^{459} + x^{458} + x^{457} + x^{455} + x^{454} - x^{453} \\
& + x^{452} - x^{451} + x^{450} + x^{449} + x^{448} + x^{447} - x^{446} - x^{445} + x^{444} - x^{443} + x^{442} \\
& + x^{441} - x^{440} + x^{439} - x^{437} + x^{436} + x^{435} + x^{434} + x^{433} - x^{432} - x^{431} + x^{428} \\
& + x^{427} - x^{426} + x^{425} - x^{422} + x^{421} + x^{419} + x^{418} - x^{417} + x^{416} + x^{415} + x^{414} \\
& - x^{412} - x^{409} + x^{408} - x^{407} - x^{406} + x^{405} + x^{404} - x^{403} - x^{402} + x^{400} - x^{399} \\
& + x^{398} + x^{397} + x^{395} + x^{394} - x^{393} + x^{392} + x^{391} + x^{390} - x^{388} + x^{387} - x^{385} \\
& - x^{383} + x^{382} - x^{381} - x^{380} - x^{379} + x^{377} + x^{376} + x^{375} + x^{374} - x^{373} - x^{372} \\
& - x^{370} + x^{369} + x^{367} + x^{366} + x^{365} - x^{363} + x^{362} - x^{361} - x^{360} - x^{359} + x^{358} \\
& - x^{357} - x^{356} - x^{355} - x^{354} + x^{353} - x^{352} + x^{351} + x^{350} - x^{349} + x^{348} - x^{347} \\
& + x^{346} + x^{345} + x^{344} + x^{342} + x^{341} - x^{340} - x^{339} - x^{338} - x^{336} - x^{335} - x^{334} \\
& + x^{333} + x^{332} + x^{331} + x^{330} + x^{329} - x^{327} - x^{326} + x^{325} - x^{324} + x^{323} + x^{322} \\
& + x^{321} + x^{320} - x^{318} - x^{317} + x^{316} + x^{315} - x^{314} + x^{313} + x^{312} + x^{311} - x^{310} \\
& - x^{309} + x^{308} + x^{307} - x^{306} - x^{305} + x^{304} - x^{303} + x^{302} - x^{301} - x^{299} - x^{298} \\
& + x^{297} - x^{296} - x^{295} + x^{294} - x^{292} - x^{291} - x^{290} - x^{288} + x^{287} - x^{286} + x^{285} \\
& - x^{284} - x^{283} - x^{282} - x^{281} + x^{280} + x^{279} - x^{278} - x^{277} + x^{276} - x^{275} - x^{274} \\
& - x^{273} - x^{272} - x^{270} + x^{269} - x^{267} - x^{266} + x^{265} + x^{264} - x^{263} - x^{262} + x^{261} \\
& - x^{260} - x^{259} - x^{258} - x^{257} - x^{256} - x^{255} + x^{254} - x^{252} + x^{251} - x^{250} - x^{249} \\
& - x^{246} + x^{245} - x^{244} - x^{243} + x^{241} + x^{239} + x^{238} + x^{236} - x^{235} + x^{234} + x^{232} \\
& - x^{231} + x^{230} + x^{228} - x^{227} + x^{226} + x^{225} - x^{224} - x^{223} + x^{222} + x^{221} - x^{219} \\
& + x^{218} - x^{217} - x^{215} - x^{214} + x^{213} + x^{212} + x^{211} + x^{210} + x^{209} - x^{208} - x^{207} \\
& + x^{206} + x^{205} + x^{204} - x^{203} + x^{202} - x^{201} - x^{200} - x^{199} + x^{198} + x^{197} - x^{194} \\
& - x^{193} + x^{192} - x^{189} + x^{188} + x^{187} - x^{186} - x^{185} - x^{184} - x^{182} - x^{181} - x^{180} \\
& + x^{179} - x^{178} + x^{176} - x^{175} - x^{173} - x^{172} + x^{171} + x^{170} - x^{169} + x^{168} - x^{166} \\
& - x^{165} - x^{164} + x^{163} - x^{162} + x^{161} + x^{160} - x^{158} + x^{157} - x^{155} + x^{154} + x^{153} \\
& + x^{152} - x^{151} - x^{150} + x^{149} + x^{148} + x^{147} + x^{145} - x^{144} - x^{143} + x^{142} - x^{141} \\
& + x^{140} + x^{138} - x^{137} + x^{136} + x^{135} + x^{134} - x^{133} + x^{132} - x^{131} + x^{130} - x^{129} \\
& + x^{128} - x^{127} - x^{126} + x^{125} - x^{124} + x^{123} - x^{122} - x^{121} + x^{120} + x^{119} - x^{118} \\
& + x^{117} - x^{116} + x^{114} - x^{113} - x^{112} - x^{110} - x^{109} + x^{108} - x^{107} - x^{106} + x^{105} \\
& - x^{104} - x^{103} - x^{102} + x^{101} + x^{100} + x^{98} + x^{97} - x^{96} - x^{95} - x^{94} + x^{93} \\
& - x^{92} + x^{91} + x^{90} + x^{89} - x^{88} + x^{87} - x^{86} + x^{85} - x^{84} - x^{83} + x^{82} + x^{81} \\
& + x^{80} - x^{79} - x^{77} - x^{75} - x^{74} - x^{73} + x^{72} - x^{70} - x^{69} + x^{68} - x^{67} + x^{66} \\
& - x^{65} + x^{64} - x^{63} - x^{62} - x^{61} + x^{58} - x^{57} - x^{56} - x^{55} - x^{54} - x^{52} + x^{51} \\
& - x^{50} - x^{49} + x^{48} - x^{47} + x^{46} - x^{44} - x^{43} + x^{41} - x^{40} + x^{39} + x^{38} + x^{37} \\
& + x^{36} - x^{35} + x^{34} - x^{33} + x^{32} - x^{31} - x^{30} + x^{28} + x^{27} - x^{26} + x^{25} - x^{24} \\
& + x^{23} - x^{21} + x^{20} - x^{19} - x^{18} + x^{17} + x^{16} + x^{15} + x^{13} + x^{12} - x^{11} + x^{10} \\
& - x^9 + x^7 - x^6 + x^4 + x^2 - x - 1
\end{aligned}
$$

This value is different from the original one (we have $f = x^{236} \circledast f'$), but it can be used to decrypt messages nonetheless.

## 4.2    Choice of Parameters and Running Times Table

Here we give estimation of the running times for the different sets of parameters and the values chosen for $m$, $n$ and $c$.

| Case | A | B | | C | | D |
|------|---|---|---|---|---|---|
| $m$ | 1 | 4 | 1 | 1 | 4 | 1 |
| $n$ | 1 | 1 | 2 | 2 | 2 | 3 |
| $c$ | 18 | 15 | 24 | 24 | 24 | 33 |
| Avg no of collisions | 3.36 | 0.712 | 1.75 | 0.832 | 0.7 | 1.27 |
| No of ciphertexts (testing 1 collision) | − | 4.5 | − | 2.2 | 2.25 | − |
| No of ciphertexts (testing 2 collisions) | 7 | − | 2 | 2 | − | 2 |
| Time to test for 1 collision | − | 1s | − | 6s | 85s | − |
| Time to test for 2 collisions | 25s | − | 135s | 4mn | − | 1h |

These running times have been obtain on a single PC, using GP/PARI CALCULATOR Version 2.0.14.

# 5    Plaintext Awareness with Our Chosen-Ciphertext Attack

The attack described in the previous sections uses the fact that one can build a ciphertext without knowing the corresponding plaintext. A cryptosystem is said to be plaintext aware if it is infeasible for an attacker to construct a valid ciphertext without knowing the corresponding plaintext (see [2] which first introduced this notion and [1] which had a corrected definition). So in [11] Silverman proposed to use a system similar to OAEP to make NTRU plaintext aware. OAEP stands for Optimal Asymmetric Encryption Padding. It has been proposed by Mihir Bellare and Phillip Rogaway in [2] and describes an embedding scheme using an hash and a generating function that achieves plaintext-aware encryption. However, since OAEP applies only to a one-way trapdoor function, it had to be adapted to work for NTRU.

## 5.1    A Description of the Embedding Scheme Proposed for NTRU

We let

$$\mathcal{P}_p(N) = \{\text{polynomials of degree at most } N - 1 \text{ with mod } p \text{ coefficients}\},$$

and we write

$$[g]_p = \begin{cases} g \text{ with its coefficients reduced} \\ \text{modulo } p \text{ into the range } ] - p/2, p/2]. \end{cases}$$

We need a generating function and a hash function

$$G : \mathcal{P}_p(N) \rightarrow \mathcal{P}_p(N) \text{ and } H : \mathcal{P}_p(N) \times \mathcal{P}_p(N) \rightarrow \mathcal{P}_p(K).$$

To encrypt a message, one chooses a plaintext $m$ from the set of plaintexts $\mathcal{P}_p(N - K)$ and a polynomial $\phi \in \mathcal{L}_\phi$. One computes

$$e \equiv p\phi \circledast h + [m + H(m, [p\phi \circledast h]_p)X^{N-K} + G([p\phi \circledast h]_p)]_p \pmod{q}. \quad (1)$$

To decrypt the message, the receiver uses his private key $f$ and the standard NTRU decryption method to recover a polynomial

$$n = [F_p \circledast [f \circledast e]_q]_p \in \mathcal{P}_p(N).$$

Next he computes

$$b \equiv e - n \pmod{p} \quad \text{and} \quad c \equiv n - G(b) \pmod{p}.$$

and he writes $c$ in the form

$$c = c' + c''X^{N-K} \text{ with } \deg(c') < N - K \text{ and } \deg(c'') < K.$$

Finally, he compares the quantities

$$c'' \text{ and } H(c', b).$$

If they are the same, he accepts $c'$ as a valid decryption. Otherwise he rejects the message as invalid.

An attacker who does not know the underlying plaintext of a cipher message will have a probability of $p^{-K}$ of producing a valid ciphertext.

We are now going to show how our attack is modified with this encapsulation.

## 5.2 Adaptation of Our Attack

**Principle.** With this embedding, an attacker can detect when a message is valid or invalid. Our goal is to produce special messages that may be either valid or invalid and learn information from their acceptance or rejection.

As in the unpadded version, this is achieved by replacing $p\phi \circledast h$ by a well chosen polynomial. We add to this polynomial the correct encapsulation of a message $m$, so that the ciphertext will be accepted when there is no collision in the polynomial and rejected otherwise.

The principle of our attack is close to what Hall, Goldberg and Schneier call a *reaction attack* in [6]. It is a chosen-ciphertext attack but does not require that the attacker sees the decrypted plaintext. He only needs to know whether the ciphertext was correctly decrypted or rejected for errors.

Such attacks have been studied on NTRU by Hoffstein and Silverman in [8] but they applied on the unpadded version of the cryptosystem.

**Choice of a Polynomial $P$.** Let

$$P \equiv x^{i_1} + \cdots + x^{i_n} + h \circledast (x^{j_1} + \cdots x^{j_m}) \pmod{q}, \qquad i_k, j_l \in \mathbb{N}.$$

and choose $n$ and $m$ such that the average number of collisions, as defined in section 3.3, in $P$ is near 1, and preferably a little smaller, so that we can expect $P$ to have no more than one collision. If there is no collision, there will be no decryption failure, and we will know we need to change $P$. We will have to try different $P$, till we found a suitable one.

Now, since multiplying by $\pm x^i$ does not change the propriety of $f$ and $h$ to act as private and public key, we can assume the collision happens at degree 0 and is a collision of 1. This will simplify the presentation of the attack.

**Information Obtained from Decryption Failure.** Now if we can ask the decryption of messages of the form $cx^i + cP$, for $i$ ranging from 0 to $N - 1$, with $c$ such that $c \equiv 0 \bmod P$, $(n + m)c < q/2$ and $(n + m + 1)c > q/2$, we can discover all coefficients equal to 1 in $f$. Indeed let us assume that we send a message of the above form and that we expect the decrypted message to be 0. If the answer of the decryption is not 0, then the decryption process will send an error since we cannot know the plaintext.

Now, as we have seen in section 3.1, decryption will be different from 0 if and if only there is collision between the $(N - i)$th coefficient of $f$ and the unique collision in $P$. So if decryption is 0, the $(N - i)$th coefficient in $f$ will be a 0 or a $-1$ and if decryption is different than 0, that is if we have a decryption error, we know that the $(N - i)$th coefficient of $f$ is a 1. Similarly, with messages $cx^i - cP$, a decryption error indicates that the $(N - i)$th coefficient of $f$ is a $-1$. By testing those $2N$ messages, we can reconstruct a key $f'$ equivalent to $f$.

**Influence of the Encapsulation.** But, as stated above, we now have to add some valid encapsulated message to our test cipher $cx^i \pm cP$ (otherwise all our test messages will be rejected and we will not learn anything), so we do not send $cx^i \pm cP$, but $cx^i \pm cP + m'$. The message $m'$ can be chosen as the correct encapsulation of any message $m$, where $p\phi \circledast h$ has been replaced by $cx^i \pm cP$ in the formula (1).

After multiplication by $f$, we obtain $cx^i \circledast f \pm cP \circledast f + m' \circledast f$. The coefficients of $m' \circledast f$ may be of size $q/4$ and thus can produce a wrong decryption where we should have had a good one according only to $cx^i \circledast f \pm cP \circledast f$. It is not possible to get rid of the influence of $m' \circledast f$, but we can reduce it. It is indeed possible to take for $m$ the value $-G([cx^i \pm cP]_p) \bmod p$ truncated to degree $N - K$, so that $m' = [m + H(m, [cx^i \pm cP]_p)X^{N-k} + G([cx^i \pm cP]_p)]_p$ will have all its coefficients of degree less than $N - K$ equal to zero. $m'$ has now only approximately $2K/3$ non zero coefficients, and $m' \circledast f$ will have coefficients whose absolute value may be less than $\min((5c - q/2), (q/2 - 4c))$. Then hopefully $cP + m'$ will have the same property than $cP$, that is produce a wrong message when added to $cx^i$ if and if only the $(N - i)$th coefficient of $f$ is 1. Note that if $cP + m'$ verify this, we can proceed exactly as described above to recover the private key. The problem is that $m'$ should be recalculated each time, for each value of $cx^i \pm cP$. But, since $m'$ comes from $[[cP]_q]_p$, let us see what happens when we add $cx^i$ to $cP$: in the majority of cases, the addition of $cx^i$ to $[cP]_q$ will not induce a new reduction modulo $q$ so that $[[cP]_q + cx^i]_p = [[cP]_q]_p$ (recall that $c \equiv 0 \bmod p$), and $m'$ will stay the same. For such $i$, we can use the system described above to determine the corresponding coefficients of $f$. For the other coefficients, we cannot be really sure of the coefficients we obtain, even if there is a good probability for them to be right. It is then possible to use either LLL algorithm to find the missing coefficients or choose another value for $P$ and repeat the process.

## 5.3   Example

**Algorithm.** We give first a brief description of the resulting algorithm to attack NTRU.

1. Choose appropriate values for $m$ and $n$ such that the heuristic number of collisions $\frac{2d_f^n d_g^m}{N^{n+m-1}}$ will be near 1.
2. Select a suitable $c$ with $c \equiv 0 \bmod p$, $(m + n)c < q/2$ and $(m + n + 1)c > q/2$.

3. Select a value of a polynomial P.

$$P = x^{i_1} + \cdots + x^{i_n} + h \circledast (x^{j_1} + \cdots + x^{j_m}) \pmod{q}$$

4. Produce $m'$ corresponding to $cP$: $m' = [m + H(m, [cP]_p)X^{N-k} + G([cP]_p)]_p$ with $m = [-G([cP]_p)]_p \bmod X^{N-K}$.
5. Ask the decryption of $cP + m'$. The answer should be $m$. If not, go back to 3.
6. For all $i$ such that $[[cx^i + cP]_q]_p = [[cP]_q]_p$, ask decryption of $cx^i + cP + m'$. If the answer is a decryption error, the (N-i)th coefficient of $f'$ is a 1, else we know it is *not* a 1. For all other $i$, the (N-i)th coefficient of $f'$ may be a 1.
7. At the same time, for all $i$ such that $[[-cx^i + cP]_q]_p = [[cP]_q]_p$, ask decryption of $-cx^i + cP + m'$. If the answer is a decryption error, the (N-i)th coefficient of $f'$ is a $-1$, else we know it is *not* a $-1$. Note that if we had $[[cx^i + cP]_q]_p \neq [[cP]_q]_p$, then $[[-cx^i + cP]_q]_p = [[cP]_q]_p$. So a coefficient can not both possibly be a 1 and $-1$.
8. Note also that if $cx^i + cP + m'$ gave a decryption error, then $-cx^i + cP + m'$ should not. If this is the case, we know that $m'$ introduced decryption errors and we go back to step 3.
9. If after a few messages there is still no decryption failure, there is no collision in $P$. Go back to step 3.
10. Count the minimal and maximal number of 1 and $-1$ in $f'$. If this number is not consistent with the value of $d_f$, go back to step 3.
11. Merge with preceeding informations obtained on $f'$. Eventually repeat with another $P$ (step 3).

**Application.** Here is an example of the attack with the following set of parameters:

- $(N, p, q) = (503, 3, 256)$
- $n_f = 216$
- $n_g = 72$
- $K = 107$

Those are the parameters proposed in [11] to offer the highest security.
For $n = 1$ and $m = 3$, we find an average number of collisions equal to 1.267.
We want $c \equiv 0 \bmod 3$, $4c < 128$, $5c > 128$. We choose $c = 27$.
We tested the following polynomials $P$:

- $P = 1 + h \circledast (x + x^2 + x^3)$
- $P = 1 + h \circledast (x + x^2 + x^4)$
- $P = 1 + h \circledast (x + x^2 + x^5)$
- $P = 1 + h \circledast (x + x^2 + x^6)$
- $P = 1 + h \circledast (x + x^2 + x^7)$

The good ones where:

- $P = 1 + h \circledast (x + x^2 + x^4)$
- $P = 1 + h \circledast (x + x^2 + x^7)$

The other ones failed at step 8 or 9.
After merging the informations gained from these two polynomials, we had only 15 possible keys left. It is then easy to find the good one by trying to decipher a ciphertext or by testing whether $h \circledast f' \equiv \pm x^i \circledast g \bmod q$ for some $i$.

We were able to recover the private key with less than $5N$ calls to the decryption oracle.

We give now a few statistics of our algorithm with the different sets of parameters.

| Case | A | B | B | C | D |
|---|---|---|---|---|---|
| Value for $K$ | 17 | 49 | 49 | 65 | 107 |
| Avg no of ciphertexts | 230 | 310 | 620 | 950 | 2100 |
| Avg running time | | 11s | 17mn | 2mn | 6mn | 36mn |

Remark: even for the highest security parameters, two successful polynomials were enough to recover sufficient information on the secret key.

## 5.4   Protection against This Attack

Hoffstein and Silverman described in [8] a similar attack but did not take into account the digital envelope. However he proposed different ways of countering it:

- Change the key very often. This solution requires that one send the actual public key to the receiver before each communication. Each time, we will need to have the new public key signed with a digital certificate, proving the origin of the key. Under these conditions, there cannot be off-line communication.
- Track decryption failure. Decryption failure should occur rarely under normal circumstances. While under a ciphertext attack, this will happens quite often. One can detect an undergoing attack and change the key. The attacker has still the power of forcing someone to change its public key when he wants.
- Induce randomness. This solution consist in adding some random $px^i$ to the message before its decryption. This can lead to produce invalid messages from goods messages when OAEP is used. It may also produce errors in our attack, but sufficient information might still be obtained.
- Coefficient distribution analysis. The number of coefficients of the polynomial $p\phi g + fm$ falling into ranges close to $q/2$ or $-q/2$ will be larger than usual when the attack takes place. So one can discover the attack by looking counting the number of coefficients in such ranges and simply not respond to inflated polynomial.

In fact, the easiest protection against this attack is to replace the padding described in [11] by the construction from [4]. This construction works in the random oracle model and provably turns any asymmetric system into a system resistant to adaptive chosen-ciphertext attacks.

## 6   Conclusion

The NTRU cryptosystem makes use of the independence of reduction modulo two relatively prime integers $p$ and $q$. This cryptosystem have proved secure against different attacks, such as the brute force attack, the meet-in-the-middle attack and lattice based attacks. Unfortunately, the structure of the private keys $f$ and $g$ opens a way to the chosen-ciphertext attack that was described here, even when the padding in [11] is used; so alternative padding/hashing methods such as those described in [4] should be used to avoid the attacks described in this paper.

# References

1. Mihir Bellare, Anand Desai, David Pointcheval, and Phillip Rogaway. Relations among notions of security for public-key encryption schemes. In Hugo Krawczyk, editor, *Advances in Cryptology — CRYPTO'98*, volume 1462 of *Lecture Notes in Computer Science*, pages 26–45. Springer, 1998.
2. Mihir Bellare and Phillip Rogaway. Optimal asymmetric encryption. In A. de Santis, editor, *Advances in Cryptology — EUROCRYPT'94*, volume 950 of *Lecture Notes in Computer Science*, pages 92–111. Springer-Verlag, 1994.
3. D. Coppersmith and A. Shamir. Lattice attacks on NTRU. In *Advances in Cryptology — EUROCRYPT'97*, volume 1233 of *Lecture Notes in Computer Science*, pages 52–61, 1997.
4. Eiichiro Fujisaki and Tatsuaki Okamoto. Secure integration of asymmetric and symmetric encryption schemes. In Michael Wiener, editor, *Advances in Cryptology — CRYPTO'99*, volume 1666 of *Lecture Notes in Computer Science*, pages 537–554. Springer-Verlag, 1999.
5. H. Gilbert, D. Gupta, A.M. Odlyzko, and J.-J. Quisquater. Attacks on shamir's 'rsa for paranoids'. *Information Processing Letters*, 68:197–199, 1998. http://www.research.att.com/~amo/doc/recent.html.
6. Chris Hall, Ian Goldberg, and Bruce Schneier. Reaction attacks against several public-key cryptosystems. In G. Goos, J. Hartmanis, and J; van Leeuwen, editors, *ICICS'99*, volume 1726 of *Lecture Notes in Computer Science*, pages 2–12. Springer-Verlag, 1999.
7. Jeffrey Hoffstein, Jill Pipher, and Joseph H. Silverman. NTRU: A ring based public key cryptosystem. In *ANTS'3*, volume 1423 of *Lecture Notes in Computer Science*, pages 267–288. Springer Verlag, 1998.
8. Jeffrey Hoffstein and Joseph H. Silverman. Reaction attacks against the NTRU public key cryptosystem. Technical Report 15, NTRU Cryptosystems, August 1999.
9. M. Joye and J.-J. Quisquater. On the importance of securing your bins: the garbage-man-in-the-middle attack. *4th ACM Conf. Computer Comm. Security*, pages 135–141, 1997.
10. A.K. Lenstra, H.W. Lenstra, and L. Lovász. Factoring polynomials with polynomial coefficients. *Math. Annalen*, 261:515–534, 1982.
11. Joseph H. Silverman. Plaintext awareness and the NTRU PKCS. Technical Report 7, NTRU Cryptosystems, July 1998.
12. Joseph H. Silverman. Estimated breaking times for NTRU lattices. Technical Report 12, NTRU Cryptosystems, March 1999.

# Privacy Preserving Data Mining

Yehuda Lindell[1] and Benny Pinkas[2]*

[1] Department of Computer Science and Applied Math, Weizmann Institute of
Science, Rehovot, ISRAEL. lindell@wisdom.weizmann.ac.il
[2] School of Computer Science and Engineering, Hebrew University of Jerusalem,
Jerusalem, ISRAEL. bpinkas@cs.huji.ac.il

**Abstract.** In this paper we introduce the concept of privacy preserving
data mining. In our model, two parties owning confidential databases
wish to run a data mining algorithm on the union of their databases,
without revealing any unnecessary information. This problem has many
practical and important applications, such as in medical research with
confidential patient records.
Data mining algorithms are usually complex, especially as the size of
the input is measured in megabytes, if not gigabytes. A generic secure
multi-party computation solution, based on evaluation of a circuit computing the algorithm on the entire input, is therefore of no practical use.
We focus on the problem of decision tree learning and use ID3, a popular and widely used algorithm for this problem. We present a solution
that is considerably more efficient than generic solutions. It demands
very few rounds of communication and reasonable bandwidth. In our
solution, each party performs by itself a computation of the same order
as computing the ID3 algorithm for its own database. The results are
then combined using efficient cryptographic protocols, whose overhead
is only logarithmic in the number of transactions in the databases. We
feel that our result is a substantial contribution, demonstrating that secure multi-party computation can be made practical, even for complex
problems and large inputs.

## 1 Introduction

We consider a scenario where two parties having private databases wish to co-
operate by computing a data mining algorithm on the union of their databases.
Since the databases are confidential, neither party is willing to divulge any of
the contents to the other. We show how the involved data mining problem of decision tree learning can be efficiently computed, with no party learning anything
other than the output itself. We demonstrate this on ID3, an algorithm widely
used and implemented in many real applications.

**Confidentiality Issues in Data Mining.** A key problem that arises in any en
masse collection of data is that of *confidentiality*. The need for secrecy is sometimes due to law (e.g. for medical databases) or can be motivated by business
interests. However, sometimes there can be mutual gain by *sharing* of data. A

---

* Supported by an Eshkol grant of the Israel Ministry of Science.

M. Bellare (Ed.): CRYPTO 2000, LNCS 1880, pp. 36–54, 2000.
© Springer-Verlag Berlin Heidelberg 2000

key utility of large databases today is research, whether it be scientific, or economic and market oriented. The medical field has much to gain by pooling data for research; as can even competing businesses with mutual interests. Despite the potential gain, this is not possible due to confidentiality issues which arise.

We address this question and show that practical solutions are possible. Our scenario is one where two parties $P_1$ and $P_2$ own databases $D_1$ and $D_2$. The parties wish to apply a data-mining algorithm to the joint database $D_1 \cup D_2$ without revealing any unnecessary information about their individual databases. That is, the only information learned by $P_1$ about $D_2$ is that which can be learned from the output, and vice versa. We do not assume any "trusted" third party who computes the joint output.

**Very Large Databases and Efficient Computation.** We have described a model which is exactly that of multi-party computation. Therefore, there exists a secure solution for *any* functionality (Goldreich et. al. in [13]). As we discuss in Section 1.1, due to the fact that these solutions are generic, they are highly inefficient. In our case where the inputs are very large and the algorithms reasonably complex, they are far from practical.

It is clear that any reasonable solution must have the individual parties do the majority of the computation independently. Our solution is based on this guiding principle and in fact, the number of bits communicated is dependent on the number of transactions by a logarithmic factor only.

**Semi-Honest Parties.** In any multi-party computation setting, a *malicious* party can always alter his input. In the data-mining setting, this fact can be very damaging as an adversarial party may define his input to be the empty database. Then, the output obtained is the result of the algorithm on the other party's database alone. Although this attack cannot be prevented, we would like to limit attacks by malicious parties to altering their input only. However, for this initial work we assume that the parties are *semi-honest* (also termed *passive*). That is, they follow the protocol as it is defined, but may record all intermediate messages sent in an attempt to later derive additional information. We leave the question of an efficient solution to the malicious party setting for future work. In any case, as was described above, malicious parties cannot be prevented from obtaining meaningful confidential information and therefore a certain level of trust is anyway needed between the parties. We remark that the semi-honest model is often a realistic one; that is, deviating from a specified program which may be buried in a complex application is a non-trivial task.

## 1.1   Related Work

Secure two party computation was first investigated by Yao [21], and was later generalized to multi-party computation in [13,2,5]. These works all use a similar methodology: the function $F$ to be computed is first represented as a combinatorial circuit, and then the parties run a short protocol for every gate in the circuit. While this approach is appealing in its generality and simplicity, the protocols it generates depend on the size of the circuit. This size depends on the size of the input (which might be huge as in a data mining application), and on the complexity of expressing $F$ as a circuit (for example, a multiplication circuit is quadratic in the size of its inputs). We stress that secure computation of small circuits with small inputs can be *practical* using the [21] protocol.[1]

There is a major difference between the protocol described in this paper and other examples of multi-party protocols (e.g. [3,11,6]). While previous protocols were efficient (polynomial) in the size of their inputs, this property does not suffice for data mining applications, as the input consists of huge databases. In the protocol presented here, most of the computation is done individually by each of the parties. They then engage in a few secure circuit evaluations on very small circuits. We obtain very few rounds of communication with bandwidth which is practical for even very large databases.

*Outline:* The next section describes the problem of classification and a widely used solution to it, decision trees. Following this, Section 3 presents the security definition and Section 4 describes the cryptographic tools used in the solution. Section 5 contains the protocol itself and its proof of security. Finally, the main subprotocol that privately computes random shares of $f(v_1, v_2) \stackrel{\text{def}}{=} (v_1 + v_2) \ln(v_1 + v_2)$ is described in Section 6.

## 2   Classification by Decision Tree Learning

This section briefly describes the machine learning and data mining problem of *classification* and ID3, a well-known algorithm for it. The presentation here is rather simplistic and very brief and we refer the reader to Mitchell [15] for an in-depth treatment of the subject. The ID3 algorithm for generating decision trees was first introduced by Quinlan in [19] and has since become a very popular learning tool.

The aim of a classification problem is to classify transactions into one of a discrete set of possible categories. The input is a structured database comprised of attribute-value pairs. Each row of the database is a *transaction* and each column is an *attribute* taking on different values. One of the attributes in the database is designated as the *class* attribute; the set of possible values for this attribute being the classes. We wish to predict the class of a transaction by viewing only the non-class attributes. This can thus be used to predict the class of new transactions for which the class is unknown.

For example, a bank may wish to conduct credit risk analysis in an attempt to identify non-profitable customers before giving a loan. The bank then defines "Profitable-customer" (obtaining values "yes" or "no") to be the class attribute. Other database attributes may include: Home-Owner, Income, Years-of-Credit, Other-Delinquent-Accounts and other relevant information. The bank is then interested in obtaining a tool which can be used to classify a *new* customer as potentially profitable or not. The classification may also be accompanied with a probability of error.

---

[1] The [21] protocol requires only two rounds of communication. Furthermore, since the circuit and inputs are small, the bandwidth is not too great and only a reasonable number of oblivious transfers need be executed.

## 2.1   Decision Trees and the ID3 Algorithm

A decision tree is a rooted tree containing nodes and edges. Each internal node is a test node and corresponds to an attribute; the edges leaving a node correspond to the possible values taken on by that attribute. For example, the attribute "Home-Owner" would have two edges leaving it, one for "Yes" and one for "No". Finally, the leaves of the tree contain the *expected* class value for transactions matching the path from the root to that leaf.

Given a decision tree, one can predict the class of a new transaction $t$ as follows. Let the attribute of a given node $v$ (initially the root) be $A$, where $A$ obtains possible values $a_1, ..., a_m$. Then, as described, the $m$ edges leaving $v$ are labeled $a_1, ..., a_m$ respectively. If the value of $A$ in $t$ equals $a_i$, we simply go to the son pointed to by $a_i$. We then continue recursively until we reach a leaf. The class found in the leaf is then assigned to the transaction.

The following notation is used: $R$: a set of *attributes*; $C$: the *class* attribute and $T$: a set of *transactions*. The ID3 algorithm assumes that each attribute is categorical, that is containing discrete data only, in contrast to continuous data such as age, height etc.

The principle of the ID3 algorithm is as follows:

The tree is constructed top-down in a recursive fashion. At the root, each attribute is tested to determine how well it alone classifies the transactions. The "best" attribute (to be discussed below) is then chosen and we partition the remaining transactions by it. We then recursively call ID3 on each partition (which is a smaller database containing only the appropriate transactions and without the splitting attribute). See Figure 1 for a description of the ID3 algorithm.

---

**ID3**$(R, C, T)$

1. If $R$ is empty, return a leaf-node with the class value of the majority of the transactions in $T$.
2. If $T$ consists of transactions with all the same value $c$ for the class attribute, return a leaf-node with the value $c$ (finished classification path).
3. Otherwise,
   (a) Find the attribute that *best* classifies the transactions in $T$, let it be $A$.
   (b) Let $a_1, ..., a_m$ be the values of attribute $A$ and let $T(a_1), ..., T(a_m)$ be a partition of $T$ s.t. every transaction in $T(a_i)$ has the attribute value $a_i$.
   (c) Return a tree whose root is labeled $A$ (this is the test attribute) and has edges labeled $a_1, ..., a_m$ such that for every $i$, the edge $a_i$ goes to the tree ID3$(R - \{A\}, C, T(a_i))$.

---

**Fig. 1.** The ID3 Algorithm for Decision Tree Learning

What remains is to explain how the *best* predicting attribute is chosen. This is the central principle of ID3 and is based on information theory. The entropy of the class attribute clearly expresses the difficulty of prediction. We know the class of a set of transactions when the class entropy for them equals zero. The idea is therefore to check which attribute reduces the information of the class-attribute by the most. This results in a greedy algorithm which searches for a small decision tree consistent with the database. As a result of this, decision trees are usually relatively small, even for large databases.

The exact test for determining the best attribute is defined as follows. Let $c_1, ..., c_\ell$ be the class-attribute values. Let $T(c_i)$ be the set of transactions with class $c_i$. Then the information needed to identify the class of a transaction in $T$ is the entropy, given by:

$$H_C(T) = \sum_{i=1}^{\ell} -\frac{|T(c_i)|}{|T|} \log \frac{|T(c_i)|}{|T|}$$

Let $A$ be a non-class attribute. We wish to quantify the information needed to identify the class of a transaction in $T$ *given* that the value of $A$ has been obtained. Let $A$ obtain values $a_1, ..., a_m$ and let $T(a_j)$ be the transactions obtaining value $a_j$ for $A$. Then, the conditional information of $T$ given $A$ is given by:

$$H_C(T|A) = \sum_{j=1}^{m} \frac{|T(a_j)|}{|T|} H_C(T(a_j))$$

Now, for each attribute $A$ the information-gain[2], is defined by

$$\text{Gain}(A) \stackrel{\text{def}}{=} H_C(T) - H_C(T|A)$$

The attribute $A$ which has the maximum gain over all attributes in $R$ is then chosen.

Since its inception there have been many extensions to the original ID3 algorithm, the most well-known being C4.5. We consider only the simpler ID3 algorithm and leave extensions to more advanced versions for future work.

## 2.2 The ID3$_\delta$ Approximation

The ID3 algorithm chooses the "best" predicting attribute by comparing entropies that are given as real numbers. If at a given point, two entropies are very close together, then the two (different) trees resulting from choosing one attribute or the other are expected to have almost the same predicting capability. Formally stated, let $\delta$ be some small value. Then, for a pair of attributes $A_1$ and $A_2$, we say that $A_1$ and $A_2$ have $\delta$-*equivalent information gains* if

$$|H_C(T|A_1) - H_C(T|A_2)| < \delta$$

This definition gives rise to an approximation of ID3. Denote by $\mathcal{ID3}_\delta$ the set of all possible trees which are generated by running the ID3 algorithm, and choosing either $A_1$ or $A_2$ in the case that they have $\delta$-equivalent information gains. We actually present a protocol for secure computation of a specific algorithm ID3$_\delta \in \mathcal{ID3}_\delta$, in which the choice of $A_1$ or $A_2$ is implicit by an approximation that is used instead of the log function. The value of $\delta$ influences the efficiency, but only by a logarithmic factor.

---

[2]  Note that the gain measure biases attributes with many values and another measure called the *Gain Ratio* is therefore sometimes used. We present the simpler version here.

# 3 Security Definition – Private Computation of Functions

The model for this work is that of general multi-party computation, more specifically between two *semi-honest* parties. Our formal definitions here are according to Goldreich in [12]. We now present in brief the definition for general two-party computation of a functionality with *semi-honest* parties only. We present a formalization based on the simulation paradigm (this is equivalent to the ideal-model definition in the semi-honest case).

*Formal Definition.* The following definitions are taken from [12]. We begin with the following notation:

- Let $f : \{0,1\}^* \times \{0,1\}^* \mapsto \{0,1\}^* \times \{0,1\}^*$ be a functionality where $f_1(x,y)$ *(resp., $f_2(x,y)$)* denotes the first *(resp., second)* element of $f(x,y)$ and let $\Pi$ be a two-party protocol for computing $f$.
- The view of the first *(resp., second)* party during an execution of $\Pi$ on $(x,y)$, denoted $\text{view}_1^{\Pi}(x,y)$ *(resp., $\text{view}_2^{\Pi}(x,y)$)*, is $(x,r,m_1,...,m_t)$ *(resp., $(y,r,m_1,...,m_t)$)* where $r$ represents the outcome of the first *(resp., second)* party's *internal* coin tosses, and $m_i$ represents the $i$'th message it has received.
- The output of the first *(resp., second)* party during an execution of $\Pi$ on $(x,y)$ is denoted $\text{output}_1^{\Pi}(x,y)$ *(resp., $\text{output}_2^{\Pi}(x,y)$)*, and is implicit in the party's view of the execution.

We note that in the case of $\text{ID3}_\delta$ itself, we have $f_1 = f_2 = \text{ID3}_\delta$ (however, in the subprotocols that we use it is often the case that $f_1 \neq f_2$).

**Definition 1** (privacy w.r.t. semi-honest behavior): *For a functionality $f$, we say that $\Pi$ privately computes $f$ if there exist probabilistic polynomial time algorithms, denoted $S_1$ and $S_2$, such that*

$$\{(S_1(x, f_1(x,y)), f_2(x,y))\}_{x,y \in \{0,1\}^*} \overset{c}{\equiv} \left\{(\text{view}_1^{\Pi}(x,y), \text{output}_2^{\Pi}(x,y))\right\}_{x,y \in \{0,1\}^*} \quad (1)$$

$$\{(f_1(x,y), S_2(y, f_2(x,y)))\}_{x,y \in \{0,1\}^*} \overset{c}{\equiv} \left\{(\text{output}_1^{\Pi}(x,y), \text{view}_2^{\Pi}(x,y))\right\}_{x,y \in \{0,1\}^*} \quad (2)$$

*where $\overset{c}{\equiv}$ denotes computational indistinguishability.*

Equations (1) and (2) state that the views of the parties can be simulated by a polynomial time algorithm given access to the party's *input and output only*. We emphasize that the parties here are semi-honest and the view is therefore exactly according to the protocol definition. We note that it is not enough for the simulator $S_1$ to generate a string indistinguishable from $\text{view}_1^{\Pi}(x,y)$. Rather, the *joint distribution* of the simulator's output and $f_2(x,y)$ must be indistinguishable from $(\text{view}_1^{\Pi}(x,y), \text{output}_2^{\Pi}(x,y))$. See [12] for a discussion on why this is essential.

*Composition of Private Protocols.* The protocol for privately computing $\text{ID3}_\delta$ is composed of many invocations of smaller private computations. In particular, we reduce the problem to that of privately computing smaller subproblems and show how to compose them together in order to obtain a complete $\text{ID3}_\delta$ solution. This composition is shown to be secure in Goldreich [12].

## 3.1   Secure Computation of Approximations

Our work takes $ID3_\delta$ as the starting point and security is guaranteed relative to the approximated algorithm, rather than to ID3 itself. We present a secure protocol for computing $ID3_\delta$. That is, $P_1$ can compute his view given $D_1$ and $ID3_\delta(D_1 \cup D_2)$ only (likewise $P_2$). However, this does *not* mean that $ID3_\delta(D_1 \cup D_2)$ reveals the "same" information as $ID3(D_1 \cup D_2)$ does. In fact, it is clear that although the computation of $ID3_\delta$ is secure, *different* information is revealed (intuitively though, no "more" information is revealed)[3].

The problem of secure distributed computation of approximations was introduced and discussed by Feigenbaum et. al. [10]. Their main motivation is a scenario in which the computation of an approximation to a function $f$ might be considerably more efficient than the computation of $f$ itself. The security definition requires that the approximation does not reveal more about the inputs than $f$ does. In addition, the paper presents several general techniques for computing approximations, and efficient protocols for computing approximations of distances.

# 4   Cryptographic Tools

## 4.1   Oblivious Transfer

The notion of 1-out-2 oblivious transfer $(OT_1^2)$ was suggested by Even, Goldreich and Lempel [8], as a generalization of Rabin's "oblivious transfer" [20]. This protocol involves two parties, the *sender* and the *receiver*. The sender has two inputs $\langle X_0, X_1 \rangle$, and the receiver has an input $\sigma \in \{0, 1\}$. At the end of the protocol the receiver should learn $X_\sigma$ and no other information, and the sender should learn nothing. Very attractive non-interactive $OT_1^2$ protocols were presented in [1]. More recent results in [17] reduce the amortized overhead of $OT_1^2$, and describe non-interactive $OT_1^2$ of strings whose security is not based on the "random oracle" assumption. Oblivious transfer protocols can be greatly simplified if the parties are assumed to be semi-honest, as they are in the application discussed in this paper.

## 4.2   Oblivious Evaluation of Polynomials

In the oblivious polynomial evaluation problem there is a sender who has a polynomial $P$ of degree $k$ over some finite field $\mathcal{F}$ and a receiver with an element $x \in \mathcal{F}$. The receiver obtains $P(x)$ without learning anything else about the polynomial $P$ and the sender learns nothing about $x$. This primitive was introduced in [16]. For our solution we use a new protocol [7] that requires $O(k)$ exponentiations in order to evaluate a polynomial of degree $k$ (where the '$O$' coefficient is very small). This is important as we work with low-degree polynomials. Following are the basic ideas of this protocol.

---

[3] Note that although our implementation approximates many invocations of the ln function, none of these approximations is revealed. The only approximation which becomes known to the parties is the final result of $ID3_\delta$.

Let $P(y) = \sum_{i=0}^{k} a_i y^i$ and $x$ be the sender and receiver's respective inputs. The following protocol enables the receiver to compute $g^{P(x)}$, where $g$ is a generator of a group in which the Decisional Diffie-Hellman assumption holds. The protocol is very simple since it is assumed that the parties are semi-honest. It can be converted to one which computes $P(x)$ using the the methods of Paillier [18], who presented a trapdoor for computing discrete logs. Security against malicious parties can be obtained using proofs of knowledge. The protocol consists of the following steps:

- The receiver chooses a secret key $s$, and sends $g^s$ to the sender.
- For $0 \leq i \leq k$, the receiver computes $c_i = (g^{r_i}, g^{s \cdot r_i} g^{x^i})$, where $r_i$ is random. The receiver sends $c_0, \ldots, c_k$ to the sender.
- The sender computes $C = \Pi_{i=0}^{k}(c_i)^{a_i} = (g^R, g^{sR} g^{P(x)})$, where $R = \sum_{i=0}^{k} r_i a_i$. It then chooses a random value $r$ and computes $C' = (g^R \cdot g^r, g^{sR} g^{P(x)} \cdot g^{sr})$ and sends it to the receiver.
- The receiver divides the second element of $C'$ by the first element of $C'$ raised to the power of $s$, and obtains $g^{P(x)}$.

By the DDH assumption, the sender learns nothing of $x^i$ from the messages $c_0, \ldots, c_k$ sent by the receiver to the sender. On the other hand, the receiver learns nothing of $P$ from $C'$.

### 4.3  Oblivious Circuit Evaluation

The two party protocol of Yao [21] solves the following problem. There are two parties, a party $A$ which has an input $x$, and a party $B$ which has as input a function $f$ and a combinatorial circuit that computes $f$. At the end of the protocol $A$ outputs $f(x)$ and learns no other information about $f$, while $B$ learns nothing at all. We employ this protocol for the case that $f$ depends on two inputs $x$ and $y$, belonging to $A$ and $B$ respectively. This is accomplished by having $B$ simply hardwire his input $y$ into the circuit (that is, $B$'s input is a function $f(\cdot, y)$ and $A$ obtains $f(x, y)$ from the circuit).[4]

The overhead of the protocol involves (1) $B$ sending to $A$ tables of size linear in the size of the circuit, (2) $A$ and $B$ engaging in an oblivious transfer protocol for every input wire of the circuit, and (3) $A$ computing a pseudo-random function a constant number of times for every gate. Therefore, the number of rounds of this protocol is constant (namely, two rounds using non-interactive oblivious transfer), and the main computational overhead is that of running the oblivious transfers.

*Computing Random Shares.* Note that by defining $r_1 = F(x, (y, r_2)) \stackrel{\text{def}}{=} f(x, y) - r_2$, $A$ and $B$ obtain random shares summing to $f(x, y)$.

## 5  The Protocol

The central idea of our protocol is that all intermediate values of the computation seen by the players are uniformly distributed. At each stage, the players obtain

---

[4] In the case that $f$ is known and the parties are semi-honest, Yao's evaluation constitutes a secure protocol for the described problem.

random shares $v_1$ and $v_2$ such that their sum equals an appropriate intermediate value. Efficiency is achieved by having the parties do most of the computation independently.

We assume that there is a known upper bound on the size of the union of the databases, and that the attribute-value names are public.[5]

*Solution Outline.* The "most difficult" step in privately computing $ID3_\delta$ reduces to oblivious evaluation of the $x \ln x$ function (Section 5.1). (A private protocol for this task is presented separately in Section 6.) Next, we show how given a protocol for computing $x \ln x$, we can privately find the next attribute in the decision tree (Section 5.2). Finally, we describe how the other steps of the $ID3_\delta$ algorithm are privately computed and show the complete private protocol for computing $ID3_\delta$ (Section 5.3).

## 5.1   A Closer Look at $ID3_\delta$

The part of $ID3_\delta$ which is hardest to implement in a private manner is step 3(a). In this step the two parties must find the attribute $A$ that best classifies the transactions $T$ in the database, namely the attribute that provides the maximum information gain. This step can be stated as: *Find the attribute $A$ which minimizes the conditional information of $T$ given $A$, $H_C(T|A)$.* Examine $H_C(T|A)$ for an attribute $A$ with $m$ possible values $a_1, \ldots, a_m$, and a class attribute $C$ with $l$ possible values $c_1, \ldots, c_\ell$.

$$H_C(T|A) = \sum_{j=1}^{m} \frac{|T(a_j)|}{|T|} H_C(T(a_j))$$

$$= \frac{1}{|T|} \sum_{j=1}^{m} |T(a_j)| \sum_{i=1}^{\ell} -\frac{|T(a_j, c_i)|}{|T(a_j)|} \cdot \log(\frac{|T(a_j, c_i)|}{|T(a_j)|})$$

$$= \frac{1}{|T|} \left( -\sum_{j=1}^{m} \sum_{i=1}^{\ell} |T(a_j, c_i)| \log(|T(a_j, c_i)|) + \sum_{j=1}^{m} |T(a_j)| \log(|T(a_j)|) \right) (3)$$

Note that since the algorithm is only interested in finding the attribute $A$ which minimizes $H_C(T|A)$, the coefficient $1/|T|$ can be ignored. Also, natural logarithms can be used instead of logarithms to the base 2.

The database is a union of two databases, $D_1$ which is known to $P_1$ and $D_2$ which is known to $P_2$. The number of transactions for which attribute $A$ has value $a_j$ can therefore be written as $|T(a_j)| = |T_1(a_j)| + |T_2(a_j)|$, where $|T_b(a_j)|$ is the number of transactions with attribute $A$ set to $a_j$ in database $D_b$ (likewise $T(a_j, c_i)$ is the number of transaction with $A = a_j$ and the class attribute set to $c_i$). The values $|T_1(a_j)|$ and $|T_1(a_j, c_i)|$ can be computed by party $P_1$ independently, and the same holds for $P_2$. Therefore the expressions that should be compared can be written as a sum of expressions of the form

$$(v_1 + v_2) \cdot \ln(v_1 + v_2),$$

---

[5]  It is clear that the databases must have the same structure with previously agreed upon attribute names.

where $v_1$ is known to $P_1$ and $v_2$ is known to $P_2$. The main task is, therefore, to privately compute $x \ln x$ and a protocol for this task is described in Section 6. The exact definition of this protocol is provided in Figure 2.

## 5.2   Finding the Attribute with Maximum Gain

Given the above protocol for privately computing shares of $x \ln x$, the attribute with the maximum information gain can be determined. This is done in two stages: first, the parties obtain shares of $H_C(T|A) \cdot |T| \cdot \ln 2$ for all attributes $A$ and second, the shares are input into a very small circuit which outputs the appropriate attribute. In this section we refer to a field $\mathcal{F}$ which is defined so that $|\mathcal{F}| > H_C(T|A) \cdot |T| \cdot \ln 2$.

*Stage 1 (computing shares)* : For every attribute $A$, for every attribute-value $a_j \in A$ and every class $c_i \in C$, $P_1$ and $P_2$ use the $x \ln x$ protocol in order to obtain $w_{A,1}(a_j)$, $w_{A,2}(a_j)$, $w_{A,1}(a_j, c_i)$ and $w_{A,2}(a_j, c_i) \in_R \mathcal{F}$ such that

$$w_{A,1}(a_j) + w_{A,2}(a_j) = |T(a_j)| \cdot \log(|T(a_j)|) \mod |\mathcal{F}|$$
$$w_{A,1}(a_j, c_i) + w_{A,2}(a_j, c_i) = |T(a_j, c_i)| \cdot \log(|T(a_j, c_i)|) \mod |\mathcal{F}|$$

Now, define $\hat{H}_C(T|A) \stackrel{\text{def}}{=} H_C(T|A) \cdot |T| \cdot \ln 2$. Then,

$$\hat{H}_C(T|A) = -\sum_{j=1}^{m}\sum_{i=1}^{\ell} |T(a_j, c_i)| \cdot \ln(|T(a_j, c_i)|) + \sum_{j=1}^{m} |T(a_j)| \cdot \ln(|T(a_j)|)$$

Then, $P_1$ (and likewise $P_2$) computes his share in $\hat{H}_C(T|A)$ as follows:

$$S_{A,1} = -\sum_{j=1}^{m}\sum_{i=1}^{\ell} w_{A,1}(a_j, c_i) + \sum_{j=1}^{m} w_{A,1}(a_j) \mod |\mathcal{F}|$$

It is clear that $S_{A,1} + S_{A,2} = \hat{H}_C(T|A) \mod |\mathcal{F}|$ and we therefore have that for every attribute $A$, $P_1$ and $P_2$ obtain shares in $\hat{H}_C(T|A)$ (this last step involves local computation only).

*Stage 2 (finding the attribute):* It remains to find the attribute with the minimum $\hat{H}_C(T|A)$ (and therefore the minimum $H_C(T|A)$). This is done via a Yao circuit evaluation [21]. We note that since $\hat{H}_C(T|A) < |\mathcal{F}|$, it holds that either $S_{A,1} + S_{A,2} = \hat{H}_C(T|A)$ or $S_{A,1} + S_{A,2} = \hat{H}_C(T|A) + |\mathcal{F}|$.

The parties run an oblivious evaluation of a circuit with the following functionality. The circuit input is the shares of both parties for each $\hat{H}_C(T|A)$. The circuit first computes each $\hat{H}_C(T|A)$ (by subtracting $|\mathcal{F}|$ if the sum is larger than $|\mathcal{F}| - 1$ or leaving it otherwise), and then compares the results to find the smallest among them. This circuit has $2|R|$ inputs of size $\log |\mathcal{F}|$ and its size is $O(|R| \log |\mathcal{F}|)$. Note that $|R| \log |\mathcal{F}|$ is a small number and thus this circuit evaluation is efficient.

*Privacy:* Stage 1 is clearly private as it involves many invocations of a private protocol that outputs random shares, followed by a local computation. Stage 2 is also private as it involves a single invocation of Yao's oblivious circuit evaluation and nothing more.

Note the efficiency achieved above. Each party has to compute the same set of values $|T(a_j, c_i)|$ as it computes in an individual computation of ID3. For each of these values it engages in the $x \ln x$ protocol. (We stress that the number of values here does not depend on the number of transactions, but rather on the number of different possible values for each attribute, which is usually smaller by orders of magnitude.) It sums the results of all these protocols together, and engages in an oblivious evaluation of a circuit whose size is linear in the number of attributes.

## 5.3   The Private ID3$_\delta$ Protocol

In the previous subsection we showed how each node can be privately computed. The complete protocol for privately computing ID3$_\delta$ can be seen below. The steps of the protocol correspond to those in the original algorithm (see Figure 1).

## Protocol 1 (Protocol for Private Computation of ID3$_\delta$:)

**Step 1:** *If $R$ is empty, return a leaf-node with the class value of the majority of the transactions in $T$.*

Since the set of attributes is known to both parties, they both publicly know if $R$ is empty. If yes, the parties do an oblivious evaluation of a circuit whose inputs are the values $\langle |T_1(c_1)|, \ldots, |T_1(c_\ell)| \rangle$ and $\langle |T_2(c_1)|, \ldots, |T_2(c_\ell)| \rangle$, and whose output is $i$ such $|T_1(c_i)| + |T_2(c_i)|$ is maximal. The size of this circuit is linear in $\ell$ and in $\log(|T|)$.

**Step 2:** *If $T$ consists of transactions with all the same class $c$, return a leaf-node with the value $c$.*

In order to compute this step privately, we must determine whether both parties remain with the same single class or not. We define a fixed symbol $\perp$ symbolizing the fact that a party has more than one remaining class. A party's input to this step is then $\perp$, or $c_i$ if it is its one remaining class. All that remains to do is check *equality* of the two inputs. The value causing the equality can then be publicly announced as $c_i$ (halting the tree on this path) or $\perp$ (to continue growing the tree from the current point).

The equality check can be executed in one of two ways: (1) Using the "comparing information without leaking it" protocols of Fagin, Naor, and Winkler [9]. This solution requires the execution of $\log(\ell + 1)$ oblivious transfers. (2) Using a protocol suggested in [16] and which involves the oblivious evaluation of linear polynomials. The overhead of this solution is $O(1)$ oblivious transfers, using the oblivious polynomial evaluation protocol of [7].

**Step 3:** *(a) Determine the attribute that best classifies the transactions in $T$, let it be $A$.*

For every value $a_j$ of every attribute $A$, and for every value $c_i$ of the class attribute $C$, the two parties run the $x \ln x$ protocol of Section 6 for $T(a_j)$ and $T(a_j, c_i)$. They then continue as described in Section 5.2 by computing independent additions and inputting the results into a small circuit. Finally, they perform an oblivious evaluation of the circuit with the result being the attribute with the highest information gain, $A$. This is public knowledge as it becomes part of the output.

*(b,c) Recursively call* $ID3_\delta$ *for the remaining attributes on the transaction sets* $T(a_1), \ldots, T(a_m)$ *(where* $a_1, \ldots, a_m$ *are the values of attribute A).*
The result of 3(a) and the attribute values of $A$ are public and therefore both parties can individually partition the database and prepare their input for the recursive calls.

Although each individual step of the above protocol has been shown to be private, we must show that the composition is also private. The central issue in the proof involves showing that the control flow can be predicted from the input and output only.

**Theorem 2** *The protocol for computing* $ID3_\delta$ *is* private.

*Proof.* In this proof the simulator is described in generic terms as it is identical for $P_1$ and $P_2$. Furthermore, we skip details which are obvious Recall that the simulator is given the output decision tree.

We need to show that any information learned by the computation can be learned directly from the input and output. This is done by showing how the views can be correctly simulated based solely on the input and output. The computation of the tree is recursive beginning at the root. For each node, a "splitting" class is chosen (due to it having the highest information gain) developing the tree to the next level. Any implementation defines the order of developing the tree and this order is used by the simulator to write the messages received in the correct order. Therefore according to this order, at any given step the computation is based on finding the highest information gain for a *known* node (for the proof we ignore optimizations which find the gain for more than one node in parallel). We differentiate between two cases: (1) a given node is a leaf node and (2) a given node is not a leaf.
    1. *The Current Node in the Computation is a Leaf-Node:* The simulator checks, by looking at the input, if the set of attributes $R$ at this point is empty or not. If it is *not* empty (this can be deduced from the tree and the attribute-list which is public), then the computation proceeds to Step (2). In this case, the simulator writes that the oracle-answer from the equality call in Step (2) is equal (or else it would not be a leaf). On the other hand, if the list of attributes *is* empty, the computation is executed in Step (1) and the simulator writes the output of the majority evaluation to be the class appearing in the leaf.
    2. *The Current Node in the Computation is not a Leaf-Node:* In this case Step (1) is skipped and the oracle-answer of Step (2) must be not-equal; this is therefore what the simulator writes. The computation then proceeds to Step (3) which involves many invocations of the $x \ln x$ protocol, returning values uniformly distributed in $\mathcal{F}$. Therefore, the simulator simply chooses the correct number of random values (based on the public list of attribute names, values and class values) and writes them. The next step of the algorithm is a local computation (not included in the view) and an oblivious circuit evaluation. The

simulator simply looks to see which class is written in the tree at this node and writes the class name as the output from the circuit evaluation.

The computation then continues to the next node in the defined order of traversal. This completes the proof.  ∎

*Remark.* It is both surprising and interesting to note that if Steps (1) and (2) of the protocol are switched (as the algorithm is in fact presented in [15]), then it is no longer private. This is due to the equality evaluation in Step (2), which may leak information about the other party's input. Consider the case of a computation in which at a certain point the list of attributes is *empty* and $P_1$ has only one class $c$ left in his remaining transactions. The output of the tree at this point is a leaf with a class, assume that the class is $c$. From the output it is impossible for $P_1$ to know if $P_2$'s transactions also have only one remaining class or if the result is because the majority of both together is $c$. The majority circuit of Step (1) covers both cases and therefore does not reveal this information. However, if $P_1$ and $P_2$ first execute the equality evaluation, this information is revealed.

*Complexity.* A detailed analysis of the complexity of the protocol is presented in Appendix A. The overhead is dominated by the $x \ln x$ protocol.

## 6   A Protocol for Computing $x \ln x$

This section describes an efficient protocol for privately computing the $x \ln x$ function, as defined in Figure 2.

---

- **Input:** $P_1$'s input is a value $v_1$; $P_2$'s input is $v_2$.
- **Auxiliary input:** A large enough field $\mathcal{F}$, the size of which will be discussed later.
- **Output:** $P_1$ obtains $w_1 \in \mathcal{F}$ and $P_2$ obtains $w_2 \in \mathcal{F}$ such that:
  1. $w_1 + w_2 = (v_1 + v_2) \cdot \ln(v_1 + v_2) \bmod |\mathcal{F}|$
  2. $w_1$ and $w_2$ are uniformly distributed in $\mathcal{F}$ when viewed independently of one another.

---

**Fig. 2.** Definition of the $x \ln x$ protocol.

There are several difficulties in the design of such a protocol. Firstly, it is not clear how to obliviously compute the natural logarithm efficiently. Furthermore, the protocol must multiply two values together. An initial idea is to use Yao's generic two party circuit evaluation protocol [21] and construct a multiplication circuit. However, the size of this circuit is of the order of the *multiplication of the sizes of its inputs*. This subprotocol is to be repeated many times throughout the complete ID3$_\delta$ protocol and its efficiency is, therefore, crucial.

The solution requires a linear size circuit and a small number of simple oblivious evaluation protocols. The problem is divided into two parts: First it is shown how to compute shares of $\ln x$ from shares of $x$. Secondly, we show how to obtain shares of the product $x \ln x$ given separate shares of $x$ and $\ln x$.

## 6.1   Computing Shares of $\ln x$

We now show how to compute random shares $u_1$ and $u_2$ such that $u_1 + u_2 = \ln x$. The starting point for the solution is the Taylor series of the natural logarithm, namely:

$$\ln(1 + \varepsilon) = \sum_{i=1}^{\infty} \frac{(-1)^{i-1} \varepsilon^i}{i} = \varepsilon - \frac{\varepsilon^2}{2} + \frac{\varepsilon^3}{3} - \frac{\varepsilon^4}{4} + \cdots \qquad \text{for} \quad -1 < \varepsilon < 1$$

It is easy to verify that the error for a partial evaluation of the series is as follows:

$$\left| \ln(1 + \varepsilon) - \sum_{i=1}^{k} \frac{(-1)^{i-1} \varepsilon^i}{i} \right| < \frac{|\varepsilon|^{k+1}}{k+1} \cdot \frac{1}{1 - |\varepsilon|} \tag{4}$$

As is demonstrated in Section 6.3, the error shrinks exponentially as $k$ grows.

Now, given an input $x$, let $2^n$ be the power of 2 which is closest to $x$ (in the ID3$_\delta$ application, note that $n < \log |T|$). Therefore, $x = 2^n(1 + \varepsilon)$ where $-1/2 \le \varepsilon \le 1/2$. Consequently,

$$\ln(x) = \ln(2^n(1 + \varepsilon)) = n \ln 2 + \varepsilon - \frac{\varepsilon^2}{2} + \frac{\varepsilon^3}{3} - \frac{\varepsilon^4}{4} + \cdots$$

Our aim is to compute this Taylor series to the $k$'th place. Let $N$ be a predetermined (public) upper-bound on the value of $n$ ($N > n$ always). Now, we use a small circuit that receives $v_1$ and $v_2$ as input (the value of $N$ is hardwired into it) and outputs shares of $2^N \cdot n \ln 2$ (for computing the first element in the series of $\ln x$) and $\varepsilon \cdot 2^N$ (for computing the remainder of the series). This circuit is easily constructed: notice that $\varepsilon \cdot 2^n = x - 2^n$, where $n$ can be determined by looking at the two most significant bits of $x$, and $\varepsilon \cdot 2^N$ is obtained simply by shifting the result by $N - n$ bits to the left. The possible values of $2^N n \ln 2$ are hardwired into the circuit. As we have described, random shares are obtained by having one of the parties input random values $\alpha_1, \beta_1 \in_R \mathcal{F}$ into the circuit and having the circuit output $\alpha_2 = \varepsilon \cdot 2^N - \alpha_1$ and $\beta_2 = 2^N \cdot n \ln 2 - \beta_1$ to the other party. The parties therefore have shares $\alpha_1, \beta_1$ and $\alpha_2, \beta_2$ such that

$$\alpha_1 + \alpha_2 = \varepsilon 2^N \quad \text{and} \quad \beta_1 + \beta_2 = 2^N n \ln 2$$

The second stage of the protocol involves computing shares of the Taylor series approximation. In fact, it computes shares of

$$\text{lcm}(2, \ldots k) \cdot 2^N \left( n \ln 2 + \varepsilon - \frac{\varepsilon^2}{2} + \frac{\varepsilon^3}{3} - \cdots \frac{\varepsilon^k}{k} \right) \approx \text{lcm}(2, \ldots k) 2^N \ln x \tag{5}$$

(where $\text{lcm}(2, \ldots, k)$ is the lowest common multiple of $\{2, \ldots, k\}$, and we multiply by it to ensure that there are no fractions). In order to do this $P_1$ defines the following polynomial:

$$Q(x) = \text{lcm}(2, \ldots, k) \cdot \sum_{i=1}^{k} \frac{(-1)^{i-1}}{2^{N(i-1)}} \frac{(\alpha_1 + x)^i}{i} - w_1$$

where $w_1 \in_R \mathcal{F}$ is randomly chosen. It is easy to see that

$$w_2 \stackrel{\text{def}}{=} Q(\alpha_2) = \text{lcm}(2,...,k) \cdot 2^N \cdot \left( \sum_{i=1}^{k} \frac{(-1)^{i-1} \varepsilon^i}{i} \right) - w_1$$

Therefore by a single oblivious polynomial evaluation of the $k$-degree polynomial $Q(\cdot)$, $P_1$ and $P_2$ obtain random shares $w_1$ and $w_2$ to the approximation in Equation (5). Namely $P_1$ defines $u_1 = w_1 + \text{lcm}(2,\ldots,k)\beta_1$ and likewise $P_2$. We conclude that

$$u_1 + u_2 \approx \text{lcm}(2,\ldots,k)2^N \cdot \ln x$$

This equation is accurate up to an approximation error which we bound, and the shares are random as required. Since $N$ and $k$ are known to both parties, the additional multiplicative factor of $2^N \cdot \text{lcm}(2,\ldots,k)$ is public and can be removed at the very end. Notice that *all* the values in the computation are integers (except for $2^N n \ln 2$ which is given as the closest integer number).

*The size of the field $\mathcal{F}$.* It is necessary that the field be chosen large enough so that the initial inputs in each evaluation and the final output be between 0 and $|\mathcal{F}| - 1$. Notice that all computation is based on $\varepsilon 2^N$. This value is raised to powers up to $k$ and multiplied by $\text{lcm}(2,\ldots,k)$. Therefore a field of size $2^{Nk+2k}$ is clearly large enough, and requires $(N+2)k$ bits for representation.

We now summarize the $\ln x$ protocol:

**Protocol 2 (Protocol $\ln x$)**

1. $P_1$ and $P_2$ input their shares $v_1$ and $v_2$ into an oblivious evaluation protocol for a circuit outputting: (1) Random shares $\alpha_1$ and $\alpha_2$ of $\varepsilon 2^N$ (i.e. $\alpha_1 + \alpha_2 = \varepsilon 2^N \mod |\mathcal{F}|$). (2) Random shares $\beta_1, \beta_2$ such that $\beta_1 + \beta_2 = 2^N \cdot n \ln 2$.
2. $P_1$ chooses $w_1 \in_R \mathcal{F}$ and defines the following polynomial

$$Q(x) = \text{lcm}(2,\ldots,k) \cdot \sum_{i=1}^{k} \frac{(-1)^{i-1}}{2^{N(i-1)}} \frac{(\alpha_1 + x)^i}{i} - w_1$$

3. $P_1$ and $P_2$ then execute an oblivious polynomial evaluation with $P_1$ inputting $Q(\cdot)$ and $P_2$ inputting $\alpha_2$, in which $P_2$ obtains $w_2 = Q(\alpha_2)$.
4. $P_1$ and $P_2$ define $u_1 = \text{lcm}(2,\ldots,k)\beta_1 + w_1$ and $u_2 = \text{lcm}(2,\ldots,k)\beta_2 + w_2$ respectively. We have that $u_1 + u_2 \approx 2^N \text{lcm}(2,\ldots,k) \cdot \ln x$

**Proposition 3** *Protocol 2 constitutes a private protocol for computing random shares of $c \cdot \ln x$ in $\mathcal{F}$, where $c = 2^N \text{lcm}(2,\ldots,k)$.*

*Proof.* We first show that the protocol correctly computes shares of $c \ln x$. In order to do this, we must show that the computation over $\mathcal{F}$ results in a correct result over the reals. We first note that *all* the intermediate values are integers. In particular, $\varepsilon 2^n$ equals $x - 2^n$ and is therefore an integer as is $\varepsilon 2^N$ (since $N > n$). Furthermore, every division by $i$ ($2 \le i \le k$) is counteracted by a multiplication by $\operatorname{lcm}(2, \ldots, k)$. The only exception is $2^N n \ln 2$. However, this is taken care of by having the original circuit output the closest integer to $2^N n \ln 2$ (although the rounding to the closest integer introduces an additional approximation error, it is negligible compared to the approximation error of the Taylor series).

Secondly, the field $\mathcal{F}$ is defined to be large enough so that all intermediate values (i.e. the sum of shares) and the final output (as a real number times $2^N \cdot \operatorname{lcm}(2, \ldots, k)$) are between 0 and $|\mathcal{F}| - 1$. Therefore the two shares uniquely identify the result, which equals the sum (over the integers) of the two random shares if it is less than $|\mathcal{F}|$, or the sum minus $|\mathcal{F}|$ otherwise.

The proof of privacy appears in the full version of the paper. ∎

## 6.2 Computing Shares of $x \ln x$

We begin by briefly describing a simple multiplication protocol that on private inputs $a_1$ and $a_2$ outputs random shares $b_1$ and $b_2$ (in some finite field $\mathcal{F}$) such that $b_1 + b_2 = a_1 \cdot a_2$.

**Protocol 3 (Protocol $Mult(a_1, a_2)$)**

The protocol is very simple and is based on an oblivious evaluation of a linear polynomial. The protocol begins by $P_1$ choosing a random value $b_1 \in \mathcal{F}$ and defining a linear polynomial $Q(x) = a_1 x - b_1$. $P_1$ and $P_2$ then engage in an oblivious evaluation of $Q$, in which $P_2$ obtains $b_2 = Q(a_2) = a_1 \cdot a_2 - b_1$. We define the respective outputs of $P_1$ and $P_2$ as $b_1$ and $b_2$ giving us that $b_1 + b_2 = a_1 \cdot a_2$.

**Proposition 4** *Protocol 3 constitutes a private protocol for computing* Mult *as defined above.*

We are now ready to present the complete $x \ln x$ protocol:

**Protocol 4 (Protocol $x \ln x$)**

1. $P_1$ and $P_2$ use Protocol 2 for privately computing shares of $\ln x$ in order to obtain random shares $u_1$ and $u_2$ such that $u_1 + u_2 = \ln x$.
2. $P_1$ and $P_2$ use two invocations of Protocol 3 in order to obtain shares of $u_1 \cdot v_2$ and $u_2 \cdot v_1$.
3. $P_1$ (resp., $P_2$) then defines his output $w_1$ (resp., $w_2$) to be the sum of the two *Mult* shares and $u_1 \cdot v_1$ (resp., $u_2 \cdot v_2$).
4. We have that $w_1 + w_2 = u_1 v_1 + u_1 v_2 + u_2 v_1 + u_2 v_2 = (u_1 + u_2)(v_2 + v_2) = x \ln x$ as required.

**Theorem 5** *Protocol 4 is a protocol for privately computing random shares of* $x \ln x$.

The correctness of the protocol is straightforward. The proof of the privacy properties appears in the full version of the paper.

*Complexity* The detailed analysis of the complexity is presented in Appendix A.

## 6.3   Choosing the Parameter $k$

Recall that the parameter $k$ defines the accuracy of the Taylor approximation of the "ln" function. Given $\delta$ and the database, we analyze which $k$ we need to take in order to ensure that the defined $\delta$-approximation is correctly estimated[6] From here on we denote an approximation of the value $z$ by $\widetilde{z}$.

The approximation definition of ID3$_\delta$ requires that for all $A_1, A_2$

$$H_C(T|A_1) > H_C(T|A_2) + \delta \Rightarrow \widetilde{H_C}(T|A_1) > \widetilde{H_C}(T|A_2)$$

This is clearly fulfilled if $\left| H_C(T|A_b) - \widetilde{H_C}(T|A_b) \right| < \frac{\delta}{2}$ for $b = 1, 2$.

We now bound the difference on each $|\ln x - \widetilde{\ln x}|$ in order that the above condition is fulfilled. By replacing $\log x$ by $\frac{1}{\ln 2}|\ln x - \widetilde{\ln x}|$ in Equation (3) computing $H_C(T|A)$, we obtain a bound on the error of $\left| H_C(T|A_1) - \widetilde{H_C}(T|A_1) \right|$. A straightforward algebraic manipulation gives us that if $\frac{1}{\ln 2}|\ln x - \widetilde{\ln x}| < \frac{\delta}{4}$, then the error is less than $\frac{\delta}{2}$ as required. As we have mentioned (Equation (4)), the $\ln x$ error is bounded by $\frac{|\varepsilon|^{k+1}}{k+1} \frac{1}{1-|\varepsilon|}$ and this is maximum at $|\varepsilon| = \frac{1}{2}$ (recall that $-\frac{1}{2} \le \varepsilon \le \frac{1}{2}$). Therefore, given $\delta$, we set $\frac{1}{2^k k+1} < \frac{\delta}{4} \cdot \ln 2$ or $k + \log(k+1) > \log\left[\frac{4}{\delta \ln 2}\right]$ (for $\delta = 0.0001$, it is enough to take $k > 12$). Notice that the value of $k$ is *not* dependent on the input database.

## References

1. M. Bellare and S. Micali, *Non-interactive oblivious transfer and applications*, Advances in Cryptology - Crypto '89, pp. 547-557, 1990.
2. M. Ben-Or, S. Goldwasser and A. Wigderson, *Completeness theorems for non cryptographic fault tolerant distributed computation*, 20th STOC, (1988), 1-9.
3. D. Boneh and M. Franklin, *Efficient generation of shared RSA keys*, Proc. of Crypto' 97, LNCS, Vol. 1233, Springer-Verlag, pp. 425–439, 1997.
4. R. Canetti, *Security and Composition of Multi-party Cryptographic Protocols*. To appear in the Journal of Cryptology. Available from the Theory of Cryptography Library at http://philby.ucsd.edu/cryptlib, 1998.
5. D. Chaum, C. Crepeau and I. Damgard, *Multiparty unconditionally secure protocols*, 20th Proc. ACM Symp. on Theory of Computing, (1988), 11-19.
6. B. Chor, O. Goldreich, E. Kushilevitz and M. Sudan, *Private Information Retrieval*, 36th FOCS, pp. 41–50, 1995.
7. R. Cramer, N. Gilboa. M. Naor, B. Pinkas and G. Poupard, *Oblivious Polynomial Evaluation*, 2000.
8. S. Even, O. Goldreich and A. Lempel, *A Randomized Protocol for Signing Contracts*, Communications of the ACM **28**, pp. 637–647, 1985.
9. R. Fagin, M. Naor and P. Winkler, *Comparing Information Without Leaking It*, Communications of the ACM, vol 39, May 1996, pp. 77-85.

---

[6] An additional error is introduced by rounding the value $2^N n \ln 2$ to the closest integer. We ignore this error as it is negligible compared to the approximation error of the ln function.

10. J. Feigenbaum, J. Fong, M. Strauss and R. N. Wright, *Secure Multiparty Computation of Approximations*, manuscript, 2000.
11. N. Gilboa, *Two Party RSA Key Generation*, Proc of Crypto '99, Lecture Notes in Computer Science, Vol. 1666, Springer-Verlag, pp. 116–129, 1999.
12. O. Goldreich, *Secure Multi-Party Computation*, 1998. (Available at http://philby.ucsd.edu)
13. O. Goldreich, S. Micali and A. Wigderson, *How to Play any Mental Game - A Completeness Theorem for Protocols with Honest Majority*. In 19th ACM Symposium on the Theory of Computing, pp. 218-229, 1987.
14. J. Kilian, **Uses of randomness in algorithms and protocols**, MIT Press, 1990.
15. T. Mitchell, **Machine Learning.** McGraw Hill, 1997.
16. M. Naor and B. Pinkas, *Oblivious Transfer and Polynomial Evaluation*, Proc. of the 31st STOC, Atlanta, GA, pp. 245-254, May 1-4, 1999.
17. M. Naor and B. Pinkas, *Efficient Oblivious Transfer Protocols*, manuscript, 2000.
18. P. Paillier, *Public-Key Cryptosystems Based on Composite Degree Residuocity Classes*. Proc. of Eurocrypt '99, LNCS Vol. 1592, pp. 223–238, 1999.
19. J. Ross Quinlan, *Induction of Decision Trees*. Machine Learning 1(1): 81-106(1986)
20. M. O. Rabin, *How to exchange secrets by oblivious transfer*, Tech. Memo TR-81, Aiken Computation Laboratory, 1981.
21. A.C. Yao, *How to generate and exchange secrets*, Proc. of the 27th IEEE Symp. on Foundations of Computer Science, 1986, pp. 162–167.

# A   Complexity

The communication complexity is measured by two parameters: the number of rounds and the bandwidth of all messages sent. As for the computation overhead, it is measured by the number of exponentiations and oblivious transfers (ignoring evaluations of pseudo-random functions, since they are more efficient by a few orders of magnitude).

**Parameters:** The overhead depends on the following parameters:

- $T$, the number of transactions.
- $k$, the length of the Taylor series, which affects the accuracy.
- $\mathcal{F}$, the field over which the computation is done. This is set as a function of the above two parameters, namely $\log|\mathcal{F}| = (k + 2)\log|T|$
- $|R|$, the number of attributes.
- $m$, the number of possible values for each attribute (to simplify the notation assume that this is equal for all attributes).
- $\ell$, the number of possible values for the class attribute.
- $|E|$, the length of an element in the group in which oblivious transfers and exponentiations are implemented. To simplify the notation we assume that $|E| > \log|\mathcal{F}| = k\log|T|$.
- $|S|$, the length of a key for a pseudorandom function used in the circuit evaluation (say, 80 or 100 bits long).
- $|D|$, the number of nodes in the decision tree.

A very detailed analysis of the complexity is given in the full version of the paper. The $\ln x$ protocol (Protocol 2) affects the complexity the most. Its overall overhead is $O(\max(\log|T|, k))$ oblivious transfers. Since $|T|$ is usually large (e.g. $\log|T| = 20$), and on the other hand $k$ can be set to small values (e.g. $k = 12$), the overhead can be defined as $O(\log |T|)$ oblivious transfers. The main communication overhead is incurred by the circuit evaluation and is $O(k \log |T| \cdot |S|)$ bits.

**Finding the best attribute for a node.** This step requires running the $\ln x$ protocol for every attribute and for every combination of attribute-value and class-value, and evaluating a small circuit. The communication overhead is $O(|R|m\ell k \log |T| \cdot |S|)$ bits and the computation overhead is $O(|R|m\ell \log |T|)$ oblivious transfers. The number of rounds is $O(1)$.

**Computing all nodes of the decision tree.** All nodes on the same level of the tree can be computed in parallel. We therefore have that the number of rounds equals $O(d)$ where $d$ is the depth of the tree. The value of $d$ is upper bound by $|R|$ but is expected to be much smaller.

**Overall complexity:**

- **Parameters:** For a concrete example, assume that there are a million transactions $|T| = 2^{20}$, $|R| = 15$ attributes, each attribute has $m = 10$ possible values, the class attribute has $\ell = 4$ values, and $k = 10$ suffices to have the desired accuracy. Say that the depth of the tree is $d = 7$, and that it uses private keys of length $|S| = 80$ bits.
- **Rounds:** There are $O(d)$ rounds.
- **Communication:** The communication overhead is $O(|D| \cdot |R|m\ell k \log |T| \cdot |S|)$. In our example, this is $|D| \cdot 15 \cdot 10 \cdot 4 \cdot 10 \cdot 20 \cdot 80 = 9,600,000|D|$ bits times a very small constant factor. We conclude that the communication per node can be transmitted in a matter of seconds using a fast communication network (e.g. a T1 line with 1.5Mbps bandwidth, or a T3 line with 35Mbps).
- **Computation:** The computation overhead is $O(|D| \cdot |R|m\ell \log |T|)$. In our example, this is an order of $|D| \cdot 15 \cdot 10 \cdot 4 \cdot 20 = 12,000|D|$ exponentiations and oblivious transfers. Assuming that a modern PC can compute 50 exponentiations per second, we conclude that the computation per node can be completed in a matter of minutes.

In the full paper we present a comparison to generic solutions that shows that our protocol achieves a considerable improvement (both in comparison to the complete ID3 protocol and to the $x \ln x$ protocol).

# Reducing the Servers Computation in Private Information Retrieval: PIR with Preprocessing

Amos Beimel[1], Yuval Ishai[2], and Tal Malkin[3]

[1] Dept. of Computer Science, Ben-Gurion University, Beer-Sheva 84105, Israel.
beimel@cs.bgu.ac.il
[2] DIMACS and AT&T Labs – Research, USA. yuval@dimacs.rutgers.edu
[3] AT&T Labs – Research, 180 Park Ave., Florham Park, NJ 07932, USA.
tal@research.att.com.

**Abstract.** Private information retrieval (PIR) enables a user to retrieve a specific data item from a database, replicated among one or more servers, while hiding from each server the identity of the retrieved item. This problem was suggested by Chor et al. [11], and since then efficient protocols with sub-linear communication were suggested. However, in all these protocols the servers' computation for *each* retrieval is at least *linear* in the size of entire database, even if the user requires just one bit.

In this paper, we study the *computational* complexity of PIR. We show that in the standard PIR model, where the servers hold only the database, linear computation cannot be avoided. To overcome this problem we propose the model of *PIR with preprocessing*: Before the execution of the protocol each server may compute and store polynomially-many information bits regarding the database; later on, this information should enable the servers to answer each query of the user with more efficient computation. We demonstrate that preprocessing can save work. In particular, we construct, for any constant $k \geq 2$, a $k$-server protocol with $O(n^{1/(2k-1)})$ communication and $O(n/\log^{2k-2} n)$ work, and for any constants $k \geq 2$ and $\epsilon > 0$ a $k$-server protocol with $O(n^{1/k+\epsilon})$ communication and work. We also prove some lower bounds on the work of the servers when they are only allowed to store a small number of extra bits. Finally, we present some alternative approaches to saving computation, by batching queries or by moving most of the computation to an off-line stage.

## 1 Introduction

In this era of the Internet and www.bigbrother.com, it is essential to protect the privacy of the small user. An important aspect of this problem is hiding the information the user is interested in. For example, an investor might want to know the value of a certain stock in the stock-market without revealing the identity of this stock. Towards this end, Chor, Goldreich, Kushilevitz, and Sudan [11] introduced the problem of Private Information Retrieval (PIR). A PIR protocol allows a user to access a database such that the server storing the database does not gain any information on the records the user read. To make the problem

M. Bellare (Ed.): CRYPTO 2000, LNCS 1880, pp. 55–73, 2000.
© Springer-Verlag Berlin Heidelberg 2000

more concrete, the database is modeled as an $n$ bit string $x$, and the user has some index $i$ and is interested in privately retrieving the value of $x_i$.

Since its introduction, PIR has been an area of active research, and various settings and extensions have been considered (e.g., [2,25,10,20,18,17,14,9,8,19,15], [21,1]). Most of the initial work on PIR has focused on the goal of minimizing the *communication*, which was considered the most expensive resource. However, despite considerable success in realizing this goal, the real-life applicability of the proposed solutions remains questionable. One of the most important practical restrictions is the *computation* required by the servers in the existing protocols; in all protocols described in previous papers, the (expected) work of the server(s) involved is at least $n$, the size of the entire database, for a single query of the user. This computation overhead may be prohibitive, since the typical scenario for using PIR protocols is when the database is big.

In this paper, we initiate the study of using preprocessing to reduce server computation.[1] We demonstrate that, while without any preprocessing linear computation is unavoidable, with preprocessing and some extra storage, computation can be reduced. Such a tradeoff between storage and computation is especially motivated today; as storage becomes very cheap, the computation time emerges as the more important resource. We also provide some lower bounds on this tradeoff, relating the amount of additional storage and the computation required. Finally, we present some alternative approaches to saving computation. While this paper is still within the theoretical realm, we hope that the approach introduced here will lead to PIR protocols which are implemented in practice.

**Previous Work.** Before proceeding, we give a brief overview of some known results on PIR. The simplest solution to the PIR problem is that of communicating the entire database to the user. This solution is impractical when the database is large. However, if the server is not allowed to gain *any* information about the retrieved bit, then the linear communication complexity of this solution is optimal [11]. To overcome this problem, Chor et al. [11] suggested that the user accesses replicated copies of the database kept on different servers, requiring that each server gets absolutely no information on the bit the user reads (thus, these protocols are called *information-theoretic* PIR protocols). The best information-theoretic PIR protocols known to date are summarized below: (1) a 2-server protocol with communication complexity of $O\left(n^{1/3}\right)$ bits [11], (2) a $k$-server protocol, for a constant $k$, with communication complexity of $O\left(n^{1/(2k-1)}\right)$ bits [2] (improving on [11], see also [19]), and (3) a protocol with $O\left(\log n\right)$ servers and communication complexity of $O\left(\log^2 n \log\log n\right)$ bits [5,6,11]. In all these protocols it is assumed that the servers do not communicate with each other.[2]

A different approach for reducing the communication is to limit the power of the servers; i.e., to relax the perfect privacy requirement into *computational indistinguishability* against computationally bounded servers (thus, these protocols are called *computational* PIR protocols). Following a 2-server construc-

---

[1] [17] have used preprocessing in a different model, allowing to move most computation to special purpose servers (though not reducing the total work). See more below.

[2] Extensions to $t$-private PIR protocols, in which the user is protected against collusions of up to $t$ servers, have been considered in [11,19,7].

tion of Chor and Gilboa [10], Kushilevitz and Ostrovsky [20] proved that in this setting one server suffices; under a standard number theoretic assumption they construct, for every constant $\epsilon > 0$, a *single* server protocol with communication complexity of $O(n^\epsilon)$ bits. Cachin, Micali, and Stadler [9] present a single server protocol with polylogarithmic communication complexity, based on a new number theoretic intractability assumption. Other works in this setting are [25,24,27,8,15,21,1].

The only previous work that has addressed the servers' computation is that of Gertner, Goldwasser, and Malkin [17] (see also [23]), who present a model for PIR utilizing special-purpose privacy servers, achieving stronger privacy guarantees and small computation for the original server holding the database. While their protocols save computation for the original server, the computation of the special-purpose servers (who do not hold the database) is still linear for every query. In contrast, our goal is to reduce the *total* computation by all servers. Di-Crescenzo, Ishai, and Ostrovsky [14] present another model for PIR using special-purpose servers. By extending their technique, it is possible to shift most of the servers' work to an off-line stage, at the expense of requiring additional off-line work for each future query. This application is discussed in Section 5.

**Our Results.** As a starting point for this work, we prove that in any $k$-server protocol the total *expected* work of the servers is at least $n$ (or negligibly smaller than $n$ in the computational setting). Consequently, we suggest the model of PIR with preprocessing: Before the first execution of the protocol each server computes and stores some information regarding the database. These bits of information are called the *extra bits* (in contrast to the original data bits). Later on, this information should enable the servers to perform less computation for each of the (possibly many) queries of the users.[3] The number of extra bits each server is allowed to store in the preprocessing stage is polynomial in $n$.

We demonstrate that preprocessing can save computation. There are three important performance measurements that we would like to minimize: communication, servers' work (i.e., computation), and storage. We describe a few protocols with different trade-offs between these parameters. We first construct, for any $\epsilon > 0$ and constant $k \geq 2$, a $k$-server protocol with $O(n^{1/(2k-1)})$ communication, $O\left(n/(\epsilon \log n)^{2k-2}\right)$ work, and $O(n^{1+\epsilon})$ extra bits (where $n$ is the size of the database). The importance of this protocol is that it saves work without increasing the communication compared to the best known information-theoretic PIR protocols. We define a combinatorial problem for which a better solution will further reduce the work in this protocol. Our second construction moderately increases the communication; however the servers' work is much smaller. For any constants $k \geq 2$ and $\epsilon > 0$, we construct a $k$-server protocol with polynomially many extra bits and $O(n^{1/k+\epsilon})$ communication and work. All the above protocols maintain information-theoretic user privacy.

We prove, on the negative side, that if the servers are only allowed to store a bounded number of bits in the preprocessing stage, then their computation in response to each query is big. In particular, we prove that if the servers are

---

[3] This problem can be rephrased in terms of Yao's cell-probe model [28]; in the full version of the paper we elaborate on the connection with this model.

allowed to store only $e$ extra bits ($e \geq 1$) in the preprocessing stage, then the expected work of the servers is $\Omega(n/e)$.

Finally, we suggest two alternative approaches for saving work. First, we suggest batching multiple queries to reduce the amortized work per query, and show how to achieve sub-linear work while maintaining the same communication. Second, we show how to shift most of the work to an off-line stage, applying a separate preprocessing procedure for each future query. While generally more restrictive than our default model, both of these alternative approaches may be applied in the single-server case as well.

**Organization.** In Section 2 we provide the necessary definitions, in Section 3 we construct PIR protocols with reduced work, and in Section 4 we prove our lower bounds. In Section 5 we present the alternative approaches of batching and off-line communication, and in Section 6 we mention some open problems.

## 2   Definitions

We first define one-round[4] information-theoretic PIR protocols. A $k$-server PIR protocol involves $k$ servers $\mathcal{S}_1, \ldots, \mathcal{S}_k$, each holding the same $n$-bit string $x$ (the database), and a user who wants to retrieve a bit $x_i$ of the database.

**Definition 1 (PIR).** *A $k$-server PIR protocol $\mathcal{P} = (\mathcal{Q}_1, \ldots, \mathcal{Q}_k, \mathcal{A}_1, \ldots, \mathcal{A}_k, \mathcal{C})$ consists of three types of algorithms: query algorithms $\mathcal{Q}_j(\cdot, \cdot)$, answering algorithms $\mathcal{A}_j(\cdot, \cdot)$, and a reconstruction algorithm $\mathcal{C}(\cdot, \cdot, \ldots, \cdot)$ (C has $k + 2$ arguments). At the beginning of the protocol, the user picks a random string $r$ and, for $j = 1, \ldots, k$, computes a query $q_j = \mathcal{Q}_j(i, r)$ and sends it to server $\mathcal{S}_j$. Each server responds with an answer $a_j = \mathcal{A}_j(q_j, x)$ (the answer is a function of the query and the database; without loss of generality, the servers are deterministic). Finally, the user computes the bit $x_i$ by applying the reconstruction algorithm $\mathcal{C}(i, r, a_1, \ldots, a_k)$. A PIR protocol is secure if:*

**Correctness.** *The user always computes the correct value of $x_i$. Formally, $\mathcal{C}(i, r, \mathcal{A}_1(\mathcal{Q}_1(i, r), x), \ldots, \mathcal{A}_k(\mathcal{Q}_k(i, r), x)) = x_i$ for every $i \in \{1, \ldots, n\}$, every random string $r$, and every database $x \in \{0, 1\}^n$.*

**Privacy.** *Each server has no information about the bit that the user tries to retrieve: For every two indices $i_1$ and $i_2$, where $1 \leq i_1, i_2 \leq n$, and for every $j$, where $1 \leq j \leq k$, the distributions $\mathcal{Q}_j(i_1, \cdot)$ and $\mathcal{Q}_j(i_2, \cdot)$ are identical.*

We next define the model proposed in this paper, PIR with preprocessing. Adding the preprocessing algorithm $\mathcal{E}$ will become meaningful when we define the work in PIR protocols.

**Definition 2 (PIR with Preprocessing).** *A PIR protocol with $e$ extra bits $\mathcal{P} = (\mathcal{E}, \mathcal{Q}_1, \ldots, \mathcal{Q}_k, \mathcal{A}_1, \ldots, \mathcal{A}_k, \mathcal{C})$ consists of 4 types of algorithms: preprocessing algorithm $\mathcal{E}$ which computes a mapping from $\{0, 1\}^n$ to $\{0, 1\}^e$, query and*

---

[4] All the protocols constructed in this paper, as well as most previous PIR protocols, are one-round. This definition may be extended to multi-round PIR in the natural way. All our results (specifically, our lower bounds) hold for the multi-round case.

*reconstruction algorithms $\mathcal{Q}_j$ and $\mathcal{C}$ which are the same as in regular PIR protocols, and the answer algorithms $\mathcal{A}_j(\cdot, \cdot, \cdot)$, which, in addition to the query $q_j$ and the database $x$, have an extra parameter – the extra bits $\mathcal{E}(x)$. The privacy is as above and the correctness includes $\mathcal{E}$:*

**Correctness with Extra Bits.** *The user always computes the correct value of $x_i$. Formally, $\mathcal{C}(i, r, \mathcal{A}_1(\mathcal{Q}_1(i,r), x, \mathcal{E}(x)), \ldots, \mathcal{A}_k(\mathcal{Q}_k(i,r), x, \mathcal{E}(x))) = x_i$ for every $i \in \{1, ..., n\}$, every random string $r$, and every database $x \in \{0,1\}^n$.*

Next we define the work in a PIR protocol. We measure the work in a simplistic way, only counting the number of bits that the servers read (both from the database itself and from the extra bits). This is reasonable when dealing with lower bounds (and might even be too conservative as the work might be higher). In general this definition is not suitable for proving upper bounds. However, in all our protocols the servers' total work is linear in the number of bits they read.

**Definition 3 (Work in PIR).** *Fix a PIR protocol $\mathcal{P}$. For a query $q$ and database $x \in \{0,1\}^n$ we denote the the number of bits that $\mathcal{S}_j$ reads from $x$ and $\mathcal{E}(x)$ in response to $q$ by $\mathrm{BITS}_j(x,q)$.[5] For a random string $r$ of the user, an index $i \in \{1, \ldots, n\}$, and a database $x$, the work of the servers is defined as the sum of the number of bits each server reads. Formally, $\mathrm{WORK}(i, x, r) \stackrel{def}{=} \sum_{j=1}^{k} \mathrm{BITS}_j(x, \mathcal{Q}_j(i, r))$. Finally, the work of the servers for an $i \in \{1, \ldots, n\}$, and a database $x$ is the expected value, over $r$, of $\mathrm{WORK}(i, x, r)$. That is, $\mathrm{WORK}(i, x) \stackrel{def}{=} \mathrm{E}_r [\mathrm{WORK}(i, x, r)]$.*

**Notation.** We let $[m]$ denote the set $\{1, \ldots, m\}$. For a set $A$ and an element $i$, define $A \oplus i$ as $A \cup \{i\}$ if $i \notin A$ and as $A \setminus \{i\}$ if $i \in A$. For a finite set $A$, define $i \in_U A$ as assigning a value to $i$ which is chosen randomly with uniform distribution from $A$ independently of any other event. We let $\mathrm{GF}(2)$ denote the finite field of two elements. All logarithms are taken to the base 2. By $H$ we denote the binary entropy function; that is, $H(p) = -p \log p - (1-p) \log(1-p)$.

## 3  Upper Bounds

We show that preprocessing can reduce the work. We start with a simple protocol which demonstrates ideas of the protocols in the rest of the section, in Section 3.1 we present a 2-server protocol with $O(n/\log^2 n)$ work, and in Section 3.2 we construct a $k$-server protocol, for a constant $k$, with $O(n/\log^{2k-2} n)$ work. In these protocols the communication is $O(n^{1/(2k-1)})$ for $k$ servers. In Section 3.3 we describe a combinatorial problem concerning spanning of cubes; a good construction for this problem will reduce the work in the previous protocols. Finally, in Section 3.4 we utilize PIR protocols with short query complexity to obtain $k$-server protocols with $O(n^{1/k+\epsilon})$ work and communication.

**A Warm-Up.** We show that using $n^2/\log n$ extra bits we can reduce the work to $n/\log n$. This is only a warm-up as the communication in this protocol is $O(n)$. We save work in a simple 2-server protocol of Chor et al. [11].

---

[5]  Technically speaking, also $\mathcal{E}(x)$ should have been a parameter of BITS. However, since $\mathcal{E}$ is a function of $x$ we can omit it.

ORIGINAL PROTOCOL [11]. The user selects a random set $A^1 \subseteq [n]$, and computes $A^2 = A^1 \oplus i$. The user sends $A^j$ to $\mathcal{S}_j$ for $j = 1, 2$. Server $\mathcal{S}_j$ answers with $a^j = \bigoplus_{\ell \in A^j} x_\ell$. The user then computes $a^1 \oplus a^2$ which equals $\bigoplus_{\ell \in A^1} x_\ell \oplus \bigoplus_{\ell \in A^1 \oplus i} x_\ell = x_i$. Thus, the user outputs the correct value. The communication in this protocol is $2(n + 1) = O(n)$, since the user needs to send $n$ bits to specify a random subset $A^j$ to $\mathcal{S}_j$, and $\mathcal{S}_j$ replies with a single bit.

OUR CONSTRUCTION. We use the same queries and answers as in the above protocol, but use preprocessing to reduce the servers' work while computing their answers. Notice that in the above protocol each server only computes the exclusive-or of a subset of bits. To save on-line work, the servers can precompute the exclusive-or of some subsets of bits. More precisely, the set $[n]$ is partitioned to $n/\log n$ disjoint sets $D_1, \ldots, D_{n/\log n}$ of size $\log n$ (e.g., $D_t = \{(t-1)\log n + 1, \ldots, t \log n\}$ for $t = 1, \ldots, n/\log n$). Each server computes the exclusive-or for every subset of these sets. That is, for every $t$, where $1 \leq t \leq n/\log n$, and every $G \subseteq D_t$, each server computes and stores $\oplus_{\ell \in G} x_\ell$. This requires $(n/\log n) \cdot 2^{\log n} = n^2/\log n$ extra bits. Once a server has these extra bits, it can compute its answer as an exclusive-or of $n/\log n$ bits; that is, $\mathcal{S}_j$ computes the exclusive-or of the pre-computed bits $\bigoplus_{\ell \in A^j \cap D_1} x_\ell, \ldots, \bigoplus_{\ell \in A^j \cap D_{n/\log n}} x_\ell$.

### 3.1  A 2-Server Protocol with Improved Work

We describe, for every constant $\epsilon$, a 2-server protocol with $O(n^{1+\epsilon})$ extra bits, $O(n^{1/3})$ communication, and $O(n/(\epsilon^2 \log^2 n))$ work. Thus, our protocol exhibits tradeoff between the number of extra bits and the work. As the best known information-theoretic 2-server protocol without extra bits requires $O(n^{1/3})$ communication, our protocol saves work without paying in the communication.

**Theorem 1.** *For every $\epsilon$, where $\epsilon > 4/\log n$, there exists a 2-server PIR protocol with $n^{1+\epsilon}$ extra bits in which the work of the servers is $O(n/(\epsilon^2 \log^2 n))$ and the communication is $O(n^{1/3})$.*

*Proof.* We describe a simpler (and slightly improved) variant of a 2-server protocol of [11], and then show how preprocessing can save work for the servers.

ORIGINAL PROTOCOL (VARIANT OF [11]). Let $n = m^3$ for some $m$, and consider the database as a 3-dimensional cube, i.e., every $i \in [n]$ is represented as $\langle i_1, i_2, i_3 \rangle$ where $i_r \in [m]$ for $r = 1, 2, 3$. This is done using the natural mapping from $\{0,1\}^{m^3}$ to $(\{0,1\}^m)^3$. In Fig. 1 we describe the protocol. It can be checked that each bit, except for $x_{i_1,i_2,i_3}$, appears an even number of times in the exclusive-or the user computes in Step 3, thus cancels itself. Therefore, the user outputs $x_{i_1,i_2,i_3}$ as required. Furthermore, the communication is $O(m) = O(n^{1/3})$.

OUR CONSTRUCTION. To save on-line work the servers pre-compute the exclusive-or of some sub-cubes of bits. Let $\alpha = 0.5\epsilon \log n$. The set $[m]$ is partitioned to $m/\alpha$ disjoint sets $D_1, \ldots, D_{m/\alpha}$ of size $\alpha$ (e.g., $D_t = \{(t-1)\alpha + 1, \ldots, t\alpha\}$ for $t = 1, \ldots, m/\alpha$). For every $\ell \in [m]$, every $t_1, t_2$, where $1 \leq t_1, t_2 \leq m/\alpha$, every $G_1 \subseteq D_{t_1}$, and every $G_2 \subseteq D_{t_2}$, each server computes and stores the three bits

---

**A Two Server Protocol with Low Communication**

1. The user selects three random sets $A_1^1, A_2^1, A_3^1 \subseteq [m]$, and computes $A_r^2 = A_r^1 \oplus i_r$ for $r = 1, 2, 3$. The user sends $A_1^j, A_2^j, A_3^j$ to $\mathcal{S}_j$ for $j = 1, 2$.
2. Server $\mathcal{S}_j$ computes for every $\ell \in [m]$

$$a_{1,\ell}^j \overset{\text{def}}{=} \bigoplus_{\ell_1 \in A_2^j, \ell_2 \in A_3^j} x_{\ell, \ell_1, \ell_2}, \qquad a_{2,\ell}^j \overset{\text{def}}{=} \bigoplus_{\ell_1 \in A_1^j, \ell_2 \in A_3^j} x_{\ell_1, \ell, \ell_2}, \qquad \text{and}$$

$$a_{3,\ell}^j \overset{\text{def}}{=} \bigoplus_{\ell_1 \in A_1^j, \ell_2 \in A_2^j} x_{\ell_1, \ell_2, \ell},$$

and sends the $3m$ bits $\left\{a_{r,\ell}^j : r \in \{1, 2, 3\}, \ell \in [m]\right\}$ to the user.
3. The user outputs $\bigoplus_{r=1,2,3}(a_{r,i_r}^1 \oplus a_{r,i_r}^2)$.

---

**Fig. 1.** A two server protocol with communication $O(n^{1/3})$.

$\bigoplus_{\ell_1 \in G_1, \ell_2 \in G_2} x_{\ell, \ell_1, \ell_2}$, $\bigoplus_{\ell_1 \in G_1, \ell_2 \in G_2} x_{\ell_1, \ell, \ell_2}$, and $\bigoplus_{\ell_1 \in G_1, \ell_2 \in G_2} x_{\ell_1, \ell_2, \ell}$. This requires $3m \cdot (m/\alpha)^2 \cdot 2^{2\alpha} \leq m^3 \cdot 2^{\epsilon \log n} = n^{1+\epsilon}$ extra bits. Once a server has these extra bits, it can compute each bit of its answer as an exclusive-or of $O(m^2/\alpha^2)$ pre-computed bits.

ANALYSIS. The answer of each server contains $O(m)$ bits, and each bit requires reading $O(m^2/\alpha^2)$ bits. Thus, the number of bits that each server reads is $O(m^3/\alpha^2) = O\left(n/(\epsilon \log n)^2\right)$.     $\square$

## 3.2   A $k$-Server Protocol with Small Communication

We present a $k$-server protocol with $O(n^{1+\epsilon})$ extra bits, $O\left(n^{1/(2k-1)}\right)$ communication, and $O\left(n/(\epsilon \log n)^{2k-2}\right)$ work for constant $k$. (The best known information-theoretic $k$-server protocol without extra bits requires the same communication and $O(n)$ work).

**Theorem 2.** *For every $k$ and $\epsilon > 4k/\log n$, there is a $k$-server PIR protocol with $n^{1+\epsilon}$ extra bits in which the work is $O\left((2k)^{4k}n/(\epsilon \log n)^{2k-2}\right)$ and the communication is $O(n^{1/(2k-1)})$. If $k$ is constant, the work is $O\left(n/(\epsilon \log n)^{2k-2}\right)$, and if $k \leq 0.5(\log n)^{1/4}$ and $\epsilon \geq 1$ then the work is $O\left(n/(\epsilon \log n)^{k-2}\right)$.*

*Proof.* We save work in a $k$-server protocol of Ishai and Kushilevitz [19].

ORIGINAL PROTOCOL [19]. As the protocol of [19] involves some notation, we only describe its relevant properties. Let $n = m^d$ for some $m$ and for $d = 2k - 1$. The database is considered as a $d$-dimensional cube. That is, every index $i \in [n]$ is represented as $\langle i_1, i_2, \ldots, i_d \rangle$ where $i_r \in [m]$ for $r = 1, 2, \ldots, d$. A sub-cube of the $d$-dimensional cube is defined by $d$ sets $A_1, \ldots, A_d$ and contains all indices $\langle i_1, i_2, \ldots, i_d \rangle$ such that $i_r \in A_r$ for every $r$. A sub-cube is a $(d-1)$-dimensional sub-cube if there exists some $r$ such that $|A_r| = 1$. In the protocol from [19] each server has to compute, for $k^d m$ sub-cubes of dimension $(d-1)$, the exclusive-or of bits of the sub-cube. The communication in the protocol is $O\left(k^3 n^{1/(2k-1)}\right)$.

OUR CONSTRUCTION. To save on-line work the servers compute in advance the exclusive-or of bits for some $(d-1)$-dimensional sub-cubes. Let $\alpha = \frac{\epsilon \log n}{d-1}$. The set $[m]$ is partitioned to $m/\alpha$ disjoint sets $D_1, \ldots, D_{m/\alpha}$ of size $\alpha$. For every $r \in \{1, \ldots, d\}$, every $\ell \in [m]$, every $t_1, t_2, \ldots, t_{d-1}$, where $1 \leq t_1, t_2, \ldots, t_{d-1} \leq m/\alpha$, every $G_1 \subseteq D_{t_1}$, every $G_2 \subseteq D_{t_2}$, $\ldots$, and every $G_{d-1} \subseteq D_{t_{d-1}}$, each server computes and stores the bit $\bigoplus_{\ell_1 \in G_1, \ldots, \ell_{d-1} \in G_{d-1}} x_{\ell_1, \ldots, \ell_{r-1}, \ell, \ell_r, \ldots, \ell_{d-1}}$. This requires $dm \cdot (m/\alpha)^{d-1} \cdot 2^{(d-1)\alpha} < m^d \cdot 2^{(d-1)\alpha} = n \cdot 2^{(d-1)\frac{\epsilon \log n}{d-1}} = n^{1+\epsilon}$ extra bits (the inequality holds since $d^{d-1} < 2$ and since $\epsilon > 4k/\log n$). Once a server has these extra bits, it can compute each exclusive-or of the bits of any $(d-1)$-dimensional sub-cube as an exclusive-or of $O(m^{d-1}/\alpha^{d-1})$ pre-computed bits.

ANALYSIS. The answer of each server requires computing the exclusive-or of the bits of a $(d-1)$-dimensional sub-cube for $O(k^d m)$ sub-cubes, and each sub-cube requires reading $O((m/\alpha)^{d-1})$ bits. Thus, the number of bits that each server reads is $O(k^d m^d/\alpha^{d-1})$. Recall that $d = 2k - 1$, thus the work reduces to $O\left((2k)^{4k} n/(\epsilon \log n)^{2k-2}\right)$. □

## 3.3   Can the Protocols Be Improved?

We now describe a combinatorial problem concerning spanning of cubes. This problem is a special case of a more general problem posed by Dodis [16]. Our protocols in Section 3.1 and Section 3.2 are based on constructions for this problem; better constructions will enable to further reduce the work in these protocols.

We start with some notation. Consider the collection of all $d$-dimensional sub-cubes $\mathcal{F}_d \stackrel{\text{def}}{=} \{G_1 \times \ldots \times G_d : G_1, \ldots, G_d \subseteq [m]\}$. The exclusive-or of subsets of $[m]^d$ is defined in the natural way: For sets $S_1, \ldots, S_t \subseteq [m]^d$, the point $\ell \in [m]^d$ is in $\bigoplus_{j=1}^{t} S_j$ if and only if $\ell$ is in an odd number of sets $S_j$.

**Definition 4 ($q$-xor basis).** $\mathcal{X} \subseteq 2^{[m]^d}$ *is a $q$-xor basis of $\mathcal{F}_d$ if every sub-cube in $\mathcal{F}_d$ can be expressed as the exclusive-or of at most $q$ sets from $\mathcal{X}$.*

For example, for $D_1, \ldots, D_{m/\log m}$ the partition of $[m]$, defined in Section 3.1, the collection $\mathcal{X}_0 \stackrel{\text{def}}{=} \{G_1 \times G_2 : \exists i, j \; G_1 \subseteq D_i, G_2 \subseteq D_j\}$ is a $m^2/\log^2 m$-xor basis of $\mathcal{F}_2$. We next show how to use a $q$-xor basis of $\mathcal{F}_2$ for 2-server PIR protocols. A similar claim holds for $q$-xor basis of $\mathcal{F}_{2k-2}$ for $k$-server PIR protocols.

**Lemma 1.** *If $\mathcal{X}$ is a $q$-xor basis of $\mathcal{F}_2$ then there exists a 2-server PIR protocol in which the communication is $O(n^{1/3})$, the work is $O(n^{1/3}q)$, and the number of extra bits is $O(n^{1/3}|\mathcal{X}|)$.*

*Proof.* We start with the protocol of [11], described in Fig. 1, in which $n = m^3$. For each set $S \in \mathcal{X}$, each server computes and stores $3m|\mathcal{X}|$ bits: for every $\ell \in [m]$ it stores the bits $\bigoplus_{(\ell_1,\ell_2) \in S} x_{\ell,\ell_1,\ell_2}$, $\bigoplus_{(\ell_1,\ell_2) \in S} x_{\ell_1,\ell,\ell_2}$, and $\bigoplus_{(\ell_1,\ell_2) \in S} x_{\ell_1,\ell_2,\ell}$. In the protocol of [11] each server has to compute the exclusive-or of the bits of a 2-dimensional sub-cube for $O(m)$ sub-cubes. Each exclusive-or requires reading at most $q$ stored bits, hence the total work per server is $O(mq) = O(n^{1/3}q)$.[6] □

---

[6]   Each server should be able to efficiently decide which $q$ bits it needs for computing each answer bit; otherwise our measurement of work may be inappropriate.

Lemma 1 suggests the following problem:

**The combinatorial Problem.** *Construct a q-xor basis of $\mathcal{F}_d$ of size* $\mathrm{poly}(m^d)$ *such that q is as small as possible.*

It can be shown that the smallest $q$ for which there is a $q$-xor basis of $\mathcal{F}_d$ whose size is $\mathrm{poly}(m^d)$ satisfies $\Omega(m/\log m) \leq q \leq O(m^d/\log^d m)$. We do not know where in this range the minimum $q$ lies. A construction with a smaller $q$ than the current upper bound will further reduce the work in PIR protocols.

### 3.4   Utilizing PIR Protocols with Logarithmic Query Length

If we have a PIR protocol with logarithmic query length and sub-linear answer length, then it is feasible for the servers to compute and store in advance the answers to *all* of the (polynomially many) possible queries. When a server receives a query it only needs to read the prepared answer bits. In general,

**Lemma 2.** *If there is a k-server PIR protocol in which the length of the query sent to each server is $\alpha$ and the length of answer of each server is $\beta$, then there is a k-server PIR protocol with $\beta$ work per server, $\alpha + \beta$ communication, and $2^\alpha \cdot \beta$ extra-bits.*

A 2-server PIR protocol with $\alpha = \log n$ and sub-linear $\beta$ is implied by communication complexity results of [26,4,3]. The most recent of those, due to Ambainis and Lokam [3], implies an upper bound of $\beta = n^{0.728...+o(1)}$.[7] We use similar techniques to construct a family of PIR protocols which provides a general tradeoff between $\alpha$ and $\beta$. In particular, our construction allows the exponent in the polynomial bounding the answer length to get arbitrarily close to $1/2$ while maintaining $O(\log n)$ query length. At the heart of the construction is the following lemma of Babai, Kimmel, and Lokam [4]. Let $\Lambda(m, w) = \sum_{h=0}^{w}\binom{m}{h}$.

**Lemma 3 ([4]).** *Let $p(Y_1, Y_2, \ldots, Y_m)$ be a degree-d m-variate polynomial[8] over* GF(2). *Let $y_\ell^h$, where $1 \leq h \leq k$ and $1 \leq \ell \leq m$, be arbitrary $km$ elements of* GF(2), *and $y_\ell = \sum_{h=1}^{k} y_\ell^h$ for $\ell = 1, \ldots, m$. Suppose that each $\mathcal{S}_j$ knows all $(k-1)m$ bits $y_\ell^h$ with $h \neq j$ and the polynomial $p$, and that the user knows all $km$ values $y_\ell^h$ but does not know $p$. Then, there exists a communication protocol in which each $\mathcal{S}_j$ simultaneously sends to the user a single message of length $\Lambda(m, \lfloor d/k \rfloor)$, and the user always outputs the correct bit value of $p(y_1, \ldots, y_m)$.*

The key idea in our construction is to apply Lemma 3 where $(y_1, y_2, \ldots, y_m)$ is a "convenient" encoding of the retrieval index $i$. Specifically, by using a low-weight encoding of $i$, the data bit $x_i$ can be expressed as a low-degree polynomial (depending on $x$) in the bits of the encoding. By letting the user secret-share the encoding of $i$ among the servers in an appropriate manner, Lemma 3 will allow the servers to communicate $x_i$ to the user efficiently. Low-weight encodings (over larger fields) have been previously used in PIR-related works [6,11,14]. However, it is the combination of this encoding with Lemma 3 which gives us the extra power.

---

[7]   This immediately implies a protocol with $n^{0.728...+o(1)}$ communication and work and $n^{1.728...+o(1)}$ extra-bits.

[8]   A degree-$d$ polynomial is a multi-linear polynomial of (total) degree *at most* $d$.

**Theorem 3.** *Let $m$ and $d$ be positive integers such that $\Lambda(m, d) \geq n$. Then, for any $k \geq 2$, there exists a $k$-server PIR protocol with $\alpha = (k-1)m$ query bits and $\beta = \Lambda(m, \lfloor d/k \rfloor)$ answer bits per server.*

*Proof.* Assign a distinct length-$m$ binary encoding $E(i)$ to each index $i \in [n]$, such that $E(i)$ contains at most $d$ ones. (Such an encoding exists since $\Lambda(m, d) \geq n$.) For each $x \in \{0, 1\}^n$, define a degree-$d$ $m$-variate polynomial $p_x$ over GF(2) such that $p_x(E(i)) = x_i$ for every $i \in [n]$.[9] Specifically, let $p_x(Y_1, \ldots, Y_m) = \sum_{i=1}^{n} x_i \cdot p^{(i)}(Y_1, \ldots, Y_m)$, where each $p^{(i)}$ is a fixed degree-$d$ polynomial such that $p^{(i)}(E(i'))$ equals 1 if $i = i'$ and equals 0 if $i \neq i'$. (The polynomials $p^{(i)}$ can be constructed in a straightforward way; details are omitted from this version.) The protocol with the specified complexity is described below. The user encodes $i$ as the $m$-bit string $y = E(i)$, and breaks each bit $y_\ell$, where $1 \leq \ell \leq m$, into $k$ additive shares $y_\ell^1, \ldots, y_\ell^k$; that is, $y_\ell^1, \ldots, y_\ell^{k-1}$ are chosen uniformly at random from GF(2), and $y_\ell^k$ is set so that the sum (i.e., exclusive-or) of the $k$ shares is equal to $y_\ell$. The user sends to each $\mathcal{S}_j$ the $(k-1)m$ shares $y_\ell^h$ with $h \neq j$. The query sent to each $\mathcal{S}_j$ consists of $(k-1)m$ uniformly random bits, guaranteeing the privacy of the protocol. By Lemma 3, each server can send $\Lambda(m, \lfloor d/k \rfloor)$ bits to the user such that the user can reconstruct $p_x(y_1, \ldots, y_m) = x_i$.     □

We note that, by using constant-weight encodings, Theorem 3 can be used to improve the communication complexity of the 2-server protocol from [11] and its $k$-server generalizations from [2,19] by constant factors. This and further applications of the technique are studied in [7]. For the current application, however, we will be most interested in denser encodings, in which the *relative* weight $d/m$ is fixed as some constant $\theta$, where $0 < \theta \leq 1/2$. In the following we rely on the approximation $2^{(H(\theta)-o(1))m} \leq \Lambda(m, \lfloor \theta m \rfloor) \leq 2^{H(\theta)m}$ (cf. [22, Theorem 1.4.5]). For $\Lambda(m, \lfloor \theta m \rfloor) \geq n$ to hold, it is sufficient to let $m = (1/H(\theta) + o(1)) \log n$. Substituting the above $m$ and $d = \lfloor \theta m \rfloor$ in Theorem 3, and applying the transformation to PIR with preprocessing described in Lemma 2, we obtain:

**Theorem 4.** *For any integer $k \geq 2$ and constant $0 < \theta \leq 1/2$, there exists a $k$-server protocol with $n^{H(\theta/k)/H(\theta)+o(1)}$ communication and work, and $n^{(k-1+H(\theta/k))/H(\theta)+o(1)}$ extra bits.*

In particular, since $H(\theta/k)/H(\theta)$ tends to $1/k$ as $\theta$ tends to 0, we have:

**Theorem 5.** *For any constants $k \geq 2$ and $\epsilon > 0$ there exists a $k$-server protocol with polynomially many extra bits and $O(n^{1/k+\epsilon})$ communication and work.*

The number of extra bits in the protocols of Theorem 4 may be quite large. By partitioning the database into small blocks, as in [11], it is possible to obtain a more general tradeoff between the storage and the communication and work. Specifically, by using blocks of size $n^\mu$, where $0 < \mu \leq 1$, we

---

[9] The *existence* of an encoding $E : [n] \to \text{GF}(2)^m$ such that $x_i$ can be expressed as a degree-$d$ polynomial in the encoding of $i$ easily follows from the fact that the space of degree-$d$ $m$-variate polynomials has dimension $\Lambda(m, d)$. We use the specific low-weight encoding for concreteness. Furthermore, the condition $\Lambda(m, d) \geq n$ is essential for the existence of such encoding.

obtain a protocol with $n^{\mu H(\theta/k)/H(\theta)+(1-\mu)+o(1)}$ communication and work and $n^{\mu(k-1+H(\theta/k))/H(\theta)+(1-\mu)+o(1)}$ extra bits.[10] It follows that for any constant $\epsilon > 0$ there exists a constant $\epsilon' > 0$ such that there is a 2-server protocol with $O(n^{1+\epsilon})$ extra bits and $O(n^{1-\epsilon'})$ communication and work.

*Remark 1.* There is a $k$-server PIR protocol with *one* extra bit (which is the exlusive-or of all bits in the database), $\frac{k}{2k-1} \cdot n$ work, and $O(n)$ communication. Thus, with 1 extra-bit we can save a constant fraction of the work. In Section 4 we show that with a constant number of bits at most a constant fraction of the computation can be saved, and if the bit is an exclusive-or of a subset of the data bits then the computation is at least $n/2$. Thus, this protocol illustrates that our lower bounds are essentially tight. The protocol will be described in the full version of this paper.

## 4   Lower Bounds

We prove that without preprocessing (namely without extra bits), the expected number of bits all servers must read is at least $n$, the size of the database. We then prove that if there are $e$ extra bits ($e \geq 1$) then the expected number of bits all servers must read is $\Omega(n/e)$. These lower bounds hold for any number of servers $k$, and regardless of the communication complexity of the protocol.

Note that we only prove that the expectation is big, since there could be specific executions where the servers read together less bits.[11] The fact that there are executions with small work should be contrasted with single-server (computational) PIR protocols without preprocessing, where the server has to read the entire database for each query, except with negligible probability: if the server does not read $x_\ell$ in response to some query, it knows that the user is not interested in $x_\ell$, violating the user's privacy.

We start with some notation crucial for the lower bound. Fix a PIR protocol, and denote the user's random input by $r$. Let $\mathcal{C} \subseteq \{0,1\}^n$ be a set of strings (databases) to be fixed later. Define $B_j(i)$ as the set of all indices that server $\mathcal{S}_j$ reads in order to answer the user's query when the database is chosen uniformly from $\mathcal{C}$. Note that the set of bits $B_j(i)$ that $\mathcal{S}_j$ reads is a function of the query and the values of the bits that the server has already read. Since the query is a function of the index and the user's random input, the set $B_j(i)$ is a random variable of the user's random input $r$ and the database $c \in_U \mathcal{C}$. Next define

$$\mathrm{P}(\ell) \stackrel{\text{def}}{=} \max_{1 \leq i \leq n} \left\{ \Pr_{r,c} \left[ \ell \in \bigcup_{j=1}^{k} B_j(i) \right] \right\}. \tag{1}$$

---

[10]  In particular, the protocol obtained by letting $k = 2$, $\theta = 1 - 1/\sqrt{2}$, and $\mu = H(\theta)$ is very similar (and is slightly superior) to the protocol implied by [3].

[11]  For example, consider the following 2-server protocol (without any extra bits). The user with probability $\frac{1}{n}$ sends $i$ to server $\mathcal{S}_1$, and nothing to server $\mathcal{S}_2$, and with probability $(1 - \frac{1}{n})$ sends a random $j \neq i$ to $\mathcal{S}_1$, and sends $[n]$ to $\mathcal{S}_2$. Server $\mathcal{S}_j$, upon reception of a set $B_j$, replies with the bits $\{x_\ell : \ell \in B_j\}$ to the user.

That is, for every index $i$ we consider the probability that at least one server reads $x_\ell$ on a query generated for index $i$, and $\mathrm{P}(\ell)$ is the maximum of these probabilities. Furthermore, define the random variable $B_j \stackrel{\text{def}}{=} B_j(1)$ (by Lemma 4 below, the random variable $B_j$ would not change if we choose another index instead of 1). Finally, for every $\ell$ define $\mathrm{P}_j(\ell) \stackrel{\text{def}}{=} \mathrm{Pr}_{r,c}[\ell \in B_j]$, that is, the probability that $x_\ell$ is read by $\mathcal{S}_j$ (again, by Lemma 4 below, this probability is the same no matter which index $i$ was used to generate the query).

## 4.1   Technical Lemmas

We start with three lemmas that will be used to establish our lower bounds. First note that, by the user's privacy, the view of $\mathcal{S}_j$, and in particular $B_j(i)$, is identically distributed for 1 and for any $i$. Thus,

**Lemma 4.** *For every $j \in \{1, \ldots, k\}$, every index $i \in [n]$, and every set $B \subseteq [n]$,*
$\mathrm{Pr}_{r,c}[B_j(i) = B] = \mathrm{Pr}_{r,c}[B_j = B]$.

**Lemma 5.** *For every $j \in \{1, \ldots, k\}$ it hold that $\mathrm{E}_{r,c}[|B_j|] = \sum_{\ell=1}^{n} \mathrm{P}_j(\ell)$.*

*Proof.* Define the random variables $Y_1, \ldots, Y_n$ where $Y_\ell = 1$ if $\ell \in B_j$ and $Y_\ell = 0$ otherwise. Clearly, $\mathrm{E}_{r,c}[Y_\ell] = \mathrm{Pr}[Y_\ell = 1] = \mathrm{P}_j(\ell)$. Furthermore, $|B_j| = \sum_{\ell=1}^{n} Y_\ell$. Thus, $\mathrm{E}_{r,c}[|B_j|] = \mathrm{E}_{r,c}[\sum_{\ell=1}^{n} Y_\ell] = \sum_{\ell=1}^{n} \mathrm{E}_{r,c}[Y_\ell] = \sum_{\ell=1}^{n} \mathrm{P}_j(\ell)$. □

Next we prove that $\sum_{\ell=1}^{n} \mathrm{P}(\ell)$ is a lower bound on the expected number of bits the servers read, namely on the expected work for a random database in $\mathcal{C}$.

**Lemma 6.** *For every $i \in [n]$, $\mathrm{E}_{r,c}\left[\sum_{j=1}^{k} |B_j(i)|\right] \geq \sum_{\ell=1}^{n} \mathrm{P}(\ell)$.*

*Proof.* First, for an index $\ell \in [n]$ let $i_\ell$ be an index that maximizes the probability in the r.h.s. of (1), that is, $\mathrm{P}(\ell) = \mathrm{Pr}_{r,c}\left[\ell \in \bigcup_{j=1}^{n} B_j(i_\ell)\right]$. Second, by Lemma 4,

$$\mathrm{P}_j(\ell) = \Pr_{r,c}\left[\ell \in B_j(i_\ell)\right]. \tag{2}$$

Therefore, using the union bound,

$$\mathrm{P}(\ell) = \Pr_{r,c}\left[\ell \in \bigcup_{j=1}^{n} B_j(i_\ell)\right] \leq \sum_{j=1}^{k} \Pr_{r,c}\left[\ell \in B_j(i_\ell)\right] = \sum_{j=1}^{k} \mathrm{P}_j(\ell). \tag{3}$$

Third, by Lemma 4,

$$\mathrm{E}_{r,c}[|B_j(i)|] = \mathrm{E}_{r,c}[|B_j|]. \tag{4}$$

Thus, by linearity of the expectation, Equation (4), Lemma 5, and Inequality (3)

$$\mathrm{E}_{r,c}\left[\sum_{j=1}^{k} |B_j(i)|\right] = \sum_{j=1}^{k} \mathrm{E}_{r,c}[|B_j(i)|] = \sum_{j=1}^{k} \mathrm{E}_{r,c}[|B_j|]$$

$$= \sum_{j=1}^{k}\left(\sum_{\ell=1}^{n} \mathrm{P}_j(\ell)\right) = \sum_{\ell=1}^{n}\left(\sum_{j=1}^{k} \mathrm{P}_j(\ell)\right) \geq \sum_{\ell=1}^{n} \mathrm{P}(\ell). \quad □$$

We express Lemma 6 as a lower bound on the work for a specific database.

**Corollary 1.** *For every PIR protocol there exists a database $c \in \{0,1\}^n$ such that for every $i \in [n]$, $\mathrm{WORK}(i,c) \geq \sum_{\ell=1}^{n} \mathrm{P}(\ell)$.*

*Proof.* By our definitions $\mathrm{E}_{r,c \in \mathcal{C}} \sum_{j=1}^{k} |B_j(i)| = \mathrm{E}_{r,c \in \mathcal{C}} \sum_{j=1}^{k} \mathrm{BITS}_j(c, \mathcal{Q}_j(i,r)) = \mathrm{E}_{c \in \mathcal{C}} \mathrm{WORK}(i,c)$. Thus, by Lemma 6, $\mathrm{E}_{c \in \mathcal{C}} \mathrm{WORK}(i,c) \geq \sum_{\ell=1}^{n} \mathrm{P}(\ell)$. Therefore, there must be some $c \in \mathcal{C}$ such that $\mathrm{WORK}(i,c) \geq \sum_{\ell=1}^{n} \mathrm{P}(\ell)$. ☐

*Remark 2.* In the full version of this paper we prove that the corollary holds (up to a negligible difference) even if we replace the perfect privacy of the $k$-server PIR protocol with computational privacy. Thus, all the lower bounds in this section hold, up to a negligible difference, for $k$-server computational PIR protocols as well.

## 4.2   Lower Bound without Extra Bits

We next prove that without extra bits the expected number of bits that the servers read is at least $n$. This lower bound holds for every database. For simplicity we prove this lower bound for the case that the database is $0^n$. The idea behind the lower bound is that one cannot obtain the value of $x_\ell$ without reading $x_\ell$, thus for every query the user generates with index $\ell$ at least one server must read $x_\ell$. This implies that $\mathrm{P}(\ell) = 1$ and the lower bound follows Corollary 1.

**Theorem 6.** *For every PIR protocol without extra bits and for every $i \in [n]$ $\mathrm{WORK}(i, 0^n) \geq n$.*

*Proof.* By Corollary 1 it is enough to prove that $\mathrm{P}(\ell) \geq 1$ for every $\ell$. (Trivially, $\mathrm{P}(\ell) \leq 1$). Define $\mathcal{C} = \{0^n\}$, i.e., the probabilities $\mathrm{P}(\ell)$ are defined when the value of the database is $0^n$. However, without reading the value of a bit, the servers do not know this value. If when the user queries about the $\ell$th bit no server reads this bit, then the answers of all the servers are the same for the databases $0^n$ and $0^{\ell-1} 1 0^{n-\ell}$, so with this query for one of these databases the user errs with probability at least $1/2$ in the reconstruction of $x_\ell$. Thus, by the correctness, for any possible query of the user generated with index $\ell$, at least one of the servers must read $x_\ell$. Thus, $\mathrm{P}(\ell) \geq \mathrm{Pr}_r \left[ \ell \in \bigcup_{j=1}^{k} B_j(\ell) \right] = 1$. ☐

## 4.3   Lower Bound with Extra Bits

In this section we show that a small number of extra bits cannot reduce the work too much. The proof uses information theory, and especially properties of the entropy function $\mathrm{H}$ (see, e.g., [12]).

To describe the ideas of the proof of the lower bound we first consider a special case where each of the $e$ extra bits is an exclusive-or of a subset of the bits of the database. That is, there is a system of $e$ linear equations over $\mathrm{GF}(2)$ that determines the values of the extra bits; the unknowns are the bits of the database. This is the case in all our protocols. (Better lower bounds for this case are presented in the end of this section.)

By Corollary 1 we need to prove that the probabilities $P(\ell)$ are big. We fix the database to be $x = 0^n$, therefore the values of the extra bits are fixed to 0 as well. Assume towards a contradiction that $P(\ell) < 1/(e + 1)$ for at least $e + 1$ indices, which, w.l.o.g., are $1, \ldots, e+1$. This implies that for every $i$, where $1 \le i \le e + 1$, when the user is retrieving the $i$th bit, the servers, with positive probability, do not read any of the bits $x_1, \ldots, x_{e+1}$.

Now let $x_{e+2}, \ldots, x_n$ and all the extra bits be zero. We have established that in this case the servers with positive probability do not read the bits $x_1, \ldots, x_{e+1}$, and the user concludes that $x_1 = 0, \ldots, x_{e+1} = 0$ from the answers of the servers. Hence, by the correctness, it must hold that $x_1 = 0, \ldots, x_{e+1} = 0$. But in this case the linear system that determines the extra bits is reduced to a system of $e$ homogeneous linear equations over $GF(2)$ where the unknowns are the bits $x_1, \ldots, x_{e+1}$. Any homogeneous system with $e$ equations and $e + 1$ unknowns has a non-trivial solution. Therefore, there is a non-zero database in which $x_{e+2}, \ldots, x_n$ and all the extra bits be zero, contradiction since at least one bit $x_i$ among $x_1, \ldots, x_{e+1}$ is not determined by $x_{e+2}, \ldots, x_n$ and the extra bits.

The above proof is only the rough idea of the proof of the general case. One problem in the general case is that we cannot fix the value of the database, and we need more sophisticated methods.

**Theorem 7.** *For every PIR protocol with $e$ extra bits, there is some database $c \in \{0,1\}^n$ such that for every $i \in [n]$, $WORK(i, c) \ge \frac{n}{4e} - \frac{1}{2}$.*

*Proof.* Since there are $e$ extra bits, there exits a value for these bits that is computed for at least $2^{n-e}$ databases. Fix such a value for the extra bits and let $\mathcal{C}$ be the set of databases with this value for the extra bits. Thus, $|\mathcal{C}| \ge 2^{n-e}$. Let $C$ be a random variable distributed uniformly over $\mathcal{C}$, and $C_i$ be the $i$th bit of $C$. By definition,

$$H(C) = \log |\mathcal{C}| \ge n - e. \tag{5}$$

We will prove that for all indices, but at most $2e$, it holds that $P(\ell) \ge 1/(4e)$. Thus, the theorem follows from Corollary 1. Without loss of generality, assume that $P(1) \le P(2) \le \ldots \le P(n)$. We start with a simple analysis of the entropies of $C_i$. First, by properties of conditional entropy,

$$H(C) = H(C_1 \ldots C_n) \le H(C_{2e+1} \ldots C_n) + \sum_{\ell=1}^{2e} H(C_\ell | C_{2e+1} \ldots C_n). \tag{6}$$

Second, since $C_{2e+1} C_{2e+2} \ldots C_n$ obtains at most $2^{n-2e}$ values,

$$H(C_{2e+1} C_{2e+2} \ldots C_n) \le n - 2e. \tag{7}$$

Combining (5), (6), and (7),

$$\sum_{\ell=1}^{2e} H(C_\ell | C_{2e+1} C_{2e+2} \ldots C_n) \ge H(C_1 \ldots C_n) - H(C_{2e+1} \ldots C_n) \ge e. \tag{8}$$

The next lemma, together with (8), shows that not too many $P(\ell)$ are small.

**Lemma 7.** *If* $P(\ell) < \frac{1}{4e}$ *for every* $\ell \in [2e]$, *then* $H(C_\ell | C_{2e+1} \ldots C_n) < 0.5$ *for every* $\ell \in [2e]$.

*Proof.* Fix $\ell$ and consider an experiment where the database $c$ is chosen uniformly from $\mathcal{C}$ and the PIR protocol is executed with the user generating a random query for index $\ell$. With probability at least half, none of the bits $x_1, \ldots, x_{2e}$ are read by any server in this execution (the probability is taken over the random input of the user and over the uniform distribution of $c \in \mathcal{C}$). Denote by $\mathcal{C}' \subseteq \mathcal{C}$ the set of all strings in $\mathcal{C}$ for which there is a positive probability that none of the bits $x_1, \ldots, x_{2e}$ is read by any server (this time the probability is taken only over the random input of the user). Thus, $|\mathcal{C}'| \geq 0.5|\mathcal{C}|$. Since the user always reconstructs the correct value of the bit $x_\ell$, then for every $c' \in \mathcal{C}'$ the values of the bits $c'_{2e}, \ldots, c'_n$ determine the value of $c'_\ell$; that is, for every $c \in \mathcal{C}$ if $c_m = c'_m$ for every $m \in \{2e+1, \ldots, n\}$, then $c_\ell = c'_\ell$. Now, define a random variable $Z$ where $Z = 1$ if the values of the bits $c_{2e}, \ldots, c_n$ determine the value of $c_\ell$ and $Z = 0$ otherwise. In particular, $Z$ must be 1 for any string in $\mathcal{C}'$. Hence,

$$\Pr_c[Z = 1] \geq 0.5. \tag{9}$$

Furthermore,

$$H(C_\ell | C_{2e+1} \ldots C_n Z = 1) = 0. \tag{10}$$

On the other hand, since $C_\ell$ obtains at most two values,

$$H(C_\ell | C_{2e+1} \ldots C_n Z = 0) \leq H(C_\ell) \leq 1. \tag{11}$$

By definition of conditional entropy, (11), (10), and (9)

$$\begin{aligned}
H(C_\ell &| C_{2e+1} \ldots C_n Z) \\
&= \Pr[Z = 0] \cdot H(C_\ell | C_{2e+1} \ldots C_n Z = 0) + \Pr[Z = 1] \cdot H(C_\ell | C_{2e+1} \ldots C_n Z = 1) \\
&\leq \Pr[Z = 0] \; < \; 0.5. 
\end{aligned} \tag{12}$$

The values of $C_{2e+1}, \ldots, C_n$ determine the value of $Z$, i.e., $H(Z | C_{2e+1} \ldots C_n) = 0$. Thus, $H(C_\ell | C_{2e+1} \ldots C_n) = H(C_\ell | C_{2e+1} \ldots C_n Z) < 0.5$. □

Lemma 7 and (8) imply that $P(\ell) \geq 1/4e$ for at least one $\ell \in \{1, \ldots, 2e\}$. Since we assume, without loss of generality, that $P(1) \leq P(2) \leq \ldots \leq P(n)$, then $\sum_{\ell=1}^n P(\ell) \geq \sum_{\ell=2e+1}^n P(\ell) \geq (n - 2e)/4e$, and by Corollary 1 the work of servers is as claimed in the theorem. □

**Better Lower Bounds for Exclusive-or Extra Bits.** If each extra bit is an exclusive-or of the bits of a subset of the database, then the lower bound of Theorem 7 can be improved by a factor of $\log n$, as stated in the following theorem (whose proof is omitted).

**Theorem 8.** *If* $e < \log n$, *then in every* $k$-*server PIR protocol with* $e$ *exclusive-or extra bits the work of the servers is at least* $(n - 2^e)/2$. *If* $\log n \leq e \leq \sqrt{n}$, *then in every* $k$-*server PIR protocol with* $e$ *exclusive-or extra bits the work is* $\Omega(n \log n/e)$.

As explained in Remark 1, the lower bound for a constant number of extra bits is essentially tight, as a matching upper bound protocol with one extra bit exists.

# 5    Alternative Approaches for Saving Work

The PIR with preprocessing model allows to reduce the on-line work in PIR protocols. In this section we discuss two alternative approaches for achieving the same goal. While both are in a sense more restrictive than our original model, in some situations they may be preferred. For instance, they both allow to substantially reduce the on-line work in *single*-server computational PIR, which is an important advantage over the solutions of Section 3.

## 5.1    Batching Queries

In the first alternative setting, we allow servers to batch together several queries before replying to all of them. By default, no preprocessing is allowed. The main performance measures of PIR protocols in this setting are: (1) the *amortized communication complexity*, defined as the average communication per query; (2) the *amortized work* per query; (3) the *batch size*, i.e., the minimum number of queries which should be processed together; and (4) the extra *space* required for storing and manipulating the batched queries. Note that in the case of a single user, the trivial PIR solution of communicating the entire database gives an optimal tradeoff between the batch size and the amortized work, namely their product is $n$.[12] However, this solution provides a poor tradeoff between the amortized communication and the batch size (their product is $n$). Moreover, as in the remainder of this paper, we are primarily interested in the general situation where different queries may originate from different users.[13]

Our main tool for decreasing the amortized work is a reduction to matrix multiplication. The savings achieved by the state-of-the-art matrix multiplication algorithms can be translated into savings in the amortized work of the PIR protocols. To illustrate the technique, consider the 2-server PIR protocol described in Fig. 1. In a single invocation of this protocol, each server has to compute the exclusive-or of $O(n^{1/3})$ two-dimensional sub-cubes. Each such computation can be expressed as evaluating a product (over $\mathrm{GF}(2)$) of the form $a^t X b$, where $a$ and $b$ are vectors in $\mathrm{GF}(2)^{n^{1/3}}$ determined by the user's query, and $X$ is an $n^{1/3} \times n^{1/3}$ matrix determined by the database $x$. It follows that the answers to $n^{1/3}$ queries can be computed by evaluating $O(n^{1/3})$ matrix products of the form $A \cdot X \cdot B$, where the $j$-th row of $A$ and the $j$-th column of $B$ are determined by the $j$-th query. The communication complexity of the protocol is $O(n^{1/3})$ per query, and its space and time requirements depend on the matrix multiplication algorithm being employed. Letting $\omega$ denote the exponent of matrix multiplication (Coppersmith and Winograd [13] prove that $\omega < 2.376$), the amortized work can be as low as $O(n^{1/3} n^{\omega/3})/n^{1/3} = O(n^{\omega/3})$, with batch size $n^{1/3}$.

Finally, we note that the same approach can also be employed towards reducing the amortized work in computational single-server PIR protocols, when batching queries of users who share the same key. In the protocols from [20,24,27], which utilize homomorphic encryption, the server's computation on multiple

---

[12] This is optimal by the lower bound of Theorem 6.

[13]    In this setting, the amortized communication complexity cannot be smaller than the communication complexity of a corresponding (single-query) PIR protocol.

queries can be reduced to evaluating several matrix products. In each product one matrix depends on the queries and is given in an encrypted form (using a key held by the user) and the other depends on the database and is given in a plain form. Now, by the definition of homomorphic encryption, an encryption of the sum of two encrypted values and an encryption of the product of an encrypted value with a non-encrypted value are both easy to compute. It turns out that these two operations are sufficient for implementing a fast matrix multiplication algorithm where one of the matrices is given in an encrypted form and the output may be encrypted as well. It follows (e.g., by modifying the protocol from [20]) that for any constant $\epsilon > 0$ there is a constant $\epsilon' > 0$, such that there exists a single-server PIR protocol with $O(n^\epsilon)$ batch size, $O(n^\epsilon)$ communication, $O(n^{1-\epsilon'})$ amortized work, and sub-linear extra space.

## 5.2  Off-Line Interaction

In the PIR with preprocessing model, a single off-line computational effort can reduce the on-line work in each of an unlimited number of future queries. It is natural to ask whether the on-line work can be further reduced if a separate off-line procedure is applied for each query. More precisely, we allow the user and the servers to engage in an off-line protocol, involving both communication and computation, so as to minimize the total on-line work associated with answering a *single* future query. (The off-line protocol may be repeated an arbitrary number of times, allowing to efficiently process many on-line queries.) During the off-line stage, the database $x$ is known to the servers but the retrieval index $i$ is unknown to the user. The goal is to obtain protocols with a small on-line work and "reasonable" off-line work.[14]

Towards achieving the above goal we extend an idea from [14]. Given any $k$-server PIR protocol ($k \geq 1$) in which the user sends $\alpha$ query bits to each server and receives $\beta$ bits in return, Di-Crescenzo et al. [14] show how to construct another $k$-server protocol where: (1) in the off-line stage the user sends $\alpha$ bits to each server and receives nothing in return; (2) in the on-line stage the user sends $\log n$ bits to each server and receives $\beta$ bits in return. Since there are only $n$ possible on-line queries made by the user, the servers can pre-compute the answers to each of these queries. Thus, with $\beta n$ storage, $O(\alpha)$ off-line communication and polynomial off-line computation, the on-line work is reduced to $O(\beta)$. Fortunately, most known PIR protocols admit variants in which the answer complexity $\beta$ is very small, as small as a single bit in the multi-server case, while $\alpha$ is still sub-linear (see [14] for a detailed account). For instance, in the 2-server computational PIR protocol of Chor and Gilboa [10], $\alpha = 2^{O(\sqrt{\log n})}$ and $\beta = 1$. Furthermore, by utilizing the structure of specific PIR protocols (including the one from [10]), the off-line computation of each server may be reduced to multiplying a length-$n$ data vector by an $n \times n$ Toeplitz matrix determined by the user's query. Thus, using the FFT algorithm, the total off-line computation can be made very close to linear.

---

[14]  This may be compared to the approach of Gertner et al. [17], where instead of shifting most computation to a "more reasonable place" (special purpose servers), here we shift most computation to a "more reasonable time" (the off-line stage).

# 6   Open Problems

We have shown that using preprocessing in PIR protocols one can obtain polynomial savings in the amount of computation without severely affecting the communication complexity. However, this work only initiates the study on PIR with preprocessing, and there are many open problems for further research. The obvious open problem is if more substantial savings are possible:

> *How much can the work be reduced using polynomially many extra bits?*
> *How much can be saved using* linearly *many extra bits?*

All the solutions provided in this work (with the exception of Section 5) are multi-server, information-theoretic PIR protocols. It is therefore natural to ask:

> *Can preprocessing substantially save work in single-server PIR protocols?*

**Acknowledgments.** We thank Oded Goldreich for suggesting the question of preprocessing in PIR protocols. Part of the work of Amos Beimel was done while in Harvard University, supported by grants ONR-N00014-96-1-0550 and ARO-DAAL03-92-G0115.

# References

1. W. Aiello, S. Bhatt, R. Ostrovsky, and S. Rajagopalan. Fast Verification of Any Remote Procedure Call: Short Witness-Indistinguishable One-Round Proofs for NP. In *ICALP 2000*.
2. A. Ambainis. Upper bound on the communication complexity of private information retrieval. In *24th ICALP*, volume 1256 of *LNCS*, pages 401–407, 1997.
3. A. Ambainis and S. Lokam. Improved upper bounds on the simultaneous messages complexity of the generalized addressing function. In *LATIN 2000*.
4. L. Babai, P. Kimmel, and S. Lokam. Simultaneous messages vs. communication. In *12th STACS*, volume 900 of *LNCS*, pages 361–372, 1995.
5. D. Beaver and J. Feigenbaum. Hiding instances in multioracle queries. In *7th STACS*, volume 415 of *LNCS*, pages 37–48. Springer-Verlag, 1990.
6. D. Beaver, J. Feigenbaum, J. Kilian, and P. Rogaway. Locally random reductions: Improvements and applications. *J. of Cryptology*, 10:17–36, 1997. Early version: Security with small communication overhead, *CRYPTO '90*.
7. A. Beimel and Y. Ishai. On private information retrieval and low-degree polynomials. Manuscript, 2000.
8. A. Beimel, Y. Ishai, E. Kushilevitz, and T. Malkin. One-way functions are essential for single-server private information retrieval. In *31th STOC*, pages 89–98, 1999.
9. C. Cachin, S. Micali, and M. Stadler. Computationally private information retrieval with polylogarithmic communication. In *EUROCRYPT '99*, volume 1592 of *LNCS*, pages 402–414. Springer, 1999.
10. B. Chor and N. Gilboa. Computationally private information retrieval. In *29th STOC*, pages 304–313, 1997.
11. B. Chor, O. Goldreich, E. Kushilevitz, and M. Sudan. Private information retrieval. In *36th FOCS*, pages 41–51, 1995. Journal version: *JACM*, 45:965–981, 1998.
12. T. M. Cover and J. A. Thomas. *Elements of Information Theory*. John Wiley & Sons, 1991.

13. D. Coppersmith and S. Winograd. Matrix multiplication via arithmetic progressions. *J. Symbolic Comput.*, 9:251-280, 1990.
14. G. Di-Crescenzo, Y. Ishai, and R. Ostrovsky. Universal service-providers for database private information retrieval. In *17th PODC*, pages 91–100, 1998.
15. G. Di-Crescenzo, T. Malkin, and R. Ostrovsky. Single-database private information retrieval implies oblivious transfer. In *EUROCRYPT 2000*, volume 1807 of *LNCS*, pages 122 –138, 2000.
16. Y. Dodis. Space-Time Tradeoffs for Graph Properties. Master's thesis, Massachusetts Institute of Technology, 1998.
17. Y. Gertner, S. Goldwasser, and T. Malkin. A random server model for private information retrieval. In *RANDOM '98, 2nd Workshop on Randomization and Approximation Techniques in CS*, vol. 1518 of *LNCS*, pages 200–217. 1998.
18. Y. Gertner, Y. Ishai, E. Kushilevitz, and T. Malkin. Protecting data privacy in private information retrieval schemes. In *30th STOC*, pages 151–160, 1998.
19. Y. Ishai and E. Kushilevitz. Improved upper bounds on information theoretic private information retrieval. In *31th STOC*, pages 79 – 88, 1999.
20. E. Kushilevitz and R. Ostrovsky. Replication is not needed: Single database, computationally-private information retrieval. In *38th FOCS*, pages 364–373, 1997.
21. E. Kushilevitz and R. Ostrovsky. One-way trapdoor permutations are sufficient for non-trivial single-server private information retrieval. In *EUROCRYPT 2000*, volume 1807 of *LNCS*, pages 104–121, 2000.
22. J. H. van Lint. *Introduction to Coding Theory*. Springer-Verlag, 1982.
23. T. Malkin. *A Study of Secure Database Access and General Two-Party Computation*. PhD thesis, MIT, 2000. http://theory.lcs.mit.edu/~cis/cis-theses.html .
24. E. Mann. Private access to distributed information. Master's thesis, Technion - Israel Institute of Technology, Haifa, 1998.
25. R. Ostrovsky and V. Shoup. Private information storage. In *29th STOC*, pages 294–303, 1997.
26. P. Pudlák and V. Rödl. Modified Ranks of Tensors and the Size of Circuits. In *25th STOC*, pages 523–531, 1993.
27. J. P. Stern. A new and efficient all-or-nothing disclosure of secrets protocol. In *ASIACRYPT '98*, volume 1514 of *LNCS*, pages 357–371. Springer, 1998.
28. A.C. Yao. Should tables be sorted? *JACM*, 28:615–628, 1981.

# Parallel Reducibility for Information-Theoretically Secure Computation

Yevgeniy Dodis[1] and Silvio Micali[1]

Laboratory for Computer Science, Massachusetts Institute of Technology, USA.
{yevgen,silvio}@theory.lcs.mit.edu

**Abstract.** Secure Function Evaluation (SFE) protocols are very hard to design, and *reducibility* has been recognized as a highly desirable property of SFE protocols. Informally speaking, reducibility (sometimes called modular composition) is the automatic ability to break up the design of complex SFE protocols into several simpler, individually secure components. Despite much effort, only the most basic type of reducibility, *sequential reducibility* (where only a single sub-protocol can be run at a time), has been considered and proven to hold for a specific class of SFE protocols. Unfortunately, sequential reducibility does not allow one to save on the number of rounds (often the most expensive resource in a distributed setting), and achieving more general notions is not easy (indeed, certain SFE notions provably enjoy sequential reducibility, but fail to enjoy more general ones).

In this paper, for information-theoretic SFE protocols, we

- Formalize the notion of *parallel reducibility*, where sub-protocols can be run at the same time;
- Clarify that there are two *distinct* forms of parallel reducibility:
  - ⋆ *Concurrent reducibility*, which applies when the order of the sub-protocol calls is not important (and which reduces the round complexity dramatically as compared to sequential reducibility); and
  - ⋆ *Synchronous reducibility*, which applies when the sub-protocols must be executed simultaneously (and which allows modular design in settings where sequential reducibility does not even apply).
- Show that a large class of SFE protocols (i.e., those satisfying a slight modification of the original definition of Micali and Rogaway [15]) provably enjoy (both forms of) parallel reducibility.

## 1   Introduction

The objective of this paper is to understand, define, and prove the implementability of the notion of parallel reducibility for information-theoretically secure multiparty computation. Let us start by discussing the relevant concepts.

M. Bellare (Ed.): CRYPTO 2000, LNCS 1880, pp. 74–92, 2000.
© Springer-Verlag Berlin Heidelberg 2000

**SFE Protocols.** A secure function evaluation (SFE) is a communication protocol enabling a network of players (say, having a specified threshold of honest players) to compute a (probabilistic) function in a way that is as correct and as private as if an uncorruptable third party had carried out the computation on the players' behalf. SFE protocols were introduced by Goldreich, Micali and Wigderson [13] in a *computational* setting (where the parties are computationally bounded, but can observe all communication), and by Ben-Or, Goldwasser and Wigderson [4] and Chaum, Crépeau and Damgård [7] in an *information-theoretic* setting (where the security is unconditional, and is achieved by means of *private channels*[1]). We focus on the latter setting.

**SFE Definitions.** Together with better SFE protocols, increasingly precise definitions for information-theoretic SFE have been proposed; in particular, those of Beaver [2], Goldwasser and Levin [11], Canetti [5], and Micali and Rogaway [15]. At a high-level, these definitions express that whatever an adversary can do in the *real model* (i.e., in the running of the actual protocol, where no trusted party exists) equals what an adversary can do in the *ideal model* (i.e., when players give their inputs to the trusted third party, who then computes the function for them). This more or less means that the most harm the adversary can do in the real model consists of changing the inputs of the faulty players (but not based on the inputs of the honest players!), and then running the protocol honestly.

All these prior definitions are adequate, in the sense that they (1) reasonably capture the desired intuition of SFE, and (2) provide for the existence of SFE protocols (in particular, the general protocol of [4] satisfies all of them). Were properties (1) and (2) all one cared about, then the most "liberal" definition of SFE might be preferable, because it would allow a greater number of reasonable protocols to be called secure. However, if one cared about satisfying *additional* properties, such as reducibility (i.e., as discussed below, the ability of designing SFE protocols in a modular fashion), then *more stringent* notions of SFE would be needed.

**Reducibility and Sequential Reducibility.** Assume that we have designed a SFE protocol, $F$, for a function $f$ in a so called *semi-ideal* model, where one can use a trusted party to evaluate some other functions $g^1, \ldots, g^k$. Assume also that we have designed a SFE protocol, $G_i$, for each function $g^i$. The reducibility property says that, by substituting the ideal calls to the $g^i$'s in $F$ with the corresponding SFE protocols $G_i$'s, we are *guaranteed* to obtain a SFE protocol for $f$ in the *real* model.

Clearly, reducibility is quite a fundamental and desirable property to have, because it allows one to break the task of designing a secure protocol for a complex function into the task of designing secure protocols for simpler functions. Reducibility, however, is not trivial to satisfy. After considerable effort, only the the most basic notion of reducibility, *sequential reducibility*, has been proved

---

[1] This means that every pair of players has a dedicated channel for communication, which the adversary can listen to only by corrupting one of the players.

to hold for some SFE notions: those of [5] and [15]. Informally, sequential reducibility guarantees that substituting the ideal calls to the $g^i$'s in $F$ with the corresponding $G_i$'s yields a SFE protocol for $f$ in the real model *only if* a single $G_i$ is executed (in its entirety!) at a time.[2] Therefore, sequential reducibility is not general enough to handle protocols like the expected $O(1)$-round Byzantine agreement protocol of [10] (which relies on the concurrent execution of $n^2$ specific SFE protocols) whose security, up to now, must be proven "from scratch".

## 1.1   Our Results

In this paper, we put forward the notion of *parallel reducibility* and show which kinds of SFE protocols satisfy it. We actually distinguish two forms of parallel reducibility:

- *Concurrent reducibility.*
  This type of reducibility applies when, in the semi-ideal model, the $g^1, \ldots, g^k$ can be executed in any order. The goal of concurrent reducibility is *improving the round-complexity* of modularly designed SFE protocols.

- *Synchronous reducibility.*
  This type of reducibility applies when, in the semi-ideal model, the $g^1, \ldots, g^k$ must be executed "simultaneously." The goal of synchronous reducibility is *enlarging the class of modularly designed SFE protocols* (while being round-efficient as well).

**Concurrent Reducibility.** There are many ways to schedule the execution of several programs $G_1, \ldots, G_k$. Each such way is called an *interleaving*. The $k!$ sequential executions of $G_1, \ldots, G_k$ are examples of interleavings. But they are very special and "very few," because interleavings may occur at a round-level. For instance, we could execute the $G_i$'s one round at a time in a round-robin manner, or we could simultaneously execute, in single round $r$, the $r$-th round (if any) of all the $G_i$'s. Saying that programs $G_1, \ldots, G_k$ are *concurrently executable*, relative to some specified goal, means that this goal is achieved *for all of their interleavings*.

Assume now that a function $f$ is securely evaluated by a semi-ideal protocol $F$ which, in a set of contiguous instructions, only makes ideal calls to functions $g^1, \ldots, g^k$, and let $G_i$ be a SFE protocol for $g^i$ (in the real model). Then, a fundamental question arise:

> *Will substituting each $g^i$ with $G_i$ yield a (real-model) SFE protocol for $f$ in which the $G_i$'s are concurrently executable?*

Let us elaborate on this question. Assume, for instance, that $F$ calls $g^2$ on inputs that include an output of $g^1$. Then we clearly cannot hope that the $G_i$'s are

---

[2] This is true even if, within $F$, one could "ideally evaluate" all or many of the $g^i$'s "in parallel."

concurrently executable. Thus, to make sense of the question, all the inputs to the $g^i$'s should be determined before any of them is ideally evaluated. Moreover, even if all the $g^i$'s are evaluated on completely unrelated and "independent" inputs, $F$ may be secure only for some orders of the $g^i$'s, but not for others, which is illustrated by the following example.

*Example 1:* Let $f$ be the coin-flipping function (that takes no inputs and outputs a joint random bit), let $g^1$ be the coin-flipping function as well, and let $g^2$ be the majority function on $n$ bits. Let now $F$ be the following semi-ideal protocol. Each player $P_j$ locally flips a random bit $b_j$. Then the players "concurrently" use ideal calls to $g^1$ and $g^2(b_1, \ldots, b_n)$, getting answers $r$ and $c$ respectively. The common output of $F$ is $r \oplus c$. We claim that $F$ is secure if we first call $g^2$ (the majority) and then $g^1$ (the coin-flip), but insecure if we do it the other way around. Indeed, irrespective of which $c$ we get in the first ordering, since $r$ is random (and independent of $c$), then so is $r \oplus c$. On the other hand, assume we first learn the random bit $r$ and assume faulty players want to bias the resulting coin-flip towards 0. Then the faulty players pretend that their (supposedly random) inputs $b_j$ for the majority are all equal to $r$. Provided there are enough faulty players, this strategy will bias the outcome $c$ of $g^2$ (the majority) towards $r$, and thus the output of $F$ towards 0.

Clearly, in the case of the above example, we cannot hope to execute the $G_i$'s concurrently: one of the possible interleavings is the one that sequentially executes the $G_i$'s in the order that is insecure even in the semi-ideal model. Thus, the example illustrates that the following condition is *necessary* for the concurrent execution of the $G_i$'s.

**Condition 1:** $F$ is secure in the semi-ideal model for any order of the $g^i$'s.

Is this necessary condition also sufficient? Of course, the answer also depends on the type of SFE notion we are using. But, if the answer were YES, then we would get the "strongest possible form of concurrent reducibility." Let us then be optimistic and put forward the following informal definition.

**Definition 1:** We say that a SFE notion satisfies *concurrent reducibility* if, whenever the protocols $F, G_1, \ldots, G_k$ satisfy this SFE notion, Condition 1 is (both necessary and) sufficient for the concurrent execution of the $G_i$'s inside $F$ (in the real model).

Our optimism is justified in view of the following

**Theorem 1:** A slight modification of the SFE notion of Micali and Rogaway [15] satisfies concurrent reducibility.

We note that the SFE notion of Micali and Rogaway is the strictest one proposed so far, and that we have been unable to prove analogous theorems for all other, more liberal notions of SFE. We conjecture that no such analogous theorems exist for those latter notions. In support of our conjecture, we shall point out in

Section 4.3 the strict properties of the definition of [15] that seem to be essential in establishing Theorem 1.

We remark that concurrent reducibility is important because it implies significant efficiency gains in the round-complexity (often the most expensive resource) of modularly designed SFE protocols. This is expressed by the following immediate Corollary of Definition 1.

> **Corollary 1:** Assume that $F, g^1, \ldots, g^k$ satisfy Condition 1, that $G_i$ is a protocol for $g^i$ taking $R_i$ rounds, and that $F, G_1, \ldots, G_k$ are SFE protocols according to a SFE notion satisfying concurrent reducibility. Then, there is a (real model) SFE implementation of $F$ executing all the $G_i$'s in $\max(R_1, \ldots, R_k)$ rounds.

This number of rounds is the smallest one can hope for, and should be contrasted with $R_1 + \cdots + R_k$, the number of rounds required by sequential reducibility.

**Synchronous Reducibility.** The need to execute several protocols in parallel does not necessarily arise from efficiency considerations or from the fact that it is nice not to worry about the order of the execution. A special type of parallel execution, *synchronous execution*, is needed for correctness itself.

*Example 2:* Let $f$ be the two-player coin-flipping function that returns a random bit to the first two players, $P_1$ and $P_2$, of a possibly larger network. That is, $f(\lambda, \lambda, \lambda, \ldots, \lambda) = (x, x, \lambda, \ldots, \lambda)$, where $x$ is a random bit (and $\lambda$ is the empty string). Consider now the following protocol $F$: player $P_1$ randomly and secretly selects a bit $x_1$, player $P_2$ randomly and secretly selects a bit $x_2$, and then $P_1$ and $P_2$ "exchange" their selected bits and both output $x = x_1 \oplus x_2$.

Clearly, $F$ is a secure function evaluation of $f$ only if the exchange of $x_1$ and $x_2$ is "simultaneous", that is, whenever $P_1$ learns $x_2$ only after it declares $x_1$, and vice versa. This requirement can be modeled as the parallel composition of two sending protocols: $g^1(x_1, \lambda, \lambda, \ldots, \lambda) = (x_1, x_1, \lambda, \ldots, \lambda)$ and $g^2(\lambda, x_2, \lambda, \ldots, \lambda) = (x_2, x_2, \lambda, \ldots, \lambda)$. That is, we can envisage a semi-ideal protocol in which players $P_1$ and $P_2$ locally flip coins $x_1$ and $x_2$, then *simultaneously* evaluate $g^1$ and $g^2$, and finally exclusive OR their outputs of $g^1$ and $g^2$. However, *no sequential order* of the ideal calls to $g^1$ and $g^2$ would result in a secure two-player coin-flipping protocol. This example motivates the introduction of a special type of parallel composition (for *security* rather than efficiency considerations).

The ability to evaluate several functions *synchronously* is very natural to define in the ideal model: the players simultaneously give all their inputs to the trusted party, who then gives them all the outputs (i.e., no output is given before all inputs are presented). We can also naturally define the corresponding semi-ideal model, where the players can ideally and simultaneously (i.e., within a single round) evaluate several functions. Assume now that we have a semi-ideal protocol $F$ for some function $f$ which simultaneously evaluates functions $g^1, \ldots, g^k$, and let $G_i$ be a secure protocol for $g^i$. Given an interleaving $I$ of the $G_i$'s, we let $F^I$ denote the (real-model) protocol where we substitute the *single* ideal call to $g^1, \ldots, g^k$ with $k$ real executions of the protocols $G_i$ interleaved according to $I$. As apparent from Example 2, we cannot hope that *every* interleaving $I$ will be "good," that is, will yield a SFE protocol $F^I$ for $f$. (For instance, in the semi-ideal coin-flipping protocol $F$ of Example 2, no matter how we design SFE protocols $G_1$ and $G_2$ for $g^1$ and $g^2$, any sequential interleaving of $G_1$ and $G_2$ yields an insecure protocol.) Actually, the guaranteed existence of even a *single* good interleaving cannot be taken for granted, therefore:

> *Can we be guaranteed that there is always an interleaving $I$*
> *of $G_1, \ldots, G_k$ such that $F^I$ is a SFE protocol for $f$?*

Of course, the answer to the above question should depend on the notion of SFE we are using. This leads us to the following informal definition.

**Definition 2:** We say that a SFE notion satisfies *synchronous reducibility* if, whenever the protocols $F, G_1, \ldots, G_k$ satisfy this SFE notion, there exists an interleaving $I$ such that $F^I$ is a SFE protocol under this notion.

Example 2 not only shows that there are bad interleavings, but also that a "liberal" enough definition of SFE will not satisfy synchronous reducibility. In particular,

**Lemma 2:** The SFE notions of [5,2,11] do not support synchronous reducibility.

Indeed, according to the SFE notions of [5,2,11], the protocol $G_1$ consisting of player $P_1$ sending $x_1$ to player $P_2$ is a secure protocol for $g^1$. Similarly, the protocol $G_2$ consisting of player $P_2$ sending $x_2$ to player $P_1$ is a secure protocol for $g^2$. However, there is *no interleaving* of $G_1$ and $G_2$ that will result in a secure coin-flip. This is because the last player to send its bit (which includes the case when the players exchange their bits in one round, due to the "rushing" ability of the adversary; see Section 2) is completely controlling the outcome.
On the positive side, we show[3]

**Theorem 2:** A slight modification of the SFE notion of Micali and Rogaway [15] satisfies synchronous reducibility.

Theorem 2 actually has quite a constructive form. Namely, the nature of the definition in [15] not only guarantees that "good" interleavings $I$ always exist, but also that there are many of them, that they are easy to find, and that some of them produce efficient protocols. We summarize the last property in the following corollary.

**Corollary 2:** With respect to a slightly modified definition of SFE of Micali and Rogaway [15], let $F$ be an ideal protocol for $f$ that simultaneously calls functions $g^1, \ldots, g^k$, and let $G_i$ be an $R_i$-round SFE protocol for $g^i$. Then there exists (an easy to find) interleaving $I$ of the $G_i$'s, consisting of at most $2 \cdot \max(R_1, \ldots, R_k)$ rounds, such that $F^I$ is a SFE protocol for $f$.

---

[3] As is illustrated in Section 4.3, the above "natural" protocols $G_1$ and $G_2$ are indeed *insecure* according to the definition of [15].

In other words, irrespective of the number of sub-protocols, we can synchronously interleave them using at most twice as many rounds as the longest of them takes.[4] Let us remark that, unlike Corollary 1 (that simply follows from the definition of concurrent reducibility), Corollary 2 crucially depends on the very notion of [15], as is discussed more in Section 4.3.

**In Sum.** We have (1) clarified the notion of parallel reducibility, (2) distilled two important flavors of it, (3) modified slightly the SFE notion of Micali and Rogaway, and (4) showed that there exist SFE notions (e.g., the modified notion of [15]) as well as general SFE protocols (e.g., the one of [4]) that satisfy (both forms of) parallel reducibility.

Enjoying (both forms of) parallel reducibility do not necessarily imply that the definition of [15] is "preferable" to others. If the protocol one is designing is simple enough or is unlikely to be composed in parallel with other protocols, other definitions are equally adequate (and may actually be simpler to use). However, understanding which SFE notions enjoy parallel reducibility is crucial in order to simplify the complex task of designing secure computation protocols.

## 2    The (Modified) Micali-Rogaway Definition of SFE

Consider a probabilistic function $f(\mathbf{x}, r) = (f_1(\mathbf{x}, r), \ldots, f_n(\mathbf{x}, r))$ (where $\mathbf{x} = (x_1, \ldots, x_n)$). We wish to define a protocol $F$ for computing $f$ that is *secure* against any *adversary* $A$ that is allowed to *corrupt* in a dynamic fashion up to $t$ (out of $n$) players.[5]

### 2.1    Protocols and Adversaries

**Protocol:** An $n$-party *protocol* $F$ is a tuple $(\hat{F}, LR, CR, \mathcal{I}, \mathcal{O}, f, t)$ where

- $\hat{F}$ is a collection of $n$ interactive probabilistic Turing machines that interact in synchronous rounds.
- $LR$ — the last round of $F$ (a fixed integer, for simplicity).
- $CR$ — the *committal round* (a fixed integer, for simplicity).
- $\mathcal{I}$ — the *effective-input function*, a function from strings to strings.
- $\mathcal{O}$ — the *effective-output function*, a function from strings to strings.
- $f$ — a probabilistic function (which $F$ is supposed to compute).
- $t$ — a positive integer less than $n$ (a bound on the number of players that may be corrupted).

---

[4]  We note that the factor of 2 is typically too pessimistic. As it will be clear from the precise statement of synchronous reducibility in Section 3, natural protocols $G_i$ (like the ones designed using a general paradigm of [4]) can be synchronously interleaved in $\max(R_1, \ldots, R_k)$ rounds.

[5]  More generally, one can have an adversary that can corrupt only certain "allowable" subsets of players. The collection of these allowable subsets is usually called the *adversary structure*. For simplicity purposes only, we consider *threshold* adversary structures, i.e. the ones containing all subsets of cardinality $t$ or less. We call any such adversary $t$-*restricted*.

**Adversary:** An *adversary* $A$ is a probabilistic algorithm.

**Executing $F$ and $A$:** Adversary $A$ interacts with protocol $F$ as a traditional adaptive adversary in the rushing model. Roughly, this is explained below.

The execution of $F$ with an adversary $A$ proceeds as follows. Initially, each player $j$ has an input $x_j$ (for $f$) and an auxiliary input $a_j$, while $A$ has an auxiliary input $\alpha$. (Auxiliary inputs represent any a-priori information known to the corresponding party like the history of previous protocol executions. An honest player $j$ should ignore $a_j$, but $a_j$ might be useful later to the adversary.) At any point during the execution of $F$, $A$ is allowed to corrupt some player $j$ (as long as $A$ corrupts no more than $t$ players overall). By doing so, $A$ learns the entire *view of $j$* (i.e., $x_j$, $a_j$, $j$'s random tape, and all the messages sent and received by $j$) up to this point input. From now on, $A$ can completely control the behavior of $j$ and thus make $j$ deviate from $F$ in any malicious way. At the beginning of each round, $A$ first learns all the messages sent from currently good players to the corrupted ones.[6] Then $A$ can adaptively corrupt several players, and only then does he send the messages from bad players to good ones. Without loss of generality, $A$ never sends a message from a bad player to another bad player.

At the end of $F$, the *view of $A$*, denoted $View(A, F)$ consists of $\alpha$, $A$'s random coins and the views of all the corrupted players. The *traffic of a player $j$ up to round $R$* consists of all the messages received and sent by $j$ up to round $R$. Such traffic is denoted $traffic_j(R)$ (or by $traffic_j(R, F[A])$ whenever we wish to stress the protocol and the adversary executing with it).

**Effective Inputs and Outputs of a Real Execution:** In an execution of $F$ with $A$, the *effective input* of player $j$ (whether good or bad), denoted $\hat{x}_j^F$, is determined at the *committal round* $CR$ by evaluating the effective-input function $\mathcal{I}$ on $j$'s traffic at round $CR$: $\hat{x}_j^F = \mathcal{I}(traffic_j(CR, F[A]))$. The *effective output* of player $j$, denoted $\hat{y}_j^F$, is determined from $j$'s traffic at the last round $LR$ via the effective output function $\mathcal{O}$: $\hat{y}_j^F = \mathcal{O}(traffic_j(LR, F[A]))$. Note that, for now, the effective inputs $\hat{\mathbf{x}}^F$ and outputs $\hat{\mathbf{y}}^F$ are unrelated to computing $f$.

**History of a Real Execution:** We let the *history of a real execution*, denoted $History(A, F)$, to be $\langle View(A, F), \hat{\mathbf{x}}^F, \hat{\mathbf{y}}^F \rangle$. Intuitively, the history contains all the relevant information of what happened when $A$ attacked the protocol $F$: the view of $A$, i.e. what he "learned", and the effective inputs and outputs of all the players.

## 2.2  Simulators and Adversaries

**Simulator:** A *simulator* is a probabilistic, oracle-calling, algorithm $S$.

---

[6] We can even let the adversary schedule the delivery of good-to-bad messages and let him adaptively corrupt a new player in the middle of this process. For simplicity, we stick to our version.

**Executing $S$ with $A$:** Let $A$ be an adversary for a protocol $F$ for function $f$. In an execution of $S$ with $A$, there are no real players and there is no real network. Instead, $S$ interacts with $A$ in a round-by-round fashion, playing the role of all currently good players in an execution of $A$ with the real network, i.e.: (1) (makes up and) sends to $A$ a view of a player $j$ immediately after $A$ corrupts $j$, (2) sends to $A$ the messages of currently good players to currently bad players[7] and (3) receives the messages sent by $A$ (on behalf of the corrupted players) to currently good players. In performing these tasks, $S$ makes use of the following *oracle* $O(\mathbf{x}, \mathbf{a})$[8]:

- *Before CR.* When a player $j$ is corrupted by $A$ before the committal round, $O$ immediately sends $S$ the input values $x_j$ and $a_j$. In particular, $S$ uses these values in making up the view of $j$.
- *At CR.* At the end of the committal round $CR$, $S$ sends $O$ the value $\hat{x}_j^S = \mathcal{I}(traffic_j(CR))$ for each corrupted player $j$.[9] In response, $O$ randomly selects a string $r$, sets $\hat{x}_j^S = x_j$ for all currently good players $j$, computes $\hat{\mathbf{y}}^S = f(\hat{\mathbf{x}}^S, r)$, and for each corrupted player $j$ sends $\hat{y}_j^S$ back to $S$.
- *After CR.* When a player $j$ is corrupted by $A$ after the committal round, $O$ immediately sends $S$ the input values $x_j$ and $a_j$, as well as the computed value $\hat{y}_j^S$. In particular, $S$ uses these values in making up the view of $j$.

We denote by $View(A, S)$ the view of $A$ when interacting with $S$ (using $O$).

**Effective Inputs and Outputs of a Simulated Execution:** Consider an execution of $S$ (using oracle $O(\mathbf{x}, \mathbf{a})$) with an adversary $A$. Then, the effective inputs of this execution consist of the above defined values $\mathbf{x}^S$. Namely, if a player $j$ is corrupted before the committal round $CR$, then its effective input is $\hat{x}_j^S = \mathcal{I}(traffic_j(CR, S[A]))$; otherwise ($j$ is never corrupted, or is corrupted after the committal round) its effective input is $\hat{x}_j^S = x_j$. The effective outputs are the values $\mathbf{y}^S$ defined above. Namely, $\hat{\mathbf{y}}^S = f(\hat{\mathbf{x}}^S, r)$, where $r$ is the random string chosen by $O$ right after the committal round.

**History of a Simulated Execution:** We let the *history of a simulated execution*, denoted $History(A, S)$, to be $\langle View(A, S), \hat{\mathbf{x}}^S, \hat{\mathbf{y}}^S \rangle$. Intuitively, the history contains all the relevant information of what happened when $A$ was communicating with $S$ (and $O$): the view of $A$, i.e. what he "learned", and the effective inputs and outputs of all the players.

---

[7] Notice that $S$ does not (and cannot) produce the messages from good players to good players.

[8] Such oracle is meant to represent the trusted party in an ideal evaluation of $f$. Given this oracle, $S$'s goal is making $A$ believe that it is executing $F$ in a real network in which the players have inputs $\mathbf{x}$ and auxiliary inputs $\mathbf{a}$.

[9] Here $traffic_j(R) = traffic_j(R, S[A])$ of a corrupted player $j$ denotes what $A$ "thinks" the traffic of $j$ after round $R$ is.

## 2.3 Secure Computation

**Definition 3:** An $n$-party protocol $F$ is a SFE protocol for a probabilistic $n$-input/$n$-output function $f(\mathbf{x}, r)$, if there exists a simulator $S$ such that for any input $\mathbf{x} = (x_1, \ldots, x_n)$, auxiliary input $\mathbf{a} = (a_1, \ldots, a_n)$, and any adversary $A$ with some auxiliary input $\alpha$, the histories of the real and the simulated executions are identically distributed:

$$History(A, F) \equiv History(A, S) \tag{1}$$

Equivalently, $\langle View(A, F), \ \hat{\mathbf{x}}^F, \ \hat{\mathbf{y}}^F \rangle \equiv \langle View(A, S), \ \hat{\mathbf{x}}^S, \ \hat{\mathbf{y}}^S \rangle$.

## 2.4 Remarks

Let us provide a minimal discussion of the above definition of SFE.

**Simulators and Oracles vs. Ideal Adversaries.** A standard benchmark in determining if a SFE notion is "reasonable" is the fact that for every real adversary $A$ there exists an "ideal adversary" $A'$ that can produce (in the ideal model with the trusted party) the same view as $A$ got from the real network.[10] We argue that the existence of a simulator $S$ in Definition 3 indeed implies the existence of such an adversary $A'$. $A'$ simply runs $A$ against the simulator $S$. If $A$ corrupts a player $j$ before the committal round, $A'$ corrupts $j$ in the ideal model, and gives the values $x_j$ and $a_j$ (that it just learned) to $S$ on behalf of the oracle $O$. Right after the committal round of $F$ has been simulated by $S$, $A'$ computes from the traffic of $A$ the effective inputs $\hat{x}_j^S$ of currently corrupted players $j$, hands them to the trusted party, and returns the outputs of the corrupted players to $S$ on behalf of $O$. Finally, if $A$ corrupts a player $j$ after the committal round, $A'$ corrupts $j$ in the ideal model, and gives the values $x_j$, $a_j$ and the output of $j$ (that it just learned) to $S$ on behalf of the oracle $O$. At the end, $A'$ simply outputs the resulting view of $A$ in the simulation.[11]

We notice, however, that the "equivalent" ideal adversary $A'$ implied by our definition is much more special than the possible ideal adversary envisaged by other definitions (e.g., [5]).[12]

**Our Modifications of the Original SFE Notion of Micali and Rogaway.** We contribute a slightly cleaner and more powerful version of the SFE notion of [15]. Their original original notion was the first to advocate and highlight the importance of *blending together* privacy and correctness, a feature inherited by all subsequent SFE notions. We actually use a stronger (and more compactly expressed) such blending by demanding the equality of the joint distributions

---

[10] In fact, this requirement is more or less the SFE definition of [5].

[11] The construction of $A'$ intuitively explains the definition of effective inputs $\hat{\mathbf{x}}^S$ and effective outputs $\hat{\mathbf{y}}^S$ of the simulated execution, as they are exactly the inputs/outputs in the run of $A'$ in the ideal model.

[12] For instance, such $A'$ is constrained to run $A$ only once and in a black-box manner.

of "view, inputs and outputs" in the real and in the simulated executions —
a suggestion of [11], which was followed by other SFE notions as well. We also
extend the original SFE notion of [15] to include *probabilistic* functions.

**Simulator Complexity.** Because we are in an information-theoretic setting, we
certainly do not want to impose any computational restrictions on the adversary.
However, even though we chose not to do it for simplicity, we could demand
that the simulator to be efficient (i.e., polynomial-time in the running time of
the protocol[13]). Indeed, (1) the natural simulator for the general protocol of
[4] is efficient, and (2) our parallel-reducibility theorems would hold even if we
required simulators to be efficient.

## 3    The Notion of Parallel Reducibility

First, let us define the *semi-ideal* model which generalizes the real model with
the ability to ideally evaluate some functions. More precisely, in addition to
regular rounds (where each player sends messages to other players), the semi-
ideal model allows players to have *ideal rounds*. In such a round, the players can
*simultaneously* evaluate several functions $g^1, \ldots, g^k$ using a trusted third party.
More specifically, at the beginning of this round each player gives the $k$-tuple of
his inputs to a trusted party. At the end of the round, each player gets back from
the trusted party the corresponding $k$-tuple of outputs. (Note, these $k$-tuples are
parts of players' traffic.)

Our definition of security of a protocol $F$ in the semi-ideal model is the same
as that of a real model protocol with the following addition:

- The simulator $S$ has to simulate all the ideal rounds as well, since they are
  part of what the adversary $A$ expects. $S$ has to do this using no special "$g$-
  oracle". In other words, given the $g$-inputs of corrupted players in an ideal
  round, $S$ has to generate the corresponding outputs of corrupted players and
  give them back to $A$. Also, when $A$ corrupts a player $j$, $S$ has to produce
  on its own the $g$-inputs/outputs of player $j$ during all the ideal rounds that
  happened so far (as these are parts of $j$'s traffic, and therefore $j$'s view).

Let $F$ be a SFE protocol for $f$ in the semi-ideal model, and let us fix our attention on any particular ideal round $R$ that evaluates some functions $g^1, \ldots, g^k$.
We say that the ideal round $R$ is *order-independent* if for any sequential ordering $\pi$ of $g^1, \ldots, g^k$, semi-ideal protocol $F$ remains secure if we replace the ideal
round $R$ with $k$ ideal rounds evaluating a single $g^i$ at a time in the order given
by $\pi$ (we denote this semi-ideal protocol by $F^\pi$).

---

[13] Some other SFE notions (e.g., that of [5]) demand that, for each adversary $A$, there
is a simulator $S_A$ that is efficient compared to the running time of $A$. Note that such
a requirement is meaningless in our definition. Indeed, our simulator is *universal*:
it must reply "properly" and "on-line" to the messages it receives, without any
knowledge of which adversary might have generated them.

Let $G_1, \ldots, G_k$ be SFE protocols for $g^1, \ldots, g^k$. We would like to substitute the ideal calls to the $g^i$'s with the corresponding protocols $G_i$'s and still get a secure protocol for $f$. As we informally argued before, there are many ways to substitute (or to *interleave*) the $G_i$'s, which is made precise by the following definition.

**Definition 4:**

- An *interleaving* of protocols $G_1, \ldots, G_k$ is any schedule $I$ of their execution. Namely, a single round of an interleaving may execute in parallel one round of one or more of the $G_i$'s with the only restriction that the rounds of each $G_i$ are executed in the same order as they are in $G_i$.

- A *synchronous interleaving* of protocols $G_1, \ldots, G_k$ with committal rounds $CR_1, \ldots, CR_k$ is any interleaving $I$ such that for any $1 \leq i, \ell \leq k$, round $CR_i$ of $G_i$ strictly precedes round $CR_\ell + 1$ of $G_\ell$. We call the place after all the "pre-committal" rounds but before all the "post-committal" rounds the *synchronization point of $I$*.

- Given an interleaving $I$ of $G_1, \ldots, G_k$, we let $F^I$ be a protocol obtained by substituting the ideal round $R$ with the execution of the protocols $G_1, \ldots, G_k$ in the order specified by $I$. The committal round of $F^I$, its effective input and output functions are defined in a straightforward manner from those of $F$ and $G_1, \ldots, G_k$. More specifically, given the traffic of player $j$ in $F^I$, we replace all $j$'s traffic inside $G_i$ (if any) with the *effective inputs and outputs* of player $j$ in $G_i$, and apply the corresponding effective input/output function of $F$ to the resulting traffic. We also remark that when we run $G_i$, we let the auxiliary input of player $j$ to be its view of the computation so far.

The fundamental question addressed by parallel reducibility is

*Assuming $F, G_1, \ldots, G_k$ are SFE protocols, under which conditions is $F^I$ a SFE protocol as well?*

We highlight two kinds of sufficient conditions: (1) special properties of the protocol $F$ making $F^I$ secure *irrespective of $I$* (which will lead us to *concurrent reducibility*), and (2) restrictions on the interleaving $I$ such that mere security of $F$ and $G_1, \ldots, G_k$ is enough (which will lead us to *synchronous reducibility*). The following Main Theorem restates Theorem 1 and 2 of the introduction.

**Parallel-Reducibility Theorem:** Consider the SFE notion of Definition 3. Let $F$ be a semi-ideal SFE protocol for $f$ evaluating $g^1, \ldots, g^k$ in an ideal round $R$; let $G_i$ be a SFE protocol for $g^i$; and let $I$ be an interleaving of $G_1, \ldots, G_k$. Then $F^I$ is a SFE protocol for $f$ if either of the following conditions holds:

1. **(Concurrent-Reducibility Theorem)** Round $R$ is order-independent.
2. **(Synchronous-Reducibility Theorem)** Interleaving $I$ is synchronous.

As we argued in the introduction, if we want $F^I$ to be secure for all $I$, round $R$ must be order-independent. Thus, the modified definition of Micali and Rogaway achieves the strongest form of concurrent reducibility. On the other, hand, we also argued that if we do not put any extra conditions on $F$ and $G_1, \ldots, G_k$ (aside from being SFE protocols), not all interleavings $I$ necessarily result in a SFE protocol. In fact, we showed in Lemma 2 that under a "too liberal" definition of SFE (which includes all SFE definitions other than Micali-Rogaway), it could be that *no interleaving* $I$ will result in a secure protocol $F^I$. The stringent definition of Micali-Rogaway (in particular, the existence of a committal round) not only shows that such an interleaving *must* exist, but also allows us to define a rich class of interleavings which guarantee the security of $F^I$: the only thing we require is that all the "pre-committal" rounds precede all the "post-committal" rounds. In other words, players should first "declare" all their inputs to $g^i$'s, and only then proceed with the "actual computation" of any of the $g^i$'s. The intuition behind this restriction is clear: this is exactly what happens in the semi-ideal model when the players simultaneously evaluate $g^1, \ldots, g^k$ in $F$.

*Remark 1:* In the parallel-reducibility theorem we do not allow the adversary choose the interleaving $I$ adaptively in the process of the computation. This is only done for simplicity. For example, synchronous reducibility will hold provided the adversary is restricted to select a synchronous interleaving $I$. And concurrent reducibility holds if the semi-ideal protocol $F$ remains secure if we allow the semi-ideal adversary adaptively order the ideal calls to $g^1, \ldots, g^k$.

## 4   Proof of the Parallel-Reducibility Theorem

For economy and clarity of presentation, we shall prove both concurrent and synchronous reducibility "as together as possible". Let $S$ be the simulator for $F$, let $\pi$ be the order of committal rounds of the $G_i$'s in the interleaving $I$ (if several committal rounds of $G_i$'s happen in one round, order them arbitrarily), and let $S_i$ be the simulator for $G_i$. We need to construct the simulator $S^I$ for $F^I$. The proofs for the concurrent and synchronous reducibility are going to be very similar, the main differences being the following:

- *Concurrent Reducibility.* Since $R$ is an order-independent round of $F$, the protocol $F^\pi$ is also secure, i.e. has a simulator $S^\pi$. We will use $S^\pi$ instead of $S$ (together with $S_1 \ldots S_k$) in constructing $S^I$. In particular, $S^\pi$ will simulate the ideal call to $g^i$ right after the committal round of $G_i$, which is exactly the order given by $\pi$.

- *Synchronous Reducibility.* Here we must use $S$ itself. In particular, at some point $S$ will have to simulate the *simultaneous* ideal call to $g^1, \ldots, g^k$, and expects to see all the inputs of the corrupted players. Since the interleaving $I$ is a synchronous interleaving, it has a synchronization point where all the effective inputs of the corrupted players are defined before any of the $G_i$'s went on "with the rest of the computation." It is at this point where we let $S$ simulate the ideal call, because we will be able to provide $S$ with all the (effective) inputs.

To simplify matters, we can assume without loss of generality that each round of $I$ executes one round of a *single* $G_i$. Indeed, if we can construct a simulator for any such interleaving, we can do it for any interleaving executing in one round a round of several $G_i$'s: arbitrarily split this round into several rounds executing a single $G_i$ and use the simulator for this new interleaving to simulate the original interleaving.[14]

## 4.1  The Simulator $S^I$

As we will see in Section 4.2, the actual proof will construct $S^I$ in $k$ stages, that is, will construct $k$ simulators $S^1, \ldots, S^k$, where $S^k$ will be $S^I$. However, we present the final $S^I$ right away because it provides a good intuition of why the proof "goes through".

For concreteness, we concentrate on the concurrent reducibility case. As one can expect, $S^I$ simply runs $S^\pi$ and uses $S_1, \ldots, S_k$ to simulate the interleaving of $G_1, \ldots, G_k$.

- Run $S^\pi$ up to round $R$ (can do it since $F^I$ and $F^\pi$ are the same up to round $R$).
- Tell each $S_i$ to corrupt all the players already corrupted by the adversary (it is irrelevant what we give to $S_i$ as their inputs).
- Assume we execute some round of protocol $G_i$ in the interleaving $I$. $S^I$ then uses $S_i$ to produce the needed messages from good-to-bad players and gives back to $S_i$ the response of the adversary.
- Right after the committal round $CR_i$ of $G_i$ has been simulated, use the *effective input function of $G_i$* and the traffic of the adversary in the simulation of $G_i$ to determine the effective input $w_j^i$ of each corrupted player $j$ to $g^i$.
- We notice that at this stage $S^\pi$ is *exactly* waiting to simulate the ideal call to $g^i$ for the adversary. So $S^I$ gives $S^\pi$ the effective inputs $w_j^i$ as the adversary's inputs to $g^i$, and learns from $S^\pi$ the output $z_j^i$ of each corrupted player $j$.
- We notice that after round $CR_i$ has been simulated, the simulator $S_i$ expects to see the outputs of all the corrupted players from the $g^i$-oracle that does not exist in our simulation. Instead, we give $S_i$ the values $z_j^i$ that we just learned from $S^\pi$.
- We keep running the above simulation up to the end of the interleaving $I$. We note that at this stage $S^\pi$ has just finished simulating the ideal calls to all the $g^i$'s, and waits to keep the simulation of $F^\pi$ starting from round $R + 1$. And we just let $S^\pi$ do it intil the end of $F^I$ (we can do it since $F^I$ and $F^\pi$ are the same again from this stage).
- It remains to describe how $S^I$ handles the corruption requests of the adversary. This will depend on where in $F^I$ the corruption request happens. But in any case $S^I$ tells $S^\pi$ that the adversary asked to corrupt player $j$ and learns from $S^\pi$ the view $V_j$ of $j$ in (the simulation of) $F^\pi$.

---

[14] Here we use the fact that non-corrupted players execute all the $G_i$'s independently from each other, so the adversary can only benefit by executing one round of a single $G_i$ at a time.

⋆ If the corruption request happens before round $R$, simply return $V_j$ to the adversary.

⋆ Otherwise, the adversary expects to see (possibly partial) transcript of $j$ inside every $G_i$, which $V_j$ does not contain. However, $V_j$ still contains the supposed inputs $w_j^i$ of player $j$ to each $g^i$.

⋆ For each $i$ we now ask the simulator $S_i$ to corrupt player $j$ in order to learn its view inside $G_i$. To answer this request, $S_i$ needs help from the $g^i$-oracle (that does not exist in our simulation), which $S^I$ provides as follows.

- If the corruption happened before the committal round $CR_i$, $S_i$ only expects to see the input and the auxiliary input of player $j$ to $g^i$. We give him $w_j^i$ as the actual input and extract from $V_j$ the view of $j$ prior to round $R$ as $j$'s auxiliary input.

- If the corruption happened after round $CR_i$,[15] $S_i$ also expects to see the output $z_j^i$ of player $j$ in $g^i$. However, in this case such an output is also contained in $V_j$, since right after the (already elapsed) round $CR_i$, we have simulated the ideal call to $g^i$ in $F^\pi$. Thus, $z_j^i$ is part of $j$'s view in $F^\pi$, and as such should be included by $S^\pi$ in $V_j$.

⋆ We see that in any of the above two cases we can provide $S_i$ with the information it expects. Therefore, we get back the view $W_j^i$ of $j$ in $G_i$ so far.

⋆ $S^I$ now simply combines $V_j$ with $W_j^1, \ldots, W_j^k$ to get the final simulated view of $j$, and gives it back to the adversary (we will argue later that the security of the $G_i$'s implies that these views "match").

We remark that the simulator for synchronous reducibility is very similar. We essentially need to replace $S^\pi$ by $S$ and let $S$ simulate the single ideal call to $g^1, \ldots, g^k$ at the synchronization point of $I$, when the traffic of the adversary will simultaneously give $S$ the (effective) inputs of the corrupted players to all the $g^i$'s.

## 4.2    Proof Outline

While we have already constructed the simulator $S^I$, in the proof we will need to use the security of some particular $G_i$. Therefore, we will need "to move slowly" from the assumed secure protocol $F$ or $F^\pi$ (evaluating all the $g^1, \ldots, g^k$ ideally) to the protocol $F^I$ (whose security we need to establish and which runs $k$ real protocols $G_1, \ldots, G_k$). Roughly, we need to "eliminate" one ideal call (to some $g^i$) at a time, by "replacing" it with the protocol $G_i$. Using the security of $G_i$, we

---

[15] This includes the case when the corruption happened "after the end" of $G_i$. We treat this corruption as having the adversary corrupt player $j$ at the very end of the computation of $G_i$. This kind of "post-executuion" corruption has caused a lot of problems preventing some other SFE notions to satisfy reducibility. In our situation, this case presents no special problems due to the universality of the simulator and the information-theoretic security.

will then argue that this "substitution" still leaves the resulting protocol a SFE protocol for $f$. To make the above idea more precise, we need some notation.[16]

First, from the interleaving $I$ of $G_1, \ldots, G_k$, we define the "projection interleaving" $I^i$ (for each $i \leq k$). This is the interleaving of the protocols $G_1, \ldots, G_i$ intermixed with the ideal calls to $g^{i+1}, \ldots, g^k$. More precisely, we remove from $I$ the rounds of all $G_\ell$ for $\ell > i$. For concurrent reducibility, we add the ideal calls to $g^\ell$ (for every $\ell > i$) right after the place where we previously had the committal round of $G_\ell$. We notice that this order of the ideal calls is consistent with the permutation $\pi$. In particular, we will identify the "base" interleaving $I^0$ of $g^1, \ldots, g^k$ with the permutation $\pi$. For synchronous reducibility, we add a *single* ideal call to $g^{i+1}, \ldots, g^k$ right at the *synchronization point* of $I$, and still call the resulting interleaving $I^i$ of $G_1, \ldots, G_i, g^{i+1}, \ldots, g^k$ a synchronous interleaving. Notice that $I^{i-1}$ is also a "projection" of $I^i$.

Slightly abusing the notation, we now define (in a straightforward way) "intermediate" semi-ideal protocols $F^i = F^{I^i}$, which essentially replace the ideal calls to $g^1, \ldots, g^i$ with $G_1, \ldots, G_i$ (but leave the ideal calls to $g^{i+1}, \ldots, g^k$). We note that $F^k = F^I$ and $F^0$ is either $F^\pi$ (the concurrent case) or $F$ (the synchronous case). We know by the assumption of the theorem that $F^0$ is secure, and need to show that $F^k$ is secure. Naturally, we show it by induction by showing that the security of $F^{i-1}$ implies that of $F^i$. Not surprisingly, this inductive step will follow from the security of $G_i$.

To summarize, the only thing we need to establish is the following. Assume $F^{i-1}$ is a SFE protocol for $f$ with the simulator $S^{i-1}$. We need to construct a simulator $S^i$ for $F^i$ such that for all inputs of the players and for any adversary $A^i$ in $F^i$, we get

$$History(A^i, F^i) \equiv History(A^i, S^i) \tag{2}$$

We construct $S^i$ from $S^{i-1}$ and the simulator $S_i$ for $G_i$. Essentially, $S^i$ will run $S^{i-1}$ in $F^i$ and use $S_i$ (together with $S^{i-1}$'s simulation of the ideal call to $g^i$) to answer the adversary inside the $G_i$. In the "other direction", given adversary $A^i$ in $F^i$, we define the adversary $A^{i-1}$ in $F^{i-1}$. This adversary will run $A^i$ in $F^{i-1}$, and will also use $S_i$ (together with the ideal call to $g^i$ in $F^{i-1}$) to interact with $A^i$ inside $G_i$. Informally, we will say that "$S^i = S^{i-1} + S_i$" and "$A^{i-1} = A^i + S_i$".

We observe that the security of $F^{i-1}$ implies that

$$History(A^{i-1}, F^{i-1}) \equiv History(A^{i-1}, S^{i-1}) \tag{3}$$

which is the same as

$$\langle View(A^{i-1}, F^{i-1}), \hat{\mathbf{x}}^{F^{i-1}}, \hat{\mathbf{y}}^{F^{i-1}} \rangle \equiv \langle View(A^{i-1}, S^{i-1}), \hat{\mathbf{x}}^{S^{i-1}}, \hat{\mathbf{y}}^{S^{i-1}} \rangle \tag{4}$$

Now, since $A^{i-1}$ essentially runs $A^i$ in the background, the view of $A^{i-1}$ (against both $F^{i-1}$ and $S^{i-1}$) will naturally "contain" the view of $A^i$. We denote these

---

[16] Below, we will try to use superscripts when talking about the notions related to computing $f$, like $F^i$, $S^i$, $A^i$. And we will use subscripts for the notions related to computing some $g^i$, like $G_i$, $S_i$, $A_i$.

views by $View(A^i, F^{i-1} + S_i)$ and $View(A^i, S^{i-1} + S_i)$, and let

$$History(A^i, F^{i-1} + S_i) \stackrel{\text{def}}{=} \langle View(A^i, F^{i-1} + S_i), \hat{\mathbf{x}}^{F^{i-1}}, \hat{\mathbf{y}}^{F^{i-1}} \rangle \qquad (5)$$

$$History(A^i, S^{i-1} + S_i) \stackrel{\text{def}}{=} \langle View(A^i, S^{i-1} + S_i), \hat{\mathbf{x}}^{S^{i-1}}, \hat{\mathbf{y}}^{S^{i-1}} \rangle \qquad (6)$$

Thus, Equation (3) (i.e., assumed security of $F^{i-1}$) implies that

$$History(A^i, F^{i-1} + S_i) \equiv History(A^i, S^{i-1} + S_i) \qquad (7)$$

However, from the definition of $S^i = S^{i-1} + S_i$ and the definitions of the effective inputs/outputs of $F^i$ based on those of $F^{i-1}$, it will immediately follow that the latter distribution is *syntactically the same* as $History(A^i, S^i)$! That is,

$$History(A^i, S^{i-1} + S_i) \equiv History(A^i, S^i) \qquad (8)$$

Therefore, Equation (7) and Equation (8) imply that what remains to prove in order to show Equation (2) is that

$$History(A^i, F^i) \equiv History(A^i, F^{i-1} + S_i) \qquad (9)$$

We remark that the "environments" $F^i$ and $F^{i-1} + S_i$ are identical except the former runs the actual protocol $G_i$, while the latter evaluates $g^i$ ideally and uses the simulator $S_i$ to deal with $A^i$ inside $G_i$. Not surprisingly, the last equality (whose verification is the main technical aspect of the proof) will follow from the security of $G_i$. Namely, assuming that the last equality is false, we will construct an adversary $A_i$ for $G_i$ such that $History(A_i, G_i) \not\equiv History(A_i, S_i)$, a contradiction. Roughly, $A_i$ will simulate the whole network of players in $F^i$ (both the adversary $A^i$ and the honest players!), except when executing $G_i$.

This completes a brief outline of the proof. Additional details can be found in [9].

## 4.3   The Definitional Support of Parallel Reducibility

Since at least synchronous reducibility provably does not hold for other SFE definitions, one may wonder what specific features of our modified definition of [15] are "responsible" for parallel reducibility. While such key features can be properly appreciated only from the full proof of the parallel-reducibility theorem, we can already informally highlight two such features on the basis of the above proof outline.

**On-Line Simulatability.** The simulator $S$ not only is universal (i.e., independent of the adversary $A$) and not only interacts with $A$ in a black-box manner, but must also interact with $A$ "on-line". In other words, $S$ runs with $A$ *only once*: each time that $S$ sends a piece of information to $A$, this piece becomes part of $A$'s *final view*. This is in contrast with traditional simulators, which would be allowed to interact with $A$ arbitrarily many times, to "rewind" $A$ in the middle of an execution, and to produce any string they want as $A$'s entire view.

The ability to generate $A$'s final view on-line is probably the most crucial for achieving any kind of parallel reducibility. For example, an adversary $A$ of the composed protocol might base it actions in sub-protocol $G_1$ depending on what it sees in sub-protocol $G_2$ and vice versa. Therefore, the resulting views of $A$ inside $G_1$ and $G_2$ are very *inter-dependent*. It thus appears crucial that, in order to simulate these inter-dependent views, the simulator $S_i$ for $G_i$ should be capable of extending $A$'s view inside $G_i$ incrementally "in small pieces" (as it happens with $A$'s view in the real execution) that should never "be taken back". If, instead, we were only guaranteed that the simulator could simulate the *entire* (as opposed to "piece-by-piece") view of $A$ in each of the $G_i$'s separately, there is no reason to expect that these separate views would be as inter-dependent as $A$ can make them in the real model. As demonstrated in Section 4.1, on the other hand, having on-line "one-pass" simulation makes it very easy to define the needed on-line simulator for $A$.

**Committal Rounds.** Intuitively, the committal round corresponds to the "synchronization point" in the ideal function evaluation: when all the players have sent their inputs to the trusted party, but have not received their corresponding outputs yet. Not surprisingly, the notion of the committal round plays such a crucial role in synchronous reducibility. In particular, the very existence of "good" interleavings (i.e., synchronous interleaving, as stated in Theorem 2) is based on the committal rounds. Committal rounds also play a crucial role in Corollary 2. Indeed, the greedy concurrent execution of all the "pre-committal" rounds of any number of sub-protocols $G_1, \ldots, G_k$ (which takes at most $\max(R_1, \ldots, R_k)$ rounds), followed by the greedy concurrent execution of all the "post-committal" rounds of $G_1, \ldots, G_k$ (which also takes at most $\max(R_1, \ldots, R_k)$ rounds), yields a *synchronous interleaving* of $G_1, \ldots, G_k$ with the claimed number of rounds.

**The Price of Parallel Reducibility.** The definitional support of parallel reducibility "comes at a price": it rules out some reasonable protocols from being called secure. For example, having $P_1$ simply send $x_1$ to $P_2$ is not a secure protocol (in the sense of [15] and Definition 3) for the function $g^1(x_1, \lambda, \lambda, \ldots, \lambda) = (x_1, x_1, \lambda, \ldots, \lambda)$ of Example 2. Indeed, assume the adversary $A$ corrupts player $P_2$ before the protocol starts and does not corrupt anyone else later on. Then $A$ will learn $x_1$ in the real execution. Therefore, for the simulator $S$ to match the view of $A$, it must also send $x_1$ to $A$ in round 1. For doing so, $S$ must learn $x_1$ from its oracle *before round* 1. Since $A$ does not corrupt player 1, this can only happen when $S$ learns the output of corrupted player $P_2$ (which is indeed $x_1$) after the committal round. Unfortunately, the committal round *is* round 1 itself, because only then does $P_1$ manifest its input $x_1$ via its own message traffic. Thus, $S$ will learn $x_1$ only *after round* 1, which is too late.

In sum, a reasonable protocol for function $g^1$ is excluded by the Definition 3 from being secure, but this "price" has a reason: Example 2 proves that such (individually) reasonable protocol is not synchronously reducible.

**Parallel Reducibility in Other Settings.** We have examined the concept of parallel reducibility in the *information-theoretic* setting. In particular, our proof of the parallel-reducibility theorem strongly uses information-theoretic security. It is a very interesting open question to examine parallel reducibility in the *statistical* and *computational* settings.

# References

1. D. Beaver, Foundations of Secure Interactive Computing. *Proc. of CRYPTO'91*, pp. 377-391, 1991.
2. D. Beaver, Secure multi-party protocols and zero-knowledge proof systems tolerating a faulty majority. *Journal of Cryptology*, 4(2), pp. 75–122, 1991.
3. D. Beaver and S. Goldwasser, Multi-party computation with faulty majority, *Proc. of the 30th FOCS*, pp. 468–473, 1989.
4. M. Ben-Or, S. Goldwasser and A. Wigderson, Completeness Theorems for Non-Cryptographic Fault-Tolerant Distributed Computation, *Proc. of the 20th STOC*, pp. 1–10, 1998.
5. R. Canetti, Security and Composition of Multi-party Cryptographic Protocols. *Journal of Cryptology*, 13(1):143–202.
6. R. Canetti, Studies in Secure Multi-party Computation and Application, *Ph.D. Thesis*, Weizmann Institute, Israel, 1995.
7. D. Chaum, C. Crépeau and I. Damgård, Multiparty unconditionally secure protocols, *Proc. of the 20th STOC*, pp. 11–19, 1988.
8. R. Cramer, U. Maurer, and I. Damgård, General secure multiparty computation from any linear secret-sharing scheme, *Proc. EUROCRYPT'00*, pp. 316–334, 2000.
9. Y. Dodis and S. Micali. Parallel Reducibility for Information-Theoretically Secure Computation. Manuscript in progress.
10. P. Feldman and S. Micali, Optimal algorithms for Byzantine agreement, *SIAM J. on Computing*, 26(4):873-933, 1997.
11. S. Goldwasser and L. Levin, Fair computation of general functions in presence of immoral majority, *Proc. CRYPTO '90*, pp. 75–84, 1990.
12. O. Goldreich, Secure Multi-Party Computation, First draft available at `http://theory.lcs.mit.edu/~oded`.
13. O. Goldreich, S. Micali and A. Wigderson, How to play any mental game, *Proc. of the 19th STOC*, pp. 218–229, 1987.
14. K. Kilian, E. Kushilevitz, S. Micali and R. Ostrovsky, Reducibility and Completeness in Private Computations, To appear in SIAM J. on Computing, preliminary versions in *Proc. of the 23rd STOC*, 1991 by Kilian and in *Proc. of the 35th FOCS*, 1994 by Kushilevitz, Micali and Ostrovsky.
15. S. Micali and P. Rogaway, Secure computation, *Proc. CRYPTO '91*, pp. 392–404, 1991. Also in Workshop On Multi-Party Secure Computation, Weizmann Institute, Israel, 1998.
16. T. Rabin and M. Ben-Or, Verifiable Secret Sharing and Multi-party Protocols with Honest Majority, *Proc. of 21st STOC*, pp. 75–83, 1989.
17. A. Yao, Protocols for secure computation, *Proc. of the 23rd FOCS*, pp. 160–164, 1982.

# Optimistic Fair Secure Computation

## (Extended Abstract)

Christian Cachin and Jan Camenisch

IBM Research, Zurich Research Laboratory
CH-8803 Rüschlikon, Switzerland
{cca,jca}@zurich.ibm.com

**Abstract.** We present an efficient and fair protocol for secure two-party computation in the optimistic model, where a partially trusted third party $T$ is available, but not involved in normal protocol executions. $T$ is needed only if communication is disrupted or if one of the two parties misbehaves. The protocol guarantees that although one party may terminate the protocol at any time, the computation remains fair for the other party. Communication is over an asynchronous network. All our protocols are based on efficient proofs of knowledge and involve no general zero-knowledge tools. As intermediate steps we describe efficient verifiable oblivious transfer and verifiable secure function evaluation protocols, whose security is proved under the decisional Diffie-Hellman assumption.

## 1  Introduction

Secure computation between distrusting parties is a fundamental problem in cryptology. Suppose two parties $A$ with input $x$ and $B$ with input $y$ wish to jointly compute a function $f(x, y)$ of their inputs without revealing anything else than the result. It is known that any function can be computed securely and with only few rounds of interaction under cryptographic assumptions [36,26,25].

However, if the computation should also be *fair* and give a guarantee that $A$ learns $f(x, y)$ if and only if $B$ learns $f(x, y)$, two-party protocols inevitably come at the cost of many rounds of interaction [36]. The reason is that a malicious party could always quit the protocol early, e.g., as soon as it obtains the information it is interested in, and the other party may not get any output at all. The only way to get around this are several rounds of interaction, in which the result is revealed verifiably and gradually bit-by-bit so that a cheating party has an unfair advantage of at most one bit [36,9,15,8].

This work presents an efficient protocol for *fair secure computation* using a third party $T$ to ensure fairness, which is not actively involved if $A$ and $B$ are honest and messages are delivered without errors. This approach has been proposed for fair exchange (e.g., of digital signatures) by Asokan, Schunter, Shoup, and Waidner [1,2] and is known as the *optimistic model*. Its main benefits are a small, constant number of rounds of interaction between $A$ and $B$, independent of the security parameter, and the minimal involvement of $T$. Our secure computation protocol maintains the privacy of one party's inputs even if $T$ should

M. Bellare (Ed.): CRYPTO 2000, LNCS 1880, pp. 93–111, 2000.
© Springer-Verlag Berlin Heidelberg 2000

collude with the other party (unlike [2]). We achieve this by combining Yao's technique for securely evaluating a circuit with efficient zero-knowledge proofs.

We consider actually a more general model of fair secure computation, in which there are two functions, $f_A(x, y)$ and $f_B(x, y)$, and $A$ should learn $f_A(x, y)$ if and only if $B$ learns $f_B(x, y)$, evaluated on the *same* inputs.

A key feature of our protocol is that it works in an *asynchronous* environment such as the Internet, where messages between $A$ and $B$ might be lost or reordered.

Our protocol is *efficient* in the sense that its complexity is directly proportional to the size of the circuit computing $f$ and does not involve large initial costs. All our zero-knowledge proofs and verifiable primitives are based on proofs of knowledge about discrete logarithms, without resorting to expensive general zero-knowledge proof techniques involving NP-reductions. Our solution is of practical relevance for cases where $A$ and $B$ want to compute $f$ with a small circuit, for example, to evaluate the predicate $x_A \geq x_B$ (the "millionaire's problem" [35]), which has applications to on-line bidding and auctions.

Baum and Waidner [3] and Micali [29] have observed before that fair two-party computation is feasible in the optimistic model. They used general tools and did not focus on efficient protocols for small circuits, however.

### 1.1   Overview

We build the fair secure computation protocol in several steps and use intermediate concepts and protocols that may be of independent interest.

Recall Yao's approach to secure function evaluation [36]: The circuit constructor $A$ scrambles the bits on the wires of the circuit by replacing each with a random token, encrypting the truth tables of all gates accordingly such that two tokens together decrypt the corresponding token on the outgoing wire, and providing the cleartext interpretation for the tokens appearing in the circuit output. It sends the encrypted circuit to $B$ (the circuit evaluator), who obtains the tokens corresponding to his input bits using one-out-of-two oblivious transfer; this ensures that he learns nothing about other tokens. $B$ is then able to evaluate the circuit and to compute the output on his own. Note that secure function evaluation is one-sided because only $B$ learns the output.

Our fair secure computation protocol, presented in Section 6, consists of two intertwined executions of *verifiable secure function evaluation* (VFE) on committed inputs between $A$ and $B$, plus recovery involving $T$. Verifiable secure function evaluation is a protocol (which we define in Section 5) extending Yao's construction that computes a given function on committed inputs of $A$ and $B$.

In order to obtain the initial tokens, $A$ and $B$ use a *verifiable oblivious transfer* (VOT) protocol that performs a one-out-of-two oblivious transfer on committed values (as defined in Section 4).

However, this solution is not sufficient for fair secure computation in the optimistic model. We need to escrow some information in the VFE construction such that a third party $T$ can open the result of the computation in case the sender refuses to continue or some of its messages are lost. (The escrow protocol is defined and described in Section 3.4.)

These protocols are based on proofs of knowledge about discrete logarithms and verifiable encryption. Our notation for proofs of knowledge is introduced in Section 3.2 and allows to describe modular composition of proofs. For verifiable encryption we use the methods of Camenisch and Damgård [10] as described in Section 3.3. Our model for optimistic fair secure two-party computation is formalized in Section 2.

### 1.2   Related Work

Beaver, Micali, and Rogaway [6] give a constant-round cryptographic protocol for multi-party computation. Its specialization to three parties is related to our three-party model in that it guarantees fairness against one malicious party, but $T$ needs to be always involved.

Fair protocols for two-party computation (and extensions to multiple parties) have previously been investigated by Chaum, Damgård, and van de Graaf [13], by Beaver and Goldwasser [5], and by Goldwasser and Levin [27]. They combine oblivious circuit evaluation with gradual release techniques to obtain fairness, but without focus on particularly efficient protocols.

Feige, Kilian, and Naor [24] study an extension of the multi-party secure computation models using a third party $T$, which receives a single message, does some computation, and outputs the function value, but does not learn anything else about the inputs. Under cryptographic assumptions, every polynomial-time computable function can be computed efficiently (i.e., in polynomial time) in their model. In our model, $T$ is not involved in regular computations and only used in case some party misbehaves.

## 2   Optimistic Fair Secure Two-Party Computation

### 2.1   Notation

The security parameter is denoted by $k$. The random choice of an element $x$ from a set $\mathcal{X}$ with uniform distribution is denoted by $x \in_R \mathcal{X}$. The concatenation of strings is denoted by $\|$.

The statistical difference between two probability distributions $P_X$ and $P_Y$ is denoted by $|P_X - P_Y|$. A quantity $\epsilon_k$ is called *negligible* (as a function of $k$) if for all $c > 0$ there exists a constant $k_0$ such that $\epsilon_k < \frac{1}{k^c}$ for all $k > k_0$. The formal security notion is defined in terms of indistinguishability of probability ensembles indexed by $k$, but extension from a single random variable to an ensemble is assumed implicitly. Two probability ensembles $X = \{X_k\}$ and $Y = \{Y_k\}$ are called *computationally indistinguishable* (written $X \overset{c}{\approx} Y$) if for every algorithm $D$ that runs in probabilistic polynomial time (in $k$), the quantity $|\text{Prob}[D(X_k) = 1] - \text{Prob}[D(Y_k) = 1]|$ is negligible.

## 2.2   Definition

The parties $A$, $B$, and $T$ are probabilistic interactive Turing Machines (PITM) that communicate via secure channels in an asynchronous environment. Let $f : \mathcal{X}_A \times \mathcal{X}_B \to \mathcal{Y}_A \times \mathcal{Y}_B$ be a deterministic function with two inputs and two outputs that $A$ and $B$ want to evaluate, possibly using $T$'s help. Suppose $f$ can be evaluated by a polynomial-sized circuit in $k$ (the extension to probabilistic functions is straightforward and omitted). Let $f_A : \mathcal{X}_A \times \mathcal{X}_B \to \mathcal{Y}_A$ denote the restriction of $f$ to $A$'s output and let $f_B : \mathcal{X}_A \times \mathcal{X}_B \to \mathcal{Y}_B$ denote the restriction of $f$ to $B$'s output. $A$ has private input $x_A$ and should output $f_A(x_A, x_B)$ and $B$ has private input $x_B$ and should output $f_B(x_A, x_B)$.

These requirements are expressed formally in terms of the simulatability paradigm for general secure multi-party computation [4,30,25,12], although we consider only three parties. In this paradigm, the requirements on a protocol are expressed in terms of an ideal process, where the parties have access to a universally trusted device that performs the actual computation. A protocol is considered secure if all an adversary may do in the real world can also happen in the ideal process; formally, for every real-world adversary there must exist some adversary in the ideal process such that the real protocol execution is indistinguishable from execution of the ideal process.

First, one has to define the real-world model and the ideal process. We assume static corruption throughout this work.

*The real-world model.* We consider an asynchronous three-party protocol as a collection $(A, B, T)$ of PITM. All parties are initialized with the public inputs of the protocol that includes the function $f$, $T$'s public key $y_T$, and possibly further parameters of the encryption schemes. The private inputs are $x_A$ for $A$, $x_B$ for $B$, and $z_T$ for $T$.

There is no global clock and the parties are linked by secure authenticated channels in the following sense. All communication is driven by the adversary in form of a scheduler $\mathcal{S}$. There exists a global set $\mathcal{M}$ of undelivered messages tagged with $(S, R)$ that denote sender $S$ and receiver $R$. $\mathcal{M}$ is initially empty. At each step, $\mathcal{S}$ chooses a party $P$, selects some message $M \in \mathcal{M}$ with receiver $P$, and activates $P$ with $M$ on its communication input tape. If $\mathcal{M}$ is empty, $P$ may also be activated with empty input. $P$ performs some computation and eventually writes a message $(R, \tau)$ to its communication output tape. The message $\tau$ is then added to $\mathcal{M}$, tagged with $(P, R)$. $\mathcal{S}$ repeats this step arbitrarily often and is not allowed to terminate as long as $\mathcal{M}$ contains messages with receiver or sender equal to $T$. (In other words, $\mathcal{S}$ must eventually deliver all messages between $T$ and any other party $P \in \{A, B\}$, but may suppress messages between $A$ and $B$.) Honest parties eventually generate an output as prescribed by the protocol and terminate by raising a corresponding flag; they will not process any more messages.

An adversary in the real world is an algorithm $C$ that controls $\mathcal{S}$ and at most two of the parties $A$, $B$, and $T$. Parties controlled by the adversary are called corrupt; we assume their output is empty. The adversary itself outputs an arbitrary function of its view, which consists of the information observed by the scheduler and all messages written to and read from communication tapes of corrupted parties. W.l.o.g. we assume the adversary is deterministic. For a fixed adversary $C$ and inputs $x_A$ and $x_B$, the joint output of $A$, $B$, $T$, and $C$, denoted by $O_{ABTC}(x_A, x_B)$, is a random variable induced by the internal coins of the honest parties.

*The ideal process.* The ideal process consists of algorithms $\bar{A}$, $\bar{B}$, and $\bar{T}$, and uses on a universally trusted party $U$ to specify all desired properties of the real protocol. $U$ is parametrized by $f$. $\bar{A}$ has input $x_A$, $\bar{B}$ has input $x_B$, and $\bar{T}$ has no input. The operation is as follows. $\bar{A}$ sends a message in $\mathcal{X}_A \cup \{\perp\}$ to $U$, and $\bar{B}$ sends a message in $\mathcal{X}_B \cup \{\perp\}$ to $U$, and $\bar{T}$ sends two distinct messages to $U$ in arbitrary order, one containing a value $b_A \in \mathcal{Y}_A \cup \{\diamond, \perp\}$ and the other one containing a value $b_B \in \mathcal{Y}_B \cup \{\diamond, \perp\}$. Messages are delivered instantly.

$U$ is a device that computes two messages, $m_A$ and $m_B$, for $\bar{A}$ and $\bar{B}$, respectively. Each message is generated as soon as all necessary inputs have arrived. The message for $\bar{A}$ depends on $x_A, x_B$, and $b_A$, and is given by

$$
m_A = \begin{cases}
f_A(x_A, x_B) & \text{if } b_A = \diamond \text{ and } x_A \neq \perp \text{ and } x_B \neq \perp \\
\perp & \text{if } b_A = \diamond, \text{ but } x_A = \perp \text{ or } x_B = \perp \\
b_A & \text{if } b_A \neq \diamond.
\end{cases}
$$

$m_B$ is computed analogously from $x_A, x_B$, and $b_B$.

Honest parties in the ideal process operate as follows. $\bar{A}$ and $\bar{B}$ just send their input to $U$ and $\bar{T}$ sends $b_A = \diamond$ and $b_B = \diamond$. $\bar{A}$ and $\bar{B}$ then wait for an answer from $U$, output the received value, and terminate. $\bar{T}$ halts as soon as it has sent two messages to $U$ and outputs nothing.

The ideal-process adversary is an algorithm $\bar{C}$ that controls the behavior of the corrupted parties in the ideal process. It sees the inputs of a corrupted party and may substitute them by an arbitrary value before sending the specified message to $U$. The adversary sees also $U$'s answer to a corrupted party. Corrupted parties output nothing, but the adversary outputs an arbitrary function of all information gathered in the protocol.

For a fixed (deterministic) adversary $\bar{C}$ and inputs $x_A$ and $x_B$, the output of the ideal process is the concatenation of all outputs, denoted by $O_{\bar{A}\bar{B}\bar{T}\bar{C}}(x_A, x_B)$.

In contrast to most of the literature using the simulation paradigm for secure computation, each party (including $U$) sends a message as soon as it is ready in this asynchronous specification. This means that an adversary may also delay the message of a corrupted party until it has obtained the output of another corrupted party.

*Simulatability.* We are now ready to state the definition of fair secure computation. Seemingly separate requirements on a protocol such as correctness, privacy, and fairness are expressed via the simulatability by an ideal process. Recall that an adversary in the real world is an algorithm $C$ that controls $\mathcal{S}$ and at most two of the three parties and that $C$'s output is arbitrary.

**Definition 1.** *Let $f : \mathcal{X}_A \times \mathcal{X}_B \to \mathcal{Y}_A \times \mathcal{Y}_B$ be a function that can be evaluated by a polynomial-sized circuit. We say that a protocol $(A, B, T)$ performs* fair secure computation *if for every real-world adversary $C$, there exists an adversary $\bar{C}$ in the ideal process such that for all $x_A \in \mathcal{X}_A$ and for all $x_B \in \mathcal{X}_B$, the joint distribution of all outputs of the ideal process is computationally indistinguishable from the outputs in the real world, i.e.,*

$$O_{ABTC}(x_A, x_B) \overset{c}{\approx} O_{\bar{A}\bar{B}\bar{T}\bar{C}}(x_A, x_B).$$

*A fair secure computation protocol is called* optimistic *if whenever all parties follow the protocol and messages between them are delivered instantly, then $T$ does not receive or send any message.*

Remarks on the above definition.

1. By the design of the ideal process, fairness is only guaranteed if $T$ is not colluding with $A$ or $B$. This is unavoidable because a cheating participant of a two-party protocol may always refuse to send the last message. Protocols to defend against such misbehavior require a number of rounds of interaction that is inverse proportional to the cheating probability [36,9].
2. Conversely, if $T$ is corrupt, then the computation may be unfair and an honest party, say $A$, may not receive its output. Moreover, $B$ and $T$ may still decide to block $A$ after seeing $f_B$ and even cause $A$ to output a value that has nothing to do with $f_A$. This occurs in the ideal process if $\bar{T}$ colluding with $\bar{B}$ delays sending $b_A$ until it has observed $\bar{B}$'s output and then decides to send $b_A \neq \diamond$. But notice that $\bar{T}$ and $\bar{B}$ together do not learn more about Alice's input than what follows from $f_B$.
3. A stronger requirement would be that $T$ is only permitted to send $\diamond$ or $\perp$, but not a substitute for $A$ or $B$'s output. The current model reflects a corresponding property of our protocol because $T$'s actions in the resolve protocols are not verifiable. However, by making all proofs non-interactive and resorting to the random oracle model, our protocol satisfies also this stronger requirement.
4. Our model applies only to an isolated three-party case (as is customary in the literature on secure computation). A multi-user model that allows for concurrent execution of multiple protocol instances can be constructed by combining our model with techniques proposed by Asokan et al. [2]. Basically, a unique transaction identifier has to be added to all messages and techniques for concurrent composition of zero-knowledge proofs have to be used [20].

## 3    Proofs of Knowledge and Verifiable Encryption

This section introduces our notation for proofs of knowledge about discrete logarithms, the notion for verifiable encryption, and our escrow scheme. It starts with a description of the underlying encryption schemes.

## 3.1   Preliminaries

A *semantically secure public-key cryptosystem* $(E_k, D_k)$ with security parameter $k$ consists of a (public) probabilistic encryption algorithm $E_k(\cdot)$ and a (secret) decryption algorithm $D_k(\cdot)$. The encryption algorithm $E_k : \mathcal{M} \to \mathcal{C}$ takes a message $m \in \mathcal{M}$ and outputs a ciphertext $c$; the corresponding decryption algorithm $D_k : \mathcal{C} \to \mathcal{M}$ computes $m$ from $c$.

Semantic security asserts that an eavesdropper cannot get partial information about the plaintext from a ciphertext [28]. More precisely, $(E_k, D_k)$ is a semantically secure public-key system if for two arbitrary messages $m_0$ and $m_1$, the random variables representing the two encryptions $E_k(m_0)$ and $E_k(m_1)$ are computationally indistinguishable.

The protocols in this paper are mostly based on ElGamal encryption [22]. Let $G$ be a group of large prime order $q$ (polynomial in $k$) and let $g \in G$ be a randomly chosen generator. An ElGamal public key is $(g, y)$ for $y = g^x$ with a randomly chosen $x \in \mathbb{Z}_q$ and the corresponding secret key is $x$. ElGamal encryption of a message $m \in G$ proceeds as follows:

**Algorithm** ElGamal$(g, y)(m)$

1.  choose a random $r \in \mathbb{Z}_q$;
2.  compute and output $(c, c') = (g^r, my^r)$.

The decryption algorithm computes $m = c'/c^x$ and outputs $m$.

Consider the two distributions over $G^4$ with $D_0 = (g, g^x, g^y, g^z)$ for $x, y, z \in_R \mathbb{Z}_q$ and $D_1 = (g, g^x, g^y, g^{xy})$ for $x, y \in_R \mathbb{Z}_q$. The *Decisional Diffie-Hellman (DDH) assumption* is that there exists no probabilistic polynomial-time (PPT) algorithm that distinguishes with non-negligible probability between $D$ and $R$. By a random self-reduction property [34,31], the DDH assumption is equivalent to assuming that there is no PPT algorithm that *decides with high probability* for all tuples $(g, g^x, g^y, g^z)$ if $z = xy \mod q$. It is well known that ElGamal encryption is semantically secure under the DDH assumption.

Using a hybrid argument, one can show that also the two distributions

$$M_0 = (g, g^{x_1}, \ldots, g^{x_n}, g^{y_1}, \ldots, g^{y_m}, g^{z_1}, \ldots, g^{z_{nm}})$$

with $x_i, y_j, z_{ij} \in_R \mathbb{Z}_q$ and

$$M_1 = (g, g^{x_1}, \ldots, g^{x_n}, g^{y_1}, \ldots, g^{y_m}, g^{x_1 y_1}, \ldots, g^{x_n y_m})$$

with $x_i, y_j \in_R \mathbb{Z}_q$ for $i = 1, \ldots, n$ and $j = 1, \ldots, m$ are computationally indistinguishable under the DDH Assumption. The argument is essentially the same as the one by Naor and Reingold [31].

## 3.2   Proofs of Knowledge about Discrete Logarithms

We introduce a notation for describing proofs of knowledge about discrete logarithms. Such three-move proofs of knowledge can be composed efficiently in parallel and in a modular way, as shown by Cramer, Damgård, and Schoemakers [17]. The notation was first used by Camenisch and Stadler [11] and subsumes several discrete logarithm-based proof techniques (see the references therein). Our extension allows to describe modular composition.

Let $G$ be a group of large prime order $q$ and let $g, g_1 \in G$ be generators such that $\log_g g_1$ is not known (e.g. provided by a trusted dealer).

The simplest example of such a proof is the proof of knowledge of a discrete logarithm of $y \in G$ [33]. For reference, we recall some of properties of this protocol between a prover $P$ and verifier $V$. Public inputs are $(g, y)$ and $P$'s private input is $x$ such that $y = g^x$. First, $P$ computes a commitment $t = g^r$ with $r \in_R \mathbb{Z}_q$ and sends it to $V$. Then $V$ sends to $P$ a random challenge $c \in \{0, 1\}^{k'}$, to which $P$ responds with $s = r - cx \mod q$, where $k'$ is a security parameter. $V$ accepts if and only if $t = g^s y^c$. We denote this protocol by

$$PK \log(g, y)$$
$$\{\xi : y = g^\xi\}.$$

The witness(es) are conventionally written in Greek letters and only known to the prover while all other parameters are known to the verifier as well.

Unlike the simplifying description above, we assume that all proofs here are actually three-move concurrent zero-knowledge protocols, i.e., carried out using trapdoor commitments for the first message $t$. Such trapdoor commitments may be constructed, for example, using an additional generator $h \in G$, which is chosen at random by a trusted dealer or is determined in a once-and-for-all setup phase; the zero-knowledge simulator can extract the trapdoor $\log_g h$ from this. It will allow the simulator to open a given commitment $t$ in an arbitrary way upon receiving a challenge $c$ because it can compute suitable $s$ from the trapdoor, without having to rewind the verifier (for more details see, e.g., [20]); this allows also arbitrarily large challenges (i.e., $k' = O(k)$).

This basic protocol can be extended in many ways. For example,

$$PK \operatorname{rep}(g, g_1, y)$$
$$\{\xi, \rho : y = g^\xi g_1{}^\rho\}$$

denotes a proof of knowledge of a representation of $y$ with respect to $g$ and $g_1$.

Proofs written in this notation may be composed in a modular way. It is known that this is sound for monotone boolean expressions from the results of Cramer et al. [17]. For instance, the prover can convince the verifier that he knows the representation of at least one of $x$ and $y$ w.r.t. bases $g$ and $g_1$ with

$$PK \operatorname{or}(g, g_1, x, y)$$
$$\{\operatorname{rep}(g, g_1, x) \vee \operatorname{rep}(g, g_1, y)\}.$$

It is also possible to prove that two discrete logarithms (or parts of representations) are equal [14]. We give an example of this technique. It shows that a commitment $z$ contains the product modulo $q$ of the two values committed to in $x$ and $y$:

$$PK \ \mathsf{mul}(g, g_1, x, y, z)$$
$$\{\alpha, \beta, \gamma, \delta, \varepsilon : x = g^\alpha g_1{}^\gamma \wedge y = g^\beta g_1{}^\delta \wedge z = y^\alpha g_1{}^\varepsilon\}.$$

This works also for $z = g^a g_1{}^r$ with $r = 0$ and arbitrary $a \in \mathbb{Z}_q$, which is needed in Section 5.

When such proofs are combined, some optimizations are often possible, just like in assembly code that is produced by a compiler from a high-level language. An example that occurs in Section 5 is that multiple parallel commitments to the same value are introduced, where only one of them is needed.

## 3.3   Verifiable Encryption

Verifiable encryption is an important building block here and has been used for publicly verifiable secret sharing [34], key escrow, and optimistic fair exchange [2]. It is a two-party protocol between a prover and encryptor $P$ and a verifier and receiver $V$. Their common inputs are a public encryption key $E$, a public value $v$, and a binary relation $\mathcal{R}$ on bit strings. As a result of the protocol, $V$ either rejects or obtains the encryption $c$ of some value $s$ under $E$ such that $(s, v) \in \mathcal{R}$. For instance, $\mathcal{R}$ could be the relation $(s, g^s) \subset \mathbb{Z}_q \times G$. The protocol should ensure that $V$ accepts an encryption of an invalid $s$ only with negligible probability and that $V$ learns nothing beyond the fact that the encryption contains some $s$ with $(s, v) \in \mathcal{R}$. The encryption key $E$ typically belongs to a third party, which is not involved in the protocol at all.

Generalizing the protocol of Asokan et al. [2], Camenisch and Damgård [10] provide a verifiable encryption scheme for all relations $\mathcal{R}$ that have an honest-verifier zero-knowledge three-move proof of knowledge where the second message is a random challenge and the witness can be computed from two transcripts with the same first message but different challenges. This includes most known proofs of knowledge, and in particular, all proofs about discrete logarithms from the previous section. The verifiable encryption scheme is itself a three-move proof of knowledge of the encrypted witness $s$ and is zero-knowledge if a semantically secure encryption scheme is used [10].

We use a similar notation as above and denote by, e.g.,

$$VE \ (\mathsf{ElGamal}, (g, y), tag)\{\xi : v = g^\xi\}$$

the verifiable encryption protocol for the ElGamal scheme, whereby $\log_g v$ along with $tag$ is encrypted under public key $y$. The $tag$, an arbitrary bit string, is needed for the composition of such protocols, as we will see later. The ciphertext $c$ is represented by (a function of) the verifier's transcript of this protocol, which we abbreviate by writing $c \leftarrow VE \ (\mathsf{ElGamal}, (g, y), tag)\{\xi : v = g^\xi\}$, and is stored by $V$.

Together with the corresponding secret key ($x = \log_g y$ in this example), transcript $c$ contains enough information to decrypt the witness efficiently. We assume that the corresponding decryption algorithm $\mathsf{VD}(\mathsf{ElGamal}, (g, x), c, string)$ is subject to the condition that a *tag* matching *string* is encrypted in $c$; $\mathsf{VD}$ outputs the witness in this case and $\perp$ in all other cases.

We refer to Camenisch and Damgård [10] for further details of the verifiable encryption scheme.

### 3.4   Escrow Schemes

A (verifiable) *escrow scheme* [2] is a protocol involving three parties: a sender $S$, a receiver $R$, and a third party $T$, whose public key $y_T$ of an encryption scheme is known to $S$ and $R$. We require that $T$'s encryption scheme is semantically secure against adaptive chosen-ciphertext attacks [21]. $S$ has a bit string $a$ as private input. $T$'s private input is $z_T$, the secret key corresponding to $y_T$. Furthermore, there is a public input string *tag* for $S$ and $R$ that controls the condition under which $T$ may resolve the escrow of $a$.

The operation of an escrow scheme consists of two phases. In the first phase, only $S$ and $R$ interact. If $R$ accepts Phase I, then he is guaranteed to receive $a$ in Phase II as long as either $S$ or $T$ is honest. That is, $R$ either receives a single message from $S$ that will allow him to compute $a$ (and hence $T$ needs not participate in the protocol at all) or, if this does not happen, $R$ sends $T$ a single request containing *tag*, to which $T$ will reply with $a$.

Several escrow schemes with different tags may be run concurrently among the same participants.

The security requirements of the escrow scheme are that a malicious $R$ cannot gain any information on $a$ before Phase II. More precisely, for all bit strings $a'$, $a''$, and *tag*, suppose $S$ runs Phase I of the escrow scheme with $R^*$ on *tag* and $a \in \{a', a''\}$ chosen at random. Subsequently $R^*$ interacts arbitrarily with $T$ subject only to the condition that it never submits a request containing *tag* to $T$; the escrow scheme is secure if such an $R^*$ cannot distinguish $a = a'$ from $a = a''$ with more than negligible probability.

A secure escrow scheme can be implemented easily using verifiable encryption and a cryptosystem for $T$ that is semantically secure against chosen-ciphertext attacks. We use the Cramer-Shoup cryptosystem [18], denoted by $\mathsf{CS}$, with public key $y_T$ and private key $z_T$.

In Phase I, $S$ chooses $u \in_R Z_q^*$, computes $A = g^a g_1{}^u$, and sends $A$ to $R$. $S$ and $R$ also carry out $PK\ \mathsf{rep}(g, g_1, A)$ and

$$out \leftarrow VE\ (\mathsf{CS}, y_T, tag)\{\alpha, \beta : A = g^\alpha g_1{}^\beta\}.$$

In Phase II, $S$ sends $a$ and $u$ to $R$ and $R$ verifies that $A = g^a g_1{}^u$. If this check fails or if $R$ did not receive a message from $S$, then $R$ sends to $T$ the message $(out, tag)$. $T$ runs $\mathsf{VD}(\mathsf{CS}, z_T, out, tag)$ and sends the output to $R$. In either case, $R$ learns $a$.

It is easy to see that this is a secure escrow scheme using the security of $\mathsf{CS}$ and the properties of $PK$ and $VE$.

## 4    Verifiable Oblivious Transfer

This section describes a variant of oblivious transfer that is needed for our fair secure computation protocol. Oblivious transfer, proposed by Rabin [32] and by Even, Goldreich, and Lempel [23], is a fundamental primitive for multi-party computation. In its basic incarnation as a one-out-of-two oblivious transfer, a sender $S$ has two input bits $b_0$ and $b_1$, and a receiver $R$ has a bit $c$. As a result of the protocol $R$ should obtain $b_c$, but should not learn anything about $b_{c\oplus 1}$ whereas $S$ should not get any information about $c$.

A *verifiable oblivious transfer* (VOT) is an oblivious transfer on committed values, where the sender $S$ has made two commitments $A_0$ and $A_1$, containing two values $a_0$ and $a_1$, and $R$ has made a commitment $C$, containing a bit $c$. The requirements are that $R$ outputs $a_c$ without learning anything about $a_{c\oplus 1}$ and that $S$ does not learn anything about $c$. (A *committed oblivious transfer* as described by Crépeau, van de Graaf, and Tapp [19] is a similar protocol that performs an oblivious transfer *of commitments* such that $R$ ends up being committed to $a_c$; Cramer and Damgård [16] give an efficient implementation for this.)

Suppose the commitments $A_0, A_1$, and $C$ are of the form $B = g^b g_1{}^r$ for a randomly chosen $r \in \mathbb{Z}_q$ and committed value $b \in \mathbb{Z}_q$. In this section, we assume that *corresponding commitments* are computed correctly from the inputs $a_0$, $a_1$, and $c$. In other words, a *commitment oracle* receives $a_0$ and $a_1$ from $S$, chooses random $t_0, t_1 \in \mathbb{Z}_q$, places $A_0 = g^{a_0} g_1{}^{t_0}$ and $A_1 = g^{a_1} g_1{}^{t_1}$ in the public input, and returns $t_0$ and $t_1$ to $S$ privately; similarly, it receives $c$ from $R$, computes $C = g^c g_1{}^r$ using a random $r \in \mathbb{Z}_q$, places $C$ in the public input and gives $r$ privately to $R$. This commitment oracle is an artificial construction for using VOT as part of a larger protocol. Alternatively, one might assume that $S$ and $R$ generated and exchanged the commitments beforehand, together with a proof that they are constructed correctly; this is indeed how VOT is used in Section 6 below.

The following protocol is based on verifiable encryption and the oblivious transfer constructions by Even et al. [23] and Bellare and Micali [7]. Our notational convention for such protocols is as follows. All inputs are written as argument lists in parentheses, grouped by the receiving party; the first list contains public inputs, the second list private inputs of the first party ($S$), the third list private inputs of the second party ($R$), and so on.

**Protocol** VOT$(g, g_1, A_0, A_1, C)(a_0, a_1, t_0, t_1)(c, r)$

1. $S$ as encryptor and $R$ as receiver engage in two verifiable encryption protocols

$$out_0 \leftarrow VE \; (\mathsf{ElGamal}, (g_1, C), \emptyset)\{\alpha, \beta : A_0 = g^\alpha g_1{}^\beta\}$$
$$out_1 \leftarrow VE \; (\mathsf{ElGamal}, (g_1, \frac{C}{g}), \emptyset)\{\alpha, \beta : A_1 = g^\alpha g_1{}^\beta\}.$$

2. If $R$ accepts both of the above protocols, he computes

$$a_c = \mathsf{VD}(\mathsf{ElGamal}, (g_1, r), out_c, \emptyset).$$

The above protocol uses $R$'s commitment $C$ directly as encryption public key and saves one round compared to the direct adoption of the Bellare-Micali scheme. The way the commitment $C$ is constructed from $c$ ensures that $R$ knows $\log_{g_1}(C/g^c) = r$ needed to decrypt $out_c$, but not the discrete logarithm needed to decipher the other encryption. (The proof of the following lemma is omitted from this extended abstract.)

**Lemma 1.** *Under the DDH assumption, Protocol* VOT *is a secure verifiable oblivious transfer.*

## 5   Verifiable Secure Function Evaluation

Verifiable secure function evaluation (VFE) is an interactive protocol between a circuit constructor $A$ and an evaluator $B$. Both parties have as common public input values $C_A$ and $C_B$, representing commitments to their inputs. $A$ has two private inputs strings: her input string $x_A$ and a string $r_A$ allowing her to open $C_A$; likewise, $B$ has two private input strings, $x_B$ and $r_B$. Their goal is to evaluate $f_B$ on the committed inputs such that $B$ learns $f_B(x_A, x_B)$.

We assume here, as already in Section 4, that all commitments are computed correctly from the inputs, which in turn may have been chosen in an arbitrary way. More precisely, assume $A$ gives $x_A$ to a commitment oracle, which computes $C_A$ according to the specified commitment scheme using the random bits $r_A$ and returns $C_A$ and $r_A$ (similarly for $B$). These are the corresponding commitments used below. (Alternatively, one might assume that $A$ and $B$ generated and exchanged correct commitments beforehand.)

Given concrete implementations of a parties $A$ and $B$, a protocol execution between $A$ and $B$ with inputs $C_A, C_B, x_A, x_B, r_A$, and $r_B$ defines naturally the views $V_A$ and $V_B$ of $A$ and $B$, respectively, which are families of random variables determined by the public input, $A$'s private input, $B$'s private input, and the internal random coins. Moreover, if $B$ is deterministic then $V_B$ is a random variable depending only on $A$'s coin flips.

**Definition 2.** *A* verifiable secure function evaluation protocol *for a function* $f_B : \mathcal{X}_A \times \mathcal{X}_B \rightarrow \mathcal{Y}_B$ *between $A$ and $B$ satisfies the following requirements:*

**Correctness:** *If $A$ and $B$ are honest and follow the protocol, then $\forall x_A \in \mathcal{X}_A, \forall x_B \in \mathcal{X}_B$ and corresponding commitments, $B$ outputs $f_B(x_A, x_B)$ except with negligible probability.*

**Soundness:** *$\forall A^*$ and $\forall x_A^* \in \mathcal{X}_A$ and corresponding commitments $C_A^*$, if the protocol starts with public inputs $C_A^*, C_B$, then, except with negligible probability, $B$ outputs $f_B(x_A^*, x_B)$ or $\perp$.*

**Privacy:** *We consider two cases, corresponding to cheating $B$ and cheating $A$.*

  *1. Privacy for A: $\forall B^*$ there exists a probabilistic polynomial-time algorithm (PPT) $SIM_{B^*}$ such that $\forall x_A \in \mathcal{X}_A$ and $\forall x_B^* \in \mathcal{X}_B$ with corresponding commitments $C_A, C_B^*$,*

$$V_{B^*}(C_A, C_B^*, x_A, r_A, x_B^*, r_B^*) \overset{c}{\approx} SIM_{B^*}(C_A, C_B^*, f_B(x_A, x_B^*), x_B^*).$$

2. *Privacy for B:* $\forall A^*$ *there exists a PPT algorithm* $SIM_{A^*}$ *such that* $\forall x_B \in$ $\mathcal{X}_B$ *and* $\forall x_A^* \in \mathcal{X}_A$ *with corresponding commitments* $C_A^*, C_B$,

$$V_{A^*}(C_A^*, C_B, x_A^*, r_A^*, x_B, r_B) \overset{c}{\approx} SIM_{A^*}(C_A^*, C_B, x_A^*).$$

The soundness condition binds $A$ to her committed inputs. The corresponding binding for $B$ is part of the privacy condition for $A$, which ensures that $B$ is committed to the value $x_B$ at which he evaluates $f_B$ before the protocol starts. This is needed to use the one-sided concept of VFE as a building block for optimistic fair secure computation below.

## 5.1   Overview of the Encrypted Circuit Construction

We give a brief description of our protocol and the "encrypted circuit construction"; it follows the approach to secure function evaluation developed by Yao [36], but uses public-key encryption instead of pseudo-random functions for the sake of verifiability. Suppose $A$'s private input is a binary string $x_A = (x_{A,1}, \ldots, x_{A,n_A})$ and $B$'s private input is a binary string $x_B = (x_{B,1}, \ldots, x_{B,n_B})$; assume further w.l.o.g. that $f_B$ is represented a binary circuit consisting of NAND gates.

**Protocol** VFE$(g, g_1, C_A, C_B, f_B)(x_A, r_A)(x_B, r_B)$

V1. $A$ produces an encrypted version of the circuit computing $f_B$. The circuit consists of gates and wires linking the gates. Except for input and output wires, each wire connects the output of one gate with the input of one or more other gate(s). For each wire, $A$ chooses two random *tokens* $s_0$ and $s_1$, representing bits 0 and 1 on this wire, and produces unconditionally hiding commitments $u_0$ and $u_1$ to these tokens.

For each gate, $A$ encrypts the truth table as follows: First, the bits are replaced by (new) commitments to the tokens representing the bits. Next, for each row, a "row public key" for encryption is computed and added to the table such that the corresponding secret key can be derived from combining the two input tokens of the row. Finally, all four rows are permuted randomly.

These tables and the commitments are sent to $B$ as an ordered list such that $B$ knows which commitment represents token 0 or 1 etc. Moreover, $A$ proves to $B$ in zero-knowledge that the commitments and the encrypted gates are consistent, ensuring (1) that the tokens of the input and output wires are the same as those committed to in the truth table, (2) that the secret key for each row of a gate is derived correctly from the input tokens of the row, and (3) that each encrypted gate implements NAND.

V2. For each row of each gate of the circuit, $A$ and $B$ engage in verifiable encryption of the output token under the row public key.

V3. For each of her input bits, $A$ sends to $B$ the corresponding token and proves to him that this is consistent with her input $x_A$ committed in $C_A$. Furthermore, $B$ obtains the tokens representing his input bits through $n_B$ verifiable oblivious transfers from $A$ to $B$ and $A$ opens all the commitments of the output wires.

V4. Once $B$ has obtained all this information, he is able to evaluate the circuit gate by gate on his own.

Suppose w.l.o.g. the circuit consists of $n$ NAND gates $\mathcal{G}_1, \ldots, \mathcal{G}_n$ and $n + n_A + n_B$ wires $\mathcal{W}_1, \ldots, \mathcal{W}_{n+n_A+n_B}$ and has $n_A + n_B$ inputs and $n_O$ outputs. Wires $\mathcal{W}_1, \ldots, \mathcal{W}_n$ are output wires of the gates $\mathcal{G}_1, \ldots, \mathcal{G}_n$. Wires $\mathcal{W}_{n+1}, \ldots, \mathcal{W}_{n+n_A}$ are input wires of $A$ and $\mathcal{W}_{n+n_A+1}, \ldots, \mathcal{W}_{n+n_A+n_B}$ are input wires of $B$. Wires $\mathcal{W}_{n-n_O+1}, \ldots, \mathcal{W}_n$ are the output wires of the circuit; except for those, any wire is an input to at least one gate.

The commitment to $A$'s input $x_A$ is $C_A = (C_{A,1}, \ldots, C_{A,n_A})$, where for $i = 1, \ldots, n_A$, a bit commitment

$$C_{A,i} = g^{x_{A,i}} g_1^{r_{A,i}}$$

has been constructed using a random $r_{A,i} \in \mathbb{Z}_q$ and $r_A = (r_{A,1}, \ldots, r_{A,n_A})$ is a private input of $A$.

Similarly, the commitment to $B$'s input $x_B$ is $C_B = (C_{B,1}, \ldots, C_{B,n_B})$, where for $i = 1, \ldots, n_B$, a bit commitment

$$C_{B,i} = g^{x_{B,i}} g_1^{r_{B,i}}$$

has been constructed using a random $r_{B,i} \in \mathbb{Z}_q$ and $r_B = (r_{B,1}, \ldots, r_{B,n_B})$ is a private input of $B$.

The details of the verifiable secure function evaluation protocol and its analysis are omitted from this extended abstract.

## 6    Optimistic Fair Secure Computation Protocol

We are now ready to describe our protocol for optimistic fair secure two-party computation. In short, the protocol consists of two intertwined executions of the verifiable secure function evaluation protocol from the previous section, where the output tokens are not directly revealed, but mutually escrowed with $T$ first and opened later. Recall that optimistic fair secure computation involves three parties $A$, $B$, and $T$, in the asynchronous communication model of Definition 1.

In the following we use Protocol VOT from Section 4 and the secure escrow scheme based on Cramer-Shoup encryption from Section 3.4.

Common inputs are a function $f : \mathcal{X}_A \times \mathcal{X}_B \to \mathcal{Y}_A \times \mathcal{Y}_B$, $T$'s public key $y_T$, and generators $g, g_1 \in G$. The private input of $A$ is $x_A \in \mathcal{X}_A$, the private input of $B$ is $x_B \in \mathcal{X}_B$, and the private input of $T$ is the secret key $z_T$ corresponding to $y_T$.

**Protocol** FAIRCOMP$(g, g_1, f, y_T)(x_A)(x_B)(z_T)$

F1. $A$ chooses $r_{A,1}, \ldots, r_{A,n_A} \in_R \mathbb{Z}_q$, computes the commitments

$$C_A = (C_{A,1}, \ldots, C_{A,n_A}) = (g^{x_{A,1}} g_1^{r_{A,1}}, \ldots, g^{x_{A,n_A}} g_1^{r_{A,n_A}}),$$

sends $C_A$ to $B$, and runs with $B$

$$PK\{\alpha_1, \beta_1, \ldots, \alpha_{n_A}, \beta_{n_A} : C_{A,1} = g^{\alpha_1}g_1^{\beta_1} \wedge \cdots \wedge C_{A,n_A} = g^{\alpha_{n_A}}g_1^{\beta_{n_A}})\}.$$

If $B$ rejects any proof, it outputs $\perp$ and halts.

F2. $B$ chooses $r_{B,1}, \ldots, r_{B,n_B} \in_R \mathbb{Z}_q$, computes the commitments

$$C_B = (C_{B,1}, \ldots, C_{B,n_B}) = (g^{x_{B,1}}g_1^{r_{B,1}}, \ldots, g^{x_{B,n_B}}g_1^{r_{B,n_B}}),$$

sends $C_B$ to $A$, and runs with $A$

$$PK\{\alpha_1, \beta_1, \ldots, \alpha_{n_B}, \beta_{n_B} : C_{B,1} = g^{\alpha_1}g_1^{\beta_1} \wedge \cdots \wedge C_{B,n_B} = g^{\alpha_{n_B}}g_1^{\beta_{n_B}})\}.$$

If $A$ rejects any proof, it outputs $\perp$ and halts.

F3. $A$ and $B$ invoke a modification of Protocol $\mathsf{VFE}(g, g_1, C_A, C_B, f_B)(x_A, r_A)$ $(x_B, r_B)$, where they replace opening the commitments of the output tokens by escrowing them with $T$. That is, in Step V3, $A$ and $B$ run Phase I of the escrow scheme for each of the values $s_{i,0}, s_{i,1}, r_{i,0}, r_{i,1}$ tagged with $C_A\|C_B\|f_B\|i$ for $i = n - n_O + 1, \ldots, n$ in the circuit computing $f_B$. They interrupt Protocol $\mathsf{VFE}$ after Step V3. (Note that $T$ has not been involved so far.)

If this fails, $B$ simply outputs $\perp$ and halts.

F4. $B$ and $A$ invoke a modification of Protocol $\mathsf{VFE}(g, g_1, C_B, C_A, f_A)(x_B, r_B)$ $(x_A, r_A)$, where they replace opening the commitments of the output tokens by escrowing them with $T$. That is, in Step V3, $B$ and $A$ run Phase I of the escrow scheme for each of the values $s_{i,0}, s_{i,1}, r_{i,0}, r_{i,1}$ tagged with $C_A\|C_B\|f_A\|i$ for $i = n - n_O + 1, \ldots, n$ in the circuit computing $f_A$. They interrupt Protocol $\mathsf{VFE}$ after Step V3.

If this fails, $A$ invokes Protocol $\mathsf{abort}$ with $T$. If $T$ answers $\mathsf{abort}$, then $A$ outputs $\perp$ and halts. If $T$ answers $\mathsf{resolve}\|\mathit{transcript}$ then $A$ completes the VFE protocol computing $f_A$ as read from $\mathit{transcript}$ (continuing with Step V3), outputs $O_A$, and halts.

F5. $A$ and $B$ continue with Phase II of the escrow protocols started in Step F3. According to this, $A$ sends $B$ the corresponding messages, $B$ checks their contents, and if a check fails or if some message does not arrive, $B$ invokes Protocol $\mathsf{B\text{-}resolve}$ with $T$. If $T$ answers $\mathsf{abort}$, then $B$ outputs $\perp$ and halts. If $T$ answers $\mathsf{resolve}\|\mathit{transcript}$ then $B$ completes the VFE protocol computing $f_B$ as read from $\mathit{transcript}$ (continuing with Step V3), outputs $O_B$, and halts.

Otherwise $B$ resumes Protocol $\mathsf{VFE}$ started in Step F3 with Step V4 and obtains $O_B$.

F6. $B$ and $A$ continue with Phase II of the escrow protocols started in Step F4. According to this, $B$ sends $A$ the corresponding messages. Then $B$ outputs $O_B$ and halts.

$A$ checks the messages received from $B$, and if a check fails or if some message does not arrive, $A$ invokes Protocol $\mathsf{A\text{-}resolve}$ with $T$. If $T$ answers $\mathsf{abort}$, $A$ outputs $\perp$ and halts.

If $T$ answers resolve$\|transcript$ then $A$ completes the VFE protocol computing $f_A$ as read from $transcript$ from Step V3, outputs $O_A$, and halts. Otherwise $A$ resumes Protocol VFE started in Step F4 with Step V4, outputs $O_A$, and halts.

We now describe the sub-protocols for aborting and resolving. They also take place in the model of Definition 1, where all parties maintain internal state (private inputs are sometimes mentioned nevertheless). In particular, $T$ maintains a list of tuples internally and processes all abort and resolve requests atomically. Recall that the transcript of a party of a protocol consists of all messages received or sent by this party.

Protocol abort is a protocol between $A$ and $T$; it is invoked by $A$ with inputs $C_A$ and $C_B$.

**Protocol** abort$(g, g_1, f, y_T)(C_A, C_B)()$

1. $A$ sends the message (abort, $C_A\|C_B\|f$) to $T$.
2. If $T$'s internal state contains an entry of the form $(C_A\|C_B\|f, string)$, then $T$ returns to $A$ the message $string$.
3. Otherwise, $T$ adds the tuple $(C_A\|C_B\|f, \text{abort})$ to its internal state and returns to $A$ the message abort.

Protocol B-resolve is a protocol between $B$ and $T$; it is invoked by $B$ with input a string $transcript$, containing $B$'s complete transcript of Steps F1–F4 in Protocol FAIRCOMP, which includes also $C_A$ and $C_B$.

**Protocol** B-resolve$(g, g_1, f, y_T)(transcript)(z_T)$

1. $B$ sends the message (B-resolve, $transcript$) to $T$.
2. If $T$'s internal state contains an entry of the form $(C_A\|C_B\|f, string)$, then $T$ returns to $B$ the message $string$ and halts.
3. Otherwise, $B$ and $T$ run Steps V1–V3 of Protocol VFE$(g, g_1, C_B, C_A, f_A)$ $(x_B, r_B)(\emptyset)$ unmodified with $B$ in the role of circuit constructor (VFE-)$A$ and $T$ in the role of circuit evaluator (VFE-)$B$. They stop after Step 1 in Protocol VOT, before $T$ would have to decrypt the tokens. (Thus, $T$'s inputs to the protocol may be empty.)
   If $T$ rejects any of the proofs by $B$, then $T$ adds the tuple $(C_A\|C_B\|f, \text{abort})$ to its internal state and returns to $B$ the message abort.
4. Otherwise, $T$ reads the $transcript$ sent by $B$ and carries out its part of Phase II for the escrows of the tokens on the output wires for $f_B$ from Step F3. $T$ opens the escrows subject to all tags matching $C_A\|C_B\|f_B\|i$. In other words, $T$ runs the decryption algorithm VD(CS, $z_T, \ldots$) and returns the outputs to $B$ if all tags match, or $\perp$ if one or more decryptions yield $\perp$. $T$ computes the transcript $t$ of Protocol B-resolve and adds $(C_A\|C_B\|f,$ resolve$\|t)$ to its internal state.

Protocol A-resolve is a protocol between $A$ and $T$; it is invoked by $A$ with input a string $transcript$, containing her complete transcript of Steps F1–F3 in Protocol FAIRCOMP, which includes also $C_A$ and $C_B$.

**Protocol** A-resolve$(g, g_1, f, y_T)(transcript)(z_T)$

1. $A$ sends the message (**A-resolve**, $transcript$) to $T$.
2. If $T$'s internal state contains an entry of the form $(C_A \| C_B \| f, string)$, then $T$ returns to $A$ the message $string$ and halts.
3. Otherwise, $A$ and $T$ run Steps V1–V3 of Protocol VFE$(g, g_1, C_A, C_B, f_B)$ $(x_A, r_A)(\emptyset)$ unmodified with $A$ in the role of circuit constructor (VFE-)$A$ and $T$ in the role of circuit evaluator (VFE-)$B$. They stop after Step 1 in Protocol VOT, before $T$ would have to decrypt the tokens. (Thus, $T$'s inputs to the protocol may be empty.)
   If $T$ rejects any of the proofs by $A$, then $T$ adds the tuple $(C_A \| C_B \| f, \texttt{abort})$ to its internal state and returns to $A$ the message **abort**.
4. Otherwise, $T$ reads the $transcript$ sent by $A$ and carries out its part of Phase II for the escrows of the tokens on the output wires for $f_A$ from Step F4. $T$ opens the escrows subject to all tags matching $C_A \| C_B \| f_A \| i$. In other words, $T$ runs the decryption algorithm VD$(\mathsf{CS}, z_T, \dots)$ and returns the outputs to $A$ if all tags match, or $\perp$ if one or more decryptions yield $\perp$. $T$ computes the transcript $t$ of Protocol A-resolve and adds $(C_A \| C_B \| f, \texttt{resolve} \| t)$ to its internal state.

Remarks about the protocol.

1. Protocol **FAIRCOMP** as described above consists of seven rounds (14 moves). By pipelining the execution of Steps F1–F4 one can reduce this to five rounds (ten moves). Using non-interactive proofs in the random oracle model, this could even be reduced further to three rounds (six moves).
2. A major difference between the resolve protocols here and those used for optimistic fair exchange of signatures [2] is that $T$ cannot directly replace the other party here. Whereas in a fair exchange of digital signatures, $T$ can verify that the party requesting to resolve supplies a correct signature, $T$ has to re-run almost the complete VFE protocol here. After $T$ has done this, the other party is able to complete VFE and its part of the computation from this transcript.
3. $T$ does not have to know any secrets of the other party for re-running VFE. For instance, in Step 3 of Protocol B-resolve, when $B$ and $T$ run Protocol VFE for $f_A$ (and $T$ plays the role of $A$), $T$ does not have to know anything about $A$'s secret input $x_A$ besides the commitments $C_A$; this follows because the VFE protocol is stopped after Step V3 and because of a special feature of the underlying Protocol VOT, in which the commitments are used for encryption.

It can be shown that under the DDH assumption, Protocol **FAIRCOMP** is an optimistic fair secure computation protocol (omitted).

## Acknowledgments

We thank Ran Canetti and Victor Shoup for helpful suggestions and discussions about modeling optimistic fair secure computation.

# References

1. N. Asokan, M. Schunter, and M. Waidner, "Optimistic protocols for fair exchange," in *Proc. 4th ACM Conference on Computer and Communications Security*, pp. 6, 8–17, 1997.
2. N. Asokan, V. Shoup, and M. Waidner, "Optimistic fair exchange of digital signatures," *IEEE Journal on Selected Areas in Communications*, vol. 18, pp. 591–610, Apr. 2000.
3. B. Baum-Waidner and M. Waidner, "Optimistic asynchronous multi-party contract signing," Research Report RZ 3078 (#93124), IBM Research, Nov. 1998.
4. D. Beaver, "Secure multiparty protocols and zero-knowledge proof systems tolerating a faulty minority," *Journal of Cryptology*, vol. 4, no. 2, pp. 75–122, 1991.
5. D. Beaver and S. Goldwasser, "Multiparty computation with faulty majority (extended announcement)," in *Proc. 30th IEEE Symposium on Foundations of Computer Science (FOCS)*, pp. 468–473, 1989.
6. D. Beaver, S. Micali, and P. Rogaway, "The round complexity of secure protocols," in *Proc. 22nd Annual ACM Symposium on Theory of Computing (STOC)*, pp. 503–513, 1990.
7. M. Bellare and S. Micali, "Non-interactive oblivious transfer and applications," in *Advances in Cryptology: CRYPTO '89* (G. Brassard, ed.), vol. 435 of *Lecture Notes in Computer Science*, pp. 547–557, Springer, 1990.
8. M. Ben-Or, O. Goldreich, S. Micali, and R. L. Rivest, "A fair protocol for signing contracts," *IEEE Transactions on Information Theory*, vol. 36, pp. 40–46, Jan. 1990.
9. E. F. Brickell, D. Chaum, I. Damgård, and J. van de Graaf, "Gradual and verifiable release of a secret," in *Advances in Cryptology: CRYPTO '87* (C. Pomerance, ed.), vol. 293 of *Lecture Notes in Computer Science*, Springer, 1988.
10. J. Camenisch and I. Damgård, "Verifiable encryption and applications to group signatures and signature sharing," Tech. Rep. RS-98-32, BRICS, Departement of Computer Science, University of Aarhus, Dec. 1998.
11. J. Camenisch and M. Stadler, "Efficient group signature schemes for large groups," in *Advances in Cryptology: CRYPTO '97* (B. Kaliski, ed.), vol. 1233 of *Lecture Notes in Computer Science*, pp. 410–424, Springer, 1997.
12. R. Canetti, "Security and composition of multi-party cryptographic protocols," *Journal of Cryptology*, vol. 13, no. 1, pp. 143–202, 2000.
13. D. Chaum, I. Damgård, and J. van de Graaf, "Multiparty computations ensuring privacy of each party's input and correctness of the result," in *Advances in Cryptology: CRYPTO '87* (C. Pomerance, ed.), vol. 293 of *Lecture Notes in Computer Science*, Springer, 1988.
14. D. Chaum and T. P. Pedersen, "Wallet databases with observers," in *Advances in Cryptology: CRYPTO '92* (E. F. Brickell, ed.), vol. 740 of *Lecture Notes in Computer Science*, pp. 89–105, Springer-Verlag, 1993.
15. R. Cleve, "Limits on the security of coin flips when half the processors are faulty," in *Proc. 18th Annual ACM Symposium on Theory of Computing (STOC)*, pp. 364–369, 1986.
16. R. Cramer and I. Damgård, "Linear zero-knowledge—a note on efficient zero-knowledge proofs and arguments," in *Proc. 29th Annual ACM Symposium on Theory of Computing (STOC)*, 1997.
17. R. Cramer, I. Damgård, and B. Schoemakers, "Proofs of partial knowledge and simplified design of witness hiding protocols," in *Advances in Cryptology: CRYPTO '94* (Y. G. Desmedt, ed.), vol. 839 of *Lecture Notes in Computer Science*, 1994.

18. R. Cramer and V. Shoup, "A practical public-key cryptosystem provably secure against adaptive chosen-ciphertext attack," in *Advances in Cryptology: CRYPTO '98* (H. Krawczyk, ed.), vol. 1462 of *Lecture Notes in Computer Science*, Springer, 1998.

19. C. Crépeau, J. van de Graaf, and A. Tapp, "Committed oblivious transfer and private multi-party computation," in *Advances in Cryptology: CRYPTO '95* (D. Coppersmith, ed.), vol. 963 of *Lecture Notes in Computer Science*, Springer, 1995.

20. I. B. Damgård, "Efficient concurrent zero-knowledge in the auxiliary string model," in *Advances in Cryptology: EUROCRYPT 2000* (B. Preneel, ed.), vol. 1087 of *Lecture Notes in Computer Science*, pp. 418–430, Springer, 2000.

21. D. Dolev, C. Dwork, and M. Naor, "Non-malleable cryptography (extended abstract)," in *Proc. 23rd Annual ACM Symposium on Theory of Computing (STOC)*, pp. 542–552, 1991.

22. T. ElGamal, "A public key cryptosystem and a signature scheme based on discrete logarithms," *IEEE Transactions on Information Theory*, vol. 31, pp. 469–472, July 1985.

23. S. Even, O. Goldreich, and A. Lempel, "A randomized protocol for signing contracts," *Communications of the ACM*, vol. 28, pp. 637–647, 1985.

24. U. Feige, J. Kilian, and M. Naor, "A minimal model for secure computation (extended abstract)," in *Proc. 26th Annual ACM Symposium on Theory of Computing (STOC)*, pp. 554–563, 1994.

25. O. Goldreich, "Secure multi-party computation." Manuscript, 1998. (Version 1.1).

26. O. Goldreich, S. Micali, and A. Wigderson, "How to play any mental game or a completeness theorem for protocols with honest majority," in *Proc. 19th Annual ACM Symposium on Theory of Computing (STOC)*, pp. 218–229, 1987.

27. S. Goldwasser and L. Levin, "Fair computation of general functions in presence of immoral majority," in *Advances in Cryptology: CRYPTO '90* (A. J. Menezes and S. A. Vanstone, eds.), vol. 537 of *Lecture Notes in Computer Science*, Springer, 1991.

28. S. Goldwasser and S. Micali, "Probabilistic encryption," *Journal of Computer and System Sciences*, vol. 28, pp. 270–299, 1984.

29. S. Micali, "Secure protocols with invisible trusted parties." Presentation at the Workshop on Multi-Party Secure Protocols, Weizmann Institute of Science, Israel, June 1998.

30. S. Micali and P. Rogaway, "Secure computation," in *Advances in Cryptology: CRYPTO '91* (J. Feigenbaum, ed.), vol. 576 of *Lecture Notes in Computer Science*, pp. 392–404, Springer, 1992.

31. M. Naor and O. Reingold, "Number-theoretic constructions of efficient pseudorandom functions," in *Proc. 38th IEEE Symposium on Foundations of Computer Science (FOCS)*, 1997.

32. M. O. Rabin, "How to exchange secrets by oblivious transfer," Tech. Rep. TR-81, Harvard University, 1981.

33. C. P. Schnorr, "Efficient signature generation by smart cards," *Journal of Cryptology*, vol. 4, pp. 161–174, 1991.

34. M. Stadler, "Publicly verifiable secret sharing," in *Advances in Cryptology: EUROCRYPT '96* (U. Maurer, ed.), vol. 1233 of *Lecture Notes in Computer Science*, pp. 190–199, Springer, 1996.

35. A. C. Yao, "Protocols for secure computation," in *Proc. 23rd IEEE Symposium on Foundations of Computer Science (FOCS)*, pp. 160–164, 1982.

36. A. C. Yao, "How to generate and exchange secrets," in *Proc. 27th IEEE Symposium on Foundations of Computer Science (FOCS)*, pp. 162–167, 1986.

# A Cryptographic Solution
# to a Game Theoretic Problem

Yevgeniy Dodis[1], Shai Halevi[2], and Tal Rabin[2]

[1] Laboratory for Computer Science, MIT,
545 Tech Square, Cambridge, MA 02139, USA.
yevgen@theory.lcs.mit.edu
[2] IBM T.J. Watson Research Center,
P.O. Box 704, Yorktown Heights, New York 10598, USA.
{shaih,talr}@watson.ibm.com

**Abstract.** In this work we use cryptography to solve a game-theoretic problem which arises naturally in the area of two party strategic games. The standard game-theoretic solution concept for such games is that of an *equilibrium*, which is a pair of "self-enforcing" strategies making each player's strategy an optimal response to the other player's strategy. It is known that for many games the expected equilibrium payoffs can be much higher when a trusted third party (a "mediator") assists the players in choosing their moves (*correlated equilibria*), than when each player has to choose its move on its own (*Nash equilibria*). It is natural to ask whether there exists a mechanism that eliminates the need for the mediator yet allows the players to maintain the high payoffs offered by mediator-assisted strategies. We answer this question affirmatively provided the players are computationally bounded and can have free communication (so-called "cheap talk") prior to playing the game.

The main building block of our solution is an efficient cryptographic protocol to the following *Correlated Element Selection* problem, which is of independent interest. Both Alice and Bob know a list of pairs $(a_1, b_1) \ldots (a_n, b_n)$ (possibly with repetitions), and they want to pick a *random* index $i$ such that Alice learns only $a_i$ and Bob learns only $b_i$. Our solution to this problem has constant number of rounds, negligible error probability, and uses only very simple zero-knowledge proofs. We then show how to incorporate our *cryptographic* protocol back into a *game-theoretic* setting, which highlights some interesting parallels between cryptographic protocols and extensive form games.

## 1 Introduction

The research areas of Game Theory and Cryptography are both extensively studied fields with many problems and solutions. Yet, the cross-over between them is surprisingly small: very rarely are tools from one area borrowed to address problems in the other. Some examples of using game-theoretic concepts to solve cryptographic problems include the works of Fischer and Wright [17] and

M. Bellare (Ed.): CRYPTO 2000, LNCS 1880, pp. 112–130, 2000.
© Springer-Verlag Berlin Heidelberg 2000

Kilian [26]. In this paper we show an example in the other direction of how cryptographic tools can be used to address a natural problem in the Game Theory world.

## 1.1   Two Player Strategic Games

The game-theoretic problem that we consider in this work belongs to the general area of *two player strategic games*, which is an important field in Game Theory (see [20,32]). In the most basic notion of a two player game, there are two players, each with a set of possible moves. The game itself consists of each player choosing a move from its set, and then both players executing their moves simultaneously. The rules of the game specify a *payoff* function for each player, which is computed on the two moves. Thus, the payoff of each player depends both on its move and the move of the other player. A *strategy* for a player is a (possibly randomized) method for choosing its move. A fundamental assumption of these games, is that each player is *rational*, i.e. its sole objective is to maximize its (expected) payoff.

A pair of players' strategies achieves an *equilibrium* when these strategies are *self-enforcing*, i.e. each player's strategy is an *optimal response* to the other player's strategy. In other words, once a player has chosen a move and believes that the other player will follow its strategy, its (expected) payoff will not increase by changing this move. This notion was introduced in the classical work of Nash [31].

In a *Nash equilibrium*, each player chooses its move *independently* of the other player. (Hence, the induced distribution over the pairs of moves is a product distribution.) Yet, Aumann [2] showed that in many games, the players can achieve much higher expected payoffs, while preserving the "self-enforcement" property, if their strategies are *correlated* (so the induced distribution over the pairs of moves is no longer a product distribution). To actually implement such a *correlated equilibrium*, a "trusted third party" (called a *mediator*) is postulated. This mediator chooses the pair of moves according to the right joint distribution and *privately* tells each player what its designated move is. Since the strategies are correlated, the move of one player typically carries some information (not known a-priori) on the move of the other player. In a correlated equilibrium, no player has an incentive to deviate from its designated move, even knowing this extra information about the other player's move.

## 1.2   Removing the Mediator

As the game was intended for two players, it is natural to ask if correlated equilibria can be implemented without actually having a mediator. In the language of cryptography, we ask if we can design a two party game to eliminate the trusted third party from the original game. It is well known that in the standard cryptographic models the answer is positive, provided that the two players can interact, that they are computationally bounded, and assuming some standard hardness assumptions ([22,34]). We show that this positive answer can be carried over to the Game Theory model as well. Specifically, we consider an *extended game*, in which the players first exchange messages (this part is called "cheap talk" by game theorists and is quite standard; see Myerson [30] for survey), and then choose their moves and execute them simultaneously as in the original game. The payoffs are still computed as a function of the moves, according to the same payoff function as in the original game.

It is very easy to see that every Nash equilibrium payoff of the extended game is also a correlated equilibrium payoff of the original game (the mediator can simulate the pre-play communication stage). Our hope would be to show that any Correlated equilibrium payoffs of the original game can always be achieved by some Nash equilibrium of the extended game. However, Barany [3] showed that this is generally not true. Namely, that Nash equilibria payoffs of the extended game are inside the convex hull of the Nash equilibria payoffs of the original game, which often does not include many correlated equilibria payoffs of the original game (see Section 2 for an example).

In this work we overcome this difficulty by considering the realistic scenario where the players are *computationally bounded*. In other words, while Game Theory typically assumes that the players have unlimited computational capabilites when they need to make their decisions, we will assume that the players are restricted to *probabilistic polynomial time*. Of independent interest to Game Theory, we will define a new concept of a *computational Nash equilibrium* as a pair of efficient strategies where no *polynomially bounded* player can gain a non-negligible advantage by not following its strategy (see Section 3 for formal definitions). Then, we prove the following:

**Theorem 1.** *Let $G$ be any two player strategic game and let $G'$ be the extended game of $G$. If secure two-party protocols exist for non-trivial functions, then for any correlated equilibrium $s$ of $G$ there exists a computational Nash equilibrium $\sigma$ of $G'$, such that the payoffs for both players are the same in $\sigma$ and $s$.*

In other words, any correlated equilibrium payoffs of $G$ can be achieved using a computational Nash equilibrium of $G'$. Thus, the mediator can be eliminated if the players are computationally bounded and can communicate prior to the game.

We stress that although this theorem seem quite natural and almost trivial from a cryptography point of view, the models of Game Theory and Cryptography are significantly different, and thus proving it in the Game Theory framework requires some care. In particular, two-party cryptographic protocols always assume that at least one player is honest, while the other player could be arbitrarily malicious. In the game-theoretic setting, on the other hand, *both players are selfish and rational*: they (certainly) deviate from the protocol if they benefit from it, and (can be assumed to) follow their protocol otherwise. Also, it is important to realize that in this setting we cannot use cryptography to "enforce" honest behavior. This is due to the fact that a "cheating player" who was "caught cheating" during the protocol, can still choose a move that would maximizes its profit. We discuss these and some other related issues further in Section 2.

## 1.3    Doing It Efficiently

Although the assumption of Theorem 1 can be proven using tools of generic two-party computations [22,34], it would be nice to obtain computational Nash equilibria (i.e. protocols) which are more efficient than the generic ones. In Section 4 we observe that for many cases, the underlying cryptographic problem reduces to a problem which we call *Correlated Element Selection*. We believe that this natural problem has other cryptographic application and is of independent interest. In this problem, two players, $A$ and $B$, know a list of pairs $(a_1, b_1), \ldots, (a_n, b_n)$ (maybe with repetitions), and they need to jointly choose a random index $i$, so that player $A$ only learns the value $a_i$ and player $B$ only learns the value $b_i$.[1] Our final protocol for this problem is very intuitive, has constant number of rounds, negligible error probability, and uses only very simple zero-knowledge proofs.

Our protocol for Correlated Element Selection uses as a tool a useful primitive which we call *blindable encryption* (which can be viewed as a counterpart of blindable signatures [10]). Stated roughly, blindable encryption is the following: given an encryption $c$ of an (unknown) message $m$, and an additional message $m'$, a random encryption of $m + m'$ can be easily computed. This should be done without knowing $m$ or the secret key. Examples of semantically secure blindable encryption schemes (under appropriate assumptions) include Goldwasser-Micali [23], ElGamal [15] and Benaloh [5]. (In fact, for our Correlated Element Selection protocol, it is sufficient to use a weaker notion of blindability, such as the one in [33].) Aside from our main application, we also observe that blindable encryption appears to be a very convenient tool for devising efficient two-party protocols and suggest that it might be used more often. (For example, in the full version of this paper we show a very simple protocol to achieve 1-*out-of-n* *Oblivious Transfer* protocol from any secure blindable encryption scheme.)

## 1.4    Related Work

*Game Theory.* Realizing the advantages of removing the mediator, various papers in the Game Theory community have been published to try and achieve this goal. Similarly to our work, Barany [3] shows that the mediator can be replaced by pre-play communication but he requires four or more players for this communication, even for a game which is intended for two players. In his protocol only two players participate as "decision makers" during the pre-play communication, and (at least two) other players help them to hide information from each other (as Barany showed, two players do not suffice). Barany's protocol works in an information-theoretic setting (which explains the need for four players; see [6].) Of course, if one is willing to use a group of players to simulate the mediator, then the general multiparty computation tools (e.g. [6,11]) can

---

[1] A special case of Correlated Element Selection when $a_i = b_i$ is just the standard *coin-flipping* problem [7]. However, this is a degenerate case of the problem, since it requires no secrecy. In particular, none of the previous coin-flipping protocols seem to extend to solve our problem.

also be used, even though the solution of [3] is simpler. Forges [18,19] extends these results to more general classes of games. The work of Lehrer and Sorin [27] describes protocols that "reduce" the role of the mediator (the mediator receives private signals from the players and makes deterministic *public* announcements). Mailath et al. [29] show that the set of correlated equilibria of the original game coincides with the set of Nash equilibria of the so called "local-interaction game" (where many players are paired up randomly and play the original game). The distinguishing feature of our work is the observation that placing realistic computational restrictions on the players allows them to achieve results which are *provably* impossible when the players are computationally unbounded.

*Cryptography.* We already mentioned the relation of our work to generic two-party secure computations [22,34]. We note that some of our techniques (in particular, the zero-knowledge proofs) are similar to those used for mixing networks (see [1,25] and the references therein), even though our usage and motivation are quite different. Additionally, encryption schemes with various "blinding properties" were used for many different purposes, including among others for secure storage [21], and secure circuit evaluations [33].

## 2   Background in Game Theory

*Two-player Games.* Although our results apply to a much larger class of two-player games, we demonstrate them on the simplest possible class of finite *strategic games* (with complete information). Such a game $G$ has two players 1 and 2, each of whom has a finite set $A_i$ of possible *actions* and a *payoff function* $u_i : A_1 \times A_2 \mapsto R$ $(i = 1, 2)$, known to both players. The players move simultaneously, each choosing an action $a_i \in A_i$. The *payoff* of player $i$ is $u_i(a_1, a_2)$. The (probabilistic) algorithm that tells player $i$ which action to take is called its *strategy*, and a pair of strategies is called a *strategy profile*. In our case, a strategy $s_i$ of player $i$ is simply a probability distribution over its actions $A_i$, and a strategy profile $s = (s_1, s_2)$ is a probability distribution over $A_1 \times A_2$. Classical Game Theory assumes that each player is *selfish and rational*, i.e. only cares about maximizing its (expected) payoff. As a result, we are interested in strategy profiles that are *self-enforcing*. In other words, even knowing the strategy of the other player, each player still has no incentive to deviate from its own strategy. Such a strategy profile is called an *equilibrium*.

*Nash equilibrium.* This is the best known notion of an equilibrium [31]. It corresponds to a strategy profile in which players' strategies are *independent*. More precisely, the induced distribution over the pairs of actions, must be a product distribution, $s(A_1 \times A_2) = s_1(A_1) \times s_2(A_2)$. Deterministic (or *pure*) strategies are a special case of such strategies, where $s_i$ assigns probability 1 to some action. For strategies $s_1$ and $s_2$, we denote by $u_i(s_1, s_2)$ the *expected* payoff for player $i$ when players independently follow $s_1$ and $s_2$.

**Definition 1.** *A* Nash equilibrium *of a game $G$ is an independent strategy profile $(s_1^*, s_2^*)$, such that for any $a_1 \in A_1$, $a_2 \in A_2$, we have $u_1(s_1^*, s_2^*) \geq u_1(a_1, s_2^*)$ and $u_2(s_1^*, s_2^*) \geq u_2(s_1^*, a_2)$.*

In other words, given that player 2 follows $s_2^*$, $s_1^*$ is an optimal response of player 1 and vice versa.

*Correlated equilibrium.* While Nash equilibrium is quite a natural and appealing notion (since players can follow their strategies independently of each other), one can wonder if it is possible to achieve higher expected payoffs if one allows *correlated* strategies.

In a correlated strategy profile [2], the induced distribution over $A_1 \times A_2$ can be an arbitrary distribution, not necessarily a product distribution. This can be implemented by having a trusted party (called *mediator*) sample a pair of actions $(a_1, a_2)$ according to some *joint* probability distribution $s(A_1 \times A_2)$, and "recommend" the action $a_i$ to player $i$. We stress that knowing $a_i$, player $i$ now knows a *conditional distribution* over the actions of the other player (which can be different for different $a_i$'s), but knows *nothing more*. We denote these distributions by $s_2(\cdot \mid a_1)$ and $s_1(\cdot \mid a_2)$.

For any $a_1' \in A_1, a_2' \in A_2$, let $u_1(a_1', s_2 \mid a_1)$ be the expected value of $u_1(a_1', a_2)$ when $a_2$ is distributed according to $s_2(\cdot \mid a_1)$ (similarly for $u_2(s_1, a_2' \mid a_2)$). In other words, $u_1(a_1', s_2 \mid a_1)$ measures the expected payoff of player 1 if his recommended action was $a_1$ (thus, $a_2$ is distributed according to $s_2(\cdot \mid a_1)$), but it decided to play $a_1'$ instead. As before, we let $u_i(s)$ be the expected value of $u_i(a_1, a_2)$ when $(a_1, a_2)$ are drawn according to $s$. Similarly to Nash equilibrium, a more general notion of a *correlated equilibrium* is defined, which ensures that players have no incentive to deviate from the "recommendation" they got from the mediator.

**Definition 2.** *A* correlated equilibrium *is a strategy profile $s^* = s^*(A_1 \times A_2) = (s_1^*, s_2^*)$, such that for any $(a_1^*, a_2^*)$ in the support of $s^*$, any $a_1 \in A_1$ and $a_2 \in A_2$, we have $u_1(a_1^*, s_2^* \mid a_1^*) \geq u_1(a_1, s_2^* \mid a_1^*)$ and $u_2(s_1^*, a_2^* \mid a_2^*) \geq u_2(s_1^*, a_2 \mid a_2^*)$.*

Given Nash (resp. Correlated) equilibrium $(s_1^*, s_2^*)$, we say that $(s_1^*, s_2^*)$ achieves *Nash (resp. Correlated) equilibrium payoffs* $[u_1(s_1^*, s_2^*), u_2(s_1^*, s_2^*)]$.

Correlated equilibria of any game form a convex set, and therefore always include the convex hull of Nash equilibria. However, it is well known that correlated equilibria can give equilibrium payoffs *outside* (and significantly better!) than anything in the convex hull of Nash equilibria payoffs. This is demonstrated in the following simple example first observed by Aumann [2], who also defined the notion of correlated equilibrium. Much more dramatic examples can be shown in larger games.[2]

*Game of "Chicken".* We consider a simple $2 \times 2$ game, the so-called game of "Chicken" shown in the table to the right. Here each player can either "dare"

---

[2] For example, there are games with a unique Nash equilibrium $s$ and many Correlated equilibria giving *both* players much higher payoffs than $s$.

($D$) or "chicken out" ($C$). The combination ($D, D$) has a devastating effect on both players (payoffs $[0, 0]$), ($C, C$) is quite good (payoffs $[4, 4]$), while each player would ideally prefer to dare while the other chickens-out (giving him 5 and the opponent 1). While the "wisest" pair of actions is ($C, C$), this is not a Nash equilibrium, since both players are willing to deviate to $D$ (believing that the other player will stay at $C$). The game is easily seen to have three Nash equilibria: $s^1 = (D, C)$, $s^2 = (C, D)$ and $s^3 = (\frac{1}{2} \cdot D + \frac{1}{2} \cdot C, \ \frac{1}{2} \cdot D + \frac{1}{2} \cdot C)$. The respective Nash equilibrium payoffs are $[5, 1]$, $[1, 5]$ and $[\frac{5}{2}, \frac{5}{2}]$. We see that the first two pure strategy Nash equilibria are "unfair", while the last mixed equilibrium has small payoffs, since the mutually undesirable outcome ($D, D$) happens with non-zero probability $\frac{1}{4}$ in the product distribution. The best "fair" strategy profile in the convex hull of the Nash equilibria is the combination $\frac{1}{2}s^1 + \frac{1}{2}s^2 = (\frac{1}{2}(C, D) + \frac{1}{2}(D, C))$, yielding payoffs $[3, 3]$. On the other hand, the profile $s^* = (\frac{1}{3}(C, D) + \frac{1}{3}(D, C) + \frac{1}{3}(C, C))$ is a correlated equilibrium, yielding payoffs $[3\frac{1}{3}, 3\frac{1}{3}]$ outside any convex combination of Nash equilibria.

|   | C | D |
|---|---|---|
| C | 4,4 | 1,5 |
| D | 5,1 | 0,0 |

*"Chicken"*

|   | C | D |
|---|---|---|
| C | 1/4 | 1/4 |
| D | 1/4 | 1/4 |

*Mixed Nash $s^3$*

|   | C | D |
|---|---|---|
| C | 1/3 | 1/3 |
| D | 1/3 | 0 |

*Correlated $s^*$*

To briefly see that this is a correlated equilibrium, consider the "row player" 1 (same works for player 2). If it is recommended to play $C$, its expected payoff is $\frac{1}{2} \cdot 4 + \frac{1}{2} \cdot 1 = \frac{5}{2}$ since, conditioned on $a_1 = C$, player 2 is recommended to play $C$ and $D$ with probability $\frac{1}{2}$ each. If player 1 switched to $D$, its expected payoff would still be $\frac{1}{2} \cdot 5 + \frac{1}{2} \cdot 0 = \frac{5}{2}$, making player 1 reluctant to switch. Similarly, if player 1 is recommended $D$, it knows that player 2 plays $C$ (as ($D, D$) is never played in $s^*$), so its payoff is 5. Since this is the maximum payoff of the game, player 1 would not benefit by switching to $C$ in this case. Thus, we indeed have a correlated equilibrium, where each player's payoff is $\frac{1}{3}(1 + 5 + 4) = 3\frac{1}{3}$, as claimed.

## 3   Implementing the Mediator

In this section we show how to remove the mediator using cryptographic means. We assume the existence of generic secure two-party protocols and show how to achieve our goal by using such protocols in the *game-theoretic* (rather than its designated cryptographic) setting. In other words, the players remain selfish and rational, even when running the cryptographic protocol. In Section 4 we give an efficient implementation for the types of cryptographic protocols that we need.

*Extended Games.* To remove the mediator, we assume that the players are (1) computationally bounded and (2) can communicate prior to playing the original game, which we believe are quite natural and minimalistic assumptions. To formally define the computational power of the players, we introduce an external security parameter into the game, and require that the strategies of both players can be computed in probabilistic polynomial time in the security parameter.[3]

To incorporate communication into the game, we consider an *extended game*, which is composed of three parts: first the players are given the security parameter and they freely exchange messages (i.e., execute any two-party protocol), then each player locally selects its move, and finally both players execute their move simultaneously. The final payoffs $u'_i$ of the extended game are just the corresponding payoffs of the original game applied to the players' simultaneous moves at the last step.

The notions of a strategy and a strategy profile are straightforwardly generalized from those of the basic game, except that they are full-fledged probabilistic algorithms telling each player what to do *in each situation*. We now define the notion of a *computational Nash equilibrium* of the extended game, where the strategies of both players are restricted to probabilistic polynomial time (PPT). Also, since we are talking about a computational model, the definition must account for the fact that the players may break the underlying cryptographic scheme with negligible probability (e.g., by guessing the secret key), thus gaining some advantage in the game. In the definition and discussion below, we denote by $negl(k)$ some function that is negligible in $k$.

**Definition 3.** *A computational Nash equilibrium of an extended game $G$ is an independent strategy profile $(\sigma_1^*, \sigma_2^*)$, such that*

*(a) both $\sigma_1^*$, $\sigma_2^*$ are PPT computable; and*
*(b) for any other PPT computable strategies $\sigma_1'$, $\sigma_2'$, we have*
$$u_1(\sigma_1', \sigma_2^*) \leq u_1(\sigma_1^*, \sigma_2^*) + negl(k) \text{ and } u_2(\sigma_1^*, \sigma_2') \leq u_2(\sigma_1^*, \sigma_2^*) + negl(k).$$

We notice that the new "philosophy" for both players is still to maximize their expected payoff, except that the players will not change their strategy if their gain is negligible.

The idea of getting rid of the mediator is now very simple. Consider a correlated equilibrium $s(A_1 \times A_2)$ of the original game $G$. Recall that the job of the mediator is to sample a pair of actions $(a_1, a_2)$ according to the distribution $s$, and to give $a_i$ to player $i$. We can view the mediator as a trusted party who securely computes a probabilistic (polynomial-time) function $s$. Thus, to remove it we can have the two players execute a cryptographic protocol $P$ that securely computes the function $s$. The strategy of each player would be to follow the protocol $P$, and then play the action $a$ that it got from $P$.

Yet, several issues have to be addressed in order to make this idea work. First, the above description does not completely specify the strategies of the players. A full specification of a strategy must also indicate what a player should do if the other player *deviates* from its strategy (in our case, does not follow the protocol $P$). While cryptography does not address this question (beyond the guarantee that the other player is likely to detect the deviation and abort the protocol), it is

---

[3] Note that the parameters of the original game (like the payoff functions, the correlated equilibrium distribution, etc.) are all independent of the security parameter, and thus can always be computed "in constant time".

crucial to resolve it in our setting, since "the game must go on": No matter what happens inside $P$, both players eventually have to take simultaneous actions, and receive the corresponding payoffs (which they wish to maximize). Hence we must explain how to implement a "punishment for deviation" within the game-theoretic framework.

*Punishment for Deviations.* We employ the standard game-theoretic solution, which is to punish the cheating player to his *minmax level.* This is the smallest payoff that one player can "force" the other player to have. Namely, the minmax level of player 2 is $\underline{v_2} = \min_{s_1} \max_{s_2} u_2(s_1, s_2)$. Similarly, minmax level of player 1 is $\underline{v_1} = \min_{s_2} \max_{s_1} u_1(s_1, s_2)$. To complete the description of our proposed equilibrium, we let each player punish the other player to its minmax level, if the other player deviates from $P$ and is "caught". Namely, if player 2 cheats, player 1 will play in the last stage of the game the strategy $\underline{s_1}$ achieving the minmax payoff $\underline{v_2}$ for player 2 and vice versa. Note that the instances where a player deviates from $P$ but this is not detected falls under the negligible probability that the protocol will fail. Note also that in "interesting" games, the minmax payoff would be strictly smaller than the correlated equilibrium payoffs. Intuitively, in this case the only potentially profitable cheating strategy is an "honest but curious" behavior, where a player follows the prescribed protocol but tries nonetheless to learn additional information about the action of the other player. Any other cheating strategy would carry an overwhelming probability of "getting caught", hence causing a real loss. Thus, we first observe the following simple fact:

**Lemma 1.** *Let $s^* = (s_1^*, s_2^*)$ be a correlated equilibrium. For any action $a_1$ of player 1 which occurs with non-zero probability in $s^*$, denote $\mu_1(a_1) = u_1(a_1, s_2^*|a_1)$. That is, $\mu(a_1)$ is the expected payoff of player 1 when its recommended action is $a_1$. Similarly, we define for player 2 $\mu_2(a_2) = u_2(s_1^*|a_2, a_2)$.*

*Let $\underline{v_i}$ be the minmax payoff of player $i$, then for every $a_1, a_2$ that occur with non-zero probability in $s^*$, it holds that $\mu_i(a_i) \geq \underline{v_i}$.*

Theorem 1 now follows almost immediately from Lemma 1 and the security of $P$. Intuitively, since (a) a cheating player that "gets caught" is going to lose by Lemma 1 and (b) the security of $P$ implies that cheating is detected with very high probability, we get that the risk of getting caught out-weighs the benefits of cheating, and players will not have an incentive to deviate from the protocol $P$. (A particular type of cheating in $P$ is "early stopping". Since the extended game must always result in players getting payoffs, early stopping is not an issue in game theory, since it will be punished by the minmax level as well.)

Somewhat more formally, let $v_1 = u_1(s_1^*, s_2^*)$, and consider that 1 is a cheating player who uses some arbitrary (but PPT computable) strategy $s_1'$ (the analysis for player 2 is similar). Let the action taken by player 1 in the extended game be considered its output of the protocol. The output of player 2 is whatever is specified in its part of the protocol $P$, which is either an action (if the protocol runs to completion) or "abort" (if some "cheating" is detected).

According to standard definitions of secure protocols (e.g., the one by Canetti [9]), $P$ is secure if the above output pair can be simulated in an "ideal model". This "ideal model" is almost exactly the model of the trusted mediator, except that player 1 may choose to have the mediator abort before it recommends an action to player 2 (in which case the output of player 2 in the ideal model is also "abort"). The security of $P$ implies that the output distribution in the execution of the protocol in the "real world" is indistinguishable from that of the "ideal world".

Consider now the function $\tilde{u}_1(\cdot, \cdot)$, which denotes the "payoff of player 1" in the extended game, given a certain output pair. That is, if the output is a pair of actions $(a_1, a_2)$ than $\tilde{u}_1(a_1, a_2) = u_1(a_1, a_2)$, and if the output of the second player is "abort" then $\tilde{u}_1(a_1, \text{"abort"}) = u_1(a_1, \underline{a_2})$, where $\underline{a_2}$ is the minmax move for player 2. Note that in the real world, the function $\tilde{u}_1$ indeed represents the payoff of player 1 using strategy $s'_1$, but note also that this function is well defined even in the ideal world. Clearly, the expected value of $\tilde{u}_1$ in the real world is at most negligibly higher than in the ideal world. Otherwise, the output distributions in the two worlds could be distinguished with a non-negligible advantage by comparing the value of this function to some properly chosen threshold, contradicting the security of the protocol $P$.

Therefore, to prove Theorem 1 it is sufficient to show that the expected value of $\tilde{u}_1$ in the ideal world is at most $v_1$ (which is equal to the correlated equilibrium payoff of player 1 in the original game $G$). This is where we use Lemma 1: this lemma tells us that in the ideal world, no matter what action that is recommended to player 1, this player cannot increase the expected value of $\tilde{u}_1$ by aborting the mediator before it recommends an action to player 2. Hence, we can upper bound the expected value of $\tilde{u}_1$ in the ideal world by considering a strategy of player 1 that never aborts the mediator. Such strategy corresponds exactly to a strategy in the original game $G$ (with the mediator), and so it cannot achieve expected payoff of more than $v_1$. This completes the proof.

*Subgame Perfect Equilibrium.* In looking at the computational Nash equilibrium we constructed, one may wonder why would a player want to carry out the "minmax punishment" when it catches the other player cheating (since this "punishment" may also hurt the "punishing player"). The answer is that the notion of Nash equilibrium only requires player's actions to be optimal *provided the other player follows its strategy*. Thus, it is acceptable to carry out the punishment even if this results in a loss for *both* players. We note that this oddity (known as an "empty threat" in the game-theoretic literature) is one of the reason the concept of Nash equilibrium is considered weak in certain situations. As a result, game theorists often consider a stricter version of a Nash equilibrium for extended games, called a *subgame perfect* equilibrium.

In the full version we show that Theorem 1 can be broadened to the case of the subgame perfect equilibrium. Generally stated, we prove that every "interesting" correlated-equilibrium payoff of the game $G$ can be achieved by a subgame perfect equilibrium of an extended game $G'$.

# 4   The Correlated Element Selection Problem

In most common games, the joint strategy of the players is described by a short list of pairs $\{(\text{move1}, \text{move2})\}$, where the strategy is to choose at random one pair from this list, and have Player 1 play move1 and Player 2 play move2. (For example, in the game of chicken the list consists of three pairs $\{(D, C), (C, D), (C, C)\}$.)[4]

Hence, to obtain an efficient solution for such games, we need an efficient cryptographic protocol for the following problem: Two players, $A$ and $B$, know a list of pairs $(a_1, b_1), \ldots, (a_n, b_n)$ (maybe with repetitions), and they need to jointly choose a random index $i$, and have player $A$ learn only the value $a_i$ and player $B$ learn only the value $b_i$. We call this problem the *Correlated Element Selection* problem. In this section we describe our efficient solution for this problem. We start by presenting some notations and tools that we use (in particular, "blindable encryption schemes"). We then show a simple protocol that solves this problem in the special case where the two players are "honest but curious", and explain how to modify this protocol to handle the general case where the players can be malicious.

## 4.1   Notations and Tools

We denote by $[n]$ the set $\{1, 2, \ldots n\}$. For a randomized algorithm $A$ and an input $x$, we denote by $A(x)$ the output distribution of $A$ on $x$, and by $A(x; r)$ we denote the output string when using the randomness $r$. If one of the inputs to $A$ is considered a "key", then we write it as a subscript (e.g., $A_k(x)$). We use $pk, pk_1, pk_2, \ldots$ to denote public keys and $sk, sk_1, sk_2, \ldots$ to denote secret keys.

The main tool that we use in our protocol is *blindable encryption schemes*. Like all public-key encryption schemes, blindable encryption schemes include algorithms for key-generation, encryption and decryption. In addition they also have a "blinding" and "combining" algorithms. We denote these algorithms by *Gen*, *Enc*, *Dec*, *Blind*, and *Combine*, respectively. Below we formally define the blinding and combining functions. In this definition we assume that the message space $M$ forms a group (which we denote as an additive group with identity 0).

**Definition 4 (Blindable encryption).** *A public-key encryption scheme $\mathcal{E}$ is blindable if there exist (PPT) algorithms Blind and Combine such that for every message $m$ and every ciphertext $c \in Enc_{pk}(m)$:*

- *For any message $m'$ (also referred to as the "blinding factor"), $Blind_{pk}(c, m')$ produces a random encryption of $m + m'$. Namely, the distribution $Blind_{pk}(c, m')$ should be equal to the distribution $Enc_{pk}(m + m')$.*

$$Enc_{pk}(m + m') \equiv Blind_{pk}(c, m') \tag{1}$$

---

[4] Choosing from the list with distribution other than the uniform can be accommodated by having a list with repetitions, where a high-probability pair appears many times.

– If $r_1, r_2$ are the random coins used by two successive "blindings", then for any two blinding factors $m_1, m_2$,

$$Blind_{pk}(Blind_{pk}(c, m_1; \ r_1), m_2; \ r_2)$$
$$= Blind_{pk}(c, m_1 + m_2; Combine_{pk}(r_1, r_2)) \qquad (2)$$

Thus, in a blindable encryption scheme anyone can "randomly translate" the encryption $c$ of $m$ into an encryption $c'$ of $m + m'$, without knowledge of $m$ or the secret key, and there is an efficient way of "combining" several blindings into one operation.

Both the ElGamal and the Goldwasser-Micali encryption schemes can be extended into blindable encryption schemes. We note that most of the components of our solution are independent of the specific underlying blindable encryption scheme, but there are some aspects that still have to be tailored to each scheme. (Specifically, proving that the key generation process was done correctly is handled differently for different schemes. See details in the full paper [13].)

## 4.2   A Protocol for the Honest-but-Curious Case

For the case of honest-but-curious players, one can present an "almost trivial" solution using any 1-out-of-$n$ oblivious transfer protocol. However, in order to be able to derive an efficient protocol also for the general case, our starting point would be a somewhat different (but still very simple) protocol.

Let us recall the Correlated Element Selection problem. Two players share a public list of pairs $\{(a_i, b_i)\}_{i=1}^n$. For reasons that will soon become clear, we call the two players the "Preparer" $(P)$ and the "Chooser" $(C)$. The players wish to pick a random index $i$ such that $P$ only learns $a_i$ and $C$ only learns $b_i$. Figure 1 describes the Correlated Element Selection protocol for the honest-but-curious players. We employ a semantically secure blindable encryption scheme and for simplicity, we assume that the keys for this scheme were chosen by a trusted party ahead of time and given to $P$, and that the public key was also given to $C$.

At the beginning of the protocol, the Preparer randomly permutes the list, encrypts it element-wise and sends the resulting list to the Chooser. (Since the encryption is semantically secure, the Chooser "cannot extract any useful information" about the permutation $\pi$.) The Chooser picks a random pair of ciphertexts $(c_\ell, d_\ell)$ from the permuted list (so the final output pair will be the decryption of these ciphertexts). It then blinds $c_\ell$ with 0 (i.e. makes a random encryption of the same plaintext), blinds $d_\ell$ with a random blinding factor $\beta$, and sends the resulting pair of ciphertexts $(e, f)$ back to the Preparer. Decryption of $e$ gives the Preparer its element $a$ (and nothing more, since $e$ is a *random* encryption of $a$ after the blinding with 0), while the decryption $\tilde{b}$ of $f$ does not convey the value of the actual encrypted message since it was blinded with a random blinding factor. The Preparer sends $\tilde{b}$ to the Chooser, who recovers his element $b$ by subtracting the blinding factor $\beta$.

It is easy to show that if both players follow the protocol then their output is indeed a random pair $(a_i, b_i)$ from the known list. Moreover, at the end of the

## Protocol CES-1

| | |
|---|---|
| *Common inputs*: List of pairs $\{(a_i, b_i)\}_{i=1}^{n}$, public key $pk$. | |
| *Preparer knows*: secret key $sk$. | |

$P$ :      **1. Permute and Encrypt.**
Pick a random permutation $\pi$ over $[n]$.
Let $(c_i, d_i) = (Enc_{pk}(a_{\pi(i)}),\ Enc_{pk}(b_{\pi(i)}))$, for all $i \in [n]$.
Send the list $\{(c_i, d_i)\}_{i=1}^{n}$ to $C$.

$C$ :      **2. Choose and Blind.**
Pick a random index $\ell \in [n]$, and a random blinding factor $\beta$.
Let $(e, f) = (Blind_{pk}(c_\ell, 0),\ Blind_{pk}(d_\ell, \beta))$.
Send $(e, f)$ to $P$.

$P$ :      **3. Decrypt and Output.**
Set $a = Dec_{sk}(e)$, $\tilde{b} = Dec_{sk}(f)$. Output $a$.
Send $\tilde{b}$ to $C$.

$C$ :      **4. Unblind and Output.**
Set $b = \tilde{b} - \beta$. Output $b$.

**Fig. 1.** Protocol for Correlated Element Selection in the honest-but-curious model.

protocol the Preparer has no information about $b$ other than what's implied by its own output $a$, and the Chooser gets "computationally no information" about $a$ other than what's implied by $b$. Hence we have:

**Theorem 2.** *Protocol* CES-1 *securely computes the (randomized) function of the Correlated Element Selection problem in the honest-but-curious model.*

*Proof omitted.*

### 4.3   Dealing with Dishonest Players

*Generic transformation.* Following the common practice in the design of secure protocols, one can modify the above protocol to deal with dishonest players by adding appropriate zero-knowledge proofs. That is, after each flow of the original protocol, the corresponding player proves in zero knowledge that it indeed followed its prescribed protocol: After Step 1, the Preparer proves that it knows the permutation $\pi$ that was used to permute the list. After Step 2 the Chooser proves that it knows the index $\ell$ and the blinding factor that was used to produce the pair $(e, f)$. Finally, after Step 3 the Preparer proves that the plaintext $\tilde{b}$ is indeed the decryption of the ciphertext $f$. Given these zero-knowledge proofs, one can appeal to general theorems about secure two-party protocols, and prove that the resulting protocol is secure in the general case of potentially malicious players.

We note that the zero-knowledge proofs that are involved in this protocol can be made very efficient, so even this "generic" protocol is quite efficient (these are essentially the same proofs that are used for mix-networks in [1], see description in the full paper). However, a closer look reveals that one does not need all the power of the generic transformation, and the protocol can be optimized in several ways. Some of the optimizations are detailed below, while protocols for the zero-knowledge proofs and issues of key generation can be found in the full paper [13]. The resulting protocol CES-2 is described in Figure 2.

**Theorem 3.** *Protocol* CES-2 *securely computes the (randomized) function of the Correlated Element Selection problem.*

*Proof omitted.*

*Proof of proper decryption.* To withstand malicious players, the Preparer $P$ must "prove" that the element $\tilde{b}$ that it send in Step 3 of CES-1 is a proper decryption of the ciphertext $f$. However, this can be done in a straightforward manner without requiring zero-knowledge proofs. Indeed, the Preparer can reveal additional information (such as the randomness used in the encryption of $f$), as long as this extra information does not compromise the semantic security of the ciphertext $e$. The problem is that $P$ may not be able to compute the randomness of the blinded value $f$ (for example, in ElGamal encryption this would require computation of discrete log). Hence, we need to devise a different method to enable the proof.

The proof will go as follows: for each $i \in [n]$, the Preparer sends the element $b_{\pi(i)}$ and corresponding random string that was used to obtain ciphertexts $d_i$ in the first step. The Chooser can then check that the element $d_\ell$ that it chose in Step 2 was encrypted correctly, and learn the corresponding plaintext.

Clearly, in this protocol the Chooser gets more information than just the decryption of $f$ (specifically, it gets the decryption of all the $d_i$'s). However, this does not affect the security of the protocol, as the Chooser now sees a decryption of a permutation of a list that he knew at the onset of the protocol. This permutation of the all $b_i$'s does not give any information about the output of the Preparer, other than what is implied by its output $b$. In particular, notice that if $b$ appears more than once in the list, then the Chooser does not know which of these occurrences was encrypted by $d_\ell$.

Next, we observe that after the above change there is no need for the Chooser to send $f$ to the Preparer; it is sufficient if $C$ sends only $e$ in Step 2, since it can compute the decryption of $d_\ell$ by itself.

*A weaker condition in the second proof-of-knowledge.* Finally, we observe that since the security of the Chooser relies on an information-theoretic argument, the second proof-of-knowledge (in which the Chooser proves that it knows the index $\ell$) does not have to be fully zero-knowledge. In fact, tracing through the proof of security, one can verify that it is sufficient for this proof to be *witness independent* in the sense of Feige and Shamir [16].

## Protocol CES-2

---

*Common inputs*: List of pairs $\{(a_i, b_i)\}_{i=1}^n$, public key $pk$.
*Preparer knows*: secret key $sk$.

$P$ :     **1. Permute and Encrypt.**
Pick a random permutation $\pi$ over $[n]$, and random strings $\{(r_i, s_i)\}_{i=1}^n$.
Let $(c_i, d_i) = (Enc_{pk}(a_{\pi(i)}; r_{\pi(i)}),\ Enc_{pk}(b_{\pi(i)}; s_{\pi(i)}))$, for all $i \in [n]$.
Send $\{(c_i, d_i)\}_{i=1}^n$ to $C$.

**Sub-protocol $\Pi_1$**: $P$ proves in zero-knowledge that it knows the randomness $\{(r_i, s_i)\}_{i=1}^n$ and permutation $\pi$ that were used to obtain the list $\{(c_i, d_i)\}_{i=1}^n$.

$C$ :     **2. Choose and Blind.**
Pick a random index $\ell \in [n]$.
Send to $P$ the ciphertext $e = Blind_{pk}(c_\ell, 0)$.

**Sub-protocol $\Pi_2$**: $C$ proves in a witness-independent manner that it knows the randomness and index $\ell$ that were used to obtain $e$.

$P$ :     **3. Decrypt and Output.**
Set $a = Dec_{sk}(e)$. Output $a$.
Send to $C$ the list of pairs $\{(b_{\pi(i)}, s_{\pi(i)})\}_{i=1}^n$ (in this order).

$C$ :     **4. Verify and Output.**
Denote by $(b, s)$ the $\ell$'th entry in this lists (i.e., $(b, s) = (b_{\pi(\ell)}, s_{\pi(\ell)})$ ).
If $d_\ell = Enc_{pk}(b; s)$ then output $b$.

---

**Fig. 2.** Protocol for Correlated Element Selection.

*Blinding by Zero.* Notice that for the modified protocol we did not use the full power of blindable encryption, since we only used "blindings" by zero. Namely, all that was used in these protocols is that we can transform any ciphertext $c$ into a *random* encryption of the same plaintext. (The zero-knowledge proofs also use only "blindings" by zero.) This is exactly the "random self-reducibility" property used by Sander et al. [33].

*Efficiency.* We note that all the protocols that are involved are quite simple. In terms of number of communication flows, the key generation step and Step 1 take at most five flows each, using techniques which appear in Appendix A. Step 2 takes three flows and Step 3 consists of just one flow. Moreover, these flows can be piggybacked on each other. Hence, we can implement the protocol with only five flows of communication, which is equal to the five steps which are required by a single proof. In terms of number of operations, the complexity of the protocol is dominated by the complexity of the proofs in Steps 1 and 2. The proof in Step 1 requires $nk$ blinding operations (for a list of size $n$ and security parameter $k$), and the proof of Step 2 can be optimized to about $nk/2$ blinding operations on the average. Hence, the whole protocol has about $\frac{3}{2}nk$ blinding operations.[5]

# 5 Epilogue: Cryptography and Game Theory

The most interesting aspect of our work is the synergy achieved between cryptographic solutions and the game-theory world. Notice that by implementing our cryptographic solution in the game-theory setting, we gain on the game-theory front (by eliminating the need for a mediator), but we also gain on the cryptography front (for example, in that we eliminate the problem of early stopping). In principle, it may be possible to make stronger use of the game theory setting to achieve improved solutions. For example, maybe it is possible to prove that in the context of certain games, a player does not have an incentive to deviate from its protocol, and so in this context there is no point in asking this player to prove that it behaves honestly (so we can eliminate some zero-knowledge proofs that would otherwise be required).

More generally, it may be the case that working in a model in which "we know what the players are up to" can simplify the design of secure protocols. It is a very interesting open problem to find interesting examples that would demonstrate such phenomena.

We conclude with the table that shows some parallels between Cryptography and Game Theory that we discussed.

| Issue | Cryptography | Game Theory |
|---|---|---|
| Incentive | None | Payoff |
| Players | Totally Honest/Malicious | Always Rational |
| Punishing Cheaters | Outside Model | Central Part |
| Solution Concept | Secure Protocol | Equilibrium |
| Early Stopping | Problem | Not an Issue |

---

[5] We note that the protocol includes just a single decryption operation, in Step 3. In schemes where encryption is much more efficient than decryption – such as the Goldwasser-Micali encryption – this may have a significant impact on the performance of the protocol.

# References

1. M. Abe. Universally Verifiable Mix-net with Verification Work Independent on the number of Mix-centers. In *Proceedings of EUROCRYPT '98*, pp. 437-447, 1998.
2. R. Aumann. Subjectivity and Correlation in Randomized Strategies. In *Journal of Mathematical Economics*, 1, pp. 67-95, 1974
3. I. Barany. Fair distribution protocols or how the players replace fortune. *Mathematics of Operations Research*, 17(2):327–340, May 1992.
4. M. Bellare, R. Impagliazzo, and M. Naor. Does parallel repetition lower the error in computationally sound protocols? In *38th Annual Symposium on Foundations of Computer Science*, pages 374–383. IEEE, 1997.
5. J. Benaloh. Dense Probabilistic Encryption. In *Proc. of the Workshop on Selected Areas in Cryptography*, pp. 120-128, 1994.
6. M. Ben-Or, S. Goldwasser, and A. Wigderson. Completeness theorems for non-cryptographic fault-tolerant distributed computation. In *Proceedings of the 20th Annual ACM Symposium on Theory of Computing*, pages 1–10, 1988.
7. M. Blum. Coin flipping by telephone: A protocol for solving impossible problems. In *CRYPTO '81*. ECE Report 82-04, ECE Dept., UCSB, 1982.
8. G. Brassard, D. Chaum, and C. Crépeau. Minimum disclosure proofs of knowledge. *JCSS*, 37(2):156–189, 1988.
9. R. Canetti, Security and Composition of Multi-parti Cryptographic Protocols. *Journal of Cryptology*, 13(1):143–202.
10. D. Chaum. Blind signatures for untraceable payment. In *Advances in Cryptology – CRYPTO '82*, pages 199–203. Plenum Press, 1982.
11. D. Chaum, C. Crépeau, and E. Damgård. Multiparty unconditionally secure protocols. In *Advances in Cryptology – CRYPTO '87*, volume 293 of *99 Lecture Notes in Computer Science*, pages 462–462. Springer-Verlag, 1988.
12. R. Cramer, I. Damgard, and P. MacKenzie. Efficient zero-knowledge proofs of knowledge without intractability assumptions. Proceedings of *PKC 2000* January 2000, Melbourne, Australia.
13. Y. Dodis and S. Halevi and T. Rabin. Cryptographic Solutions to a Game Theoretic Problem. http://www.research.ibm.com/security/DHR00.ps.
14. C. Dwork, M. Naor, and A. Sahai. Concurrent zero knowledge. In *Proceedings of the 30th Annual ACM STOC* , pages 409–418. ACM Press, 1998.
15. T. ElGamal. A public key cryptosystem and a signature scheme based on discrete logarithms. In *CRYPTO '84, LNCS 196*, pages 10–18. Springer-Verlag, 1985.
16. U. Feige and A. Shamir. Witness indistinguishable and witness hiding protocols. In *Proceedings of the 22nd Annual ACM STOC* , pages 416–426. ACM Press, 1990.
17. M. Fischer, R. Wright. An Application of Game-Theoretic Techniques to Cryptography. In *Advances in Computational Complexity Theory*, DIMACS Series in Discrete Mathematics and Theoretical Computer Science, vol. 13, pp. 99–118, 1993.
18. F. Forges. Can sunspots repalce the mediator? In J. of Math. Economics, 17:347–368, 1988.
19. F. Forges. Universal Mechanisms, In Econometrica, 58:1341–1364, 1990.
20. D. Fudenberg, J. Tirole. Game Theory. MIT Press, 1992.
21. J. Garay, R. Gennaro, C. Jutla, and T. Rabin. Secure distributed storage and retrieval. In *Proc. 11th International Workshop on Distributed Algorithms (WDAG '97), LNCS 1320*, pages 275–289. Springer-Verlag, 1997.
22. O. Goldreich, S. Micali, and A. Wigderson. How to play any mental game. In *Proceedings of the 19th Annual ACM Symposium on Theory of Computing*, pages 218–229, 1987.

23. S. Goldwasser and S. Micali. Probabilistic encryption. *Journal of Computer and System Sciences*, 28(2):270–299, April 1984.
24. S. Goldwasser, S. Micali, and C. Rackoff. The knowledge complexity of interactive proof systems. *SIAM Journal on Computing*, 18(1):186–208, 1989.
25. M. Jakobsson. A Practical Mix. In *Proceedings of EUROCRYPT '98*, pp. 448–461, 1998.
26. J. Kilian. (More) Completeness Theorems for Secure Two-Party Computation In *Proc. of STOC*, 2000.
27. E. Lehrer and S. Sorin. One-shot public mediated talk. Discussion Paper 1108, Northwestern University, 1994.
28. P. MacKenzie. Efficient ZK Proofs of Knowledge. Unpublished manuscript, 1998.
29. G. Mailath, L. Samuelson and A. Shaked. Correlated Equilibria and Local Interaction In *Economic Theory*, 9, pp. 551-556, 1997.
30. R. Myerson. Communication, correlated equilibria and incentive compatibility. In *Handbook of Game Theory*, Vol. II, Elsevier, Amsterdam, pp. 827-847, 1994.
31. J.F. Nash. Non-Cooperative Games. *Annals of Mathematics*, 54 pages 286–295.
32. M. Osborne, A. Rubinstein. A Course in Game Theory. The MIT Press, 1994.
33. T. Sander, A. Young, and M. Yung. Non-interactive CryptoComputing for NC1. In *40th Annual Symposium on Foundations of Computer Science*, pages 554–567. IEEE, 1999.
34. A. C. Yao. Protocols for secure computations (extended abstract). In *23rd Annual Symposium on Foundations of Computer Science*, pages 160–164. IEEE, Nov. 1982.

# A    Reducing the Error in a Zero-Knowledge Proof-of-Knowledge

Below we describe a known transformation from any 3-round, constant-error zero-knowledge proof-of-knowledge into a 5-round, negligible error zero knowledge proof-of-knowledge, that uses trapdoor commitment schemes. We were not able to trace the origin of this transformation, although related ideas and techniques can be found in [14,28,12].

Assume that you have some 3-round, constant-error zero-knowledge proof-of-knowledge protocol, and consider the 3-round protocol that you get by running the constant-error protocol many times in parallel. Denote the first prover message in the resulting protocol by $\alpha$, the verifier message by $\beta$, and the last prover message by $\gamma$. Note that since the original protocol was 3-round, then parallel repetition reduces the error exponentially (see proof in [4]). However, this protocol is no longer zero-knowledge.

To get a zero-knowledge protocol, we use a trapdoor (or *Chameleon*) commitment schemes [8]. Roughly, this is a commitment scheme which is computationally binding and unconditionally secret, with the extra property that there exists a trapdoor information, knowledge of which enables one to open a commitment in any way it wants.

In the zero-knowledge protocol, the prover sends to the verifier in the first round the public-key of the trapdoor commitment scheme. The verifier then commits to $\beta$, the prover sends $\alpha$, the verifier opens the commitment to $\beta$, and the prover sends $\gamma$ and also *the trapdoor for the commitment*. The zero-knowledge simulator follows the one for the standard 4-round protocol. The knowledge extractor, on the other hand, first runs one instance of the proof to get the trapdoor, and then it can effectively ignore the commitment in the second round, so you can use the extractor of the original 3-round protocol.

# Differential Fault Attacks
# on Elliptic Curve Cryptosystems
## (Extended Abstract)

Ingrid Biehl[1], Bernd Meyer[2], and Volker Müller[3]

[1] University of Technology, Computer Science Department,
Alexanderstraße 10, 64283 Darmstadt, Germany,
biehl@informatik.tu-darmstadt.de
[2] Siemens AG, Corporate Technology,
81730 München, Germany,
bernd.meyer@mchp.siemens.de
[3] Universitas Kristen Duta Wacana,
Jl. Dr. Wahidin 5–19, Yogyakarta 55224, Indonesia, vmueller@ukdw.ac.id

**Abstract.** In this paper we extend the ideas for differential fault attacks on the RSA cryptosystem (see [4]) to schemes using elliptic curves. We present three different types of attacks that can be used to derive information about the secret key if bit errors can be inserted into the elliptic curve computations in a tamper-proof device. The effectiveness of the attacks was proven in a software simulation of the described ideas.

**Key words:** Elliptic Curve Cryptosystem, Differential Fault Attack.

## 1 Introduction

Elliptic curves have gained especially much attention in public key cryptography in the last few years. Standards for elliptic curve cryptosystems (ECC) and signature schemes were developed [7]. The security of ECC is usually based on the (expected) difficulty of the discrete logarithm problem in the group of points on an elliptic curve. In many practical applications of ECC the secret key (the solution to a discrete logarithm problem) is stored inside a *tamper-proof* device, usually a smart card. It is considered to be impossible to extract the key from the card without destroying the information. For security reasons the decryption or signing process is usually also done inside the card.

Three years ago a new kind of attack on smart card implementations of cryptosystems became public, the so called *differential fault attack (DFA)*, which has been successful in attacking RSA [4], DES [3], and even helps reverse-engineering unknown cryptosystems. The basic idea of DFA is the enforcement of bit errors into the decryption or signing process which is done inside the smart card. Then information on the secret key can leak out of the card. In RSA implementations for example this information can be used to factor the RSA modulus (at least with some non-negligible probability), which is equivalent to computing the secret RSA key. So far there is no method known to extend the ideas of [4] to

M. Bellare (Ed.): CRYPTO 2000, LNCS 1880, pp. 131–146, 2000.
© Springer-Verlag Berlin Heidelberg 2000

cryptosystems based on the discrete logarithm problem over elliptic curves. In this paper we investigate how DFA techniques can be used to compute the secret key of an ECC smart card implementation. Our attacks can be used for elliptic curves defined over arbitrary finite fields.

We consider the following scenario: a cryptographically strong elliptic curve is publicly known as part of the public key. The secret key $d \in \mathbb{Z}$ is stored inside a tamper-proof device, unreadable for outside users. On input of some point $P$ on the chosen elliptic curve, the device computes and outputs the point $d \cdot P$. We assume that we have access to the tamper-proof device such that we can compute $d \cdot P$ for arbitrary input points $P$.

The main common idea behind the attacks in Sect. 4 is the following: by inserting (in the first mentioned attack) or by disturbing the representation of a point by means of a random register fault we enforce the device to apply its point addition resp. multiplication algorithm to a value which is not a point on the given but on some different curve. It is a crucial observation as we will show in Sect. 3 that the result of this computation is a point on the new probably cryptographically less strong curve which can be exploited to compute $d$. Thus these attacks work by misusing the tamper-proof device to execute its computation steps on group structures not originally intended by the designer of the cryptosystem. Similar ideas have been previously described in [10] where small order subgroups in $(\mathbb{Z}/p\mathbb{Z})^*$ are exploited to compute part of the secret key and in [5] for attacks against identification schemes. It is shown in [5] how identification schemes can be used to prove knowledge of logarithms and roots which do not even exist in the subgroup where the cryptosystem should make its computations.

Moreover, we present a DFA-like attack in Sect. 5 which is similar to attacks against RSA in [4]. There so called *register faults* are used to attack RSA smart card implementations. Register faults are transient faults that affect current data inside a register. All the circuitry is not influenced by these faults and works properly. For a more detailed discussion of that fault model, we refer to [4, Sect. 3]. We use the same fault model and assume that we can enforce random register faults in the decryption or signing process. Incorrect output values caused by random register faults are used to compute possible intermediate values of the computation and parts of the secret key. The intermediate values are not necessarily unique and one has to repeat the attack to get successively all bits of the secret key. The analysis of the probability of non-uniqueness and so of the costs of the computation of the secret key is the technically most complicated part of the analysis in the considered ECC case and cannot be based on the ideas presented in [4]. We sketch it in the appendix.

We know no widespread applications of smart cards for signature generation or decryption where complete points are the output of the used tamper-proof device. Therefore, we consider additionally as a more realistic scenario the situation that the tamper-proof device implements El-Gamal decryption. For El-Gamal decryption we can show that the attacks from Sect. 4.1 and 5 have expected polynomial running time. Furthermore, it is shown that the attack of Sect. 4.2

can be used against El-Gamal decryption and the elliptic curve digital signature scheme in expected subexponential running time.

The fault models of DFA attacks have been criticized for being purely theoretical. In [2] it is argued that a random one-bit error would be more likely to crash the processor of the tamper-proof device or yield an uninformative error than to produce a faulty ciphertext. Instead, *glitch attacks* which have already been used in the pay-TV hacking community, are presented in [2,1,8] as a more practical approach for differential fault analysis. The attacker applies a rapid transient in the clock or the power supply of the chip. Due to different delays in various signal paths, this affects only some signals and by varying the parameters of the attack, the CPU can be made to execute a number of wrong instructions. By carefully choosing the timing of the glitch, the attacker can possibly enforce register faults in the decryption or signing process and apply our attacks.

The paper is structured as follows: Section 2 gives an introduction to the well known theory of elliptic curves. Section 3 examines *pseudo-addition*, an operation which will play a crucial part in the DFA attacks. Sections 4 and 5 describe three different attacks on ECC systems and show how faults can be used to determine the secret key $d$. We close with comments on possible countermeasures.

## 2   Elliptic Curves

In this section we review several well known facts about elliptic curves. Let $K$ be a finite field of arbitrary characteristic, and let $a_1, a_2, a_3, a_4, a_6 \in K$ be elements such that the discriminant of the polynomial given in (1) is not zero (the formula for the discriminant can be found in, e.g., [6]). Then the group of points $E(K)$ on the elliptic curve $E = (a_1, a_2, a_3, a_4, a_6)$ is given as

$$\left\{ (x,y) \in K^2 : y^2 + a_1 xy + a_3 y = x^3 + a_2 x^2 + a_4 x + a_6 \right\} \cup \left\{ \mathcal{O} \right\}, \qquad (1)$$

where $\mathcal{O} := (\infty, \infty)$. Pairs of elements of $K^2$ which satisfy the polynomial equation (1) are denoted as *points on E*. In the following we use subscripts like $P_E$ to show that $P$ is a point on the elliptic curve $E$. We define the following operation:

- for all $P_E \in E(K)$, set $P_E + \mathcal{O}_E = \mathcal{O}_E + P_E := P_E$,                    (2)
- for $P_E = (x,y)_E$, set $-P_E := (x, -y - a_1 x - a_3)_E$,
- for $x_1 = x_2$ and $y_2 = -y_1 - a_1 x_1 - a_3$, set $(x_1, y_1) + (x_2, y_2) := \mathcal{O}_E$,
- in all other situations, set $(x_1, y_1)_E + (x_2, y_2)_E := (x_3, y_3)_E$, where

$$x_3 = \lambda^2 + a_1 \lambda - a_2 - x_1 - x_2$$

$$y_3 = -y_1 - (x_3 - x_1)\lambda - a_1 x_3 - a_3$$

with

$$\lambda = \begin{cases} \dfrac{3 x_1^2 + 2 a_2 x_1 + a_4 - a_1 y_1}{2 y_1 + a_1 x_1 + a_3} & \text{if } x_1 = x_2 \text{ and } y_1 = y_2, \\[2ex] \dfrac{y_1 - y_2}{x_1 - x_2} & \text{otherwise.} \end{cases}$$

As shown in [6], this operation makes $E(K)$ to an abelian (additive) group with zero element $\mathcal{O}_E$. For any positive integer $m$ we define $m \cdot P_E$ to be the result of adding $P_E$ $m - 1$ times to itself. A crucial point that we will use in further sections is the fact that the curve coefficient $a_6$ is not used in any of the addition formulas given above, but follows implicitly from the fact that the point $P_E$ is assumed to be on the curve $E$.

In almost all practical ECC systems the discrete logarithm (DL) problem in the group of points on an elliptic curve is used as a trapdoor one-way function. The DL problem is defined as follows: given an elliptic curve $E$ and two points $P_E$, $d \cdot P_E$ on $E$, compute the minimal positive multiplier $d$. A *cryptographically strong elliptic curve* is an elliptic curve such that the discrete logarithm problem in the group of points is expected (up to current knowledge) to be difficult. ECC system implementations should always use cryptographically strong curves.

We will show in the following sections that random register faults can be used to compute information about a secret key $d$ which is stored inside a tamper-proof device that computes $d \cdot P$ for some input point $P$. Thus our scenario becomes applicable if the device is used for the computation of the trapdoor one-way function $d \cdot P$ in a larger protocol. In practice however neither EC signature generation nor EC cryptosystems use tamper-proof devices which output complete points. Consider for example the following EC El-Gamal cryptosystem (without point compression):

Let $E$ be a cryptographically strong elliptic curve. Given a point $P \in E$ assume that $Q = d \cdot P$ is the public key and $1 \leq d < \operatorname{ord}(P)$ the secret key of some user. For a point $R$ let $x(R)$ denote the $x$-coordinate. The EC El-Gamal cryptosystem (without point compression) is given as follows:

| Encryption | Decryption |
|---|---|
| Input: message $m$, public key | Input: $(H, m')$, secret key $d$ |
| choose $1 < k < \operatorname{ord}(P)$ randomly | compute $d \cdot H$ |
| return $(k \cdot P, x(k \cdot Q) \oplus m)$ | return $m' \oplus x(d \cdot H)$ |

If we combine the input and the output of the decryption process, then we can consider El-Gamal decryption as a black box that computes on input of some point $H$ the $x$-coordinate of $d \cdot H$. Using the curve equation corresponding to the input point $H$ we can determine the points $d \cdot H$ and $-(d \cdot H)$. But we have to stress that one cannot distinguish which one of this pair of points is $d \cdot H$.

## 3    Pseudo-addition and Pseudo-multiplication

Let $E$ be a fixed cryptographically strong elliptic curve defined over a finite field $K$. We start with the following question: what happens when we use the operation defined in (2) for arbitrary pairs in $K^2$ instead for points on $E$? In this section we will answer this question and deduce some properties of this new operation.

Let $a_1, a_2, a_3, a_4 \in K$ be the coefficients of $E$ with the exception of $a_6$. It should be noted that $a_6$ does not occur in the addition formulas (2) and is therefore not needed. Then it is easy to see that the operation (2) is also well-defined for arbitrary elements in $\mathcal{P} := K^2 \cup \{(\infty, \infty)\}$ (assuming that division by zero has the result $\infty$). For two arbitrary pairs $P_i \in \mathcal{P}$, $i = 1, 2$, we denote this operation as *pseudo-addition* and write $P_1 \oplus P_2$. *Pseudo-subtraction* is defined as pseudo-addition with the negative point and denoted with $P_1 \ominus P_2 = P_1 \oplus (-P_2)$. Moreover, for any positive integer $n \in \mathbb{N}$ and any pair $P_1 \in \mathcal{P}$, we define a *pseudo-multiplication* $n \otimes P_1$ as the result of $(\cdots ((P_1 \oplus P_1) \oplus P_1) \oplus \cdots) \oplus P_1$, where pseudo-addition $\oplus$ is used exactly $n - 1$ times.

We present a few facts on the operation $\oplus$. Testing a few random example pairs in $\mathcal{P}$, it becomes obvious that pseudo-addition $\oplus$ is in general no longer associative. We can however prove the following weaker results on pseudo-addition.

**Theorem 1.** *Let two elements $(x_i, y_i) \in \mathcal{P}$, $i = 1, 2$, be given. Pseudo-addition is*

1. *commutative, i.e. $(x_1, y_1) \oplus (x_2, y_2) = (x_2, y_2) \oplus (x_1, y_1)$,*
2. *"weakly associative": if $x_1 \neq x_2$ or $(x_1, y_1) = \pm(x_2, y_2)$*

$$\Big((x_1, y_1) \oplus (x_2, y_2)\Big) \ominus (x_2, y_2) = (x_1, y_1).$$

*Proof.* The first assertion of the theorem follows directly from the symmetry of the formulas given in (2), testing all cases for the second assertion is a minor exercise for a computer algebra system. $\square$

The discrete logarithm problem for elliptic curves is defined after multiplication of a point with a scalar. The following theorem describes a property of pseudo-multiplication.

**Theorem 2.** *Let the number of elements in the field $K$ be $q$. For at least $q^2 + 1 - 4q$ elements $P \in \mathcal{P}$ and all positive integers $n, m$, pseudo-multiplication satisfies*

1. $n \otimes (m \otimes P) = (n \cdot m) \otimes P$,
2. $(n \otimes P) \oplus (m \otimes P) = (n + m) \otimes P$.

*Proof.* Note first that the assertions are trivial for the pair $\mathcal{O}$. Let therefore $P = (x, y) \in \mathcal{P}$. Define $a_6' = y^2 + a_1 xy + a_3 y - x^3 - a_2 x^2 - a_4 x$. If $(a_1, a_2, a_3, a_4, a_6')$ defines an elliptic curve, then obviously $P$ is a point on this curve, and the result of the theorem follows directly from the associativity of point addition. The number of exceptional pairs $(x, y)$ that do not lead to elliptic curves can easily be bounded by $4q$ since for given coefficients $a_1, a_2, a_3, a_4$ there are only two possibilities for $a_6$ such that the discriminant becomes zero. $\square$

Finally, we examine how a fast multiplication algorithm behaves when used with pseudo-addition instead of ordinary point addition. A direct consequence of Theorem 2 is the following theorem.

**Theorem 3.** *Given a pair* $P = (x, y) \in \mathcal{P}$ *and a positive integer* $m$. *Assume that the tuple* $(a_1, a_2, a_3, a_4, y^2 + a_1 xy + a_3 y - x^3 - a_2 x^2 - a_4 x)$ *defines an elliptic curve* $E'$ *over* $K$. *Then any fast multiplication type algorithm with input* $(m, P, a_1, a_2, a_3, a_4)$ *computes the result* $m \otimes P$ *accordingly to the addition defined in Sect. 2. Moreover, we have the equality* $m \otimes P = m \cdot P_{E'}$, *where* $P_{E'} = P$ *and* $m \cdot P_{E'}$ *are points on* $E'$ *and the latter is computed with "ordinary" point additions.*

*Remark 1.* The crucial idea of pseudo addition is the fact that one of the curve coefficients is not used in the addition formulas. However a different point representation, so called *projective coordinates*, is also often used in practice. The addition formulas for such representations (see, e.g., [7, A.10.4]) have the same property. Therefore, the ideas presented in this paper can be adapted to other point representations typically used in practical applications.

## 4    Faults at the Beginning of the Multiplication

We start with the description of elliptic curve fault attacks. The first type of attacks however does not need the generation of any fault; it is an attack on "bad" implementations of ECC systems.

### 4.1    No Correctness Check for Input Points

The first attack is applicable when the device neither explicitly checks whether an input point $P$ nor the result of the computation really is a point on the cryptographically strong elliptic curve $E$ which is a parameter of the system. The attack is simple and should not be applicable to a well designed system, but nevertheless such a "bug" might happen in practice.

Let $E = (a_1, a_2, a_3, a_4, a_6)$ be a given cryptographically strong elliptic curve, which is part of the setup of the ECC system. In this situation we input a pair $P \in \mathcal{P}$ into the tamper-proof device which is not a point on $E$, but a point on some other elliptic curve $E'$. We choose the input pair $P = (x, y)$ carefully, such that with $a_6' = y^2 + a_1 xy + a_3 y - x^3 - a_2 x^2 - a_4 x$ the tuple $(a_1, a_2, a_3, a_4, a_6')$ defines an elliptic curve $E'$ whose order has a small divisor $r$ and such that $\text{ord}(P) = r$. With Theorem 3 we know that the output of the tamper-proof device with input $P$ is then $d \cdot P$ on $E'$. Therefore, we end up with a discrete logarithm problem in the subgroup of order $r$ generated by $P \in E'$, namely given points $P$, $d \cdot P$ on $E'$, find $d \mod \text{ord}(P)$. We can repeat this procedure with a different choice of $P$ and use the Chinese Remainder Theorem to compute the correct value of $d$.

This algorithm is quite efficient if we do not choose $P$, but the curve $E'$ first and compute $P$. The construction of such an elliptic curve $E'$ can be done in essentially the same way as in the elliptic curve construction method described in [7]. First we try to find an integer $m$ in the Hasse interval such that $(q + 1 - m)^2 - 4q$ has a large square factor and $m$ a small factor. Then we can determine the $j$-invariant of an elliptic curve defined over $K$ which has group order $m$. Finally, we have to check whether there exists an elliptic curve with coefficients $a_1, \dots, a_4, a_6'$ that has the given $j$-invariant. The latter test can be solved by factoring a polynomial of degree 2 and yields $a_6'$. We check for a few random values of $x$ whether $y^2 + a_1 xy + a_3 y - x^3 - a_2 x^2 - a_4 x - a_6' = 0$ is solvable for $y$. The pair $P_{E'} = (x, y)$ is chosen as input. Since $m$ has a small divisor, given $d \cdot P_{E'}$ we can then determine the secret key modulo this small divisor (at least when this small divisor divides the order of $P_{E'}$ on $E'$).

If we apply this attack to the device computing the El-Gamal decryption as described in Sect. 2 we cannot determine the $y$-coordinate of the resulting point uniquely. Given its $x$-coordinate $w$ we can compute values $z, z'$ such that $(w, z), (w, z') \in E'$ and $(w, z_1) = -(w, z_2)$, but we cannot decide which of these points is $d \cdot P$ on $E'$. By computation of the discrete logarithms of $(w, z)$ and $(w, z')$ we therefore get values $c, c'$ with $c \equiv -c' \mod \mathrm{ord}(P)$ and either $d \equiv c \mod \mathrm{ord}(P)$ or $d \equiv c' \mod \mathrm{ord}(P)$. Thus we get $d^2 \equiv c^2 \mod \mathrm{ord}(P)$. To compute $d$ we have to choose sufficiently many points $P_i$ with small order such that $\mathrm{lcm}(\mathrm{ord}(P_1), \dots, \mathrm{ord}(P_s)) \geq d^2$. Then we get equations $d^2 \equiv c_i^2 \mod \mathrm{ord}(P_i)$ for $1 \leq i \leq s$ and can compute the value $d^2$ as an integer using the Chinese Remainder Theorem. The integer square root is the secret key $d$.

## 4.2   Placing Register Faults Properly

In the second attack we assume that we can enforce register faults inside the "tamper-proof" device at some precise moment at the beginning of the multiplication process. If the "tamper-proof" device checks whether the given input point is a point in the group of points of the cryptographically strong elliptic curve $E$, the attack of Sect. 4.1 is no more applicable. Assume however that we can produce one register fault inside the tamper-proof device right after this test is finished. Then the device computes internally with a pair $P'$ which differs in exactly one bit from the input point $P$. Therefore, the device computes and – if it does not check whether the output is a point on $E$ – outputs $d \otimes P'$. With Theorem 3 we deduce that $d \otimes P'$ lies on the same elliptic curve $E'$ as $P'$. We determine $a_6'$ such that the output pair $d \otimes P'$ satisfies the curve equation with coefficients $(a_1, a_2, a_3, a_4, a_6')$. If these coefficients define an elliptic curve $E'$, we have reduced the original DL problem on $E$ to a DL problem on $E'$: check for all possible candidates $P'$ ($P'$ is unknown outside the device, but remember that $P'$ differs in only one bit from the known point $P$) whether this candidate is a point on $E'$ and – if so – try to solve the DL problem on $E'$. First, we compute $\mathrm{ord}(E')$ the number of points on $E'$ using algorithms for point counting. If $\mathrm{ord}(E')$ has a small divisor $r$, we solve the DL problem for the points $(\mathrm{ord}(E')/r) \cdot P_{E'}'$ and $d \cdot ((\mathrm{ord}(E')/r) \cdot P_{E'}')$. This gives an equation $d \equiv c \mod r$ for some value $c$. Repeating this step with different divisors $r$ we can compute $d$ with the Chinese Remainder Theorem.

As described in Sect. 2, we can consider El-Gamal decryption as a black box that on input of some point $P$ computes $x(d \cdot P)$ where $d$ is the secret key stored inside the tamper-proof device. Note however that we cannot apply directly the attack from this section since we do not know the $y$-coordinate of the output point. Without the $y$-coordinate we cannot determine the curve $E'$ to which the output $P'$ belongs. In general there are many possible curves. It is however possible to solve the DL problem with non-negligible probability if there exists a curve $E'$ corresponding to a base point $P'$ resulting from a one-bit error such that the order of $E'$ is smooth. Then we use the algorithm of Pohlig-Hellman (see [12]) to compute $d$.

Similar to the analysis of Lenstra's *Elliptic Curve Factoring Method* [9], it follows that we have to consider subexponentially many random elliptic curves until one of them has (subexponentially) smooth order. Thus the expected number of trials of the attack with random points $P \in E$ until we find such a smooth curve and can determine the secret multiplier $d$ is subexponential again.

A similar situation occurs in the elliptic curve DSA signature scheme. In EC DSA, we have two primes $p, q$ which are about the same size, an elliptic curve $E$ over $\mathbb{F}_q$, and a point $P$ on $E$ of order $p$. The public key is $(p, q, E, P, Q)$ where $Q = d \cdot P$ for some secret value $d$. To sign a message $m$ with hash value $h$, the signer randomly chooses an integer $1 < k < p - 1$, computes $k \cdot P = (x_1, y_1)$, $r \equiv x_1 \mod p$, and $s \equiv k^{-1}(h + dr) \mod p$. The signature is $(r, s)$.

Please note that we cannot input a point here but a publicly known point $P$ is used as base point for the computation. We again disturb the computation of $k \cdot P$ by a register fault right at the beginning, i.e. $P$ is replaced by some $P'$. The tamper-proof device then computes the signature $r' \equiv x(k \cdot P') \mod p$, and $s' \equiv k^{-1}(h + dr') \mod p$. Knowing this signature, we can use the following algorithm for all possible candidates $\tilde{P}$ for $P'$:

- compute the curve $\tilde{E}$ corresponding to $\tilde{P}$ – if it exists –,
- derive from $r'$ a small set of possible values for the $x$-coordinate of $x(k \cdot P')$ (since $p, q$ are of about the same size),
- compute two candidates for the corresponding $y$-coordinate by means of the equation for $\tilde{E}$.

In case $\tilde{P}$ was correctly chosen and $\tilde{E}$ is a weak curve with respect to the discrete logarithm problem and $\mathrm{ord}(\tilde{P}) > p - 1$, one can first find $k$, and then the secret key $d$ as $d \equiv r'^{-1}(s'k - h) \mod p$.

If we disturb the base point $P$ in such a way that $P'$ and $P$ differ only in one bit, we have only $2\log(q)$ possible choices for the curve $E'$ and it is very unlikely that we get a curve with subexponentially smooth order and that the attack succeeds. But if we manage to change $o(\log(q))$ many bits at once such that we get subexponentially many different choices for $E'$ then there is with high probability at least one curve with smooth order among them and we can compute the one-time key $k$ and so the secret key $d$, i.e. the signature scheme is completely broken. The expected number of trials to get such a curve $E'$ is subexponential again.

## 5  Faults at Random Moments of the Multiplication

In this section we sketch an attack that works even if we cannot influence the exact position in the computation process, at which the enforced random register fault happens.

In [4], the authors show how to attack RSA smart card implementations by enforcing register faults at random time in the decryption or signing process. The most important operation in RSA is fast exponentiation. For elliptic curves, the situation is similar and we can use some of the ideas of [4].

In the following we assume that the used elliptic curve is cryptographically strong, especially we assume that $E(\mathbb{F}_q)$ contains a subgroup of prime order $p$ with $p > q/\log(q)$. The operation $Q = d \cdot P$ is usually done with either a "right-to-left" or a "left-to-right" multiplication algorithm. Since the ideas for the attacks in both cases are very similar we restrict ourselves here to the "right-to-left" multiplication algorithm and show: if one can enforce a fault randomly in a register at a random state of the computation than one can recover the secret key in expected polynomial time.

We start with a result for a fault model where we can introduce register faults during the computation of an a-priory chosen *specific block* of multiplier bits, e.g. we assume that we can repeatedly input some point $P_E$ on $E$ into the tamper-proof device and enforce a register fault during $m$ successive iterations of the fast multiplication algorithm. Then we will show that we can relax this condition, i.e. even if one cannot influence at which block the register fault happens one can deduce the secret key after an expected number of polynomially many enforced random register faults. We will present a rather informal description of the attack which abstracts from some less important details.

The right-to-left multiplication algorithm works as follows (we denote by $(d_{n-1} \, d_{n-2} \, \ldots \, d_0)_2$ the binary representation of a positive integer $d$, where $d_0$ is the least significant bit):

```
H = P; Q = 0;
for i = 0 , ... , n-1 do
   if (d_i == 1) then Q = Q + H;
   H = 2 * H;
output Q;
```

To simplify the notation assume that we know the binary length $n$ of the unknown multiplier $d$ (note that an attacker can "guess" the length of $d$). Denote by $Q^{(i)}$, $H^{(i)}$ the value stored in the variable $Q$, $H$ in the algorithm description before iteration $i$.

The basic attack operation works as follows: we use the tamper-proof device with some input point $P_E$ to get the correct result $Q^{(n)} = d \cdot P_E$ and moreover we restart it with input $P_E$ but enforce a random register fault to get a faulty result $\tilde{Q}^{(n)}$. Assume that we enforce the register fault in iteration $n - m \le j < n$, and that this fault flips one bit in a register holding the variable $Q$ (the case that a bit in $H$ is flipped can be handled similarly). Then $\tilde{Q}^{(j)}$ is a *disturbed Q-value*, i.e. a pair in $\mathcal{P}^2$ that differs in exactly one bit from $Q^{(j)}$.

Next we try to find the index of the first iteration $j'$ with $j' > j$ and $d_{j'} = 1$ given $Q^{(n)}$ and $\tilde{Q}^{(n)}$. For simplicity reasons we assume that there is at least one non-zero bit among the $m$ most significant bits of $d$, i.e. $j'$ exists (we omit the technically more difficult case of $m$ zero bits here for reasons of readability). We can find a candidate for the disturbed $Q$-value $\tilde{Q}^{(j')}$ with the following method: successively, we check each $i$ with $n - m \leq i < n$ as candidate for $j'$, each $x \in \{0,1\}^{n-i}$ with least significant bit 1 as candidate for the $i$ most significant bits of $d$, and each $Q_x^{(i)} = Q^{(n)} - x \cdot 2^i \cdot P_E$ as candidate for $Q^{(j)}$. For each choice of $x$ and $i$ we consider all disturbed $Q$-values $\tilde{Q}_x^{(i)}$ which we can derive from $Q_x^{(i)}$ by flipping one bit. Then we check whether this may be the disturbed value which appeared in the device, i.e. we simulate the computation of the device, compute the corresponding result value and check whether it is identical with the found value $\tilde{Q}^{(n)}$. More precisely: we use pseudo-additions with points $x_\ell 2^{i+\ell} \cdot P_E$ for $\ell = 0, \dots, n - i - 1$ where $x = (x_{n-i-1} \dots x_0)_2$ with $x_0 = 1$ is the binary representation of $x$ to get for candidates $i, x, Q_x^{(i)}$, and $\tilde{Q}_x^{(i)}$ the corresponding faulty result

$$\tilde{Q}_x^{(n)} = (\cdots ((\tilde{Q}_x^{(i)} \oplus x_0 2^i \cdot P_E) \oplus x_1 2^{i+1} \cdot P_E) \oplus \cdots) \oplus x_{n-i-1} 2^{n-1} \cdot P_E.$$

If $\tilde{Q}_x^{(n)}$ is equal to the faulty result $\tilde{Q}^{(n)}$ output by the device, then we have found $i$ as a candidate for $j'$, $\tilde{Q}_x^{(i)}$ as a candidate for $\tilde{Q}^{(j')}$, and the binary representation of $x$ as a candidate for the upper $n - j'$ bits of $d$.

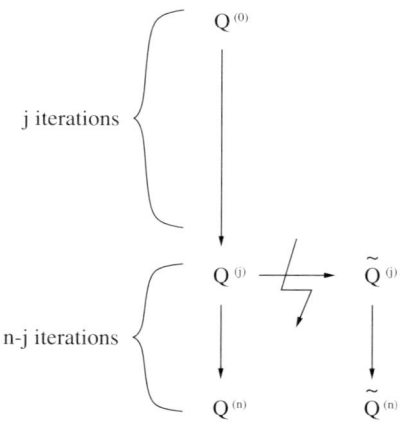

By trying faults on $Q$ and on $H$ and all $m$ possibilities for $i$ and corresponding integers $x$ we can make sure that this procedure outputs at least one candidate for $\tilde{Q}^{(j')}$ (or for $\tilde{H}^{(j)}$, in the case the fault occurs in $H$). In case there is only one candidate suitable for $P_E$, $Q^{(n)}$, $m$, and for $\tilde{Q}^{(n)}$ we have computed the $n - j'$ upper bits of the secret key $d$. One can show that the probability is small that more than one candidate survives (more details can be found in the appendix).

To reveal step by step all bits of $d$ we start to compute the most significant bits as explained above and work downwards to the least significant bits by iterating the same procedure with new random register faults in blocks of at most $m$ iterations. In each step we use the information that we already know about $d$ to restrict the range of test integers $x$ which have to be considered.

**Theorem 4.** *Let* $m = o(\log\log\log q)$ *and let* $n$ *be the binary length of the secret multiplier. Assume that we can generate a register fault in an a-priory chosen block of* $m$ *iterations of the multiplication algorithm. Using an expected number of* $O(n)$ *register faults we can determine the secret key* $d$ *in expected* $O(nm2^m(\log q)^3)$ *bit operations.*

Finally, we consider the more general situation in which we cannot induce register faults in small blocks, but only at random moments during the multiplication. As in [4] one can show that for a large enough number $\ell$ of disturbed computations we get a reasonable probability that errors happen in each block of $m$ iterations.

**Theorem 5.** *Let* $E$ *be an elliptic curve defined over a finite field with* $q$ *elements, let* $m = o(\log\log\log q)$, *and let* $n$ *be the binary length of the secret multiplier. Given* $\ell = O((n/m)\log(n))$ *faults, the secret key can be extracted from a device implementing the "right-to-left" multiplication algorithm in expected* $O(n\,2^m\,(\log q)^3\,\log(n))$ *bit operations.*

Thus this theorem can be summarized as follows: if we consider the size of the used finite field as a constant then we need $O(n\log(n))$ accesses to the tamper-proof device to compute in $O(n\log(n))$ bit operations the secret key of bit length $n$. Please notice that the block size $m$ we used as parameter of our algorithm reflects the tradeoff between the number of necessary register faults and the running time to analyse the output values influenced by these faults. It depends on the attackers situation whether more accesses to the tamper-proof device or more time for the analysis can be spent.

*Remark 2.* We have implemented a software simulation of the algorithm given above and attacked several hundred randomly chosen elliptic curves. Obviously, one can find easily non-unique solutions for the indices $j$ and the parts of $x$ of the corresponding discrete logarithms if the order of the base point $P_E$ is small in comparison with $2^m$ where $m$ is the length of the block containing the error. Also, if the size of the field is very small ($< 1000$) the algorithm often finds contradicting solutions. Both cases are not relevant for a cryptographically strong elliptic curve. In all tested examples with size of the field bigger than $2^{64}$, randomly chosen curve, and random point on the curve we determined the complete secret multiplier $d$ without problems.

If we apply this attack to the device computing the El-Gamal decryption as described in Sect. 2 we cannot determine the $y$-coordinate of the resulting point uniquely. Since we know the equation of the curve we can compute points $Q$ and $-Q$ such that the correct result $Q^{(n)}$ of the device is one of these points. We start the described attack on both points $Q$ and $-Q$ and compare only the $x$-coordinate of the disturbed results of the attack with the $x$-coordinate of the faulty result $x(\tilde{Q}^{(n)})$ of the device. Using this procedure we find at least one candidate for some point $\tilde{Q}^{(j)}$ (or for some point $\tilde{H}^{(j)}$, in the case the fault occurs in $H$) and can determine the upper bits of the secret multiplier $d$ if the candidate is unique.

## 6  Countermeasures

It became obvious in the preceding sections that DFA techniques for elliptic curves depend mainly on the ability to disturb a point on $E$ to "leave" the group of points and become an ordinary pair in $\mathcal{P}$. Countermeasures against all attacks presented in this paper are therefore obvious. Although it is part of the protocols of most cryptosystems based on elliptic curves to check whether input points indeed belong to a given cryptographically strong elliptic curve it follows from the described attacks that it is even more important for the tamper-proof device to check the output point or any point which serves as basis for the computation of some output values. If any of these points, input points or computed points, do not satisfy this condition, no output is allowed to leave the device. This countermeasure for ECC is similar to the countermeasures proposed against DFA for RSA where the consistency of the output also has to be checked by the device.

### Acknowledgements

We would like to thank the unknown referees for several suggestions which improved the quality and readability of the paper. Moreover, we would like to thank Susanne Wetzel and Erwin Heß for discussions. Our thank belongs especially to Arjen K. Lenstra who gave us a lot of support to improve the paper and pointed out to us the subexponential time attacks against El-Gamal decryption and EC DSA in Sect. 4.2.

## References

1. R. J. Anderson and M. G. Kuhn: *Tamper Resistance – a Cautionary Note*, Proceedings of Second USENIX Workshop on Electronic Commerce 1996, pp. 1–11.
2. R. J. Anderson and M. G. Kuhn: *Low Cost Attacks on Tamper Resistant Devices*, Lecture Notes in Computer Science 1361, Proceedings of International Workshop on Security Protocols 1997, Springer, pp. 125–136.
3. E. Biham and A. Shamir: *Differential Fault Analysis of Secret Key Cryptosystems*, Lecture Notes of Computer Science 1294, Proceedings of CRYPTO'97, Springer, pp. 513–525.
4. D. Boneh, R. A. DeMillo, and R. J. Lipton: *On the Importance of Checking Cryptographic Protocols for Faults*, Lecture Notes of Computer Science 1233, Proceedings of EUROCRYPT'97, Springer, pp. 37–51.
5. M. Burmester: *A Remark on the Efficiency of Identification Schemes*, Lecture Notes of Computer Science 473, Proceedings of EUROCRYPT'90, Springer, pp. 493–495.
6. I. Connell: *Elliptic Curve Handbook*, Preprint, 1996.

7. IEEE P1363 Draft Version 12: *Standard Specifications for Public Key Cryptography*, available on the Homepage of the IEEE.
8. O. Kömmerling and M. G. Kuhn: *Design Principles for Tamper-Resistant Smart-card Processors*, Proceedings of USENIX Workshop on Smartcard Technology 1999, pp. 9–20.
9. H. W. Lenstra: *Factoring Integers with Elliptic Curves*, Annals of Mathematics, **126** (1987), pp. 649–673.
10. C. H. Lim and P. J. Lee: *A Key Recovery Attack on Discrete Log-based Schemes Using a Prime Order Subgroup*, Lecture Notes of Computer Science 1294, Proceedings of CRYPTO'97, Springer, pp. 249–263.
11. A. Menezes: *Elliptic Curve Public Key Cryptosystems*, Kluwer Academic Publishers, 1993.
12. S. Pohlig and M. Hellman: *An Improved Algorithm for Computing Logarithms over* GF($p$) *and its Cryptographic Significance*, IEEE Transactions on Information Theory, vol. 24 (1978), pp. 106–110.
13. J. H. Silverman: *The Arithmetic of Elliptic Curves*, Graduate Texts in Mathematics 106, Springer 1986.

# Appendix: Success Probability of the Attack in Sect. 5

We denote by $Q^{(i)}$ resp. $H^{(i)}$ the value stored in the variable $Q$ resp. $H$ before iteration $i$ of the right-to-left multiplication algorithm described in Sect. 5. We know also the correct result $Q^{(n)} = d \cdot P_E$ and a faulty result $\tilde{Q}^{(n)}$ for a given base point $P_E$ on $E$.

We define a *disturbed Q-value* with respect to $P_E$, $Q^{(n)}$, $m$ to be a pair in $\mathcal{P}^2$ that differs in exactly one bit from some $Q^{(i)}$ for $n - m \leq i \leq n$. Assume that we enforce a register fault in iteration $n - m \leq j < n$, and that this fault flips one bit in a register holding the variable $Q$. Denote by $\tilde{Q}^{(j)}$ the resulting disturbed $Q$-value. According to the right-to-left multiplication algorithm we try all possible indices $n - m \leq i < n$ and all integers $x$ with exactly $n - i$ bits (least significant bit 1) to compute candidates $\tilde{Q}_x^{(i)}$ for disturbed $Q$-values that lead to the faulty result $\tilde{Q}^{(n)}$. The second place where a register fault can happen is the register holding the variable $H$ in the algorithm. The procedure for this case is quite similar. Again, we try all possible indices $n - m \leq i < n$ and all integers $x$ of exactly $n - i$ bits (least significant bit 1). If the fault is now introduced in the variable $H$ (i.e. into one of the points $H^{(i)} = 2^i \cdot P_E$), this results in some disturbed $H$-value $\tilde{H}^{(i)}$ and is then propagated by the loop of the algorithm. By trying both $Q$- and $H$-case and all $m$ possibilities for $i$ and corresponding integers $x$ we can make sure that this procedure outputs at least one candidate for $\tilde{Q}^{(j)}$ or for $\tilde{H}^{(j)}$. In case there is only one candidate suitable for $P_E$, $Q^{(n)}$, $m$ and for $\tilde{Q}^{(n)}$ we call this candidate a *uniquely determined disturbed value* with respect to $P_E$, $Q^{(n)}$, $m$. Otherwise, a candidate is called *non-uniquely determined disturbed value*.

In Lemma 2 we will prove that for $m = o(\log\log\log q)$, all $d$ and almost all points $P_E$ there are at most three different non-uniquely disturbed values. Thus the expected number of necessary repetitions of attacks (i.e. choosing a point

$P_E$ and causing a random register fault in the last $m$ iterations), until one finds a uniquely determined disturbed value, is constant. Next we give an estimate for the probability that an attack allows us to find a uniquely determined disturbed value. For background on elliptic curve theory, we recommend [6] or [13].

**Lemma 1.** *Let $m = o(\log \log \log q)$ and assume that we can generate register faults in the last $m$ iterations of the algorithm. The number of points $P_E$ for which there exist more than three different non-uniquely determined disturbed values with respect to $P_E$, $Q^{(n)}$, $m$ is bounded by $O((\log \log q)(\log q)^5)$.*

*Proof.* We want to bound the number of points $P_E$ for which there exists at least four different non-uniquely determined disturbed values with respect to $P_E$, $Q^{(n)}$, $m$. Thus there are at least two pairs of disturbed values where each pair leads under the secret key $d$ with $Q^{(n)} = d \cdot P_E$ to the same faulty multiplication result. Since these disturbed values are either $H$- or $Q$-values the following cases must be considered: each such pair either consists of two disturbed $Q$-values, or two disturbed $H$-values or is a pair consisting of one disturbed $Q$- and one disturbed $H$-value. We show for all nine cases that the number of points $P_E$ for which there exists four different non-uniquely determined disturbed values can be bounded by $O((\log \log q)(\log q)^5)$.

We consider the first case that all four non-uniquely disturbed values are $Q$-values. Then there exist integers $x_i$ of binary length at most $m$, points $P_i \in E(\mathbb{F}_q)$, and bit locations $r_i$ for $1 \le i \le 4$ such that

1. $\quad P_1 + x_1 \cdot R = P_2 + x_2 \cdot R = Q^{(n)}$,
2. $\quad P_{1,(r_1)} \otimes x_1 R = P_{2,(r_2)} \otimes x_2 R$,
3. $\quad P_3 + x_3 \cdot R = P_4 + x_4 \cdot R = Q^{(n)}$,
4. $\quad P_{3,(r_3)} \otimes x_3 R = P_{4,(r_4)} \otimes x_4 R$,

where $n$ is the binary length of the secret multiplier, $R = 2^{n-m} \cdot P_E$, $P_{i,(j)}$ denotes a pair which is obtained by switching bit $j$ of point $P_i$ (numbering the bits of $x$- and $y$-coordinate appropriately), and the notation $P \otimes w \cdot R$ serves as abbreviation for the computation $(\cdots((P \oplus w_0 \cdot R) \oplus w_1 \cdot 2 \cdot R) \oplus \cdots) \oplus w_{k-1} \cdot 2^{k-1} \cdot R$ for an integer $w = (w_{k-1} \ldots w_0)_2$. (The values $P_{i,(j)}$ are the non-uniquely determined disturbed values to the faulty results $P_{1,(r_1)} \otimes x_1 R$, and $P_{3,(r_3)} \otimes x_3 R$.)

We translate the four conditions above into polynomial equations using the concept of formal points. Assume that $P_1$ is given formally as $(X_1, Y_1)$ and $R$ as $(X_2, Y_2)$. Using the theory of division polynomials (see [6]), it follows directly that the $X_2$-degree of the numerator, denominator, of points $x \cdot R$ for arbitrary $m$-bit integers $x$ is $O(2^{2m})$. Combining the first and third equation (note that $Q^{(n)}$ occurs in both equations), we see with the addition formulas that the $x$-coordinates of all the points $P_i, i \ge 2$, can be written as rational functions of constant degree in $X_1, Y_1, Y_2$ and of degree $O(2^{cm})$ in $X_2$ for some small constant $c$ (both numerator and denominator). The essentially same idea can be used to find an equation from the second and the fourth equation: we compute the left hand side as rational functions (using the representation of $P_1, P_3$ in $X_1, Y_1, X_2, Y_2$, respectively), introducing new variables for the faults $r_1, r_3$. Similarly, we transform the right hand side of the second and fourth equation into a rational function, introducing new variables for the faults $r_2, r_4$ and using the representation of $P_2, P_4$ as function in $X_1, Y_1, X_2, Y_2$. Then we can derive a polynomial of $X_1, X_2$-degree $O(2^{c'm})$ for some small constant $c'$. Using the fact that both $P_1$ and $R$ are points on $E$, we can remove the variables $Y_1, Y_2$ with the help of the curve equation, increasing the exponent in the degree formula by a constant. Finally, we determine the resultant in the variable $X_2$ of both these equations, thereby removing $X_2$ and getting an equation of $X_1$-degree $O(2^{2^{c''m}})$ for some constant $c''$ (the resultant can be determined by computing the determinant of the so called Sylvester matrix). By substituting all possible values for $r_i, 1 \leq i \leq 4$, (note that $r_i$ are bit faults) and substituting all possible values for $x_i, 1 \leq i \leq 4$, (note that $0 \leq x_i \leq 2^m$), and observing that $m = o(\log \log \log q)$ and so $O(2^{2^{c''m}}) = O(\log q)$, we get $O(2^{4m}(\log q)^4) = O((\log \log q)(\log q)^4)$ equations of $X_1$-degree $O(\log q)$ each. Therefore, the total number of possibilities for $X_1$ and the number of possible points $P_E$ is at most $O((\log \log q)(\log q)^5)$.

The number of points $P_E$ for the other cases can be analyzed analogously.     □

**Lemma 2.** *Let* $m = o(\log \log \log q)$ *and* $q$ *be sufficiently large. The expected number of attacks, i.e. random choices of a point* $P_E$ *of* $E(\mathbb{F}_q)$ *and random register faults in the last* $m$ *iterations of the right-to-left multiplication algorithm, until one finds a uniquely determined disturbed value, is 2.*

*Proof.* Since $E(\mathbb{F}_q)$ is cryptographically strong, it contains a subgroup of prime order $p$ with $p > \frac{q}{\log q}$. Using the Hasse theorem it follows from the previous lemma that we will get a point $P_E$ with probability $1 - c(\log \log q)(\log q)^5/q$ (for some constant $c$) of order at least $p$ and for which there exists at most three different non-uniquely determined disturbed values with respect to $P_E, Q^{(n)}, m$. Since at least all $H^{(i)}$ for $i = n - m, \dots, n$ are different and consist of $2 \log q$ bits, there are $O(m \log q)$ bit positions in the computation process which could be disturbed. Since there are less than four non-uniquely determined disturbed values for $P_E$ the probability to disturb the computation in a way which will lead to one of these non-uniquely determined disturbed values is bounded by $3/(m \log q)$. It follows that with probability more than $1/2$ each attack will lead to a uniquely determined disturbed value.     □

**Lemma 3.** *Let* $m = o(\log \log \log q)$. *Assume that we can generate random register faults in the last* $m$ *iterations of the algorithm of the attack. Then the expected number of applications of the algorithm with independent random register faults is* $O(m)$ *until we can compute the* $m$ *most significant bits of the secret key* $d$. *Thus the expected number of bit operations is* $O(m^2 2^m (\log q)^3)$.

*Proof.* For a running time analysis we note that the number of fault positions is at most $4 \log q$ in each iteration (there are at most 2 points that can be disturbed, the $x$- and $y$-coordinate of each point have at most $\log q$ bits). For each of the $2^{m+1}$ different integers $x$ we have at most $m$ pseudo-additions which can be done in $O(m(\log q)^2)$ bit operations each. In addition we have to compute for all indices $n - m \leq j < n$ the corresponding values $Q^{(j)}$ and $H^{(j)}$ which can be done in $O((\log q)^3)$ bit operations.

We learn all $m$ most significant bits of $d$ if the error changes the value $H^{(n-m)}$. The probability that a random error during the last $m$ iterations disturbs a bit of $H^{(n-m)}$ is $\frac{1}{2m}$. Therefore, we can lower bound the probability of success if we have $k$ independent randomly disturbed results by $1 - (1 - \frac{1}{2m})^k$. Since the register faults are induced at random places, we derive that we expect to need $k = O(m)$ many faulty applications before we have found all the upper $m$ bits of $d$. Combining all the partial results, we get the expected $O(m^2 2^m (\log q)^3)$ bit operations.                                                                    □

Lemma 2 is the basis for an algorithm to determine the complete multiplier $d$. The basic idea is the usage of Lemma 2 successively on blocks of size $m$. Note the fact that we can "compute backwards" once we know the upper $m$ bits of $d$ to generate a DL problem with a smaller multiplier. Computing backwards from the correct output of the device for a given base point to get a correct intermediate point is trivial. For the disturbed output of the device it follows from Theorem 1 that we can compute backwards the faulty result with high probability too to get a faulty intermediate result. Then we can apply Lemma 2 again on the pair of correct intermediate point and faulty intermediate result. Thus we get:

**Theorem 6.** *Let $m = o(\log \log \log q)$ and let $n$ be the binary length of the secret multiplier. Assume that we can generate a register fault in a block of $m$ iterations of the right-to-left multiplication algorithm. Using an expected number of $O(n)$ register faults we can determine the secret key $d$ in expected $O(nm2^m (\log q)^3)$ bit operations.*

# Quantum Public-Key Cryptosystems

Tatsuaki Okamoto, Keisuke Tanaka, and Shigenori Uchiyama

NTT Laboratories
1-1 Hikari-no-oka Yokosuka-shi, Kanagawa-ken 239-0847, Japan
{okamoto, keisuke, uchiyama}@isl.ntt.co.jp
Tel: +81-468-59-2511
Fax: +81-468-59-3858

**Abstract.** This paper presents a new paradigm of cryptography, *quantum public-key cryptosystems*. In quantum public-key cryptosystems, all parties including senders, receivers and adversaries are modeled as *quantum* (probabilistic) poly-time Turing (QPT) machines and only classical channels (i.e., no quantum channels) are employed. A *quantum trapdoor one-way function*, $f$, plays an essential role in our system, in which a QPT machine can compute $f$ with high probability, any QPT machine can invert $f$ with negligible probability, and a QPT machine with trapdoor data can invert $f$. This paper proposes a concrete scheme for quantum public-key cryptosystems: a quantum public-key encryption scheme or quantum trapdoor one-way function. The security of our schemes is based on the computational assumption (over QPT machines) that a class of subset-sum problems is intractable against any QPT machine. Our scheme is very efficient and practical if Shor's discrete logarithm algorithm is efficiently realized on a quantum machine.

## 1 Introduction

### 1.1 Background and Problem

The concept of public-key cryptosystems (PKCs) introduced by Diffie and Hellman [18] and various theories for proving the security of public-key cryptosystems and related protocols (e.g., [22]) have been constructed on the Turing machine (TM) model. In other words, public-key cryptosystems and related theories are founded on Church's thesis, which asserts that any reasonable model of computation can be efficiently simulated on a probabilistic Turing machine. However, a new model of computing, the quantum Turing machine (QTM), has been investigated since the 1980's. It seems reasonable to consider a computing model that makes use of the quantum mechanical properties as our world behaves quantum mechanically. Several recent results provide informal evidence that QTMs violate the feasible computation version of Church's thesis [17,38,37]. The most successful result in this field was Shor's (probabilistic) polynomial time algorithms for integer factorization and discrete logarithm in the QTM model [37], since no (probabilistic) polynomial time algorithm for these problems has been found in the classical Turing machine model.

M. Bellare (Ed.): CRYPTO 2000, LNCS 1880, pp. 147–165, 2000.
© Springer-Verlag Berlin Heidelberg 2000

Although Shor's result demonstrates the positive side of the power of QTMs, other results indicate the limitation of the power of QTMs. Bennett, Bernstein, Brassard, and Vazirani [5] show that relative to an oracle chosen uniformly at random, with probability 1, class NP cannot be solved on a QTM in time $o(2^{n/2})$. Although this result does not rule out the possibility that NP $\subseteq$ BQP, many researchers consider that it is hard to find a probabilistic polynomial time algorithm to solve an NP-complete problem even in the QTM model, or conjecture that NP $\not\subseteq$ BQP.

Shor's result, in particular, greatly impacted practical public-key cryptosystems such as RSA, (multiplicative group/elliptic curve versions of) Diffie–Hellman and ElGamal schemes, since almost all practical public-key cryptosystems are constructed on integer factoring or the discrete logarithm problem. Therefore, if a QTM is realized in the future, we will lose almost all practical public-key cryptosystems. Since public-key cryptosystems are becoming one of the infrastructures of our information network society, we should resolve this technical and social crisis before a QTM is realized.

## 1.2   Our Results

This paper proposes a solution to this problem. First we show a natural extension of the concept of public-key cryptosystems to the QTM model, the *quantum public-key cryptosystem (QPKC)*. The classical model, TM in PKC, is replaced by the quantum model, QTM in QPKC. That is, in QPKC, all parties in QPKC are assumed to be (probabilistic) polynomial time QTMs. All channels are classical (i.e., not quantum) in our model of QPKC. We can naturally extend the definitions of one-way functions, trapdoor one-way functions, public-key encryption, digital signatures, and the related security notions.

We then show a concrete practical scheme to realize the concept of QPKC. The proposed scheme is a quantum public-key encryption (QPKE) scheme, or quantum trapdoor one-way function. The security of our scheme is based on the computational assumption (over QPT machines) that a class of subset-sum problems (whose density is at least 1) is intractable against QTM adversaries[1]. In this scheme, the underlying quantum (not classical) mechanism is only Shor's discrete logarithm algorithm, which is employed in the key generation stage (i.e., off-line stage). Encryption and decryption (i.e., on-line stage) require only classical mechanisms and so are very efficient.

## 1.3   Related Works

**1 [Quantum cryptography (QC)]** The concept of *quantum cryptography* (QC), which utilizes a quantum channel and classical TMs (as well as a classical channel), was proposed by Bennet et al. [7,6,10,9], and some protocols such as oblivious transfer based on this concept have also been presented [12,8,16,29].

---

[1]   We can also défine adversaries based on the non-uniform model as quantum circuits [1,19].

QC is one of the solutions to the above-mentioned problem when a QTM is realized in the future: that is, QC will be used for key-distribution in place of public-key encryption if a QTM is realized. The major difference between QC and QPKC is that QC employs a quantum channel (and classical channel) while QPKC employs only a classical channel. The security assumption for a QC scheme is quantum mechanics (believed by most physicists), while that for a QPKC scheme is a computational assumption (e.g., existence of a one-way function) in the QTM model.

Although several experimental QC systems have been already realized in the current technologies, recently reported security flaws of these systems are due to their realistic restrictions of quantum channels such as channel losses, realistic detection process, modifications of the qubits through channels, and fixed dark count error over long distance channels [11]. In addition, it is likely that much more complicated communication networks will be utilized in the future, and it seems technically very hard and much costly to realize a quantum channel from end to end through such complicated networks even in the future.

Accordingly, the QPKC approach seems much more promising, since in many applications encryption and key-distribution should be realized by end-to-end communication through (classical) complicated communication networks.

QC provides no solution to the problem of digital signatures when a QTM is realized: that is, QC cannot be used in digital signatures. Hence, our QPKC approach may be the only possible solution to the problem of digital signatures when a QTM is realized.

## 2 [Traditional public-key cryptosystems based on NP-hard problems]

Many public-key cryptosystems based on NP-hard problems have been presented. These schemes were designed under the traditional public-key cryptosystem model (i.e., all parties are assumed to be classical Turing machines). If, however, such a scheme is also secure against QTM adversaries, it can be an example of our model, QPKC. This is because: the QPKC model allows us to employ the quantum mechanism for key generation, encryption, and decryption, but a PKC model, in which all parties but adversaries are classical TMs and only adversaries are QTMs, is still included in the QPKC model as a special case, since the classical TM is covered by QTM. Unfortunately, however, almost all existing public-key cryptosystems based on NP-hard problems have been broken, and the security of the unbroken systems often seems suspicious due to the lack of simplicity in the trapdoor tricks.

The advantage of our new paradigm, QPKC, over the traditional approach based on NP-hard problems is that quantum mechanisms are employed for key-generation, encryption, or decryption as well as adversaries. That is, we obtain new freedom in designing PKC because we can utilize a quantum mechanism for key-distribution and encryption/decryption. Actually, this paper shows a typical example, a knapsack-type scheme; its trapdoor trick is very simple and it looks much more secure than any knapsack-type scheme based on the traditional approach.

As for digital signatures, we can theoretically construct a concrete signature scheme based on any one-way function. This means that the scheme can be as secure as an NP-hard problem, if inverting the underlying one-way function is an NP-hard problem. Since such a construction is usually impractical, we believe that the QPKC approach will provide a way to construct an efficient signature scheme secure against QPT adversaries.

**3 [Knapsack-type cryptosystems]** The subset-sum (or subset-product) problems are typical NP-hard problems. Knapsack-type cryptosystems are based on theses problems.

The proposed scheme is a knapsack-type cryptosystem, and is closely related to the Merkle–Hellman "multiplicative" trapdoor knapsack scheme [30][2], and the Chor–Rivest scheme [13].

The Merkle–Hellman scheme was broken by Odlyzko [33] under some condition and has also been broken due to its low-density (asymptotically its density is zero). Typical realizations of the Chor–Rivest scheme were also cryptanalyzed by Schnorr–Hoerner and Vaudenay [36,39], because of the known low cardinality of the subset-sum and the symmetry of the trapdoor information.

Note that these two schemes already use the trick of computing the discrete logarithm in the key-generation stage. Since they do not assume a quantum mechanism, the recommendation was to use a specific class of the discrete logarithm that could be easily computed by a classical machine.

Since we have freedom for selecting the underlying discrete logarithm problem, our scheme enjoys the use of more general mathematical tools than these two schemes. The proposed scheme employs the ring of integers, $\mathcal{O}_K$, of an algebraic number field, $K$, while the Merkle–Hellman scheme employs the ring of rational integer, $\mathbb{Z}$; the Chor–Rivest scheme employs the ring of polynomials over a finite field, $\mathbb{F}_p[x]$. The discrete logarithm in $\mathcal{O}_K/\mathfrak{p}$ should be computed in our scheme, while the discrete logarithms in $\mathbb{Z}/p\mathbb{Z}$ and $\mathbb{F}_p[x]/(g(x))$ are computed, where $\mathfrak{p}$, $p$, and $g(x)$ are prime ideal, rational prime, and irreducible polynomial, respectively. All of them are discrete logarithms in finite fields.

Our scheme offers many advantages over these two schemes:

- No information on the underlying algebraic number field, $K$, in our scheme is revealed in the public-key, while it is publicly known that $\mathbb{Z}$ and $\mathbb{F}_p[x]$ are employed in the Merkle–Hellman and the Chor–Rivest schemes respectively. Here, note that there are exponentially many candidates from which $K$ can be selected.
- No information on the underlying finite field is revealed in our scheme, while the underlying finite field is revealed in the Chor–Rivest scheme.

---

[2] Note that this scheme is different from the famous Merkle–Hellman knapsack scheme based on the super-increasing vector. Morii–Kasahara [31] and Naccache–Stern [32] also proposed a different type of multiplicative knapsack scheme, but their idea does not seem useful for our purpose since the scheme is vulnerable if the discrete logarithm is tractable.

− The density of a subset-sum problem in our scheme is at least 1, while that
for the Merkle–Hellman scheme is asymptotically 0. If the parameters are
chosen appropriately, the information rate in our scheme is asymptotically 1.

## 2   Quantum Public-Key Cryptosystems

This section defines quantum public-key cryptosystems (QPKCs) and related
notions. These definitions are straightforwardly created from the classical definitions just by replacing a classical Turing machine (or classical circuits) with a
quantum Turing machine (QTM) (or quantum circuits). Accordingly, this section defines only typical notions regarding QPKCs such as quantum one-way
functions, quantum public-key encryption, and quantum digital signatures. We
can easily extend the various classical security notions to the QPKC model.

**Definition 1.** *A function $f$ is called* quantum one-way *(QOW) if the following
two conditions hold:*

1. *[Easy to compute]   There exists a polynomial time QTM, $A$, so that, on
   input $x$, $A$ outputs $f(x)$ (i.e., $A(x)=f(x)$).*
2. *[Hard to invert]   For every probabilistic polynomial time QTM, Adv, every
   polynomial poly, and all sufficiently large $n$,*

$$\Pr[Adv(f(x)) \in f^{-1}(f(x))] < 1/poly(n).$$

*The probability is taken over the distribution of $x$, the (classical) coin flips
of Adv, and quantum observation of Adv.*

*Note that all variables in this definition are classical strings, and no quantum
channel between any pair of parties is assumed.*

**Remark:** We can also define the *non-uniform* version of this notion, in which
*Adv* is defined as polynomial size quantum circuits.[3] [1,19]

**Definition 2.** *A* quantum public-key encryption *(QPKE) scheme consists of
three probabilistic polynomial time QTMs, $(G, E, D)$, as follows:*

1. *$G$ is a probabilistic polynomial time QTM for generating keys. That is, $G$,
   on input $1^n$, outputs $(e, d)$ with overwhelming probability in $n$ (taken over
   the classical coin flips and quantum observation of $G$), where $e$ is a public-
   key, $d$ is a secret-key, and $n$ is a security parameter. (W.o.l.g., we suppose
   $|e| = |d| = n$.)*

---

[3] The concept of a quantum one-way function has been also presented by [19] independently from us. Our paper solves one of their open problems: find a candidate
one-way function that is not classical one-way. The key generation function of our
proposed scheme with input of a secret-key to output the corresponding public-key
is such a candidate one-way function.

2. *E is an encryption function that produces ciphertext c, and D is a decryption function. For every message m of size $|m| = n$, every polynomial poly, and all sufficiently large n,*

$$\Pr[D(E(m, e), d) = m] > 1 - 1/poly(n).$$

*The probability is taken over the (classical) coin flips and quantum observation of $(G, E, D)$.*

*Note that all variables in this definition are classical strings, and no quantum channel between any pair of parties is assumed.*

**Remark:** We omit the description of *security* in the above-mentioned definition, since we can naturally and straightforwardly extend the definitions of *one-wayness* (i.e., hard to invert $E(\cdot, e)$), *semantic security* [23] and *non-malleability* [4] to the QPKE model. In addition, passive and active adversaries (*adaptively chosen ciphertext attackers*) can be introduced for QPKE in the same manner as done for the classical PKE models [4]. The only difference between the classical and quantum security definitions is just that all adversaries are assumed to be probabilistic polynomial time QTMs (or polynomial size quantum circuits) in QPKC.

In addition, we can employ the random oracle model [2] to prove the security of QPKC schemes, since the random oracle model is generic and independent of the computation model. So, the conversions by Bellare–Rogaway [3] and Fujisaki–Okamoto [20,21] are useful to enhance the security of the QPKE scheme proposed in this paper.

**Definition 3.** *A* quantum digital signature *(QDS) scheme consists of three probabilistic polynomial time QTMs, $(G, S, V)$, as follows:*

1. *G is a probabilistic polynomial time QTM for generating keys. That is, G, on input $1^n$, outputs $(s, v)$ with overwhelming probability in n (taken over the classical coin flips and quantum observation of G), where s is a (secret) signing-key, v is a (public) verification-key, and n is a security parameter. (W.o.l.g., we suppose $|s| = |v| = n$.)*
2. *S is a signing function that produces signature $\sigma$, and V is a verification function. For every message m of size $|m| = n$, every polynomial poly, and all sufficiently large n,*

$$\Pr[(V(m, S(m, s), v) = 1] > 1 - 1/poly(n).$$

*The probability is taken over the (classical) coin flips and quantum observation of $(G, S, V)$.*

*Note that all variables in this definition are classical strings, and no quantum channel between any pair of parties is assumed.*

**Remark:** Similarly to QPKE, we can naturally and straightforwardly extend the security definitions of *universal/existential unforgeability* and active adversaries (*adaptively chosen message attackers*) [24] to the QDS model.

# 3   Proposed Scheme

## 3.1   Basic Idea

The basic idea to realize QPKC is to employ an appropriate NP-hard problem as an intractable primitive problem, since the concept of QPKC is based on the assumption, NP-complete $\not\subseteq$ BQP. What is the most suitable NP-hard problem? We believe that the subset-sum (or subset-product) problem is one of the most suitable problems, since the algorithms to solve the subset-sum (or subset-product) problem and the ways to realize public-key cryptosystems based on this problem have been extensively studied for the last 20 years. Another promising candidate is the lattice problem, which seems to be closely related to the subset-sum problem.

There are two typical trapdoor tricks for subset-sum or subset-product problems. One is to employ super-increasing vectors for the subset-sum and prime factorization for the subset-product. Such a tractable trapdoor vector is transformed into a public-key vector, which looks intractable. However, almost all transformation tricks from a trapdoor subset-sum (or subset-product, resp.) vector to another subset-sum (or subset-product, resp.) vector have been cryptanalyzed due to their linearity and low density.

One promising idea for the transformation is, if computing a logarithm is feasible, to employ a non-linear transformation, exponentiation (and logarithm), that bridges the subset-sum and subset-product problems. To the best of our knowledge, two schemes have been proposed on this type of transformation: One is the Merkle–Hellman "multiplicative" trapdoor knapsack scheme [30], and the other is the Chor–Rivest scheme [13]. Unfortunately, typical realizations of these schemes have been cryptanalyzed.

To overcome the weakness of these schemes, the proposed scheme employs the ring of integers, $\mathcal{O}_K$, of an algebraic number field, $K$, which is randomly selected from exponentially many candidates. See Section 1.3 for a comparison with these two schemes.

## 3.2   Notation and Preliminaries

This section introduces notations and propositions on the algebraic number theory employed in this paper. Refer to some textbooks (e.g., [27,28,14]) for more details.

We denote an algebraic number field by $K$, the ring of integers of $K$ by $\mathcal{O}_K$, and the norm of $I$ by $\mathcal{N}(I)$. (In this paper, $I$ is an integer or ideal of $\mathcal{O}_K$). We also denote the logarithm of $n$ to the base 2 by $\log n$, and that to the base $e$ by $\ln n$.

Before going to the description of our scheme, we present two propositions.

**Proposition 1.** *If $K$ is a number field and $\mathfrak{p}$ is a prime ideal of $\mathcal{O}_K$, then $\mathcal{O}_K/\mathfrak{p}$ is a finite field, $\mathbb{F}_{p^f}$, and $\mathcal{N}(\mathfrak{p}) = p^f$. There exists an integral basis, $[\omega_1, \ldots, \omega_l]$, such that each residue class of $\mathcal{O}_K/\mathfrak{p}$ is uniquely represented by*

$$a_1\omega_1 + \cdots + a_l\omega_l,$$

*where $l$ is the degree of $K$, $0 \leq a_i < e_i (i = 1, \ldots, l)$, and $[e_1\omega_1, \cdots, e_n\omega_l]$ is an integral basis of $\mathfrak{p}$. Note that $\prod_{i=1}^{l} e_i = p^f$.*

Here we note that, by using the HNF (Hermite Normal Form) representation of prime ideals, we can always assume that $\omega_1 = 1$, and $e_1 = p$. (For more detail, see Section 4.7 and Exercise 17 of [14].)

Note that $\mathcal{O}_K/\mathfrak{p}$ has some properties of integral domain and norm in addition to the structure of $\mathbb{F}_q$. These properties are specified by $K$ and $\mathfrak{p}$. (In our scheme, the variety of the properties characterized by $K$ and $\mathfrak{p}$ is utilized to enhance the security, since $K$ and $\mathfrak{p}$ are concealed against adversaries and there are exponentially many candidates for $K$ and $\mathfrak{p}$.)

The following proposition is a generalized version of Fermat's little theorem (obtained from the fact that $\mathcal{O}_K/\mathfrak{p}$ is a finite field, $\mathbb{F}_q$).

**Proposition 2 (Fermat's little theorem).** *Let $\mathfrak{p}$ be a prime ideal of $\mathcal{O}_K$, and a non-zero element $g$ from $\mathcal{O}_K \backslash \mathfrak{p}$. Then we have*

$$g^{\mathcal{N}(\mathfrak{p})-1} \equiv 1 \pmod{\mathfrak{p}}.$$

Here, note that $\mathcal{O}_K$ is not always a unique factorization domain, although our decryption algorithm utilizes factorization of an element (integer) of $\mathcal{O}_K$.

### 3.3   Proposed Scheme

### Key Generation

1. Fix a set $\mathcal{K}$ of algebraic number fields, available to the system.
2. Randomly choose an algebraic number field, $K$, from $\mathcal{K}$. Let $\mathcal{O}_K$ be its ring of integers.
3. Fix size parameters $n$, $k$ from $\mathbb{Z}$.
4. Choose a prime ideal, $\mathfrak{p}$, of $\mathcal{O}_K$, and randomly choose an element, $g$, of $\mathcal{O}_K$ such that $g$ is a generator of the multiplicative group of finite field $\mathcal{O}_K/\mathfrak{p}$. Here, an element in $\mathcal{O}_K/\mathfrak{p}$ is uniquely represented by basis $[1, \omega_2, \ldots, \omega_l]$ and integer tuple $(e_1, e_2, \ldots, e_l)$ (where $e_1 = p$) defined by Proposition 1. That is, for any $x \in \mathcal{O}_K$, there exist rational integers $x_1, x_2, \ldots, x_l \in \mathbb{Z}$ $(0 \leq x_i < e_i)$ such that $x \equiv x_1 + x_2\omega_2 + \cdots + x_l\omega_l \pmod{\mathfrak{p}}$. Note that $p$ is the rational prime below $\mathfrak{p}$.
5. Choose $n$ integers $p_1, \ldots, p_n$ from $\mathcal{O}_K/\mathfrak{p}$ with the condition that $\mathcal{N}(p_1), \ldots, \mathcal{N}(p_n)$ are co-prime, and for any subset $\{p_{i_1}, p_{i_2}, \ldots, p_{i_k}\}$ from $\{p_1, p_2, \ldots, p_n\}$, there exist rational integers $a_1, a_2, \ldots, a_l$ $(0 \leq a_i < e_i)$ such that $\prod_{j=1}^{k} p_{i_j} = a_1 + a_2\omega_2 + \cdots + a_l\omega_l$.
6. Use Shor's algorithm for finding discrete logarithms to get $a_1, \ldots, a_n$ such that

$$p_i \equiv g^{a_i} \pmod{\mathfrak{p}},$$

where $a_i \in \mathbb{Z}/(\mathcal{N}(\mathfrak{p}) - 1)\mathbb{Z}$, and $1 \leq i \leq n$.
7. Randomly choose a rational integer, $d$, in $\mathbb{Z}/(\mathcal{N}(\mathfrak{p}) - 1)\mathbb{Z}$.
8. Compute $b_i = (a_i + d) \bmod (\mathcal{N}(\mathfrak{p}) - 1)$ for each $1 \leq i \leq n$.
9. The public key is $(\mathcal{K}, n, k, b_1, b_2, \ldots, b_n)$, and the private key is $(K, g, d, \mathfrak{p}, p_1, p_2, \ldots, p_n)$.

**Encryption**

1. Fix the length of plaintext $M$ to $\lfloor \log \binom{n}{k} \rfloor$.
2. Encode $M$ into a binary string $m = (m_1, m_2, \ldots, m_n)$ of length $n$ and of Hamming weight $k$ (i.e., of having exactly $k$ 1's) as follows:
   (a) Set $l \leftarrow k$.
   (b) For $i$ from 1 to $n$ do the following:
       If $M \geq \binom{n-i}{l}$ then set $m_i \leftarrow 1$, $M \leftarrow M - \binom{n-i}{l}$, $l \leftarrow l - 1$. Otherwise, set $m_i \leftarrow 0$. (Notice that $\binom{l}{0} = 1$ for $l \geq 0$, and $\binom{0}{l} = 0$ for $l \geq 1$.)
3. Compute ciphertext $c$ by
$$c = \sum_{i=1}^{n} m_i b_i.$$

**Decryption**

1. Compute $r = (c - kd) \bmod (\mathcal{N}(\mathfrak{p}) - 1)$.
2. Compute
$$u \equiv g^r \pmod{\mathfrak{p}}.$$
3. Find $m$ as follows: If $p_i \mid u$ then set $m_i \leftarrow 1$. Otherwise, set $m_i \leftarrow 0$. After completing this procedure for all $p_i$'s ($1 \leq i \leq n$), set $m = (m_1, \ldots, m_n)$.
4. Decode $m$ to plaintext $M$ as follows:
   (a) Set $M \leftarrow 0$, $l \leftarrow k$.
   (b) For $i$ from 1 to $n$ do the following:
       If $m_i = 1$, then set $M \leftarrow M + \binom{n-i}{l}$ and $l \leftarrow l - 1$.

### 3.4   Correctness and Remarks

**1 [Decryption]** We show that decryption works. We observe that

$$u \equiv g^r \equiv g^{c-kd} \equiv g^{(\sum_{i=1}^{n} m_i b_i) - kd} \equiv g^{\sum_{i=1}^{n} m_i a_i} \pmod{\mathfrak{p}}$$

$$\equiv \prod_{i=1}^{n} (g^{a_i})^{m_i} \pmod{\mathfrak{p}}$$

$$\equiv \prod_{i=1}^{n} p_i^{m_i} \pmod{\mathfrak{p}}$$

$$= \prod_{i=1}^{n} p_i^{m_i},$$

since, from the condition of $(p_1, \ldots, p_n)$, $\prod_{i=1}^{n} p_i^{m_i}$ can be represented by $a_1 + a_2\omega_2 + \cdots + a_l\omega_l$ for some rational integers $a_1, a_2, \ldots, a_l$ ($0 \leq a_i < e_i$).

Since $\mathcal{O}_K$ is not always a unique factorization domain, we select $p_1, \ldots, p_n$ so that $\mathcal{N}(p_1), \ldots, \mathcal{N}(p_n)$ are co-prime. It follows that a product of $p_1, \ldots, p_n$ is uniquely factorized if we use only these elements as factors. Thus, a ciphertext is uniquely deciphered if a product of $p_1, \ldots, p_n$ is correctly recovered.

**2 [Number fields]** Considering efficiency and security, a typical example for $\mathcal{K}$ is the set of quadratic fields, $\{\mathbb{Q}(\sqrt{D})\}$. Especially, the set of imaginary quadratic fields is strongly recommended as $\mathcal{K}$ (see Appendix 2 for how to select parameters). Even in this set $\mathcal{K} = \{\mathbb{Q}(\sqrt{-D})\}$ of fields, there are exponentially many candidates.

**3 [Special Parameters]** Although the parameters of the imaginary quadratic fields for our scheme are described in Appendix 2, we will now show another way to select parameters, $(p_1, p_2, \ldots, p_n)$, for more general fields. Rational primes, $p_1, p_2, \ldots, p_n$, are selected such that $\prod_{j=1}^{k} p_{i_j} < e_1 = p$ for any subset $\{p_{i_1}, p_{i_2}, \ldots, p_{i_k}\}$ from $\{p_1, p_2, \ldots, p_n\}$. In that case, for any subset $\{p_{i_1}, p_{i_2}, \ldots, p_{i_k}\}$ from $\{p_1, p_2, \ldots, p_n\}$, there exists a rational integer $a_1$ $(0 < a_1 < e_1)$ such that $\prod_{j=1}^{k} p_{i_j} = a_1$. That is, $\prod_{j=1}^{k} p_{i_j}$ can be represented by $\prod_{j=1}^{k} p_{i_j} = a_1 \omega_1 + \cdots + a_l \omega_l$, where $a_2 = \cdots = a_l = 0$. Here we note that, by using the HNF (Hermite Normal Form) representation of prime ideals, we can always assume that $\omega_1 = 1$, and $e_1 > 1$ (For more detail, see Section 4.7 and Exercise 17 of [14]).

The shortcoming of this case is that $\prod_{j=1}^{k} p_{i_j} < e_1 = p = (\mathcal{N}(\mathfrak{p}))^{1/f}$, when $\mathcal{N}(\mathfrak{p})) = p^f$. Then the density and rate should be smaller (the density is about $\frac{n}{fk \log n}$, and the rate is about $\frac{k \log n - k \log k}{fk \log n}$) when $f$ is greater than 1. (Note that $\prod_{j=1}^{k} p_{i_j} \approx \mathcal{N}(\mathfrak{p})$, in the case of Appendix 2: imaginary quadratic fields.) See below for a discussion of density and rate.

**4 [Density and rate]** Here we estimate the density and rate in the case of Appendix 2. (see Section 3.5 for the definition of density and information rate.) The size of $b_i$ (i.e., $|b_i|$) is $|\mathcal{N}(\mathfrak{p})|$, $k \times |\mathcal{N}(p_i)| \approx |\mathcal{N}(\mathfrak{p})|$, and $|\mathcal{N}(p_i)| \approx 2 \log n$. Accordingly, ignoring a minor term, we obtain $|b_i| \approx |\mathcal{N}(\mathfrak{p})| \approx 2k \log n$. Hence the density $D$ of our scheme is estimated by $\frac{n}{2k \log n}$, and the rate $R$ by $\frac{k \log n - k \log k}{2k \log n}$. If $k = 2^{(\log n)^c}$ for a constant $c < 1$, the information rate, $R$, is asymptotically $1/2$, and density, $D$, is asymptotically $\infty$.

**5 [Shor's algorithm]** Key generation uses Shor's algorithm for finding discrete logarithms. The scope of Shor's original algorithm is for multiplicative cyclic groups. In particular, given a rational prime $p$, a generator $g$ of the group $(\mathbb{Z}/p\mathbb{Z})^\times$, and a target rational integer $x$ from $(\mathbb{Z}/p\mathbb{Z})^\times$, Shor's algorithm can find a rational integer $a$ from $\mathbb{Z}/(p-1)\mathbb{Z}$ such that

$$g^a = x \bmod p.$$

Shor's algorithm basically uses three registers. The first and the second registers are for all of the rational integers from 0 to $q$ where $q$ is, roughly, a large rational integer, and the third is for $g^a x^{-b} \bmod p$. Our scheme only needs to change the contents in the third register to

$$g^a p_i^{-b} \bmod \mathfrak{p}.$$

Since each of these contents can be computed efficiently even by classical computers, we can find the discrete logarithms in our scheme.

**6 [Coding]** We next mention about the encoding scheme used in encryption and decryption. This scheme is well known in combinatorial literature. (see [15]. This scheme is also employed by the Chor–Rivest cryptosystem.) This encoding scheme is used mainly for avoiding the low-density attacks mentioned later.

**7 [Complexity]** Here we mention about the time complexity needed for the key generation as well as the encryption and the decryption. The most difficult part in the key generation is the computation of discrete logarithms at line 6. In particular, we compute $n$ discrete logarithms $a_1, \ldots, a_n$ in field $\mathcal{O}_K/\mathfrak{p}$. For the encryption, once we get the encoded string by line 2 in the encryption, all we need is to add $k$ integer, each smaller than $\mathcal{N}(\mathfrak{p})$. For the decryption, we perform the modular exponentiation $g^r \bmod \mathfrak{p}$ in line 2. This dominates the running time of the decryption. Raising a generator $g$ to a power in the range up to $\mathcal{N}(\mathfrak{p})$ takes at most $2 \times \log \mathcal{N}(\mathfrak{p})$ modular multiplications by using a standard multiplication technique.

### 3.5  Security Consideration

We provide an initial analysis for the security of our scheme by considering several possible attack approaches.

We can use quantum computers also for attacks in our setting. As far as we know, despite recent attempts at designing efficient quantum algorithms for problems where no efficient classical probabilistic algorithm is known, all known such quantum algorithms are for some special cases of the hidden subgroup problem. Let $f$ be a function from a finitely generated group $G_1$ to a finite set such that $f$ is constant on the cosets of a subgroup $G_2$. Given a way of computing $f$, a hidden subgroup problem is to find $G_2$ (i.e., a generating set for $G_2$). The problems of factoring and finding discrete logarithms can be formulated as instances of the hidden subgroup problems.

There is also a result by Grover [25] for database search. He shows that the problem of finding an entry with the target value can be searched in $O(\sqrt{N})$ time, where $N$ is the number of entries in the database. This result implies NP-complete problems can be solved in $O(\sqrt{N})$ time.

However, if we do not put a structure in the database, i.e., we need to ask oracles for the contents in the database, it is known that we cannot make algorithms whose time complexity is $o(\sqrt{N})$. Thus, it is widely believed that NP-complete problems cannot be solved in polynomial time even with quantum computers.

**Finding secret keys from public keys.** Recall that we have the public key $(\mathcal{K}, n, k, b_1, b_2, \ldots, b_n)$, and the secret key $(K, g, d, \mathfrak{p}, p_1, p_2, \ldots, p_n)$, where

$$p_i \equiv g^{a_i} \pmod{\mathfrak{p}},$$

and $b_i = (a_i + d) \mod (\mathcal{N}(\mathfrak{p}) - 1)$. In a passive attack setting, the attacker has only information on the public key. The information on $n$ and $k$ only exposes problem size.

Assume we choose exponentially large $\mathcal{K}$. First, $K$ seems to be impossible to guess, since we have exponentially large possibilities for $K$. Second, $g$ and $d$ would be hard to guess, even if $K$ is revealed. This is again because we have exponentially large possibilities for them.

Third, if $K$ is revealed, we could guess only a subset of $p_i$'s, since we have chosen roughly $n$ prime elements out of $cn$ ones, where $c$ is a constant. Suppose we find a subset of $p_i$'s. In order to use them in the attack by Odlyzko for multiplicative-knapsack [33], the size of the subset must be fairly large. In addition, it is necessary to find the correspondences between the elements of the subset and $b_i$'s. Here we observe that $b_i$'s seem to be random because of the discrete-log relation in our function. Thus, it seems impossible for any reasonable relation between public keys and private keys to be made without knowing $K$, $g$, $d$, and $\mathfrak{p}$, so the critical attacks of directly finding public keys from secret keys seem to be difficult.

Notice that, in contrast to our scheme, the Chor–Rivest cryptosystem exposes the information corresponding to $K$, $p_i$'s, and $q$ of the underlying $\mathbb{F}_q$ in the public key, which enables the attackers to make use of the symmetry of the secret keys (see [39]).

**Finding plaintexts from ciphertexts.** For many knapsack-type cryptosystems, low-density attacks are known to be effective. Thus, they might be effective against our scheme. A low-density attack finds plaintexts from ciphertexts by directly solving feasible solutions to the subset sum problems that the cryptosystem is based on.

The subset-sum problem is, given positive rational integers $c$ and $a_1, \ldots, a_n$ to solve the equation $c = \sum_{i=1}^n m_i a_i$ with each $m_i \in \{0, 1\}$. Let $a = \{a_1, \ldots, a_n\}$. The density $d(a)$ of a knapsack system is defined to be $d(a) = \frac{n}{\log(\max_i a_i)}$. Density is an approximate measure of the information rate for knapsack-type cryptosystems. The shortest vector in a lattice solves almost all subset sum problems whose density is less than 0.9408 [35]. If we choose appropriate parameters for our scheme, the density is at least 1 (see Section 3.4).

It is known that the algorithms for finding the shortest vector in a lattice can be used to find the solutions to the subset sum problems. The LLL algorithm plays an important role in this kind of attack. However, it is not known that the LLL algorithm can be improved with the quantum mechanism. Incidentally, as far as we know, for any approximation algorithm, it is not known that its approximation ratio can be improved by the addition of the quantum mechanism.

Information rate $R$ is defined to be $\frac{\log |M|}{N}$, where $|M|$ is the size of message space and $N$ is the number of bits in a cipher text. If we select appropriate parameters, the information rate of our scheme is about $1/2$ (see Section 3.4).

Notice again here that it is widely believed that NP-complete problems cannot be solved efficiently even with quantum computers. Since the subset-sum problem is a typical NP-complete problem, our scheme with appropriate parameters does not seem to be open to successful crucial attacks that find plaintexts from ciphertexts even if quantum computers are used.

## 4   Extensions

We can extend our QPKC model to more general ones. One possible extension is to relax the restriction of variables employed inside QTMs to quantum strings. For example, a secret key of QPKE or QDS can be a quantum string (qubits) stored in a quantum register. The other possible extension is to use quantum channels as well as QTMs and classical channels. However, these extensions are beyond the scope of this paper.

Another direction in extension is to extend the computational model to other non-classical models such as DNA computers.

## 5   Conventional PKC Version

Our techniques to construct QPKC schemes using knapsack problems can be also employed to realize standard (non-quantum) public-key encryption based on conventional (non-quantum) algorithms [34]. We utilize the Chinese remainder theorem technique in the key generation procedure to compute the discrete logarithm very efficiently even if conventional (non-quantum) algorithms are used. In our construction, the secrecy of the underlying field, $K$, still guarantees its security.

## 6   Concluding Remarks

This paper presented a new paradigm of cryptography, quantum public-key cryptosystems (QPKCs), which consist of quantum public-key encryption (QPKE) and quantum digital signatures (QDSs). It also proposed a concrete scheme for quantum public-key cryptosystems, that will be very efficient if a QTM is realized.

The situation of this paper is comparable to that in the late 1970's, when many new ideas were proposed to realize Diffie–Hellman's paradigm. Almost all trials such as the so-called knapsack cryptosystems based on subset-sum and subset-product problems failed, and only the schemes based on integer factoring and discrete logarithm problems are still alive and widely employed.

The main purpose of this paper is to explicitly raise the concept of quantum public-key cryptosystems and to encourage researchers to create and cryptanalyze concrete QPKC schemes to investigate the feasibility of this concept.

There are many open problems regarding this concept as follows:

1. Find attacks on our QPKE scheme. (In particular, as an initial trial, cryptanalyze a restricted version of our scheme, where the underlying algebraic number field, $K$, is published and limited to the rational number field, $\mathbb{Q}$ (See Appendix 1)).

2. Find (indirect) evidence that a one-way function exists in the QTM model, or show that NP $\not\subseteq$ BQP under a reasonable assumption.
3. Realize a concrete quantum digital signature (QDS) scheme.
4. Extend the concept of QPKC (see Section 4).
5. Realize QPKC schemes based on various NP-hard problems.
6. Realize QPKC schemes that employ Shor's factoring algorithm or Grover's database search algorithm.

# References

1. BARENCO, A., BENNETT, C. H., CLEVE, R., DiVINCENZO, D. P., MARGOLUS, N., SHOR, P., SLEATOR, T., SMOLIN, J., AND WEINFURTER, H. Elementary Gates for Quantum Computation. *Physical Review A 52*, 5 (Nov. 1995), 3457–3467.
2. BELLARE, M., AND ROGAWAY, P. Entity authentication and key distribution. In *Advances in Cryptology—CRYPTO '93* (22–26 Aug. 1993), D. R. Stinson, Ed., vol. 773 of *Lecture Notes in Computer Science*, Springer-Verlag, pp. 232–249.
3. BELLARE, M., AND ROGAWAY, P. Optimal Asymmetric Encryption—How to Encrypt with RSA. In *Advances in Cryptology—EUROCRYPT'94* (1994), pp. 92–111.
4. BELLARE, M., DESAI, A., POINTCHEVAL, D., AND ROGAWAY, P. Relations among Notions of Security for Public-Key Encryption Schemes. In *Advances in Cryptology—CRYPTO'98* (1998), pp. 26–45.
5. BENNETT, C. H., BERNSTEIN, E., BRASSARD, G., AND VAZIRANI, U. Strengths and weaknesses of quantum computing. *SIAM J. Comput. 26*, 5 (Oct. 1997), 1510–1523.
6. BENNETT, C. H., BESSETTE, F., BRASSARD, G., SALVAIL, L., AND SMOLIN, J. Experimental quantum cryptography. *Journal of Cryptology 5*, 1 (1992), 3–28.
7. BENNETT, C. H., AND BRASSARD, G. An update on quantum cryptography. In *Advances in Cryptology: Proceedings of CRYPTO 84* (19–22 Aug. 1984), G. R. Blakley and D. Chaum, Eds., vol. 196 of *Lecture Notes in Computer Science*, Springer-Verlag, 1985, pp. 475–480.
8. BENNETT, C. H., BRASSARD, G., CRÉPEAU, C., AND SKUBISZEWSKA, M.-H. Practical quantum oblivious transfer. In *Advances in Cryptology—CRYPTO '91* (11–15 Aug. 1991), J. Feigenbaum, Ed., vol. 576 of *Lecture Notes in Computer Science*, Springer-Verlag, 1992, pp. 351–366.
9. BENNETT, C. H., BRASSARD, G., AND EKERT, A. K. Quantum cryptography. *Scientific America 262*, 10 (Oct. 1992), 26–33.
10. BENNETT, C. H., BRASSARD, G., AND MERMIN, N. D. Quantum cryptography without Bell's theorem. *Physical Review Letters 68*, 5 (Feb. 1992), 557–559.
11. BRASSARD, G., LÜTKENHAUS, N., TAL, M., AND SANDERS, B. C. Security Aspects of Practical Quantum Cryptography. In *Advances in Cryptology—EUROCRYPT2000* (2000), pp. 289–299.
12. BRASSARD, G., AND CRÉPEAU, C. Quantum bit commitment and coin tossing protocols. In *Advances in Cryptology—CRYPTO '90* (11–15 Aug. 1990), A. J. Menezes and S. A. Vanstone, Eds., vol. 537 of *Lecture Notes in Computer Science*, Springer-Verlag, 1991, pp. 49–61.
13. CHOR, B., AND RIVEST, R. L. A knapsack-type public key cryptosystem based on arithmetic in finite fields. *IEEE Trans. on Information Theory 34* (1988), 901–909.
14. COHEN, H. *A Course in Computational Algebraic Number Theory*. Springer, 1993.

15. COVER, T. M. Enumerative source encoding. *IEEE Trans. on Information Theory IT-19* (1973), 901–909.

16. CRÉPEAU, C., AND SALVAIL, L. Quantum oblivious mutual identification. In Guillou and Quisquater [26], pp. 133–146.

17. DEUTSCH, D., AND JOZSA, R. Rapid solution of problems by quantum computation. *Proc. R. Soc. Lond. A 439* (1992), 553–558.

18. DIFFIE, W., AND HELLMAN, M. New directions in cryptography. *IEEE Trans. on Information Theory IT-22*, 6 (1976), 644–654.

19. DUMAIS, P., MAYERS, D., AND SALVAIL, L. Perfectly Concealing Quantum Bit Commitment from any Quantum One-Way Permutation. In *Advances in Cryptology—EUROCRYPT2000* (2000), pp. 300–315.

20. FUJISAKI, E. AND OKAMOTO, T. How to Enhance the Security of Public-Key Encryption at Minimum Cost. In *PKC'99* (1999), pp. 53–68.

21. FUJISAKI, E. AND OKAMOTO, T. Secure Integration of Asymmetric and Symmetric Encryption Schemes. In *Advances in Cryptology—CRYPTO'99* (1999), pp. 537–554.

22. GOLDREICH, O. On the foundations of modern cryptography. In *Advances in Cryptology—CRYPTO '97* (17–21 Aug. 1997), B. S. Kaliski Jr., Ed., vol. 1294 of *Lecture Notes in Computer Science*, Springer-Verlag, pp. 46–74.

23. GOLDWASSER, S., AND MICALI, S. Probabilistic encryption. *J. Comput. Syst. Sci. 28*, 2 (Apr. 1984), 270–299.

24. GOLDWASSER, S., MICALI, S., AND RIVEST, R. L. A digital signature scheme secure against adaptive chosen-message attacks. *SIAM J. Comput. 17*, 2 (Apr. 1988), 281–308.

25. GROVER, L. K. A fast quantum mechanical algorithm for database search. In *Proceedings of the Twenty-Eighth Annual ACM Symposium on the Theory of Computing* (Philadelphia, Pennsylvania, 22–24 May 1996), pp. 212–219.

26. GUILLOU, L. C., AND QUISQUATER, J.-J., Eds. *Advances in Cryptology—EUROCRYPT 95* (21–25 May 1995), vol. 921 of *Lecture Notes in Computer Science*, Springer-Verlag.

27. LANG, S. *Algebraic Number Theory, Second Edition*, Springer, 1994.

28. MARCUS, D. A. *Number Fields*, Springer, 1977.

29. MAYERS, D. Quantum key distribution and string oblivious transfer in noisy channels. In *Advances in Cryptology—CRYPTO '96* (18–22 Aug. 1996), N. Koblitz, Ed., vol. 1109 of *Lecture Notes in Computer Science*, Springer-Verlag, pp. 343–357.

30. MERKLE, R. C., AND HELLMAN, M. E. Hiding information and signatures in trapdoor knapsacks. *IEEE Trans. on Information Theory 24* (1978), 525–530.

31. MORII, M., AND KASAHARA, M. New Public Key Cryptosystem Using Discrete Logarithms over $GF(p)$. *Trans. of the IEICE J71-D*, 2 (Feb. 1988), 448–453 (In Japanese).

32. NACCACHE, D., AND STERN, J. A New Public-Key Cryptosystem. In *Advances in Cryptology—EUROCRYPT'97* (1997), pp. 27–36.

33. ODLYZKO, A. M. Cryptanalytic attacks on the multiplicative knapsack cryptosystem and on Shamir's fast signature scheme. *IEEE Trans. on Information Theory IT-30* (1984), 594–601.

34. OKAMOTO, T., AND TANAKA, K. A New Approach to Knapsack Cryptosystems. manuscript (2000).

35. ORTON, G. A Multiple-Iterated Trapdoor for Dense Compact Knapsacks. In *Advances in Cryptology—EUROCRYPT'94* (1994), pp. 112–130.

36. SCHNORR, C. P., AND HÖRNER, H. H. Attacking the Chor–Rivest cryptosystem by improved lattice reduction. In Guillou and Quisquater [26], pp. 1–12.

37. SHOR, P. W. Polynomial-time algorithms for prime factorization and discrete logarithms on a quantum computer. *SIAM J. Comput.* 26, 5 (Oct. 1997), 1484–1509.
38. SIMON, D. R. On the power of quantum computation. *SIAM J. Comput.* 26, 5 (Oct. 1997), 1474–1483.
39. VAUDENAY, S. Cryptanalysis of the Chor–Rivest cryptosystem. In *Advances in Cryptology—CRYPTO'98* (1998), pp. 243–256.

# Appendix 1:  Restricted Version for Explaining Our Scheme

This section presents a very restricted version of our scheme in order to help readers to understand our scheme more easily. Since this version seems to be much less secure than the full version, we do not recommend this version for practical usage, although we have not found any effective attack even against this restricted version.

Suppose that we set $\mathcal{K} = \{\mathbb{Q}\}$. i.e., we have only the field $\mathbb{Q}$ of rational numbers for the system. Then, the ring $\mathcal{O}_K$ of integers of $\mathbb{Q}$ is $\mathbb{Z}$. In this section, we use a *prime* to refer to a rational prime, and an *integer* a rational integer. The restricted version of our scheme is as follows:

## Key Generation

1. Fix size parameters $n$, $k$ from $\mathbb{Z}$.
2. Randomly choose a prime $p$, a generator $g$ of the group $(\mathbb{Z}/p\mathbb{Z})^{\times}$, and $n$ co-primes $p_1, \ldots, p_n \in \mathbb{Z}/p\mathbb{Z}$ such that $\prod_{j=1}^{k} p_{i_j} < p$ for any subset $\{p_{i_1}, p_{i_2}, \ldots, p_{i_k}\}$ from $\{p_1, p_2, \ldots, p_n\}$.
3. Use Shor's algorithm for finding discrete logarithms to get integers $a_1, \ldots, a_n \in \mathbb{Z}/(p-1)\mathbb{Z}$ satisfying $p_i \equiv g^{a_i} \pmod{p}$, for each $1 \le i \le n$.
4. Randomly choose a integer $d \in \mathbb{Z}/(p-1)\mathbb{Z}$.
5. Compute $b = (a_i + d) \bmod (p-1)$, for each $1 \le i \le n$.
6. The public key is $(n, k, b_1, b_2, \ldots, b_n)$, and the secret key is $(g, d, p, p_1, p_2, \ldots, p_n)$.

## Encryption

1. Fix the length of plaintext $M$ to $\lfloor \log \binom{n}{k} \rfloor$.
2. Encode $M$ into a binary string $m = (m_1, m_2, \ldots, m_n)$ of length $n$ and Hamming weight $k$ (i.e., having exactly $k$ 1's) as follows:
   (a) Set $l \leftarrow k$.
   (b) For $i$ from 1 to $n$ do the following:
       If $M \ge \binom{n-i}{l}$ then set $m_i \leftarrow 1$, $M \leftarrow M - \binom{n-i}{l}$, $l \leftarrow l - 1$. Otherwise, set $m_i \leftarrow 0$. (Notice that $\binom{0}{0} = 1$ for $l \ge 0$, and $\binom{0}{l} = 0$ for $l \ge 1$.)
3. Compute the ciphertext $c$ by $c = \sum_{i=1}^{n} m_i b_i$.

**Decryption**

1. Compute $r = (c - kd) \bmod (p - 1)$.
2. Compute $u = g^r \bmod q$.
3. Find the factors of $u$. If $p_i$ is a factor, then set $m_i \leftarrow 1$. Otherwise, $m_i \leftarrow 0$.
4. Decode $m$ to the plaintext $M$ as follows:
   (a) Set $M \leftarrow 0, l \leftarrow k$.
   (b) For $i$ from 1 to $n$ do the following:
       If $m_i = 1$, then set $M \leftarrow M + \binom{n-i}{l}$ and $l \leftarrow l - 1$.

# Appendix 2:  Imaginary Quadratic Field Version of Our Scheme

This section presents the imaginary quadratic field version of our scheme. Before describing the proposed scheme, we will briefly review basic results of the arithmetic on imaginary quadratic fields and present a proposition.

## Imaginary Quadratic Fields

Let $K = \mathbb{Q}(\sqrt{-D})$ be an imaginary quadratic field of discriminant $-D$. Here we note that the ring of integers, $\mathcal{O}_K$, of $K$ has an integral basis $[1, \omega]$, where $\omega = \sqrt{-D/4}$ (if $-D \equiv 0 \pmod 4$), $\omega = \frac{1+\sqrt{-D}}{2}$ (otherwise), and this called the standard basis of $\mathcal{O}_K$. Let $\mathfrak{p}$ be a prime ideal of $\mathcal{O}_K$ of residue degree $f$, namely, $\mathcal{N}(\mathfrak{p}) = p^f$, where $p$ is a rational prime integer below $\mathfrak{p}$. Then we can take an integral basis of $\mathfrak{p}$ as $[p, e_2\omega_2]$, where $e_2 = p^{f-1}$, and $\omega_2 = b + \omega$ with some rational integer $b$ (e.g. if $-D \equiv 0 \pmod 4$ and $-D$ is a quadratic residue $\bmod\, p$, then $b$ is a root of $b^2 \equiv -D \pmod p$). We also call this basis the standard basis of $\mathfrak{p}$. Then, each residue class of $\mathcal{O}_K/\mathfrak{p}$ is uniquely represented by $x_1 + x_2\omega_2$, where $-p/2 < x_1 < p/2$ and $-e_2/2 < x_2 < e_2/2$ (cf. Proposition 1). From here, we fix a complete representative system of $\mathcal{O}_K/\mathfrak{p}$ as follows:

$$R(\mathfrak{p}) = \{x_1 + x_2\omega_2 \in \mathcal{O}_K \mid -p/2 < x_1 < p/2, -e_2/2 < x_2 < e_2/2\}.$$

We then have the following proposition.

**Proposition 3.** *Let* $K = \mathbb{Q}(\sqrt{-D})$ *be an imaginary quadratic field of discriminant* $-D$, $\mathfrak{p}$ *a prime ideal of* $\mathcal{O}_K$ *with* $\mathcal{N}(\mathfrak{p}) = p^f$, *where* $f = 1, 2$. *Let* $[1, \omega]$ *and* $[p, e_2\omega_2]$ *be the standard basis of* $\mathcal{O}_K$ *and* $\mathfrak{p}$, *respectively.*
*Then*

1. *Case:* $\mathcal{N}(\mathfrak{p}) = p$ $(f = 1)$
   *For any integer* $x = x_1 + x_2\omega_2 \in \mathcal{O}_K$, *if it satisfies* $\mathcal{N}(x) < p^2/4$ *and* $x_2 = 0$, *then we have* $x \in R(\mathfrak{p})$. *In this case, we can take* $R(\mathfrak{p})$ *as* $\{x \in \mathbb{Z} \mid -p/2 < x < p/2\}$.
2. *Case:* $\mathcal{N}(\mathfrak{p}) = p^2$ $(f = 2)$
   *For any integer* $x = x_1 + x_2\omega_2 \in \mathcal{O}_K$, *if* $-D \equiv 0 \pmod 4$ *and* $\mathcal{N}(x) < p^2/4$, *then we have* $x \in R(\mathfrak{p})$, *while if* $-D \equiv 1 \pmod 4$ *and* $\mathcal{N}(x) < \frac{(p-1)^2 D}{4(1+D)}$, *then we have* $x \in R(\mathfrak{p})$.

*Proof.* In the case of $f = 1$, $x = x_1 + x_2\omega_2 \in R(\mathfrak{p})$ if and only if $x_1^2 < p^2/4$ and $x_2 = 0$, namely, $\mathcal{N}(x) < p^2/4$ and $x_2 = 0$. For the case of $f = 2$ and $-D \equiv 0 \pmod 4$, it is sufficient to show that $\{(x_1, x_2) \in \mathbb{Z}^2 \mid x_1 + x_2\omega_2 \in R(\mathfrak{p})\}$ contains $\{(x_1, x_2) \in \mathbb{Z}^2 \mid x_1^2 + \frac{D}{4}x_2^2 < \frac{p^2}{4}\}$. Note that we can take $b = 0$ in the standard basis, namely, $\omega_2 = \omega = \sqrt{-D/4}$ and $e_2 = p$. Similarly, for the case of $f = 2$ and $-D \equiv 2, 3 \pmod 4$, it is sufficient to show that $\{(s, t) \in \mathbb{Z}^2 \mid -3p/2 < s < 3p/2, -p/2 < t < p/2, s \equiv t \pmod 2\}$ contains $\{(s, t) \in \mathbb{Z}^2 \mid s^2 + Dt^2 < \frac{D(p-1)^2}{1+D}, s \equiv t \pmod 2\}$. By drawing pictures, we can easily show these relationships.

## Proposed Scheme

We will now present our proposed scheme using imaginary quadratic fields.

## Key Generation

1. Fix a set $\mathcal{K}$ of imaginary quadratic fields, available to the system.
2. Randomly choose an imaginary quadratic field, $K = \mathbb{Q}(\sqrt{-D})$, where $-D$ is the discriminant of $K$, from $\mathcal{K}$. Let $\mathcal{O}_K$ be the ring of integers of $K$.
3. Fix size parameters $n, k$ from $\mathbb{Z}$.
4. Choose a prime ideal, $\mathfrak{p}$, of degree 2 from $\mathcal{O}_K$, and randomly choose an element, $g$, of $\mathcal{O}_K$ such that $g$ is a generator of the multiplicative group of finite field $\mathcal{O}_K/\mathfrak{p}$. Here, an element in $\mathcal{O}_K/\mathfrak{p}$ is uniquely represented by basis $[1, \omega_2]$ and integer pair $(p, p)$. That is, for any $x \in \mathcal{O}_K$, there exist integers $x_1, x_2 \in \mathbb{Z}$, $-p/2 < x_1, x_2 < p/2$ such that $x \equiv x_1 + x_2\omega_2 \pmod{\mathfrak{p}}$.
5. Choose $n$ integers $p_1, \ldots, p_n$ from $\mathcal{O}_K/\mathfrak{p}$ with the condition that $\mathcal{N}(p_1), \ldots, \mathcal{N}(p_n)$ are co-prime, and for any subset $\{p_{i_1}, p_{i_2}, \ldots, p_{i_k}\}$ from $\{p_1, p_2, \ldots, p_n\}$, if $-D \equiv 0 \pmod 4$, $\prod_{j=1}^{k} \mathcal{N}(p_{i_j}) < \frac{p^2}{4}$, otherwise, $\prod_{j=1}^{k} \mathcal{N}(p_{i_j}) < \frac{(p-1)^2 D}{4(1+D)}$.
6. Use Shor's algorithm for finding discrete logarithms to get $a_1, \ldots, a_n$ such that
$$p_i \equiv g^{a_i} \pmod{\mathfrak{p}},$$
where $a_i \in \mathbb{Z}/(\mathcal{N}(\mathfrak{p}) - 1)\mathbb{Z}$, and $1 \leq i \leq n$.
7. Randomly choose a rational integer, $d$, in $\mathbb{Z}/(\mathcal{N}(\mathfrak{p}) - 1)\mathbb{Z}$.
8. Compute $b_i = (a_i + d) \bmod (\mathcal{N}(\mathfrak{p}) - 1)$ for each $1 \leq i \leq n$.
9. The public key is $(\mathcal{K}, n, k, b_1, b_2, \ldots, b_n)$, and the private key is $(K = \mathbb{Q}(\sqrt{-D}), g, d, \mathfrak{p}, p_1, p_2, \ldots, p_n)$.

## Encryption

1. Fix the length of plaintext $M$ to $\lfloor \log \binom{n}{k} \rfloor$.
2. Encode $M$ into a binary string $m = (m_1, m_2, \ldots, m_n)$ of length $n$ and of Hamming weight $k$ (i.e., of having exactly $k$ 1's) as follows:
   (a) Set $l \leftarrow k$.

(b) For $i$ from 1 to $n$ do the following:
   If $M \geq \binom{n-i}{l}$ then set $m_i \leftarrow 1$, $M \leftarrow M - \binom{n-i}{l-1}$, $l \leftarrow l - 1$. Otherwise, set $m_i \leftarrow 0$. (Notice that $\binom{l}{0} = 1$ for $l \geq 0$, and $\binom{0}{l} = 0$ for $l \geq 1$.)

3. Compute ciphertext $c$ by $c = \sum_{i=1}^{n} m_i b_i$.

## Decryption

1. Compute $r = (c - kd) \bmod (\mathcal{N}(\mathfrak{p}) - 1)$.
2. Compute $u \equiv g^r \pmod{\mathfrak{p}}$.
3. Find $m$ as follows:
   (a) Let $[1, \omega_2]$ and $(p, p)$ be the basis and integer pair defined by Proposition 1. From the selection of $p_1, \ldots, p_n$, $u$ can be represented by $u = a_1 + a_2\omega_2$ for some integers $a_1, a_2 \in \mathbb{Z}$ with $-p/2 < a_i < p/2$ $(i = 1, 2)$.
   (b) Do the following:
   If $p_i \mid u$ then set $m_i \leftarrow 1$. Otherwise, set $m_i \leftarrow 0$. After completing this procedure for all $p_i$'s $(1 \leq i \leq n)$, set $m = (m_1, \cdots, m_n)$.
4. Decode $m$ to plaintext $M$ as follows:
   (a) Set $M \leftarrow 0$, $l \leftarrow k$.
   (b) For $i$ from 1 to $n$ do the following:
   If $m_i \leftarrow 1$, then set $M \leftarrow M + \binom{n-i}{l}$ and $l \leftarrow l - 1$.

**Remark 1:** Note that we can easily choose a prime ideal, $\mathfrak{p}$, of degree 2 as follows: choose any rational prime, $p$, such that $-D$ is a quadratic non-residue $\bmod p$, then set $\mathfrak{p} = p\mathcal{O}_K$. In other words, $p$ is also a prime element in $\mathcal{O}_K$. Furthermore, it can be efficiently checked whether $p$ is a prime in $\mathcal{O}_K$ or not by computing the Legendre symbol, $\left(\frac{-D}{p}\right)$, namely $p$ is a prime element in $\mathcal{O}_K$ if and only if $\left(\frac{-D}{p}\right) = -1$, and always selected such $p$ from the set of all rational primes with probability about $1/2$.

**Remark 2:** Note that, in Step 5 of the key generation stage, for any subset $\{p_{i_1}, p_{i_2}, \ldots, p_{i_k}\}$ from $\{p_1, p_2, \ldots, p_n\}$, $\prod_{j=1}^{k} \mathcal{N}(p_{i_j}) = \mathcal{N}(\prod_{j=1}^{k} p_{i_j}) < \frac{p^2}{4}$, ( or $\frac{(p-1)^2 D}{4(1+D)}$ ), so $\prod_{j=1}^{k} p_{i_j} \in R(\mathfrak{p})$ by Proposition 3. That is, there exist integers $a_1, a_2 \in \mathbb{Z}$ $(-p/2 < a_1, a_2 < p/2)$ such that $u = a_1 + a_2\omega_2$ in Step 3(a) of the Decryption. The typical selection of $p_1, \ldots, p_n$ presented in Section 3.4 may be restricted, in fact, we take $p_1, \ldots, p_n$ from the rational integers, but the selection introduced above is more general than the typical one. That is, we can take $p_i$'s from $\mathbb{Z}$ as well as $\mathcal{O}_K$ by using such a characterization with the norm in the imaginary quadratic field case.

# New Public-Key Cryptosystem Using Braid Groups

Ki Hyoung Ko[1], Sang Jin Lee[1], Jung Hee Cheon[2],
Jae Woo Han[3], Ju-sung Kang[3], and Choonsik Park[3]

[1] Department of Mathematics, Korea Advanced Institute of Science and Technology,
Taejon, 305-701, Korea
{knot,sjlee}@knot.kaist.ac.kr
[2] Department of Mathematics, Brown university, Providence, RI 02912, USA
and Securepia, Korea
jhcheon@math.brown.edu
[3] Section 8100, Electronics and Telecommunications Research Institute,
Taejon, 305-600, Korea
{jwhan,jskang,csp}@etri.re.kr

**Abstract.** The braid groups are infinite non-commutative groups naturally arising from geometric braids. The aim of this article is twofold. One is to show that the braid groups can serve as a good source to enrich cryptography. The feature that makes the braid groups useful to cryptography includes the followings: (i) The word problem is solved via a fast algorithm which computes the canonical form which can be efficiently manipulated by computers. (ii) The group operations can be performed efficiently. (iii) The braid groups have many mathematically hard problems that can be utilized to design cryptographic primitives. The other is to propose and implement a new key agreement scheme and public key cryptosystem based on these primitives in the braid groups. The efficiency of our systems is demonstrated by their speed and information rate. The security of our systems is based on topological, combinatorial and group-theoretical problems that are intractible according to our current mathematical knowledge. The foundation of our systems is quite different from widely used cryptosystems based on number theory, but there are some similarities in design.

**Key words:** public key cryptosystem, braid group, conjugacy problem, key exchange, hard problem, non-commutative group, one-way function, public key infrastructure

## 1 Introduction

### 1.1 Background and Previous Results

Since Diffie and Hellman first presented a public-key cryptosystem(PKC) in [11] using a trapdoor one-way function, many PKC's have been proposed and broken.

Most of successful PKC's require large prime numbers. The difficulty of factorization of integers with large prime factors forms the ground of RSA [29] and

M. Bellare (Ed.): CRYPTO 2000, LNCS 1880, pp. 166–183, 2000.
© Springer-Verlag Berlin Heidelberg 2000

its variants such as Rabin-Williams [28,36], LUC's scheme [32] or elliptic curve versions of RSA like KMOV [20]. Also the difficulty of the discrete logarithm problem forms the ground of Diffie-Hellman type schemes like ElGamal [12], elliptic curve cryptosystem, DSS, McCurley [23].

There have been several efforts to develop alternative PKC's that are not based on number theory. The first attempt was to use NP-hard problems in combinatorics like Merkle-Hellman Knapsack [24] and its modifications. Though many cryptographers have been pessimistic about combinatorial cryptography after the breakdown of the Knapsack-type PKC's by Shamir [30], Brickell [9], Lagarias [22], Odlyzko [26], Vaudenay [35] and others, and after the appearance of Brassard theorem [8], there may still be some hopes as Koblitz has noted in [21]. The other systems that are worth to mention are the quantum cryptography proposed by Bennet and Brassard [4] and the lattice cryptography proposed by Goldreich, Goldwasser and Halevi [18].

Another approach is to use hard problems in combinatorial group theory such as the word problem [1,37,17] or using the Lyndon words [31]. Recently Anshel-Anshel-Goldfeld proposed in [2] a key agreement system and a PKC using groups where the word problem is easy but the conjugacy problem is intractible. And they noted that the usage of braid groups is particularly promising. Our proposed systems is based on the braid groups but is independent from their algebraic key establishment protocol on monoids in [2].

Most of cryptosystems derived from combinatorial group theory are mainly theoretical or have certain limitations in wide and general practice. This is perhaps due to the rack of efficient description of group elements and operations or due to the difficulty of implementing cryptosystems themselves.

## 1.2   The Features of Braid Groups

The $n$-braid group $B_n$ is an infinite noncommutative group of $n$-braids defined for each positive integer $n$. There is a natural projection from $B_n$ to the group $\Sigma_n$ of all $n!$ $n$-permutations and so $B_n$ can be thought as a resolution of $\Sigma_n$. In this article, we first show that the braid groups have the following nice properties, unlike the usual combinatorial groups, so that one can build cryptosystems satisfying both security and efficiency requirements.

1. There is a natural way to describe group elements as data which can be handled by computers: Theorem 1 shows that there is a canonical form for a braid, which can be described as an ordered tuple $(u, \pi_1, \pi_2, \ldots, \pi_p)$, where $u$ is an integer and $\pi_i$'s are $n$-permutations. The canonical form can remove the difficulties in using words in the description of the group elements.

2. There are fast algorithms to perform the group operations: The product of two words $U$ and $V$ is just the concatenation $UV$ and therefore the group operation for the purpose of cryptography really means hiding the factors $U$ and $V$. This can be achieved by converting $UV$ into its canonical form. For a group whose element has no canonical form, this can be achieved only by rewriting via defining relations and a retrieval must be done by a solution

to the word problem. Let $U$ and $V$ be $n$-braids whose canonical forms are represented by $p$ and $q$ permutations respectively. Theorem 2 shows that the canonical form of the product $UV$ can be computed in time $\mathcal{O}(pqn \log n)$ and the canonical form of the inverse of $U$ can be computed in time $\mathcal{O}(pn)$

3. There are many hard problems based on topological or group-theoretical open problems and one can sometimes design (trap-door) one-way functions based on these problems that can be described basically by group operations.

4. As $n$ grows in the braid groups $B_n$, the computation of group operations become harder in $\mathcal{O}(n \log n)$. On the other hand, a naive computation of the inverses of one-way functions are seem to be at least $\mathcal{O}(n!)$. Consequently, $n$ plays a reliable role of a security parameter.

### 1.3   Our Results

After exploring cryptographic aspects of the braid groups in §2, we propose a trapdoor one-way function that is based on one of the hard problems in §2.3 and construct a key exchange scheme and a public-key cryptosystem in §3. A theoretic operating characteristics and implementation of our PKC will be given in §4 and so they are readily available in practice. Our PKC has the following features.

1. Our key exchange scheme is based on a variation of the conjugacy problem similar to the Diffie-Hellman problem and our PKC is constructed from this key exchange scheme. Therefore our PKC and behaves somewhat similarly to ElGamal PKC.

2. Our PKC is non-deterministic: The ciphertext depends on both of the plaintext and the braid chosen randomly at each session.

3. The message expansion is at most 4-1.

4. There are two parameters $p, n$ in our PKC so that the message length becomes $pn \log n$. The encryption and decryption are $\mathcal{O}(p^2 n \log n)$ operations. The security level against brute force attacks is $\mathcal{O}((n!)^p) = \mathcal{O}(\exp(pn \log n))$. Thus the parameter $n$ rapidly increases the security level without sacrificing the speed.

Our cryptosystems are efficient enough, comparing to other widely used cryptosystems. The security of our scheme is discussed and a possible attack based on a mathematical knowledge is introduced in §5. As a further study, possible improvements of our cryptosystems and possible replacements of the braid groups are discussed in §6.

## 2   A Cryptographic Aspect of the Braid Groups

The braid group was first introduced by Artin in [3]. Because these groups play important roles in low dimensional topology, combinatorial group theory and representation theory, considerable research has been done on these groups. In this section, we will briefly introduce the notion of braids and give evidence that

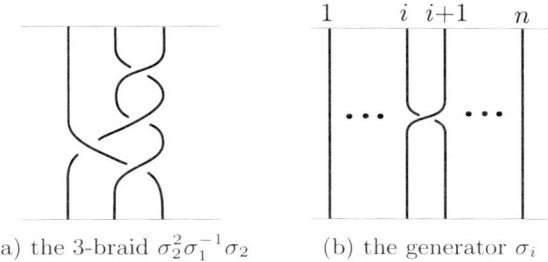

(a) the 3-braid $\sigma_2^2\sigma_1^{-1}\sigma_2$          (b) the generator $\sigma_i$

**Fig. 1.** An example of braid and the generator

the braid groups can also play important roles in cryptography. The general reference for braid theory is the Birman's book [5] and for the word problem and conjugacy problem, see [6,13,14,16].

This section is composed as follows: §2.1 is the definition of the braid groups. In §2.2 we first summarize the known results on the word problem (or the canonical form problem). Theorem 1 is important since it enables one to encode a braid into a data format that can be handled easily by computers. The remains are supplementary to this theorem.

In §2.3 we list hard problems that are potential sources to develop primitives in cryptography.

### 2.1 Definition of the $n$-Braid Group

The $n$-braid group $B_n$ is defined by the following group presentation.

$$B_n = \left\langle \sigma_1, \ldots, \sigma_{n-1} \mid \begin{array}{l} \sigma_i\sigma_j\sigma_i = \sigma_j\sigma_i\sigma_j \text{ if } |i - j| = 1 \\ \sigma_i\sigma_j = \sigma_j\sigma_i \quad\quad \text{ if } |i - j| \geq 2 \end{array} \right\rangle \qquad (*)$$

The integer $n$ is called the *braid index* and each element of $B_n$ is called an $n$-*braid*. Braids have the following geometric interpretation: an $n$-braid is a set of disjoint $n$ strands all of which are attached to two horizontal bars at the top and at the bottom such that each strand always heads downward as one walks along the strand from the top to the bottom. The braid index is the number of strings. See Figure 1(a) for an example. Two braids are *equivalent* if one can be deformed to the other continuously in the set of braids. In this geometric interpretation, $\sigma_i$ is the elementary braid as in Figure 1(b).

The multiplication $ab$ of two braids $a$ and $b$ is the braid obtained by positioning $a$ on the top of $b$. The identity $e$ is the braid consisting of $n$ straight vertical strands and the inverse of $a$ is the reflection of $a$ with respect to a horizontal line. So $\sigma_i^{-1}$ can be obtained from $\sigma_i$ by switching the over-strand and under-strand. See [5] for details.

Note that if we add the relation $\sigma_i^2 = 1$, $i = 1, \ldots, n - 1$, to the presentation $(*)$, it becomes the group presentation of the $n$-permutation group $\Sigma_n$, where $\sigma_i$ corresponds to the transition $(i, i + 1)$. So there is a natural surjective homomorphism $\rho: B_n \to \Sigma_n$. Let's denote a permutation $\pi \in \Sigma_n$, $\pi(i) = b_i$, by

$\pi = b_1 b_2 \cdots b_n$. In terms of geometric braids, the homomorphism $\rho \colon B_n \to \Sigma_n$ can be described as follows: given a braid $a$, let the strand starting from the $i$-th upper position ends at the $b_i$-th lower position. Then $\rho(a)$ is the permutation $b_1 b_2 \cdots b_n$.

## 2.2   Describing Braids Using Permutations

The easiest way to describe a braid is to write it as a word on $\sigma_i$'s. But there is no unique way to do this. For example, all the words in the following formula represent the same braid $\Delta_4$ in Figure 2(a), where the defining relations, $\sigma_1 \sigma_2 \sigma_1 = \sigma_2 \sigma_1 \sigma_2$, $\sigma_2 \sigma_3 \sigma_2 = \sigma_3 \sigma_2 \sigma_3$ and $\sigma_1 \sigma_3 = \sigma_3 \sigma_1$, are applied to the underlined subwords.

$$\Delta_4 = \sigma_1 \sigma_2 \sigma_3 \underline{\sigma_1 \sigma_2 \sigma_1} = \sigma_1 \underline{\sigma_2 \sigma_3 \sigma_2} \sigma_1 \sigma_2 = \underline{\sigma_1 \sigma_3} \sigma_2 \sigma_3 \sigma_1 \sigma_2$$
$$= \sigma_3 \underline{\sigma_1 \sigma_2 \sigma_1} \sigma_3 \sigma_2 = \sigma_3 \sigma_2 \sigma_1 \underline{\sigma_2 \sigma_3 \sigma_2} = \sigma_3 \sigma_2 \sigma_1 \sigma_3 \sigma_2 \sigma_3$$

In 1947, Artin proved that a braid can be described uniquely as an automorphism of the free group of rank $n$ [3]. In late sixties, Garsides solved the word problem after exploring the properties of the semigroup of positive words in [16] and his idea was improved by Thurston [14], Elrifai-Morton [13] and Birman-Ko-Lee [6]. They showed that there is a fast algorithm to compute the canonical form, which is unique for their results briefly.

Before stating the theorem, we introduce the notions of the *permutation braid* and the *fundamental braid*. To each permutation $\pi = b_1 b_2 \cdots b_n$, we associate an $n$-braid $A$ that is obtained by connecting the upper $i$-th point to the lower $b_i$-th point by a straight line and then making each crossing positive, i.e. the line between $i$ and $b_i$ is under the line between $j$ and $b_j$ if $i < j$. For example if $\pi = 4213$, then $A = \sigma_1 \sigma_2 \sigma_1 \sigma_3$ as in Figure 2(b).

The braids made as above is called a *permutation braid* or a *canonical factor* and $\tilde{\Sigma}_n$ denotes the set of all permutation braids. The correspondence from a permutation $\pi$ to a canonical factor $A$ is a right inverse of $\rho \colon B_n \to \Sigma_n$ as a set function. So the cardinality of $\tilde{\Sigma}_n$ is $n!$. The permutation braid can be characterized by the property that every crossing is positive and any pair of strands crosses at most once [13,14].

The permutation braid corresponding to the permutation $\Omega_n = n(n-1) \cdots \cdots (2)1$ is called the *fundamental braid* and denoted by $\Delta_n$. If there is no confusion on the braid index, we drop the subscript to write just $\Delta$. See Figure 2(a) for an example.

The following theorem gives a method to describe a braid. It is Theorem 2.9 of [13], where they proved the theorem for the positive words, and then discuss the general words in the next section. The notion of left-weightedness will be explained right after.

**Theorem 1.** *For any $W \in B_n$, there is a unique representation, called the* left-*canonical form, as*

$$W = \Delta^u A_1 A_2 \cdots A_p, \qquad u \in \mathbb{Z}, \ A_i \in \tilde{\Sigma}_n \setminus \{e, \Delta\},$$

(a) the fundamental braid $\Delta_4$     (b) the braid $A \in \tilde{\Sigma}_n$ corresponding to $\pi = 3124$

**Fig. 2.** The permutation braid and the fundamental braid

*where $A_i A_{i+1}$ is left-weighted for $1 \leq i \leq p - 1$.*

So we can describe a braid $W = \Delta^u A_1 A_2 \cdots A_p$ by a tuple $(u, \pi_1, \pi_2, \ldots, \pi_p)$, where the canonical factor $A_i$ corresponds to the permutation $\pi_i$. Here $p$ is called the *canonical length*, denoted by $\text{len}(W)$, of $W$. We will use this description when implementing cryptosystems. Now we explain briefly the idea of Garside, Thurston, Elrifai-Morton and Birman-Ko-Lee, following the paper of Elrifai-Morton [13].

1. Note that the relations in the group presentation $(*)$ relate two positive words with same word length. Let $B_n^+$ be the semigroup defined by the generators and relations in the presentation. Garside proved that the natural homomorphism $: B_n^+ \rightarrow B_n$ is injective [16]. Thus two positive words $P$ and $Q$ are equivalent in $B_n$ if and only if $P$ and $Q$ are equivalent in $B_n^+$.
2. For a positive word $P$, the *starting set* $S(P)$ and the *finishing set* $F(P)$ is defined by

$$S(P) = \{i \mid P = \sigma_i P' \text{ for some } P' \in B_n^+\},$$
$$F(P) = \{i \mid P = Q'\sigma_i \text{ for some } Q' \in B_n^+\}.$$

For a canonical factor $A$ corresponding to a permutation $\pi \in \Sigma_n$, $S(A) = \{i \mid \pi(i) > \pi(i+1)\}$ [14] and similarly $F(A) = \{i \mid \pi^{-1}(i) > \pi^{-1}(i+1)\}$. We note that $S(A)$ is just the descent set [33] defined in the combinatorics. So for $\Delta_4$ in Figure 2(a), $S(\Delta_4) = F(\Delta_4) = \{1, 2, 3\}$ and for $A = \sigma_1\sigma_2\sigma_1\sigma_3$ in Figure 2(b), $S(A) = \{1, 2\}$ and $F(A) = \{1, 3\}$.
3. The fundamental braid $\Delta$ has the following two properties.
   (a) For each $1 \leq i \leq n - 1$, $\Delta = \sigma_i A_i = B_i \sigma_i$ for some permutation braids $A_i$ and $B_i$.
   (b) For any $1 \leq i \leq n - 1$, $\sigma_i \Delta = \Delta \sigma_{n-i}$.
   For an arbitrary word $W$ on $\sigma_i$'s, we can replace each occurrence of $\sigma_i^{-1}$ by the formula $\sigma_i^{-1} = \Delta^{-1} B_i$ from the first property and collect $\Delta^{-1}$'s to the left by using the second property to get the expression $W = \Delta^u P$, $P \in B_n^+$.
4. For any positive word $P$, there is a unique decomposition, which is called the *left-weighted decomposition* as follows:

$$P = A_1 P_1, \qquad A_1 \in \tilde{\Sigma}_n, \ \ P_1 \in B_n^+, \ \ F(A_1) \supset S(P_1).$$

By iterating the left-weighted decomposition $P = A_1 P_1$, $P_1 = A_2 P_2$, ...,
and then collecting $\Delta$'s to the left, we have the *left-canonical form*

$$P = \Delta^u A_1 A_2 \cdots A_p, \qquad u \in \mathbb{Z}, \ A_i \in \tilde{\Sigma}_n \setminus \{e, \Delta\},$$

where $A_i A_{i+1}$ is left-weighted. (In fact, $A_i A_{i+1}$ is left-weighted for all $1 \leq i < p$ if and only if $A_i(A_{i+1} \cdots A_p)$ is left-weighted for all $1 \leq i < p$.) *This left canonical form is unique.*

5. By combining 3 and 4, we have the left-canonical form for arbitrary braids as in Theorem 1.

**Theorem 2.** *1. Let $W$ be a word on $\sigma_i$'s with word length $\ell$. Then the left-canonical form of $W$ can be computed in time $\mathcal{O}(\ell^2 n \log n)$.*

*2. Let $U = \Delta^u A_1 \cdots A_p$ and $V = \Delta^v B_1 \cdots B_q$ be the left-canonical forms of $n$-braids. Then we can compute the left-canonical form of $UV$ in time $\mathcal{O}(pqn \log n)$.*

*3. If $U = \Delta^u A_1 \cdots A_p$ is the left-canonical form of $U \in B_n$, then we can compute the left-canonical form of $U^{-1}$ in time $\mathcal{O}(pn)$.*

*Proof.* The proofs of 1 and 2 are in [14]. The left-canonical form of $U^{-1}$ is given by $U^{-1} = \Delta^{-u-p} A_p' \cdots A_1'$, where $A_i'$ is the permutation braid such that $\Delta = A_i(\Delta^{u+i} A_i' \Delta^{-u-i})$ [13]. Let $\pi_i$ and $\Omega$ be the permutations corresponding to $A_i$ and $\Delta$. Then $A_i'$ is the permutation braid corresponding to the permutation $\pi' = \Omega^{-u-i}(\pi_i^{-1}\Omega)\Omega^{u+i}$. Since $\Omega^2$ is the identity, we can compute $\pi_i'$ in $\mathcal{O}(n)$. Thus we can compute the whole left-canonical form of $U^{-1}$ in time $\mathcal{O}(pn)$ as desired.    □

In order to analyze the security against brute force attacks, we will need a lower bound for the number of $n$-braids of a given canonical length. The estimate given in the following theorem has some room to improve.

**Theorem 3.** *The number of $n$-braids of canonical length $p$ is at least $(\lfloor \frac{n-1}{2} \rfloor !)^p$.*

*Proof.* Since $\lfloor \frac{n-1}{2} \rfloor = r$ for $n = 2r + 1$ and $n = 2r + 2$, and clearly there are more $(2r + 2)$-braids than $(2r + 1)$-braids of a fixed canonical length, we may assume that $n = 2r + 1$. Consider the two subsets of the $\tilde{\Sigma}_n$.

$$\mathcal{S} = \left\{ A \in \tilde{\Sigma}_n \mid S(A) \subset \{1, 2, \ldots, r\} \text{ and } F(A) \supset \{2, 4, \ldots, 2r\} \right\}$$

$$\mathcal{T} = \left\{ A \in \tilde{\Sigma}_n \mid S(A) \subset \{2, 4, \ldots, 2r\} \text{ and } F(A) \supset \{1, 2, \ldots, r\} \right\}$$

We will show that there are injective functions from $\Sigma_r \to \mathcal{S}$ and $\Sigma_r \to \mathcal{T}$. It is easy to see that the functions are not surjective and so $|\mathcal{S}| > r!$ and $|\mathcal{T}| > r!$. Since for any $A \in \mathcal{S}$ and $B \in \mathcal{T}$, $AB$ and $BA$ are left-weighted, there are at least $(r!)^p$ $n$-braids of canonical length $p$ and so we are done.

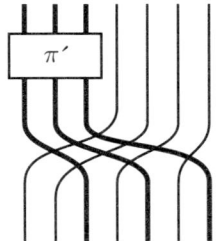                    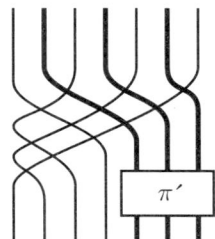

(a) The construction of element of $\mathcal{S}$    (b) The construction of element of $\mathcal{T}$

**Fig. 3.**

For a permutation $\pi' \in \Sigma_r$, we construct a canonical factor $A$ in $\mathcal{S}$ by defining the corresponding permutation $\pi \in \Sigma_{2r+1}$ by

$$\pi(i) = \begin{cases} 2\pi'(i) + 1 & \text{for } 1 \leq i \leq r, \\ 1 & \text{if } i = r + 1, \\ 2(i - r - 1) & \text{for } r + 2 \leq i \leq 2r + 1. \end{cases}$$

See Figure 3(a) for an illustration for $r = 3$. For $r + 1 \leq i \leq 2r$, $\pi(i) < \pi(i+1)$ so that $i \notin S(A)$. Thus $S(A) \subset \{1, 2, \ldots, r\}$. And since $\pi^{-1}(2i) \geq r + 1$ and $\pi^{-1}(2i+1) \leq r$, $2i \in F(A)$ for any $1 \leq i \leq r$. So $F(A) \supset \{2, 4, \ldots, 2r\}$.

Similarly, for a permutation $\pi' \in \Sigma_r$, we construct a canonical factor $A$ in $\mathcal{T}$ by defining the corresponding permutation $\pi \in \Sigma_{2r+1}$ by

$$\begin{aligned} \pi(2i-1) &= (r+2) - i & \text{for } 1 \leq i \leq r + 1, \\ \pi(2i) &= (r+1) + \pi'(i) & \text{for } 1 \leq i \leq r. \end{aligned}$$

See Figure 3(b) for an illustration for $r = 3$. Since $\pi(2i-1) = (r+2) - i \leq r + 1$ and $\pi(2i) = (r+1) + \pi'(i) > r + 1$, $(2i-1) \notin S(A)$ for $1 \leq i \leq r$. Thus $S(A) \subset \{2, 4, \ldots, 2r\}$. And since $\pi^{-1}(i) = 2r - 2i + 3 > 2r - 2i + 1 = \pi^{-1}(i+1)$ for $1 \leq i \leq r$, $F(A) \supset \{1, 2, \ldots, r\}$.                                        □

### 2.3   Hard Problems in the Braid Group

We describe some of problems in braid groups that are mathematically hard and may be interesting in cryptography.

We say that $x$ and $y$ are *conjugate* if there is an element $a$ such that $y = axa^{-1}$. And for $m < n$, $B_m$ can be considered as a subgroup of $B_n$ generated by $\sigma_1, \ldots, \sigma_{m-1}$.

1. CONJUGACY DECISION PROBLEM
   **Instance:** $(x, y) \in B_n \times B_n$.
   **Objective:** Determine whether $x$ and $y$ are conjugate or not.

2. CONJUGACY SEARCH PROBLEM
   **Instance:** $(x, y) \in B_n \times B_n$ such that $x$ and $y$ are conjugate.
   **Objective:** Find $a \in B_n$ such that $y = axa^{-1}$.

3. GENERALIZED CONJUGACY SEARCH PROBLEM
   **Instance:** $(x, y) \in B_n \times B_n$ such that $y = bxb^{-1}$ for some $b \in B_m$, $m \leq n$.
   **Objective:** Find $a \in B_m$ such that $y = axa^{-1}$.

4. CONJUGACY DECOMPOSITION PROBLEM
   **Instance:** $(x, y) \in B_n \times B_n$ such that $y = bxb^{-1}$ for some $b \in B_m$, $m < n$.
   **Objective:** Find $a', a'' \in B_m$ such that $y = a'xa''$.

5. $p$-TH ROOT PROBLEM
   **Instance:** $(y, p) \in B_n \times \mathbb{Z}$ such that $y = x^p$ for some $x \in B_n$
   **Objective:** Find $z \in B_n$ such that $y = z^p$.

6. CYCLING[1] PROBLEM
   **Instance:** $(y, r) \in B_n \times \mathbb{Z}$ such that $y = \mathbf{c}^r(x)$ for some $x \in B_n$
   **Objective:** Find $z \in B_n$ such that $y = \mathbf{c}^r(z)$.

7. MARKOV PROBLEM
   **Instance:** $y \in B_n$ such that $y$ is conjugate to a braid of the form $w\sigma_{n-1}^{\pm 1}$ for some $w \in B_{n-1}$
   **Objective:** Find $(z, x) \in B_n \times B_{n-1}$ such that $zyz^{-1} = x\sigma_{n-1}^{\pm 1}$.

The CONJUGACY DECISION PROBLEM and the CONJUGACY SEARCH PROBLEM are very important because there are many topologically important problems defined up to conjugacy. But they are so difficult that there is no known polynomial time algorithm to solve this problem. The GENERALIZED CONJUGACY SEARCH PROBLEM is a generalized version of the conjugacy problem, which has a restriction on the braid that conjugates $x$. We will use this problem to propose an one-way function in §3. The CONJUGACY DECOMPOSITION PROBLEM is trivial for $m = n$ and is easier than GENERALIZED CONJUGACY SEARCH PROBLEM. We conjecture that GENERALIZED CONJUGACY SEARCH PROBLEM is equivalent to CONJUGACY DECOMPOSITION PROBLEM for some choices of $x$. The security of the key agreement scheme proposed in §3 are in fact based on CONJUGACY DECOMPOSITION PROBLEM.

It seems that one can write down potential one-way functions from the above problems. Furthermore since we can always extract a fixed number of factors at the designated position in a canonical form of a braid, it may be possible to design potential (keyed) hash functions from a combination of the above problems.

Finally we remark that the MARKOV PROBLEM is closely related to the study of knots and links via braids. As every knot theorist dreams of a complete classification of knots and links, this problem should be hard.

---

[1] For an $n$-braid $x = \Delta^u A_1 \cdots A_p$ in the left-canonical form, the *cycling* of $x$ is defined by $\mathbf{c}(x) = \Delta^u A_2 \cdots A_p \tau^u(A_1)$, where the automorphism $\tau \colon B_n \to B_n$ is defined by $\tau(\sigma_i) = \sigma_{n-i}$ for $i = 1, \ldots, n-1$.

# 3   The Cryptosystem Using Braid Groups

In this section, we propose a one-way function based on the difficulty of the GENERALIZED CONJUGACY SEARCH PROBLEM. Also we propose a key agreement scheme and a PKC using the proposed one-way function. But we don't have a digital signature scheme based on the braid groups yet.

Note: All the braids in this section are supposed to be in the left-canonical form. For example, for $a, b \in B_n$, $ab$ means the left-canonical form of $ab$ and so it is hard to guess its factors $a$ or $b$ from $ab$.

## 3.1   Proposed One-Way Function

We consider two subgroups $LB_\ell$ and $RB_r$ of $B_{\ell+r}$. $LB_\ell$ (resp. $RB_r$) is the subgroup of $B_{\ell+r}$ consisting of braids made by braiding left $\ell$ strands(resp. right $r$ strands) among $\ell+r$ strands. Thus $LB_\ell$ is generated by $\sigma_1, \ldots, \sigma_{\ell-1}$ and $RB_r$ is generated by $\sigma_{\ell+1}, \ldots, \sigma_{\ell+r-1}$. An important role is played by the commutative property that for any $a \in LB_\ell$ and $b \in RB_r$, $ab = ba$.

Now we propose the following one-way function

$$f: LB_\ell \times B_{\ell+r} \to B_{\ell+r} \times B_{\ell+r}, \qquad f(a, x) = (axa^{-1}, x).$$

It is a one-way function because given a pair $(a, x)$, it is easy to compute $axa^{-1}$ but all the known attacks need exponential time to compute $a$ from the data $(axa^{-1}, x)$. This one-way function is precisely based on the GENERALIZED CONJUGACY SEARCH PROBLEM.

The securities of our key agreement scheme and PKC are based on the difficulty of the following problem.

[**Base Problem**]
**Instance:** The triple $(x, y_1, y_2)$ of elements in $B_{\ell+r}$ such that $y_1 = axa^{-1}$ and $y_2 = bxb^{-1}$ for some hidden $a \in LB_\ell$ and $b \in RB_r$.
**Objective:** Find $by_1b^{-1}(= ay_2a^{-1} = abxa^{-1}b^{-1})$.

We do not know whether this problem is equivalent to the GENERALIZED CONJUGACY SEARCH PROBLEM, even though the latter problem implies the former problem. The two problems seem to have the almost same complexity and this phenomenon is similar to the case of the Diffie-Hellman problem and the discrete logarithm problem.

The role of $x$ is similar to that of $g$ in the Diffie-Hellman problem to find $g^{xy}$ from $g^x$ and $g^y$. In order to make our base problem hard, $x$ must be sufficiently complicated by avoiding the "reducible" braids $x_1x_2z$ where $x_1 \in LB_\ell$, $x_2 \in RB_r$ and $z$ is a $(\ell+r)$-braid that commutes with both $LB_\ell$ and $RB_r$ as depicted in Figure 4 for $\ell = r = 3$. If $x$ were decomposed into $x_1x_2z$, then $by_1b^{-1} = (ax_1a^{-1})(bx_2b^{-1})z$ would be obtained from $y_1 = (ax_1a^{-1})x_2z$ and $y_2 = x_1(bx_2b^{-1})z$ without knowing $a$ and $b$. It is shown by Fenn-Rolfsen-Zhu in [15] that $(\ell+r)$-braids that commute with $RB_r$ (or $LB_\ell$) are of the form $x_1z$ (or $x_2z$, respectively) up to full twists $\Delta_\ell^2$ and $\Delta_r^2$ of left $\ell$ strands and right $r$

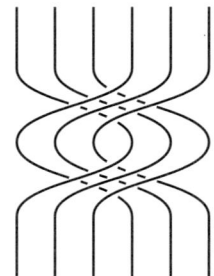

**Fig. 4.** An example of reducible braid

strands. The probability for a randomly chosen $(\ell + r)$-braid of canonical length $q$ to be reducible is small, that is, roughly $(\ell! r!/(\ell + r)!)^q$.

CONJUGACY DECOMPOSITION PROBLEM also implies the base problem since for $a', a'' \in LB_\ell$ such that $a' x a'' = y_1$, we have $a' y_2 a'' = a' b x b^{-1} a'' = b a' x a'' b^{-1} = b y_1 b^{-1}$. We note that CONJUGACY DECOMPOSITION PROBLEM is trivial if $x$ is reducible. Thus a necessary condition on $x$ for which CONJUGACY DECOMPOSITION PROBLEM becomes equivalent to GENERALIZED CONJUGACY SEARCH PROBLEM is that $x$ is not reducible. A sufficient condition on $x$ is that $x c x^{-1}$ is not in $B_m$ for each nontrivial $c \in B_m$. But we do not have a good characterization of this sufficient condition yet and further study on the choice of $x$ may be required to maintain the soundness of our base problem.

Recall the surjection $\rho \colon B_n \to \Sigma_n$ into the permutation group. In order to prevent adversaries from computing $\rho(a)$ and $\rho(b)$ by looking at $\rho(x)$, $\rho(y_1)$ and $\rho(y_2)$, $x$ should be a pure braid so that $\rho(y_1)$ and $\rho(y_2)$ as well as $\rho(x)$ are the identity.

## 3.2   Key Agreement

Now we propose a key agreement system between A(lice) and B(ob). This is the braid group version of the Diffie-Hellman key agreement system.

1. **Preparation step:** An appropriate pair of integers $(\ell, r)$ and a sufficiently complicated $(\ell + r)$-braid $x \in B_{\ell+r}$ are selected and published. The requirement to be sufficiently complicated has been discussed in §3.1.
2. **Key agreement:** Perform the following steps each time a shared key is required.
   (a) A chooses a random secret braid $a \in LB_\ell$ and sends $y_1 = a x a^{-1}$ to B.
   (b) B chooses a random secret braid $b \in RB_r$ and sends $y_2 = b x b^{-1}$ to A.
   (c) A receives $y_2$ and computes the shared key $K = a y_2 a^{-1}$.
   (d) B receives $y_1$ and computes the shared key $K = b y_1 b^{-1}$.

Since $a \in LB_\ell$ and $b \in RB_r$, $ab = ba$. Thus

$$ a y_2 a^{-1} = a(b x b^{-1}) a^{-1} = b(a x a^{-1}) b^{-1} = b y_1 b^{-1} $$

and so Alice and Bob obtain the same braid.

Since the Anshel-Anshel-Goldfeld's key agreement in [2] is also based on combinatorial groups and conjugacy problems, it seems necessary to discuss the difference between our key agreement and the Anshel-Anshel-Goldfeld's key agreement. The points of their algebraic key establishment protocol in [2] are the homomorphic property of an one-way function and the public key of multiple arguments. But our key agreement relies neither on the homomorphic property nor on the public key of multiple arguments.

The group theoretic application in [2] uses the following generalization of the usual conjugacy search problem:

Given words $t_1, \ldots, t_k$, and $at_1a^{-1}, \ldots, at_ka^{-1}$ in a group $G$, find such a word $a$ in $G$.

On the other hand our key agreement is based on another generalization of the conjugacy search problem as follows:

Given words $x$ and $axa^{-1}$ in a group $G$ and given a subgroup $H$ of $G$, find such a word $a$ in $H$.

We believe that the two generalizations are independent each other, especially for the braid group $B_n$ and its subgroup $LB_\ell$ under our current mathematical knowledge.

It should be also noted that the trapdoors of two key agreements are distinct. Our scheme uses the commutativity between two subgroups $LB_\ell$ and $RB_r$. On the other hand the Anshel-Anshel-Goldfeld's scheme uses the homomorphic property of conjugations, that is, $(asa^{-1})(ata^{-1}) = asta^{-1}$.

### 3.3   Public-Key Cryptosystem

By using the key agreement system in §3.2, we construct a new PKC. Let $H: B_{\ell+r} \to \{0,1\}^k$ be an ideal hash function from the braid group to the message space.

1. **Key generation:**
   (a) Choose a sufficiently complicated $(\ell + r)$-braid $x \in B_{\ell+r}$.
   (b) Choose a braid $a \in LB_\ell$.
   (c) Public key is $(x, y)$, where $y = axa^{-1}$; Private key is $a$.
2. **Encryption:** Given a message $m \in \{0,1\}^k$ and the public key $(x, y)$.
   (a) Choose a braid $b \in RB_r$ at random.
   (b) Ciphertext is $(c, d)$, where $c = bxb^{-1}$ and $d = H(byb^{-1}) \oplus m$.
3. **Decryption:** Given a ciphertext $(c, d)$ and private key $a$, compute $m = H(aca^{-1}) \oplus d$.

Since $a$ and $b$ commute, $aca^{-1} = abxb^{-1}a^{-1} = baxa^{-1}b^{-1} = byb^{-1}$. So $H(aca^{-1}) \oplus d = H(byb^{-1}) \oplus H(byb^{-1}) \oplus m = m$ and the decryption recovers the original braid $m$.

Because $H$ is an ideal hash function, our PKC is semantically secure relative to the decisional version of our base problem; if the adversary can compute some information of the message from the public key and the ciphertext, (s)he can also compute some information of $byb^{-1} = abxa^{-1}b^{-1}$ from $x$, $axa^{-1}$ and $bxb^{-1}$.

We hope that one can make a semantically secure PKC using the proposed one-way function without assuming the hash function to be an ideal hash function.

# 4  The Theoretic Operating Characteristics and Implementation

In this section, we discuss the theoretic operating characteristics of our PKC and the security/message length parameters for future implementations. Because our PKC has not been fully implemented yet as a computer program, we can not compare its speed with other PKC's. But we can report the speed of a conversion algorithm into left canonical forms that is the essential part of our PKC

Recall that our PKC uses three braids $x \in B_{\ell+r}$, $a \in LB_\ell$ and $b \in RB_r$, and the ciphertext is $(bxb^{-1}, H(abxa^{-1}b^{-1}) \oplus m)$. When we work with braids, we should consider two parameters, the braid index and the canonical length. For simplicity, we assume that the braid indexes in our PKC are $\ell = r = \frac{n}{2}$ and the canonical lengths are $\text{len}(x) = \text{len}(a) = \text{len}(b) = p$. The followings are the discussions about the operating characteristics of our PKC, which is summarized in Table 1.

1. An $n$-permutation can be represented by an integer $0 \leq N < n!$. Since $n! \sim \exp(n \log n)$, a braid with $p$ canonical factors can be represented by a bit string of size $pn \log n$.
2. For braids $y_1, y_2 \in B_n$, $\text{len}(y_1 y_2) \leq \text{len}(y_1) + \text{len}(y_2)$ and for $y_1 \in LB_\ell$, $y_2 \in RB_r$, $\text{len}(y_1 y_2) = \max\{\text{len}(y_1), \text{len}(y_2)\}$. So $\text{len}(bxb^{-1})$ and $\text{len}(abxa^{-1}b^{-1})$ are at most $3p$. For generic choices of $a, b$, and $x$, they are no less than $2p$. Thus we assume that $\text{len}(bxb^{-1})$ and $\text{len}(abxa^{-1}b^{-1})$ are between $2p$ and $3p$.
3. The size of the private key $a$ is $p\ell \log \ell \sim p\frac{n}{2} \log \frac{n}{2} \sim \frac{1}{2}pn \log n$.
4. The size of the public key $bxb^{-1}$ is at most $3pn \log n$.
5. By Theorem 3, the number of $n$ braids with $2p$ canonical factors is at least the exponential of

$$\log\left(\lfloor \tfrac{n-1}{2} \rfloor!\right)^{2p} = 2p \log\left(\lfloor \tfrac{n-1}{2} \rfloor!\right) \sim 2p \log\left(\tfrac{n}{2}!\right) \sim 2p \cdot \tfrac{n}{2} \cdot \log \tfrac{n}{2} \sim pn \log n.$$

Thus we may let the bit length of $H(abxa^{-1}b^{-1})$ equal $pn \log n$ and so the message length is also $pn \log n$. Since the bit size of $bxb^{-1}$ is at most $3pn \log n$, the size of ciphertext is at most $3pn \log n + pn \log n = 4pn \log n$. Hence the message expansion is less than 4-1.
6. As noted earlier, the encryption/decryption speed is $\mathcal{O}(p^2 n \log n)$.
7. The hardness of the brute force attack to compute $a$ from $axa^{-1}$, or equivalently to compute $b$ from $bxb^{-1}$, is proportional to $(\ell!)^p = (\frac{n}{2}!)^p \sim \exp(\frac{1}{2}pn \log n)$.

| Plaintext block | $pn \log n$ bits |
| --- | --- |
| Ciphertext block | $4pn \log n$ bits |
| Encryption speed | $\mathcal{O}(p^2 n \log n)$ operation |
| Decryption speed | $\mathcal{O}(p^2 n \log n)$ operation |
| Message expansion | 4-1 |
| Private key length | $\frac{1}{2}pn \log n$ bits |
| Public key length | $3pn \log n$ bits |
| Hardness of brute force attack | $(\frac{n}{2}!)^p \sim \exp(\frac{1}{2}pn \log n)$ |

**Table 1.** The operating characteristics of our PKC

| $s$ | $p$ | Braid index $n$ | Expected message length $k$ | Ellapsed time $t$ (sec) | Kbits/(sec) | Hardness of brute force attack |
| --- | --- | --- | --- | --- | --- | --- |
| 11 | 5 | 50 | 1071 | 0.0112 | 93.4 | 251 |
| 11 | 5 | 70 | 1662 | 0.0210 | 77.3 | 399 |
| 11 | 5 | 90 | 2294 | 0.0344 | 65.2 | 559 |
| 17 | 7 | 50 | 1499 | 0.0173 | 84.6 | 418 |
| 17 | 7 | 70 | 2327 | 0.0325 | 69.9 | 665 |
| 17 | 7 | 90 | 3212 | 0.0532 | 59.0 | 931 |
| 32 | 12 | 50 | 2570 | 0.0326 | 77.0 | 837 |
| 32 | 12 | 70 | 3989 | 0.0611 | 63.8 | 1329 |
| 32 | 12 | 90 | 5507 | 0.1037 | 51.9 | 1863 |

**Table 2.** The performance of the algorithm converting into the left canonical forms.

Both the security level and the message length are affected at the same extent by $p$ and $n \log n$, but the speed is quadratic in $p$ and linear in $n \log n$. Thus it is better to increase $n$ rather than $p$ in order to increase the security level.

Table 2 shows speed of the canonical form algorithm in the braid group. It is the total ellapsed time of our C-program which takes a pair of integers $(n, s)$, and then generates $s$ random canonical factors $A_1, \ldots, A_s$, and then computes the left-canonical form of $A_1 \cdots A_s$ in Pentium III 450MHz. The table shows that the multiplication in braid groups is efficient.

The C-program inputs pairs $(n, s)$ and outputs the ellasped time $t$. The other entries in Table 2 are computed from $(n, s)$ and $t$ as follows:

1. Let $p = \lceil \frac{s}{3} \rceil + 1$ be the canonical length of the fixed braid $x$ of our PKC. Let $q = \frac{n-p}{2}$. Then we consider the PKC, where $\text{len}(x) = p$, $\text{len}(a) = \text{len}(b) = q$.
2. The expected message length $k$ is computed by $p \log n!$.
3. Kbits/(sec) is computed by $k/(2^{10}t)$.
4. Hardness of brute force attack is computed by $\log(\frac{n}{2}!)^q$.

# 5  Security Analysis

In this section, we analyze the security of the proposed encryption scheme.

## 5.1  Similarity with ElGamal Scheme

Our PKC is similar to the ElGamal PKC in design and it has the following properties.

1. The problem of breaking our PKC is equivalent to solving the base problem, as breaking the ElGamal PKC is equivalent to solving the Diffie-Hellman problem. In the proposed scheme, the ciphertext is

$$(c, d) = (bxb^{-1}, H(abxa^{-1}b^{-1}) \oplus m)$$

   and decrypting the ciphertext into a plaintext $m$ is equivalent to knowing $abxa^{-1}b^{-1}$.

2. Like any other probabilistic PKC's, it is critical to use different key $b$ for each session: If the same session key $b$ is used to encrypt both of $m_1$ and $m_2$ whose corresponding ciphertexts are $(c_1, d_1)$ and $(c_2, d_2)$, then $m_2$ can be easily computed from $(m_1, d_1, d_2)$ because $H(byb^{-1}) = m_1 \oplus d_1 = m_2 \oplus d_2$.

## 5.2  Brute Force Attack

A possible brute force attack is to compute $a$ from $axa^{-1}$ or $b$ from $bxb^{-1}$, which is just an attack to GENERALIZED CONJUGACY SEARCH PROBLEM. Assume that we are given a pair $(x, y)$ of braids in $B_{\ell+r}$, such that $y = axa^{-1}$ for some $a \in LB_\ell$. The braid $a$ can be chosen from an infinite group $LB_\ell$ in theory. But in a practical system, the adversary can generate all braids $a = \Delta^u A_1 \ldots A_p$ in the canonical form with some reasonable bound for $p$ and check whether $y = axa^{-1}$ holds. The necessary number is at least $(\frac{\ell-1}{2}!)^p$ by Theorem 3. If $\ell = 45$ and $p = 2$, then $(\frac{\ell-1}{2}!)^p > 2^{139}$, which shows that the brute force attack is of no use.

We note that there might be another $a' \in LB_\ell$ such that $y = a'xa'^{-1}$. Then $a^{-1}a'$ must be a member of the centralizer $C(x)$ of $x$. For a generic $x$ and a fixed canonical length, the probability for a braid in $LB_\ell$ to be a member of $C(x) \cap LB_\ell$ seems negligible, that is, it is hard to find such an $a'$ different from $a$.

Another possible brute force attack is to find $a' \in LB_\ell$ such that $x^{-1}a^{-1}a'x \in LB_\ell$, which is an attack to CONJUGACY DECOMPOSITION PROBLEM. As we conjectured earlier, there are choices of $x$ so that CONJUGACY DECOMPOSITION PROBLEM implies GENERALIZED CONJUGACY SEARCH PROBLEM, that is, such an $a'$ must equals to $a$. Thus we need to concern only about an attack to GENERALIZED CONJUGACY SEARCH PROBLEM.

## 5.3   Attack Using the Super Summit Set

The adversary may try to use a mathematical solution to the conjugacy problem by Garside [16], Thurston [14], Elrifai-Morton [13] and Birman-Ko-Lee [6]. But the known algorithms find an element $a \in B_{\ell+r}$, not in $LB_\ell$. Hence the attack using the super summit set will not be successful.

# 6   Further Study

1. We think that further primitives and cryptosystems can be found by using hard problems in the braid groups. For example, a new digital signature scheme is waiting for our challenge.
2. GENERALIZED CONJUGACY SEARCH PROBLEM implies CONJUGACY DE-COMPOSITION PROBLEM that in turn implies the base problem in §3.1. We would like to know what choice of $x$ makes these three problems equivalent. If this question is too challenging, it is nice to know a practical sufficient condition on $x$ that makes the first two problems equivalent. It seems hard to prove directly that CONJUGACY DECOMPOSITION PROBLEM or the base problem are intractible. On the other hand, GENERALIZED CONJUGACY SEARCH PROBLEM seems mathematically more interesting and so it could attract more research.
3. We may try to use other groups with an one-way function based on the conjugacy problem and so on. To support our ideas, the group must have the following properties.
   - The word problem should be solved by a fast algorithm. It would be much better if the word problem is solved by a fast algorithm which computes a canonical form. For example, the automatic groups may be good candidates [14,25].
   - The conjugacy problem must be hard. The permutation group does not satisfy this requirement.
   - It should be easy to digitize the group element.
   Our idea can be applied to matrix groups. In particular, for an $n$-braid, we can compute its image of Burau representation [5], which is an $n \times n$ matrix in $GL_n(\mathbb{Z}[t, t^{-1}])$. One might expect that the conjugacy problem in this matrix group is easier than in the braid groups. But it does not seem so. And it is not easy to encode the message into a matrix and vice versa.

**Acknowledgement**

We wish to thank Dan Boneh, Mike Brenner and anonymous referees for their valuable comments and suggestions. The first two authors also wish to thank Joan Birman for her kind introduction of the braid theory and her continuing encouragement. The first author was supported in part by the National Research Laboratory Grant from the Ministry of Science and Technology in the program year of 1999.

# References

1. I. Anshel and M. Anshel, *From the Post-Markov theorem through decision problems to public-key cryptography*, Amer. Math. Monthly **100** (1993), no. 9, 835–844.
2. I. Anshel, M. Anshel and D. Goldfeld, *An algebraic method for public-key cryptography*, Mathematical Research Letters **6** (1999) 287–291.
3. E. Artin, *Theory of braids*, Annals of Math. **48** (1947), 101–126.
4. C. H. Bennet and G. Brassard, *Quantum cryptography: Public key distribution and coin tossing*, Proc. IEEE Int. Conf. Computers, Systems and Signal Processing (Bangalore, India, 1984), 175–179.
5. J. S. Birman, *Braids, links and mapping class groups*, Annals of Math. Study, no. **82**, Princeton University Press (1974).
6. J. S. Birman, K. H. Ko and S. J. Lee, *A new approach to the word and conjugacy problems in the braid groups*, Advances in Math. **139** (1998), 322-353.
7. D. Boneh, *Twenty years of attacks on the RSA cryptosystem*, Notices Amer. Math. Soc. **46** (1999), 203–213.
8. G. Brassard, *A note on the complexity of cryptography*, IEEE Transactions on Information Theory **25** (1979), 232-233.
9. E. F. Brickell, *Breaking iterated knapsacks*, Advances in Cryptology, Proceedings of Crypto '84, Lecture Notes in Computer Science **196**, ed. G. R. Blakley and D. Chaum, Springer-Verlag (1985), 342–358.
10. P. Dehornoy, *A fast method for comparing braids*, Advances in Math. **125** (1997), 200-235.
11. W. Diffie and M. E. Hellman, *New directions in cryptography*, IEEE Transactions on Informaton Theory **22** (1976), 644–654.
12. T. ElGamal, *A public key cryptosystem and a signature scheme based on discrete logarithms*, IEEE Transactions on Information Theory **31** (1985), 469–472.
13. E. A. Elrifai and H. R. Morton, *Algorithms for positive braids*, Quart. J. Math. Oxford **45** (1994), no. 2, 479–497.
14. D. Epstein, J. Cannon, D. Holt, S. Levy, M. Paterson and W. Thurston, *Word processing in groups*, Jones & Bartlett, 1992.
15. R. Fenn, D. Rolfsen and J. Zhu *Centralisers in the braid group and singular braid monoid*, Enseign. Math. (2) **42** (1996), no. 1-2, 75–96.
16. F. A. Garside, *The braid group and other groups*, Quart. J. Math. Oxford **20** (1969), no. 78, 235–254.
17. M. Garzon and Y. Zalcstein, *The complexity of Grigorchuk groups with application to cryptography*, Theoretical Computer Sciences **88** (1991) 83–98.
18. O. Goldreich, S. Goldwasser and S. Halevi, *Public-key cryptosystems from lattice reduction problems*, Advances in Cryptology, Proceedings of Crypto '97, Lecture Notes in Computer Science **1294**, ed. B. Kaliski, Springer-Verlag (1997), 112–131.
19. E. S. Kang, K. H. Ko and S. J. Lee, *Band-generator presentation for the 4-braid group*, Topology Appl. **78** (1997), 39-60.
20. K. Komaya, U. Maurer, T. Okamoto and S. Vanston, *New public-key schemes bases on elliptic curves over the ring* $\mathbf{Z}_n$, Advances in Cryptology, Proceedings of Crypto '91, Lecture Notes in Computer Science **576**, ed. J. Feigenbaum, Springer-Verlag (1992), 252–266
21. N. Koblitz, *Algebraic aspects of cryptography*, Algorithms and Computations in Mathematics **3** (1998) Springer-Verlag, Berlin.
22. J. C. Lagarias, *Knapsack public key cryptosystems and Diophantine approximation*, Advances in Cryptology: Proceedings of Crypto '83, ed. by D. Chaum, Plenum Publishing (1984), 3–24.

23. K. McCurley, *A key distribution system equivalent to factoring*, Journal of Cryptology **1** (1988), 95–105.
24. R. C. Merkle and M. E. Hellman, *Hiding information and signatures in trapdoor knapsacks*, IEEE Transactions on Information Theory **24** (1978), 525–530.
25. L. Mosher, *Mapping class groups are automatic*, Ann. Math. **142** (1995), 303–384.
26. A. M. Odlyzko, *The rise and fall of knapsack cryptosystems*, Cryptology and Computational Number Theory, Proc. Symp. App. Math. **42** (1990), 75–88.
27. M. S. Paterson and A. A. Rasborov, *The set of minimal braids is co-NP-complete*, J. Algorithms. **12** (1991), 393–408.
28. M. O. Rabin, *Digitized signatures and public-key functions as intractible as factorization*, MIT Laboratory for Computer Science Technical Report, LCS/TR-212 (1979).
29. R. L. Rivest, A. Shamir and L. Adleman, *A method for obtaining digital signatures and public key cryptosystems*, Communications of the ACM **21** (1978), 120–126.
30. A. Shamir, *A polynomial time algorithm for breaking the basis Merkle-Hellman cryptosystem*, Advances in Cryptology: Proceedings of Crypto '82, ed. by D. Chaum et al., Plenum Publishing (1983), 279–288.
31. R. Siromoney and L. Mathew, *A public key cryptosystem based on Lyndon words*, Information Proceeding Letters **35** (1990) 33-36.
32. P. Smith and M. Lennon, *LUC: A new public key system*, Proceedings of the IFIP TC11 Ninth International Conference on Information Security, ed. E. Dougall, IFIP/Sec 93, 103–117, North-Holland, 1993.
33. R. P. Stanley, *Enumerative combinatorics*, Wadsworth and Brooks/Cole, 1986.
34. Y. Tsiounis and M. Yung, *On the security of Elgamal based encryption*, In PKC '98, Lecture Notes in Computer Science **1431**, Springer-Verlag (1998), 117–134.
35. S. Vaudenay, *Cryptanalysis of the Chor-Rivest Cryptosystem*, Advances in Cryptology: Proceedings of Crypto '98, Lecture Notes in Computer Science **1462**, ed. Krawczyk, Springer-Verlag (1998), 243–256.
36. H. Williams, *Some public-key crypto-funtions as intractible as factorization*, Advances in Cryptology, Proceedings of Crypto '84, Lecture Notes in Computer Science **196**, ed. G. R. Blakley and D. Chaum, Springer-Verlag (1985), 66–70.
37. N. R. Wagner and M. R. Magyarik, *A public-key cryptosystem based on the word problem*, Advances in Cryptology, Proceedings of Crypto '84, Lecture Notes in Computer Science **196**, ed. G. R. Blakley and D. Chaum, Springer-Verlag (1985), 19–36.

# Key Recovery and Forgery Attacks on the MacDES MAC Algorithm

Don Coppersmith[1], Lars R. Knudsen[2], and Chris J. Mitchell[3]

[1] IBM Research, T.J. Watson Research Center, Yorktown Heights, NY 10598, USA
copper@watson.ibm.com
[2] Department of Informatics, University of Bergen, N-5020, Bergen, Norway
lars.knudsen@ii.uib.no, http://www.ii.uib.no/~larsr
[3] Information Security Group, Royal Holloway, University of London, Egham, Surrey
TW20 0EX, UK
c.mitchell@rhbnc.ac.uk, http://isg.rhbnc.ac.uk/cjm

**Abstract.** We describe a series of new attacks on a CBC-MAC algorithm due to Knudsen and Preneel including two key recovery attacks and a forgery attack. Unlike previous attacks, these techniques will work when the MAC calculation involves prefixing the data to be MACed with a 'length block'. These attack methods provide new (tighter) upper bounds on the level of security offered by the MacDES technique.

**Key words.** Message Authentication Codes. Cryptanalysis. CBC-MAC.

## 1   Introduction

CBC-MACs, i.e. Message Authentication Codes (MACs) based on a block cipher in Cipher Block Chaining (CBC) mode, have been in wide use for many years for protecting the integrity and origin of messages. A variety of minor modifications to the 'basic' CBC-MAC have been devised and adopted over the years, in response to various cryptanalytic attacks (for a survey see [7]). The latest version of the international standard for CBC-MACs, ISO/IEC 9797–1, [4], which was recently published, contains a total of six different MAC algorithms.

This paper is concerned with one of these algorithms, namely MAC Algorithm 4. This algorithm has only recently been added to the draft international standard, and was intended to offer a higher degree of security than previous schemes at a comparable computational cost. It was originally proposed by Knudsen and Preneel, [6] and, when used with the DES block cipher, was given the name 'MacDES'.

M. Bellare (Ed.): CRYPTO 2000, LNCS 1880, pp. 184–196, 2000.
© Springer-Verlag Berlin Heidelberg 2000

Some key recovery attacks against this scheme have recently been described, [3], but these do not work when 'Padding method 3' is used, which involves prefixing the data to be MACed with a length block. The key recovery and forgery attacks described below are designed specifically to work in this case.

## 2    Preliminaries

MAC algorithm 4 uses three block cipher keys, $K$, $K'$ and $K''$, where either $K''$ is derived from $K'$, or $K'$ and $K''$ are both derived from a single key. However, for the attacks here we make no assumptions about how $K'$ and $K''$ are related. We assume that the block cipher uses $k$-bit keys. We denote the block cipher encryption operation by $Y = e_K(X)$, where $Y$ is the $n$-bit ciphertext block corresponding to the $n$-bit plaintext block $X$, and $K$ is the $k$-bit key. We denote the corresponding decryption operation by $X = d_K(Y)$.

The MAC is computed on a data string by first padding the data string so that it contains an integer multiple of $n$ bits, and then breaking it into a series of $n$-bit blocks. If the $n$-bit blocks derived from the padded data string are $D_1, D_2, \ldots, D_q$, then the MAC computation is as follows.

$$
\begin{aligned}
H_1 &= e_{K''}(e_K(D_1)), \\
H_i &= e_K(D_i \oplus H_{i-1}), \quad (2 \le i \le q-1), \quad \text{and} \\
M &= e_{K'}(e_K(D_q \oplus H_{q-1})),
\end{aligned}
$$

for some $H_1, H_2, \ldots, H_{q-1}$. Finally, $M$ is truncated as necessary to form the $m$-bit MAC.

ISO/IEC FDIS 9797–1 provides three different padding methods. Padding Method 1 simply involves adding between 0 and $n-1$ zeros, as necessary, to the end of the data string. Padding Method 2 involves the addition of a single 1 bit at the end of the data string followed by between 0 and $n-1$ zeros. Padding Method 3 involves prefixing the data string with an $n$-bit block encoding the bit length of the data string, with the end of the data string padded as in Padding Method 1.

When using one of the six MAC algorithms from ISO/IEC FDIS 9797–1, it is necessary to choose one of the three padding methods, and the degree of truncation to be employed. We consider the case where Padding Method 3 is used, and where there is no truncation (as already noted, the case where either of the other two Padding Methods is used is dealt with in [3]). Hence, given that the block cipher in use has an $n$-bit block length, the MAC has $m = n$ bits. E.g., in the case of DES we have $m = n = 64$ and $k = 56$.

Finally note that we assume that all MACs are computed and verified using one particular triple of keys $(K, K', K'')$. The two key recovery attacks described below are designed to discover these keys. The forgery attack also described in this paper enables the observer of messages and corresponding MACs to obtain a new (valid) message/MAC pair which had not been observed. Of course, in general, a successful key recovery attack enables arbitrary numbers of forgeries to be constructed in a trivial way.

# 3    A Key Recovery Attack

We start by describing a chosen plaintext attack for key recovery which is more efficient than any previously known attacks on this MAC scheme. In a later section we describe a still more efficient key recovery attack.

## 3.1    Outline of Attack

The attack operates in two stages. In stage 1 we find a pair of $n$-bit block-pairs: $(D_1, D_2)$ and $(D_1', D_2')$, which should be thought of as the first pair of blocks of longer padded messages, with the property that the 'partial MACs' for these two pairs are equal, i.e. so that if

$$H_1 = e_{K''}(e_K(D_1)),$$
$$H_2 = e_K(D_2 \oplus H_1),$$
$$H_1' = e_{K''}(e_K(D_1')), \quad \text{and}$$
$$H_2' = e_K(D_2' \oplus H_1'),$$

then $H_2 = H_2'$. This is what is usually referred to as an 'internal collision', although we call this particular case a 'hidden internal collision' since it will not be evident from a complete MAC calculation.

Given that $D_1$ and $D_1'$ are to be thought of as the first blocks of padded messages (and given we are assuming Padding Method 3 is in use), $D_1$ and $D_1'$ will be encodings of the length of the unpadded messages. The attack relies on $D_1'$ being an encoding of a message for which the padded version has $q+1$ blocks ($q \geq 4$) and $D_1$ being an encoding of a message for which the padded version has $q$ blocks.

In stage 2 we will use this pair of block-pairs to launch an attack which is essentially the same as Attack 1 of [3].

## 3.2    Stage 1 — Finding the Hidden Internal Collision

We first choose $D_1$ and $D_1'$. Typically one might choose $D_1$ to be an encoding of $3n$ and $D_1'$ to be an encoding of $4n$, which will mean that $D_1$ will be the first block of a 4-block padded message and $D_1'$ will be the first block of a 5-block padded message. For the purposes of the discussion here we suppose that $D_1$ encodes a bit-length resulting in a $q$-block padded message ($q \geq 4$).

The attacker chooses (arbitrary) values for $n$-bit blocks labeled $D_2$ and $D_5, D_6, \ldots, D_q$. The attacker then generates $2^{n/2}$ messages which, when padded using Padding Method 3, will have the form

$$D_1, D_2, X, Y, D_5, D_6, \ldots, D_q$$

where $X$ and $Y$ are arbitrary $n$-bit blocks, and, by some means, obtains the MACs for these messages. By routine probabilistic arguments (called the 'birthday attack', see [7]), there is a good chance that two of the messages will have the same MAC.

Suppose the two pairs of blocks $(X, Y)$ involved are $(X_1, Y_1)$ and $(X_2, Y_2)$. Then if

$$H_1 = e_{K''}(e_K(D_1)),$$
$$H_2 = e_K(D_2 \oplus H_1),$$
$$H_3 = e_K(X_1 \oplus H_2),$$
$$H_4 = e_K(Y_1 \oplus H_3),$$
$$H_3' = e_K(X_2 \oplus H_2), \quad \text{and}$$
$$H_4' = e_K(Y_2 \oplus H_3'),$$

we know that $H_4 = H_4'$. We call the two pairs $(X_1, Y_1)$ and $(X_2, Y_2)$ a 'diagnostic pair'. These can be used to find the desired hidden internal collision.

The attacker now constructs $2^n$ message pairs. Each pair will, when padded using Padding Method 3, have the form:

$$D_1', W, X_1, Y_1, D_5', D_6', \ldots, D_{q+1}'$$

and

$$D_1', W, X_2, Y_2, D_5', D_6', \ldots, D_{q+1}'$$

where $W$ varies over all possible $n$-bit blocks, $(X_1, Y_1)$ and $(X_2, Y_2)$ are as above, and $D_5', D_6', \ldots, D_{q+1}'$ are arbitrary (if desired, they could be different for each message pair). The attacker now, by some means, discovers whether or not the two messages within each pair have the same MAC; this will typically require $2^n$ chosen MACs, together with $2^n$ MAC verifications.

First consider the case where the 'partial MAC' from the block-pair $(D_1', W)$ is the same as the partial MAC from the block pair $(D_1, D_2)$. That is, if $H_2$ is as above, and if

$$H_1' = e_{K''}(e_K(D_1')), \quad \text{and}$$
$$H_2' = e_K(W \oplus H_1'),$$

then suppose that $H_2 = H_2'$. Note that there will always exist a value $W$ giving this property, since as $W \oplus H_1'$ ranges over all possible $n$-bit blocks, then so will $H_2'$. But, given the discussion above regarding the diagnostic pairs $(X_1, Y_1)$ and $(X_2, Y_2)$, this immediately means that the pair of messages involving this value of $W$ will yield the same MACs as each other.

Second consider the pairs of messages for all the other $2^n - 1$ values of $W$. In these cases we know that $H_2 \neq H_2'$. There will be at least one 'false positive', namely when $H_2' = H_2 \oplus X_1 \oplus X_2$. For the remaining cases, assuming that the block cipher behaves in a random way, the probability of the MACs from the message pair being identical is $2^{-n}$.

Hence, as the search proceeds through the entire set of message pairs, we would expect approximately three 'positives' — one corresponding to the case we desire, i.e. where $H_2 = H_2'$, one where $H_2' = H_2 \oplus X_1 \oplus X_2$, and one random 'false positive'. If a second diagnostic pair is available then this can be used to immediately rule out any false positives. If not then we can proceed to stage 2 with all the 'positives' from stage 1, and all but the genuine positive will yield inconsistent results.

Note that the cost of this part of the attack is $2^{n/2}$ 'chosen MACs' (to find the diagnostic pair), and $2^n$ chosen MAC calculations and $2^n$ MAC verifications to find the hidden internal collision.

## 3.3  Stage 2 — Recovering the Key

From the previous stage we have a pair of $n$-bit block-pairs: $(D_1, D_2)$ and $(D'_1, D'_2)$, with the property that the 'partial MACs' for these two pairs are equal, i.e. so that if

$$H_1 = e_{K''}(e_K(D_1)),$$
$$H_2 = e_K(D_2 \oplus H_1),$$
$$H'_1 = e_{K''}(e_K(D'_1)), \quad \text{and}$$
$$H'_2 = e_K(D'_2 \oplus H'_1),$$

then $H_2 = H'_2$. In addition $D'_1$ encodes the length of a message, which, when padded, contains $(q+1)$ blocks, and $D_1$ encodes the length of a message, which, when padded, contains $q$ blocks.

The attacker now, by some means, obtains the MAC for a set of $2^{n/2}$ padded messages of the form

$$D'_1, D'_2, E_3, E_4, \ldots, E_{q+1}$$

where $E_3, E_4, \ldots, E_{q+1}$ are arbitrary. By the usual 'birthday attack' arguments, there is a good chance that two of the messages will have the same MAC. Suppose the two padded strings $D'_1, D'_2, E_3, E_4, \ldots, E_{q+1}$ and $D'_1, D'_2, E'_3, E'_4, \ldots, E'_{q+1}$ yield the same MAC, and suppose that the common MAC is $M$. Note that the cost of this part of the attack is $2^{n/2}$ 'chosen MACs'. (Note also that we need to ensure that $E_{q+1} \neq E'_{q+1}$).

Next submit two chosen padded strings for MACing, namely

$$D_1, D_2, E_3, E_4, \ldots, E_q$$

and

$$D_1, D_2, E'_3, E'_4, \ldots, E'_q$$

namely the strings one obtains by deleting the last block from each of the above two messages and replacing the first two blocks by $D_1$ and $D_2$ (these remain 'valid' padded messages because $D_1$ encodes the length of a message which, when padded, contain $q$ blocks). If we suppose that the MACs obtained are $M'$ and $M''$ respectively, then we know immediately that

$$d_{K'}(M') \oplus E_{q+1} = d_K(d_{K'}(M)) = d_{K'}(M'') \oplus E'_{q+1}$$

since $(D_1, D_2)$ and $(D'_1, D'_2)$ yield the same 'partial MAC'.

Now run through all possibilities $L$ for the unknown key $K'$, and set $x(L) = d_L(M')$ and $y(L) = d_L(M'')$. For the correct guess $L = K'$ we will have $x(L) = d_{K'}(M')$ and $y(L) = d_{K'}(M'')$, and hence $E_{q+1} \oplus x(L) = E'_{q+1} \oplus y(L)$. This will hold for $L = K'$ and probably not for any other value of $L$, given that $k < n$ (if $k \geq n$ then either a second 'collision' or a larger brute force search will probably be required).

Having recovered $K'$, we do an exhaustive search for $K$ using the relation $d_{K'}(M') \oplus E_{q+1} = d_K(d_{K'}(M))$ (which requires $2^k$ block cipher encryptions). Finally we can recover $K''$ by exhaustive search on any known text/MAC pair, e.g. from the set of $2^{n/2}$, which again will require $2^k$ block cipher encryptions.

### 3.4   Complexity of the Attack

We start by introducing a simple way of quantifying the effectiveness of an attack. Following the approach used in [4], we use a four-tuple $[a, b, c, d]$ to specify the size of the resources needed by the attacker, where

- $a$ denotes the number of off-line block cipher encipherments (or decipherments),
- $b$ denotes the number of known data string/MAC pairs,
- $c$ denotes the number of chosen data string/MAC pairs, and
- $d$ denotes the number of on-line MAC verifications.

The reason for distinguishing between the numbers $c$ and $d$ is that, in some environments, it may be easier for the attacker to obtain MAC verifications (i.e. to submit a data string/MAC pair and receive an answer indicating whether or not the MAC is valid) than to obtain the genuine MAC value for a chosen message.

Using this notation, the complexity of Stage 1 of the attack is $[0, 0, 2^n, 2^n]$ and the complexity of stage 2 is $[2^{k+2}, 0, 2^{n/2}, 0]$ (note that we ignore lower order terms). Hence the overall attack complexity is $[2^{k+2}, 0, 2^n, 2^n]$, i.e. in the case of DES the attack complexity is $[2^{58}, 0, 2^{64}, 2^{64}]$.

This is sufficiently high to rule out most practically conceivable attacks. However it is substantially less than the complexity of the best previously known attack as given in [6], which was $[2^{89}, 0, 2^{65}, 2^{55}]$.

## 4   A More Efficient Key Recovery Attack

We now present a second key recovery attack which is considerably more efficient than the attack just described. The attack is in two stages. The second stage is identical to the second stage of the first attack; the improvement is in the first stage. We find a pair of $n$-bit blocks $(D_1, D'_1)$ with the following properties.

- $D'_1$ encodes the length of a message which, when padded, contains $q + 1$ blocks.
- $D_1$ encodes the length of a message which, when padded, contains $q$ blocks.
- $e_{K''}(e_K(D_1)) \oplus e_{K''}(e_K(D'_1)) = V$, for some *known* $n$-bit block $V$.

Then choosing $D_2' = D_2 \oplus V$ yields a pair of $n$-bit block-pairs $(D_1, D_2)$ and $(D_1', D_2')$ for which the partial MACs are equal.

For fixed values $D_1, D_4, D_5, \ldots, D_q$, a total of $2^{n/2}$ different values of $X$, and a set of $t$ different values of $Y$, by some means obtain the MACs of the padded messages $(D_1, X, Y, D_4, D_5, \ldots, D_q)$. Choose the $2^{n/2}$ values of $X$ to cover all the $n$-bit blocks with most significant $n/2$ bits set to zero (for simplicity we assume $n$ is even). If $t$ is sufficiently large, then for most values of $X$ there will exist at least one tuple $(Y, X', Y')$ such that

$$\mathrm{MAC}(D_1, X, Y, D_4, D_5, \ldots, D_q) = \mathrm{MAC}(D_1, X', Y', D_4, D_5, \ldots, D_q).$$

In such a case we know that

$$e_K(X \oplus e_{K''}(e_K(D_1))) \oplus e_K(X' \oplus e_{K''}(e_K(D_1))) = Y \oplus Y'. \qquad (1)$$

If the $Y$ values are fixed for every value of $X$, each such match results in an additional match, since a match for messages with blocks $X, Y$ and $X', Y'$ also gives a match for messages with blocks $X, Y'$ and $X', Y$. To avoid this (it doesn't help us) for each value of $X$ we choose the $Y$ values randomly depending on $X$.

Now consider the graph $G$ whose vertices are the messages $X$, and with an edge between $X$ and $X'$ when a relationship of the type (1) is known. This graph has $2^{n/2}$ vertices and a number of edges dependent on $t$. With $t = 2^{n/4}t'$ we have a total of $T = 2^{3n/4}t'$ messages and about $T^2/2 = 2^{3n/2}t'^2/2$ pairs of messages. Assuming that the underlying block cipher behaves randomly this results in about $2^{n/2}t'^2/2$ pairs of messages with colliding MAC values. Thus the graph $G$ will have $2^{n/2}$ vertices and approximately $2^{n/2}t'^2/2$ edges. If we view the graph as a random graph [9], then a "giant component" will arise when the number of edges is sufficiently larger than half the number of vertices. With $t' = 2$ it can be shown that with high probability there is a component containing 98% of all vertices.[1]

We know the value

$$e_K(X \oplus e_{K''}(e_K(D_1))) \oplus e_K(X' \oplus e_{K''}(e_K(D_1)))$$

whenever $X$ and $X'$ are in the same connected component, by adding the appropriate equations together. So for most values of $X$ we know the value

$$f(X) = e_K(X \oplus e_{K''}(e_K(D_1))) \oplus e_K((X \oplus 1) \oplus e_{K''}(e_K(D_1)))$$

where 1 denotes the $n$-bit block with least significant bit 1 and all other bits set to zero.

---

[1] With $s$ vertices and $cs/2$ randomly placed edges, with $c > 1$, there is a single "giant component" whose size is almost exactly $(1 - t(c))s$, where [1]

$$t(c) = (1/c)\sum_{k=1}^{\infty}(k^{k-1}(ce^{-c})^k)/k!.$$

For $c = 4$, $1 - t(c) = 0.98$.

Now repeat the above process but for a set of padded messages with $q + 1$ rather than $q$ blocks. In this case label the fixed values $D_1', D_4', D_5', \ldots, D_{q+1}'$, and obtain the MACs for messages

$$D_1', Z, Y, D_4', D_5', \ldots, D_{q+1}'$$

for $2^{n/2+1}$ values of $Z$ and $t$ values of $Y$, where the values of $Z$ cover all $n$-bit blocks whose least significant $n/2$ bits are forced to 0 (except the single least significant bit which covers both values). As previously, if $t$ is sufficiently large then for many (most) values of $Z$ we will have an equation of the form

$$g(Z) = e_K(Z \oplus e_{K''}(e_K(D_1'))) \oplus e_K((Z \oplus 1) \oplus e_{K''}(e_K(D_1'))).$$

Now find values $X$, $Z$ such that $f(X) = g(Z)$. Then we know with a high probability that either

$$X \oplus e_{K''}(e_K(D_1)) = Z \oplus e_{K''}(e_K(D_1')),$$

or

$$X \oplus e_{K''}(e_K(D_1)) = Z \oplus 1 \oplus e_{K''}(e_K(D_1')).$$

This (almost) gives us the desired relationship between $D_1$ and $D_1'$. In fact the next stage of the attack can be carried out for both possible relationships. This will not significantly affect the overall attack complexity, since the complexity of the second stage is much less than that of the first stage.

### Complexity of attack

It should be clear that finding the relationship between the desired pair of blocks $(D_1, D_1')$ requires $3 \times 2^{n/2} \times t$ 'chosen MACs', where $t = 2^{n/4}t'$. Hence the complexity of the first stage of the attack is $[0, 0, 3t' \times 2^{3n/4}, 0]$ for small $t' \geq 1$. The complexity of the second stage of the attack is $[2^{k+2}, 2^{n/2}, 0, 0]$. Thus, assuming that the second stage of the attack is performed twice, we get an overall attack complexity of $[2^{k+3}, 2^{n/2+1}, 3t' \times 2^{3n/4}, 0]$. In the case of DES this gives an attack complexity of $[2^{59}, 2^{33}, s \times 2^{48}, 0]$ for a small $s \geq 3$.

## 5   A MAC Forgery Attack

We next consider a forgery attack against the same MAC scheme, and which uses a similar method of attack. What is particularly significant about this attack is that, analogously to the attack in [6], it is based almost exclusively on 'MAC verifications' rather than 'chosen MACs'. As mentioned above, in certain circumstances it may be substantially easier for the attacker to obtain MAC verifications (i.e. to submit a data string/MAC pair and receive an answer indicating whether or not the MAC is valid) than to obtain the genuine MAC value for a chosen message. The attack also requires almost no memory.

### 5.1   Details of Attack

By some means suppose the attacker obtains the MAC, $M$, for the padded message

$$D_1, D_2, D_3, ..., D_q$$

(where $D_1$ encodes the length of a $(q-1)n$-bit message). Suppose next that the attacker submits the $2^n$ messages

$$D_1', W, D_2, D_3, ..., D_q$$

for MAC verification with candidate MAC $M$, where $W$ takes on all possible values, and where $D_1'$ encodes the length of a $qn$-bit message. Precisely one of the $2^n$ messages will have valid MAC $M$.

Armed with the correct $W$ it is now possible to forge the MAC of any padded message of $q$ blocks by requesting the MAC of a padded message of $q+1$ blocks or vice versa. This is because we know that

$$\mathrm{MAC}(D_1', W, E_2, ..., E_q) = \mathrm{MAC}(D_1, E_2, ..., E_q)$$

for any blocks $E_2, E_3, \ldots, E_q$.

There are variants of this attack which allow the block $W$ to be inserted between any pair of blocks. Also the attack is only dependent on the block length, and will also work against Triple DES CBC-MAC and other iterated MAC schemes of similar structure.

### 5.2   Complexity

It is straightforward to see that the complexity of the above attack is simply $[0, 0, 1, 2^n]$. In addition, once the $2^n$ verifications have been performed, each additional MAC forgery only requires one 'chosen MAC'.

This compares with the best previously known forgery attack for this MAC scheme, namely the Preneel-van Oorschot attack, [8], which has complexity $[0, 0, 2^{n/2}, 0]$.

## 6   Preventing the Attacks

Before considering countermeasures against these attacks it is first important to note that if $k$ (the bit length of the key) and $n$ (the cipher block length) are chosen to be sufficiently large, then all these attacks become infeasible. In particular, with the lengths envisaged for the emerging Advanced Encryption Standard (AES) these attacks are of academic interest only. However, for the time being many systems are reliant on ciphers such as DES, for which $k$ and $n$ are both uncomfortably small. Thus, finding countermeasures which do not involve too many additional encryption operations remains of practical importance.

## 6.1  Using Serial Numbers

Probably the simplest way of preventing all the attacks described above is to use *Serial Numbers*, as described in [4]. The basic idea is to prepend a unique serial number to a data string prior to computing a MAC. That is, every time a MAC is generated by a device, that device ensures that the data to be MACed is prepended with a number which has never previously been used for this purpose (at least within the lifetime of the key).

Although it is not stated explicitly in [4], it would seem that it is intended that the serial number should be prepended to the message prior to padding, and this is the assumption we make here. Note also that it will be necessary to send the serial number with the message, so that the intended recipient can use it to help recompute the MAC (as is necessary to verify it).

It is fairly simple to see why this approach foils the attacks described in this paper. All attacks require the forger to obtain the MAC for a *chosen* data string. However, the attacker is now no longer in a position to choose the data string, since the MAC generator will insert a serial number (previously unused) as part of the MAC computation process. Note that an attacker can still verify MACs on particular messages using serial numbers of his own choice.

Note that the effectiveness of Serial numbers against forgery attacks on the MAC scheme considered here is discussed in more detail in [2].

## 6.2  A Further MacDES Variant

Another possible way to defeat the attacks described previously is to modify the MAC scheme to introduce an extra key. The key recovery attacks exploit the fact that one key of the CBC-chaining equals the first of the two keys in the final double encryption. We can therefore preclude such attacks by using the key $K$ only in the middle step of the MAC calculation and not in the first and final steps. I.e. we can introduce a fourth key, $K'''$, which could be derived from $K$ (with $K''$ derived from $K'$), and put:

$$H_1 = e_{K''}(e_{K'''}(D_1)),$$
$$H_i = e_K(D_i \oplus H_{i-1}), (2 \leq i \leq q-1), \text{ and}$$
$$M = e_{K'}(e_{K'''}(D_q \oplus H_{q-1})).$$

However, this modified scheme can still be attacked with about $2^{65}$ chosen plaintexts and $2^{64}$ work. The number $2^{65}$ refers to the number of different messages for which MACs are required, and not the number of different plaintext blocks (which obviously cannot exceed $2^{64}$). We now sketch the attack.

Select $2^{60}$ values of $X_i$ that are 0 in the most significant 4 bits (say), and $2^5$ arbitrary values of $Y_j$. Fix words $D_1$ and $D_4$ corresponding to a padded message of eventual length $4 \times 64$ bits. (Thus $D_1$ encodes a message length of $3 \times 64$ bits.) By some means obtain MACs for the $2^{65}$ messages

$$(D_1, X_i, Y_j, D_4).$$

We are guaranteed to have at least $2^{64}$ coincidences of the form

$$\text{MAC}(D_1, X_i, Y_j, D_4) = \text{MAC}(D_1, X_k, Y_m, D_4).$$

Recalling that $H_1$ is constant throughout this exercise (but unknown), each coincidence gives the knowledge that

$$e_K(X_i \oplus H_1) \oplus Y_j = e_K(X_k \oplus H_1) \oplus Y_m = d_K(H_3),$$

whence

$$e_K(X_i \oplus H_1) \oplus e_K(X_k \oplus H_1) = Y_j \oplus Y_m.$$

Construct a graph of $2^{60}$ vertices, each vertex representing an allowable value of $X_i$. An edge joins two vertices when we have knowledge of the type given in the last equation. That is, vertices $X_i$ and $X_k$ are joined when we know the value of

$$e_K(X_i \oplus H_1) \oplus e_K(X_k \oplus H_1).$$

Because we have $2^{64}$ edges and $2^{60}$ vertices, routine arguments about random graphs (see, for example [9]) predicts that we will have one 'giant connected component' in this graph, which contains most of the vertices. If two vertices $X_i, X_p$ lie in the same connected component, we can (by chasing edges and adding the corresponding equations) find the corresponding sum

$$e_K(X_i \oplus H_1) \oplus e_K(X_p \oplus H_1).$$

In particular for most 'acceptable' values of $X_i$ we know the sum

$$f(X_i) = e_K(X_i \oplus H_1) \oplus e_K((X_i \oplus 1) \oplus H_1).$$

However, we still do not know $K$ or $H_1$.

Now for each guess $k$ for $K$, and each of $2^4$ choices of $z$, compute $e_k(z) \oplus e_k(z \oplus 1)$ and see whether this equals $f(X_i)$ for some value of $X_i$. If so, compute also $e_k(z \oplus 2) \oplus e_k(z \oplus 3)$ and check whether that matches $f(X_i \oplus 2)$ for the same $X_i$. If so, then $K$ and $H_1$ can be determined. This last step takes time $2^{56} \times 2^4 \times 4 = 2^{62}$.

Having recovered $K$ and $H_1$, it is then necessary to break double-DES to recover $K'$ and $K'''$ (complexity approximately $[2^{56+t}, 0, 0, 0]$ with $2^{56-2t}$ space [10]). This also involves doing the above attack twice, since two pairs of input and output of double-DES are need to determine the secret keys. Finally, break single DES to find $K''$.

## 6.3    Truncation

Perhaps the most obvious countermeasure to the attacks described here is to choose the MAC length $m$ such that $m < n$, i.e. to truncate the Output Block of the MAC calculation. However, Knudsen [5] has shown that, even when truncation is employed, in some cases the same attacks can still be made at the cost of a modest amount of additional effort. Moreover, if $m$ is made smaller, then certain trivial forgery attacks become easier to mount (see, for example, [2]).

We now consider how truncation affects the key recovery and forgery attacks described above.

**Key recovery attacks.** The key recovery attacks no longer work as described. However, as we now sketch, a $2^{64}$ chosen text attack can still work, even with truncated (32-bit) MACs and even with the four-key variant (as described in Section 6.2).

In fact for most of the attack it is possible to make a trade-off between chosen texts and computation. Also, it is possible to recover all keys, or to build a 'dictionary'. The attack complexity is approximately $2^{64+3-p}$ chosen texts and $2^{56+2p}$ computation, for $0 < p < 32$, and with full key-recovery we need an additional amount of $2^{56+t}$ computation using $2^{56-2t}$ space and about $2^{57}$ MAC verifications.

Pick a block of $x = 2^{64-2p}$ words $X$, with the low order $2p$ bits set to 0. For each $X$, pick $y = 2^p$ words $Y$. Optimally these depend on $X$ in a random fashion (to avoid some duplications). Pick four messages $Z$, fixed throughout. Fix the initial block $D_1$. Obtain the MACs of the $2^{64-2p} \times 2^p \times 4$ blocks $(D_1, X, Y, Z)$. Let $h(X, Y)$ be the concatenation of the four MACs of

$$(D_1, X, Y, Z_1), (D_1, X, Y, Z_2), (D_1, X, Y, Z_3), (D_1, X, Y, Z_4).$$

The coincidence $h(X, Y) = h(X', Y')$ is (essentially) equivalent to $H_3(D_1, X, Y) = H_3(D_1, X', Y')$ (the equality after three rounds).

Now use the same random graph idea as previously, with about $x$ edges among the $x$ vertices. Throw in a slop factor so that there are $2x$ edges. Then nearly everything is in the giant component. Evaluate, for each $X$, the function

$$g(X) = H_2(D_1, X) \oplus H_2(D_1, X \oplus 1) = e_K(H_1 \oplus X) \oplus e_K(H_1 \oplus X \oplus 1).$$

Now for each of $w = 2^{2p}$ trial values $W$, and each of $2^{56}$ trial values $k$ for $K$, evaluate $f(k, W) = e_k(W) \oplus e_k(W \oplus 1)$. The $W$s are 0 in the high order $64 - 2p$ bits. For each false positive $g(X) = f(k, W)$, do a bit more sleuthing. Eventually you find the right setting: $K = k$, $H_1 = X \oplus W$.

Then (if $p = 0$ and we had initially $2^{67}$ chosen texts) we have enough information to get the truncated 32 bits of $e_{K'}(e_{K'''}(U))$ for each 64-bit input $U$, although we don't have enough information to easily get $K'$ and $K'''$ themselves. If $p$ is larger than 0 then we will only build a partial dictionary.

Alternatively, once $K$ and $H_1$ are recovered, one can break double-DES to recover $K''$ and $K'''$. At this point, $K$, $K''$, and $K'''$ are known and one finds $K'$ using about $2^{57}$ MAC verifications.

**Forgery attacks.** Truncating the MAC values does not substantially increase the complexity of the forgery attack described in Section 5. For example, if the MAC length $m = 32$, two known-text MACs (of equal lengths) will be required. In the verification step, a check is performed on the second message (only) if the first verification succeeds.

## 7  Conclusions

We have seen that the most effective key recovery attack against the MacDES scheme (with Padding Method 3) has complexity $[2^{59}, 2^{33}, s \times 2^{48}, 0]$ for a small

$s \geq 3$. This compares with the previously best known attack which has complexity $[2^{89}, 0, 2^{65}, 2^{55}]$. This means that this scheme is still better than the ANSI retail MAC, i.e. MAC algorithm 3 from [4], but not as much as previously thought. In addition a new forgery attack against this scheme (and others) has been described, requiring just one 'chosen MAC'. The use of MAC truncation makes the attacks considerably more difficult. As an example, when used with DES and a MAC value of 32 bits we outlined a key-recovery attack of complexity $[2^{64}, 0, 2^{63}, 2^{57}]$ (with possible trade-offs between chosen texts and computation). If Serial Numbers are employed, then the attacks appear to become even more infeasible.

# References

1. B. Bollobás. *Random graphs*. Academic Press, 1985.
2. K. Brincat and C.J. Mitchell. A taxonomy of CBC-MAC forgery attacks. Submitted, January 2000.
3. D. Coppersmith and C.J. Mitchell. Attacks on MacDES MAC algorithm. *Electronics Letters*, **35**:1626–1627, 1999.
4. International Organization for Standardization, Genève, Switzerland. *ISO/IEC 9797–1, Information technology — Security techniques — Message Authentication Codes (MACs) — Part 1: Mechanisms using a block cipher*, December 1999.
5. L.R. Knudsen. Chosen-text attack on CBC-MAC. *Electronics Letters*, **33**:48–49, 1997.
6. L.R. Knudsen and B. Preneel. MacDES: MAC algorithm based on DES. *Electronics Letters*, **34**:871–873, 1998.
7. A.J. Menezes, P.C. van Oorschot, and S.A. Vanstone. *Handbook of Applied Cryptography*. CRC Press, Boca Raton, 1997.
8. B. Preneel and P.C. van Oorschot. On the security of iterated Message Authentication Codes. *IEEE Transactions on Information Theory*, **45**:188–199, 1999.
9. J. Spencer. *Ten lectures on the probabilistic method*. Society for Industrial and Applied Mathematics, Philadelphia, PA, second edition, 1994.
10. P.C. van Oorschot and M.J. Wiener. Parallel collision search with cryptanalytic applications. *Journal of Cryptology*, 12(1):1–28, 1999.

# CBC MACs for Arbitrary-Length Messages: The Three-Key Constructions

John Black[1] and Phillip Rogaway[2]

[1] Dept. of Computer Science, University of Nevada, Reno NV 89557, USA,
blackj@cs.ucdavis.edu
[2] Dept. of Computer Science, University of California at Davis, Davis, CA 95616,
USA, rogaway@cs.ucdavis.edu, http://www.cs.ucdavis.edu/~rogaway

**Abstract.** We suggest some simple variants of the CBC MAC that let you efficiently MAC messages of arbitrary lengths. Our constructions use three keys, $K1$, $K2$, $K3$, to avoid unnecessary padding and MAC any message $M \in \{0,1\}^*$ using $\max\{1, \lceil |M|/n \rceil\}$ applications of the underlying $n$-bit block cipher. Our favorite construction, XCBC, works like this: if $|M|$ is a positive multiple of $n$ then XOR the $n$-bit key $K2$ with the last block of $M$ and compute the CBC MAC keyed with $K1$; otherwise, extend $M$'s length to the next multiple of $n$ by appending minimal $10^i$ padding ($i \geq 0$), XOR the $n$-bit key $K3$ with the last block of the padded message, and compute the CBC MAC keyed with $K1$. We prove the security of this and other constructions, giving concrete bounds on an adversary's inability to forge in terms of her inability to distinguish the block cipher from a random permutation. Our analysis exploits new ideas which simplify proofs compared to prior work.

## 1 Introduction

This paper describes some simple variants of CBC MAC. These algorithms correctly and efficiently handle messages of any bit length. In addition to our schemes, we introduce new techniques to prove them secure. Our proofs are much simpler than prior work. We begin with some background.

THE CBC MAC. The CBC MAC [6,8] is the simplest and most well-known way to make a message authentication code (MAC) out of a block cipher. Let's recall how it works. Let $\Sigma = \{0,1\}$ and let $E : \mathsf{Key} \times \Sigma^n \to \Sigma^n$ be a block cipher: it uses a key $K \in \mathsf{Key}$ to encipher an $n$-bit block $X$ into an $n$-bit ciphertext $Y = E_K(X)$. The message space for the CBC MAC is $(\Sigma^n)^+$, meaning binary strings whose lengths are a positive multiple of $n$. So let $M = M_1 \cdots M_m$ be a string that we want to MAC, where $|M_1| = \cdots = |M_m| = n$. Then $\mathrm{CBC}_{E_K}(M)$, the CBC MAC of $M$ under key $K$, is defined as $C_m$, where $C_i = E_K(M_i \oplus C_{i-1})$ for $i = 1, \ldots, m$ and $C_0 = 0^n$.

Bellare, Kilian, and Rogaway proved the security of the CBC MAC, in the sense of reduction-based cryptography [2]. But their proof depends on the assumption that it is only messages of one fixed length, $mn$ bits, that are being

M. Bellare (Ed.): CRYPTO 2000, LNCS 1880, pp. 197–215, 2000.
© Springer-Verlag Berlin Heidelberg 2000

MACed. Indeed when message lengths can vary, the CBC MAC is *not* secure. This fact is well-known. As a simple example, notice that given the CBC MAC of a one-block message $X$, say $T = \mathrm{CBC}_{E_K}(X)$, the adversary immediately knows the CBC MAC for the two-block message $X \parallel (X \oplus T)$, since this is once again $T$.

Thus the CBC MAC (in the "raw" form that we have described) has two problems: it can't be used to MAC messages outside of $(\Sigma^n)^+$, and all messages must have the same fixed length.

DEALING WITH VARIABLE MESSAGE LENGTHS: EMAC. When message lengths vary, the CBC MAC must be embellished. There have been several suggestions for doing this. The most elegant one we have seen is to encipher $\mathrm{CBC}_{E_{K1}}(M)$ using a new key, $K2$. That is, the domain is still $(\Sigma^n)^+$ but one defines EMAC (for encrypted MAC) by $\mathrm{EMAC}_{E_{K1}, E_{K2}}(M) = E_{K2}(\mathrm{CBC}_{E_{K1}}(M))$. This algorithm was developed for the RACE project [3]. It has been analyzed by Petrank and Rackoff [10] who show, roughly said, that an adversary who obtains the MACs for messages which total $\sigma$ blocks cannot forge with probability better than $2\sigma^2/2^n$.

Among the nice features of EMAC is that one need not know $|M|$ prior to processing the message $M$. All of our suggestions will retain this feature.

OUR CONTRIBUTIONS. EMAC has a domain limited to $(\Sigma^n)^+$ and uses $1+|M|/n$ applications of the block cipher $E$. In this paper we refine EMAC in three ways: (1) we extend the domain to $\Sigma^*$; (2) we shave off one application of $E$; and (3) we avoid keying $E$ by multiple keys. Of course we insist on retaining provable security (across all messages lengths).

In Section 2 we introduce three refinements to EMAC, which we call ECBC, FCBC, and XCBC. These algorithms are natural extensions of the CBC MAC. We would like to think that this is an asset. The point here is to strive for economy, in terms of both simplicity and efficiency.

Figure 1 summarizes the characteristics of the CBC MAC variants mentioned in this paper. The top three rows give known constructions (two of which we have now defined). The next three rows are our new constructions. Note that our last construction, XCBC, retains essentially all the efficiency characteristics of the CBC MAC, but extends the domain of correct operation to all of $\Sigma^*$. The cost to save one invocation of the block cipher and extend our domain to $\Sigma^*$ is a slightly longer key. We expect that in most settings the added overhead to create and manage the longer key is minimal, and so XCBC may be preferred.

For each of the new schemes we give a proof of security. Rather than adapt the rather complex proof of [10], or the even more complicated one of [2], we follow a new tack, viewing EMAC as an instance of the Carter-Wegman paradigm [5,12]: with EMAC one is enciphering the output of a universal-2 hash function. This universal-2 hash function is the CBC MAC itself. Since it is not too hard to upper bound the collision probability of the CBC MAC (see Lemma 3), this approach leads to a simple proof for EMAC, and ECBC as well. We then use the security of ECBC to prove security for FCBC, and then we use the security of FCBC to prove security for XCBC. In passing from FCBC to XCBC we use

| Construct | Domain | #E Appls | #E Keys | Key Length |
|-----------|--------|----------|---------|------------|
| CBC | $\Sigma^{nm}$ | $\|M\|/n$ | 1 | $k$ |
| EMAC | $(\Sigma^n)^+$ | $1 + \|M\|/n$ | 2 | $2k$ |
| EMAC* | $\Sigma^*$ | $1 + \lceil(\|M\|+1)/n\rceil$ | 2 | $2k$ |
| ECBC | $\Sigma^*$ | $1 + \lceil\|M\|/n\rceil$ | 3 | $3k$ |
| FCBC | $\Sigma^*$ | $\lceil\|M\|/n\rceil$ | 3 | $3k$ |
| XCBC | $\Sigma^*$ | $\lceil\|M\|/n\rceil$ | 1 | $k + 2n$ |

**Fig. 1.** *The CBC MAC and five variants. Here $M$ is the message to MAC and $E :$ $\Sigma^k \times \Sigma^n \to \Sigma^n$ is a block cipher. The third column gives the number of applications of $E$, assuming $\|M\| > 0$. The fourth column is the number of different keys used to key $E$. For CBC the domain is actually $(\Sigma^n)^+$, but the scheme is secure only on messages of some fixed length, $nm$.*

a general lemma (Lemma 4) which says, in effect, that you can always replace a pair of random independent permutations $\pi_1(\cdot), \pi_2(\cdot)$ by a pair of functions $\pi(\cdot), \pi(\cdot \oplus K)$, where $\pi$ is a random permutation and $K$ is a random constant.

NEW STANDARDS. This work was largely motivated by the emergence of the Advanced Encryption Standard (AES). With the AES should come a next-generation standard for using it to MAC. Current CBC MAC standards handle this, for example, in an open-ended informational appendix [8]. We suggest that the case of variable message lengths is the *usual* case in applications of MACs, and that a modern MAC standard should specify only algorithms which will correctly MAC any sequence of bit strings. The methods here are simple, efficient, provably sound, timely, and patent-free—all the right features for a contemporary standard.

## 2   Schemes ECBC, FCBC, and XCBC

ARBITRARY-LENGTH MESSAGES WITHOUT OBLIGATORY PADDING: ECBC. We have described the algorithm $\text{EMAC}_{E_{K1},E_{K2}}(M) = E_{K2}(\text{CBC}_{E_{K1}}(M))$. One problem with EMAC is that its domain is limited to $(\Sigma^n)^+$. What if we want to MAC messages whose lengths are *not* a multiple of $n$?

The simplest approach is to use obligatory $10^i$ padding: always append a "1" bit and then the minimum number of "0" bits so as to make the length of the padded message a multiple of $n$. Then apply EMAC. We call this method EMAC*. Formally, $\text{EMAC}^*_{E_{K1},E_{K2}}(M) = \text{EMAC}_{E_{K1},E_{K2}}(M \| 10^{n-1-\|M\| \bmod n})$. This construction works fine. In fact, it is easy to see that this form of padding *always* works to extend the domain of a MAC from $(\Sigma^n)^+$ to $\Sigma^*$.

One unfortunate feature of EMAC* is this: if $\|M\|$ is already a multiple of $n$ then we are appending an entire extra block of padding, and seemingly "wasting" an application of $E$. People have worked hard to optimize new block ciphers—it seems a shame to squander some of this efficiency with an unnecessary application of $E$. Moreover, in practical settings we often wish to MAC very short

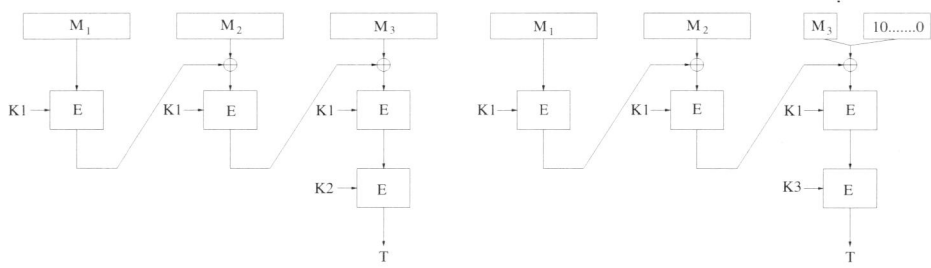

**Fig. 2.** *The ECBC construction using a block cipher $E$ : Key $\times \Sigma^n \rightarrow \Sigma^n$. The construction uses three keys, $K1, K2, K3 \in$ Key. On the left is the case where $|M|$ is a positive multiple of $n$, while on the right is the case where it isn't.*

messages, where saving one invocation of the block cipher can be a significant performance gain.

Our first new scheme lets us avoid padding when $|M|$ is a nonzero multiple of $n$. We simply make two cases: one for when $|M|$ is a positive multiple of $n$ and one for when it isn't. In the first case we compute $\text{EMAC}_{E_{K1}, E_{K2}}(M)$. In the second case we append minimal $10^i$ padding $(i \geq 0)$ to make a padded message $P$ whose length is divisible by $n$, and then we compute $\text{EMAC}_{E_{K1}, E_{K3}}(P)$. Notice the different second key—$K3$ instead of $K2$—in the case where we've added padding. Here, in full, is the algorithm. It is also shown in Figure 2.

---

**Algorithm** $\text{ECBC}_{E_{K1}, E_{K2}, E_{K3}}(M)$
**if** $M \in (\Sigma^n)^+$
    **then return** $E_{K2}(\text{CBC}_{E_{K1}}(M))$
    **else return** $E_{K3}(\text{CBC}_{E_{K1}}(M \parallel 10^i))$, where $i = n - 1 - |M| \bmod n$

---

In Section 4 we prove that ECBC is secure. We actually show that it is a good pseudorandom function (PRF), not just a good MAC. The security of ECBC does not seem to directly follow from Petrank and Rackoff's result [10]. At issue is the fact that there is a relationship between the key $(K1, K2)$ used to MAC messages in $(\Sigma^n)^+$ and the key $(K1, K3)$ used to MAC other messages.

IMPROVING EFFICIENCY: FCBC. With ECBC we are using $\lceil |M|/n \rceil + 1$ applications of the underlying block cipher. We can get rid of the $+1$ (except when $|M| = 0$). We start off, as before, by padding $M$ when it is outside $(\Sigma^n)^+$. Next we compute the CBC MAC using key $K1$ for all but the final block, and then use either key $K2$ or $K3$ for the final block. Which key we use depends on whether or not we added padding. The algorithm follows, and is also shown in Figure 3. In Section 5 we prove the security of this construction. Correctness follows from the result on the security of ECBC.

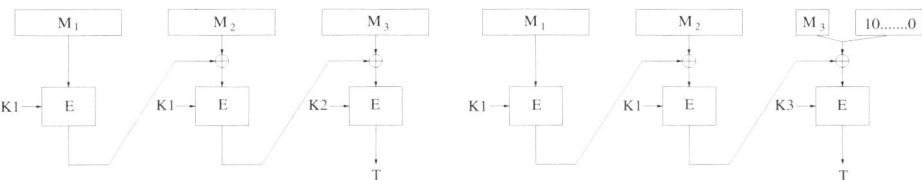

**Fig. 3.** *The FCBC construction with a block cipher $E : \mathsf{Key} \times \Sigma^n \to \Sigma^n$. The construction uses three keys, $K1, K2, K3 \in \mathsf{Key}$. On the left is the case where $|M|$ is a positive multiple of $n$, while on the right is the case where $|M|$ is not a positive multiple of $n$.*

---

**Algorithm** $\mathrm{FCBC}_{E_{K1}, E_{K2}, E_{K3}}(M)$
**if** $M \in (\Sigma^n)^+$
    **then** $K \leftarrow K2$, and $P \leftarrow M$
    **else** $K \leftarrow K3$, and $P \leftarrow M \parallel 10^i$, where $i \leftarrow n - 1 - |M| \bmod n$
Let $P = P_1 \cdots P_m$, where $|P_1| = \cdots = |P_m| = n$
$C_0 \leftarrow 0^n$
**for** $i \leftarrow 1$ **to** $m - 1$ **do**
    $C_i \leftarrow E_{K1}(P_i \oplus C_{i-1})$
**return** $E_K(P_m \oplus C_{m-1})$

---

AVOIDING MULTIPLE ENCRYPTION KEYS: XCBC. Most block ciphers have a key-setup cost, when the key is turned into subkeys. The subkeys are often larger than the original key, and computing them may be expensive. So keying the underlying block cipher with multiple keys, as is done in EMAC, ECBC, and FCBC, is actually not so desirable. It would be better to use the same key for all of the block-cipher invocations. The algorithm XCBC does this.

---

**Algorithm** $\mathrm{XCBC}_{E_{K1}, K2, K3}(M)$
**if** $M \in (\Sigma^n)^+$
    **then** $K \leftarrow K2$, and $P \leftarrow M$
    **else** $K \leftarrow K3$, and $P \leftarrow M \parallel 10^i$, where $i \leftarrow n - 1 - |M| \bmod n$
Let $P = P_1 \cdots P_m$, where $|P_1| = \cdots = |P_m| = n$
$C_0 \leftarrow 0^n$
**for** $i \leftarrow 1$ **to** $m - 1$ **do**
    $C_i \leftarrow E_{K1}(P_i \oplus C_{i-1})$
**return** $E_{K1}(P_m \oplus C_{m-1} \oplus K)$

---

We again make two cases. If $M \in (\Sigma^n)^+$ we CBC as usual, except that we XOR in an $n$-bit key, $K2$, before enciphering the last block. If $M \notin (\Sigma^n)^+$ then append minimal $10^i$ padding ($i \geq 0$) and CBC as usual, except that we XOR in a different $n$-bit key, $K3$, before enciphering the last block. Here, in full, is the algorithm. Also see Figure 4. The proof of security can be found in Section 6.

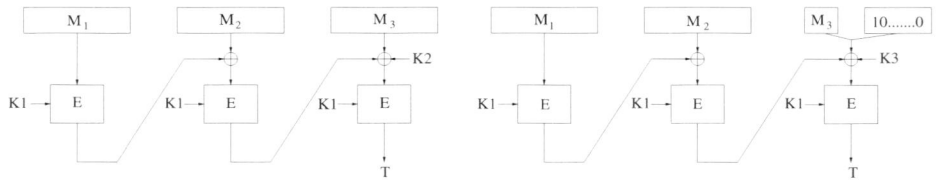

**Fig. 4.** *The XCBC construction with a block cipher* $E : \mathsf{Key} \times \Sigma^n \to \Sigma^n$. *We use keys* $K1 \in \mathsf{Key}$ *and* $K2, K3 \in \Sigma^n$. *On the left is the case where* $|M|$ *is a positive multiple of* $n$; *on the right is the case where* $|M|$ *is not a positive multiple of* $n$.

SUMMARY. We have now defined $\mathrm{CBC}_{\rho_1}(\cdot)$, $\mathrm{EMAC}_{\rho_1,\rho_2}(\cdot)$, $\mathrm{ECBC}_{\rho_1,\rho_2,\rho_3}(\cdot)$, $\mathrm{FCBC}_{\rho_1,\rho_2,\rho_3}(\cdot)$, and $\mathrm{XCBC}_{\rho_1,k_2,k_3}(\cdot)$ where $\rho_1,\rho_2,\rho_3 \colon \Sigma^n \to \Sigma^n$ and $k_2,k_3 \in \Sigma^n$. We emphasize that the definitions make sense for any $\rho_1, \rho_2, \rho_3 \colon \Sigma^n \to \Sigma^n$; in particular, we don't require $\rho_1, \rho_2, \rho_3$ be permutations. Notice that we interchangeably use notation such as $\rho_1$ and $E_{K1}$; the key $K1$ is simply naming a function $\rho_1 = E_{K1}$.

# 3   Preliminaries

NOTATION. If $A$ and $B$ are sets then $\mathrm{Rand}(A, B)$ is the set of all functions from $A$ to $B$. If $A$ or $B$ is a positive number, $n$, then the corresponding set is $\Sigma^n$. Let $\mathrm{Perm}(n)$ be the set of all permutations from $\Sigma^n$ to $\Sigma^n$. By $x \xleftarrow{R} A$ we denote the experiment of choosing a random element from $A$.

A function family is a multiset $F = \{f \colon A \to B\}$, where $A, B \subseteq \Sigma^*$. Each element $f \in F$ has a name $K$, where $K \in \mathsf{Key}$. So, equivalently, a function family $F$ is a function $F \colon \mathsf{Key} \times A \to B$. We call $A$ the domain of $F$ and $B$ the range of $F$. The first argument to $F$ will be written as a subscript. A block cipher is a function family $F \colon \mathsf{Key} \times \Sigma^n \to \Sigma^n$ where $F_K(\cdot)$ is always a permutation.

An adversary is an algorithm with an oracle. The oracle computes some function. Adversaries are assumed to never ask a query outside the domain of the oracle, and to never repeat a query.

Let $F \colon \mathsf{Key} \times A \to B$ be a function family and let $\mathcal{A}$ be an adversary. We say that $\mathcal{A}^f$ forges if $\mathcal{A}$ outputs $(x, f(x))$ where $x \in A$ and $\mathcal{A}$ never queried its oracle $f$ at $x$. We let

$$\mathbf{Adv}_F^{\mathrm{mac}}(\mathcal{A}) \stackrel{\mathrm{def}}{=} \Pr[f \xleftarrow{R} F \colon \mathcal{A}^{f(\cdot)} \text{ forges }]$$

$$\mathbf{Adv}_F^{\mathrm{prf}}(\mathcal{A}) \stackrel{\mathrm{def}}{=} \Pr[f \xleftarrow{R} F \colon \mathcal{A}^{f(\cdot)} = 1] - \Pr[R \xleftarrow{R} \mathrm{Rand}(A, n) \colon \mathcal{A}^{R(\cdot)} = 1] \,,$$

and when $A = \Sigma^n$

$$\mathbf{Adv}_F^{\mathrm{prp}}(\mathcal{A}) \stackrel{\mathrm{def}}{=} \Pr[f \xleftarrow{R} F \colon \mathcal{A}^{f(\cdot)} = 1] - \Pr[\pi \xleftarrow{R} \mathrm{Perm}(n) \colon \mathcal{A}^{\pi(\cdot)} = 1] \,.$$

We overload this notation and write $\mathbf{Adv}_F^{\mathrm{xxx}}(R)$ (where $\mathrm{xxx} \in \{\mathrm{mac}, \mathrm{prf}, \mathrm{prp}\}$) for the maximal value of $\mathbf{Adv}_F^{\mathrm{xxx}}(\mathcal{A})$ among adversaries who use resources $R$.

The resources we will deal with are: $t$, the running time of the adversary; $q$, the number of queries the adversary makes; and $\mu$, the maximal bit length of each query. Omitted arguments are unbounded or irrelevant. We fix a few conventions. To measure time, $t$, one assumes some fixed model of computation. This model is assumed to support unit-time operations for computing $f \overset{R}{\leftarrow} \mathsf{Key}$ and $F_K(P)$. Time is understood to include the description size of the adversary $\mathcal{A}$. In the case of $\mathbf{Adv}_F^{\mathrm{mac}}$ the number of queries, $q$, includes one "invisible" query to test if the adversary's output is a valid forgery.

BASIC FACTS. It is often convenient to replace random permutations with random functions, or vice versa. The following proposition lets us easily do this. For a proof see Proposition 2.5 in [2].

**Lemma 1. [PRF/PRP Switching]**   *Fix $n \geq 1$. Let $\mathcal{A}$ be an adversary that asks at most $p$ queries. Then*

$$\left| \Pr[\pi \overset{R}{\leftarrow} \mathrm{Perm}(n) : \mathcal{A}^{\pi(\cdot)} = 1] - \Pr[\rho \overset{R}{\leftarrow} \mathrm{Rand}(n,n) : \mathcal{A}^{\rho(\cdot)} = 1] \right| \leq \frac{p(p-1)}{2^{n+1}} \quad \blacksquare$$

As is customary, we will show the security of our MACs by showing that their information-theoretic versions approximate random functions. As is standard, this will be enough to pass to the complexity-theoretic scenario. Part of the proof is Proposition 2.7 of [2].

**Lemma 2. [Inf. Th. PRF $\Rightarrow$ Comp. Th. PRF]**   *Fix $n \geq 1$. Let CONS be a construction such that $\mathrm{CONS}_{\rho_1,\rho_2,\rho_3}(\cdot) \colon \Sigma^* \to \Sigma^n$ for any $\rho_1, \rho_2, \rho_3 \in \mathrm{Rand}(n,n)$. Suppose that if $|M| \leq \mu$ then $\mathrm{CONS}_{\rho_1,\rho_2,\rho_3}(M)$ depends on the values of $\rho_i$ on at most $p$ points (for $1 \leq i \leq 3$). Let $E \colon \mathsf{Key} \times \Sigma^n \to \Sigma^n$ be a family of functions. Then*

$$\mathbf{Adv}_{\mathrm{CONS}[E]}^{\mathrm{prf}}(t,q,\mu) \leq \mathbf{Adv}_{\mathrm{CONS}[\mathrm{Perm}(n)]}^{\mathrm{prf}}(q,\mu) + 3 \cdot \mathbf{Adv}_E^{\mathrm{prp}}(t',p), \text{ and}$$

$$\mathbf{Adv}_{\mathrm{CONS}[E]}^{\mathrm{mac}}(t,q,\mu) \leq \mathbf{Adv}_{\mathrm{CONS}[\mathrm{Perm}(n)]}^{\mathrm{prf}}(q,\mu) + 3 \cdot \mathbf{Adv}_E^{\mathrm{prp}}(t',p) + \frac{1}{2^n} ,$$

*where $t' = t + O(pn)$.*   $\blacksquare$

## 4   Security of ECBC

In this section we prove the security of the ECBC construction. See Section 2 for a definition of $\mathrm{ECBC}_{\rho_1,\rho_2,\rho_3}(M)$, where $\rho_1, \rho_2, \rho_3 \in \mathrm{Rand}(n,n)$.

Our approach is as follows: we view the CBC MAC as an almost universal-2 family of hash functions and prove a bound on its collision probability. Then we create a PRF by applying one of two random functions to the output of CBC MAC and claim that this construction is itself a good PRF. Applying a PRF to a universal-2 hash function is a well-known approach for creating a PRF or MAC [5,12,4]. The novelty here is the extension to three keys and, more significantly, the treatment of the CBC MAC as an almost universal-2 family of hash functions. The latter might be against one's instincts because the CBC MAC is a much stronger object than a universal hash-function family. Here we ignore that extra strength and study only the collision probability.

We wish to show that if $M, M' \in (\Sigma^n)^+$ are distinct then $\Pr_\pi[\mathrm{CBC}_\pi(M) = \mathrm{CBC}_\pi(M')]$ is small. By "small" we mean a slowly growing function of $m = |M|/n$ and $m' = |M'|/n$. Formally, for $n, m, m' \geq 1$, let the *collision probability* of the CBC MAC be

$$V_n(m, m') \stackrel{\mathrm{def}}{=}$$

$$\max_{M \in \Sigma^{nm}, \, M' \in \Sigma^{nm'}, \, M \neq M'} \left\{ \Pr[\pi \stackrel{R}{\leftarrow} \mathrm{Perm}(n) : \mathrm{CBC}_\pi(M) = \mathrm{CBC}_\pi(M')] \right\}.$$

(The shape of the character "$V$" is meant to suggest collisions.) It is not hard to infer from [10] a collision bound of $V_n(m, m') \leq 2.5 \, (m + m')^2/2^n$ by using their bound on EMAC and realizing that collisions in the underlying CBC MAC must show through to the EMAC output. A direct analysis easily yields a slightly better bound.

**Lemma 3. [CBC Collision Bound]** *Fix $n \geq 1$ and let $N = 2^n$. Let $M, M' \in (\Sigma^n)^+$ be distinct strings having $m = |M|/n$ and $m' = |M'|/n$ blocks. Assume that $m, m' \leq N/4$. Then*

$$V_n(m, m') \leq \frac{(m + m')^2}{2^n}$$

*Proof.* Although $M$ and $M'$ are distinct, they may share some common prefix. Let $k$ be the index of the last block in which $M$ and $M'$ agree. (If $M$ and $M'$ have unequal first blocks then $k = 0$.)

Each particular permutation $\pi$ is equally likely among all permutations from $\Sigma^n$ to $\Sigma^n$. In our analysis, we will view the selection of $\pi$ as an incremental procedure. This will be equivalent to selecting $\pi$ uniformly at random.

In particular, we view the computation of $\mathrm{CBC}_\pi(M)$ and $\mathrm{CBC}_\pi(M')$ as playing the game given in Figure 5. Here the notation $M_i$ indicates the $i$th block of $M$. We initially set each range point of $\pi$ as `undefined`; the notation $\mathrm{Domain}(\pi)$ represents the set of points $x$ where $\pi(x)$ is no longer `undefined`. We use $\mathrm{Range}(\pi)$ to denote the set of points $\pi(x)$ which are no longer `undefined`; we use $\overline{\mathrm{Range}(\pi)}$ to denote $\Sigma^n - \mathrm{Range}(\pi)$.

During the game, the $X_i$ are those values produced after XORing with the current message block, $M_i$, and the $Y_i$ values are $\pi(X_i)$. See Figure 6.

We examine the probability that $\pi$ will cause $\mathrm{CBC}_\pi(M) = \mathrm{CBC}_\pi(M')$, which will occur in our game iff $Y_m = Y'_{m'}$. Since $\pi$ is invertible, this occurs iff $X_m = X'_{m'}$. As we shall see, this condition will cause $bad = \mathtt{true}$ in our game. However, we actually set $bad$ to $\mathtt{true}$ in many other cases in order to simplify the analysis.

The idea behind the variable $bad$ is as follows: throughout the program (lines 5, 12, and 17) we randomly choose a range value for $\pi$ at some `undefined` domain point. Since $\pi$ has not yet been determined at this point, the selection of our range value will be an independent uniform selection: there is no dependence

```
1:   bad ← false;    for all x ∈ Σⁿ do π(x) ← undefined
2:   X₁ ← M₁;    X'₁ ← M'₁;   BAD ← {X₁, X'₁}

3:   for i ← 1 to k do
4:       if Xᵢ ∈ Domain(π) then Yᵢ ← Y'ᵢ ← π(Xᵢ) else
5:           Yᵢ ← Y'ᵢ ←ᴿ Range(π)‾;   π(Xᵢ) ← Yᵢ
6:           if i < m then begin Xᵢ₊₁ ← Yᵢ ⊕ Mᵢ₊₁
7:               if Xᵢ₊₁ ∈ BAD then bad ← true else BAD ← BAD ∪ {Xᵢ₊₁} end
8:           if i < m' then begin X'ᵢ₊₁ ← Y'ᵢ ⊕ M'ᵢ₊₁
9:               if X'ᵢ₊₁ ∈ BAD then bad ← true else BAD ← BAD ∪ {X'ᵢ₊₁} end

10:  for i ← k + 1 to m do
11:      if Xᵢ ∈ Domain(π) then Yᵢ ← π(Xᵢ) else
12:          Yᵢ ←ᴿ Range(π)‾;   π(Xᵢ) ← Yᵢ
13:          if i < m then begin Xᵢ₊₁ ← Yᵢ ⊕ Mᵢ₊₁
14:              if Xᵢ₊₁ ∈ BAD then bad ← true else BAD ← BAD ∪ {Xᵢ₊₁} end

15:  for i ← k + 1 to m' do
16:      if X'ᵢ ∈ Domain(π) then Y'ᵢ ← π(X'ᵢ) else
17:          Y'ᵢ ←ᴿ Range(π)‾;   π(X'ᵢ) ← Y'ᵢ
18:          if i < m then begin X'ᵢ₊₁ ← Y'ᵢ ⊕ M'ᵢ₊₁
19:              if X'ᵢ₊₁ ∈ BAD then bad ← true else BAD ← BAD ∪ {X'ᵢ₊₁} end
```

**Fig. 5.** *Game used in the proof of Lemma 3. The algorithm gives one way to compute the CBC MAC of distinct messages $M = M_1 \cdots M_m$ and $M' = M'_1 \cdots M'_{m'}$. These messages are identical up to block $k$. The computed MACs are $Y_m$ and $Y_{m'}$, respectively.*

on any prior choice. If the range value for $\pi$ were already determined by some earlier choice, the analysis would become more involved. We avoid the latter condition by setting *bad* to `true` whenever such interdependencies are detected. The detection mechanism works as follows: throughout the processing of $M$ and $M'$ we will require $\pi$ be evaluated at $m + m'$ domain points $X_1, \cdots, X_m$ and $X'_1, \cdots, X'_{m'}$. If all of these domain points are distinct (ignoring duplications due to any common prefix of $M$ and $M'$), we can rest assured that we are free to assign their corresponding range points without constraint. We maintain a set $BAD$ to track which domain points have already been determined; initially $X_1$ and $X'_1$ are the only such points, since future values will depend on random choices not yet made. Of course if $k > 0$ then $X_1 = X'_1$ and $BAD$ contains only one value. Next we begin randomly choosing range points; if ever any such choice leads to a value already contained in the $BAD$ set, we set the flag *bad* to `true`.

We now bound the probability of the event that $bad = $ `true` by analyzing our game. The variable *bad* can be set `true` in lines 7, 9, 14, and 19. In each case it is required that some $Y_i$ was selected such that $Y_i \oplus M_{i+1} \in BAD$ (or possibly that some $Y'_i$ was selected such that $Y'_i \oplus M'_{i+1} \in BAD$). The set $BAD$ begins with at most 2 elements and then grows by 1 with each random choice of $Y_i$ or $Y'_i$. We know that on the $i$th random choice in the game the $BAD$ set will contain at most $i + 1$ elements. And so each random choice of $Y_i$ (resp. $Y'_i$)

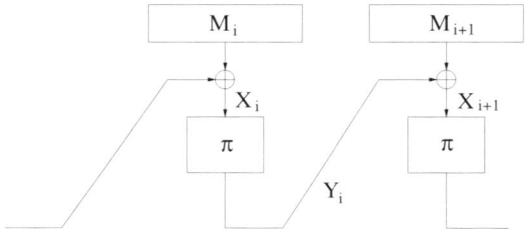

**Fig. 6.** *The labeling convention used in the proof of Lemma 3.*

from the co-range of $\pi$ will cause $Y_i \oplus M_{i+1}$ (resp. $Y_i' \oplus M_{i+1}'$) to be in $BAD$ with probability at most $(i+1)/(N-i+1)$. We have already argued that in the absence of $bad = \texttt{true}$ each of the random choices we make are independent. We make $m-1$ choices of $Y_i$ to produce $X_2$ through $X_m$ and $m'-1$ choices of $Y_i'$ to determine $X_2'$ through $X_{m'}'$ and so we can compute

$$\Pr[bad = \texttt{true}] \leq \sum_{i=1}^{m-1+m'-1} \frac{i+1}{N-i+1}.$$

Using the fact that $m, m' \leq N/4$, we can bound the above by

$$\sum_{i=1}^{m+m'-2} \frac{i+1}{N-i} \leq \frac{2}{N} \sum_{i=1}^{m+m'-2} i+1 \leq \frac{(m+m')^2}{N}.$$

This completes the proof. ∎

Fix $n \geq 1$ and let $F \colon \mathsf{Key} \times \Sigma^n \to \Sigma^n$ be a family of functions. Let $\mathrm{ECBC}[F]$ be the family of functions $\mathrm{ECBC}_{f1,f2,f3}(\cdot)$ indexed by $\mathsf{Key} \times \mathsf{Key} \times \mathsf{Key}$. We now use the above to show that $\mathrm{ECBC}[\mathrm{Perm}(n)]$ is close to being a random function.

**Theorem 1. [ECBC $\approx$ Rand]**  *Fix $n \geq 1$ and let $N = 2^n$. Let $\mathcal{A}$ be an adversary which asks at most $q$ queries each of which is at most $mn$-bits. Assume $m \leq N/4$. Then*

$$\Pr[\pi_1, \pi_2, \pi_3 \xleftarrow{R} \mathrm{Perm}(n) \colon \mathcal{A}^{\mathrm{ECBC}_{\pi_1, \pi_2, \pi_3}(\cdot)} = 1] \,-$$
$$\Pr[R \xleftarrow{R} \mathrm{Rand}(\Sigma^*, n) \colon \mathcal{A}^{R(\cdot)} = 1] \leq \frac{q^2}{2} V_n(m, m) + \frac{q^2}{2N} \leq \frac{(2m^2+1)q^2}{N}$$

*Proof.* We first compute a related probability where the final permutation is a random function; this will simplify the analysis. So we are interested in the quantity

$$\Pr[\pi_1 \xleftarrow{R} \mathrm{Perm}(n); \rho_2, \rho_3 \xleftarrow{R} \mathrm{Rand}(n, n) \colon \mathcal{A}^{\mathrm{ECBC}_{\pi_1, \rho_2, \rho_3}(\cdot)} = 1]$$
$$- \Pr[R \xleftarrow{R} \mathrm{Rand}(\Sigma^*, n) \colon \mathcal{A}^{R(\cdot)} = 1].$$

For economy of notation we encapsulate the initial parts of our experiments into the probability symbols $\mathrm{Pr}_1$ and $\mathrm{Pr}_2$, rewriting the above as

$$\mathrm{Pr}_1[\mathcal{A}^{\mathrm{ECBC}_{\pi_1,\rho_2,\rho_3}(\cdot)} = 1] - \mathrm{Pr}_2[\mathcal{A}^{R(\cdot)} = 1].$$

We condition $\mathrm{Pr}_1$ on whether a collision occurred within $\pi_1$, and we differentiate between collisions among messages whose lengths are a nonzero multiple of $n$ (which are not padded) and other messages (which are padded). Let UnpadCol be the event that there is a collision among the unpadded messages and let PadCol be the event that there is a collision among the padded messages. We rewrite the above as

$$\mathrm{Pr}_1[\mathcal{A}^{\mathrm{ECBC}_{\pi_1,\rho_2,\rho_3}(\cdot)} = 1 \mid \mathsf{UnpadCol} \vee \mathsf{PadCol}]\,\mathrm{Pr}_1[\mathsf{UnpadCol} \vee \mathsf{PadCol}]$$
$$+ \mathrm{Pr}_1[\mathcal{A}^{\mathrm{ECBC}_{\pi_1,\rho_2,\rho_3}(\cdot)} = 1 \mid \overline{\mathsf{UnpadCol} \vee \mathsf{PadCol}}]\,\mathrm{Pr}_1[\overline{\mathsf{UnpadCol} \vee \mathsf{PadCol}}]$$
$$- \mathrm{Pr}_2[\mathcal{A}^{R}(\cdot) = 1].$$

Observe that in the absence of event $(\mathsf{UnpadCol} \vee \mathsf{PadCol})$, the adversary sees the output of a random function on distinct points; that is, she is presented with random, uncorrelated points over the range $\Sigma^n$. Therefore we know

$$\mathrm{Pr}_1[\mathcal{A}^{\mathrm{ECBC}_{\pi_1,\rho_2,\rho_3}(\cdot)} = 1 \mid \overline{\mathsf{UnpadCol} \vee \mathsf{PadCol}}] \;=\; \mathrm{Pr}_2[\mathcal{A}^{R(\cdot)} = 1].$$

Bounding $\mathrm{Pr}_1[\overline{\mathsf{UnpadCol} \vee \mathsf{PadCol}}]$ and $\mathrm{Pr}_1[\mathcal{A}^{\mathrm{ECBC}_{\pi_1,\rho_2,\rho_3}(\cdot)} = 1 \mid \mathsf{UnpadCol} \vee \mathsf{PadCol}]$ by 1, we reduce the bound on the obtainable advantage to $\mathrm{Pr}_1[\mathsf{UnpadCol} \vee \mathsf{PadCol}] \leq \mathrm{Pr}_1[\mathsf{UnpadCol}] + \mathrm{Pr}_1[\mathsf{PadCol}]$. To bound this quantity, we break each event into the disjoint union of several other events; define $\mathsf{UnpadCol}_i$ to be the event that a collision in $\pi_1$ occurs for the first time as a result of the $i$th unpadded query. Let $q_u$ be the number of unpadded queries made by the adversary, and let $q_p$ be the number of padded queries. Then

$$\mathrm{Pr}_1[\mathsf{UnpadCol}] = \sum_{i=1}^{q_u} \mathrm{Pr}_1[\mathsf{UnpadCol}_i].$$

Now we compute each of the probabilities on the righthand side above. If we know no collisions occurred during the first $i - 1$ queries we know the adversary has thus far seen only images of distinct inputs under a random function. Any adaptive strategy she adopts in this case could be replaced by a non-adaptive strategy where she pre-computes her queries under the assumption that the first $i - 1$ queries produce random points. In other words, any adaptive adversary can be replaced by a non-adaptive adversary which does at least as well. A non-adaptive strategy would consist of generating all $q$ inputs in advance, and from the collision bound on the hash family we know $\mathrm{Pr}_1[\mathsf{UnpadCol}_i] \leq (i - 1)V_n(m,m)$. Summing over $i$ we get

$$V_n(m,m) \sum_{i=1}^{q_u} (i-1) \;\leq\; \frac{q_u^2}{2} V_n(m,m).$$

Repeating this analysis for the padded queries we obtain an overall bound of

$$\frac{q_u^2}{2}V_n(m,m) + \frac{q_p^2}{2}V_n(m,m) \; \le \; \frac{q^2}{2}V_n(m,m)\,.$$

Finally we must replace the PRFs $\rho_2$ and $\rho_3$ with the PRPs $\pi_2$ and $\pi_3$. Using Lemma 1 this costs an extra $q_u^2/2N + q_p^2/2N \le q^2/2N$, and so the bound becomes

$$\frac{q^2}{2}V_n(m,m) + \frac{q^2}{2N}\,,$$

and if we apply the bound on $V_n(m,m)$ from Lemma 3 we have

$$\frac{q^2}{2}\frac{4m^2}{N} + \frac{q^2}{2N} \le \frac{(2m^2+1)q^2}{N}\,. \quad \blacksquare$$

TIGHTNESS, AND NON-TIGHTNESS, OF OUR BOUND. Employing well-known techniques (similar to [11]), it is easy to exhibit a known-message attack which forges with high probability ($> 0.3$) after seeing $q = 2^{n/2}$ message/tag pairs. In this sense the analysis looks tight. The same statement can be made for the FCBC and XCBC analyses. But if we pay attention not only to the number of messages MACed, but also their lengths, then none of these analyses is tight. That is, because the security bound degrades quadratically with the total number of message blocks, while the the attack efficiency improves quadratically with the total number of messages. Previous analyses for the CBC MAC and its variants all shared these same characteristics.

COMPLEXITY-THEORETIC RESULT. In the usual way we can now pass from the information-theoretic result to a complexity-theoretic one. For completeness, we state the result, which follows using Lemma 2.

**Corollary 1. [ECBC is a PRF]**  *Fix $n \ge 1$ and let $N = 2^n$. Let $E \colon$ Key $\times$ $\Sigma^n \to \Sigma^n$ be a block cipher. Then*

$$\mathbf{Adv}^{\mathrm{prf}}_{\mathrm{ECBC}[E]}(t,q,mn) \; \le \; \frac{2m^2q^2 + q^2}{N} + 3 \cdot \mathbf{Adv}^{\mathrm{prp}}_E(t',q'), \quad and$$

$$\mathbf{Adv}^{\mathrm{mac}}_{\mathrm{ECBC}[E]}(t,q,mn) \; \le \; \frac{2m^2q^2 + q^2 + 1}{N} + 3 \cdot \mathbf{Adv}^{\mathrm{prp}}_E(t',q')$$

*where $t' = t + O(mq)$ and $q' = mq$.* $\quad \blacksquare$

It is worth noting that using a very similar argument to Theorem 1 we can easily obtain the same bound for the EMAC construction [3]. This yields a proof which is quite a bit simpler than that found in [10]. We state the theorem here, but omit the proof which is very similar to the preceding.

**Theorem 2. [EMAC $\approx$ Rand]**  *Fix $n \ge 1$ and let $N = 2^n$. Let $\mathcal{A}$ be an adversary which asks at most $q$ queries, each of which is at most $m$ $n$-bit blocks,* *where $m \leq N/4$. Then*

$$\Pr[\pi, \sigma \xleftarrow{R} \mathrm{Perm}(n) : \mathcal{A}^{\mathrm{EMAC}_{\pi,\sigma}(\cdot)} = 1] - \Pr[\rho \xleftarrow{R} \mathrm{Rand}((\Sigma^n)^+, n) : \mathcal{A}^{\rho(\cdot)} = 1]$$

$$\leq \frac{q^2}{2} V_n(m, m) + \frac{q^2}{2N} \leq \frac{(2m^2 + 1)q^2}{N} \quad \blacksquare$$

## 5  Security of FCBC

In this section we prove the security of FCBC, obtaining the same bound we had for ECBC. See Section 2 for a definition of $\mathrm{FCBC}_{\rho_1,\rho_2,\rho_3}(M)$, where $\rho_1, \rho_2, \rho_3 \in \mathrm{Rand}(n, n)$.

**Theorem 3. [FCBC $\approx$ Rand]**  *Fix $n \geq 1$ and let $N = 2^n$. Let $\mathcal{A}$ be an adversary which asks at most $q$ queries, each of which is at most $mn$ bits. Assume $m \leq N/4$. Then*

$$\Pr[\pi_1, \pi_2, \pi_3 \xleftarrow{R} \mathrm{Perm}(n) : \mathcal{A}^{\mathrm{FCBC}_{\pi_1,\pi_2,\pi_3}(\cdot)} = 1] -$$

$$\Pr[R \xleftarrow{R} \mathrm{Rand}(\Sigma^*, n) : \mathcal{A}^{R(\cdot)} = 1] \leq \frac{q^2}{2} V_n(m, m) + \frac{q^2}{2N} \leq \frac{(2m^2 + 1)q^2}{N}$$

*Proof.* Let us compare the distribution on functions

$$\{\mathrm{ECBC}_{\pi_1,\pi_2,\pi_3}(\cdot) \mid \pi_1, \pi_2, \pi_3 \xleftarrow{R} \mathrm{Perm}(n)\} \quad \text{and}$$

$$\{\mathrm{FCBC}_{\pi_1,\sigma_2,\sigma_3}(\cdot) \mid \pi_1, \sigma_2, \sigma_3 \xleftarrow{R} \mathrm{Perm}(n)\} .$$

We claim that these are the *same* distribution, so, information theoretically, the adversary has no way to distinguish a random sample drawn from one distribution from a random sample from the other. The reason is simple. In the ECBC construction we compose the permutation $\pi_1$ with the random permutation $\pi_2$. But the result of such a composition is just a random permutation, $\sigma_2$. Elsewhere in the ECBC construction we compose the permutation $\pi_1$ with the random permutation $\pi_3$. But the result of such a composition is just a random permutation, $\sigma_3$. Making these substitutions—$\sigma_2$ for $\pi_2 \circ \pi_1$, and $\sigma_3$ for $\pi_3 \circ \pi_1$, we recover the definition of ECBC. Changing back to the old variable names we have

$$\Pr[\pi_1, \pi_2, \pi_3 \xleftarrow{R} \mathrm{Perm}(n) : \mathcal{A}^{\mathrm{FCBC}_{\pi_1,\pi_2,\pi_3}(\cdot)} = 1]$$

$$= \Pr[\pi_1, \pi_2, \pi_3 \xleftarrow{R} \mathrm{Perm}(n) : \mathcal{A}^{\mathrm{ECBC}_{\pi_1,\pi_2,\pi_3}(\cdot)} = 1]$$

So the bound of our theorem follows immediately from Theorem 1.    $\blacksquare$

Since the bound for FCBC exactly matches the bound for ECBC, Corollary 1 applies to FCBC as well.

# 6   Security of XCBC

In this section we prove the security of the XCBC construction. See Section 2 for a definition of $\mathrm{XCBC}_{\rho_1,\rho_2,\rho_3}(M)$, where $\rho_1, \rho_2, \rho_3 \in \mathrm{Rand}(n, n)$ and $M \in \Sigma^*$.

We first give a lemma which bounds an adversary's ability to distinguish between a pair of random permutations, $\pi_1(\cdot)$, $\pi_2(\cdot)$, and the pair $\pi(\cdot)$, $\pi(K \oplus \cdot)$, where $\pi$ is a random permutation and $K$ is a random $n$-bit string. This lemma, and ones like it, may make generally useful tools.

**Lemma 4. [Two permutations from one]**  *Fix $n \geq 1$ and let $N = 2^n$. Let $\mathcal{A}$ be an adversary which asks at most $p$ queries. Then*

$$\left| \Pr[\pi \stackrel{R}{\leftarrow} \mathrm{Perm}(n); K \stackrel{R}{\leftarrow} \Sigma^n : \mathcal{A}^{\pi(\cdot), \pi(K \oplus \cdot)} = 1] \right.$$

$$\left. - \Pr[\pi_1, \pi_2 \stackrel{R}{\leftarrow} \mathrm{Perm}(n) : \mathcal{A}^{\pi_1(\cdot), \pi_2(\cdot)} = 1] \right| \leq \frac{p^2}{N}.$$

*Proof.* We use the game shown in Figure 7 to facilitate the analysis. Call the game in that figure Game 1. We play the game as follows: first, the initialization procedure is executed once before we begin. In this procedure we set each range point of $\pi$ as undefined; the notation $\mathrm{Domain}(\pi)$ represents the set of points $x$ where $\pi(x)$ is no longer undefined. We use $\mathrm{Range}(\pi)$ to denote the set of points $\pi(x)$ which are no longer undefined. We use $\overline{\mathrm{Range}}(\pi)$ to denote $\Sigma^n - \mathrm{Range}(\pi)$. Now, when the adversary makes a query $X$ to her left oracle, we execute the code in procedure $\pi(X)$. When she makes a query $X$ to her right oracle, we execute procedure $\pi(K \oplus X)$. We claim that in this setting we perfectly simulate a pair of oracles where the first is a random permutation $\pi(\cdot)$ and the second is $\pi(K \oplus \cdot)$. To verify this, let us examine each of the procedures in turn.

For procedure $\pi(X)$, we first check to see if $X$ is in the domain of $\pi$. We are assuming that the adversary will not repeat a query, but it is possible that $X$ is in the domain of $\pi$ if the adversary has previously queried the *second* procedure with the string $K \oplus X$. In this case we faithfully return the proper value $\pi(X)$. If her query is not in the domain of $\pi$ we choose a random element $Y$ from $S$ which we hope to return in response to her query. However, it may be that $Y$ is already in the range of $\pi$. Although we always remove from $S$ any value returned from this procedure, it may be that $Y$ was placed in the range of $\pi$ by the *second* procedure. If this occurs, we choose a random element from the co-range of $\pi$ and return it. For procedure $\pi(K \oplus \cdot)$ we behave analogously.

Note during the execution of Game 1, we faithfully simulate the pair of functions $\pi(\cdot)$ and $\pi(K \oplus \cdot)$. That is, the view of the adversary would be exactly the same if we had selected a random permutation $\pi$ and a random $n$-bit string $K$ and then let her query $\pi(\cdot)$ and $\pi(K \oplus \cdot)$ directly.

Now consider the game we get by removing the shaded statements of Game 1. Call that game Game 2. We claim that Game 2 exactly simulates two random permutations. In other words, the view of the adversary in this game is exactly as if we had given her two independent random permutations $\pi_1$ and $\pi_2$. Without

---

**Initialization:**

1:   $S, T \leftarrow \Sigma^n$;    $K \xleftarrow{R} \Sigma^n$;    **for all** $X \in \Sigma^n$ **do** $\pi(X) \leftarrow \texttt{undefined}$

**Procedure** $\pi(X)$ :

2:   **if** $X \in \mathrm{Domain}(\pi)$ **then** $bad \leftarrow \texttt{true}$, **return** $\pi(X)$

3:   $Y \xleftarrow{R} S$

4:   **if** $Y \in \mathrm{Range}(\pi)$ **then** $bad \leftarrow \texttt{true}$, $Y \xleftarrow{R} \overline{\mathrm{Range}}(\pi)$

5:   $\pi(X) \leftarrow Y$;    $S \leftarrow S - \{Y\}$;    **return** $Y$

**Procedure** $\pi(K \oplus X)$ :

6:   **if** $(K \oplus X) \in \mathrm{Domain}(\pi)$ **then** $bad \leftarrow \texttt{true}$, **return** $\pi(K \oplus X)$

7:   $Y \xleftarrow{R} T$

8:   **if** $Y \in \mathrm{Range}(\pi)$ **then** $bad \leftarrow \texttt{true}$, $Y \xleftarrow{R} \overline{\mathrm{Range}}(\pi)$

9:   $\pi(K \oplus X) \leftarrow Y$;    $T \leftarrow T - \{Y\}$;    **return** $Y$

---

**Fig. 7.** *Game used in the proof of Lemma 4. With the shaded text in place the game behaves like a pair of functions $\pi(\cdot)$, $\pi(K \oplus \cdot)$. With the shaded text removed the game behaves like a pair of independent random permutations $\pi_1(\cdot)$, $\pi_2(\cdot)$.*

loss of generality, we assume the adversary never repeats a query. Then examining procedure $\pi(X)$, we see each call to procedure $\pi(X)$ returns a random element from $\Sigma^n$ which has not been previously returned by this procedure. This clearly is a correct simulation of a random permutation $\pi_1(\cdot)$. For procedure $\pi(K \oplus X)$, we offset the query value $X$ with some hidden string $K$, and then return a random element from $\Sigma^n$ which has not been previously returned by this procedure. Note in particular that the offset by $K$ has no effect on the behavior of this procedure relative to the previous one. Therefore this procedure perfectly simulates a random independent permutation $\pi_2(\cdot)$.

In both games we sometimes set a variable $bad$ to $\texttt{true}$ during the execution of the game. In neither case, however, does this have any effect on the values returned by the game.

We name the event that $bad$ gets set to $\texttt{true}$ as B. This event is well-defined for both Games 1 and 2. Notice that Games 1 and 2 behave identically prior to $bad$ becoming $\texttt{true}$. One can imagine having two boxes, one for Game 1 and one for Game 2, each box with a red light that will illuminate if $bad$ should get set to $\texttt{true}$. These two boxes are defined as having identical behaviors until the red light comes on. Thus, collapsing the initial parts of the probabilities into the formal symbols $\mathrm{Pr}_1$ and $\mathrm{Pr}_2$, we may now say that $\mathrm{Pr}_1[\mathcal{A}^{\pi(\cdot),\pi(K \oplus \cdot)} = 1 \mid \overline{\mathsf{B}}] = \mathrm{Pr}_2[\mathcal{A}^{\pi_1(\cdot),\pi_2(\cdot)} = 1 \mid \overline{\mathsf{B}}]$. And since Games 1 and 2 behave identically until $bad$ becomes $\texttt{true}$, we know $\mathrm{Pr}_1[\mathsf{B}] = \mathrm{Pr}_2[\mathsf{B}]$. We have that

$$\left| \mathrm{Pr}_1[\mathcal{A}^{\pi(\cdot),\pi(K \oplus \cdot)} = 1] - \mathrm{Pr}_2[\mathcal{A}^{\pi_1(\cdot),\pi_2(\cdot)} = 1] \right| =$$

$$\left| \Pr_1[\mathcal{A}^{\pi(\cdot),\pi(K\oplus\cdot)} = 1 \mid \overline{\mathsf{B}}] \cdot \Pr_1[\overline{\mathsf{B}}] + \Pr_1[\mathcal{A}^{\pi(\cdot),\pi(K\oplus\cdot)} = 1 \mid \mathsf{B}] \cdot \Pr_1[\mathsf{B}] - \right.$$

$$\left. \Pr_2[\mathcal{A}^{\pi_1(\cdot),\pi_2(\cdot)} = 1 \mid \overline{\mathsf{B}}] \cdot \Pr_2[\overline{\mathsf{B}}] - \Pr_2[\mathcal{A}^{\pi_1(\cdot),\pi_2(\cdot)} = 1 \mid \mathsf{B}] \cdot \Pr_2[\mathsf{B}] \right|.$$

From the previous assertions we know this is equal to

$$\left| \Pr_1[\mathcal{A}^{\pi(\cdot),\pi(K\oplus\cdot)} = 1 \mid \mathsf{B}] \cdot \Pr_1[\mathsf{B}] - \Pr_2[\mathcal{A}^{\pi_1(\cdot),\pi_2(\cdot)} = 1 \mid \mathsf{B}] \cdot \Pr_2[\mathsf{B}] \right| =$$

$$\left| \Pr_2[\mathsf{B}] \cdot \left( \Pr_1[\mathcal{A}^{\pi(\cdot),\pi(K\oplus\cdot)} = 1 \mid \mathsf{B}] - \Pr_2[\mathcal{A}^{\pi_1(\cdot),\pi_2(\cdot)} = 1 \mid \mathsf{B}] \right) \right| \leq \Pr_2[\mathsf{B}].$$

Therefore we may bound the adversary's advantage by bounding $\Pr_2[\mathsf{B}]$.

We define $p$ events in Game 2: for $1 \leq i \leq p$, let $\mathsf{B}_i$ be the event that *bad* becomes `true`, for the first time, as a result of the $i$th query. Thus $\mathsf{B}$ is the disjoint union of $\mathsf{B}_1, \ldots, \mathsf{B}_p$ and

$$\Pr_2[\mathsf{B}] = \sum_{i=1}^{p} \Pr_2[\mathsf{B}_i] .$$

We now wish to bound $\Pr_2[\mathsf{B}_i]$. To do this, we first claim that adaptivity does not help the adversary to make $\mathsf{B}$ happen. In other words, the optimal adaptive strategy for making $\mathsf{B}$ happen is no better than the optimal non-adaptive one. Why is this? Adaptivity could help the adversary if Game 2 released information associated to $K$. But Game 2 has no dependency on $K$—the variable is not used in computing return values to the adversary. Thus Game 2 never provides the adversary any information that is relevant for creating a good $i$th query. So let $\mathcal{A}$ be an optimal adaptive adversary for making $\mathsf{B}$ happen. By the standard averaging argument there is no loss of generality to assume that $\mathcal{A}$ is deterministic. We can construct an optimal non-adaptive adversary $\mathcal{A}'$ by running $\mathcal{A}$ and simulating two independent permutation oracles. Since Game 2 returns values of the same distribution, for all $K$, the adversary $\mathcal{A}'$ will do no better or worse in getting $\mathsf{B}$ to happen in Game 2 if $\mathcal{A}'$ asks the sequence of questions that $\mathcal{A}$ asked in the simulated run. Adversary $\mathcal{A}'$ can now be made deterministic by the standard averaging argument. The resulting adversary, $\mathcal{A}''$, asks a fixed sequence of queries and yet does just as well as the adaptive adversary $\mathcal{A}$.

Now let us examine the chance that *bad* is set `true` in line 2, assuming the adversary has chosen her queries in advance. As we noted above, this can occur when the adversary asks a query $X$ after having previously issued a query $K \oplus X$ to the *second* procedure. What is the chance that this occurs? We can view the experiment as follows: at most $i-1$ queries were asked of the second procedure; let's name those queries $Q = \{X_1, \cdots, X_{i-1}\}$. We wish to bound the chance that a randomly chosen $X$ will cause $K \oplus X \in Q$. But this is simply asking the chance that $K \in \{X \oplus X_1, \cdots, X \oplus X_{i-1}\}$ which is at most $(i-1)/N$.

What is the chance that *bad* becomes `true` in line 4? This occurs when the randomly chosen $Y$ is already in the range of $\pi$. Since $Y$ is removed from $S$ each time the first procedure returns it, this occurs when the *second* procedure returned such a $Y$ and it was then subsequently chosen by the first procedure.

Since at most $i - 1$ values were returned by the second procedure, the chance is at most $(i - 1)/N$ that this could occur.

The same arguments apply to lines 6 and 8. Therefore we have $\Pr_2[\mathsf{B}_i] \leq 2(i - 1)/N$, and

$$\Pr_2[\mathsf{B}] \leq \sum_{i=1}^{p} \frac{2(i-1)}{N} \leq \frac{p^2}{N}$$

which completes the proof.   ∎

We may now state and prove the theorem for the security of our XCBC construction. The bound follows quickly from the the bound on FCBC and the preceding lemma.

**Theorem 4. [XCBC ≈ Rand]**  *Fix $n \geq 1$ and let $N = 2^n$. Let $\mathcal{A}$ be an adversary which asks at most $q$ queries, each of which is at most $mn$ bits. Assume $m \leq N/4$. Then*

$$\left| \Pr[\pi_1 \xleftarrow{R} \mathrm{Perm}(n); K2, K3 \xleftarrow{R} \Sigma^n : \mathcal{A}^{\mathrm{XCBC}_{\pi_1,K2,K3}(\cdot)} = 1] \right.$$

$$\left. - \Pr[R \xleftarrow{R} \mathrm{Rand}(\Sigma^*, n) : \mathcal{A}^{R(\cdot)} = 1] \right|$$

$$\leq \frac{q^2}{2} V_n(m, m) + \frac{(2m^2 + 1)q^2}{N} \leq \frac{(4m^2 + 1)q^2}{N}$$

*Proof.* By the triangle inequality, the above difference is at most

$$\left| \Pr[\pi_1, \pi_2, \pi_3 \xleftarrow{R} \mathrm{Perm}(n) : \mathcal{A}^{\mathrm{FCBC}_{\pi_1,\pi_2,\pi_3}(\cdot)} = 1] \right.$$

$$\left. - \Pr[R \xleftarrow{R} \mathrm{Rand}(\Sigma^*, n) : \mathcal{A}^{R(\cdot)} = 1] \right|$$

$$+ \left| \Pr[\pi_1, \pi_2, \pi_3 \xleftarrow{R} \mathrm{Perm}(n) : \mathcal{A}^{\mathrm{FCBC}_{\pi_1,\pi_2,\pi_3}(\cdot)} = 1] \right.$$

$$\left. - \Pr[\pi_1 \xleftarrow{R} \mathrm{Perm}(n); K2, K3 \xleftarrow{R} \Sigma^n : \mathcal{A}^{\mathrm{XCBC}_{\pi_1,K2,K3}(\cdot)} = 1] \right|$$

and Theorem 3 gives us a bound on the first difference above. We now bound the second difference. Clearly this difference is at most

$$\left| \Pr[\pi_1, \pi_2, \pi_3 \xleftarrow{R} \mathrm{Perm}(n) : \mathcal{A}^{\pi_1(\cdot), \pi_2(\cdot), \pi_3(\cdot)} = 1] \right.$$

$$\left. - \Pr[\pi_1 \xleftarrow{R} \mathrm{Perm}(n); K2, K3 \xleftarrow{R} \Sigma^n : \mathcal{A}^{\pi_1(\cdot), \pi_1(K2 \oplus \cdot), \pi_1(K3 \oplus \cdot)} = 1] \right|$$

since any adversary which does well in the previous setting could be converted to one which does well in this setting. (Here we assume that $\mathcal{A}$ makes at most $mq$ total queries of her oracles). Applying Lemma 4 twice we bound the above by $2m^2 q^2/N$. Therefore our overall bound is

$$\frac{q^2}{2} V_n(m, m) + \frac{q^2}{2N} + \frac{2m^2 q^2}{N} \leq \frac{q^2}{2} V_n(m, m) + \frac{(2m^2 + 1)q^2}{N} \ .$$

And if we apply the bound on $V_n(m, m)$ from Lemma 3 we have

$$\frac{q^2}{2}\frac{4m^2}{N} + \frac{(2m^2 + 1)q^2}{N} \leq \frac{(4m^2 + 1)q^2}{N}$$

as required. ∎

XOR-AFTER-LAST-ENCIPHERING DOESN'T WORK. The XCBC-variant that XORs the second key just *after* applying the final enciphering does *not* work. That is, when $|M|$ is a nonzero multiple of the blocksize, we'd have $\mathrm{MAC}_{\pi,K}(M) = \mathrm{CBC}_\pi(M) \oplus K$. This is no good. In the attack, the adversary asks for the MACs of three messages: the message $\mathbf{0} = 0^n$, the message $\mathbf{1} = 1^n$, and the message $\mathbf{1} \parallel \mathbf{0}$. As a result of these three queries the adversary gets tag $T_0 = \pi(\mathbf{0}) \oplus K$, tag $T_1 = \pi(\mathbf{1}) \oplus K$, and tag $T_2 = \pi(\pi(\mathbf{1})) \oplus K$. But now the adversary knows the correct tag for the (unqueried) message $\mathbf{0} \parallel (T_0 \oplus T_1)$, since this is just $T_2$: namely, $\mathrm{MAC}_{\pi,K}(\mathbf{0} \parallel (T_0 \oplus T_1)) = \pi(\pi(\mathbf{0}) \oplus (\pi(\mathbf{0}) \oplus K) \oplus (\pi(\mathbf{1}) \oplus K)) \oplus K = \pi(\pi(\mathbf{1})) \oplus K = T_2$. Thanks to Mihir Bellare for pointing out this attack.

ON KEY-SEARCH ATTACKS. If FCBC or XCBC is used with an underlying block cipher (like DES) which is susceptible to exhaustive key search, then the MACs inherit this vulnerability. (The same can be said of ECBC and EMAC, except that the double encryption which these MACs employ would seem to necessitate a meet-in-the-middle attack.) It was such considerations that led the designers of the retail MAC, ANSI X9.19, to suggest triple encryption for enciphering the last block [1]. It would seem to be possible to gain this same exhaustive-key-search strengthening by modifying XCBC to *again* XOR the second key ($K2$ or $K3$) with the result of the last encipherment. (If one is using DES, this amounts to switching to DESX for the last encipherment [9].) We call this variant XCBCX. Likely one could prove good bounds for it in the Shannon model. However, none of this is necessary or relevant if one simply starts with a strong block cipher.

COMPLEXITY-THEORETIC RESULT. We can again pass from the information-theoretic result to the complexity-theoretic one:

**Corollary 2. [XMAC is a PRF]**  *Fix* $n \geq 1$ *and let* $N = 2^n$. *Let* $E \colon \mathsf{Key} \times \Sigma^n \to \Sigma^n$ *be a block cipher. Then*

$$\mathbf{Adv}^{\mathrm{prf}}_{\mathrm{XCBC}[E]}(t, q, mn) \leq \frac{4m^2q^2 + q^2}{N} + 3 \cdot \mathbf{Adv}^{\mathrm{prp}}_E(t', q'), \quad \text{and}$$

$$\mathbf{Adv}^{\mathrm{mac}}_{\mathrm{XCBC}[E]}(t, q, mn) \leq \frac{4m^2q^2 + q^2 + 1}{N} + 3 \cdot \mathbf{Adv}^{\mathrm{prp}}_E(t', q')$$

*where* $t' = t + O(mq)$ *and* $q' = mq$. ∎

## Acknowledgments

Shai Halevi proposed the elegant idea of using three keys to extend the domain of the CBC MAC to $\Sigma^*$, nicely simplifying an approach used in an early version of UMAC [4]. Thanks to Shai and Mihir Bellare for their comments on an early draft, and to anonymous Crypto '00 reviewers for their useful comments.

The authors were supported under Rogaway's NSF CAREER Award CCR-962540, and under MICRO grants 98-129 and 99-103, funded by RSA Data Security and ORINCON. This paper was written while Rogaway was on sabbatical at the Department of Computer Science, Faculty of Science, Chiang Mai University. Thanks for their always-kind hospitality.

# References

1. ANSI X9.19. American national standard — Financial institution retail message authentication. ASC X9 Secretariat – American Bankers Association, 1986.
2. BELLARE, M., KILIAN, J., AND ROGAWAY, P. The security of the cipher block chaining message authentication code. See www.cs.ucdavis.edu/~rogaway. Older version appears in *Advances in Cryptology – CRYPTO '94* (1994), vol. 839 of *Lecture Notes in Computer Science*, Springer-Verlag, pp. 341–358.
3. BERENDSCHOT, A., DEN BOER, B., BOLY, J., BOSSELAERS, A., BRANDT, J., CHAUM, D., DAMGÅRD, I., DICHTL, M., FUMY, W., VAN DER HAM, M., JANSEN, C., LANDROCK, P., PRENEEL, B., ROELOFSEN, G., DE ROOIJ, P., AND VANDE-WALLE, J. *Final Report of Race Integrity Primitives*, vol. 1007 of *Lecture Notes in Computer Science*. Springer-Verlag, 1995.
4. BLACK, J., HALEVI, S., KRAWCZYK, H., KROVETZ, T., AND ROGAWAY, P. UMAC: Fast and secure message authentication. In *Advances in Cryptology – CRYPTO '99* (1999), Lecture Notes in Computer Science, Springer-Verlag.
5. CARTER, L., AND WEGMAN, M. Universal hash functions. *J. of Computer and System Sciences*, 18 (1979), 143–154.
6. FIPS 113. Computer data authentication. Federal Information Processing Standards Publication 113, U.S. Department of Commerce/National Bureau of Standards, National Technical Information Service, Springfield, Virginia, 1994.
7. GOLDREICH, O., GOLDWASSER, S., AND MICALI, S. How to construct random functions. *Journal of the ACM 33*, 4 (1986), 210–217.
8. ISO/IEC 9797-1. Information technology – security techniques – data integrity mechanism using a cryptographic check function employing a block cipher algorithm. International Organization for Standards, Geneva, Switzerland, 1999. Second edition.
9. KILIAN, J., AND ROGAWAY, P. How to protect DES against exhaustive key search. In *Advances in Cryptology – CRYPTO '96* (1996), vol. 1109 of *Lecture Notes in Computer Science*, Springer-Verlag, pp. 252–267.
10. PETRANK, E., AND RACKOFF, C. CBC MAC for real-time data sources. Manuscript 97-10 in http://philby.ucsd.edu/cryptolib.html, 1997.
11. PRENEEL, B., AND VAN OORSCHOT, P. On the security of two MAC algorithms. In *Advances in Cryptology — EUROCRYPT '96* (1996), vol. 1070 of *Lecture Notes in Computer Science*, Springer-Verlag, pp. 19–32.
12. WEGMAN, M., AND CARTER, L. New hash functions and their use in authentication and set equality. In *J. of Comp. and System Sciences* (1981), vol. 22, pp. 265–279.

# L-collision Attacks against Randomized MACs

Michael Semanko

Department of Computer Science & Engineering,
University of California at San Diego,
9500 Gilman Drive,
La Jolla, California 92093, USA.
msemanko@cs.ucsd.edu

**Abstract.** In order to avoid birthday attacks on message authentication schemes, it has been suggested that one add randomness to the scheme. One must be careful about how randomness is added, however. This paper shows that prefixing randomness to a message before running the message through an iterated MAC leads to an attack that takes only $O\left(2^{(l+r)/3} + \max\{2^{l/2}, 2^{r/2}\}\right)$ queries to break, where $l$ is the size of the MAC iteration output and $r$ is the size of the prefixed randomness.

**Keywords:** MACs, message authentication codes, randomness, L-collision, birthday attacks

## 1    Introduction

### 1.1    Problem

A message authentication scheme allows people to ensure that messages can travel between them without being altered. These schemes are basic cryptographic primitives, and thus they are widely used. Message authentication schemes consist of three major algorithms: one for computing a message's authentication tag, one for checking that a message's authentication tag is valid, and one for picking a key. By tagging a message before it is transmitted, one can protect the message from being maliciously altered. One popular form of message authentication is the iterated message authentication scheme.

Iterated message authentication schemes are those message authentication schemes which break up the message into smaller blocks and then make repeated use of a small keyed function on these blocks. The keyed function takes as input a block of the message and the previous result of the function to yield the next result. This is repeated for each block of the message, and the final result of the function is then output. CBC-MAC [1], HMAC [2], NMAC [2], and EMAC [7] are all examples of iterated MACs. All of these iterated MACs suffer from the same security flaw: internal collision attacks [8], referred to more generally as birthday attacks. Internal collision attacks can break any deterministic, stateless, iterated MAC in $O\left(2^{l/2}\right)$ queries, where $l$ is the size of the internal chaining value.

M. Bellare (Ed.): CRYPTO 2000, LNCS 1880, pp. 216–228, 2000.
© Springer-Verlag Berlin Heidelberg 2000

## 1.2    Possible Solutions

In an attempt to make more efficient MACs, people have come up with methods to try to avoid birthday attacks. There are two obvious possibilities: either make the MAC function stateful or make it probabilistic. We briefly discuss each.

STATEFUL ITERATED MACS. Stateful iterated MACs are iterated MACs which maintain $s$ bits of state between queries, and access this state in the tagging algorithm. The state is then changed so that a different state is always presented as the input to the transformation. The authentication tag then is the $l$-bit output of the MAC prepended to the $s$ bits of state. These iterated MACs can actually completely avoid the internal collision attack, as shown in [4]. The counter-based XOR-MAC, XMACC, requires approximately $O\left(2^l\right)$ authenticated messages to create a forgery, for instance [4]. XMACC achieves this by prepending a counter to messages before they are run through the iteration.

Stateful MACs have problems of their own which make them less used. It is often not convenient to maintain those $s$ bits of state between authentications; for example, you cannot use a stateful iterated MAC if you want multiple senders to share an authentication key. All of the systems that should be able to authenticate would have to share the same $s$ bits of state in order to maintain the security of the signatures. This is very difficult to do if there are many machines or if those machines may not always be connected.

RANDOMIZED ITERATED MACS. Using randomized iterated MACs is another approach that could be taken. A randomized iterated MAC is a iterated MAC which has a probabilistic tagging algorithm. The authentication tag in this case is the $l$-bit output of the MAC prepended to the $r$ bits of randomness.

A randomized scheme seems like it would be the answer to all of these problems. It does not need to maintain state between authentications. It can be used on multiple machines with the same key. What is missing? The problem with randomized schemes is that no one is really sure how secure randomized schemes are. Security proofs of randomized MAC schemes become difficult if one tries to prove security beyond the birthday bound. Designing a randomized authentication scheme that provably avoids birthday attacks was the topic of [3]. This paper succeeded only by tripling to quadrupling the size of the authentication tag. No one yet has designed a randomized MAC scheme with a proof of security that beats the birthday bound without a significant increase in authentication tag size or computation time.

Two approaches which are often suggested are to either prepend or append a random message block to the message before the keyed function iterations are performed. These approaches are promising because they only add a small amount of computation time and do not require a lot of randomness. There have not yet been results showing the security of either of these schemes.

## 1.3    Background and Related Work

Bellare, Killian, and Rogaway's paper [5], provides a formal analysis of the security of CBC-MAC. This paper is a good starting point from which the effects of birthday attacks can be seen on message authentication schemes in use. Later, Preneel and van Oorschot described the internal collision attack on MACs in [8]. This attack works upon all deterministic iterated MACs. They propose using keyed hash functions with large outputs to mitigate the effects of this attack. Although this works, it leads to MACs which may be less efficient than possible.

In Bellare, Guerin, and Rogaway's paper [4], the XOR-MAC schemes were introduced. One of these schemes, XMACC, was the first MAC which provably avoided the birthday attack problem. XMACC is a stateful MAC, however, and thus has all of the problems associated with a stateful MAC scheme. For a while, the only way to not be affected by birthday attacks was to use a stateful counter-based scheme.

More recently, Bellare, Goldreich, and Krawczyk provided a randomized, stateless scheme in [3] which is secure beyond the birthday bound in terms of the size of the hashed message; however, is not nearly as secure in terms of the size of the entire authentication tag because they had to transmit multiple random numbers in the tag.

Part of the problem with the term "birthday bound" lays in the fact that there are many parameters to a MAC, and even more to a randomized iterated MAC. An increase in any of these parameters can generally lead to an increase in the security of the MAC. In the best case, we would like to be able to increase the security of MACs, without a significant increase in the key size, computation time, authentication tag size, blocksize, or randomness.

### 1.4   Results

This paper shows a new form of birthday attack, called the L-collision attack. This attack applies to schemes that prepend random data before MACing. The L-collision attack allows an adversary to create a forgery with a constant probability in only $O\left(2^{(l+r)/3} + \max\{2^{l/2}, 2^{r/2}\}\right)$ queries where $l$ is the size of the internal chaining value of the randomized MAC, and $r$ is the size of the randomness.

Section 2 provides basic definitions that allow us to examine iterated message authentication schemes formally. Section 3 describes how the internal collision attack works, as this is the basis for our new attack. Section 4 presents a variation upon MAC schemes, which is based upon the idea of prepending a random block. This scheme is often secure against birthday attacks, but it falls to our new attack.

## 2   Definitions

This section presents the basic definitions of message authentication schemes, their security, and iterated message authentication schemes.

## 2.1    Function Families

Function families are vital components to MACs, and thus we wish to treat them formally. We do so in the format of [5].

**Definition 1.** *Let $h\colon \{0,1\}^k \times \{0,1\}^i \to \{0,1\}^j$ be a map. We say that $h$ is a function family. We define $\{0,1\}^k$ as the keyspace of $h$, $\{0,1\}^i$ as the domain of $h$, and $\{0,1\}^j$ as the range of $h$. Then for each particular key $K$, we define a map $h_K\colon \{0,1\}^i \to \{0,1\}^j$ such that $\forall u \in \{0,1\}^i, h_k(u) = h(k,u)$. We say that $h_K$ is a particular instance of $h$.*

Function families can be seen in such primitives as keyed-hash functions and block ciphers. Often, when proving security, we do not want to consider the effects of a particular block cipher or other keyed function on the security of a scheme. Thus, we replace the function with a function drawn at random from the set of all functions.

## 2.2    Message Authentication Schemes

We first define the syntax of message authentication schemes so that we may examine their security. This definition is a combination of the schemes used in [5] and [4].

**Definition 2.** *A message authentication scheme $\mathcal{MA} = (\mathsf{Tag}, \mathsf{Vf}, \mathsf{Key})$ consists of three algorithms as follows:*

- *A randomized key generation algorithm, $\mathsf{Key}$, which returns a key from the set $\{0,1\}^k$.*
- *A tagging algorithm, $\mathsf{Tag}$ which can be either randomized or stateful. It takes a key $K$ and a message $M$, and returns a tag $\sigma$ from $\{0,1\}^t$.*
- *A deterministic verification algorithm, $\mathsf{Vf}$, which takes a key $K$, a message $M$, and a tag $\sigma$ to return a bit $v$. We say that $\sigma$ is a valid tag for a message $M$ under a key $K$ if $\mathsf{Vf}(M, \sigma, K) = 1$.*

*We require that $\mathsf{Vf}(M, \mathsf{Tag}(M, K), K) = 1$ for all $M \in \{0,1\}^*$. The scheme is said to be deterministic if the tagging algorithm is deterministic.*

The security of a message authentication scheme is based upon the probability that an adversary without the secret key can create a forgery, which is a correct message/tag pair such that the verification procedure considers the pair valid.

## 2.3    Iterated Message Authentication Schemes

There are no interesting known general attacks on message authentication schemes as they are given above. We need to define a more specific message authentication scheme in order to be able to talk about actual attacks. For this reason, we now define iterated message authentication schemes. Figure 1 shows the general layout of an iterated MAC tagging algorithm, and below we present a definition based upon [8].

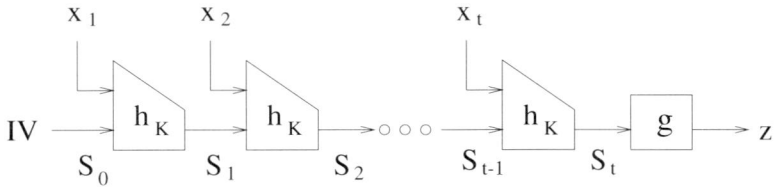

**Fig. 1.** A general scheme for iterated MACs

**Definition 3.** *Let $h$: $\{0,1\}^k \times \{0,1\}^b \times \{0,1\}^l \to \{0,1\}^l$ be a function family with domain $\{0,1\}^b \times \{0,1\}^l$, and let $g$: $\{0,1\}^l \to \{0,1\}^o$ be an output transformation. We associate to $h$ and $g$ the following iterative message authentication scheme $\mathcal{IMA} = (\mathsf{Tag}_h, \mathsf{Vf}_h, \mathsf{Key})$:*

- *$\mathsf{Tag}_h$ first divides a message $M$ to be tagged into b-bit blocks, labeled $x_1, \ldots$ $\ldots, x_t$, where $M = x_1 || \ldots || x_t$ (where $||$ denotes concatenation). We call $k$ the keysize of the MAC, $b$ the blocksize of the iterated MAC, $l$ the chaining variable size, and $o$ the output size of the MAC.*
- *$\mathsf{Tag}_h$ makes use of the keyed function $h$ iteratively to retrieve a value according to the algorithm $h_K^*(x_1 || \ldots || x_t)$, which we define below.*

```
function h*_K(x_1 || ... || x_t)
    S_0 ← IV
    for i = 1, ..., t do
        S_i ← h_K(x_i, S_{i-1})
    endfor
    return S_t
```

Note that this definition of iterated MACs includes the well-known CBC-MAC by making $h_k(x_i, S_{i-1}) = f_k(x_i \oplus S_{i-1})$, where $f_k$ is the block cipher to be used in CBC-MAC. For this paper, we shall be set the output transformation $g$ to be the identity function. Thus, the chaining variable size and the output size of the iterated MACs in this paper shall both be $l$. Changing the attack to deal with different output transformations can be done in the same way as in [8].

## 3    Birthday Attacks

If we wish to avoid birthday attacks with a message authentication scheme, we must be able to define what a birthday attack is. Usually, the type of birthday attacks that MAC designers attempt to avoid are known as internal collision attacks. The general idea behind an internal collision attack is that if we find two messages with the same $n$ block suffix but different $m$ block prefixes that lead to the same output, then we are likely to be able to change the suffix and still have the two messages have the same output. Such collisions may be found because iterated MACs have an internal state. A collision on the output means that at some point, the internal state of the two MACs was probably the same. Given the same internal state, and the same suffixes, the MAC will be likely to output the same tags.

For completeness, we define the internal collision attack, as given in [8], more formally in the Appendix.

## 4    A New Attack

We first present a basic modification to the iterated MAC scheme, having a random message prefix. Many random prefix versions of MACs seem to be secure against internal collision attacks. We then present the L-collision attack, a new attack which breaks these schemes. Finally, we find the probability that the attack produces a valid forgery of an unqueried message for any number of queries.

### 4.1    Random Prefix Versions of Message Authentication Schemes

The iterated message authentication schemes that we examine in this paper are modified by adding a random prefix to the message to be tagged before the tagging algorithm is run. We denote a random prefix version of a scheme by prepending it with RP-. For example, CBC-MAC with a random prefix added to the message can be refered to as RP-CBC-MAC. We shall use the phrase RP-MAC to denote a general iterated MAC with a random prefix.

One thing to note about RP-MACs is that internal collision attacks do not seem to work. The reason for this is that although one may get two messages to collide on some output, one cannot get any further queries with one of the two random numbers used in the collision. These random numbers are prepended to the messages, and thus become part of the prefix. We are thus unable to generate a query to the third message in the internal collision attack because we cannot get the same prefix again.

### 4.2    The L-collision Attack

We now describe a new attack that can be used to defeat a random prefix scheme. This is not a length-based attack, and any MAC which would normally fall to an internal collision attack would fall to this new attack if random data were first prepended.

We first define the idea of a collision formally so that the similarities can be seen in the definition of a collision and an L-collision.

**Definition 4.** *Let A be a set, and let C be a list of elements from the set A. We say that the list C contains a collision if there exist at least two elements of the list C that are the same.*

The idea of a collision can be extended by thinking about what sort of collisions can occur when we take the list $C$ from the Cartesian product of two sets. An

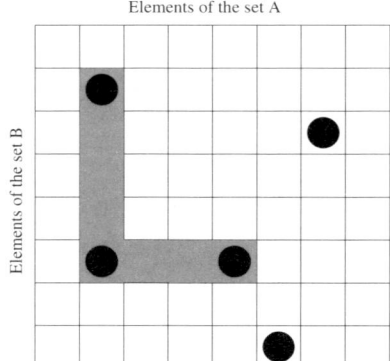

Elements of the set A

Elements of the set B

**Fig. 2.** An L-collision given 5 queries, where there are only 8 possible random values and 8 possible output values.

L-collision, intuitively, is a chain of two collisions in this Cartesian product. One collision is in the first set, and the second collision is in the second set. These two collisions contain a single common point. A picture of this can be seen in Figure 2. In the figure, the x-axis is indexed by the elements of the first set, the y-axis is indexed by the elements of the second set, and the black balls represent the elements of the list $C$. An L-collision is highlighted in the figure. The L-shape that is formed by the collision chain is the reason we refer to these chains as L-collisions. We now state the formal definition for an L-collision.

**Definition 5.** *Let $A, B$ be two sets, and let $C$ be a list of elements from the set $A \times B$. We say that the set $C$ contains an L-collision over $A$ and $B$ if there exist distinct elements $(a, b), (a, y), (x, b) \in C$. We call these elements the collision points. We call $(a, b)$ the pivot of the L-collision, $(a, y)$ the A-collision end, and $(x, b)$ the B-collision end.*

For the L-collision attack, we let one of the sets be the possible random values that could be prefixed to the message, and we let the other set be the possible output values. By getting a certain L-collision over these sets, we can forge a message.

The messages queried have two $m$ block message prefixes, and two $n$ block message suffixes. There are four different ways these prefixes and suffixes can be combined into complete messages. By generating enough queries on three of these combinations, we hope to forge the fourth combination.

**Definition 6.** *Let $m, n$ be any positive integers. Then we define an L-collision attack as an attack performed by an adversary $A$ that, given an oracle $g = \mathsf{Tag}(K, \cdot)$ attacks a MAC scheme as follows:*

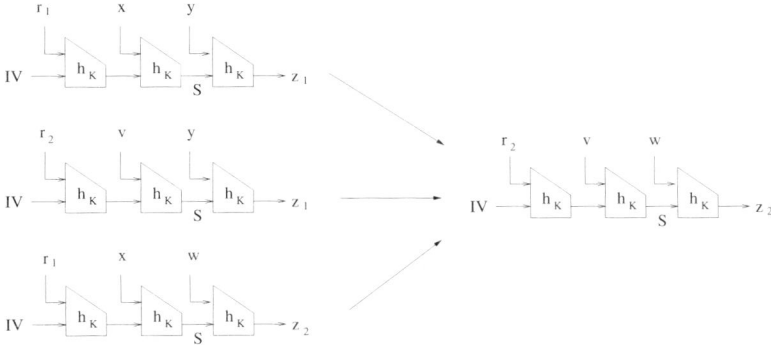

**Fig. 3.** An L-collision on the three messages $M_1$, $M_2$, and $M_3$ allows us to forge the fourth message $M_4$. Notice that the internal state $S$ is the same before the third block in all of the messages. For clarity, $m = 1$ and $n = 1$ in this example.

---

**function $A^g$**
$\quad x \overset{R}{\leftarrow} \{0,1\}^{mb}$
$\quad v \overset{R}{\leftarrow} \{0,1\}^{mb} - \{x\}$
$\quad y \overset{R}{\leftarrow} \{0,1\}^{nb}$
$\quad w \overset{R}{\leftarrow} \{0,1\}^{nb} - \{y\}$
$\quad M_1 \leftarrow x \parallel y; \; M_2 \leftarrow v \parallel y$
$\quad M_3 \leftarrow x \parallel w; \; M_4 \leftarrow v \parallel w$
$\quad$**for** $i = 1$ **to** 3 **do**
$\quad\quad$**for** $j = 1$ **to** $\lfloor \frac{q}{3} \rfloor$ **do**
$\quad\quad\quad (r_{ij}, z_{ij}) \leftarrow g(M_i)$
$\quad\quad$**endfor**
$\quad$**endfor**
$\quad$**if** $\exists\, e, d, f$ such that $r_{1d} = r_{3e}$ and $z_{1d} = z_{2f}$, **then**
$\quad\quad$**return** $(M_4, (r_{2f}, z_{3e}))$

---

Figure 3 shows us why the L-collision attack works. Because the internal state $S$ is the same in all of the messages at some point, we know that the outputs depend only upon the final block(s) which occur after this point. Thus, we know that the outputs of the fourth message and the second message are going to be the same. The theorem below specifies the parameters for the L-collision attack.

**Theorem 1.** *For any deterministic, non-stateful iterated message authentication scheme MAC, there is a constant $c > 0$ such that an adversary $A$ using the above strategy requires only*

- *$q_s$ tagging queries,*
- *$q_v$ verification queries, and*
- *$c(b + l)(q_s + q_v)$ time*

*to be able to forge a message for the random prefix version of MAC, RP-MAC,*
*with probability $\epsilon \approx \left(\frac{q_s^3}{162 \cdot 2^{l+r}}\right)\left(1 - \frac{n}{n+1}\right)$, where $n$ is the number of blocks in the*
*suffixes $y$ and $w$, and $q_s$ satisfies $\max\{2^{l/2}, 2^{r/2}\} < q_s < 2^{(l+r)/3}$.*

If we assume that the MAC is CBC-MAC or has a similar combine-then-permute structure, the above result is similar except that $\epsilon \approx \left(\frac{q_s^3}{162 \cdot 2^{l+r}}\right)$. We leave the proof of these theorems for later. In order to be able to complete this proof, we must first be able to find the probability that a randomly chosen set contains L-collisions over two sets.

### 4.3   L-collision Probabilities

We now can turn our attention away from the cryptographic problem for a while, to examine the probabilistic problem of the frequency of L-collisions. We can represent L-collisions in a rectangle, with rows representing the elements of set $A$ and columns representing the elements of the set $B$.

**Definition 7.** *Suppose we pick $q$ elements randomly and independently from the set $A \times B$ where $|A| = M$ and $|B| = N$. Then $L(M, N, q)$ denotes the probability of at least one L-collision over $A$ and $B$ within these elements.*

We now provide a lemma that gives us the approximate probability of finding such an L-collision.

**Lemma 1.** *Let $L(M, N, q)$ be defined as above. Then,*

$$L(M, N, q) \approx \frac{q^3}{6MN} \tag{1}$$

*for all $q$ such that $\max\{\sqrt{M}, \sqrt{N}\} < q < (MN)^{\frac{1}{3}}$.*

The results of this lemma can be seen intuitively through the following argument. The probability that any given triple, $(a, b), (c, d), (e, f)$ taken randomly from $A \times B$, is an L-collision is equal to the probability that $a = c$ times the probability that $b = f$, since these events are independent. Thus, a triple is an L-collision with probability of $\frac{1}{MN}$. If we chose three elements from the $q$ random queries, we have $\frac{q^3}{6}$ ways of choosing the elements without regard to order. Putting this together, we expect to see an L-collision in the queries with probability $\frac{q^3}{6MN}$.

### 4.4   Analysis of L-collision Attack

Now that we know bounds on the probability that there is an L-collision, we can prove Theorem 1.

*Proof (Theorem 1).* We first show that the algorithm provides a valid forgery in the case where the **if** statement is true. So, suppose there exists an $e, d, f$ such that $r_{1d} = r_{3e}$ and $z_{1d} = z_{2f}$. Then we know that $(r_{1d}, z_{1d}) = \mathsf{Tag}(K, x \parallel y)$ and $(r_{2f}, z_{2f}) = \mathsf{Tag}(K, v \parallel y)$ collide on the iteration outputs since $z_{1d} = z_{2f}$. The message prefixes which are likely to have led to this collision are $r_{1d} \parallel x$ and $r_{2f} \parallel v$ (note that the randomness is prefixed to the message). Because the suffixes of both messages are the same, we have a birthday style collision here. According to [8], the collision occurs before the $n$ block suffix $y$ with probability $\left(1 - \frac{n}{n+1}\right)$ for a random function and with probability 1 for a CBC-style iterative function with the block cipher replaced by a random permutation. Once we have this collision, all we need is one query of $r_{1d} \parallel x$ with a different suffix, and we will be able to forge the message $r_{2f} \parallel v$ with that same suffix.

We get this additional query by the fact that we know $(r_{3e}, z_{3e}) = \mathsf{Tag}(K, x \parallel w)$. Since $r_{3e} = r_{1d}$, we have the same prefix, $r_{1d} \parallel x$ with a different suffix, $w$. Because it is likely that the output will be the same, we replace the prefix of the third message with the colliding prefix of the second message. This gives us to $r_{2f} \parallel v \parallel w$, a valid forgery under the output $z_{3e}$. It is clear that $v \parallel w$ has never previously been queried, since the only queries were to $x \parallel y$, $x \parallel w$, and $v \parallel y$.

We now need a lower bound on the probability that the **if** statement is be true. Notice that the if statement corresponds to there being an L-collision over $\{0,1\}^r$ and $\{0,1\}^l$, where we set the pivot to be $(r_{1d}, z_{1d})$, we set the $\{0,1\}^r$-collision end to be $(r_{3e}, z_{3e})$, and we set the $\{0,1\}^l$-collision end to be $(r_{2f}, z_{2f})$.

Lemma 1 shows an approximation of the probability that an L-collision occurs over two sets, given that there are $q$ queries made. We let one of these sets be $\{0,1\}^r$, and the other be $\{0,1\}^l$, and we set the number of collisions made to be $q_s$. Then, we get a lower bound on the probability of an L-collision in this case to be $L(2^r, 2^l, q_s) \approx \frac{q_s^3}{6 \cdot 2^{l+r}}$. Just the existence of an L-collision is not enough, however. We need the L-collision to be such that the points $(r_{1d}, z_{1d})$, $(r_{3e}, z_{3e})$, and $(r_{2f}, z_{2f})$ are the pivot, the $\{0,1\}^r$-collision end, and the $\{0,1\}^l$-collision end, respectively. The probability that this happens is $\frac{\lfloor \frac{1}{3} q_s \rfloor}{q_s} \cdot \frac{\lfloor \frac{1}{3} q_s \rfloor}{q_s - 1} \cdot \frac{\lfloor \frac{1}{3} q_s \rfloor}{q_s - 2} \approx \frac{1}{27}$. Combining the two above probabilities, we get the approximate probability that the **if** statement is true, $\frac{q_s^3}{27 \cdot 6 \cdot 2^{l+r}} = \frac{q_s^3}{162 \cdot 2^{l+r}}$.

The only thing left to do is show that we can find this collision in time proportional to $q_s$, and thus finding the collision takes no more time than computing the MAC in the first place. We can do this by keeping a hash table of outputs and randomness. Then we also have hash tables of output collisions and randomness collisions. Whenever we find that we have an output collision by the output hash table, we can add this collision to the output collision hash table. Similarly for randomness collisions. We then search through all of the $M_2$ query results. If any of them have elements in both the row collision and column collision hash table, we are done. This whole procedure takes linear time in $q_s$, concluding the proof. ∎

Given Theorem 1, we can see that only $O\left(2^{(l+r)/3} + \max\{2^{l/2}, 2^{r/2}\}\right)$ queries are required to get a constant probability of the collision chain occurring. While this attack requires more queries than normal birthday attacks, it can turn out to cause almost as many problems. We can illustrate this by looking at CBC-MAC with a 64-bit block cipher. Without the random bits prepended, we saw that we could get a forgery in about $2^{32}$ queries. After prepending 64 random bits, we now require about $2^{43}$ queries. This is almost as unacceptable, given the computing power of today's machines. Given the doubling in size of our message authentication tag, we would hope that the security would also be doubled. The L-collision attack shows that this is not the case.

## 5    Conclusion

Now that we know an attack against message authentication schemes with a random prefix, there is still the open question of whether there are any non-stateful schemes which avoid collision-based attacks on tagging algorithm queries. The problem with such a request is that there are many tradeoffs which can be made to achieve better security. Most schemes involve either greatly increased computation time or greatly increased key size. What we would like to see is a simple scheme which requires just one block cipher key, that has near ideal security.

This paper has shown that there is an inherent upper bound to the security that can be achieved by prepending randomness to messages in an iterated MAC scheme. Because this bound is less than ideal, it may be in the best interest of MAC cryptanalysts to examine other ways of combining randomness with the message.

This paper has also shown that the design of randomized MAC schemes is more difficult than first imagined. The birthday bound seems to be one of many bounds between current message authentication schemes and ideal schemes. Usually, when a MAC scheme is designed, one can immediately see how birthday attacks apply. L-collision attacks are much more difficult to see, and when they apply, they can be much trickier to avoid.

### Acknowledgments

This paper would not have been possible without the guidance of my advisor, Mihir Bellare. I would like to give additional thanks to David Wagner for sheparding my paper, and for providing his comments on my paper drafts. His willingness to assist has taken a stressful load off of my shoulders.

The work done on this paper was supported in part by Mihir Bellare's 1996 Packard Foundation Fellowship in Science and Engineering and NSF CAREER Award CCR-9624439.

## References

1. ANSI X9.9. American National Standard for Financial Institution Message Authentication (Wholesale), American Bankers Association, 1981. Revised 1986.

2. M. Bellare, R. Canetti, and H. Krawczyk. Keying Hash Functions for Message Authentication. *Advances in Cryptology – Crypto 96 Proceedings*, Lecture Notes in Computer Science Vol. 1109, N. Koblitz ed., Springer-Verlag, 1996.
3. M. Bellare, O. Goldreich, and H. Krawczyk. Stateless Evaluation of Pseudorandom Functions: Security beyond the Birthday Barrier. *Advances in Cryptology – Crypto 99 Proceedings*, Lecture Notes in Computer Science Vol. 1666, M. Wiener ed., Springer-Verlag, 1999.
4. M. Bellare, R. Guerin, and P. Rogaway. XOR MACs: New Methods for Message Authentication Using Finite Pseduorandom Functions. *Advances in Cryptology – Crypto 95 Proceedings*, Lecture Notes in Computer Science Vol. 963, D. Coppersmith ed., Springer-Verlag, 1995.
5. M. Bellare, J. Killian, and P. Rogaway. The security of cipher block chaining. *Advances in Cryptology – Crypto 94 Proceedings*, Lecture Notes in Computer Science Vol. 839, Y. Desmedt ed., Springer-Verlag, 1994.
6. A. Menezes, P. van Oorschot, and S. Vanstone. Handbook of Applied Cryptography. CRC Press. 1996.
7. E. Petrank and C. Rackoff. CBC-MAC for Real-Time Data Sources. Dimacs Technical Report, 97-26, 1997.
8. B. Preneel and P. van Oorschot. MDx-MAC and Building Fast MACs from Hash Functions. *Advances in Cryptology – Crypto 95 Proceedings*, Lecture Notes in Computer Science Vol. 963, D. Coppersmith ed., Springer-Verlag, 1995.

# 6 Appendix

## 6.1 Internal Collision Attacks

We formally describe internal collision attacks below.

**Definition 8.** *Let $\mathcal{MA}$ be an iterated message authentication scheme with block-size $b$, and let $m, n$ be any positive integers. Then the internal collision attack is performed by an adversary $B$, given an oracle $g = \mathsf{Tag}(K, \cdot)$, according the following algorithm.*

$$
\begin{array}{|l|}
\hline
\textit{function } B^g \\
\quad y \xleftarrow{R} \{0,1\}^{nb}; \ w \xleftarrow{R} \{0,1\}^{nb} - \{y\} \\
\quad \textbf{for } i = 1 \textbf{ to } q \textbf{ do} \\
\quad\quad x_i \xleftarrow{R} \{0,1\}^{mb} \\
\quad\quad M_i \leftarrow x_i \parallel y \\
\quad\quad z_i \leftarrow g(M_i) \\
\quad \textbf{endfor} \\
\quad \textbf{if } \exists \, d, e \text{ such that } z_d = z_e, \textbf{ then} \\
\quad\quad M \leftarrow x_d \parallel w; \ M' \leftarrow x_e \parallel w \\
\quad\quad z \leftarrow g(M) \\
\quad\quad \textbf{return } (M', z) \\
\quad \textbf{else return null} \\
\hline
\end{array}
$$

When we get two different messages that give us the same output, we say that there is a *collision*. This adversary succeeds with probability $\left(1 - \frac{n}{n+1}\right)$

whenever there is a collision (see [8]). This is because any collision that occurs is likely to occur in the prefixes $x_d$ and $x_e$, and thus regardless of what we append to them (as long as they maintain the message length) we still have a collision. A well-known result of [8] is that these collisions are expected to occur when there are $\sqrt{2} \cdot 2^{(l/2)}$ queries. This is generalized by the following theorem:

**Theorem 2.** *For any deterministic, non-stateful iterated message authentication scheme MAC, there is a constant $c > 0$ such that an adversary $B$ using the internal collision attack requires only*

- $q_s$ *tagging queries,*
- $q_v$ *verification queries, and*
- $c(b + l)(q_s + q_v)$ *time*

*to be able to forge a message with probability* $\epsilon = \left(1 - e^{-\frac{q_s^2(n+1)}{2^{l+1}}}\right)\left(1 - \frac{n}{n+1}\right)$
*where $n$ is the number of blocks in the suffixes $y$ and $w$ and where $q_s$ satisfies* $q_s < \frac{2^{l/2}}{n}$.

This theorem is implied by the results of [8].

Notice that the internal collision attack queries only messages of the same length. This is important because collisions that occur when messages are of one length, might not occur when messages are another length. Consider for instance the case where the length of the message is prepended before querying. It is also important to notice that this attack only takes $\Theta(q)$ time. This is because for each query, we can immediately detect whether the output collided with a previous output or not using a hash table keyed to the outputs.

# On the Exact Security of Full Domain Hash

Jean-Sébastien Coron

Ecole Normale Supérieure      Gemplus Card International

45 rue d'Ulm      34 rue Guynemer

Paris, F-75230, France      Issy-les-Moulineaux, F-92447, France

coron@clipper.ens.fr      jean-sebastien.coron@gemplus.com

**Abstract.** The Full Domain Hash (FDH) scheme is a RSA-based signature scheme in which the message is hashed onto the full domain of the RSA function. The FDH scheme is provably secure in the random oracle model, assuming that inverting RSA is hard. In this paper we exhibit a slightly different proof which provides a tighter security reduction. This in turn improves the efficiency of the scheme since smaller RSA moduli can be used for the same level of security. The same method can be used to obtain a tighter security reduction for Rabin signature scheme, Paillier signature scheme, and the Gennaro-Halevi-Rabin signature scheme.

## 1 Introduction

Since the discovery of public-key cryptography by Diffie and Hellman [3], one of the most important research topics is the design of practical and provably secure cryptosystems. A proof of security is usually a computational reduction from solving a well established problem to breaking the cryptosystem. Well established problems include factoring large integers, computing the discrete logarithm modulo a prime $p$, or extracting a root modulo a composite integer. The RSA cryptosystem [9] is based on this last problem.

A very common practice for signing with RSA is to first hash the message, add some padding, and then exponentiate it with the decryption exponent. This "hash and decrypt" paradigm is the basis of numerous standards such as PKCS #1 v2.0 [10]. In this paradigm, the simplest scheme consists in taking a hash function, the output size of which is exactly the size of the modulus : this is the Full Domain Hash scheme (FDH), introduced by Bellare and Rogaway in [1]. The FDH scheme is provably secure in the random oracle model, assuming that inverting RSA, *i.e.* extracting a root modulo a composite integer, is hard. The random oracle methodology was introduced by Bellare and Rogaway in [1] where they show how to design provably secure signature schemes from any trapdoor permutation. In the random oracle model, the hash function is seen as an oracle which produces a random value for each new query.

The seminal work of Bellare and Rogaway in [1] and [2] highlights the importance, for practical applications of provable security, of taking into account the tightness of the security reduction. A security reduction is tight when breaking

M. Bellare (Ed.): CRYPTO 2000, LNCS 1880, pp. 229–235, 2000.
© Springer-Verlag Berlin Heidelberg 2000

the signature scheme leads to solving the well established problem with sufficient probability, ideally with probability one. In this case, the signature scheme is almost as secure as the well established problem. On the contrary, if the reduction is "loose", *i.e.* the above probability is too small, the guarantee on the signature scheme can be quite weak.

In this paper, we exhibit a better security reduction for the FDH signature scheme, which gives a tighter security bound. The reduction in [2] bounds the probability $\epsilon$ of breaking FDH in time $t$ by $\epsilon' \cdot (q_{hash} + q_{sig})$ where $\epsilon'$ is the probability of inverting RSA in time $t'$ comparable to $t$ and $q_{hash}$ and $q_{sig}$ are the number of hash queries and signature queries requested by the forger. The new reduction bounds the probability $\epsilon$ of breaking FDH by roughly $\epsilon' \cdot q_{sig}$ with the same running time $t$ and $t'$. This is significantly better in practice since $q_{sig}$ is usually much less than $q_{hash}$. Full domain hash is thus more secure than originally foreseen. With a tighter provable security one can safely use a smaller modulus size, which in turn improves the efficiency of the scheme.

## 2    Definitions

### 2.1    Signature Schemes

A digital signature of a message is a bit string dependent on some secret known only to the signer, and on the content of the message being signed. Signatures must be verifiable : anyone can check the validity of the signature. The following definitions are based on [5].

**Definition 1 (signature scheme).** *A signature scheme is defined by the following :*

*- The key generation algorithm Gen is a probabilistic algorithm which given $1^k$, outputs a pair of matching public and secret keys, $(pk, sk)$.*

*- The signing algorithm Sign takes the message $M$ to be signed and the secret key $sk$ and returns a signature $x = Sign_{sk}(M)$. The signing algorithm may be probabilistic.*

*- The verification algorithm Verify takes a message $M$, a candidate signature $x'$ and the public key $pk$. It returns a bit $Verify_{pk}(M, x')$, equal to 1 if the signature is accepted, and 0 otherwise. We require that if $x \leftarrow Sign_{sk}(M)$, then $Verify_{pk}(M, x) = 1$.*

Signature schemes most often use hash functions. In the following, the hash function is seen as a random oracle : the output of the hash function $h$ is a uniformly distributed point in the range of $h$. Of course, if the same input is invoked twice, identical outputs are returned.

### 2.2    Security of Signature Schemes

The security of signature schemes was formalized in an asymptotic setting by Goldwasser, Micali and Rivest [5]. Here we use the definitions of [1] and [2] which take into account the presence of an ideal hash function, and give a concrete security analysis of digital signatures. Resistance against adaptive chosen-message attacks is considered : a forger $\mathcal{F}$ can dynamically obtain signatures of messages of its choice and attempts to output a valid forgery. A *valid forgery* is a message/signature pair $(M, x)$ such that $Verify_{pk}(M, x) = 1$ but the signature of $M$ was never requested by $\mathcal{F}$.

**Definition 2.** *A forger $\mathcal{F}$ is said to $(t, q_{sig}, q_{hash}, \epsilon)$-break the signature scheme $(Gen, Sign, Verify)$ if after at most $q_{hash}(k)$ queries to the hash oracle, $q_{sig}(k)$ signatures queries and $t(k)$ processing time, it outputs a valid forgery with probability at least $\epsilon(k)$ for all $k \in \mathbb{N}$.*

**Definition 3.** *A signature scheme $(Gen, Sign, Verify)$ is $(t, q_{sig}, q_{hash}, \epsilon)$-secure if there is no forger who $(t, q_{sig}, q_{hash}, \epsilon)$-breaks the scheme.*

## 2.3   The RSA Cryptosystem

The RSA cryptosystem [9] is the most widely used public-key cryptosytem. It can be used to provide both secrecy and digital signatures.

**Definition 4 (The RSA cryptosystem).** *The RSA cryptosystem is a family of trapdoor permutations. It is specified by :*
   *- The RSA generator $\mathcal{RSA}$, which on input $1^k$, randomly selects 2 distinct $k/2$-bit primes $p$ and $q$ and computes the modulus $N = p \cdot q$. It randomly picks an encryption exponent $e \in \mathbb{Z}_{\phi(N)}^*$ and computes the corresponding decryption exponent $d$ such that $e \cdot d = 1 \bmod \phi(N)$. The generator returns $(N, e, d)$.*
   *- The encryption function $f : \mathbb{Z}_N^* \to \mathbb{Z}_N^*$ defined by $f(x) = x^e \bmod N$.*
   *- The decryption function $f^{-1} : \mathbb{Z}_N^* \to \mathbb{Z}_N^*$ defined by $f^{-1}(y) = y^d \bmod N$.*

## 2.4   Quantifying the Security of RSA

We follow the definitions of [2]. An *inverting algorithm* $\mathcal{I}$ for RSA gets input $N, e, y$ and tries to find $f^{-1}(y)$. Its success probability is the probability to output $f^{-1}(y)$ when $N, e, d$ are obtained by running $\mathcal{RSA}(1^k)$ and $y$ is set to $f(x)$ for an $x$ chosen at random in $\mathbb{Z}_N^*$.

**Definition 5.** *An inverting algorithm $\mathcal{I}$ is said to $(t, \epsilon)$-break RSA if after at most $t(k)$ processing time its success probability is at least $\epsilon(k)$ for all $k \in \mathbb{N}$.*

**Definition 6.** *RSA is said to be $(t, \epsilon)$ secure if there is no inverter which $(t, \epsilon)$-breaks RSA.*

# 3   The Full Domain Hash Signature Scheme

## 3.1   Definition

The Full Domain Hash $(GenFDH, SignFDH, VerifyFDH)$ signature scheme
[1] is defined as follows. The key generation algorithm, on input $1^k$, runs $\mathcal{RSA}(1^k)$
to obtain $(N, e, d)$. It outputs $(pk, sk)$, where $pk = (N, e)$ and $sk = (N, d)$. The
signing and verifying algorithm have oracle access to a hash function $H_{FDH}$ :
$\{0, 1\}^* \to \mathbb{Z}_N^*$. Signature generation and verification are as follows :

$SignFDH_{N,d}(M)$
   $y \leftarrow H_{FDH}(M)$
   return $y^d \bmod N$

$VerifyFDH_{N,e}(M, x)$
   $y \leftarrow x^e \bmod N; y' \leftarrow H_{FDH}(M)$
   if $y = y'$ then return 1 else return 0.

The concrete security analysis of the FDH scheme is provided by the following
theorem [1] :

**Theorem 1.** *Suppose RSA is $(t', \epsilon')$-secure. Then the Full Domain Hash signature scheme is $(t, \epsilon)$-secure where $t = t' - (q_{hash} + q_{sig} + 1) \cdot \mathcal{O}(k^3)$ and
$\epsilon = (q_{hash} + q_{sig}) \cdot \epsilon'$.*

As stated in [2], the disadvantage of this result is that $\epsilon'$ could be much
smaller than $\epsilon$. For example, if we assume like in [2] that the forger is allowed
to request $q_{sig} = 2^{30}$ signatures and computes hashes on $q_{hash} = 2^{60}$ messages,
even if the RSA inversion probability is as low as $2^{-61}$, then all we obtain is that
the forging probability is at most $1/2$, which is not satisfactory. To obtain an
acceptable level of security, we must use a larger modulus, which will affect the
efficiency of the scheme.

To obtain a better security bound, Bellare and Rogaway designed a new
scheme, the *probabilistic signature scheme* (PSS), which achieves a tight security
reduction : the probability of forging a signature is almost equally low as inverting
RSA ($\epsilon \simeq \epsilon'$). Instead, we show in the next section that a better security bound
can be obtained for the original FDH scheme.

## 3.2   The New Security Reduction

We exhibit a different reduction which gives a better security bound for FDH.
Namely, we prove the following theorem :

**Theorem 2.** *Suppose RSA is $(t', \epsilon')$-secure. Then the Full Domain Hash signature scheme is $(t, \epsilon)$-secure where*

$$t = t' - (q_{hash} + q_{sig} + 1) \cdot \mathcal{O}(k^3)$$

$$\epsilon = \frac{1}{(1 - \frac{1}{q_{sig}+1})^{q_{sig}+1}} \cdot q_{sig} \cdot \epsilon'$$

*For large $q_{sig}$,*

$$\epsilon \simeq \exp(1) \cdot q_{sig} \cdot \epsilon'$$

*Proof.* Let $\mathcal{F}$ be a forger that $(t, q_{sig}, q_{hash}, \epsilon)$-breaks FDH. We assume that $\mathcal{F}$ never repeats a hash query or a signature query. We build an inverter $\mathcal{I}$ which $(t', \epsilon')$-breaks RSA.

The inverter $\mathcal{I}$ receives as input $(N, e, y)$ where $(N, e)$ is the public key and $y$ is chosen at random in $\mathbb{Z}_N^*$. The inverter $\mathcal{I}$ tries to find $x = f^{-1}(y)$ where $f$ is the RSA function defined by $N, e$. The inverter $\mathcal{I}$ starts running $\mathcal{F}$ for this public key. Forger $\mathcal{F}$ makes hash oracle queries and signing queries. $\mathcal{I}$ will answer hash oracle queries and signing queries itself. We assume for simplicity that when $\mathcal{F}$ requests a signature of the message $M$, it has already made the corresponding hash query on $M$. If not, $\mathcal{I}$ goes ahead and makes the hash query itself. $\mathcal{I}$ uses a counter $i$, initially set to zero.

When $\mathcal{F}$ makes a hash oracle query for $M$, the inverter $\mathcal{I}$ increments $i$, sets $M_i = M$ and picks a random $r_i$ in $\mathbb{Z}_N^*$. $\mathcal{I}$ then returns $h_i = r_i^e \bmod N$ with probability $p$ and $h_i = y \cdot r_i^e \bmod N$ with probability $1 - p$. Here $p$ is a fixed probability which will be determined later.

When $\mathcal{F}$ makes a signing query for $M$, it has already requested the hash of $M$, so $M = M_i$ for some $i$. If $h_i = r_i^e \bmod N$ then $\mathcal{I}$ returns $r_i$ as the signature. Otherwise the process stops and the inverter has failed.

Eventually, $\mathcal{F}$ halts and outputs a forgery $(M, x)$. We assume that $\mathcal{F}$ has requested the hash of $M$ before. If not, $\mathcal{I}$ goes ahead and makes the hash query itself, so that in any case $M = M_i$ for some $i$. Then if $h_i = y \cdot r_i^e \bmod N$ we have $x = h_i^d = y^d \cdot r_i \bmod N$ and $\mathcal{I}$ outputs $y^d = x/r_i \bmod N$ as the inverse for $y$. Otherwise the process stops and the inverter has failed.

The probability that $\mathcal{I}$ answers to all signature queries is at least $p^{q_{sig}}$. Then $\mathcal{I}$ outputs the inverse of $y$ for $f$ with probability $1 - p$. So with probability at least $\alpha(p) = p^{q_{sig}} \cdot (1 - p)$, $\mathcal{I}$ outputs the inverse of $y$ for $f$. The function $\alpha(p)$ is maximal for $p_{max} = 1 - 1/(q_{sig} + 1)$ and

$$\alpha(p_{max}) = \frac{1}{q_{sig}} \left(1 - \frac{1}{q_{sig} + 1}\right)^{q_{sig}+1}$$

Consequently we obtain :

$$\epsilon(k) = \frac{1}{(1 - \frac{1}{q_{sig}+1})^{q_{sig}+1}} \cdot q_{sig} \cdot \epsilon'(k)$$

and for large $q_{sig}$, $\epsilon(k) \simeq \exp(1) \cdot q_{sig} \cdot \epsilon'(k)$.

The running time of $\mathcal{I}$ is the running time of $\mathcal{F}$ added to the time needed to compute the $h_i$ values. This is essentially one *RSA* computation, which is cubic time (or better). This gives the formula for $t$.

□

### 3.3  Discussion

In many security proofs in the random oracle model (including [2]), the inverter has to "guess" which hash query will be used by the adversary to produce its forgery, resulting in a factor of $q_{hash}$ in the success probability. This paper shows that a better method is to include the challenge $y$ in the answer of many hash queries so that the forgery is useful to the inverter with greater probability. This observation also applies to the Rabin signature scheme [8], the Paillier signature scheme [7] and also the Gennaro-Halevi-Rabin signature scheme [4], for which the $q_{hash}$ factor in the random oracle security proof can also be reduced to $q_{sig}$.

## 4  Conclusion

We have improved the security reduction of the FDH signature scheme in the random oracle model. The quality of the new reduction is independent from the number of hash calls performed by the forger, and depends only on the number of signatures queries. This is of practical significance since in real-world applications, the number of hash calls is only limited by the computational power of the forger, whereas the number of signature queries can be deliberately limited : the signer can refuse to sign more tha n $2^{20}$ or $2^{30}$ messages. However, the reduction is still not tight and there remains a sizable gap between the exact security of FDH and the exact security of PSS.

### Acknowledgements

I would like to thank Jacques Stern, David Pointcheval and Alexey Kirichenko for helpful discussions and the anonymous referees for their constructive comments.

## References

1. M. Bellare and P. Rogaway, *Random oracles are practical : a paradigm for designing efficient protocols*. Proceedings of the First Annual Conference on Computer and Commmunications Security, ACM, 1993.

2. M. Bellare and P. Rogaway, *The exact security of digital signatures - How to sign with RSA and Rabin*. Proceedings of Eurocrypt'96, LNCS vol. 1070, Springer-Verlag, 1996, pp. 399-416.

3. W. Diffie and M. Hellman, *New directions in cryptography*, IEEE Transactions on Information Theory, IT-22, 6, pp. 644-654, 1976.

4. R. Gennaro, S. Halevi, T. Rabin, *Secure hash-and-sign signatures without the random oracle*, proceedings of Eurocrypt'99, LNCS vol. 1592, Springer-Verlag, 1999, pp. 123-139.

5. S. Goldwasser, S. Micali and R. Rivest, *A digital signature scheme secure against adaptive chosen-message attacks*, SIAM Journal of computing, 17(2):281-308, april 1988.

6. A. Lenstra and H. Lenstra (eds.), *The development of the number field sieve*, Lecture Notes in Mathematics, vol 1554, Springer-Verlag, 1993.

7. P. Paillier, *Public-key cryptosystems based on composite degree residuosity classes*. Proceedings of Eurocrypt'99, Lecture Notes in Computer Science vol. 1592, Springer-Verlag, 1999, pp. 223-238.

8. M.O. Rabin, *Digitalized signatures and public-key functions as intractable as factorization*, MIT/LCS/TR-212, MIT Laboratory for Computer Science, 1979.

9. R. Rivest, A. Shamir and L. Adleman, *A method for obtaining digital signatures and public key cryptosystems*, CACM 21, 1978.

10. RSA Laboratories, PKCS #1 : *RSA cryptography specifications*, version 2.0, September 1998.

# Timed Commitments

## (Extended Abstract)

Dan Boneh[1] and Moni Naor[2]

[1] Stanford University, dabo@cs.stanford.edu
[2] Weizmann institute, naor@wisdom.weizmann.ac.il

**Abstract.** We introduce and construct *timed commitment* schemes, an extension to the standard notion of commitments in which a potential forced opening phase permits the receiver to recover (with effort) the committed value without the help of the committer. An important application of our timed-commitment scheme is contract signing: two mutually suspicious parties wish to exchange signatures on a contract. We show a two-party protocol that allows them to exchange RSA or Rabin signatures. The protocol is *strongly fair*: if one party quits the protocol early, then the two parties must invest comparable amounts of time to retrieve the signatures. This statement holds even if one party has many more machines than the other. Other applications, including honesty preserving auctions and collective coin-flipping, are discussed.

## 1    Introduction

This paper introduces *timed commitments*. A timed commitment is a commitment scheme in which there is an optional forced opening phase enabling the receiver to recover (with effort) the committed value without the help of the committer. A regular commitment scheme consists of two phases: (i) The commit phase at the end of which the sender is bound to some value $b$, and (ii) The reveal phase, where the sender reveals $b$ to the receiver. Following the commit phase the sender should be bound to $b$, but the receiver should be unable to learn anything about $b$. A timed commitment has an additional forced opening phase, where the receiver computes a moderately hard function and recovers the value $b$ without the participation of the sender. As a result, the value $b$ remains hidden from the receiver for only a limited amount of time.

Timed commitments satisfy the following properties: (1) *verifiable recovery:* if the commit phase ends successfully, the receiver is convinced that forced opening will yield $b$. (2) *recovery with proof:* the receiver not only recovers $b$ but also a proof of its value, so that anyone who has the commitment (or the transcript of the commit phase) can verify that $b$ is the value committed *without going through a recovery process.* (3) *commitment immune against parallel attacks:* even if a receiver has many more processors than assumed, it cannot recover $b$ much faster than a single-processor receiver.

An important application of our timed-commitment scheme is contract signing: two mutually suspicious parties wish to exchange signatures on a contract.

M. Bellare (Ed.): CRYPTO 2000, LNCS 1880, pp. 236–254, 2000.
© Springer-Verlag Berlin Heidelberg 2000

We show a two-party protocol that allows them to exchange RSA, Rabin or Fiat-Shamir signatures (or any scheme where the signature is a power of a pre-determined value mod a composite). Therefore, no special "infrastructure" is needed. One can use existing PKI and there is no need to modify the "semantics" of a signature.

Our contract signing protocol is based on gradual release of information (however, we do not release the signature "bit-by-bit"). Our protocol enjoys **strong fairness**: whatever one party may do (e.g. sending false information, stopping prematurely, etc.), the time it takes the other party to recover the signature is within a small constant of the time that it takes the misbehaving party to do so. Furthermore, our protocol is the first to resist parallel attacks. Even if one party has many more machines than the other, both parties need comparable amounts of time to recover the signature when one party aborts before the other.

Another important application of our scheme is honesty-preserving auctions. The auction participants who submit bids in sealed envelopes can have the sealing done using timed-commitments. This allows the auctioneer to convince all the participants that all the bids were considered, even if some of the parties wish to withdraw their bids.

Our timed-commitments can be used to obtain zero-knowledge in various settings, including the concurrent and the resettable settings. The most interesting application that requires all the properties of our commitment (provable recoverability and immunity to parallelization) is a three-round zero-knowledge protocol for NP in the timing model of [23].

We also confirm the folklore belief that two-party contract signing protocols require many rounds. We prove that if a protocol is to have bounded unfairness for the two parties (the ratio between the time it takes to recover a signature after early stopping) then it must take a number of rounds that is proportional to the security of the forging.

**Related Work.** The problem of contract signing was the impetus of much of the early research on cryptographic protocols. Roughly speaking contract signing protocols can be partitioned into:

- Protocols that employ a trusted third-party as a referee or judge. The third party intervenes only if things go wrong. Examples include [2,3,7,26,32,33]
- Protocols that are pure two-party (do not require a third party referee). Such protocols are based on the gradual release of signatures/secrets. Examples are [8,24,15,18,29].

Our method falls into the second category. The best scheme in this category is due to Damgard [18]. However, this scheme releases the actual signature bit-by-bit and hence is *not* immune to parallel exhaustive search attacks. If one party has access to more machines then the other then the protocol becomes unfair. This deficiency is common to all previously proposed schemes in this category.

Several recent papers focus on the trusted third party model with the goal of making it more fair, abuse free and accountable [26]. These goals are achieved by the strong fairness of our protocol.

**Timed Primitives.** Timed primitives previously came up in several contexts: Dwork and Naor [20] suggested moderately hard functions for "pricing via processing" in order to deter abuse of resources, such as spamming. Bellare and Goldwasser [4,5] suggested "time capsules" for key escrowing in order to deter widespread wiretapping. A major issue there is to verify at escrow-time that the right key is escrowed. Similar issues arise in our work, where the receiver should make sure that at the end of the commit phase the value is recoverable.

Rivest, Shamir and Wagner [35] suggested "time-locks" for encrypting data so that it is released only in the future. This is the only scheme we are aware of that took into account the parallel power of the attacker. We employ a function similar to the one they suggested. However in their setting no measures are taken to verify that the puzzle can be unlocked in the desired time.

## 2   Timed Commitments and Timed Signatures

We begin by defining our notions of *timed commitments* and *timed signatures*. We give efficient constructions for these primitives in the next section.

A $(T, t, \epsilon)$ *timed commitment* scheme for a string $S \in \{0,1\}^n$ enables Alice to give Bob a commitment to the string $S$. At a later time Alice can prove to Bob that the committed string is $S$. However, if Alice refuses to reveal $S$, Bob can spend time $T$ to forcibly retrieve $S$. Alice is assured that within time $t$ on a parallel machine with polynomially many processors, where $t < T$, Bob will succeed in obtaining $S$ with probability at most $\epsilon$. Formally, a $(T, t, \epsilon)$ timed commitment scheme consists of three phases:

**Commit phase:** To commit to a string $S \in \{0,1\}^n$ Alice and Bob execute a protocol whose outcome is a *commitment string* $C$ which is given to Bob.

**Open phase:** At a later time Alice may reveal the string $S$ to Bob. They execute a protocol so that at the end of the protocol Bob has a proof that $S$ is the committed value.

**Forced open phase:** Suppose Alice refuses to execute the open phase and does not reveal $S$. Then there exists an algorithm, called forced-open, that takes the commitment string $C$ as input and outputs $S$ and a proof that $S$ is the committed value. The algorithm's running time is at most $T$.

The commitment scheme must satisfy a number of security constraints:

BINDING: during the open phase Alice cannot convince Bob that $C$ is a commitment to $S' \neq S$.

SOUNDNESS: At the end of the commit phase Bob is convinced that, given $C$, the forced open algorithm will produce the committed string $S$ in time $T$.

PRIVACY: every PRAM algorithm $\mathcal{A}$ whose running time is at most $t$ for $t < T$ on polynomially many processors, will succeed in distinguishing $S$ from a random string, given the transcript of the commit protocol as input, with advantage at most $\epsilon$. In other words,

$$\left| \Pr[\mathcal{A}(\text{transcript}, S) = \text{"yes"}] - \Pr[\mathcal{A}(\text{transcript}, R) = \text{"yes"}] \right| < \epsilon$$

where the probability is over the random choice of $S$ and $R$ and the random bits used to create $C$ from $S$ during the commit phase.

Note that the privacy constraint measures the adversary's run time using a parallel computing model (a PRAM). Consequently, the privacy requirement ensures that even an adversary equipped with a highly parallel machine must spend at least time $t$ to forcibly open the commitment (with high probability). In other words, even an adversary with thousands of machines at his disposal cannot extract $S$ from $C$ in less than time $t$.

We define *timed signatures* analogously to timed commitments. A $(T, t, \epsilon)$ timed signature scheme enables Alice to give Bob a signature $S$ on a message $M$ in two steps. In the first step Alice commits to the signature $S$ and Bob accepts the commitment. At a later time Alice can completely reveal the signature $S$ to Bob. However, if Alice does not reveal the signature, Bob can spend time $T$ to forcibly retrieve the signature from the commitment. As before, Alice is assured that after time $t$, where $t < T$, Bob will not be able to retrieve the signature with probability more than $\epsilon$.

As before, a timed signature consists of three phases commit, open, and forced-open. The commit phase is identical to the commit phase of timed commitments. It results in Bob accepting a commitment string $C$. At a later time Alice may execute the open phase. At the end of the open phase Bob obtains a standard (message,signature) pair satisfying all the requirements of a digital signature. If the open phase is never executed Bob can run an algorithm whose run time is at most $T$ to forcibly extract the signature $S$ from the commitment $C$. In addition to soundness, a $(T, t, \epsilon)$ timed signature scheme must satisfy the following privacy requirement: all PRAM algorithms whose running time is at most $t$ will succeed in extracting $S$ from the commitment $C$ with probability at most $\epsilon$.

# 3   A Timed Commitment and Timed Signature Scheme

As one can imagine, there are two main difficulties in building a timed commitment scheme. First, during the commit phase, the committer (Alice) must convince the verifier (Bob) that the forced open algorithm will successfully retrieve the committed value. This must be done without actually running the forced open algorithm. Second, we must ensure that even an adversary with thousands of machines cannot forcibly open the commitment much faster than a legitimate party with only one machine. To solve the later issue we base our scheme on a problem that appears to be inherently sequential: modular exponentiation. We use the fact that the best known algorithm for computing $g^{(2^m)} \bmod N$ takes $m$ *sequential* squarings. The surprising fact is that there is an efficient zero-knowledge protocol, with a running time of $O(\log m)$, enabling the committer (Alice) to prove to the verifier (Bob) that the result of these $m$ squarings will produce the committed message $M$. This proof is done during the commit phase and is at the heart of our timed commitment scheme.

Let $T = 2^k$ be an integer. We build a timed commitment scheme where it takes $T = 2^k$ modular multiplications to forcibly retrieve the committed string. The commit phase takes $O(k)$ modular exponentiations. We envision $k$ as typically being in the range $[30, \ldots, 50]$. This way the forced open algorithm can take a few hours, or a few days depending on the requirements.

**Setup:** Let $n$ be a positive integer representing a certain security parameter. The committer generates two random $n$-bit primes $p_1$ and $p_2$ such that $p_1 = p_2 = 3 \bmod 4$. He computes $N = p_1 p_2$. The committer publishes $\langle N \rangle$ as a public key (alternatively he could send $N$ along with every commitment). He keeps the factors $\langle p_1, p_2 \rangle$ secret. The same modulus is used for all commitments.

**Commit phase:** The committer wishes to commit to a message $M$ of length $\ell$. The committer (Alice) and the verifier (Bob) perform the following steps:

*Step 1:* The committer picks a random $h \in \mathbb{Z}_N$. Next, the committer computes $g = h^{(\prod_{i=1}^r q_i^n)} \bmod N$ where $q_1, q_2, \ldots, q_r$ is the set of all primes less than some bound $B$. For example, one could take $B = 128$. When the verifier receives $h$ and $g$ he verifies that $g$ is constructed properly from $h$. At this point the verifier is assured that the order of $g$ in $\mathbb{Z}_N^*$ is not divisible by any primes less than $B$.

*Step 2:* The committer computes the value $u = g^{2^{2^k}} \bmod N$. She computes $u$ by first computing $a = 2^{2^k} \bmod \varphi(N)$ and then computing $u = g^a \bmod N$.

*Step 3:* Next, the committer hides the message $M$ using a pseudo random sequence generated by the BBS generator [9] whose tail is $u$. In other words, the committer hides the bits of $M$ by Xoring them with the LSB's of successive square roots of $u$ modulo $N$. More precisely, for $i = 1, \ldots, \ell$ we set $S_i = M_i \oplus \mathrm{lsb}(g^{2^{(2^k-i)}} \bmod N)$. Let $S = S_1 \ldots S_\ell \in \{0,1\}^\ell$. The commitment string is defined as $C = \langle h, g, u, S \rangle$. The committer sends $C$ to the verifier.

*Step 4:* The committer must still convince the verifier that $u$ is constructed properly, i.e. $u = g^{2^{2^k}} \bmod N$. To do so the committer constructs the following vector $W$ of length $k$:

$$W = \left\langle g^2, g^4, g^{16}, g^{256}, \ldots, g^{2^{2^i}}, \ldots, g^{2^{2^k}} \right\rangle \pmod{N}$$

She sends $W$ to the verifier. Let $W = \langle b_0, \ldots, b_k \rangle$. For each $i = 1, \ldots, k$ the committer proves in zero-knowledge to the verifier that the triple $(g, b_{i-1}, b_i)$ is a triple of the form $(g, g^x, g^{x^2})$ for some $x$. This convinces the verifier that $W$ is constructed properly. By verifying that the last element in $W$ is equal to $u$ the verifier is assured that indeed $u = g^{2^{2^k}} \bmod N$.

Each of these $k$ proofs take four rounds and they can all be done in parallel. These proofs are based on a classic zero-knowledge proof that a tuple $\langle g, A, B, C \rangle$ is a Diffie-Hellman tuple [13]. Let $q$ be the order of $g$ in $\mathbb{Z}_N^*$, and let $R$ be a security parameter. The complete protocol for proving integrity of $W$ is as follows: (unless otherwise specified, all arithmetic is done modulo $N$)

**Step 1:** The verifier picks random $c_1, \ldots, c_k \in \{0, \ldots, R\}$ and uses a regular commitment scheme to commit these values to the committer. For security against an infinitely powerful committer the verifier could use a commitment scheme that is information theoretically secure towards the committer.

**Step 2:** The committer picks random $\alpha_1, \ldots, \alpha_k \in \mathbb{Z}_q$ and computes $z_i = g^{\alpha_i}$ and $w_i = b_{i-1}^{\alpha_i}$ for $i = 1, \ldots, k$. She sends all pairs $\langle z_i, w_i \rangle_{i=1}^{k}$ to the verifier.

**Step 3:** The verifier opens the commitment in step 1 and reveals $c_1, \ldots, c_k$ to the committer.

**Step 4:** The committer responds with $\quad y_i = c_i \cdot 2^{2^{i-1}} + \alpha_i \bmod q \quad$ for all $i = 1, \ldots, k$.

**Step 5:** The verifier checks that for all $i = 1, \ldots, k$:

$$g^{y_i} \cdot b_{i-1}^{-c_i} = z_i \qquad \text{and} \qquad b_{i-1}^{y_i} \cdot b_i^{-c_i} = w_i$$

and rejects if any of these equalities does not hold.

The next two lemmas state the soundness and zero-knowledge properties of the above protocol.

**Lemma 1.** *Let $q$ be the order of $g$ in $\mathbb{Z}_N^*$ and let $d$ be the smallest prime divisor of $q$. If $W$ is constructed incorrectly then the committer will succeed in fooling the verifier with probability at most*[1] $k \cdot [\frac{1}{\min(d,R)} + o(\frac{1}{R})]$.

**Lemma 2.** *The above protocol is zero-knowledge. That is, there exists a simulator that produces a perfect simulation of the transcript for any verifier.*

The proofs of the two lemmas follow standard techniques and can be found in [13]. Recall that in Step 1 of the commitment protocol the verifier is convinced that the smallest prime divisor of $q$ (the order of $g$ in $\mathbb{Z}_N^*$) is larger than $B$. Hence, with each invocation of the protocol, the committer has a chance of at most $1/B$ in fooling the verifier. In this context, security of $2^{-70}$ is sufficient. Hence, taking $B = 128$, this level of security is obtained if the committer and verifier execute this protocol 10 times. These executions can be done in parallel.

Note that in Step 5, the verifier computes $4k$ exponentiations. However, using simultaneous multiple exponentiation [30, p. 618] the two exponentiations on the left hand side of each equality can be done for approximately the cost of one exponentiation. Hence, in reality, the verifier does work equivalent to $2k$ exponentiations. Recall that $k$ is typically in the range [30, 50]. Thus, counting 10 repetitions of the proof, the commit protocol requires at most a total of 1000 exponentiations on the verifier.

**Open phase:** Recall that the commitment string is $C = \langle h, g, u, S \rangle$. We know that $g = h^{(\prod_{i=1}^{r} q_i^n)} \bmod N$ where $q_1, q_2, \ldots, q_r$ is the set of all primes less than some bound $B$. In the open phase the committer (Alice) and verifier (Bob) execute the following protocol:

*Step 1:* Alice sends $v' = h^{2^{(2^k - \ell)}} \bmod N$ to the verifier (Bob). Bob computes

---

[1] The exact error bound is $k \cdot \left[\frac{1}{d} + \frac{\beta(d-\beta)}{d \cdot R^2}\right]$ where $\beta = R \bmod d$, $0 \le \beta \le d$. This expression is the maximum probability that a malicious prover succeeds in guessing $c_i \bmod d$ where $c_i$ is random in $\{0, \ldots, R\}$.

$v = (v')^{\prod_{i=1}^{r} q_i^n} \bmod N$. This ensures that $v$ has odd order. Bob then verifies that $v^{2^\ell} = u \bmod N$. At this point Bob has a $2^\ell$'th root of $u$. Being a $2^\ell$'th root of $u$ and having odd order ensures that $v$ is in the subgroup generated by $g$.

*Step 2:* Bob constructs the $\ell$-bit BBS pseudo-random sequence $R \in \{0,1\}^\ell$ starting at $v$. That is, for $i = 1, \ldots, \ell$ Bob sets $R_i$ to be the least significant bit of $v^{2^{(\ell-i)}}$. The message $M$ is then $M = R \oplus S$.

With an honest committer the open protocol clearly produces the committed message $M$. We show that the commitment is binding by showing that $M$ is the only possible outcome of the open protocol.

**Lemma 3.** *The commitment is binding. In other words, given a commitment $\langle h, g, u, S \rangle$ the committer can open the commitment in one way only.*

*Proof.* Due to the test in Step 1 Bob obtains from Alice $v'$ which leads to $v \in \mathbb{Z}_N$ satisfying $v^{2^\ell} = u \bmod N$. Furthermore, $v$ has odd order in $\mathbb{Z}_N^*$. Recall that during the commit phase the verifier is assured that $g$ has odd order in $\mathbb{Z}_N^*$ and that $u$ is in the subgroup generate by $g$. Since the subgroup has odd order, $u$ has a unique $2^\ell$'th root *in the subgroup*. Denote this unique $2^\ell$'th root by $v_0$. Then $v_0 = g^{2^{(2^k-\ell)}} \bmod N$. Now, observe that in $\mathbb{Z}_N^*$ there can be at most one $2^\ell$'th root of $u$ of odd order. Since both $v$ and $v_0$ are such roots we must have $v = v_0$. Consequently, there is a unique $v' \in \mathbb{Z}_N^*$ that will pass the test in Step 1. Hence, Alice is bound to a unique message $M$. $\qquad\square$

**Forced open phase:** In case the committer never executes the open phase, the verifier can retrieve the committed value $M$ himself by computing $v$ as $v = g^{2^{(2^k-\ell)}} \bmod N$ using $(2^k - \ell)$ squarings mod $N$.

We now prove that the above scheme satisfies the security properties of a timed commitment scheme. The only property that remains to be proved is *privacy:* no PRAM algorithm can obtain information about the committed string in time significantly less than the time it takes to compute $2^k$ squarings. The proof of security relies on the following complexity assumption:

$(n, n', \delta, \epsilon)$ **generalized BBS assumption:**

For $g \in \mathbb{Z}$ and a positive integer $k > n'$ let $W_{g,k} = \langle g^2, g^4, \ldots, g^{2^{2^i}}, \ldots, g^{2^{2^k}} \rangle$. Then for any integer $n' < k < n$ and any PRAM algorithm $\mathcal{A}$ whose running time is less than $\delta \cdot 2^k$ we have that

$$\left| \Pr\left[ \mathcal{A}(N, g, k, \; W_{g,k} \bmod N, g^{2^{2^{k+1}}}) = \text{``yes''} \right] - \right.$$

$$\left. \Pr\left[ \mathcal{A}(N, g, k, \; W_{g,k} \bmod N, R^2) = \text{``yes''} \right] \right| < \epsilon$$

where the probability is taken over the random choice of an $n$-bit RSA modulus $N = p_1 p_2$ where $p_1, p_2$ are equal size primes satisfying $p_1 = p_2 = 3 \bmod 4$, an element $g \in \mathbb{Z}_N$, and $R \in \mathbb{Z}_N$.

The assumption states that given $W_{g,k}$, the element $g^{2^{2^{k+1}}} \bmod N$ is indistinguishable from a random quadratic residue for any PRAM algorithm whose running time is much less than $2^k$. The parallel complexity of exponentiation modulo a composite has previously been studied by Adleman and Kompella [1] and Sorenson [38]. These results either require a super polynomials number of processors (larger than the time to factor), or give small speed ups that do not affect the generalized BBS assumption. We note that the sequential nature of exponentiation modulo a composite was previously used for time-lock cryptography [35] and benchmarking [10].

The generalized BBS assumption is sufficient for proving privacy of the scheme. This is stated in the next lemma whose proof is omitted due to lack of space.

**Theorem 1.** *Suppose the $(n, n', \delta, \epsilon)$ generalized BBS assumption holds for certain $\delta, \epsilon > 0$. Then, for $k > n'$, the above scheme is a $(T, t, \epsilon)$ timed commitment scheme with $t = \delta \cdot 2^k$ and $T = M(n) \cdot 2^k$ where $M(n)$ is the time it takes to square modulo an n-bit number.*

**Efficiency Improvements.** The timed commitment scheme described above can be made more efficient as follows:

- Rather than use a random element $g \in \mathbb{Z}_N$ whose order is close to $N$ we can use a $g$ of much smaller order. To do so, we choose $N = p_1 p_2$ where $q_1$ divides $p_1$ and $q_2$ divides $p_2$ where $q_1, q_2$ are distinct $m$ bit primes, and $m \ll \log_2 N$. We then use an element $g$ in $\mathbb{Z}_N^*$ of order $q = q_1 q_2$. Since $g$ has small order the exponentiations in Step 4 of the commitment protocol take far less time.

- Suppose the committer needs to repeatedly commit a message to the same verifier Bob. In this case the protocol can be improved significantly. During the Setup phase, the committer picks a modulus $N = p_1 p_2$ where $p_1$ and $p_2$ are strong primes, i.e. $\frac{p_1 - 1}{2}, \frac{p_2 - 1}{2}$ are prime. The committer and the verifier then execute a protocol due to Camenisch and Michels [11] to convince the verifier in zero-knowledge that $N$ is the product of two strong primes and that the smallest prime factor of $N$ is at least $m$ bits long for some predetermined $m$ (e.g. $m = 70$). This protocol is only executed once in order to validate the public key $N$. Step 1 of the commitment protocol is now replaced by the committer picking a random $h \in \mathbb{Z}_N^*$ and setting $g = h^2 \bmod N$. Let $q$ be the order of $g$ in $\mathbb{Z}_N^*$. Since $N$ is the product of strong primes greater than $2^m$ the verifier is assured that the smallest prime factor of $q$ is greater than $2^{m-1}$. Hence, by Lemma 1 the protocol for verifying integrity of $W$ in Step 4 of the commitment protocol need only be run once (rather than multiple times as discussed above).

## 3.1   A Timed Signature Scheme

A timed signature scheme can be easily built out of our timed commitment scheme and a regular signature scheme. Let $(\sigma, V, G)$ be a signature scheme ($\sigma$ takes a message and a private key and generates a signature, $V$ takes a signature and a public key and verifies the signature, and $G$ is a public/private key pair generator). The $(T, t, \epsilon)$ timed signature scheme is as follows:

**Setup:** The signer generates a public/private key pair $\langle Pub, Pr \rangle$ using algorithm $G$. The signer's public key is $\langle Pub \rangle$. He keeps $\langle Pr \rangle$ secret.

**A valid signature:** A valid signature on a message $M$ is a tuple $SIG = \langle S, C, Sig \rangle$ where (1) $C$ is a commitment string generated by the timed commitment scheme when committing to the string $S$, and (2) $Sig$ verifies using the public key $Pub$ as a valid signature on $\langle M, C \rangle$.

**Commit phase:** The signer picks a random secret string $S$ and uses the timed commitment scheme to commit $S$ to the verifier. Let $C$ be the resulting commitment string given to the verifier. The signer uses her private key $Pr$ to sign the message $\langle M, C \rangle$. Let $Sig$ be the resulting signature. The full commitment string given to the verifier is $\langle C, Sig \rangle$.

**Open phase:** The signer reveals $S$. The verifier obtains a complete valid signature as $\langle S, C, Sig \rangle$.

**Forced open:** Use the forced open algorithm provided by the timed commitment to retrieve $S$.

This signature scheme has the feature (or bug) that once the commit phase is done, the verifier can convince a third party that the signer is about to give him a signature on $M$. Indeed, the value $Sig$ in the commitment string given to the verifier could have only come from the signer. This is called an *abuse* of the protocol [26]. In Section 4.2 we construct a timed signature scheme so that after the commit phase is done the verifier cannot convince a third party that he has been talking to the signer. This is a desirable property when using timed signatures for contract signing.

# 4   Contract Signing

In this section we show how to use our timed primitives for contract signing and fair exchange of signatures. We show how to enable two untrusting parties, Alice and Bob, to exchange a signature on a joint contract. Neither party is willing to sign the contract before the other. Any timed-signature scheme can be used to solve the problem without relying on any infrastructure beyond a standard CA. The main difference between previous gradual-disclosure solutions and ours is that by using timed signatures Alice need not worry that Bob has ten times more machines than her. Timed signatures are resistant to parallel attacks, whereas previous proposals for gradual release of secrets become unfair as soon as one party has more machines than the other.

We begin by specifying the desired properties of a contract signing protocol. We then show that the timed commitment scheme of the previous section gives an especially efficient solution.

## 4.1   Contract Signing Definitions

A contract signing protocol allows two parties, $A_0$ and $A_1$, to exchange signatures on a contract $C$. Assume that $A_0$ and $A_1$ have established public keys $P_0$ and $P_1$ respectively. For a given contract $C$ the two parties exchange messages. At the end of the protocol each party $A_b$ (for $b \in \{0,1\}$) has a signature $S_{1-b}(C)$ such that a third party (contract verifier) that is given $S_{1-b}(C)$ as well as $P_0$ and $P_1$ can verify the signature. More precisely, for a contract signing protocol to be reasonable at all we need the following two conditions, which are standard in signature schemes:

**Completeness.** With overwhelming probability the *signature verification algorithm* outputs "accept" on a signature that is the result of a correct (by both parties) execution of the protocol.

**Unforgeability.** For a properly generated $P_b$ and for any contract $C$, unless $A_b$ participated in the contract signing protocol or $C$, the probability that any probabilistic polynomial-time adversary $\mathcal{A}$ succeeds in finding a signature $S$ and key $P_{1-b}$ such that verifier accepts $(C, S, P_b, P_{1-b})$ is negligible.

In addition to the normal operations (when no cheating by the other party occurs) the protocol designer should also provide to each party $A_b$ a **forced signature opening** algorithm $R_b$: given the private information of $A_b$ plus the messages exchanged with $A_{1-b}$ algorithm $R_b$ tries to produce $S_{1-b}(C)$. The time that $R_b$ is allowed to perform the recovery may be given as a parameter.

In order to define fairness we view a contract signing protocol as a game, where the goal of each party is to obtain a signature on a contract that is considered valid by the *signature verification algorithm* (for simplicity and wlog we assume that the verification algorithm is deterministic), without revealing its own signature.

**Definition 1.** *We say that a protocol is $(c, \epsilon)$-fair if the following holds: for any time $t$ smaller than some security parameter and any adversary $\mathcal{A}$ working in time $t$ as party $A_b$: let $\mathcal{A}$ choose a contract $C$ and run the contract signing protocol with party $A_{1-b}$. At some point $\mathcal{A}$ aborts the protocol and attempts to recover a valid signature $S_{1-b}(C)$. Denote $\mathcal{A}$'s probability of success by $q_1$. Suppose now that party $A_{1-b}$ runs the forced signature opening $R_{1-b}$ for time $c \cdot t$ and let $q_2$ be the probability of recovering $S_b(C)$. Then $q_1 - q_2 \leq \epsilon$.*

The protocols in this section are $(2, \epsilon)$ fair for a negligible $\epsilon$. That is, suppose $A_b$ aborts the protocol at some point and then recovers $S_{1-b}(C)$ in time $t$. Then $A_{1-b}$ can recover the signature $S_b(C)$ in time $2t$. Smaller values of $c$ can be achieved as discussed in Section 4.3.

**Strong Fairness:** One problem with the above definition is that it leaves open the possibility that one party $A_b$ would be able to convince a third party that $A_{1-b}$ is in the process of signing the contract $C$ (even without the given signature verification algorithm), whereas $A_b$ does not have the signature $S_b(C)$. This type of unfairness is referred to as "abusing" [26,37]. We propose a stronger requirement of fairness: if the adversary $\mathcal{A}$ invests time much smaller than the threshold $T$ allowing both parties to recover the desired signatures, then the transcript of the conversation is useless: there is simulator (operating in time proportional to the running-time of $\mathcal{A}$) that can output a transcript that is indistinguishable to any machine operating in time less than the threshold $T$. In other words, the protocol is zero-knowledge to an adversary not willing to invest the recovery time (in which case the full signature is extractable.)

## 4.2   The Strongly Fair Contract Signing Protocol

We now describe a signature exchange protocol that enables two parties, Alice and Bob, to exchange Rabin signatures on a contract $C$. The protocol works as follows:

**Setup:** Alice generates an $N_a$ as in the timed commitment scheme of Section 3. She will use $\langle N_a \rangle$ for all her contract signings. Similarly Bob generates $\langle N_b \rangle$. They agree on $k$ as a security parameter, e.g. $k = 40$.

**Valid signature:** Alice's signature on $C$ is a regular Rabin signature modulo $N_a$. That is, let $H \in \mathbb{Z}_{N_a}$ be the hash of $C$ properly padded prior to signing (e.g. according to PKCS1 or Bellare-Rogaway [6]). Alice's signature on $C$ is $S = H^{1/2} \bmod N_a$. Bob's signature on $C$ is defined analogously modulo $N_b$.

**Init:** To sign a contract $C$ the protocol begins with Alice picking a random $g_a \in \mathbb{Z}_N$ and generating a vector

$$W_{alice} = \left\langle g_a^2, g_a^4, g_a^{16}, g_a^{256}, \ldots, g_a^{2^{2^i}}, \ldots, g_a^{2^{2^k}} \right\rangle \pmod{N_a}$$

Alice can constructs this vector efficiently by first reducing all the exponents modulo the order of $g_a$. She sends $W_{alice}$ to Bob. Next, she convinces Bob that $W_{alice}$ is constructed correctly, i.e. $W_{alice} = \langle u_0, \ldots, u_k \rangle$ where $u_i = g_a^{2^{2^i}}$. She does so using the zero-knowledge protocol described in the commit phase in Section 3.

Let $\langle v_0, \ldots, v_k \rangle$ be the square roots modulo $N_a$ of the elements in $W_{alice}$. Alice computes the Rabin signature on $C$, namely Alice computes $S = H^{1/2} \bmod N_a$. She sends $V = S \cdot (v_0 \cdots v_k) \bmod N_a$ to Bob. Bob verifies validity of $V$ by checking that $V^2 = H \cdot (u_0 \cdots u_k) \bmod N_a$.

Bob initializes his contribution to the protocol by doing the analogous operations modulo $N_b$.

**Iteration:** from now on Alice and Bob take turns in revealing the square roots of elements in $W_{alice}$ and $W_{bob}$. Alice begins by revealing $v_k^{(alice)}$, the square root modulo $N_a$ of the last element in $W_{alice}$ (namely $u_k^{(alice)}$). Bob responds by revealing $v_k^{(bob)}$, the square root modulo $N_b$ of the last element in $W_{bob}$ (namely $u_k^{(bob)}$). Next, Alice reveals the square root of $u_{k-1}^{(alice)}$ and Bob responds by revealing the square root of $u_{k-1}^{(bob)}$. This continues for $2k$ rounds until the square roots of all elements in $W_{alice}$ and $W_{Bob}$ are revealed. At this point Bob can easily obtain Alice's signature on $C$ by computing $V/(v_0 \cdots v_k) \bmod N_a$, where $V, v_0, \ldots, v_k$ are the values sent from Alice. Alice obtains Bob's signature on $C$ by doing the same on the values sent from Bob.

**Forced Signature Opening:** suppose Bob aborts the protocol after Alice reveals only $m < k$ square roots. Then Bob has $v_k, \ldots, v_{k-m+1} \in \mathbb{Z}_{N_a}$ that are the square roots of $u_k^{(alice)}, \ldots, u_{k-m+1}^{(alice)}$. He can compute the remaining square roots by computing $v_i = g^{2^{2^i - 1}} \bmod N_a$ for all $i = 0, \ldots, m$. This requires approximately $2^m$ modular multiplications. He then obtains Alice's signature on $C$. Based on the generalized BBS assumption one can show that Bob cannot produce the signature any faster: this will imply that he can distinguish between $u_{k-m}$ and a random quadratic residue in $\mathbb{Z}_{N_a}$. Fortunately, Alice can also obtain Bob's signature on $C$ in roughly the same amount of time. Alice has to compute the square root of one more element than Bob did. Namely, she has to compute the square roots of $u_0^{(bob)}, \ldots, u_{m+1}^{(bob)}$. She does so by computing $v_i = g^{2^{2^i - 1}} \bmod N_b$ for all $i = 0, \ldots, m + 1$. This requires approximately $2^{m+1}$ modular multiplications. Thus her work load is roughly twice that of Bob's. Consequently, Bob does not gain much from prematurely aborting the protocol.

The following lemma shows that the above protocol is fair. The proof is along the lines of the above discussion, and is omitted due to space limitations.

**Lemma 4.** *Suppose the generalized BBS assumption holds for some parameters $(n, n', \delta, \epsilon)$. Say Bob aborts the protocol after only $n' < m < 2k$ rounds. He then recovers the complete signature in time $T$. Then Alice can obtain the complete signature in at most expected time $2T \cdot \epsilon/\delta$. The same holds if Alice aborts the protocol first.*

Note that even if Bob has many more machines than Alice he cannot gain much since, by assumption, parallelism does not speed up modular exponentiation. As long as Bob's machines run at approximately the same speed as Alice's machines, fairness is preserved.

The protocol also satisfies the strong fairness properties defined in the previous section. By the generalized BBS assumption, the value $u_k^{(alice)}$ looks random to a third party whose whose running time is much less than $2^k$. Since the proof of validity of $W_{alice}$ is zero-knowledge, Bob cannot use it to convince a third party that $W_{alice}$ is well formed. As a result, Bob can easily simulate the value $V$ by picking $V \in \mathbb{Z}_{N_a}$ at random, and setting $u_k = V^2/(H \cdot u_0 \cdots u_{k-1})$. Hence, Bob can simulate himself all the information he got from Alice during the commit phase. Consequently, suppose Bob misbehaves during the protocol (given the zero-knowledge proofs of consistency the only real bad behavior on Bob's part is early stopping), but Bob is able to convince with non-negligible probability a third party verifier that he is executing a contract signing protocol with Alice. Then Alice can produce a signature on the contract in time proportional to the running time of Bob and the third party verifier.

*Remark 1.* Although the above description requires $2k$ rounds, it is easy to cut it down to $k$ rounds. Simply make each party send two square roots per turn (rather than one). The advantage of early stopping is unchanged and rotates from party to party.

We also point out that any square root based signature can be used, e.g. the Fiat-Shamir method [25], or a non-oracle based one such as a variant of the Dwork-Naor scheme [21]. In the next section we show how to incorporate RSA signatures into the fair exchange scheme.

## 4.3  Extensions

**RSA Signatures.** The scheme in Section 4.2 enables two parties to exchange Rabin signatures. We describe a simple extension enabling fair exchange of RSA signatures. Let $(N_a, e)$ be the public key of Alice. The difference with respect to Rabin signatures is that now

$$W_{alice} = \left\langle g_a^{e \cdot 2}, g_a^{e \cdot 4}, g_a^{e \cdot 16}, g_a^{e \cdot 256}, \ldots, g_a^{e \cdot 2^{2^i}}, \ldots, g_a^{e \cdot 2^{2^k}} \right\rangle \pmod{N_a}$$

Note that the generalized BBS assumption implies that the next element in this sequence is indistinguishable from random in time $\delta 2^k$, since it is easy to transform a sequence

$$\left\langle g_a^2, g_a^4, g_a^{16}, g_a^{256}, \ldots, g_a^{2^{2^i}}, \ldots, g_a^{2^{2^k}} \right\rangle \pmod{N_a}$$

into $W_{alice}$ by $k$ parallel exponentiations in $e$. Showing that the vector $W_{alice}$ is constructed correctly is similar to what was done previously: for any $1 \leq i \leq k$ Alice should prove that the triple $(g, b_{i-1}, b_i)$ is of the form $(g, g^{ex}, g^{ex^2})$ for a given $e$. This can be done by the same protocol that proves triples of the form $(g, g^x, g^{x^2})$. Simply run the protocol on the triple $(g, g^{ex}, g^{e^2 x^2})$ and then verify that $b_i$ is the $e$th root of $g^{e^2 x^2}$.

The contract signing protocol proceeds along the same lines as before. For $0 \leq i \leq k$ let $u_i = g_a^{e \cdot 2^{2^i}}$, let $v_i = u_i^{1/e} \bmod N_a = g_a^{2^{2^i}}$, i.e. the $v_i$'s are defined as $e$th roots of the $u_i$'s (instead of square roots, as above). Let $H$ be the value to be signed by the RSA signature, i.e. the goal of the recipient it to obtain $H^{1/e} \bmod N_a$. The $v_i$'s mask $H^{1/e} \bmod N_a$ — Alice gives Bob $V = H^{1/e} \cdot v_0 \cdot v_1 \cdots v_k$. As before, the $v_i$'s are released one-by-one. The validity of each $v_i$ is easy to verify by comparing $v_i^e$ to $u_i$.

To argue the security of the scheme we assume the generalized BBS Assumption as well as the usual RSA one (that it is hard to extract $e$th roots). Based on these two assumptions, given $u_0, u_1, \ldots u_i$ it is hard to distinguish between $u_{i+1}$ and a random value and it is hard to compute $v_{i+1}$. Therefore in case of early stopping $i$ steps from the end it is impossible to find the RSA signature, $H^{1/e} \bmod N_a$, more efficiently than to compute $g^{2^{2^i}}$. Furthermore, a simulator can efficiently create an indistinguishable conversation.

**Other Ratios**. The signature exchange protocol of Section 4.2 achieves a fairness ratio of $c = 2$. That is, if Alice aborts the protocol, Bob has to do *twice* as must work as Alice to obtain the signature. The protocol easily generalizes to provide smaller fairness ratios as well, at the cost of increasing the number of rounds. For example, one can define $W_{alice}$ as

$$W_{alice} = \left\langle g_a^{2^{c_0}}, g_a^{2^{c_1}}, g_a^{2^{c_2}}, \ldots g_a^{2^{c_k}} \right\rangle \pmod{N_a}$$

where $c_0 = 1$ and $c_i = c_{i-1} + c_{i-2}$ for $i = 1, \ldots, k$. The proof of validity of $W_{alice}$ given in Section 3 must be changed accordingly. This approach gives a fairness ratio of $\alpha = \frac{1+\sqrt{5}}{2} \approx 1.618$. Even smaller values can be obtained by other such recurrences. The downside is that to obtain an initial security of $2^k$ the protocol must take $\log_\alpha 2^k$ rounds, as opposed to only taking $k$ rounds as in Section 4.2.

## 5   More Applications

We now describe several other applications of timed commitments. In all applications, we assume that the parties involved use clocks, but the adversary has control over the scheduling and the clocks. However, the adversary must satisfy the $(\alpha, \beta)$ constraint (for $\alpha < \beta$) of clocks [23]: for any two (possibly the same) non-faulty parties $P_1$ and $P_2$, if $P_1$ measures $\alpha$ elapsed time on its local clock and $P_2$ measures $\beta$ elapsed time on its local clock, and $P_2$ begins its measurement in real time after $P_1$ begins, then $P_2$ will finish after $P_1$ does. An $(\alpha, \beta)$ constraint is implied by many reasonable assumptions on the behavior of clocks in a system (*e.g.* the linear drift assumption). We assume that $\alpha$ is large enough that one party can send a message and expect a response in time $\alpha$, and $\beta$ is smaller than the security parameter of the commitment scheme.

**Collective Coin-Flipping.** We have two parties $A$ and $B$ who want to flip a coin in such a matter that (i) the value of the coin is unbiased and well defined even if one of the parties does not follow the protocol (if both of them don't follow, then it is a lost case) (ii) if both parties follow the protocol, then they agree on the same value for the coin.

Consider the protocol where one party commits to a bit and the other guesses it and the result is the Xor of the two bits. The problem with this simple protocol is that one party knows the results before the other and can quit early (if it doesn't like the result), thus biasing the result. Indeed Cleve[14] has shown that for any $k$-round protocol one of the parties can bias the coin with at least $1/k$ (this bound was improved in [16] to $1/\sqrt{k}$).

What we show now is a protocol that works in the $(\alpha, \beta)$ timing model using our timed commitment and (i) the coin has only negligible bias. (ii) the number of rounds is constant.
1. $A$ to $B$: pick a random bit $b_A \in \{0, 1\}$ and time-commit to $b_A$.
2. $B$ to $A$: pick a random bit $b_B \in \{0, 1\}$ and send to A
3. $A$ to $B$: open $b_A$.
The collective coin is $b_A \oplus b_B$.
*Timing:* $A$ makes sure that $B$'s message arrives within time $\alpha$ from the beginning of Step 1.
*Forced opening:* if $A$ does not open $b_A$ at Step 3, then $B$ uses the forced opening procedure to extract $b_A$ and sets the coin to $b_A \oplus b_B$.

It is easy to verify that if $B$ can bias the coin, then it can guess the value of a timed-committed bit in time smaller than the security parameter of the scheme. On the other hand $A$ can try to influence the bit only by not opening $b_A$; however this is defeated by the forced opening.

**Honesty-Preserving Auctions.** In a second-price (Vickrey) auction participants submit their bids to an item. The winner is the highest bidder but pays the second highest bid. A problem with running these auctions is establishing the trust in the auctioneer: how can the winner be assured that the auctioneer hasn't introduced a 2nd highest bid which is just $\epsilon$ less than the winning one?

Recent work focuses on many aspects of uncheatable auctions [31]. Here we are interested mostly in the *honesty-preserving* aspect of the protocol, i.e. making sure that the auctioneer is not changing some bids as a function of others. At the end of the protocol all the participants will know all the bids.
One simple solution is as follows:

1. The participants submit their bids by committing to their values. Here care must be taken to make the commitments *non-malleable* (see [19]).
2. When all of the commitments are in, the auctioneer posts them on a bulletin board. The participants should verify that their hidden bids were posted.
3. The participants then open their commitments. The auctioneer posts the results and everyone can verify that the auction was conducted properly

However, there is a problem with this solution: what if one (or more) participant refuses to open their commitment? If they are simply ignored, an auctioneer can plant several bogus bids with different values and open only those lower then the winning bid.

The timed commitment of Section 3 solves the problem. In Step 1 participants commit to their bid using a timed commitment. The bidders verify that that within time $\alpha$ the bid is posted, i.e. that from Step 1 to Step 2 no more than $\alpha$ time elapses. If in Step 3 a participant does not open its commitment, the auctioneer can "force open" the commitment using the forced opening algorithm.

Note that it is important for the commitment to have the soundness property (i.e. that at the end of the commit phase it is clear that forced open would work), otherwise the other bidders would not be convinced that it was properly opened.

**Zero-Knowledge.** We now briefly discuss the application of our timed commitments for achieving zero-knowledge in various settings. Consider *concurrent zero-knowledge*: Several parties who are *not* mutually aware of each other and may lack coordination are simultaneously engaged in zero-knowledge protocols. This settings received much attention recently (see [23,28,34,17].)The problem is in showing that the composed protocols are zero-knowledge in total. If the adversary controls the scheduling then it can create nested interactions that make the simulator's life difficult[2]. However, as suggested in [23], if the adversary is $(\alpha, \beta)$-restricted, then it is possible to obtain a constant round zero-knowledge protocol. Our timed commitments can be used for the verifier to commit to its queries. The verifier should accept only if the prover send its own commitments within time $\alpha$. The simulator force opens them and knows what to send as the prover.

---

[2]    Indeed, Kilian, Petrank and Rackoff [28] showed that no *black-box* simulation is possible for 4-round protocols and this was recently improved to 7-round [36].

Very recently Dwork and Naor [22] constructed *Zaps* - two-round witness indistinguishable proof systems - for any language in NP. Using timed-commitments with properties such as ours (verifiable recovery) they were able to show a three-round (concurrent) zero-knowledge proof system for all langauges in NP (in the $(\alpha, \beta)$-timing model.) The zap was used to prove that the commitment is proper and it is possible to perform forced opening. Note that this stands in contrast to the impossibility result even in the standard (non-concurrent) model [27]. They were also able to construct a three-round zero-knowledge protocol in the *resettable* setting [12], where prover is executed by a device that has no independent source of randomness and cannot record history.

## 6    Lower Bound on the Number of Rounds for Contract Signing

We now show that any protocol for signing contracts must take a number of rounds linearly proportional to the *advantage* one side has over the other. While contract signing has been investigated extensively we haven't quite found a similar statement in the literature. The closest that we are aware of is Cleve's [15] lower bound regarding *gradual disclosure of secrets*, a task that can be used for contract signing. However, it does not directly imply lower bounds for contract signing.

Let $A_0$ and $A_1$ be the two parties as in Section 4.1. The key to understanding the limitations of contract signing algorithms is to have a *single*-dimensional notion of progress. We choose (computing time)/(prob. of success) as the measure of progress, though there are other reasonable possibilities.

This gives us the definition of unfairness. Fix a model of computation (say a specific Turing Machine, or, for nonuniform results, Boolean circuits over the complete $\{0,1\}^2 \mapsto \{0,1\}$ basis). Fix an adversary $\mathcal{A}$ that produces a contract $C$, plays the role of $A_{1-b}$ and attempts to come up with $S_b$. Let $\gamma(\mathcal{A})$ be the running time of $\mathcal{A}$ over its probability of success. Similarly, for $\mathcal{A}$ let $\delta(\mathcal{A})$ be the running time of the forced signature opening algorithm $R_b$ over probability that $R_b$ succeeds in finding an accepting $S_{1-b}$.

**Definition 2.** *The unfairness of a protocol is the worst case of all adversaries $\mathcal{A}$ of the quantity $\gamma(\mathcal{A})/\delta(\mathcal{A})$.*

*Remark 2.* Note that this notion of unfairness is more forgiving then the one in Section 4.1, in the sense that it ignores a case where $R_b$ retrieves the signatures in the same amount of time as $\mathcal{A}$, but with slightly smaller (but not negligible) probability. Nevertheless our lower bound is applicable to this definition.

**Definition 3.** *For every signing protocol we say that the protocol has security gap $W$ if the ratio between the running time an adversary needs in order forge a signature on a message without the signers agreement and the running time of the signing and verification algorithm is at least $W$.*

For a protocol to be useful the security gap must be large, say at least $2^{50}$.

**Theorem 2.** *For every contract signing protocol and forced opening algorithm, if the protocol consists of $k$ rounds and has security gap $W$, then the unfairness of the protocol is at least $W^{1/k}$.*

All our adversary $\mathcal{A}$ will do is to stop early. For each round $i$ of the protocol we consider all programs that get the transcript of the first $i$ rounds plus the secret of one of the sides and attempt to produce a valid signature on the contract. For such a program we are interested in its running time divided by the probability it succeeds in producing a valid signature where the probability is over the coin flips of all the participants. We assume that none was cheating except for early withdrawal.

Let $T_i$ be the minimum over all machines of the above product. If the contract signing protocol is secure at all, then $T_0$ should be large (super-polynomial). From completeness (i.e. that the protocol ends with valid signatures if the participants follow the protocol) for a $k$-round protocol $T_k$ should be small and by the assumption on the security gap we have $T_0/T_k \geq W$.

If $k$ is a constant, then there must be a large gap between $T_{j-1}$ and $T_j$ for some $1 < j \leq k$, more specifically

$$T_0 = T_k \cdot \frac{T_{k-1}}{T_k} \cdot \frac{T_{k-2}}{T_{k-1}} \cdots \frac{T_0}{T_1}.$$

Therefore for at least one $1 \leq j \leq k$ we have $\frac{T_{j-1}}{T_j} \geq (\frac{T_0}{T_k})^{1/k} \geq W^{1/k}$. This gives a way for one of the participants to make the protocol unfair - the one who receives a message at step $j$ stops afterwards and tries to create the signature. For this party the product of the time and probability of success is $T_j$, whereas for the other party it is at most $T_{j-1}$. Hence the unfairness of this protocol is at least $W^{1/k}$.

Note that if $W$ is at least $2^{50}$ and we want the unfairness to be at most 2, then the number of rounds must be at least 50.

This lower bound is applicable to a wider model than had been considered previously, In particular it applies to scenarios where the participants are timed (by real clocks) and their partner expect a response with in a certain amount of time (e.g. the model of [23]). In contrast, as we have seen in Section 5, the problem of collective coin-flipping has a constant round protocol in a timed model, as opposed to the untimed one.

## 7  Conclusions

We introduced the concepts of *timed commitments* and *timed signatures*, which are useful for contract signing, auctions and other applications. In auctions timed commitments ensure that users cannot make a sealed bid and then refuse to open the bid. We emphasized the importance of defending against parallel attacks in all applications. Our constructions for timed commitments and timed signatures resist parallel attacks by relying on modular exponentiation which is believed to be an inherently serial operation.

It would be interesting to try to construct timed commitments and timed signatures based on other primitives. For example, all lattice basis reduction algorithms are sequential. Can one build timed primitives based on lattice basis reductions? It is also interesting to see what other areas in cryptography can benefit from the ability to delay one's capabilities by a fixed time period.

## Acknowledgments

We thank Vitali Shmatikov and John Mitchell for discussions regarding abuse freeness that motivated this work. We thank Victor Shoup for many helpful comments and corrections.

The first author is supported by NSF and a grant from the Packard foundation. This work was done while the second author was visiting Stanford University and IBM Almaden Research Center. Partly supported by DOD Muri grant administered by ONR and DARPA contract F30602-99-1-0530.

## References

1. L. Adleman, and K. Kompella, "Using smoothness to achieve parallelism", proc. of STOC 1988, pp. 528–538.
2. N. Asokan, V. Shoup and M. Waidner, "Optimistic fair exchange of digital signatures", in proc. Eurocrypt'98, pp. 591-606, 1998.
3. G. Ateniese, "Efficient protocols for verifiable encryption and fair exchange of digital signatures", in proc. of the 6th ACM CCS, 1999.
4. M. Bellare and S. Goldwasser, "Verifiable Partial Key Escrow", in proc. of ACM CCS, pp. 78–91, 1997.
5. M. Bellare and S. Goldwasser, "Encapsulated key escrow". MIT Laboratory for Computer Science Technical Report 688, April 1996.
6. M. Bellare and P. Rogaway, "The Exact Security of Digital Signatures - How to Sign with RSA and Rabin", EUROCRYPT 1996, pp. 399-416
7. M. Ben-Or, O. Goldreich, S. Micali and R. L. Rivest, "A Fair Protocol for Signing Contracts", IEEE Transactions on Information Theory 36/1 (1990) 40-46
8. M. Blum, "How to Exchange (Secret) Keys", STOC 1983: 440-447 and ACM TOCS 1(2): 175-193 (1983)
9. L. Blum, M. Blum and M. Shub, A Simple Unpredictable Pseudo-Random Number Generator. SIAM J. Comput. 15(2): 364-383 (1986).
10. J. Y. Cai, R. J. Lipton, R. Sedgwick, A. C. Yao, "Towards uncheatable benchmarks, Structures in Complexity", in proc. Structures in Complexity, pp. 2–11, 1993.
11. J. Camenisch, M. Michels, "Proving in Zero-Knowledge that a Number Is the Product of Two Safe Primes", EUROCRYPT 1999, pp. 107-122.
12. R. Canetti, O. Goldreich, S. Goldwasser and S. Micali. "Resettable Zero-Knowledge", ECCC Report TR99-042, Oct 27, 1999 and STOC 2000.
13. D. Chaum, and T. Pederson, "Wallet databases with observers", in Proceedings of Crypto '92, 1992, pp. 89–105.

14. R. Cleve, "Limits on the Security of Coin Flips when Half the Processors Are Faulty", in proc. STOC 1986, pp. 364–369.
15. R. Cleve, "Controlled gradual disclosure schemes for random bits and their applications", in proc. Crypto'89, 1990, pp. 573–588.
16. R. Cleve, R. Impagliazzo, "Martingales, collective coin flipping and discrete control processes", manuscript, 1993. Available:
    http://www.cpsc.ucalgary.ca/~cleve/papers.html
17. I. Damgård, "Concurrent Zero-Knowledge in the auxilary string model", in proc. EUROCRYPT '2000.
18. I. Damgård, "Practical and Provably Secure Release of a Secret and Exchange of Signatures," J. of Cryptology 8(4): 201-222 (1995)
19. D. Dolev, C. Dwork and M. Naor, "Non-malleable Cryptography", Preliminary version: Proc. 21st STOC, 1991. Full version: to appear, Siam J. on Computing. Available: http://www.wisdom.weizmann.ac.il/~naor
20. C. Dwork and M. Naor, "Pricing via Processing -or- Combatting Junk Mail", Advances in Cryptology – CRYPTO'92, pp. 139–147.
21. C. Dwork and M. Naor, "An Efficient Existentially Unforgeable Signature Scheme and Its Applications", J. of Cryptology 11, 1998, pp. 187–208,
22. C. Dwork and M. Naor, "Zaps and their applications", manuscript.
23. C. Dwork, M. Naor and A. Sahai, "Concurrent Zero-Knowledge", STOC, 1998.
24. S. Even, O. Goldreich and A. Lempel, "A Randomized Protocol for Signing Contracts," CACM 28(6): 637-647 (1985).
25. A. Fiat and A. Shamir, "How to Prove Yourself: Practical Solutions to Identification and Signature Problems", CRYPTO'86, pp. 186–194.
26. J. A. Garay, M. Jakobsson and P. D. MacKenzie: "Abuse-Free Optimistic Contract Signing," CRYPTO 1999, pp. 449-466.
27. O. Goldreich and H. Krawczyk. "On the Composition of Zero Knowledge Proof Systems," SIAM J. on Computing, Vol. 25, No. 1, pp. 169–192, 1996.
28. J. Kilian, E. Petrank and C. Rackoff, "Lower Bounds for Zero Knowledge on the Internet", FOCS 1998, pp. 484–492.
29. M. Luby, S. Micali and C. Rackoff, "How to Simultaneously Exchange a Secret Bit by Flipping a Symmetrically-Biased Coin," FOCS 1983, pp. 11-21
30. A. Menezes, P. van Oorschot and S. Vanstone, "Handbook of Applied Cryptography," CRC Press, 1996.
31. M. Naor, B. Pinkas and R. Sumner, "Privacy Preserving Auctions and Mechanism Design," Proc. of the 1st ACM conference on E-Commerce, November 1999, pp. 129 – 139.
32. B. Pfitzmann, M. Schunter, M. Waidner, "Optimal Efficiency of Optimistic Contract Signing," PODC 1998: 113-122
33. M. O. Rabin, "Transaction Protection by Beacons," JCSS 27(2): 256-267 (1983)
34. R. Richardson and J. Kilian, "On the Concurrent Composition of Zero-Knowledge Proofs," EUROCRYPT '99, pp. 415–431, 1999.
35. R. Rivest, A. Shamir and D. Wagner, "Time lock puzzles and timed release cryptography," Technical report, MIT/LCS/TR-684
36. A. Rosen, "A note on the round-complexity of concurrent zero-knowledge", these proceedings.
37. V. Shmatikov and J. C. Mitchell, "Analysis of Abuse-Free Contract Signing," in proc. of 4th Annual Conference on Financial Cryptography, 2000.
38. J. P. Sorenson, "A Sublinear-Time Parallel Algorithm for Integer Modular Exponentiation," manuscript.

# A Practical and Provably Secure Coalition-Resistant Group Signature Scheme

Giuseppe Ateniese[1], Jan Camenisch[2], Marc Joye[3], and Gene Tsudik[4]

[1] Department of Computer Science, The Johns Hopkins University
3400 North Charles Street, Baltimore, MD 21218, USA
ateniese@cs.jhu.edu
[2] IBM Research, Zurich Research Laboratory
Säumerstrasse 4, CH-8803 Rüschlikon, Switzerland
jca@zurich.ibm.com
[3] Gemplus Card International, Card Security Group
Parc d'Activités de Gémenos, B.P. 100, F-13881 Gémenos, France
marc.joye@gemplus.com
[4] Department of Information and Computer Science,
University of California, Irvine, Irvine, CA 92697-3425, USA
gts@ics.uci.edu

**Abstract.** A group signature scheme allows a group member to sign messages anonymously on behalf of the group. However, in the case of a dispute, the identity of a signature's originator can be revealed (only) by a designated entity. The interactive counterparts of group signatures are identity escrow schemes or group identification scheme with revocable anonymity. This work introduces a new provably secure group signature and a companion identity escrow scheme that are significantly more efficient than the state of the art. In its interactive, identity escrow form, our scheme is proven secure and coalition-resistant under the strong RSA and the decisional Diffie-Hellman assumptions. The security of the non-interactive variant, i.e., the group signature scheme, relies additionally on the Fiat-Shamir heuristic (also known as the random oracle model).

**Keywords:** Group signature schemes, revocable anonymity, coalition-resistance, strong RSA assumption, identity escrow, provable security.

## 1  Introduction

Group signature schemes are a relatively recent cryptographic concept introduced by Chaum and van Heyst [CvH91] in 1991. In contrast to ordinary signatures they provide anonymity to the signer, i.e., a verifier can only tell that a member of some group signed. However, in exceptional cases such as a legal dispute, any group signature can be "opened" by a designated group manager to reveal unambiguously the identity of the signature's originator. At the same time, no one — including the group manager — can misattribute a valid group signature.

The salient features of group signatures make them attractive for many specialized applications, such as voting and bidding. They can, for example, be

M. Bellare (Ed.): CRYPTO 2000, LNCS 1880, pp. 255–270, 2000.
© Springer-Verlag Berlin Heidelberg 2000

used in invitations to submit tenders [CP95]. All companies submitting a tender form a group and each company signs its tender anonymously using the group signature. Once the preferred tender is selected, the winner can be traced while the other bidders remain anonymous. More generally, group signatures can be used to conceal organizational structures, e.g., when a company or a government agency issues a signed statement. Group signatures can also be integrated with an electronic cash system whereby several banks can securely distribute anonymous and untraceable e-cash. This offers concealing of the cash-issuing banks' identities [LR98].

A concept dual to group signature schemes is identity escrow [KP98]. It can be regarded as a group-member identification scheme with revocable anonymity. A group signature scheme can be turned into an identity escrow scheme by signing a random message and then proving the knowledge of a group signature on the chosen message. An identity escrow scheme can be turned into a group signature scheme using the Fiat-Shamir heuristic [FS87]. In fact, most group signature schemes are obtained in that way from 3-move honest-verifier proof of knowledge protocols.

This paper presents a new group signature / identity escrow scheme that is provably secure. In particular, the escrow identity scheme is provably coalition-resistant under the strong RSA assumption. Other security properties hold under the decisional Diffie-Hellman or the discrete logarithm assumption. Our group signature scheme is obtained from the identity escrow scheme using the Fiat-Shamir heuristic, hence it is secure in the random oracle model.

Our new (group signature) scheme improves on the state-of-the-art exemplified by the scheme of Camenisch and Michels [CM98a] which is the only known scheme whose coalition-resistance is provable under a standard cryptographic assumption. In particular, our scheme's registration protocol (JOIN) for new members is an order of magnitude more efficient. Moreover, our registration protocol is statistically zero-knowledge with respect to the group member's secrets. In contrast, in [CM98a] the group member is required to send the group manager the product of her secret, a prime of special form, and a random prime; such products are in principle susceptible to an attack due to Coppersmith [Cop96]. Moreover, our scheme is provably coalition-resistance against an adaptive adversary, whereas for the scheme by Camenisch and Michels [CM98a] this holds only for a static adversary.

The rest of this paper is organized as follows. The next section presents the formal model of a secure group signature scheme. Section 3 overviews cryptographic assumptions underlying the security of our scheme and introduces some basic building blocks. Subsequently, Section 4 presents the new group signature scheme. The new scheme is briefly contrasted with prior work in Section 5. The security properties are considered in Section 6. Finally, the paper concludes in Section 7.

## 2    The Model

Group-signature schemes are defined as follows. (For an in-depth discussion on this subject, we refer the reader to [Cam98].)

**Definition 1.** *A* group-signature scheme *is a digital signature scheme comprised of the following five procedures:*

SETUP: *On input a security parameter $\ell$, this probabilistic algorithm outputs the initial group public key $\mathcal{Y}$ (including all system parameters) and the secret key $\mathcal{S}$ for the group manager.*

JOIN: *A protocol between the group manager and a user that results in the user becoming a new group member. The user's output is a membership certificate and a membership secret.*

SIGN: *A probabilistic algorithm that on input a group public key, a membership certificate, a membership secret, and a message $m$ outputs group signature of $m$.*

VERIFY: *An algorithm for establishing the validity of an alleged group signature of a message with respect to a group public key.*

OPEN: *An algorithm that, given a message, a valid group signature on it, a group public key and a group manager's secret key, determines the identity of the signer.*

A secure group signature scheme must satisfy the following properties:

Correctness: Signatures produced by a group member using SIGN must be accepted by VERIFY.

Unforgeability: Only group members are able to sign messages on behalf of the group.

Anonymity: Given a valid signature of some message, identifying the actual signer is computationally hard for everyone but the group manager.

Unlinkability: Deciding whether two different valid signatures were computed by the same group member is computationally hard.

Exculpability: Neither a group member nor the group manager can sign on behalf of other group members.[1] A closely related property is that of *non-framing* [CP95]; it captures the notion of a group member not being made responsible for a signature she did not produce.

Traceability: The group manager is always able to open a valid signature and identify the actual signer.

Note that the last property is also violated if a subset of group members, pooling together their secrets, can generate a valid group signature that cannot be opened by the group manager. Because this was ignored in many papers we state it explicitly as an additional property.

Coalition-resistance: A colluding subset of group members (even if comprised of the entire group) cannot generate a valid signature that the group manager cannot link to one of the colluding group members.

We observe that many group signature schemes (e.g., [CM98a,CM99b,CS97]) can be viewed as making use of two different ordinary signature schemes: one to

---

[1] Note that the above does not preclude the group manager from creating fraudulent signers (i.e., nonexistent group members) and then producing group signatures.

generate membership certificates as part of JOIN and another to actually generate group signatures as part of SIGN (cf. [CM99b]). Consequently, the properties of any secure group signature scheme must include the **Unforgeability** property (as defined in [GMR88]) for each of the two ordinary signature schemes. It is easy to see that each of: **Traceability** and **Exculpability** map into the **Unforgeability** property for the two respective signature schemes. Furthermore, together they ensure that a group signature scheme is unforgeable, i.e., that only group members are able to sign messages on behalf of the group.

The model of identity escrow schemes [KP98] is basically the same as the one for group signature schemes; the only difference being that the SIGN algorithm is replaced by an interactive protocol between a group member and a verifier.

# 3    Preliminaries

This section reviews some cryptographic assumptions and introduces the building blocks necessary in the subsequent design of our group signature scheme. (It can be skipped with no significant loss of continuity.)

## 3.1    Number-Theoretic Assumptions

The *Strong-RSA Assumption* (SRSA) was independently introduced by Barić and Pfitzmann [BF97] and by Fujisaki and Okamoto [FO97]. It strengthens the widely accepted RSA Assumption that finding $e^{\text{th}}$-roots modulo $n$ — where $e$ is the public, and thus fixed, exponent — is hard to the assumption that finding an $e^{\text{th}}$-root modulo $n$ for any $e > 1$ is hard. We give hereafter a more formal definition.

**Definition 2 (Strong-RSA Problem).** *Let $n = pq$ be an RSA-like modulus and let $G$ be a cyclic subgroup of $\mathbb{Z}_n^*$ of order $\#G$, $\lceil \log_2(\#G) \rceil = \ell_G$. Given $n$ and $z \in G$, the* Strong-RSA Problem *consists of finding $u \in G$ and $e \in \mathbb{Z}_{>1}$ satisfying $z \equiv u^e \pmod{n}$.*

**Assumption 1 (Strong-RSA Assumption).** *There exists a probabilistic polynomial-time algorithm $\mathcal{K}$ which on input a security parameter $\ell_G$ outputs a pair $(n, z)$ such that, for all probabilistic polynomial-time algorithms $\mathcal{P}$, the probability that $\mathcal{P}$ can solve the Strong-RSA Problem is negligible.*

The *Diffie-Hellman Assumption* [DH76] appears in two "flavors": (i) the Computational Diffie-Hellman Assumption (CDH) and (ii) the Decisional Diffie-Hellman Assumption (DDH). For a thorough discussion on the subject we refer the reader to [Bon98].

**Definition 3 (Decisional Diffie-Hellman Problem).** *Let $G = \langle g \rangle$ be a cyclic group generated by $g$ of order $u = \#G$ with $\lceil \log_2(u) \rceil = \ell_G$. Given $g$, $g^x$, $g^y$, and $g^z \in G$, the* Decisional Diffie-Hellman Problem *consists of deciding whether the elements $g^{xy}$ and $g^z$ are equal.*

This problem gives rise to the *Decisional Diffie-Hellman Assumption*, which was first explicitly mentioned in [Bra93] by Brands although it was already implicitly assumed in earlier cryptographic schemes.

**Assumption 2 (Decisional Diffie-Hellman Assumption).** *There is no probabilistic polynomial-time algorithm that distinguishes with non-negligible probability between the distributions $D$ and $R$, where $D = (g, g^x, g^y, g^z)$ with $x, y, z \in_R \mathbb{Z}_u$ and $R = (g, g^x, g^y, g^{xy})$ with $x, y \in_R \mathbb{Z}_u$.*

The Decisional Diffie-Hellman Problem is easier than the *(Computational) Diffie-Hellman Problem* which involves finding $g^{uv}$ from $g^u$ and $g^v$; the Decisional Diffie-Hellman Assumption is, thus, a *stronger* assumption. Both are stronger assumptions than the assumption that computing discrete logarithms is hard.

If $n$ is a *safe* RSA modulus (i.e., $n = pq$ with $p = 2p' + 1$, $q = 2q' + 1$, and $p, q, p', q'$ are all prime), it is a good habit to restrict operation to the subgroup of quadratic residues modulo $n$, i.e., the cyclic subgroup $QR(n)$ generated by an element of order $p'q'$. This is because the order $p'q'$ of $QR(n)$ has no small factors.

The next corollary shows that it is easy to find a generator $g$ of $QR(n)$: it suffices to choose an element $a \in \mathbb{Z}_n^*$ satisfying $\gcd(a \pm 1, n) = 1$ and then to take $g = a^2 \bmod n$. We then have $QR(n) = \langle g \rangle$. (By convention, $\gcd(0, n) := n$.)

**Proposition 1.** *Let $n = pq$, where $p \neq q$, $p = 2p' + 1$, $q = 2q' + 1$, and $p, q, p', q'$ are all prime. The order of the elements in $\mathbb{Z}_n^*$ are one of the set $\{1, 2, p', q', 2p', 2q', p'q', 2p'q'\}$. Moreover, if the order of $a \in \mathbb{Z}_n^*$ is equal to $p'q'$ or $2p'q' \iff \gcd(a \pm 1, n) = 1$.*     □

**Corollary 1.** *Let $n$ be as in Proposition 1. Then, for any $a \in \mathbb{Z}_n^*$ s.t. $\gcd(a \pm 1, n) = 1$, $\langle a^2 \rangle \subset \mathbb{Z}_n^*$ is a cyclic subgroup of order $p'q'$.*     □

*Remark 1.* Notice that $4 \,(= 2^2)$ always generates $QR(n)$ whatever the value of a safe RSA modulus $n$. Notice also that the Jacobi symbol $(g|n) = +1$ does *not* necessarily imply that $g$ is a quadratic residue modulo $n$ but merely that $(g|p) = (g|q) = \pm 1$, where $(g|p)$ (resp. $(g|q)$) denotes the Legendre symbol[2] of $g$ modulo $p$ (resp. $q$). For example, $(2|55) = (2|5)(2|11) = (-1)(-1) = +1$; however, there is no integer $x$ such that $x^2 \equiv 2 \pmod{55}$.

Deciding whether some $y$ is in $QR(n)$ is generally believed infeasible if the factorization of $n$ is unknown.

### 3.2   Signatures of Knowledge

So-called zero-knowledge proofs of knowledge allow a prover to demonstrate the knowledge of a secret w.r.t. some public information such that no other

---

[2] By definition, the Legendre symbol $(g|p) = +1$ if $g$ is a quadratic residue modulo $p$, and $-1$ otherwise.

information is revealed in the process. The protocols we use in the following are all 3-move protocols and can be proven zero-knowledge in an honest-verifier model. Such protocols can be performed non-interactively with the help of an ideal hash function $\mathcal{H}$ (à la Fiat-Shamir [FS87]). Following [CS97], we refer to the resulting constructs as *signatures of knowledge*. One example is the Schnorr signature scheme [Sch91] where a signature can be viewed as a proof of knowledge of the discrete logarithm of the signer's public key made non-interactive.

In the following, we consider three building blocks: signature of knowledge of (i) a discrete logarithm; (ii) equality of two discrete logarithms; and (iii) a discrete logarithm lying in a given interval. All of these are constructed over a cyclic group $G = \langle g \rangle$ the order of which $\#G$ is unknown; however its bit-length $\ell_G$ (i.e., the integer $\ell_G$ s.t. $2^{\ell_G-1} \leq \#G < 2^{\ell_G}$) is publicly known. Fujisaki and Okamota [FO97] show that, under the SRSA, the standard proofs of knowledge protocols that work for a group of known order are also proofs of knowledge in this setting. We define the *discrete logarithm* of $y \in G$ w.r.t. base $g$ as any integer $x \in \mathbb{Z}$ such that $y = g^x$ in $G$. We denote $x = \log_g y$. We assume a collision-resistant hash function $\mathcal{H} : \{0,1\}^* \to \{0,1\}^k$ which maps a binary string of arbitrary length to a $k$-bit hash value. We also assume a security parameter $\epsilon > 1$.

Showing the knowledge of the discrete logarithm of $y = g^x$ can be done easily in this setting as stated by the following definition (cf. [Sch91]).

**Definition 4.** *Let $y, g \in G$. A pair $(c, s) \in \{0,1\}^k \times \pm\{0,1\}^{\epsilon(\ell_G+k)+1}$ verifying $c = \mathcal{H}(y\|g\|g^s y^c\|m)$ is a signature of knowledge of the discrete logarithm of $y = g^x$ w.r.t. base $g$, on a message $m \in \{0,1\}^*$.*

The party in possession of the secret $x = \log_g y$ is able to compute the signature by choosing a random $t \in \pm\{0,1\}^{\epsilon(\ell_G+k)}$ and then computing $c$ and $s$ as:

$$c = \mathcal{H}(y\|g\|g^t\|m) \quad \text{and} \quad s = t - cx \quad (\text{in } \mathbb{Z}) \ .$$

A slight modification of the previous definition enables to show the knowledge and equality of two discrete logarithms of, say $y_1$ and $y_2$, with bases $g$ and $h$, i.e., knowledge of an integer $x$ satisfying $y_1 = g^x$ and $y_2 = h^x$.

**Definition 5.** *Let $y_1, y_2, g, h \in G$. A pair $(c, s) \in \{0,1\}^k \times \pm\{0,1\}^{\epsilon(\ell_G+k)+1}$ verifying $c = \mathcal{H}(y_1\|y_2\|g\|h\|g^s y_1{}^c\|h^s y_2{}^c\|m)$ is a signature of knowledge of the discrete logarithm of both $y_1 = g^x$ w.r.t. base $g$ and $y_2 = h^x$ w.r.t. base $h$, on a message $m \in \{0,1\}^*$.*

The party in possession of the secret $x$ is able to compute the signature, provided that $x = \log_g y_1 = \log_h y_2$, by choosing a random $t \in \pm\{0,1\}^{\epsilon(\ell_G+k)}$ and then computing $c$ and $s$ as:

$$c = \mathcal{H}(y_1\|y_2\|g\|h\|g^t\|h^t\|m) \quad \text{and} \quad s = t - cx \quad (\text{in } \mathbb{Z}) \ .$$

In Definition 4, a party shows the knowledge of the discrete logarithm of $y$ w.r.t. base $g$. The order of $g$ being unknown, this means that this party knows an integer $x$ satisfying $y = g^x$. This latter condition may be completed in the sense that the party knows a discrete logarithm $x$ lying in a given interval. It is a slight modification of a protocol appearing in [FO98].

**Definition 6.** *Let* $y, g \in G$. *A pair* $(c, s) \in \{0, 1\}^k \times \pm\{0, 1\}^{\epsilon(\ell+k)+1}$ *verifying* $c = \mathcal{H}(y \,\|g\, \|g^{s-cX}\, y^c \,\|m)$ *is a signature of knowledge of the discrete logarithm* $\log_g y$ *that lies in* $]X - 2^{\epsilon(\ell+k)}, X + 2^{\epsilon(\ell+k)}[$, *on a message* $m \in \{0, 1\}^*$.

From the knowledge of $x = \log_g y \in \,]X - 2^\ell, X + 2^\ell[$, this signature is obtained by choosing a random $t \in \pm\{0, 1\}^{\epsilon(\ell+k)}$ and computing $c$ and $s$ as:

$$c = \mathcal{H}(y\|g\|g^t\|m), \quad s = t - c(x - X) \quad (\text{in } \mathbb{Z}) \ .$$

*Remark 2.* Note that, although the party knows a secret $x$ in $]X - 2^\ell, X + 2^\ell[$, the signature only guarantees that $x$ lies in the extended interval $]X - 2^{\epsilon(\ell+k)}, X + 2^{\epsilon(\ell+k)}[$.

The security of all these building blocks has been proven in the random oracle model [BR93] under the strong RSA assumption in [CM98b,FO97,FO98]. That is, if $\epsilon > 1$, then the corresponding interactive protocols are statistical (honest-verifier) zero-knowledge proofs of knowledge.

# 4   The New Group Signature and Identity Escrow Schemes

This section describes our new group signature scheme and tells how an identity escrow scheme can be derived.

As mentioned in Section 2, many recent group signature schemes involve applying two types of non-group signature schemes: one for issuing certificates and one for actual group-signatures, respectively. The security of the former, in particular, is of immediate relevance because it assures, among other things, the coalition-resistance property of a group signature scheme. The reasoning for this assertion is fairly intuitive:

Each group member obtains a unique certificate from the group manager as part of JOIN where each certificate is actually a signature over a secret random message chosen by each member. As a coalition, all group members can be collectively thought of as a single adversary mounting an adaptive chosen message attack consisting of polynomially many instances of JOIN.

The main challenge in designing a practical group signature scheme is in finding a signature scheme for the certification of membership that allows the second signature scheme (which is used to produce actual group signatures) to remain efficient. Typically, the second scheme is derived (using the Fiat-Shamir heuristic) from a proof of knowledge of a membership certificate. Hence, the certification signature scheme must be such that the latter proof can be realized efficiently.

Recall that proving knowledge of a hash function pre-image is, in general, not possible in an efficient manner. Therefore, a candidate signature scheme must replace the hash function with another suitable function. However, because JOIN is an interactive protocol between the new member and the group manager, the latter can limit and influence what he signs (e.g., assure that it signs a *random* message).

## 4.1   The Group Signature Scheme

Let $\epsilon > 1$, $k$, and $\ell_p$ be security parameters and let $\lambda_1$, $\lambda_2$, $\gamma_1$, and $\gamma_2$ denote lengths satisfying $\lambda_1 > \epsilon(\lambda_2+k)+2$, $\lambda_2 > 4\ell_p$, $\gamma_1 > \epsilon(\gamma_2+k)+2$, and $\gamma_2 > \lambda_1+2$. Define the integral ranges $\Lambda = ]2^{\lambda_1} - 2^{\lambda_2}, 2^{\lambda_1} + 2^{\lambda_2}[$ and $\Gamma = ]2^{\gamma_1} - 2^{\gamma_2}, 2^{\gamma_1} + 2^{\gamma_2}[$. Finally, let $\mathcal{H}$ be a collision-resistant hash function $\mathcal{H} : \{0,1\}^* \to \{0,1\}^k$. (The parameter $\epsilon$ controls the tightness of the statistical zero-knowledgeness and the parameter $\ell_p$ sets the size of the modulus to use.)

The initial phase involves the group manager $(GM)$ setting the group public and his secret keys: $\mathcal{Y}$ and $\mathcal{S}$.

---

SETUP:

1. Select random secret $\ell_p$-bit primes $p', q'$ such that $p = 2p' + 1$ and $q = 2q' + 1$ are prime. Set the modulus $n = pq$.
2. Choose random elements $a, a_0, g, h \in_R \mathrm{QR}(n)$ (of order $p'q'$).
3. Choose a random secret element $x \in_R \mathbb{Z}^*_{p'q'}$ and set $y = g^x \bmod n$.
4. The group public key is: $\mathcal{Y} = (n, a, a_0, y, g, h)$.
5. The corresponding secret key (known only to $GM$) is: $\mathcal{S} = (p', q', x)$.

---

*Remark 3.* The group public key $\mathcal{Y}$ is made available via the usual means (i.e., embedded in some form of a public key certificate signed by a trusted authority). We note that, in practice, components of $\mathcal{Y}$ must be verifiable to prevent framing attacks. In particular, Proposition 1 provides an efficient way to test whether an element has order at least $p'q'$. Then it is sufficient to square this element to make sure it is in $\mathrm{QR}(n)$, with order $p'q'$. $GM$ also needs to provide a proof that $n$ is the product of two safe primes ([CM99a] shows how this can be done).

Suppose now that a new user wants to join the group. We assume that communication between the user and the group manager is secure, i.e., private and authentic. The selection of per-user parameters is done as follows:

---

JOIN:

1. User $P_i$ generates a secret exponent $\tilde{x}_i \in_R ]0, 2^{\lambda_2}[$, a random integer $\tilde{r} \in_R ]0, n^2[$ and sends $C_1 = g^{\tilde{x}_i} h^{\tilde{r}} \bmod n$ to $GM$ and proves him knowledge of the representation of $C_1$ w.r.t. bases $g$ and $h$.
2. $GM$ checks that $C_1 \in \mathrm{QR}(n)$. If this is the case, $GM$ selects $\alpha_i$ and $\beta_i \in_R ]0, 2^{\lambda_2}[$ at random and sends $(\alpha_i, \beta_i)$ to $P_i$.

3. User $P_i$ computes $x_i = 2^{\lambda_1} + (\alpha_i \tilde{x}_i + \beta_i \bmod 2^{\lambda_2})$ and sends $GM$ the value $C_2 = a^{x_i} \bmod n$. The user also proves to $GM$:
   (a) that the discrete log of $C_2$ w.r.t. base $a$ lies in $\Lambda$, and
   (b) knowledge of integers $u$, $v$, and $w$ such that
       i. $u$ lies in $]-2^{\lambda_2}, 2^{\lambda_2}[$,
       ii. $u$ equals the discrete log of $C_2/a^{2^{\lambda_1}}$ w.r.t. base $a$, and
       iii. $C_1^{\alpha_i} g^{\beta_i}$ equals $g^u (g^{2^{\lambda_2}})^v h^w$ (see Definition 6).
   (The statements (i–iii) prove that the user's membership secret $x_i = \log_a C_2$ is correctly computed from $C_1$, $\alpha_i$, and $\beta_i$.)
4. $GM$ checks that $C_2 \in \mathrm{QR}(n)$. If this is the case and all the above proofs were correct, $GM$ selects a random prime $e_i \in_R \Gamma$ and computes $A_i := (C_2 a_0)^{1/e_i} \bmod n$. Finally, $GM$ sends $P_i$ the new membership certificate $[A_i, e_i]$. (Note that $A_i = (a^{x_i} a_0)^{1/e_i} \bmod n$.)
5. User $P_i$ verifies that $a^{x_i} a_0 \equiv A_i^{e_i} \pmod{n}$.

*Remark 4.* As part of JOIN, $GM$ creates a new entry in the membership table and stores $\{[A_i, e_i], \text{JOIN transcript}\}$ in the new entry. (JOIN transcript is formed by the messages received from and sent to the user in the steps above. It is assumed to be signed by the user with some form of a long-term credential.)

Armed with a membership certificate $[A_i, e_i]$, a group member can generate anonymous and unlinkable group signatures on a generic message $m \in \{0, 1\}^*$:

SIGN:

1. Generate a random value $w \in_R \{0, 1\}^{2\ell_p}$ and compute:

$$T_1 = A_i y^w \bmod n, \quad T_2 = g^w \bmod n, \quad T_3 = g^{e_i} h^w \bmod n \ .$$

2. Randomly choose $r_1 \in_R \pm\{0, 1\}^{\epsilon(\gamma_2 + k)}$, $r_2 \in_R \pm\{0, 1\}^{\epsilon(\lambda_2 + k)}$, $r_3 \in_R \pm\{0, 1\}^{\epsilon(\gamma_1 + 2\ell_p + k + 1)}$, and $r_4 \in_R \pm\{0, 1\}^{\epsilon(2\ell_p + k)}$ and compute:
   (a) $d_1 = T_1^{r_1}/(a^{r_2} y^{r_3}) \bmod n$, $d_2 = T_2^{r_1}/g^{r_3} \bmod n$, $d_3 = g^{r_4} \bmod n$, and $d_4 = g^{r_1} h^{r_4} \bmod n$;
   (b) $c = \mathcal{H}(g\|h\|y\|a_0\|a\|T_1\|T_2\|T_3\|d_1\|d_2\|d_3\|d_4\|m)$;
   (c) $s_1 = r_1 - c(e_i - 2^{\gamma_1})$, $s_2 = r_2 - c(x_i - 2^{\lambda_1})$, $s_3 = r_3 - c e_i w$, and $s_4 = r_4 - c w$ (all in $\mathbb{Z}$).
3. Output $(c, s_1, s_2, s_3, s_4, T_1, T_2, T_3)$.

A group signature can be regarded as a signature of knowledge of (1) a value $x_i \in \Lambda$ such that $a^{x_i} a_0$ is the value that is ElGamal-encrypted in $(T_1, T_2)$ under $y$ and of (2) an $e_i$-th root of that encrypted value, where $e_i$ is the first part of the representation of $T_3$ w.r.t. $g$ and $h$ and that $e_i$ lies in $\Gamma$.

A verifier can check the validity of a signature $(c, s_1, s_2, s_3, s_4, T_1, T_2, T_3)$ of the message $m$ as follows:

---

**VERIFY:**

1. Compute:

$$c' = \mathcal{H}\big(g\|h\|y\|a_0\|a\|T_1\|T_2\|T_3\|a_0{}^c\,T_1{}^{s_1-c2^{\gamma_1}}/(a^{s_2-c2^{\lambda_1}}\,y^{s_3})\bmod n\,\|$$
$$T_2{}^{s_1-c2^{\gamma_1}}/g^{s_3}\bmod n\,\|T_2{}^c\,g^{s_4}\bmod n\,\|T_3{}^c\,g^{s_1-c2^{\gamma_1}}\,h^{s_4}\bmod n\,\|\,m\big)\ .$$

2. Accept the signature if and only if $c = c'$, and $s_1 \in \pm\{0,1\}^{\epsilon(\gamma_2+k)+1}$, $s_2 \in \pm\{0,1\}^{\epsilon(\lambda_2+k)+1}$, $s_3 \in \pm\{0,1\}^{\epsilon(\lambda_1+2\ell_p+k+1)+1}$, $s_4 \in \pm\{0,1\}^{\epsilon(2\ell_p+k)+1}$.

---

In the event that the actual signer must be subsequently identified (e.g., in case of a dispute) $GM$ executes the following procedure:

---

**OPEN:**

1. Check the signature's validity via the VERIFY procedure.
2. Recover $A_i$ (and thus the identity of $P_i$) as $A_i = T_1/T_2{}^x \bmod n$.
3. Prove that $\log_g y = \log_{T_2}(T_1/A_i \bmod n)$ (see Definition 5).

---

### 4.2  Deriving an Identity Escrow Scheme

Only minor changes are necessary to construct an identity escrow scheme out of the proposed group signature scheme. Specifically, the SIGN and VERIFY procedures must be replaced by an interactive protocol between a group member (prover) and a verifier. This protocol can be derived from SIGN by replacing the call to the hash function $\mathcal{H}$ by a call to the verifier. That is, the prover sends to the verifier all inputs to the hash function $\mathcal{H}$ and gets back a value $c \in \{0,1\}^{\ell_c}$ randomly chosen by the verifier, with $\ell_c = O(\log \ell_p)$. Then, the prover computes the $s_i$'s and sends these back to the verifier. The verification equation that the verifier uses to check can be derived from the argument to $\mathcal{H}$ in VERIFY. Depending on the choice of the security parameters, the resulting protocol must be repeated sufficiently many times to obtain a small enough probability of error.

## 5   Related Work

Previously proposed group signature schemes can be divided into two classes: (I) schemes where the sizes of the group public key and/or of group signatures (linearly) depend on the number of group members and (II) schemes where the sizes of the group public key and of group signatures are constant. Most of the early schemes belong to the first class. Although many of those have been proven secure with respect to some standard cryptographic assumption (such as the hardness of computing discrete logarithms) they are inefficient for large groups.

Numerous schemes of Class II have been proposed, however, most are either insecure (or of dubious security) or are grossly inefficient. The only notable and efficient group signature scheme is due to Camenisch and Michels [CM98b].

Our scheme differs from the Camenisch/Michels scheme mainly in the membership certificate format. As a consequence, our JOIN protocol has two important advantages:

(1) Our JOIN protocol is an order of magnitude more efficient since all proofs that the new group member must provide are efficient proofs of knowledge of discrete logarithms. This is in contrast to the Camenisch/Michels scheme where the group member must prove that some number is the product of two primes. The latter can be realized only with binary challenges.

(2) Our JOIN protocol is more secure for the group members, i.e., it is statistical zero-knowledge with respect to the group member's membership secret. The JOIN protocol in the Camenisch-Michels scheme is not; in fact, it requires the group member to expose the product of her secret, a prime of special form, and a random prime; such products are in principle susceptible to an attack due to Coppersmith [Cop96]. (Although, the parameters of their scheme can be set such that this attack becomes infeasible.)

Furthermore, the proposed scheme is provably coalition-resistant against an *adaptive* adversary. This offers an extra advantage:

(3) Camenisch and Michels prove their scheme coalition-resistant against a static adversary who is given all certificates as input, whereas our scheme can handle a much more powerful and realistic adversary that is allowed to *adaptively* run the JOIN protocol.

# 6   Security of the Proposed Schemes

In this section we assess the security of the new group signature scheme and the companion escrow identity scheme. We first need to prove that the following theorems hold.

**Theorem 1 (Coalition-resistance).** *Under the strong RSA assumption, a group certificate $[A_i = (a^{x_i} a_0)^{1/e_i} \bmod n, e_i]$ with $x_i \in \Lambda$ and $e_i \in \Gamma$ can be generated only by the group manager provided that the number $K$ of certificates the group manager issues is polynomially bounded.*

*Proof.* Let $\mathcal{M}$ be an attacker that is allowed to adaptively run the JOIN and thereby obtain group certificates $[A_j = (a^{x_j} a_0)^{1/e_j} \bmod n, e_j]$, $j = 1, \ldots, K$. Our task is now to show that if $\mathcal{M}$ outputs a tuple $(\hat{x}; [\hat{A}, \hat{e}])$, with $\hat{x} \in \Lambda$, $\hat{e} \in \Gamma$, $\hat{A} = (a^{\hat{x}} a_0)^{1/\hat{e}} \bmod n$, and $(\hat{x}, \hat{e}) \neq (x_j, e_j)$ for all $1 \leq j \leq K$ with non-negligible probability, then the strong RSA assumption does not hold.

Given a pair $(n, z)$, we repeatedly play a random one of the following two games with $\mathcal{M}$ and hope to calculate a pair $(u, e) \in \mathbb{Z}_n^* \times \mathbb{Z}_{>1}$ satisfying $u^e \equiv z \pmod{n}$ from $\mathcal{M}$'s answers. The first game goes as follows:

1. Select $x_1, \ldots, x_K \in \Lambda$ and $e_1, \ldots, e_K \in \Gamma$.
2. Set $a = z^{\prod_{1 \leq l \leq K} e_l} \bmod n$.

3. Choose $r \in_R \Lambda$ and set $a_0 = a^r \mod n$.
4. For all $1 \leq i \leq K$, compute $A_i = z^{(x_i + r)} \prod_{1 \leq l \leq K; l \neq i} e_l \mod n$.
5. Select $g, h \in_R \mathrm{QR}(n)$, $x \in \{1, \ldots, n^2\}$, and set $y = g^x \mod n$.
6. Run the JOIN protocol $K$ times with $\mathcal{M}$ on input $(n, a, a_0, y, g, h)$. Assume we are in protocol run $i$. Receive the commitment $C_1$ from $\mathcal{M}$. Use the proof of knowledge of a representation of $C_1$ with respect to $g$ and $h$ to extract $\tilde{x}_i$ and $\tilde{r}_i$ such that $C_1 = g^{\tilde{x}_i} h^{\tilde{r}_i}$ (this involves rewinding of $\mathcal{M}$). Choose $\alpha_i$ and $\beta_i \in ]0, 2^{\lambda_2}[$ such that the prepared $x_i = 2^{\lambda_1} + (\alpha_i \tilde{x}_i + \beta_i \mod 2^{\lambda_2})$ and send $\alpha_i$ and $\beta_i$ to $\mathcal{M}$. Run the rest of the protocol as specified until Step 4. Then send $\mathcal{M}$ the membership certificate $[A_i, e_i]$.

   After these $K$ registration protocols are done, $\mathcal{M}$ outputs $\left(\hat{x}; [\hat{A}, \hat{e}]\right)$ with $\hat{x} \in \Lambda$, $\hat{e} \in \Gamma$, and $\hat{A} = (a^{\hat{x}} a_0)^{1/\hat{e}} \mod n$.

7. If $\gcd(\hat{e}, e_j) \neq 1$ for all $1 \leq j \leq K$ then output $\perp$ and quit. Otherwise, let $\tilde{e} := (\hat{x} + r) \prod_{1 \leq l \leq K} e_l$. (Note that $\hat{A}^{\hat{e}} \equiv z^{\tilde{e}} \pmod{n}$.) Because $\gcd(\hat{e}, e_j) = 1$ for all $1 \leq j \leq K$, we have $\gcd(\hat{e}, \tilde{e}) = \gcd(\hat{e}, (\hat{x} + r))$. Hence, by the extended Euclidean algorithm, there exist $\alpha, \beta \in \mathbb{Z}$ s.t. $\alpha \hat{e} + \beta \tilde{e} = \gcd(\hat{e}, (\hat{x} + r))$. Therefore, letting $u := z^\alpha \hat{A}^\beta \mod n$ and $e := \hat{e}/\gcd(\hat{e}, (\hat{x} + r)) > 1$ because $\hat{e} > (\hat{x} + r)$, we have $u^e \equiv z \pmod{n}$. Output $(u, e)$.

The previous game is only successful if $\mathcal{M}$ returns a new certificate $[A(\hat{x}), \hat{e}]$, with $\gcd(\hat{e}, e_j) = 1$ for all $1 \leq j \leq K$. We now present a game that solves the strong RSA problem in the other case when $\gcd(\hat{e}, e_j) \neq 1$ for some $1 \leq j \leq K$. (Note that $\gcd(\hat{e}, e_j) \neq 1$ means $\gcd(\hat{e}, e_j) = e_j$ because $e_j$ is prime.)

1. Select $x_1, \ldots, x_K \in \Lambda$ and $e_1, \ldots, e_K \in \Gamma$.
2. Choose $j \in_R \{1, \ldots, K\}$ and set $a = z^{\prod_{1 \leq l \leq K; l \neq j} e_l} \mod n$.
3. Choose $r \in_R \Lambda$ and set $A_j = a^r \mod n$ and $a_0 = A_j^{e_j}/a^{x_j} \mod n$.
4. For all $1 \leq i \leq K$ $i \neq j$, compute $A_i = z^{(x_i + e_j r - x_j)} \prod_{1 \leq l \leq K; l \neq i, j} e_l \mod n$.
5. Select $g, h \in_R \mathrm{QR}(n)$, $x \in \{1, \ldots, n^2\}$, and set $y = g^x \mod n$.
6. Run the JOIN protocol $K$ times with $\mathcal{M}$ on input $(n, a, a_0, y, g, h)$. Assume we are in protocol run $i$. Receive the commitment $C_1$ from $\mathcal{M}$. Use the proof of knowledge of a representation of $C_1$ with respect to $g$ and $h$ to extract $\tilde{x}_i$ and $\tilde{r}_i$ such that $C_1 = g^{\tilde{x}_i} h^{\tilde{r}_i} \mod n$ (this involves rewinding of $\mathcal{M}$). Choose $\alpha_i$ and $\beta_i \in ]0, 2^{\lambda_2}[$ such that the prepared $x_i = 2^{\lambda_1} + (\alpha_i \tilde{x}_i + \beta_i \mod 2^{\lambda_2})$ and send $\alpha_i$ and $\beta_i$ to $\mathcal{M}$. Run the rest of the protocol as specified until Step 4. Then send $\mathcal{M}$ the membership certificate $[A_i, e_i]$.

   After these $K$ registration protocols are done, $\mathcal{M}$ outputs $\left(\hat{x}; [\hat{A}, \hat{e}]\right)$ with $\hat{x} \in \Lambda$, $\hat{e} \in \Gamma$, and $\hat{A} = (a^{\hat{x}} a_0)^{1/\hat{e}} \mod n$.

7. If $\gcd(\hat{e}, e_j) \neq e_j$ output $\perp$ and quit. Otherwise, we have $\hat{e} = t e_j$ for some $t$ and can define $Z := \hat{A}^t/A_j \mod n$ if $\hat{x} \geq x_j$ and $Z := A_j/\hat{A}^t \mod n$ otherwise. Hence, $Z \equiv (a^{|\hat{x} - x_j|})^{1/e_j} \equiv (z^{|\tilde{e}|})^{1/e_j} \pmod{n}$ with $\tilde{e} := (\hat{x} - x_j) \prod_{1 \leq l \leq K; l \neq j} e_l$. Because $\gcd(e_j, \prod_{\substack{1 \leq l \leq K \\ l \neq j}} e_l) = 1$, it follows that $\gcd(e_j, |\tilde{e}|) = \gcd(e_j, |\hat{x} - x_j|)$. Hence, there exist $\alpha, \beta \in \mathbb{Z}$ s.t. $\alpha e_j + \beta |\tilde{e}| = \gcd(e_j, |\hat{x} - x_j|)$. So, letting $u := z^\alpha Z^\beta \mod n$ and $e := e_j/\gcd(e_j, |\hat{x} - x_j|) > 1$ because $e_j > |\hat{x} - x_j|$, we have $u^e \equiv z \pmod{n}$. Output $(u, e)$.

Consequently, by playing randomly one of the Games 1 or 2 until the result is not $\perp$, an attacker getting access to machine $\mathcal{M}$ can solve the strong RSA problem in expected running-time polynomial in $K$. Because the latter is assumed to be infeasible, it follows that no one but the group manager can generate group certificates.     $\square$

**Theorem 2.** *Under the strong RSA assumption, the interactive protocol underlying the group signature scheme (i.e., the identification protocol of the identity escrow scheme) is a statistical zero-knowledge (honest-verifier) proof of knowledge of a membership certificate and a corresponding membership secret key.*

*Proof.* The proof that the interactive protocol is statistical zero-knowledge is quite standard. We restrict our attention the proof of knowledge part.

We have to show that the knowledge extractor is able to recover the group certificate once it has found two accepting tuples. Let $(T_1, T_2, T_3, d_1, d_2, d_3, d_4, c, s_1, s_2, s_3, s_4)$ and $(T_1, T_2, T_3, d_1, d_2, d_3, d_4, \tilde{c}, \tilde{s}_1, \tilde{s}_2, \tilde{s}_3, \tilde{s}_4)$ be two accepting tuples. Because $d_3 \equiv g^{s_4} T_2^{c} \equiv g^{\tilde{s}_4} T_2^{\tilde{c}} \pmod{n}$, it follows that $g^{s_4 - \tilde{s}_4} \equiv T_2^{\tilde{c} - c} \pmod{n}$. Letting $\delta_4 = \gcd(s_4 - \tilde{s}_4, \tilde{c} - c)$, by the extended Euclidean algorithm, there exist $\alpha_4, \beta_4 \in \mathbb{Z}$ s.t. $\alpha_4 (s_4 - \tilde{s}_4) + \beta_4 (\tilde{c} - c) = \delta_4$. Hence,

$$g \equiv g^{(\alpha_4 (s_4 - \tilde{s}_4) + \beta_4 (\tilde{c} - c))/\delta_4} \equiv (T_2^{\alpha_4} g^{\beta_4})^{\frac{\tilde{c} - c}{\delta_4}} \pmod{n} \ .$$

Note that we cannot have $\tilde{c} - c < \delta_4$ because otherwise $T_2^{\alpha_4} g^{\beta_4}$ is a $(\frac{\tilde{c} - c}{\delta_4})^{\text{th}}$ root of $g$, which contradicts the strong RSA assumption. Thus, we have $\tilde{c} - c = \delta_4 = \gcd(s_4 - \tilde{s}_4, \tilde{c} - c)$; or equivalently, there exists $\tau_4 \in \mathbb{Z}$ s.t. $s_4 - \tilde{s}_4 = \tau_4(\tilde{c} - c)$. So, because $s_4 + c w = \tilde{s}_4 + \tilde{c} w$, we have $\tau_4 = w$ and thus obtain

$$A_i = \frac{T_1}{y^{\tau_4}} \bmod n \ .$$

Moreover, because $d_4 \equiv g^{s_1} h^{s_4} (T_3 g^{-2^{\gamma_1}})^c \equiv g^{\tilde{s}_1} h^{\tilde{s}_4} (T_3 g^{-2^{\gamma_1}})^{\tilde{c}} \pmod{n}$, we have $g^{s_1 - \tilde{s}_1} \equiv (T_3 g^{-2^{\gamma_1}})^{\tilde{c} - c} h^{\tilde{s}_4 - s_4} \equiv (T_3 g^{-2^{\gamma_1}} h^{-\tau_4})^{\tilde{c} - c} \pmod{n}$. Let $\delta_1 = \gcd(s_1 - \tilde{s}_1, \tilde{c} - c)$. By the extended Euclidean algorithm, there exist $\alpha_1, \beta_1 \in \mathbb{Z}$ s.t. $\alpha_1 (s_1 - \tilde{s}_1) + \beta_1 (\tilde{c} - c) = \delta_1$. Therefore, $g \equiv g^{(\alpha_1 (s_1 - \tilde{s}_1) + \beta_1 (\tilde{c} - c))/\delta_1} \equiv [(T_3 g^{-2^{\gamma_1}} h^{-\tau_4})^{\alpha_1} g^{\beta_1}]^{\frac{\tilde{c} - c}{\delta_1}} \pmod{n}$. This, in turn, implies by the strong RSA assumption that $\tilde{c} - c = \delta_1 = \gcd(s_1 - \tilde{s}_1, \tilde{c} - c)$; or equivalently that there exists $\tau_1 \in \mathbb{Z}$ s.t. $s_1 - \tilde{s}_1 = \tau_1(\tilde{c} - c)$. Consequently, because $s_1 + c(e_i - 2^{\gamma_1}) = \tilde{s} + \tilde{c}(e_i - 2^{\gamma_1})$, we find

$$e_i = 2^{\gamma_1} + \tau_1 \ .$$

Likewise, from $d_2 \equiv T_2^{s_1} g^{-s_3} (T_2^{-2^{\gamma_1}})^c \equiv T_2^{\tilde{s}_1} g^{-\tilde{s}_3} (T_2^{-2^{\gamma_1}})^{\tilde{c}} \pmod{n}$, it follows that $g^{s_3 - \tilde{s}_3} \equiv (T_2^{\tau_1 + 2^{\gamma_1}})^{\tilde{c} - c} \pmod{n}$. Therefore, by the extended Euclidean algorithm, we can conclude that there exists $\tau_3 \in \mathbb{Z}$ s.t. $s_3 - \tilde{s}_3 = \tau_3(\tilde{c} - c)$. Finally, from $d_1 \equiv T_1^{s_1} a^{-s_2} y^{-s_3} (T_1^{-2^{\gamma_1}} a^{2^{\lambda_1}} a_0)^c \equiv T_1^{\tilde{s}_1} a^{-\tilde{s}_2} y^{-\tilde{s}_3} (T_1^{-2^{\gamma_1}} a^{2^{\lambda_1}} a_0)^{\tilde{c}} \pmod{n}$, we obtain $a^{\tilde{s}_2 - s_2} \equiv (T_1^{\tau_1 + 2^{\gamma_1}} y^{-\tau_3} a^{-2^{\lambda_1}} a_0^{-1})^{\tilde{c} - c}$

(mod $n$) and similarly conclude that there exists $\tau_2 \in \mathbb{Z}$ s.t. $s_2 - \tilde{s}_2 = \tau_2(\tilde{c} - c)$. Because $s_2 + c(x_i - 2^{\lambda_1}) = \tilde{s}_2 + \tilde{c}(x_i - 2^{\lambda_1})$, we recover

$$x_i = 2^{\lambda_1} + \tau_2,$$

which concludes the proof. □

**Corollary 2.** *The* JOIN *protocol is zero-knowledge w.r.t. the group manager. Furthermore, the user's membership secret key $x_i$ is a random integer from $\Lambda$.*

*Proof.* Straight-forward. □

**Corollary 3.** *In the random oracle model the group signature scheme presented in Section 4 is secure under the strong RSA and the decisional Diffie-Hellman assumption.*

*Proof.* We have to show that our scheme satisfies all the security properties listed in Definition 1.

**Correctness:** By inspection.

**Unforgeability:** Only group members are able to sign messages on behalf of the group: This is an immediate consequence of Theorem 2 and the random oracle model, that is, if we assume the hash function $\mathcal{H}$ behaves as a random function.

**Anonymity:** Given a valid signature $(c, s_1, s_2, s_3, s_4, T_1, T_2, T_3)$ identifying the actual signer is computationally hard for everyone but the group manager: Because of Theorem 2 the underlying interactive protocol is statistically zero-knowledge, no information is statistically revealed by $(c, s_1, s_2, s_3, s_4)$ in the random oracle model. Deciding whether some group member with certificate $[A_i, e_i]$ originated requires deciding whether the three discrete logarithms $\log_y T_1/A_i$, $\log_g T_2$, and $\log_g T_3/g^{e_i}$ are equal. This is assumed to be infeasible under the decisional Diffie-Hellman assumption and hence anonymity is guaranteed.

**Unlinkability:** Deciding if two signatures $(T_1, T_2, T_3, c, s_1, s_2, s_3, s_4)$ and $(\tilde{T}_1, \tilde{T}_2, \tilde{T}_3, \tilde{c}, \tilde{s}_1, \tilde{s}_2, \tilde{s}_3, \tilde{s}_4)$ were computed by the same group member is computationally hard. Simiarly as for **Anonymity**, the problem of linking two signatures reduces to decide whether the three discrete logarithms $\log_y T_1/\tilde{T}_i$, $\log_g T_2/\tilde{T}_2$, and $\log_g T_3/\tilde{T}_3$ are equal. This is, however, impossible under Decisional Diffie-Hellman Assumption.

**Exculpability:** Neither a group member nor the group manager can sign on behalf of other group members: First note that due to Corollary 2, $GM$ does not get any information about a user's secret $x_i$ apart from $a^{x_i}$. Thus, the value $x_i$ is computationally hidden from $GM$. Next note that $T_1$, $T_2$, and $T_3$ are an unconditionally binding commitments to $A_i$ and $e_i$. One can show that, if the factorization of $n$ would be publicly known, the interactive proof underlying the group signature scheme is a proof of knowledge of the discrete log of $A_i^{e_i}/a_0$ (provided that $\ell_p$ is larger than twice to output length of the hash function / size of the challenges). Hence, not even the group manager can sign on behalf of $P_i$ because computing discrete logarithms is assumed to be infeasible.

Traceability: The group manager is able to open any valid group signature and *provably* identify the actual signer: Assuming that the signature is valid, this implies that $T_1$ and $T_2$ are of the required form and so $A_i$ can be uniquely recovered. Due to Theorem 1 a group certificate $[A_i = A(x_i), e_i]$ with $x_i \in \Lambda$ and $e_i \in \Gamma$ can only be obtained from via the JOIN protocol. Hence, the $A_i$ recovered can be uniquely be linked to an instance of the JOIN protocol and thus the user $P_i$ who originated the signature can be identified.

Coalition-resistance: Assuming the random oracle model, this follows from Theorems 1 and 2.

□

**Corollary 4.** *The identity escrow scheme derived from our group signature scheme is secure under the strong RSA and the decisional Diffie-Hellman assumption.*

*Proof.* The proof is essentially the same as for Corollary 3, the difference being that we do not need the random oracle model but can apply Theorems 1 and 2 directly.

□

## 7  Conclusions

This paper presents a very efficient and provably secure group signature scheme and a companion identity escrow scheme that are based on the strong RSA assumption. Their performance and security appear to significantly surpass those of prior art. Extending the scheme to a blind group-signature scheme or to split the group manager into a membership manager and a revocation manager is straight-forward (cf. [CM98a,LR98]).

## References

BF97.    N. Barić and B. Pfitzmann. Collision-free accumulators and fail-stop signature schemes without trees. In *Advances in Cryptology — EUROCRYPT '97*, vol. 1233 of *LNCS*, pp. 480–494, Springer-Verlag, 1997.

BR93.    M. Bellare and P. Rogaway. Random oracles are practical: A paradigm for designing efficient protocols. In *1st ACM Conference on Computer and Communication Security*, pp. 62–73, ACM Press, 1993.

Bon98.   D. Boneh. The decision Diffie-Hellman problem. In *Algorithmic Number Theory (ANTS-III)*, vol. 1423 of *LNCS*, pp. 48–63, Springer-Verlag, 1998.

Bra93.   S. Brands. An efficient off-line electronic cash system based on the representation problem. Technical Report CS-R9323, Centrum voor Wiskunde en Informatica, April 1993.

CM98a.   J. Camenisch and M. Michels. A group signature scheme with improved efficiency. In *Advances in Cryptology — ASIACRYPT '98*, vol. 1514 of *LNCS*, pp. 160–174, Springer-Verlag, 1998.

CM98b.   _____. A group signature scheme based on an RSA-variant. Technical Report RS-98-27, BRICS, University of Aarhus, November 1998. An earlier version appears in [CM98a].

CM99a.     ———. Proving in zero-knowledge that a number is the product of two safe primes. In *Advances in Cryptology — EUROCRYPT '99*, vol. 1592 of *LNCS*, pp. 107–122, Springer-Verlag, 1999.

CM99b.     ———. Separability and efficiency for generic group signature schemes. In *Advances in Cryptology — CRYPTO '99*, vol. 1666 of *LNCS*, pp. 413–430, Springer-Verlag, 1999.

CP95.      L. Chen and T. P. Pedersen. New group signature schemes. In *Advances in Cryptology — EUROCRYPT '94*, vol. 950 of *LNCS*, pp. 171–181, 1995.

CS97.      J. Camenisch and M. Stadler. Efficient group signature schemes for large groups. In *Advances in Cryptology — CRYPTO '97*, vol. 1296 of *LNCS*, pp. 410–424, Springer-Verlag, 1997.

Cam98.     J. Camenisch. Group signature schemes and payment systems based on the discrete logarithm problem. PhD thesis, vol. 2 of *ETH Series in Information Security an Cryptography*, Hartung-Gorre Verlag, Konstanz, 1998. ISBN 3-89649-286-1.

Cop96.     D. Coppersmith. Finding a small root of a bivariatre interger equation; factoring with high bits known. In *Advances in Cryptology — EUROCRYPT '96*, volume 1070 of *LNCS*, pages 178–189. Springer Verlag, 1996.

CvH91.     D. Chaum and E. van Heyst. Group signatures. In *Advances in Cryptology — EUROCRYPT '91*, vol. 547 of *LNCS*, pp. 257–265, Springer-Verlag, 1991.

DH76.      W. Diffie and M. E. Hellman. New directions in cryptography. *IEEE Transactions on Information Theory*, IT-22(6): 644–654, 1976.

FO97.      E. Fujisaki and T. Okamoto. Statistical zero knowledge protocols to prove modular polynomial relations. In *Advances in Cryptology — CRYPTO '97*, vol. 1297 of *LNCS*, pp. 16–30, Springer-Verlag, 1997.

FO98.      ———. A practical and provably secure scheme for publicly verifiable secret sharing and its applications. In *Advances in Cryptology — EUROCRYPT '98*, vol. 1403 of *LNCS*, pp. 32–46, Springer-Verlag, 1998.

FS87.      A. Fiat and A. Shamir. How to prove yourself: practical solutions to identification and signature problems. In *Advances in Cryptology — CRYPTO '86*, vol. 263 of *LNCS*, pp. 186–194, Springer-Verlag, 1987.

GMR88.     S. Goldwasser, S. Micali, and R. Rivest. A digital signature scheme secure against adaptive chosen-message attacks. *SIAM Journal on Computing*, 17(2):281–308, 1988.

KP98.      J. Kilian and E. Petrank. Identity escrow. In *Advances in Cryptology — CRYPTO '98*, vol. 1642 of *LNCS*, pp. 169–185, Springer-Verlag, 1998.

LR98.      A. Lysyanskaya and Z. Ramzan. Group blind digital signatures: A scalable solution to electronic cash. In *Financial Cryptography (FC '98)*, vol. 1465 of *LNCS*, pp. 184–197, Springer-Verlag, 1998.

Sch91.     C. P. Schnorr. Efficient signature generation by smart cards. *Journal of Cryptology*, 4(3):161–174, 1991.

# Provably Secure Partially Blind Signatures

Masayuki Abe and Tatsuaki Okamoto

NTT Laboratories
Nippon Telegraph and Telephone Corporation
1-1 Hikari-no-oka Yokosuka-shi Kanagawa-ken, 239-0847 Japan
{abe,okamoto}@isl.ntt.co.jp

**Abstract.** Partially blind signature schemes are an extension of blind signature schemes that allow a signer to explicitly include necessary information (expiration date, collateral conditions, or whatever) in the resulting signatures under some agreement with the receiver. This paper formalizes such a notion and presents secure and efficient schemes based on a widely applicable method of obtaining witness indistinguishable protocols. We then give a formal proof of security in the random oracle model. Our approach also allows one to construct secure fully blind signature schemes based on a variety of signature schemes.

**Keywords:** Partially Blind Signatures, Blind Signatures, Witness Indistinguishability

## 1 Introduction

### 1.1 Background

Digital signature schemes are essential for electronic commerce as they allow one to authorize digital documents that are moved across networks. Typically, a digital signature comes with not just the document body but also attributes such as "date of issue" or "valid until", which may be controlled by the signer rather than the receiver. One can find more about those attributes in PKCS #9 [23], for instance.

Blind signature schemes, first introduced by Chaum in [5], are a variant of digital signature schemes. They allow a receiver to get a signature without giving the signer any information about the actual message or the resulting signature. This blindness property plays a central role in applications such as electronic voting (e.g. [6,12]) and electronic cash schemes (e.g. [5,7,4]) where anonymity is of great concern.

One particular shortcoming is that, since the singer's view is perfectly shut off from the resulting signatures, the signer has no control over the attributes except for those bound by the public key. For instance, if a signer issues blind signatures that are valid until the end of the week, the signer has to change his public key every week! This will seriously impact availability and performance. A similar shortcoming can be seen in a simple electronic cash system where a bank issues a blind signature as an electronic coin. Since the bank cannot inscribe the value

M. Bellare (Ed.): CRYPTO 2000, LNCS 1880, pp. 271–286, 2000.
© Springer-Verlag Berlin Heidelberg 2000

on the blindly issued coins, it has to use different public keys for different coin values. Hence the shops and customers must always carry a list of those public keys in their electronic wallet, which is typically a smart card whose memory is very limited. Some electronic voting schemes also face the same problem when an administrator issues blind signatures to authorize ballots. Since he can not include the vote ID, his signature may be used in an unintended way. This means that the public key of the administrator must be disposable. Accordingly, each voter must download a new public key for each vote.

A *partially* blind signature scheme allows the signer to explicitly include common information in the blind signature under some agreement with the receiver. For instance, the signer can attach the date of issue to his blind signatures as an attribute. If the signer issues a huge number of signatures in a day, including the date of issue will not violate anonymity. Accordingly, the attributes of the signatures can be decided independently from those of the public key.

By fixing common information to a single string, one can easily transform partially blind signature schemes into fully blind ones. However, the reverse is not that easy. One can now see that partially blind signatures are a generalized notion of blind signatures. The main subject of this paper is to consider the security of partially blind signatures and present the first secure and efficient schemes together with a formal proof of their security.

## 1.2   Related Work

In [15], Juels, Luby and Ostrovsky gave a formal definition of blind signatures. They proved the existence of secure blind signatures assuming the one-way trapdoor permutation family. Their construction was, however, only theoretical, not practical. Before [15], Pointcheval and Stern showed the security of a certain type of efficient blind signature in the random oracle model [20]. Namely, they showed that Okamoto-Schnorr and Okamoto-Guillou-Quisquater signatures [18] are secure as long as the number of issued signatures are bounded logarithmically in the security parameter. Later, in [19], Pointcheval developed a generic approach that converts logarithmically secure schemes into polynomially secure ones at the cost of two more data transmissions between the signer and the receiver. Unfortunately, his particular construction, that based on Okamoto signatures, does not immediately lead to partially blind signature schemes.

The notion of partially blind signatures was introduced in [2]. Their construction, based on RSA, was analyzed in [1]. It also showed a construction based on Schnorr signatures that withstands a particular class of attacks. There are some other heuristic constructions in the literature. One of the authors was informed that Cramer and Pedersen independently considered the same notion and constructed a scheme, which remains unavailable in public due to an embargo [8]. All in all, no provably secure and practical partially blind signature scheme has been publicly released.

## 1.3   Our Contribution

This paper first gives a formal definition of partially blind signature schemes. As partially blind signatures can be regarded as ones lying between ordinary non-blind digital signatures and fully blind signatures, they should satisfy the security requirements assigned to ordinary digital signatures and those of blind signatures.

We then present efficient partially blind signature schemes with a rigorous proof of security in the random oracle model [3] under the standard number theoretic intractability assumptions such as discrete-log or the RSA assumption. Since the technique developed by Pointcheval and Stern for proving the one-more-unforgeability [20] is not applicable to our scheme, we provide a new technique to prove the security of our scheme. The technique shown in this paper is more generic than that of [20] and applicable to variety of schemes based on the witness indistinguishable protocols including the ones that the technique of [20] is applicable to. As well as the result of [20,22], our proof guarantees that the proposed scheme is secure as long as only a logarithmic number of signatures are issued. So plugging our scheme into the generic, but yet practical scheme of [19] will yield a scheme secure up to polynomial number of signatures.

For the sake of simplicity, we put off the generic description of our approach and concentrate on describing one particular scheme based on the original (i.e. not Okamoto version of) Schnorr signature scheme. One can, however, construct a scheme in a similar way based on Guillou-Quisquater signatures [14] or variants of modified ElGamal signatures [10,21,16] at the cost of doubling the computation and communication compared to the underlying schemes.

Although our primary goal is partially blind signatures, our approach also yields secure fully blind signatures. Thus, from a different angle, our result can be seen as a widely applicable approach that turns several secure signature schemes into secure blind signatures.

## 1.4   Organization

Section 2 defines the security of partially blind signatures. In Section 3 we show a partially blind signature scheme based on Schnorr signatures. Section 4 gives a proof of security.

## 2   Definitions

In the scenario of issuing a partially blind signature, the signer and the user are assumed to agree on a piece of common information, denoted as info. In some applications, info may be decided by the signer, while in other applications it may just be sent from the user to the signer. Anyway, this negotiation is done outside of the signature scheme, and we want the signature scheme to be secure regardless of the process of agreement. We formalize this notion by introducing function $Ag(\ )$ which is defined outside of the scheme. Function $Ag$ is a polynomial-time deterministic algorithm that takes two arbitrary strings $\mathsf{info}_s$ and $\mathsf{info}_u$ that belong to the signer and the user, respectively, and outputs info. To compute $Ag$, the signer and the user will exchange $\mathsf{info}_s$ and $\mathsf{info}_u$ with each other. However, if an application allows the signer to control info, then $Ag$ is defined such that it depends only on $\mathsf{info}_s$. In such a case, the user does not need to send $\mathsf{info}_u$.

Some part of the following definitions refers to [15]. In the following, we will use the term "polynomial-time" to mean a certain period bounded by a polynomial in security parameter $n$.

**Definition 1.** *(Partially Blind Signature Scheme) A Partially blind signature scheme is a four-tuple $(\mathcal{G}, \mathcal{S}, \mathcal{U}, \mathcal{V})$.*

- $\mathcal{G}$ *is a probabilistic polynomial-time algorithm that takes security parameter $n$ and outputs a public and secret key pair $(pk, sk)$.*
- $\mathcal{S}$ *and $\mathcal{U}$ are a pair of probabilistic interactive Turing machines each of which has a public input tape, a private input tape, a private random tape, a private work tape, a private output tape, a public output tape, and input and output communication tapes. The random tape and the input tapes are read-only, and the output tapes are write-only. The private work tape is read-write. The public input tape of $\mathcal{U}$ contains $pk$ generated by $\mathcal{G}(1^n)$, the description of $Ag$, and $\mathsf{info}_u$. The public input tape of $\mathcal{S}$ contains the description of $Ag$ and $\mathsf{info}_s$. The private input tape of $\mathcal{S}$ contains $sk$, and that for $\mathcal{U}$ contains message $\mathsf{msg}$. The lengths of $\mathsf{info}_s$, $\mathsf{info}_u$, and $\mathsf{msg}$ are polynomial in $n$. $\mathcal{S}$ and $\mathcal{U}$ engage in the signature issuing protocol and stop in polynomial-time. When they stop, the public output tape of $\mathcal{S}$ contains either completed or not-completed. If it is completed, then its private output tape contains common information $\mathsf{info}^{(s)}$. Similarly, the private output tape of $\mathcal{U}$ contains either $\perp$ or $(\mathsf{info}, \mathsf{msg}, sig)$.*
- $\mathcal{V}$ *is a (probabilistic) polynomial-time algorithm that takes $(pk, \mathsf{info}, \mathsf{msg}, sig)$ and outputs either accept or reject.*

**Definition 2.** *(Completeness) If $\mathcal{S}$ and $\mathcal{U}$ follow the signature issuing protocol, then, with probability at least $1 - 1/n^c$ for sufficiently large $n$ and some constant $c$, $\mathcal{S}$ outputs completed and $\mathsf{info} = Ag(\mathsf{info}_s, \mathsf{info}_u)$ on its proper tapes, and $\mathcal{U}$ outputs $(\mathsf{info}, \mathsf{msg}, sig)$ that satisfies $\mathcal{V}(pk, \mathsf{info}, \mathsf{msg}, sig) = accept$. The probability is taken over the coin flips of $\mathcal{G}$, $\mathcal{S}$ and $\mathcal{U}$.*

We say a message-signature tuple $(\mathsf{info}, \mathsf{msg}, sig)$ is valid with regard to $pk$ if it leads $\mathcal{V}$ to *accept*.

To define the blindness property, let us introduce the following game.

**Definition 3.** *(Game A) Let $\mathcal{U}_0$ and $\mathcal{U}_1$ be two honest users that follow the signature issuing protocol.*

1. $(pk, sk) \leftarrow \mathcal{G}(1^n)$.
2. $(\mathsf{msg}_0, \mathsf{msg}_1, \mathsf{info}_{u0}, \mathsf{info}_{u1}, Ag) \leftarrow \mathcal{S}^*(sk)$.

3. *Set up the input tapes of* $\mathcal{U}_0, \mathcal{U}_1$ *as follows:*
   - *Select* $b \in_R \{0,1\}$ *and put* msg$_b$ *and* msg$_{\bar{b}}$ *on the private input tapes of* $\mathcal{U}_0$ *and* $\mathcal{U}_1$, *respectively (*$\bar{b}$ *denotes* $1 - b$ *hereafter).*
   - *Put* info$_{u0}$ *and* info$_{u1}$ *on the public input tapes of* $\mathcal{U}_0$ *and* $\mathcal{U}_1$, *respectively. Also put pk and Ag on their public input tapes.*
   - *Randomly select the contents of the private random tapes.*
4. $\mathcal{S}^*$ *engages in the signature issuing protocol with* $\mathcal{U}_0$ *and* $\mathcal{U}_1$.
5. *If* $\mathcal{U}_0$ *and* $\mathcal{U}_1$ *outputs* (info$_0$, msg$_0$, sig$_b$) *and* (info$_1$, msg$_1$, sig$_b$), *respectively, on their private tapes, and* info$_0$ = info$_1$ *holds, then give those outputs to* $\mathcal{S}^*$. *Give* $\perp$ *to* $\mathcal{S}^*$ *otherwise.*
6. $\mathcal{S}^*$ *outputs* $b' \in \{0,1\}$.

*We say that* $\mathcal{S}^*$ *wins if* $b' = b$.

**Definition 4.** *(Partial Blindness) A signature scheme is partially blind if, for all probabilistic polynomial-time algorithm* $\mathcal{S}^*$, $\mathcal{S}^*$ *wins in game A with probability at most* $1/2 + 1/n^c$ *for sufficiently large n and some constant c. The probability is taken over the coin flips of* $\mathcal{G}$, $\mathcal{U}_0$, $\mathcal{U}_1$, *and* $\mathcal{S}^*$.

As usual, one can go for stronger notion of blindness depending on the power of the adversary and its success probability. Our scheme provides *perfect* partial blindness where any infinitely powerful adversary wins with probability exactly $1/2$.

Forgery of partially blind signatures is defined in the similar way as [15] with special care for the various pieces of common information. At first look, the forgery of a partially blind signature might be considered as forging the common information, or producing $\ell_{info} + 1$ signatures with regard to info provided $\ell_{info}$ successful execution of the signature issuing protocol for that info. Forging the common information is actually the same as producing one-more signature with info where $\ell_{info} = 0$. We define unforgeability through the following game.

**Definition 5.** *(Game B)*

1. $(pk, sk) \leftarrow \mathcal{G}(1^n)$.
2. $Ag \leftarrow \mathcal{U}^*(pk)$.
3. *Put* $sk, Ag$ *and randomly taken* inf$_s$ *on proper tapes of* $\mathcal{S}$.
4. $\mathcal{U}^*$ *engages in the signature issuing protocol with* $\mathcal{S}$ *in a concurrent and interleaving way. For each* info, *let* $\ell_{info}$ *be the number of executions of the signature issuing protocol where* $\mathcal{S}$ *outputs completed and* info *on its output tapes. (For* info *that has never appeared on the private output tape of* $\mathcal{S}$, *define* $\ell_{info} = 0$.)
5. $\mathcal{U}^*$ *outputs a single piece of common information,* info, *and* $\ell_{info}+1$ *signatures* (msg$_1$, sig$_1$), ..., (msg$_{\ell_{info}+1}$, sig$_{\ell_{info}+1}$).

**Definition 6.** *(Unforgeability) A partially blind signature scheme is unforgeable if, for any probabilistic polynomial-time algorithm* $\mathcal{U}^*$ *that plays game B, the probability that the output of* $\mathcal{U}^*$ *satisfies* $\mathcal{V}(pk, info, msg_j, sig_j) = accept$ *for all* $j = 1, ..., \ell_{info} + 1$ *is at most* $1/n^c$ *for sufficiently large n and some constant c. The probability is taken over the coin flips of* $\mathcal{G}$, $\mathcal{S}$, *and* $\mathcal{U}^*$.

# 3    Construction

## 3.1    Key Idea

The security of signature schemes is defined so that they are secure against adaptive attacks [13]. To prove the security against such attacks, one has to simulate the signer without knowing the private signing key. Introducing a random oracle allows the simulation for ordinary signatures but does not help in the case of blind signatures. So, the simulator has to have a real signing key. Accordingly, we need to separate the signing key from the witness of the embedding intractable problem, such as the discrete logarithm problem, that we attempt to solve by using an attacker of the signature scheme. For this to be done, Pointcheval and Stern used the blind Okamoto signature scheme where the existence of a successful attacker implied extraction of the discrete logarithm of bases rather than the signing key. They also exploited the witness indistinguishable property of Okamoto signatures in a crucial way in their proof of security. Unfortunately, we do not know how to achieve partial blindness with their construction.

In [9], Cramer, Damgård and Schoenmakers presented an efficient method of constructing witness indistinguishable protocols. With their adaptation, one can turn a wide variety of signature schemes derived from public-coin honest verifier zero-knowledge into witness indistinguishable ones. Intuitively, the signer has one private key $x$ but uses two different public keys, $y$ and $z$, together to sign a message in such a way that the user can not distinguish which private key he has. By blinding the signing procedure, one can get fully blind witness indistinguishable signature schemes.

Our idea to achieve partial blindness is to put common information, say info, into one of those public keys. Suppose that $z = \mathcal{F}(\mathsf{info})$ where $\mathcal{F}$ is a sort of public hash function that transforms an arbitrary string to a random public key whose private key is not known to anybody. The signer then signs with private key $x$ of $y$. Since the resulting signatures are bound to public keys $y, z$, the common information info is also bound to the signature. Since blinding will not cover public keys, info (i.e. $z$) remains unblind. This adaptation preserves witness indistinguishability which we need in our proof of security.

## 3.2    Preliminaries

Let $\mathcal{G}_{DL}$ be a discrete logarithm instance generator that takes security parameter $n$ and outputs a triple $(p, q, g)$ where $p, q$ are large primes that satisfy $q|p-1$, and $g$ is an element in $\mathbb{Z}_p^*$ whose order is $q$. Let $\langle g \rangle$ denote a subgroup in $\mathbb{Z}_p^*$ generated by $g$. We assume that any polynomial-time algorithm solves $\log_g h$ in $\mathbb{Z}_q$ only with negligible probability (in the size of $q$ and coin flips of $\mathcal{G}_{DL}$ and the algorithm) when $h$ is selected randomly from $\langle g \rangle$. All arithmetic operations are done in $\mathbb{Z}_p$ hereafter unless otherwise noted.

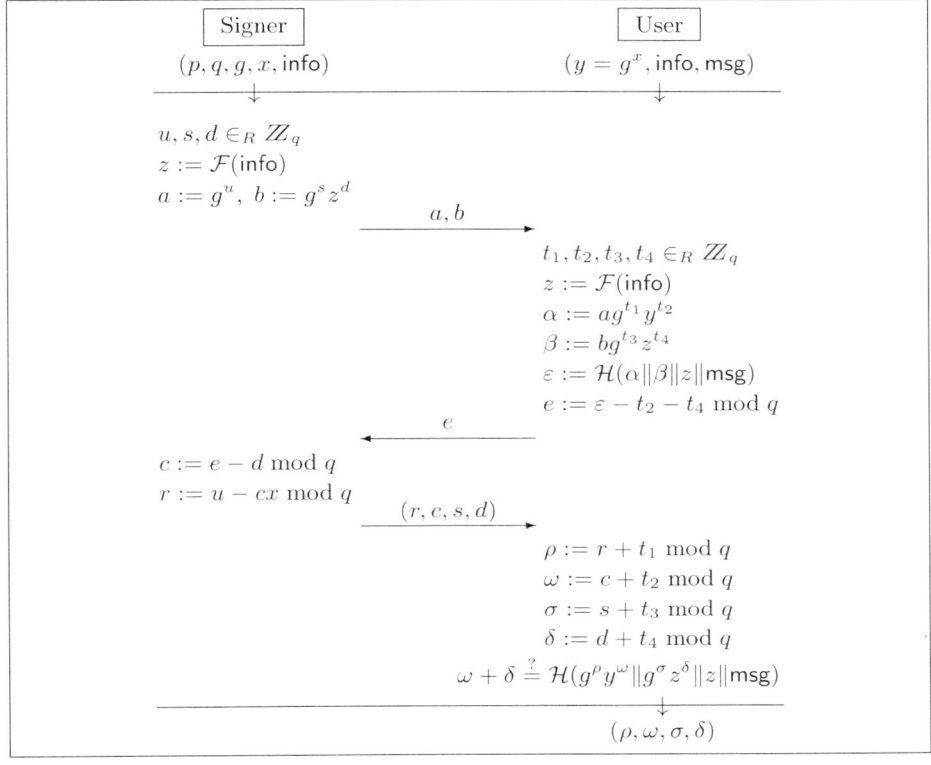

**Fig. 1.** Partially blind WI-Schnorr signature issuing protocol. The signer and the user are assumed to agree on info beforehand outside of the protocol. The signer can omit sending either $c$ or $d$ as the user can compute it himself from $e$.

### 3.3   A Partially Blind WI-Schnorr Signature Scheme

Let $\mathcal{H} : \{0,1\}^* \to \mathbb{Z}_q$ and $\mathcal{F} : \{0,1\}^* \to \langle g \rangle$ be public hash functions. Let $x \in \mathbb{Z}_q$ be a secret key and $y := g^x$ be a corresponding public key.

Signer $\mathcal{S}$ and user $\mathcal{U}$ first agree on common information info in an predetermined way. They then execute the signature issuing protocol illustrated in Figure 1. The resulting signature for message msg and common information info is a four-tuple $(\rho, \omega, \sigma, \delta)$. A signature is valid if it satisfies

$$\omega + \delta \equiv \mathcal{H}(g^\rho y^\omega \| g^\sigma \mathcal{F}(\mathsf{info})^\delta \| \mathcal{F}(\mathsf{info}) \| \mathsf{msg}) \pmod{q}.$$

Observe that the signature issuing protocol is witness indistinguishable. That is, the user's view has exactly the same distribution even if $\mathcal{S}$ executes the protocol with witness $w(= \log_g z)$ instead of $x$ computing as $v, r, c \in_R \mathbb{Z}_q$, $a := g^r y^c$, $b := g^v$, $d = e - c \bmod q$, and $s := v - dw \bmod q$.

In the above description, we assumed the use of hash function $\mathcal{F}$ that maps an arbitrary string to an element of $\langle g \rangle$. This, however, would be problematic in practice because currently available hash functions, say $\mathcal{D}$, such as SHA-1 and MD5, are of $\mathcal{D} : \{0,1\}^* \to \{0,1\}^{len}$ for some fixed $len$. An immediate thought would be to repeat $\mathcal{D}$ with random suffixes until the output eventually falls in $\langle g \rangle$. However, such a probabilistic strategy makes the running-time *expected* polynomial-time rather than *strict* polynomial-time. Furthermore, in practice, if $q$ is much smaller than $p$ as in ordinary Schnorr signatures, such a strategy is hopeless. We show two deterministic constructions of $\mathcal{F}$ assuming the use of hash function $\mathcal{D}$ with $len = |p|$.

**Construction 1** Take $p, q$ that satisfy $p = 2q + 1$. Define $\mathcal{F}$ as

$$\mathcal{F}(\mathsf{info}) \triangleq \left( \frac{\mathcal{D}(\mathsf{info})}{p} \right) \mathcal{D}(\mathsf{info}) \bmod p$$

where $\left( \dfrac{\mathcal{D}(\mathsf{info})}{p} \right)$ is the Jacobi symbol of $\mathcal{D}(\mathsf{info})$.

**Construction 2** Take $p, q$ that satisfy $q | p - 1$ and $q^2 \nmid p - 1$. Define $\mathcal{F}$ as

$$\mathcal{F}(\mathsf{info}) \triangleq \mathcal{D}(\mathsf{info})^{\frac{p-1}{q}} \bmod p.$$

The second construction is better in terms of computation as we can choose smaller $q$ such as $|q| \approx 2^{160}$. If $\mathcal{D}$ behaves as an ideal hash function, both constructions meet our requirement for the proof of security (that is, we can assign an arbitrary element of $\langle g \rangle$ as an output of $\mathcal{F}$). For simplicity, we set aside that detail and assume $\mathcal{F}$ be an atomic function in our proof of security in section 4.

## 4    Security

This section proves the security of our scheme assuming the intractability of the discrete logarithm problem and ideal randomness of hash functions $\mathcal{H}$ and $\mathcal{F}$.

**Lemma 1.** *The proposed scheme is partially blind.*

*Proof.* Let $\mathcal{S}^*$ be a player of game A. For $i = 0, 1$, let $a_i$, $b_i$, $e_i$, $r_i$, $c_i$, $s_i$, $d_i$, $\mathsf{info}_i$ be data appearing in the view of $\mathcal{S}^*$ during the execution of the signature issuing protocol with $\mathcal{U}_i$ at step 4.

When $\mathcal{S}^*$ is given $\perp$ in step 6 of the game, it is not hard to see that $\mathcal{S}^*$ wins game A with probability exactly the same as random guessing of $b$.

Suppose that $\mathsf{info}_1 = \mathsf{info}_0$, and $\{(\rho_0, \omega_0, \sigma_0, \delta_0)\}$ and $\{(\rho_1, \omega_1, \sigma_1, \delta_1)\}$ are given to $\mathcal{S}^*$. It is sufficient to show that there exists a tuple of random factors $(t_1, t_2, t_3, t_4)$ that maps $a_i, b_i, r_i, c_i, s_i, d_i$ to $\rho_j, \omega_j, \sigma_j, \delta_j$ for each $i, j \in \{0, 1\}$. ($e_i$ and $\mathsf{info}_i$ can be omitted as $c_i, d_i$ determines $e_i$, and $\mathsf{info}_i$ is common.) Define $t_1 := \rho_j - r_i$, $t_2 := \omega_j - c_i$, $t_3 := \sigma_j - s_i$, and $t_4 := \delta_j - d_i$. As $a_i = g^{r_i} y^{c_i}$ and $b_i = g^{s_i} z^{d_i}$ holds, we see that

$$
\begin{aligned}
\omega_j + \delta_j &= \mathcal{H}(g^{\rho_j} y^{\omega_j} \| g^{\sigma_j} z^{\delta_j} \| \mathcal{F}(\mathsf{info}) \| \mathsf{msg}) \\
&= \mathcal{H}(a_i g^{-r_i} y^{-c_i} g^{\rho_j} y^{\omega_j} \| b_i g^{-s_i} z^{-d_i} g^{\sigma_j} z^{\delta_j} \| \mathcal{F}(\mathsf{info}) \| \mathsf{msg}) \\
&= \mathcal{H}(a_i g^{\rho_j - r_i} y^{\omega_j - c_i} \| b_i g^{\sigma_j - s_i} z^{\delta_j - d_i} \| \mathcal{F}(\mathsf{info}) \| \mathsf{msg}) \\
&= \mathcal{H}(a_i g^{t_1} y^{t_2} \| b_i g^{t_3} y^{t_4} \| \mathcal{F}(\mathsf{info}) \| \mathsf{msg}).
\end{aligned}
$$

Thus, $a_i, b_i, r_i, c_i, s_i, d_i$ and $\rho_j, \omega_j, \sigma_j, \delta_j$ have exactly the same relation defined by the signature issuing protocol. Such $t_1, t_2, t_3, t_4$ always exist regardless of the values of $r_i, c_i, s_i, d_i$ and $\rho_j, \omega_j, \sigma_j, \delta_j$. Therefore, even an infinitely powerful $\mathcal{S}^*$ wins game A of our scheme with probability exactly $1/2$.                □

**Lemma 2.** *The proposed scheme is unforgeable if $\ell_{\mathsf{info}} < poly(\log n)$ for all* $\mathsf{info}$.

*Proof.* The proof is done in three steps. We first treat the common-part forgery where an attacker forges a signature with regard to common information $\mathsf{info}$ that has not appeared while Game B (i.e., $\ell_{\mathsf{info}} = 0$). Next we treat one-more forgery where $\ell_{\mathsf{info}} \neq 0$. For this case, we first prove the security with restricted signer $\mathcal{S}$ that issues signatures only for a fixed $\mathsf{info}$. We then eliminate the restriction by showing the reduction from the unrestricted signer model to the restricted one.

We first deal with successful common-part forger $\mathcal{U}^*$ who plays game B and produces, with probability $\mu > 1/n^c$, a valid message-signature tuple ($\mathsf{info}, \mathsf{msg}$, $\rho, \omega, \sigma, \delta$) such that $\ell_{\mathsf{info}} = 0$. This part of the proof follows that used for ID-reduction [17]. By using $\mathcal{U}^*$, we construct a machine $\mathcal{M}$ that forges a non-blind version of the WI-Schnorr signature in a passive environment (i.e. without talking with signer $\mathcal{S}$). We then use $\mathcal{M}$ to solve the discrete logarithm problem by exploiting the collision property.

Let $q_F$ and $q_H$ be the maximum number of queries asked from $\mathcal{U}^*$ to $\mathcal{F}$ and $\mathcal{H}$, respectively. Similarly, let $q_S$ be the maximum number of invocation of signer $\mathcal{S}$ in game B. All those parameters are limited by a polynomial in $n$. For simplicity, we assume that all queries are different. (For all duplicated queries to $\mathcal{F}$ and $\mathcal{H}$, return formerly defined values.) Let $(y, g, p, q)$ be the problem that we want to solve $\log_g y(= x)$ in $\mathbb{Z}_q$. Machine $\mathcal{M}$ simulates game B as follows.

1. Select $I \in_U \{1, \ldots, q_F + q_S\}$ and $J \in_U \{1, \ldots, q_H + q_S\}$.
2. Run $\mathcal{U}^*$ with $pk := (y, g, p, q)$ simulating $\mathcal{H}, \mathcal{F}$ and $\mathcal{S}$ as follows.
   - For $i$-th query to $\mathcal{F}$, return $z$ such that
     - $z := \mathcal{F}(\mathsf{info}_I)$ (i.e. ask oracle $\mathcal{F}$) if $i = I$, or
     - $z := g^{w_i}$ where $w_i \in_U \mathbb{Z}_q$, otherwise.
   - For $j$-th query to $\mathcal{H}$,
     - ask $\mathcal{H}$ if $j = J$, or
     - randomly select the answer from $\mathbb{Z}_q$, otherwise.
   - For requests to $\mathcal{S}$, first negotiate the common information. Let $\mathsf{info}_k$ be the result of the negotiation. If $\mathcal{F}(\mathsf{info}_k)$ is not defined yet, define it as mentioned above. Then,
     - if $\mathsf{info}_k \neq \mathsf{info}_I$, simulate $\mathcal{S}$ by using witness $w_k$, or
     - if $\mathsf{info}_k = \mathsf{info}_I$, we expect that $\mathcal{U}^*$ aborts the session before it receives $(r, c, s, d)$. (If $\mathcal{U}^*$ tries to complete the session, the simulation fails.) Just to simulate the state of abortion, send random $(a, b)$ to $\mathcal{U}^*$.
3. If $\mathcal{U}^*$ eventually outputs signature $(\rho, \omega, \sigma, \delta)$ with regard to $\mathsf{info}_I$ and $\mathsf{msg}_J$, output them.

Note that the queries to $\mathcal{F}$ and $\mathcal{H}$ may include the ones inquired during the simulation of $\mathcal{S}$. So, $\mathcal{F}$ and $\mathcal{H}$ are defined at at most $q_F + q_S$ and $q_H + q_S$ points during the simulation, respectively. The simulation of $\mathcal{S}$ for $\mathsf{info}_k \neq \mathsf{info}_I$ can be perfectly done with $w_k$ due to witness indistinguishability. The probability that $\mathcal{U}^*$ is successful without asking $\mathcal{F}, \mathcal{H}$ in a proper way is negligible because of the unpredictability of those hash functions. Thus, the success probability of $\mathcal{M}$ is only negligibly worse than $\frac{\mu}{(q_H + q_S)(q_F + q_S)}$ which is not negligible in $n$. By $\mu'$, we denote the success probability of $\mathcal{M}$.

Now we use $\mathcal{M}$ to solve $\log_g y$. The trick is to simulate $\mathcal{F}$ by responding to the query from $\mathcal{M}$ with $y g^\gamma$ where $\gamma$ is chosen randomly from $\mathbb{Z}_q$. Note that $\mathcal{M}$ asks each of $\mathcal{F}$ and $\mathcal{H}$ only once. Furthermore, the query to $\mathcal{F}$ happens *before* the query to $\mathcal{H}$ with overwhelming probability when $\mathcal{M}$ is successful because $\mathcal{F}(\mathsf{info})$ is contained in the inputs of $\mathcal{H}$. Next, we apply the standard replay technique [11]. That is, run $\mathcal{M}$ with a random tape and a random choice of $\mathcal{H}$. $\mathcal{M}$ then outputs a valid signature, say $(\rho, \omega, \sigma, \delta)$, with probability at least $1 - e^{-1}$ (here, $e$ is base of natural logarithms) after $1/\mu'$ trials. We then rewind $\mathcal{M}$ with the same random tape and run it with a different choice of $\mathcal{H}$. By repeating this rewind-trial $2/\mu'$ times, we get another valid signature, say $(\rho', \omega', \sigma', \delta')$, with probability at least $(1 - e^{-1})/2$. After all, with constant probability and polynomial running time, we have two valid signatures whose first messages $(a, b)$ are the same. Thus, $\rho + \omega x = \rho' + \omega' x$, $\sigma + \delta(x + \gamma) = \sigma' + \delta'(x + \gamma)$, and $\omega + \delta \neq \omega' + \delta'$ holds. Since at least $\omega \neq \omega'$ or $\delta \neq \delta'$ happens, one can get $x$ as $x = (\rho - \rho')/(\omega' - \omega) \bmod q$ or $x = (\sigma - \sigma')/(\delta' - \delta) - \gamma \bmod q$.

Next we consider the case where the forgery is attempted against $\mathsf{info}$ such that $\ell_{\mathsf{info}} \neq 0$. As the first step, we consider Game B with a single $\mathsf{info}$. Hence $z$ is common for all executions of the signature issuing protocol. Accordingly, we prove the security of fully blind version of our scheme. Let $\ell = \ell_{\mathsf{info}}$.

### Reduction Algorithm

Assume a single-info adversary, $\mathcal{U}_F^*$, which is a probabilistic polynomial time algorithm that violates unforgeability for infinitely many sizes, $n$'s, with the attack defined as Game B. (Let $n_0$ be such a size, and the success probability of $\mathcal{U}_F^*$ is at least $\eta$). Then we construct an algorithm, $\mathcal{M}$, that utilizes $\mathcal{U}_F^*$ as black-box and breaks the intractability assumption of the discrete logarithm for infinitely many $n$'s. That is, the input to $\mathcal{M}$ is $(p, q, g, z_0)$, and $\mathcal{M}$ tries to compute $w_0$ such that $z_0 = g^{w_0}$, provided $\mathcal{U}_F^*$.

First, $\mathcal{M}$ selects $b \in_U \{0, 1\}$ and assigns $(y, z)$ as $(y, z) = (g^x, z_0 g^\gamma)$ if $b = 0$, or $(y, z) = (z_0 g^\gamma, g^w)$ if $b = 1$ by choosing $\gamma$ and $x$ (or $w$) randomly from $\mathbb{Z}_q$. $\mathcal{F}$ is defined so that it returns appropriate value of $z$ according to the choice. Hereafter, without loss of generality, we assume that $b = 0$ is chosen and $(y, z) = (g^x, z_0 g^\gamma)$ is set. $\mathcal{M}$ can then simulate signer $\mathcal{S}$, since the protocol between $\mathcal{S}$ and $\mathcal{U}_F^*$ is witness indistinguishable and having $x = \log_g y$ is sufficient for $\mathcal{S}$ to complete the protocol. Let $\hat{\mathcal{S}}$ denote the signer simulated by $\mathcal{M}$.

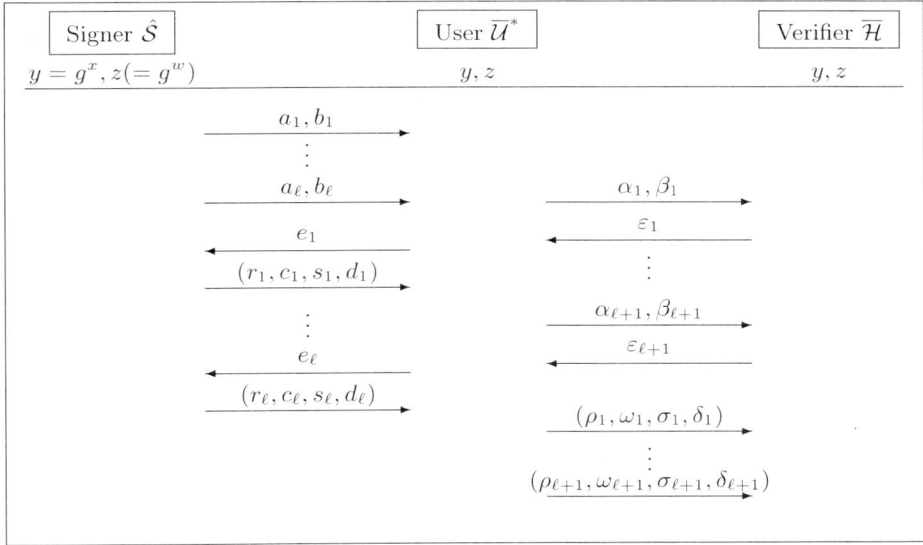

**Fig. 2.** Corresponding Divertible Identification Protocol.

If $\mathcal{U}_F^*$ is successful with probability at least $\eta$, we can find a random tape string for $\mathcal{U}_F^*$ and $\hat{\mathcal{S}}$ with probability at least $1/2$ such that $\mathcal{U}_F^*$ with $\hat{\mathcal{S}}$ succeeds with probability at least $\eta/2$.

By employing $\mathcal{U}_F^*$ as a black-box, we can construct $\overline{\mathcal{U}}^*$ which has exactly the same interface with $\hat{\mathcal{S}}$ as $\mathcal{U}_F^*$ has, and plays the role of an impersonator in the interactive identification protocol with verifier $\overline{\mathcal{H}}$ (see Fig. 2). When $\mathcal{U}_F^*$ asks at most $q_F$ queries to random oracle $\mathcal{H}$, $\overline{\mathcal{U}}^*$ is successful in completing the identification protocol with verifier $\overline{\mathcal{H}}$ with probability at least $\eta/2q_H^{\ell+1}$, since, with probability greater than $1/2q_H^{\ell+1}$, $\overline{\mathcal{U}}^*$ can guess a correct selection of $\ell+1$ queries that $\mathcal{U}^*$ eventually uses in the forgery.

$\mathcal{M}$ then use the standard replay technique for an interactive protocol to compute the discrete logarithm. $\mathcal{M}$ first runs $\overline{\mathcal{U}}^*$ with $\hat{\mathcal{S}}$ and $\overline{\mathcal{H}}$, and find a successful challenge tuple $(\varepsilon_1, \ldots, \varepsilon_{\ell+1})$. $\mathcal{M}$ then randomly chooses an index, $i \in \{1, \ldots, \ell+1\}$, and replay with the same environments and random tapes except different challenge tuple $(\varepsilon_1, \ldots, \varepsilon_{i-1}, \varepsilon_i', \ldots, \varepsilon_{\ell+1}')$ where the first $i-1$ challenges are unchanged. Since $\varepsilon_i \neq \varepsilon_i'$, at least either $\delta_i \neq \delta_i'$ or $\omega_i \neq \omega_i'$ happens. If $\delta_i \neq \delta_i'$, then $\mathcal{M}$ can compute $w(= \log_g z)$ as $w = (\sigma_i - \sigma_i')/(\delta_i' - \delta_i) \bmod q$. $\mathcal{M}$ then obtain $w_0 = w - \gamma \bmod q$ such that $z_0 = g^{w_0}$.

**Evaluation of the Success Probability**

Let $\Omega$ and $\Theta$ be random tape strings of $\mathcal{M}$ and $\mathcal{U}^*$, respectively. Note that $\Omega$ includes the random selection of $b$ and random factors in the simulation of $\mathcal{S}$. $\Omega$ and $\Theta$ are assumed to be fixed throughout this evaluation. Let $\vec{\varepsilon} = (\varepsilon_1, \ldots, \varepsilon_{\ell+1})$, and $\vec{e} = (e_1, \ldots, e_\ell)$. $\mathcal{E}$ denotes the set of all $\vec{\varepsilon}$'s (hence $\#\mathcal{E} = q^{\ell+1}$). The first

$i - 1$ elements of $\vec{\varepsilon}$, i.e. $(\varepsilon_1, \ldots, \varepsilon_{i-1})$, is denoted by $\vec{\varepsilon}_i$, and the $i$-th element of $\vec{\varepsilon}$ is denoted by $\vec{\varepsilon}_{[i]}$. We define Succ a set of successful $\vec{\varepsilon}$ such that $\vec{\varepsilon} \in$ Succ iff $\vec{\varepsilon}$ is an accepted sequence of challenges between $\overline{\mathcal{U}}^*$ and $\overline{\mathcal{H}}$.

Observe that there exists different $\vec{\varepsilon}$ and $\vec{\varepsilon}'$ that yield the same transcript between $\overline{\mathcal{U}}^*$ and $\mathcal{S}$ because $\vec{e}$ is uniquely determined from $\vec{\varepsilon}$ as $\overline{\mathcal{U}}^*$ and $\mathcal{S}$ are deterministic when $\Omega$ and $\Theta$ are fixed, and $\vec{\varepsilon}$ has more variation than $\vec{e}$. We classify elements in Succ into classes so that elements in the same class yield the same transcript between $\overline{\mathcal{U}}^*$ and $\mathcal{S}$. Precisely, we introduce a mapping, $\lambda : \vec{\varepsilon} \mapsto \vec{e}$, i.e., $\lambda(\vec{\varepsilon}) = \vec{e}$, and define an equivalence relation between elements in Succ as $\vec{\varepsilon} \sim \vec{\varepsilon}'$ iff $\lambda(\vec{\varepsilon}) = \lambda(\vec{\varepsilon}')$. Let $E(\vec{\varepsilon})$ denote the equivalence class where $\vec{\varepsilon}$ belongs.

Next we classify Succ in a different way. Let $Br(\vec{\varepsilon}, \vec{\varepsilon}') = i \in \{0, \ldots, \ell+1\}$ denote the 'branching' index such that $\vec{\varepsilon}_i = \vec{\varepsilon}'_i$ and $\vec{\varepsilon}_{[i]} \neq \vec{\varepsilon}'_{[i]}$ (define $Br(\vec{\varepsilon}, \vec{\varepsilon}') = 0$ if $\vec{\varepsilon} = \vec{\varepsilon}'$ ). For $\vec{\varepsilon} \in$ Succ, let $Br_{\max}(\vec{\varepsilon}) = i$ denote an index where $\vec{\varepsilon}$ is most likely to branch compared with randomly taken element of $E(\vec{\varepsilon})$. Formally, for $\vec{\varepsilon} \in$ Succ, $Br_{\max}(\vec{\varepsilon}) = i$ iff

$$\#\{\vec{\varepsilon}' \in E(\vec{\varepsilon}) \mid Br(\vec{\varepsilon}, \vec{\varepsilon}') = i\} = \max_{j \in \{1, \ldots, \ell+1\}} (\#\{\vec{\varepsilon}' \in E(\vec{\varepsilon}) \mid Br(\vec{\varepsilon}, \vec{\varepsilon}') = j\})$$

(if two $j$'s happen to give the same maximal value, define $i$ with the larger $j$). Now, the elements in Succ is classified by $Br_{\max}$. Let $\mathcal{E}_{i^*}$ denotes the largest class among them. Formally, $\mathcal{E}_{i^*} = \{\vec{\varepsilon} \mid Br_{\max}(\vec{\varepsilon}) = i^*\}$ where $i^* \in \{1, \ldots, \ell+1\}$ is defined so that it satisfies $\#\{\vec{\varepsilon} \mid Br_{\max}(\vec{\varepsilon}) = i^*\} = \max_{j \in \{1, \ldots, \ell+1\}} (\#\{\vec{\varepsilon}' \mid Br_{\max}(\vec{\varepsilon}') = j\})$. Note that $i^* = 0$ does not happen since $Br_{\max}(\vec{\varepsilon}) = 0$ happens only if $\#E(\vec{\varepsilon}) = 1$ and such $\vec{\varepsilon} \in$ Succ is at most $q^\ell - 1$. From the definition, it is clear that

$$\frac{\#\mathcal{E}_{i^*}}{\#\mathcal{E}} \geq \eta_1/(\ell+2).$$

Note that $\#\mathcal{E} = q^{l+1}$.

For $\vec{\varepsilon} \in \mathcal{E}_{i^*}$, define $\Gamma_{i^*}$ and $\xi_{i^*}(\vec{\varepsilon})$ as

$$\Gamma_{i^*}(\vec{\varepsilon}) = \{\varepsilon \mid {}^{\exists}\vec{\varepsilon}' \in \text{Succ} \; ; \; \vec{\varepsilon}'_{i^*} = \vec{\varepsilon}_{i^*} \; \wedge \; \vec{\varepsilon}'_{[i^*]} = \varepsilon\},$$

$$\xi_{i^*}(\vec{\varepsilon}) = \frac{\#\Gamma_{i^*}(\vec{\varepsilon})}{q}.$$

Intuitively, $\Gamma_{i^*}(\vec{\varepsilon})$ is the number of good (potentially successful) choices as the $i$-th challenge when first $i^* - 1$ challenges are fixed according to $\vec{\varepsilon}$. And $\xi_{i^*}$ is its fraction. We can obtain the following claim using the standard heavy low lemma technique [11]. Note that if $\vec{\varepsilon}$ is randomly selected from $\mathcal{E}$, the probability that $\vec{\varepsilon} \in \mathcal{E}_{i^*}$ is at least $\eta_1/(\ell+2)$, where $\eta_1 = \eta/2q_H^{\ell+1}$.

*Claim.* $\Pr_{\vec{\varepsilon} \in \mathcal{E}_{i^*}} [\xi_{i^*}(\vec{\varepsilon}) \geq \eta_1/2(\ell+2)] > 1/2$.

*Proof.* Assume that there exits a fraction, $F$, of $\mathcal{E}_{i^*}$ such that $\#F \geq \#\mathcal{E}_{i^*}/2$ and $\forall \vec{\varepsilon} \in F, \xi_{i^*}(\vec{\varepsilon}) < \eta_1/2(\ell+2)$. We then obtain, for each $\vec{\varepsilon} \in F$,

$$\#\{\vec{\varepsilon}' \in \text{Succ} \mid \vec{\varepsilon}'_{i^*} = \vec{\varepsilon}_{i^*}\} < q \times (\eta_1/2(\ell+2)) \times q^{\ell-i^*+1} = q^{\ell-i^*+2}\eta_1/2(\ell+2).$$

Since $\sum_{\vec{\varepsilon} \in F} \#\{\vec{\varepsilon}' \in \text{Succ} \mid \vec{\varepsilon}'_{i^*} = \vec{\varepsilon}_{i^*}\} \geq \#F \geq \#\mathcal{E}_{i^*}/2 = \frac{q^{\ell+1}\eta_1}{2(\ell+2)}$, the variation of the first $(i^*-1)$ challenges of the elements in $F$, i.e. $\#\{\vec{\varepsilon}_{i^*} \mid \vec{\varepsilon} \in F\}$, is strictly greater than

$$\frac{q^{\ell+1}\eta_1/2(\ell+2)}{q^{\ell-i^*+2}\eta_1/2(\ell+2)} = q^{i^*-1}.$$

As $i^* - 1$ challenges have at most $q^{i^*-1}$ variations, this is contradiction.

For each $\vec{\varepsilon} \in \mathcal{E}_{i^*}$, we arbitrarily fix a partner of $\vec{\varepsilon}$, denoted as $\vec{\varepsilon}' = Prt(\vec{\varepsilon})$, that satisfies $\vec{\varepsilon}' \neq \vec{\varepsilon}$ and $\vec{\varepsilon}' \in E(\vec{\varepsilon})$. Let $\hat{\mathcal{E}}_{i^*}$ be a set that consists of all elements of $\mathcal{E}_{i^*}$ and their partners. That is, $\hat{\mathcal{E}}_{i^*} = \mathcal{E}_{i^*} \cup \{\vec{\varepsilon}' \mid \vec{\varepsilon}' = Prt(\vec{\varepsilon})\}$. We then call a triple, $(\vec{\varepsilon}, \vec{\varepsilon}', \vec{\varepsilon}'')$, a triangle, iff $\vec{\varepsilon} \in \mathcal{E}_{i^*}$, $\vec{\varepsilon}' = Prt(\vec{\varepsilon})$, $\vec{\varepsilon}'' \in \text{Succ}$, $\vec{\varepsilon}_{i^*} = \vec{\varepsilon}''_{i^*}$, $\vec{\varepsilon}_{[i^*]} \neq \vec{\varepsilon}''_{[i^*]}$, and $\vec{\varepsilon}'_{[i^*]} \neq \vec{\varepsilon}''_{[i^*]}$. For a triangle, $(\vec{\varepsilon}, \vec{\varepsilon}', \vec{\varepsilon}'')$, we call $(\vec{\varepsilon}, \vec{\varepsilon}'')$ and $(\vec{\varepsilon}', \vec{\varepsilon}'')$ a side of the triangle, and call $(\vec{\varepsilon}, \vec{\varepsilon}')$ the base of the triangle. The number of triangles is at least

$$\#\mathcal{E}_{i^*}/3 \geq q^{\ell+1}\eta_1/(6(\ell+2)).$$

Here w.o.l.g., we assume that $y = g^x$ is chosen according to $\Omega$. Clearly, from the definition, at least one of $x$ and $w$ can be calculated from $\mathcal{M}$'s view regarding a side of a triangle, $(\vec{\varepsilon}, \vec{\varepsilon}'')$ (and $(\vec{\varepsilon}', \vec{\varepsilon}'')$). We now denote $(\vec{\varepsilon}, \vec{\varepsilon}'') \rightarrow w$ iff $w$ is extracted from $\mathcal{M}$'s view regarding $\vec{\varepsilon}$ and $\vec{\varepsilon}''$, otherwise $(\vec{\varepsilon}, \vec{\varepsilon}'') \nrightarrow w$. It is easy to see that the following claim holds.

*Claim.* Let $(\vec{\varepsilon}, \vec{\varepsilon}', \vec{\varepsilon}'')$ be a triangle. Suppose that $(\vec{\varepsilon}, \vec{\varepsilon}'') \nrightarrow w$ and $(\vec{\varepsilon}', \vec{\varepsilon}'') \nrightarrow w$. Then $(\vec{\varepsilon}, \vec{\varepsilon}') \nrightarrow w$.

*Proof.* Let $\delta, \delta'$, and $\delta''$ correspond to $\vec{\varepsilon}, \vec{\varepsilon}'$, and $\vec{\varepsilon}''$. If $(\vec{\varepsilon}, \vec{\varepsilon}'') \nrightarrow w$, then $\delta = \delta''$. If $(\vec{\varepsilon}', \vec{\varepsilon}'') \nrightarrow w$, then $\delta' = \delta''$. Therefore, $\delta = \delta'$. It follows that $(\vec{\varepsilon}, \vec{\varepsilon}') \nrightarrow w$.

We then obtain the following claim:

*Claim.* For at least $1/5$ fraction of sides, $w$ is extracted with probability at least $1/3$ over $\Omega$.

*Proof.* If $x$ ($w$ resp.) is included in $\Omega$, then $w$ ($x$ resp.) is called a *good* witness, which we want to extract. Suppose that a good witness is not obtained from at least $4/5$ fraction of sides with probability at least $2/3$ over $\Omega$. It then follows from Claim 4 that a good witness is not obtained from at least $3/5$ fraction of base, $(\vec{\varepsilon}, \vec{\varepsilon}')$, with probability at least $2/3$ over $\Omega$. When a good witness is not obtained from at least $3/5$ fraction of base, $(\vec{\varepsilon}, \vec{\varepsilon}')$, the result is (non-negligibly) biased by the witness with $\Omega$. That is, the biased result occurs with probability at least $2/3$ over $\Omega$. Since the information of a base, $(\vec{\varepsilon}, \vec{\varepsilon}')$, is independent of the witness the simulator already has as a part of $\Omega$, this contradicts that a biased result should occur with probability (over $\Omega$) less than $1/2 + 1/poly(n)$ for any polynomial *poly*.

Finally we will evaluate the total success probability of $\mathcal{M}$. The probability that $i^*$ is correctly guessed is at least $\frac{1}{\ell+1}$. When $\vec{\varepsilon}$ is randomly selected, $\vec{\varepsilon} \in \hat{\mathcal{E}}_{i^*}$

and $\xi_{i^*}(\vec{\varepsilon}) \geq \eta_1/2(\ell+2)$ with probability at least $\frac{\eta_1}{2(\ell+2)}$. $\vec{\varepsilon}''_{[i^*]} \in \Gamma_{i^*}(\vec{\varepsilon})$ is selected with probability at least $\xi_{i^*}(\vec{\varepsilon}) \geq \eta_1/2(\ell+2)$. Then $(\vec{\varepsilon}, \vec{\varepsilon}''_{[i^*]}) \to w$ with probability greater than $1/15 \ (= (1/3) \times (1/5))$. Thus, in total, the success probability of $\mathcal{M}$ is $\frac{\eta_1^2}{60(\ell+1)(\ell+2)^2}$, where $\eta_1 = \eta/2q_H^{\ell+1}$.

Now we consider the case where the common information is not all the same. Given successful forger $\mathcal{U}_B^*$ of game B, we construct successful forger $\mathcal{U}_F^*$ of the fixed-info version of game B.

The basic strategy of constructing machine $\mathcal{U}_F^*$ is to screen the conversation between $\mathcal{U}_B^*$ and $\mathcal{S}$ except for the ones involving info that $\mathcal{U}_B^*$ will output as a result of forgery. $\mathcal{U}_F^*$ simulates $\mathcal{S}$ with regard to the blocked conversations by assigning $g^w$ to $z$ with randomly picked $w$. The simulation works perfectly thanks to the witness indistinguishability of the signature issuing protocol.

Now, we describe $\mathcal{U}_F^*$ in detail. Let $q_F$ be the maximum number of queries for $\mathcal{F}$ from $\mathcal{U}_B^*$. Similarly, let $q_S$ be the maximum number of queries for $\mathcal{S}$. Observe that $\mathcal{F}$ is defined at most at $q_F+q_S$ points while $\mathcal{U}_B^*$ plays game B. For simplicity, we assume that all queries to $\mathcal{F}$ are different.

1. Select $J$ randomly from $\{1, \ldots, q_F + q_S\}$.
2. Run $\mathcal{U}_B^*$ simulating $\mathcal{F}, \mathcal{H}$ and signer $\mathcal{S}$ as follows.
   - For $j$-th query to $\mathcal{F}$, return $z$ such that
     - $z := g_j^w$ where $w_j \in_R \mathbb{Z}_q$ for $j \neq J$, or
     - $z := \mathcal{F}(\mathsf{info}_J)$ (i.e. ask $\mathcal{F}$) if $j = J$.
     If $z$ has been already defined at query point $\mathsf{info}_j$, return that value.
   - For all queries to $\mathcal{H}$, ask $\mathcal{H}$.
   - If $\mathcal{U}_B^*$ initiates the signature issuing protocol with regard to $\mathsf{info}_J$, $\mathcal{U}_F^*$ negotiates with $\mathcal{S}$ in such a way that they agree on $\mathsf{info}_J$ (this is possible because $Ag$ is deterministic). $\mathcal{U}_F^*$ then behaves transparently so that $\mathcal{U}_B^*$ can talk with $\mathcal{S}$.
   - If $\mathcal{U}_B^*$ initiates the signature issuing protocol with regard to $\mathsf{info}_j$ where $j \neq J$, $\mathcal{U}_F^*$ simulates $\mathcal{S}$ by using $w_j$.
3. Output what $\mathcal{U}_B^*$ outputs.

Note that $Ag$ is decided by $\mathcal{U}_B^*$ at the beginning of step 2. $\mathcal{U}_F^*$ is successful if $\mathcal{U}_B^*$ is successful and correct $J$ is chosen so that the final output of $\mathcal{U}_B^*$ contains $\mathsf{info}_J$. Therefore, the success probability of $\mathcal{U}_F^*$ is $\frac{\mu}{q_F+q_S}$ where $\mu$ is the success probability of $\mathcal{U}_B^*$. □

## 5    Conclusion

We have presented a formal definition of partially blind signature schemes and constructed an efficient scheme based on the Schnorr signature scheme. We then gave a proof of security in the random oracle model assuming the intractability of the discrete logarithm problem.

Although we have shown a particular construction based on Schnorr signature, the basic approach of constructing WI protocols and the proof of security do not substantially rely on the particular structure of the underlying signature scheme. Accordingly, a signature scheme derived from public-coin honest verifier zero-knowledge can be plugged into our scheme if it can be blinded. It covers, for instance, Guillou-Quisquater signature and some variants of modified ElGamal signature schemes.

As we mentioned, one can easily transform fully blind signature schemes from partially blind ones. We have shown that the reverse is possible; partially blind signature schemes can be derived from fully blind witness indistinguishable signature schemes.

# References

1. M. Abe and J. Camenisch. Partially blind signatures. In the 1997 Symposium on Cryptography and Information Security, 1997.
2. M. Abe and E. Fujisaki. How to date blind signatures. In K. Kim and T. Matsumoto, editors, *Advances in Cryptology – ASIACRYPT '96*, volume 1163 of *Lecture Notes in Computer Science*, pages 244–251. Springer-Verlag, 1996.
3. M. Bellare and P. Rogaway. Random oracles are practical: a paradigm for designing efficient protocols. In *First ACM Conference on Computer and Communication Security*, pages 62–73. Association for Computing Machinery, 1993.
4. S. Brands. Untraceable off-line cash in wallet with observers. In D. Stinson, editor, *Advances in Cryptology — CRYPTO '93*, volume 773 of *Lecture Notes in Computer Science*, pages 302–318. Springer-Verlag, 1993.
5. D. Chaum. Blind signatures for untraceable payments. In D. Chaum, R. Rivest, and A. Sherman, editors, *Advances in Cryptology — Proceedings of Crypto '82*, pages 199–204. Prenum Publishing Corporation, 1982.
6. D. Chaum. Elections with unconditionally-secret ballots and disruption equivalent to breaking RSA. In C. G. Günther, editor, *Advances in Cryptology — EUROCRYPT '88*, volume 330 of *Lecture Notes in Computer Science*, pages 177–189. Springer-Verlag, 1988.
7. D. Chaum, A. Fiat, and M. Naor. Untraceable electronic cash. In S. Goldwasser, editor, *Advances in Cryptology — CRYPTO '88*, volume 403 of *Lecture Notes in Computer Science*, pages 319–327. Springer-Verlag, 1990.
8. R. Cramer. personal communication, 1997.
9. R. Cramer, I. Damgård, and B. Schoenmakers. Proofs of partial knowledge and simplified design of witness hiding protocols. In Y. G. Desmedt, editor, *Advances in Cryptology — CRYPTO '94*, volume 839 of *Lecture Notes in Computer Science*, pages 174–187. Springer-Verlag, 1994.
10. T. ElGamal. A public key cryptosystem and a signature scheme based on discrete logarithms. In G. R. Blakley and D. Chaum, editors, *Advances in Cryptology — CRYPTO '84*, volume 196 of *Lecture Notes in Computer Science*, pages 10–18. Springer-Verlag, 1985.
11. U. Feige, A. Fiat, and A. Shamir. Zero-knowledge proofs of identity. *Journal of Cryptology*, 1:77–94, 1988.
12. A. Fujioka, T. Okamoto, and K. Ohta. A practical secret voting scheme for large scale elections. In J. Seberry and Y. Zheng, editors, *Advances in Cryptology — AUSCRYPT '92*, volume 718 of *Lecture Notes in Computer Science*, pages 244–251. Springer-Verlag, 1993.

13. S. Goldwasser, S. Micali, and R. Rivest. A digital signature scheme secure against adaptive chosen-message attacks. *SIAM Journal of Computing*, 17(2):281–308, April 1988.
14. L. C. Guillou and J.-J. Quisquater. A practical zero-knowledge protocol fitted to security microprocessor minimizing both transmission and memory. In C. G. Günther, editor, *Advances in Cryptology — EUROCRYPT '88*, volume 330 of *Lecture Notes in Computer Science*, pages 123–128. Springer-Verlag, 1988.
15. A. Juels, M. Luby, and R. Ostrovsky. Security of blind digital signatures. In B. S. Kaliski Jr., editor, *Advances in Cryptology — CRYPTO '97*, volume 1294 of *Lecture Notes in Computer Science*, pages 150–164. Springer-Verlag, 1997.
16. A. Menezes, P. Oorschot, and S. Vanstone. *Handbook of Applied Cryptography*. CRC Press, 1997.
17. K. Ohta and T. Okamoto. On concrete security treatment of signatures derived from identification. In H. Krawczyk, editor, *Advances in Cryptology — CRYPTO '98*, volume 1462 of *Lecture Notes in Computer Science*, pages 354–369. Springer-Verlag, 1998.
18. T. Okamoto. Provably secure and practical identification schemes and corresponding signature schemes. In E. F. Brickell, editor, *Advances in Cryptology — CRYPTO '92*, volume 740 of *Lecture Notes in Computer Science*, pages 31–53. Springer-Verlag, 1993.
19. D. Pointcheval. Strengthened security for blind signatures. In K. Nyberg, editor, *Advances in Cryptology — EUROCRYPT '98*, Lecture Notes in Computer Science, pages 391–405. Springer-Verlag, 1998.
20. D. Pointcheval and J. Stern. Provably secure blind signature schemes. In K. Kim and T. Matsumoto, editors, *Advances in Cryptology - ASIACRYPT '96*, volume 1163 of *Lecture Notes in Computer Science*, pages 252–265. Springer-Verlag, 1996.
21. D. Pointcheval and J. Stern. Security proofs for signature schemes. In U. Maurer, editor, *Advances in Cryptology — EUROCRYPT '96*, volume 1070 of *Lecture Notes in Computer Science*, pages 387–398. Springer-Verlag, 1996.
22. D. Pointcheval and J. Stern. Security arguments for digital signatures and blind signatures. *Journal of Cryptology*, 2000.
23. RSA Laboratories. *PKCS #9: Selected Object Classes and Attribute Types*, 2.0 edition, February 2000.

# Weaknesses in the $\mathrm{SL}_2(\mathbb{F}_{2^n})$ Hashing Scheme

Rainer Steinwandt, Markus Grassl, Willi Geiselmann, and Thomas Beth

Institut für Algorithmen und Kognitive Systeme,
Fakultät für Informatik, Universität Karlsruhe,
Am Fasanengarten 5, 76 128 Karlsruhe, Germany,
{steinwan,grassl,geiselma,EISS_Office}@ira.uka.de.

**Abstract.** We show that for various choices of the parameters in the $\mathrm{SL}_2(\mathbb{F}_{2^n})$ hashing scheme, suggested by Tillich and Zémor, messages can be modified without changing the hash value. Moreover, examples of hash functions "with a trapdoor" within this family are given. Due to these weaknesses one should impose at least certain restrictions on the allowed parameter values when using the $\mathrm{SL}_2(\mathbb{F}_{2^n})$ hashing scheme for cryptographic purposes.

## 1 Introduction

At CRYPTO '94 Tillich and Zémor [11] have proposed a class of hash functions based on the group $\mathrm{SL}_2(\mathbb{F}_{2^n})$, the group of $2 \times 2$-matrices with determinant 1 over $\mathbb{F}_{2^n}$. The hash functions are parameterized by the degree $n$ and the defining polynomial $f(X)$ of $\mathbb{F}_{2^n}$. The hash value $H(m) \in \mathrm{SL}_2(\mathbb{F}_{2^n})$ of some message $m$ is a $2 \times 2$-matrix.

At ASIACRYPT '94 a first "attack" on this hash function was proposed by Charnes and Pieprzyk [3]. They showed that the hash function is weak for some particular choices of the defining polynomial $f(X)$. However, for any chosen hash function it is easy to check if it is resistant against this attack—the order of the generators of $\mathrm{SL}_2(\mathbb{F}_{2^n})$ has to be large. This can easily be calculated. Moreover, Abdukhalikov and Kim [1] have shown that an *arbitrary* choice of $f(X)$ results in a scheme vulnerable to Charnes' and Pieprzyk's attack only with a probability of approximately $10^{-27}$.

Some additional structure of the group $\mathrm{SL}_2(\mathbb{F}_{2^n})$ was used by Geiselmann [5] to reduce the problem of finding collisions to the calculation of discrete logarithms in $\mathbb{F}_{2^n}$ or $\mathbb{F}_{2^{2n}}$ (which is feasible for the proposed values of $n \in \{130, \dots, 170\}$). The drawback of this "attack" is the extremely long message required for such a collision. (E. g., the collision given in [5] for $n = 21$ has a length of about $237\,000$ bits.).

The main advantage of the $\mathrm{SL}_2(\mathbb{F}_{2^n})$ hashing scheme to other schemes is the algebraic background that yields some proven properties about distribution, shortest length of collisions [11,15,16], and allows a parallelization of the calculation: it holds $H(m_1|m_2) = H(m_1) \cdot H(m_2)$, where $m_1|m_2$ denotes the concatenation of the two messages $m_1$, $m_2$. This parallelization property is very helpful in some applications so that Quisquater and Joye have suggested to use

M. Bellare (Ed.): CRYPTO 2000, LNCS 1880, pp. 287–299, 2000.
© Springer-Verlag Berlin Heidelberg 2000

this hash function for the authentication of video sequences [10], despite the weaknesses already known (more information on parallelizable hash functions can be found in [2,4]).

In this paper we describe some weaknesses in the $\mathrm{SL}_2(\mathbb{F}_{2^n})$ hashing scheme that also affect the generalization of this scheme to arbitrary finite fields suggested in [1]: as shown in [11], any collision in the $\mathrm{SL}_2(\mathbb{F}_{2^n})$ hashing scheme involves at least one bitstring of length $\ell \geq n$. Hence it is infeasible to search directly for a collision among the more than $2^n$ bitstrings of length $\ell \geq n$. But using some structural properties of the group $\mathrm{SL}_2(\mathbb{F}_{p^n})$ we show that for several choices of the parameters it is possible to find short bitstrings that hash to an element of small order in $\mathrm{SL}_2(\mathbb{F}_{p^n})$. Repeating such a bitstring several times, a message can be modified without changing its hash value. In the case where adequate subfields of $\mathbb{F}_{p^n}$ exist, this approach works quite efficiently. Cases where $n$ is prime are left to be resistant to this kind of attack. However, we show that—independent of $n$ being prime or not—for all suggested values $130 \leq n \leq 170$ in the $\mathrm{SL}_2(\mathbb{F}_{2^n})$ hashing scheme one can find a defining polynomial of $\mathbb{F}_{2^n}$ with a prescribed collision.

## 2   Preliminaries

### 2.1   The $\mathrm{SL}_2(\mathbb{F}_{2^n})$ Hashing Scheme

By $\mathbb{F}_{p^n}$ we denote the finite field with $p^n$ elements. The hash function $H(m)$ of Tillich and Zémor [11] is based on the group $\mathrm{SL}_2(\mathbb{F}_{2^n})$:

**Definition 1.** *Let* $A := \left(\begin{smallmatrix} \alpha & 1 \\ 1 & 0 \end{smallmatrix}\right)$, $B := \left(\begin{smallmatrix} \alpha & \alpha+1 \\ 1 & 1 \end{smallmatrix}\right)$ *be elements of* $\mathrm{SL}_2(\mathbb{F}_{2^n})$, *the group of* $2 \times 2$-*matrices with determinant* 1 *over* $\mathbb{F}_{2^n}$. *Here* $\alpha \in \mathbb{F}_{2^n}$ *is a root of a generating polynomial* $f(X)$ *of the field* $\mathbb{F}_{2^n} \simeq \mathbb{F}_2[X]/f(X)$.

*Then, according to [11], the hash value* $H(b_1 \ldots b_r) \in \mathbb{F}_{2^n}^{2 \times 2}$ *of a binary stream* $b_1 \ldots b_r$ *is defined as the product* $M_1 \cdot \ldots \cdot M_r$ *with*

$$M_i := \begin{cases} A \text{ if } b_i = 0 \; ; \\ B \text{ if } b_i = 1 \; . \end{cases}$$

The straightforward generalization of this hashing scheme is to switch to $\mathbb{F}_{p^n}$, i.e., replacing $\mathrm{SL}_2(\mathbb{F}_{2^n})$ by $\mathrm{SL}_2(\mathbb{F}_{p^n})$, the group of $2 \times 2$-matrices with determinant 1 over $\mathbb{F}_{p^n}$, generated by $A = \left(\begin{smallmatrix} \alpha & -1 \\ 1 & 0 \end{smallmatrix}\right)$ and $B = \left(\begin{smallmatrix} \alpha & \alpha-1 \\ 1 & 1 \end{smallmatrix}\right)$ (see Proposition 2).

### 2.2   Some Properties of $\mathrm{SL}_2(\mathbb{F}_{p^n})$

As stated before, we use some properties of the group $\mathrm{SL}_2(\mathbb{F}_{p^n})$ to find collisions of relatively short length. First we recall the structure of the projective special linear group $\mathrm{PSL}_2(\mathbb{F}_{p^n})$ which will prove useful for analyzing $\mathrm{SL}_2(\mathbb{F}_{p^n})$. Denoting the cyclic group with $r$ elements by $C_r$ we have (see, e.g., [7, Kapitel II, Satz 8.5])

**Proposition 1.** *Any non-identity element of* $\mathrm{PSL}_2(\mathbb{F}_{p^n})$ *is either contained in an elementary abelian p-Sylow group* $C_p^n \cong \mathfrak{P}$, *in a cyclic subgroup* $C_{(p^n-1)/k} \cong \mathfrak{U}$, *or in a cyclic subgroup* $C_{(p^n+1)/k} \cong \mathfrak{S}$, *where* $k = \gcd(p-1, 2)$. *Thus the group can be written as a disjoint union of sets*

$$\mathfrak{G} := \mathrm{PSL}_2(\mathbb{F}_{p^n}) = \mathfrak{E} \uplus \mathcal{P} \uplus \mathcal{U} \uplus \mathcal{S}$$

*where the sets* $\mathcal{E}, \mathcal{P}, \mathcal{U},$ *and* $\mathcal{S}$ *are defined by the disjoint unions*

$$\mathcal{E} := \mathfrak{E} = \{id\} \qquad\qquad \mathcal{U} := \biguplus_{g \in \mathfrak{G}} (\mathfrak{U}^g \setminus \mathfrak{E})$$

$$\mathcal{P} := \biguplus_{g \in \mathfrak{G}} (\mathfrak{P}^g \setminus \mathfrak{E}) \qquad \mathcal{S} := \biguplus_{g \in \mathfrak{G}} (\mathfrak{S}^g \setminus \mathfrak{E}) \ .$$

*(Here* $\mathfrak{H}^g := \{g^{-1}hg \colon h \in \mathfrak{H}\}$ *is the conjugate of* $\mathfrak{H}$ *with respect to* $g$.*)*

This yields immediately the structure of the group $\mathrm{SL}_2(\mathbb{F}_{p^n})$:

**Corollary 1.** *As matrix, any element of* $\mathrm{SL}_2(\mathbb{F}_{p^n})$ *can be written as* $\pm M$ *where* $M \in \mathfrak{E} \cup \mathcal{P} \cup \mathcal{U} \cup \mathcal{S}$.

*Proof.* By definition, $\mathrm{PSL}_2(\mathbb{F}_{p^n}) = \mathrm{SL}_2(\mathbb{F}_{p^n})/(\mathrm{SL}_2(\mathbb{F}_{p^n}) \cap \mathfrak{Z})$ where $\mathfrak{Z} = \{a \cdot I_2 \colon a \in \mathbb{F}_{p^n}^{\times}\}$ as matrix group (with $I_2$ denoting the $2 \times 2$ identity matrix). For $p = 2$, $\mathrm{SL}_2(\mathbb{F}_{p^n}) \cap \mathfrak{Z} = \mathfrak{E}$ and hence $\mathrm{SL}_2(\mathbb{F}_{p^n}) \cong \mathrm{PSL}_2(\mathbb{F}_{p^n})$. For $p > 2$, $\mathrm{SL}_2(\mathbb{F}_{p^n}) \cap \mathfrak{Z} = \{\pm I_2\} \cong C_2$. So we may conclude that $\mathrm{SL}_2(\mathbb{F}_{p^n}) \cong C_2 \times \mathrm{PSL}_2(\mathbb{F}_{p^n})$. $\qquad\square$

Further properties of the group $\mathrm{SL}_2(\mathbb{F}_{p^n})$ are summarized as follows:

*Remark 1.* For any subfield $\mathbb{F}_{p^m} \le \mathbb{F}_{p^n}$ of $\mathbb{F}_{p^n}$, $\mathrm{SL}_2(\mathbb{F}_{p^m})$ and all its conjugates are subgroups of $\mathrm{SL}_2(\mathbb{F}_{p^n})$.

*Remark 2.* The group $\mathrm{SL}_2(\mathbb{F}_{p^n})$ has $p^n(p^n + 1)(p^n - 1)$ elements. There are $p^{2n} - 1$ elements of order $p$ (cf. [7, Kapitel II, Satz 8.2]). Furthermore for all factors $f$ of $(p^n - 1)/\gcd(p-1, 2)$ and $(p^n + 1)/\gcd(p-1, 2)$ there are elements of order $f$. In particular, for $p = 2$ there are elements of order 3.

**Lemma 1.** *The group* $\mathrm{SL}_2(\mathbb{F}_{2^n})$ *has* $2^{2n} + (-2)^n$ *elements of order 3.*

*Proof.* Let $M = \left(\begin{smallmatrix} a & b \\ c & d \end{smallmatrix}\right) \in \mathrm{SL}_2(\mathbb{F}_{2^n})$ be of order 3. Then $M^3 = I_2$ and $M \ne I_2$, hence the characteristic polynomial $f_M(X) = X^2 + (a + d)X + (ad + bc)$ of $M$ equals $X^2 + X + 1$. The number of common solutions of the equations $a + d = 1$ and $ad + bc = 1$ is computed as follows: for $b = 0$, $c$ is unconstrained, and $a$ and $d$ are specified by $a + d = 1$ and $a^2 + a + 1 = 0$. The latter equation has two solutions iff $n$ is even. For $b \ne 0$, $a$ can take any value, and $c$ and $d$ are given by $d = a + 1$ and $c = (a^2 + a + 1)/b$. $\qquad\square$

Finally, we present relations between bitstrings and their hash value.

**Proposition 2.** *Let $\mathbb{F}_{p^n} = \mathbb{F}_p(\alpha)$. Then, as a matrix group, $\mathrm{SL}_2(\mathbb{F}_{p^n})$ is generated by*

$$A = \begin{pmatrix} \alpha & -1 \\ 1 & 0 \end{pmatrix} \quad \text{and} \quad B = \begin{pmatrix} \alpha & \alpha - 1 \\ 1 & 1 \end{pmatrix} \ .$$

*Furthermore, the hash value of a bitstring $m = b_1 \ldots b_\ell$ of length $\ell$ is of the form $M_{b_\ell}$ (where $b_\ell$ is the last bit of $m$) with*

$$M_0 = \begin{pmatrix} c_\ell(\alpha) & c_{\ell-1}(\alpha) \\ d_{\ell-1}(\alpha) & c_{\ell-2}(\alpha) \end{pmatrix} \quad \text{and} \quad M_1 = \begin{pmatrix} c_\ell(\alpha) & d_\ell(\alpha) \\ c_{\ell-1}(\alpha) & d_{\ell-1}(\alpha) \end{pmatrix} \ .$$

*Here $c_i, d_i \in \mathbb{F}_p[X]$ are polynomials of degree $i$.*

*If $M \in \mathrm{SL}_2(\mathbb{F}_{p^n})$ is the hash value of a bitstring $m$ of length $\ell < n$, the representation in the form $M_i$ is unique and the bitstring $m$ can be obtained by successively stripping the factors.*

*Proof.* (cf. also [11, proof of Lemma 3.5]) For a proof that $A$ and $B$ generate $\mathrm{SL}_2(\mathbb{F}_{p^n})$ see, e. g., [1]. Defining the degree of the zero polynomial to be $-1$, the statement is true for $H(0) = A = M_0$ and $H(1) = B = M_1$, i. e., bitstrings of length 1.

Assuming that $H(m|0)$ is of the form $M_0$, the hash value $H(m|00) = H(m|0) \cdot A$ is computed as

$$\begin{pmatrix} c_\ell(\alpha) & c_{\ell-1}(\alpha) \\ d_{\ell-1}(\alpha) & c_{\ell-2}(\alpha) \end{pmatrix} \cdot A = \begin{pmatrix} \alpha \cdot c_\ell(\alpha) + c_{\ell-1}(\alpha) & -c_\ell(\alpha) \\ \alpha \cdot d_{\ell-1}(\alpha) + c_{\ell-2}(\alpha) & -d_{\ell-1}(\alpha) \end{pmatrix} \ ,$$

i. e., $H(m|00)$ is of the form $M_0$ where the degrees of all polynomials are increased by one. Analogously, we can show that $H(m|10)$ is of the form $M_0$ and that $H(m|01)$ and $H(m|11)$ are of the form $M_1$.

Note that the minimal polynomial $f_\alpha(X) \in \mathbb{F}_p[X]$ of $\alpha$ over $\mathbb{F}_p$ has degree $n$. Hence, in polynomial representation of $\mathbb{F}_{p^n}$, no reduction occurs when $\ell < n$ and the representation is unique. Furthermore, by inspection of the degrees of the polynomials, it is easy to decide if a given matrix $M$ is of the form $M_0$ or $M_1$. This yields the final bit $b_\ell$ of a bitstring $m = m'|b_\ell$ hashing to $M$. Using the identities $H(m') = H(m'|0) \cdot A^{-1}$ and $H(m') = H(m'|1) \cdot B^{-1}$, we can strip off one factor and proceed similarly to determine the bitstring $m'$.     □

## 3     Finding Elements of Small Order

If we know a bitstring that hashes to the identity matrix then this bitstring can be inserted into a given message at arbitrary positions without changing the hash value of that message. For practical purposes we are of course particularly interested in *short* bitstrings that hash to the identity matrix. In order to find such bitstrings we want to exploit Proposition 1 and Remark 2 which imply that in case of $(p^n - 1)/\gcd(p - 1, 2)$ or $(p^n + 1)/\gcd(p - 1, 2)$ having several small factors, the group $\mathrm{SL}_2(\mathbb{F}_{p^n})$ contains various elements of small order: instead of looking for arbitrary bitstrings hashing to the identity matrix $I_2$ we try to find very short bitstrings (say less than 50 bits) which hash to a matrix of small order (say less than 300).

One family of matrices which are promising candidates for being of small order is formed by the elements $M \in \mathrm{SL}_2(\mathbb{F}_{p^n})$ whose coefficients are contained in a proper subfield $\mathbb{F}_{p^m} \lneq \mathbb{F}_{p^n}$ already, because according to Proposition 1 the order of such a matrix is bounded by $p^m + 1$. Moreover, as the orders of similar matrices coincide, we are also interested in matrices $M \in \mathrm{SL}_2(\mathbb{F}_{p^n})$ that are similar to some $M' \in \mathrm{SL}_2(\mathbb{F}_{p^m})$ with $\mathbb{F}_{p^m} \lneq \mathbb{F}_{p^n}$ (i.e. $M = N^{-1} \cdot M' \cdot N$ for some non-singular matrix $N$ with coefficients in an extension field of $\mathbb{F}_{p^m}$). By means of the trace operation (which computes the sum of the diagonal entries of a matrix) we can give the following characterization:

**Proposition 3.** *Let $M \in \mathrm{SL}_2(\mathbb{F}_{p^n})$ and $\mathbb{F}_{p^m} \leq \mathbb{F}_{p^n}$. Then*

$$M \text{ is similar to a matrix } M' \in \mathrm{SL}_2(\mathbb{F}_{p^m}) \iff \mathrm{Trace}(M) \in \mathbb{F}_{p^m} \ .$$

*Proof.* $\Longrightarrow$: Trivial.
$\Longleftarrow$: If the minimal polynomial of $M$ is linear then $M = \pm I_2 \in \mathrm{SL}_2(\mathbb{F}_p) \leq \mathrm{SL}_2(\mathbb{F}_{p^m})$. So we assume w.l.o.g. that the minimal polynomial $m(X)$ of $M$ is quadratic, i.e., $m(X) = X^2 - \mathrm{Trace}(M) \cdot X + 1 \in \mathbb{F}_{p^n}[X]$. Let $\lambda_1, \lambda_2$ be the (not necessarily distinct) eigenvalues of $M$. Then the Jordan normal form of $M$ (as a matrix in the general linear group $\mathrm{GL}_2(\mathbb{F}_{p^n}(\lambda_1, \lambda_2))$) is either $\left(\begin{smallmatrix} \lambda_1 & 1 \\ 0 & \lambda_1 \end{smallmatrix}\right)$ or $\left(\begin{smallmatrix} \lambda_1 & 0 \\ 0 & \lambda_2 \end{smallmatrix}\right)$.

In the former case we have $\lambda_1 = \lambda_2$ and $\det(M) = 1$ implies $\lambda_1 = \pm 1$, i.e., the Jordan normal form of $M$ is contained in $\mathrm{SL}_2(\mathbb{F}_p) \leq \mathrm{SL}_2(\mathbb{F}_{p^m})$. In the latter case $M$ has two different eigenvalues and is similar to the matrix $\left(\begin{smallmatrix} \lambda_1 & 0 \\ 0 & \lambda_2 \end{smallmatrix}\right)$, which is also similar to $M' := \left(\begin{smallmatrix} \mathrm{Trace}(M) & -1 \\ 1 & 0 \end{smallmatrix}\right) \in \mathrm{SL}_2(\mathbb{F}_{p^m})$, as the characteristic polynomials of $M$ and $M'$ coincide. $\qquad\square$

**Corollary 2.** *For $M \in \mathrm{SL}_2(\mathbb{F}_{p^n})$ with $\mathrm{Trace}(M) \in \mathbb{F}_{p^m} \leq \mathbb{F}_{p^n}$ we have* $\mathrm{ord}(M) \leq p^m + 1$.

*Proof.* Immediate from Proposition 1 and Proposition 3. $\qquad\square$

So if the trace $\theta$ of a matrix $M \in \mathrm{SL}_2(\mathbb{F}_{p^n})$ generates only a small subfield $\mathbb{F}_p(\theta) \lneq \mathbb{F}_{p^n}$ then $M$ is of small order. Of course, it is not sufficient to know a matrix $M \in \mathrm{SL}_2(\mathbb{F}_{p^n})$ of small order—we also need a short bitstring which hashes to $M$. Subsequently we want to verify that for certain choices of $\mathbb{F}_{p^n}$ such matrices and corresponding bitstrings can indeed be found.

### 3.1   Elements of Small Order, Functional Decomposition, and Intermediate Fields

Let $\mathbb{F}_{p^n} = \mathbb{F}_p[X]/f(X)$ where $f(X) \in \mathbb{F}_p[X]$ is an irreducible polynomial with a root $\alpha$. Moreover, assume that $f(X)$ can be expressed as a functional composition $f(X) = (g \circ h)(X) = g(h(X))$ with (non-linear) "composition factors" $g(X), h(X) \in \mathbb{F}_p[X]$—such decompositions can be found efficiently (for more information about the problem of computing functional decompositions of polynomials cf., e.g., [6,13,14]):

**Proposition 4.** *(see [8, Theorem 9]) An irreducible polynomial of degree n over* $\mathbb{F}_p$ *can be tested for the existence of a nontrivial decomposition* $g \circ h$ *in NC (parallel time* $\log^{O(1)} np$ *on* $(np)^{O(1)}$ *processors). If such a decomposition exists, the coefficients of g and h can be computed in NC.*

If the trace of a matrix $M \in \mathrm{SL}_2(\mathbb{F}_{p^n})$ equals $h(\alpha)$, then we know that the extension $\mathbb{F}_p(\mathrm{Trace}(M))/\mathbb{F}_p$ is of degree $\deg(g(Y))$—note that irreducibility of $f(X)$ implies irreducibility of $g(Y)$. So according to Corollary 2 we have $\mathrm{ord}(M) \leq p^{\deg(g(Y))} + 1$. Consequently, if $M$ can be expressed as a product in the generators $A, B \in \mathrm{SL}_2(\mathbb{F}_p(\alpha))$ with $\ell$ factors, then we obtain a bitstring of length $\leq \ell \cdot (p^{\deg(g(Y))} + 1)$ hashing to the identity $I_2 \in \mathrm{SL}_2(\mathbb{F}_2(\alpha))$.

So the idea for exploiting a decomposition $f(X) = (g \circ h)(X) \in \mathbb{F}_p[X]$ is to construct a bitstring that hashes to a matrix with trace $h(\alpha)$. As for practical purposes we are only interested in short bitstrings, we restrict ourselves to bitstrings of length $\leq n$ (for the $\mathrm{SL}_2(\mathbb{F}_{2^n})$ hashing scheme suggested in [11] we have $130 \leq n \leq 170$). In order to obtain a matrix with trace $h(\alpha)$, Proposition 2 suggests to choose the length of our bitstring equal to $\deg(h(X))$, and if $\deg(h(X))$ is not too large we can simply use an exhaustive search over all $2^{\deg(h(X))}$ bitstrings of length $\deg(h(X))$ to check whether a product of this length with the required trace exists.

## 3.2   Application to the $\mathrm{SL}_2(\mathbb{F}_{2^n})$ Hashing Scheme of Tillich and Zémor

To justify the relevance of the above discussion we apply these ideas to the $\mathrm{SL}_2(\mathbb{F}_{2^n})$ hashing scheme suggested in [11] (i.e., we choose $130 \leq n \leq 170$). As irreducible trinomials are of particular interest when implementing an $\mathbb{F}_{2^n}$ arithmetic in hardware, we first give an example of a decomposable irreducible trinomial:

*Example 1.* For the representation $\mathbb{F}_{2^{147}} \cong \mathbb{F}_2(\alpha)$ with $\alpha$ being a root of $X^{147} + X^{98} + 1 = (Y^3 + Y^2 + 1) \circ (X^{49}) \in \mathbb{F}_2[X]$ the orders of $A$ and $B$ compute to $\mathrm{ord}(A) = 2^{147} - 1, \mathrm{ord}(B) = 2^{147} + 1$, and the bitstring

$$1111101111111000100011111001010010101000111110110$$

(of length 49) hashes to a matrix with trace $\alpha^{49}$ and order 7. So we obtain a bitstring of length $7 \cdot 49 = 343$ that hashes to the identity.

Some more examples based on decomposable polynomials are listed in Table 2 in the appendix. Here we continue with an example in characteristic 3 which demonstrates that using characteristic 2 is not vital:

*Example 2.* For the representation $\mathbb{F}_{3^{90}} \cong \mathbb{F}_3(\alpha)$ with $\alpha$ being a root of

$$X^{90} + X^{78} + X^{75} - X^{69} + X^{66} - X^{63} + X^{57} + X^{56} + X^{55} + X^{54} - X^{53} - X^{48} + X^{45} -$$
$$X^{44} - X^{43} - X^{40} - X^{39} + X^{38} + X^{35} + X^{34} - X^{32} + X^{30} + X^{26} + X^{25} - X^{23} + X^{22} - X^{21} +$$
$$X^{20} + X^{19} + X^{18} + X^{16} + 1 = (Y^6 - Y^4 + Y^2 + 1) \circ (X^{15} - X^{11} - X^{10} + X^8) \in \mathbb{F}_3[X]$$

the orders of $A$ and $B$ are $\mathrm{ord}(A) = (3^{90} - 1)/4, \mathrm{ord}(B) = (3^{90} - 1)/52$, and the bitstring $000101111111100$ (of length 15) hashes to a matrix with trace

$\alpha^{15} - \alpha^{11} - \alpha^{10} + \alpha^8$ and order 56. So we obtain a bitstring of length $15 \cdot 56 = 840$ that hashes to the identity.

All of the examples mentioned make use of the existence of a nontrivial intermediate field $\mathbb{F}_p[Y]/g(Y)$ of the extension $\mathbb{F}_p \leq \mathbb{F}_p[X]/f(X)$ where $f(X) = (g \circ h)(X)$. But even in the case when $f(X)$ is indecomposable, there may exist short bitstrings where the trace of the hash value lies in a small intermediate field:

*Example 3.* The polynomial
$f(X) = X^{140} + X^{139} + X^{137} + X^{135} + X^{133} + X^{132} + X^{127} + X^{122} + X^{120} + X^{119} +$
$X^{116} + X^{114} + X^{113} + X^{112} + X^{111} + X^{106} + X^{104} + X^{101} + X^{100} + X^{94} + X^{93} + X^{91} +$
$X^{90} + X^{88} + X^{87} + X^{84} + X^{83} + X^{82} + X^{80} + X^{79} + X^{73} + X^{71} + X^{69} + X^{67} + X^{65} +$
$X^{63} + X^{62} + X^{60} + X^{59} + X^{57} + X^{56} + X^{55} + X^{53} + X^{52} + X^{51} + X^{49} + X^{47} + X^{46} +$
$X^{45} + X^{43} + X^{40} + X^{39} + X^{38} + X^{37} + X^{35} + X^{34} + X^{33} + X^{32} + X^{31} + X^{30} + X^{28} +$
$X^{27} + X^{25} + X^{23} + X^{17} + X^{16} + X^{15} + X^8 + X^7 + X^6 + X^4 + X + 1$
is indecomposable. For $\mathbb{F}_{2^{140}} \cong \mathbb{F}_2(\alpha)$ with $\alpha$ a root of $f(X)$ we have $\mathrm{ord}(A) = \mathrm{ord}(B) = 2^{140} + 1$, and the bitstring $m := 1111111110101110$ hashes to a matrix $H(m)$ with $\mathrm{Trace}(H(m)) \in \mathbb{F}_{2^{10}}$ and $\mathrm{ord}(H(m)) = 25$.

To prevent the above attack we may choose $\mathbb{F}_{p^n} \cong \mathbb{F}_p[X]/f(X)$ in such a way that $n$ is prime. Then $f$ does not permit a nontrivial functional decomposition, and there are no nontrivial intermediate fields of $\mathbb{F}_{p^n}/\mathbb{F}_p$. Moreover, in order to make the search for elements of small order not unnecessarily easy, we may also try to fix $n$ in such a way that the orders $(p^n \mp 1)/\gcd(p-1,2)$ of the cyclic groups $\mathfrak{U}$, $\mathfrak{S}$ (cf. Proposition 1) do not have many small factors. Ideally, for $p = 2$ we have the following conditions fulfilled: $n$ is prime and $(2^n - 1) \cdot (2^n + 1) = 3 \cdot p_1 \cdot p_2$ for some prime numbers $p_1$, $p_2$.

Using a computer algebra system like MAGMA one easily checks that for $\mathbb{F}_{p^n} = \mathbb{F}_{2^n}$ with $120 \leq n \leq 180$ the only possible choice for satisfying these conditions is $n = 127$; in particular none of the parameter values $130 \leq n \leq 170$ suggested in [11] meets these requirements. Furthermore, the next section shows that independent of the degree of the extension $\mathbb{F}_{p^n}/\mathbb{F}_p$ one should be careful about who is allowed to fix the actual representation of $\mathbb{F}_{p^n}$ as $\mathbb{F}_p[X]/f(X)$ used for hashing.

## 4   Deriving "Hash Functions with a Trapdoor"

In [16, Section 5.3] it is pointed out that for the $\mathrm{SL}_2(\mathbb{F}_p)$ hashing scheme discussed in [16] " ... some care should be taken in the choice of the prime number $p$, because finding simultaneously two texts and a prime number $p$ such that those two texts collide for the hash function associated to $p$, is substantially easier than finding a collision for a given $p$ ... "

In the sequel we shall discuss a related problem with the $\mathrm{SL}_2(\mathbb{F}_{2^n})$ hashing scheme of [11]—being allowed to choose a representation of $\mathbb{F}_{2^n}$ we can select a hash function "with a trapdoor":

*Example 4.* For the representation $\mathbb{F}_{2^{167}} \cong \mathbb{F}_2(\alpha)$ with $\alpha$ being a root of

$$X^{167} + X^{165} + X^{161} + X^{160} + X^{158} + X^{157} + X^{156} + X^{155} + X^{154} + X^{152} + X^{150} +$$
$$X^{148} + X^{145} + X^{143} + X^{142} + X^{140} + X^{138} + X^{137} + X^{134} + X^{131} + X^{130} + X^{126} + X^{125} +$$
$$X^{123} + X^{119} + X^{118} + X^{117} + X^{116} + X^{115} + X^{113} + X^{112} + X^{111} + X^{107} + X^{105} + X^{104} +$$
$$X^{99} + X^{96} + X^{93} + X^{91} + X^{89} + X^{88} + X^{86} + X^{85} + X^{82} + X^{81} + X^{80} + X^{77} + X^{76} +$$
$$X^{74} + X^{73} + X^{71} + X^{65} + X^{64} + X^{62} + X^{61} + X^{58} + X^{57} + X^{54} + X^{53} + X^{51} + X^{49} +$$
$$X^{48} + X^{47} + X^{45} + X^{40} + X^{38} + X^{37} + X^{35} + X^{34} + X^{33} + X^{30} + X^{29} + X^{27} + X^{26} +$$
$$X^{25} + X^{21} + X^{19} + X^{17} + X^{14} + X^{13} + X^{12} + X^{10} + X^6 + X^5 + X^2 + X + 1 \in \mathbb{F}_2[X]$$

the orders of $A$ and $B$ are $\mathrm{ord}(A) = 2^{167} + 1, \mathrm{ord}(B) = (2^{167}+1)/3$. Moreover, 167 is prime, and the prime factorizations of $(2^{167} \mp 1)$ compute to

$$2^{167} - 1 = 2349023 \cdot 79638304766856507377786162960874484906956649 \ ,$$
$$2^{167} + 1 = 3 \cdot 6235740319278519117669055286256140883865312183643 \ .$$

At first glance these parameters look reasonable. However, the bitstring

|  | 01010100 | 01101000 | 01101001 | 01110011 | 00100000 | 01101001 |
|---|---|---|---|---|---|---|
| (ASCII) | T | h | i | s |  | i |
|  | 01110011 | 00100000 | 01110100 | 01101000 | 01100101 | 00100000 |
|  | s |  | t | h | e |  |
|  | 01110111 | 01100001 | 01111001 | 00100000 | 01100001 | 00100000 |
|  | w | a | y |  | a |  |
|  | 01110100 | 01110010 | 01100001 | 01110000 | 01100100 | 01101111 |
|  | t | r | a | p | d | o |
|  | 01101111 | 01110010 | 00100000 | 01100011 | 01100001 | 01101110 |
|  | o | r |  | c | a | n |
|  | 00100000 | 01101100 | 01101111 | 01101111 | 01101011 | 00100000 |
|  |  | l | o | o | k |  |
|  | 01101100 | 01101001 | 01101011 | 01100101 | 00101110 | 00100000 |
|  | l | i | k | e | . |  |

(of length $42 \cdot 8 = 336$) hashes to a matrix with trace 0 and order 2. So we obtain a bitstring of length $336 \cdot 2 = 672$ that hashes to the identity.

The phenomenon of hash functions with a trapdoor is well-known (see, e. g., [9,12]). For the $\mathrm{SL}_2(\mathbb{F}_{2^n})$ hashing scheme deriving parameters with a trapdoor as in Example 4, is comparatively easy—the basic idea is to exploit the fact that, independent of the value of $n$, the group $\mathrm{SL}_2(\mathbb{F}_{2^n})$ always contains elements of order 2 and 3 (see Proposition 1 and Lemma 1): we start by fixing a bitstring which consists of two or three (depending on whether we want to have an element of order two or three) repetitions of an arbitrary bit sequence $m$. Then we compute the "generic hash value" $H = H(X)$ of this bitstring, i. e., instead of using the matrices $A$ and $B$ we use the matrices

$$A_X = \begin{pmatrix} X & 1 \\ 1 & 0 \end{pmatrix} \quad \text{and} \quad B_X = \begin{pmatrix} X & X+1 \\ 1 & 1 \end{pmatrix} \ ,$$

where the generator $\alpha$ is replaced by an indeterminate $X$. Next, we compute the irreducible factors $f_1, \ldots, f_r$ of the greatest common divisor of the entries of the matrix $H - I_2$. Then choosing the field $\mathbb{F}_{2^n}$ as $\mathbb{F}_{2^n} \cong \mathbb{F}_2(\alpha)$ with $\alpha$ a root of some $f_i$ guarantees that $H - I_2$ is in the kernel of the specialization $X \mapsto \alpha$. In other words, the bit sequence $m$ hashes to a matrix of order at most two resp. three. Experiments show that it is quite easy to derive weak parameters for the $\mathrm{SL}_2(\mathbb{F}_{2^n})$ hashing scheme in this way for all $127 \leq n \leq 170$ (see Table 1 in the appendix for some examples).

## 5    Constructing "Real" Collisions

We conclude by a simple example that illustrates the use of elements of small order for deriving "real" collisions, i.e. (short) non-empty bitstrings $m_1 \neq m_2$ that hash to the same value.

*Remark 3.* Let $m$ be a bitstring hashing to a matrix of order ord($m$). Then also each bitstring rot($m$) derived from $m$ through bit-wise left- or right-rotation hashes to a matrix of order ord($m$).

*Proof.* Rotating the bitstring simply translates into conjugating the hash value with a non-singular matrix. Hence, as the order of a matrix is invariant under conjugation, the claim follows.                                                               $\square$

In the following example Remark 3 is used for deriving a collision:

*Example 5.* We use the representation of $\mathbb{F}_{2^{140}}$ of Example 3. Applying a brute-force approach for constructing short products of $A$ and $B$ of small order one can derive the identities

$$(B^9 A B A B^3 A)^{25} = I_2 = (B^3 A B A B^9 A)^{25} \ .$$

As $B^3 A B A B^9 A$ is similar to $B^9 A B^3 A B A$ we get

$$(B^9 A B A B^3 A)^{25} = I_2 = (B^9 A B^3 A B A)^{25}$$

resp. after multiplication with $(B^9 A B)^{-1}$ from the left and $(B A)^{-1}$ from the right

$$A B^3 A (B^9 A B A B^3 A)^{23} B^9 A B A B^2 = B^2 A B A (B^9 A B^3 A B A)^{23} B^9 A B^3 A \ .$$

So we obtain two different bitstrings of length $5 + 16 \cdot 23 + 14 = 387$ that hash to the same value.

## 6    Summary and Conclusion

We have shown that for various choices of the parameters in the $\mathrm{SL}_2(\mathbb{F}_{2^n})$ hashing scheme, suggested in [11], messages can be modified without changing the hash value. Moreover, we have given several examples of hash functions "with a trapdoor" within this family.

In order to avoid the attacks based on functional decomposition and intermediate fields presented in Section 3, one should choose $n$ being prime. We dissuade from using the $SL_2(\mathbb{F}_{2^n})$ hashing scheme or its generalization to $SL_2(\mathbb{F}_{p^n})$ in case of $n$ being composite. Moreover, Section 4 demonstrates that even in the case of $n$ being prime it is fairly easy to find defining polynomials yielding hash functions with a trapdoor. Consequently, appropriate care should be taken in fixing the representation of $\mathbb{F}_{2^n}$ which is used for hashing (concerning the problem of avoiding trapdoors in hash functions cf., e.g., [12]).

## Acknowledgments

The first author is supported by grant DFG - GRK 209/3-98 "Beherrschbarkeit komplexer Systeme". Moreover, we would like to thank the referees for several useful remarks and references.

## References

1. K. S. ABDUKHALIKOV AND C. KIM, *On the Security of the Hashing Scheme Based on* SL₂, in Fast Software Encryption – FSE '98, S. Vaudenay, ed., vol. 1372 of Lecture Notes in Computer Science, Springer, 1998, pp. 93–102.

2. M. BELLARE AND D. MICCIANCIO, *A New Paradigm for Collision-Free Hashing: Incrementality at Reduced Cost*, in Advances in Cryptology – EUROCRYPT '97, W. Fumy, ed., vol. 1233 of Lecture Notes in Computer Science, Springer, 1997, pp. 163–192.

3. C. CHARNES AND J. PIEPRZYK, *Attacking the* SL₂ *hashing scheme*, in Advances in Cryptology – ASIACRYPT '94, J. Pieprzyk and R. Safavi-Naini, eds., vol. 917 of Lecture Notes in Computer Science, Springer, 1995, pp. 322–330.

4. I. B. DAMGÅRD, *A Design Principle for Hash Functions*, in Advances in Cryptology – CRYPTO '89, G. Brassard, ed., vol. 435 of Lecture Notes in Computer Science, Springer, 1989, pp. 416–427.

5. W. GEISELMANN, *A Note on the Hash Function of Tillich and Zemor*, in Cryptography and Coding, C. Boyd, ed., vol. 1025 of Lecture Notes in Computer Science, Springer, 1995, pp. 257–263.

6. J. GUTIÉRREZ, T. RECIO, AND C. RUIZ DE VELASCO, *Polynomial decomposition algorithm of almost quadratic complexity*, in Applied Algebra, Algebraic Algorithms and Error-Correcting Codes (AAECC-6), Rome, Italy, 1988, T. Mora, ed., vol. 357 of Lecture Notes in Computer Science, Springer, 1989, pp. 471–475.

7. B. HUPPERT, *Endliche Gruppen I*, vol. 134 of Grundlehren der mathematischen Wissenschaften, Springer, 1967. Zweiter Nachdruck der ersten Auflage.

8. D. KOZEN AND S. LANDAU, *Polynomial Decomposition Algorithms*, Journal of Symbolic Computation, 7 (1989), pp. 445–456.

9. B. PRENEEL, *Design principles for dedicated hash functions*, in Fast Software Encryption, R. Anderson, ed., vol. 809 of Lecture Notes in Computer Science, Springer, 1994, pp. 71–82.

10. J.-J. QUISQUATER AND M. JOYE, *Authentication of sequences with the* SL₂ *hash function: Application to video sequences.*, Journal of Computer Security, 5 (1997), pp. 213–223.

11. J.-P. TILLICH AND G. ZÉMOR, *Hashing with* $SL_2$, in Advances in Cryptology – CRYPTO '94, Y. Desmedt, ed., vol. 839 of Lecture Notes in Computer Science, Springer, 1994, pp. 40–49.

12. S. VAUDENAY, *Hidden Collisions on DSS*, in Advances in Cryptology – CRYPTO '96, N. Koblitz, ed., vol. 1109 of Lecture Notes in Computer Science, Springer, 1996, pp. 83–88.

13. J. VON ZUR GATHEN, *Functional Decomposition of Polynomials: The Tame Case*, Journal of Symbolic Computation, 9 (1990), pp. 281–300.

14. ———, *Functional Decomposition of Polynomials: The Wild Case*, Journal of Symbolic Computation, 10 (1990), pp. 437–452.

15. G. ZÉMOR, *Hash Functions and Graphs With Large Girths*, in Advances in Cryptology – EUROCRYPT '91, D. W. Davies, ed., vol. 547 of Lecture Notes in Computer Science, Springer, 1991, pp. 508–511.

16. ———, *Hash Functions and Cayley Graphs*, Designs, Codes and Cryptography, 4 (1994), pp. 381–394.

# Appendix: Examples

For each $127 \leq n \leq 170$ we easily found representations of $\mathbb{F}_{2^n} \cong \mathbb{F}_2[X]/f(X)$ together with a bitstring $m$ of length $n$ that hashes to a matrix of order 3. In Table 1, we list for each prime number in this range such a representation together with the corresponding bitstring. To illustrate that neither $A$ nor $B$ is of small order, the orders of $A$ and $B$ are also included in the table.

Note that all the examples have been derived by means of the computer algebra system MAGMA on usual SUN workstations at a university institute; neither specialized hard- or software nor extraordinary computational power have been used.

In Table 2 some representations of $\mathbb{F}_{2^n} \cong \mathbb{F}_2[X]/f(X)$ are listed, where the defining polynomial $f(X) = (g \circ h)(X)$ allows a nontrivial decomposition. In addition to a bitstring $m$ that hashes to a matrix $H(m)$ of small order $\mathrm{ord}(H(m))$, the orders of $A$ and $B$ are also included. In the last column, we list the total length of the resulting bitstring hashing to the identity. As low weight polynomials are of particular interest for hardware implementations of $\mathbb{F}_{2^n}$ arithmetic, a main focus is on decomposable trinomials.

**Table 1.** Weak parameters (bitstrings $m$ of length $n$ and $\operatorname{ord}(H(m)) = 3$)

| $n$ | $f(X)$ | $\operatorname{ord}(A)$ | $\operatorname{ord}(B)$ | $m$ |
|---|---|---|---|---|
| 127 | $X^{127} + X^{125} + X^{117} + X^{113} + X^{109} + X^{103} + X^{87} + X^{85} + X^{81} + X^{77} + X^{71} + X^{23} + X^{21} + X^{17} + X^{13} + X^7 + 1$ | $2^{127}-1$ | $2^{127}-1$ | $1^2 0 1^3 0 1^2 0^3 1 0 1^{114}$ |
| 131 | $X^{131} + X^{123} + X^{121} + X^{115} + X^{113} + X^{109} + X^{99} + X^{97} + X^{77} + X^{67} + X^{65} + X^{13} + X^3 + X + 1$ | $2^{131}+1$ | $2^{131}-1$ | $1^5 0^4 1 0^{121}$ |
| 137 | $X^{137} + X^{135} + X^{125} + X^{119} + X^{113} + X^{105} + X^{103} + X^{97} + X^{73} + X^{71} + X^{65} + X^9 + X^7 + X + 1$ | $2^{137}+1$ | $2^{137}-1$ | $10101010^{130}$ |
| 139 | $X^{139} + X^{133} + X^{127} + X^{123} + X^{121} + X^{117} + X^{113} + X^{107} + X^{101} + X^{97} + X^{75} + X^{69} + X^{65} + X^{11} + X^5 + X + 1$ | $2^{139}+1$ | $2^{139}-1$ | $1^3 0^2 1^3 0^{131}$ |
| 149 | $X^{149} + X^{143} + X^{137} + X^{133} + X^{129} + X^{113} + X^{111} + X^{105} + X^{101} + X^{97} + X^{85} + X^{79} + X^{73} + X^{69} + X^{65} + X^{21} + X^{15} + X^9 + X^5 + X + 1$ | $\dfrac{2^{149}+1}{3}$ | $2^{149}-1$ | $10^3 1 0^{144}$ |
| 151 | $X^{151} + X^{147} + X^{139} + X^{135} + X^{131} + X^{107} + X^{103} + X^{99} + X^{87} + X^{83} + X^{75} + X^{71} + X^{67} + X^{23} + X^{19} + X^{11} + X^7 + X^3 + 1$ | $2^{151}-1$ | $2^{151}-1$ | $1010^{148}$ |
| 157 | $X^{157} + X^{155} + X^{153} + X^{147} + X^{141} + X^{137} + X^{131} + X^{109} + X^{105} + X^{99} + X^{93} + X^{91} + X^{89} + X^{83} + X^{77} + X^{73} + X^{67} + X^{29} + X^{27} + X^{25} + X^{19} + X^{13} + X^9 + X^3 + 1$ | $2^{157}-1$ | $2^{157}-1$ | $101^2 0^2 10^{150}$ |
| 163 | $X^{163} + X^{162} + X^{157} + X^{156} + X^{141} + X^{140} + X^{109} + X^{108} + X^{99} + X^{98} + X^{93} + X^{92} + X^{77} + X^{76} + X^{35} + X^{34} + X^{29} + X^{28} + X^{13} + X^{12} + 1$ | $2^{163}-1$ | $2^{163}-1$ | $1^5 0 1 0^2 1 0^{153}$ |
| 167 | $X^{167} + X^{165} + X^{163} + X^{153} + X^{149} + X^{133} + X^{129} + X^{113} + X^{103} + X^{99} + X^{97} + X^{89} + X^{85} + X^{69} + X^{65} + X^{39} + X^{37} + X^{35} + X^{25} + X^{21} + X^5 + X + 1$ | $2^{167}+1$ | $2^{167}-1$ | $101^3 0^2 1^2 0 1 0^2 10^{153}$ |

**le 2.** Decomposable trinomials and non-monomial decompositions

| $f(X)$ | ord($A$) | ord($B$) | $m$ | or‹ |
|---|---|---|---|---|
| $+ X^{49} + 1 = (Y^3 + Y + 1) \circ (X^{49})$ | $2^{147}-1$ | $2^{147}-1$ | $0^3 101^4 0^3 10101^5 0^3 1^7 0^3 1^3 0^2 1010^5 1^2$ | |
| $+ X^{98} + 1 = (Y^3 + Y^2 + 1) \circ (X^{49})$ | $2^{147}-1$ | $2^{147}+1$ | $0^3 101^4 0^3 10101^5 0^3 1^7 0^3 1^3 0^2 1010^5 1^2$ | |
| $+ X^{62} + 1 = (Y^5 + Y^2 + 1) \circ (X^{31})$ | $2^{155}-1$ | $\frac{2^{155}+1}{11}$ | $1^7 0^4 110100001^7 010^4$ | |
| $+ X^{93} + 1 = (Y^5 + Y^3 + 1) \circ (X^{31})$ | $2^{155}-1$ | $2^{155}-1$ | $1^7 0^4 110100001^7 010^4$ | |
| $+ X^{91} + 1 = (Y^{12} + Y^7 + 1) \circ (X^{13})$ | $\frac{2^{156}-1}{2^{12}-1}$ | $2^{156}+1$ | $0110000011011$ | |
| $+ X^{81} + 1 = (Y^6 + Y^3 + 1) \circ (X^{27})$ | $2^{162}-1$ | $2^{162}+1$ | $0^2 10^6 10^2 101^2 010^2 1^2 0^3 1^2$ | |
| $+ X^{135} + 1 = (Y^6 + Y^5 + 1) \circ (X^{27})$ | $2^{162}-1$ | $2^{162}+1$ | $0^2 10^6 10^2 101^2 010^2 1^2 0^3 1^2$ | |
| $- Y^8 + Y^4 + Y^3 + 1) \circ (X^{13} + X^9)$ | $2^{130}-1$ | $2^{130}-1$ | $0010011001101$ | |
| $Y + 1) \circ (X^{19} + X^{17} + X^7)$ | $2^{133}-1$ | $2^{133}-1$ | $1110101011100000000$ | |
| $Y^8 + Y^7 + Y^6 + Y^5 + Y + 1) \circ (X^{15} + X^{11})$ | $2^{135}-1$ | $2^{135}-1$ | $010101000001011$ | |
| $Y + 1) \circ (X^{20} + X^{19} + X^{12} + X^{11} + X^4 + X^3)$ | $2^{140}-1$ | $2^{140}+1$ | $10000000000000000000$ | |
| $Y + 1) \circ (X^{21} + X^{15} + X^5)$ | $2^{147}-1$ | $2^{147}-1$ | $11110000101000000000$ | |
| $Y^4 + Y^3 + Y^2 + 1) \circ (X^{19} + X^{13} + X^{11} + X^5)$ | $2^{152}-1$ | $2^{152}-1$ | $1101101000001000000$ | |
| $Y^8 + Y^6 + Y^5 + Y^4 + Y + 1) \circ$ <br> $+ X^{11} + X^7 + X^5)$ | $2^{153}-1$ | $2^{153}-1$ | $00000010000101111$ | |
| $Y^4 + Y^3 + Y + 1) \circ (X^{20} + X^{19} + X^{10} + X^9)$ | $2^{160}-1$ | $2^{160}-1$ | $11000010100100000000$ | |
| $Y^8 + Y^6 + Y^5 + Y^4 + Y + 1) \circ$ <br> $+ X^{17} + X^{12} + X^{11} + X^8 + X^7 + X^6 + X^5)$ | $2^{162}-1$ | $2^{162}-1$ | $1010100101000000000$ | |
| $Y^7 + Y^5 + Y^3 + 1) \circ (X^{21} + X^{17} + X^{11})$ | $2^{168}-1$ | $2^{168}+1$ | $11001000110100000000$ | |
| $- Y^6 + Y^5 + Y^3 + Y^2 + Y + 1) \circ (X^{17} + X^7 + X)$ | $2^{170}+1$ | $2^{170}+1$ | $00000100011110001$ | |

# Fast Correlation Attacks
# through Reconstruction of Linear Polynomials

Thomas Johansson and Fredrik Jönsson

Dept. of Information Technology
Lund University, P.O. Box 118, 221 00 Lund, Sweden
{thomas, fredrikj}@it.lth.se

**Abstract.** The task of a fast correlation attack is to efficiently restore the initial content of a linear feedback shift register in a stream cipher using a detected correlation with the output sequence. We show that by modeling this problem as the problem of learning a binary linear multivariate polynomial, algorithms for polynomial reconstruction with queries can be modified through some general techniques used in fast correlation attacks. The result is a new and efficient way of performing fast correlation attacks.

**Keywords.** Stream ciphers, correlation attacks, learning theory, reconstruction of polynomials.

## 1   Introduction

Consider a binary additive stream cipher, i.e., a synchronous stream cipher in which the keystream, the plaintext, and the ciphertext are sequences of binary digits. The output sequence of the keystream generator, $z_1, z_2, \ldots$ is added bitwise to the plaintext sequence $m_1, m_2, \ldots$, producing the ciphertext $c_1, c_2, \ldots$. The keystream generator is initialized through a secret key $K$, and hence, each key $K$ will correspond to an output sequence. Since the key is shared between the transmitter and the receiver, the receiver can decrypt by adding the output of the keystream generator to the ciphertext and obtain the message sequence, see Figure 1.

The design goal is to efficiently produce random-looking sequences that are as "indistinguishable" as possible from truly random sequences. For a synchronous stream cipher, a known-plaintext attack is equivalent to the problem of finding the key $K$ that produced a given keystream $z_1, z_2, \ldots, z_N$. We assume that a given output sequence from the keystream generator, $z_1, z_2, \ldots, z_N$, is known to the cryptanalyst and that his task is to restore the secret key $K$.

It is common to use linear feedback shift registers, LFSRs, as building blocks in different ways. Furthermore, the secret key $K$ is usually chosen to be the initial state of the LFSRs. The feedback polynomials of the LFSRs are considered to be known.

Several cryptanalytic attacks against stream ciphers can be found in the literature [14]. One very important class of attacks on LFSR-based stream ciphers

M. Bellare (Ed.): CRYPTO 2000, LNCS 1880, pp. 300–315, 2000.
© Springer-Verlag Berlin Heidelberg 2000

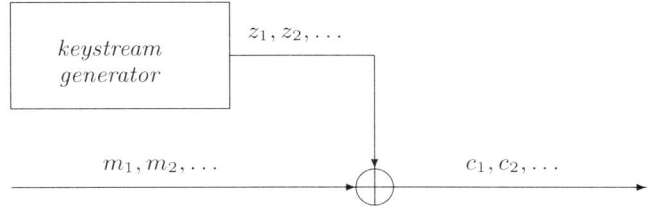

**Fig. 1.** A binary additive stream ciphers

is *correlation attacks*. The idea is that if one can detect a correlation between the known output sequence and the output of one individual LFSR, it is possible to mount a "divide-and-conquer" attack on the individual LFSR [18,19,12,13], i.e., we try to restore the individual LFSR independently from the other LFSRs. By a correlation we mean that, if $u_1, u_2, \ldots$ denotes the output of the particular LFSR, we have

$$P(u_i = z_i) \neq 1/2, \quad i \geq 1.$$

Other types of correlations may also apply.

A common methodology for producing random-like sequences from LFSRs is to combine the output of several LFSRs by a nonlinear Boolean function $f$ with desired properties [14]. The purpose of $f$ is to destroy the linearity of the LFSR sequences and hence provide the resulting sequence with a large linear complexity [14]. Note that for such a stream cipher, there is always a correlation between the output $z_n$ and either one or a set of output symbols from different LFSRs.

Finding a low complexity algorithm that successfully uses the existing correlations in order to determine a part of the secret key can be a very efficient way of attacking stream ciphers for which a correlation is identified. After the initializing ideas of Siegenthaler [18,19], Meier and Staffelbach [12,13] found a very interesting way to explore the correlation in what was called a *fast* correlation attack. A necessary condition is that the feedback polynomial of the LFSR has a very low weight. This work was followed by several papers, providing improvements to the initial results of Meier and Staffelbach, see [16,4,5,17]. However, the algorithms are efficient (good performance and low complexity) only if the feedback polynomial is of low weight. More recently, steps in other directions were taken, and in [9] it was suggested to use convolutional codes in order to improve performance [9]. This was followed by a generalization in [10], applying the use of iterative decoding and turbo codes. One main advantage compared to previous results was the fact that these algorithms now applied to a feedback polynomial of arbitrary form. Very recently, several other suggested methods have appeared, see [3,15,2].

The purpose of this paper is to show that the initial state recovery problem in a fast correlation attack can be modeled as the problem of learning a binary

linear multivariate polynomial. We show that algorithms for polynomial reconstruction with queries can be modified through some general techniques used
in fast correlation attacks. The result is a new and efficient way of performing
fast correlation attacks. Actually, two algorithms are presented, one based on a
direct search and one based on a sequential procedure. Both provide very good
simulation results as well as a theoretical platform.

The paper is organized as follows. In Section 2 we give the preliminaries
on the standard model that is used for cryptanalysis and reformulate this into
a polynomial reconstruction problem. In Section 3 we review an algorithm by
Goldreich, Rubinfeld and Sudan [7] that solves the polynomial reconstruction
problem with queries in polynomial time. In Section 4 we derive a new algorithm
for fast correlation attacks, inspired by the previous section. In Section 5 we
present a sequential version of the new algorithm, i.e, this algorithm builds a
tree of possible candidates and searches through it. In Section 6 we present
simulation results and a comparison with other algorithms, and in Section 7 a
sketch of a theoretical platform for the two algorithms is presented. We show
among other things that the central test in the algorithms is statistically optimal.

## 2   Preliminaries and Model

Most authors [19,12,13,16,4,9,10] use the approach of viewing our cryptanalysis
problem as a decoding problem over the binary symmetric channel. However,
in this section we show that it can equivalently be viewed as the problem of
learning a linear multivariate polynomial.

Let the target LFSR have length $l$ and feedback polynomial $g(x)$. Clearly, the
number of possible LFSR sequences is $2^l$. Furthermore, assume that the known
keystream sequence $\mathbf{z} = z_1, z_2, \ldots, z_N$ is of length $N$.

The assumed correlation between $u_i$ and $z_i$ is described by the correlation
probability $1/2 + \epsilon$, defined by $1/2 + \epsilon = P(u_i = z_i)$, where $0 < \epsilon < 1/2$. The
problem of cryptanalysis is the following. Given the received word $(z_1, z_2, \ldots, z_N)$
as the output of the stream cipher, find the initial state (or at least a part of it)
of the target LFSR.

It is known that the length $N$ should be at least around $N_0 = l/(1 - h(p))$
for a unique solution to the above problem [4], where $h(p)$ is the binary entropy
function. Throughout this paper we assume that $N \gg N_0$. For this case, *fast
correlation attacks* are applicable. Although this notation was initially used as
the notion for the algorithms developed by Meier and Staffelbach [12,13], we
adopt this terminology for any algorithm that finds the correct initial state
of the target LFSR significantly faster than exhaustively searching through all
initial states.

Let us now consider the unknown initial state of the target LFSR, denoted

$$\mathbf{u} = (u_1, u_2, \ldots, u_l). \tag{1}$$

Clearly, since the LFSR sequence is generated through the recursion

$$u_i = \sum_{j=1}^{l} g_j u_{i-j}, \quad i > l, \tag{2}$$

where $g(x) = 1 + g_1 x + \ldots g_l x^l$, we can express each $u_i$ as some known linear combination of the initial state $\mathbf{u}$, i.e.,

$$u_i = \sum_{j=1}^{l} w_{ij} u_j, \quad \forall i \geq 1, \tag{3}$$

where $w_{ij}, i \geq 1, 1 \leq j \leq l$ are known constants that can be calculated provided $g(x)$ is known (This is essentially the error correcting code one gets by truncating the set of LFSR sequences).

Define the *initial state polynomial*, denoted $U(\mathbf{x})$, to be

$$U(\mathbf{x}) = U(x_1, x_2, \ldots, x_l) = u_1 x_1 + u_2 x_2 + \cdots + u_l x_l. \tag{4}$$

With this notation, we can express each $u_i$ as being the initial state polynomial evaluated in some known point $\mathbf{x}_i = (w_{i1}, w_{i2}, \ldots, w_{ij})$, i.e.,

$$u_i = U(\mathbf{x}_i), \quad i \geq 1. \tag{5}$$

The correlation between $u_i$ and $z_i$ can be described by introducing a noise vector

$$\mathbf{e} = (e_1, e_2, \ldots, e_N), \tag{6}$$

where $e_i \in \mathbb{F}_2$ are independent random variables for $1 \leq i \leq N$ and $P(e_i = 0) = 1/2 + \epsilon$. Then we model the correlation by writing $\mathbf{z} = \mathbf{u} + \mathbf{e}$, giving

$$\mathbf{z} = (U(\mathbf{x}_1) + e_1, U(\mathbf{x}_2) + e_2, \ldots, U(\mathbf{x}_N) + e_N), \tag{7}$$

where $\mathbf{x}_i$ are known $l$-tuples for all $1 \leq i \leq N$. In conclusion, we have reformulated our problem into the following.

**The output vector z consists of a number of noisy observations of an unknown polynomial $U(\mathbf{x})$ evaluated in different known points $\{\mathbf{x}_1, \mathbf{x}_2, \ldots \mathbf{x}_N\}$. The task of the attacker is to determine the unknown polynomial $U(\mathbf{x})$.**

## 3   Learning Polynomials with Queries

In computational learning theory (see e.g., [7] and its references), one might want to consider the following general *reconstruction problem*:

**Given:** An oracle (black box) for an arbitrary unknown function $f : F^l \to F$, a class of functions $\mathcal{F}$ and a parameter $\delta$.

**Problem:** Provide a list of all functions $g \in \mathcal{F}$ that agree with $f$ on at least a $\delta$ fraction of the inputs.

The general reconstruction problem can be interpreted in several ways. We consider only the paradigm of learning with persistent noise. Here we assume that the output of the oracle is derived by evaluating some specific function in $\mathcal{F}$ and then adding noise to the result. A lot of work on different settings for this problem can be found.

We will now pay special attention to the work of Goldreich, Rubinfeld and Sudan in [7]. They consider a case of the reconstruction problem when the hypothesis class $\mathcal{F}$ is the set of linear polynomials in $l$ variables (actually, any polynomial degree $d$ was considered in [7], but we are only interested in the linear case). In the binary case ($F = \mathbb{F}_2$), they demonstrate an algorithm that given $\epsilon > 0$ and provided oracle access to an arbitrary function $f : F^l \to F$, runs in time poly($l/\epsilon$) and outputs a list of all linear functions in $l$ variables that agree with $f$ on at least $\delta = 1/2 + \epsilon$ of the output.

Let us immediately describe the procedure. First, the problem description can be as follows. On a selected input $\mathbf{x}$, the oracle evaluates an unknown linear function $p(\mathbf{x})$, adds a noise value $e$, and outputs the result $p(\mathbf{x}) + e$. On the next oracle access, the function is evaluated in a new point and a new noise value is added.

The algorithm for solving the above problem given in [7] is a generalization of an algorithm given in [6] (in the binary case that we consider they coincide). Consider all polynomials of the form

$$p(\mathbf{x}) = \sum_{i=1}^{l} c_i x_i.$$

The algorithm uses the concept of $i$-prefixes, which is defined to be all polynomials that can be expressed in the form $p(x_1, x_2, \ldots, x_i, 0, 0, \ldots, 0)$. This means that an $i$-prefix is a polynomial in $l$ variables in which only the first $i$ variables appear.

The algorithm proceeds in $l$ rounds, so that in the $i$th round we have a list of candidates for the $i$-prefixes of $p(\mathbf{x})$. The list of $i$-prefixes is generated by extending the list of $(i-1)$-prefixes from the previous round in all possible ways, i.e., by adding or not adding the $x_i$ variable to each of the members of the $(i-1)$-prefixes. Hence the list is doubled in cardinality. After the extension, a screening process takes place. The screening process guarantees that the $i$-prefix of the correct solution passes with high probability and that not too many other prefixes pass.

The screening process is done by testing each candidate prefix, denoted $(c_1, c_2, \ldots, c_i)$, as follows. Pick $n = $ poly($l/\epsilon$) sequences uniformly from $\mathbb{F}_2^{l-i}$. For each such sequence, denoted $(s_{i+1}, \ldots, s_l)$, and for every $\xi \in \mathbb{F}_2$, estimate the quantity

$$P(\xi) = Pr_{r_1, \ldots, r_i \in \mathbb{F}_2} \left[ f(\mathbf{r}, \mathbf{s}) = \sum_{j=1}^{i} c_j r_j + \xi \right].$$

Here $(\mathbf{r}, \mathbf{s})$ denotes the vector $(r_1, \ldots, r_i, s_{i+1}, \ldots, s_l)$. All these probabilities can be approximated simultaneously by using a sample of poly($l/\epsilon$) sequences

$(r_1, \ldots, r_i)$. A candidate is considered to pass the test if for at least one sequence $(s_{i+1}, \ldots, s_l)$ there exists $\xi$ such that the estimate $P(\xi)$ is greater than $1/2 + \epsilon/3$. It is shown in [7] that the correct candidate passes the test with overwhelming probability, and that not too many other candidates do. For more details on this algorithm, we refer to [7].

## 4  Fast Correlation Attacks Based on Algorithms for Learning Polynomials

We observe the similarities between our correlation attack problem described as a polynomial reconstruction problem as in Section 2, and the problem of learning polynomials with queries as described in the previous section.

Note that the polynomial time algorithm of the previous section can not be applied directly to the correlation attack problem, since queries are essential. *In the query case, sample points given to the oracle can be chosen, whereas for correlation attacks the sample points are simply randomly selected.* The latter problem is actually a well-known problem also in learning theory, called "learning parity with noise", and it is commonly believed to be hard, see [11,1].

Nevertheless, we are interested in finding as efficient correlation attacks as possible, and we will now derive an algorithm that is inspired by the results presented in the previous section.

Let us first briefly review our problem formulation. The recovery of the initial state of the target LFSR is viewed as the problem of recovering an unknown binary linear polynomial $U(\mathbf{x})$ in $l$ variables. To our disposal, we have a number $N$ of noisy observations of this polynomial (the output sequence), denoted

$$\mathbf{z} = (z_1, z_2, \ldots, z_N).$$

The noise is such that

$$P(z_i = U(\mathbf{x}_i)) = 1/2 + \epsilon, \quad 1 \leq i \leq N,$$

where $\mathbf{x}_i$ are known random $l$-tuples for all $1 \leq i \leq N$.

Our problem in applying the algorithm described in Section 3 is the fact that we are not able to select the points $\mathbf{x}_i$ ourselves. This can to some extent be compensated for by the following observation [9].

Assume that we have noisy observations $z_i$ and $z_j$ of the polynomial $U(\mathbf{x})$ in two points $\mathbf{x}_i$ and $\mathbf{x}_j$, respectively, i.e., $P(z_i = U(\mathbf{x}_i)) = 1/2 + \epsilon$ and $P(z_j = U(\mathbf{x}_j)) = 1/2 + \epsilon$. Since $U(\mathbf{x})$ is a linear polynomial, the sum of these two noisy observations will give rise to an even more noisy observation in the point $\mathbf{x}_i + \mathbf{x}_j$, since

$$
\begin{aligned}
P(z_i + z_j = U(\mathbf{x}_i + \mathbf{x}_j)) &= P(z_i + z_j = U(\mathbf{x}_i) + U(\mathbf{x}_j)) \\
&= P(z_i = U(\mathbf{x}_i))P(z_j = U(\mathbf{x}_j)) \\
&\quad + P(z_i \neq U(\mathbf{x}_i))P(z_j \neq U(\mathbf{x}_j)) \\
&= (1/2 + \epsilon)^2 + (1/2 - \epsilon)^2 \\
&= 1/2 + 2\epsilon^2.
\end{aligned}
$$

Next, observe that we do not have to restrict ourselves to addition of just two sample points, but can consider any sum of $t$ points. Hence, any $\sum_{j=1}^{t} z_{a_j}$, $a_1, \ldots a_t \in \{1, 2, \ldots, N\}$, will be a noisy observation of $U(\sum_{j=1}^{t} \mathbf{x}_{a_j})$ with noise level

$$P(\sum_{j=1}^{t} z_{a_j} = U(\sum_{j=1}^{t} \mathbf{x}_{a_j})) = 1/2 + 2^{t-1}\epsilon^t. \tag{8}$$

For convenience, we introduce the notation $\hat{\mathbf{x}} = \sum_{j=1}^{t} \mathbf{x}_{a_j}$ and $\hat{z} = \sum_{j=1}^{t} z_{a_j}$ and write

$$U(\hat{\mathbf{x}}) = \hat{z} + e,$$

where now $e$ is a binary random variable with $P(e = 0) = 1/2 + 2^{t-1}\epsilon^t$, from (8).

If we want to use the algorithm in Section 3 we must feed the oracle with $\hat{\mathbf{x}}$ points of a special form. An idea in the algorithm to be described is to construct such points by adding suitable vectors $\mathbf{x}_{a_j}$ in such a way that their sum is of the required form. Clearly, the noise level increases with the number of vectors in the sum, so we are interested in having as few vectors as possible summing to the desired form. On the other hand, allowing only very few vectors in the sum will give us only very few $\hat{\mathbf{x}}$ vectors of the desired form. Hence, there is a tradeoff for the value of the constant $t$. We return to this issue in the theoretical analysis.

Also, we introduce a slightly modified version of the algorithm from Section 3. The new version includes a squared distance used in the test in the screening procedure. We will later show that this is a statistically optimal distance measure. We first consider a version in which the idea of $i$-prefixes is removed. In the next section we elaborate on the idea of $i$-prefixes. A description of the basic algorithm is given in Figure 2.

Let us give an intuitive explanation of the algorithm. We first note that the algorithm recovers the first $k$ bits of the initial state, namely $u_1, \ldots, u_k$. The remaining part of the initial state can be recovered in a similar way, if desired.

Now consider the case of one hypothesized value of $(u_1, \ldots, u_k)$. We want to check whether this value, denoted $(\hat{u}_1, \ldots, \hat{u}_k)$, is correct or not. This is done by first selecting a certain $(l-k)$-tuple $\mathbf{s}_i$, and then by finding all linear combinations of $t$ vectors in $\{\mathbf{x}_1, \mathbf{x}_2, \ldots \mathbf{x}_N\}$,

$$\hat{\mathbf{x}}(i) = \sum_{j=1}^{t} \mathbf{x}_{a_j}, \tag{9}$$

having the special form

$$\hat{\mathbf{x}}(i) = (\hat{x}_1, \ldots, \hat{x}_k, \mathbf{s}_i), \tag{10}$$

for arbitrary values of $\hat{x}_1, \ldots, \hat{x}_k$ (not all zero). The complexity of this precomputation step depends on $t$, and by using some simple birthday-paradox arguments, one can show that the computation can be done in $O(N^{\lceil t/2 \rceil})$ using $O(N^{\lfloor t/2 \rfloor})$ storage.

In: $\mathbf{z} = (z_1, \ldots, z_N)$, $[\mathbf{x}_1, \mathbf{x}_2, \ldots, \mathbf{x}_N]$, and constants $t$, $k$ and $n$.

1. (Precomputation) Select $n$ different $(l-k)$-tuples $\mathbf{s}_1, \mathbf{s}_2, \ldots, \mathbf{s}_n$. For each $\mathbf{s}_i$, find all linear combinations of the form $\hat{\mathbf{x}}(i) = \sum_{j=1}^{t} \mathbf{x}_{a_j}$ which are of the special form

$$\hat{\mathbf{x}}(i) = (\hat{x}_1, \ldots, \hat{x}_k, \mathbf{s}_i),$$

   for arbitrary values of $\hat{x}_1, \ldots, \hat{x}_k$.
   Store all $\hat{\mathbf{x}}(i)$ together with all $\hat{z}(i) = \sum_{j=1}^{t} z_{a_j}$. Let the set of all such pairs have cardinality $S_i$.

2. Run through all $2^k$ values of the constants $(u_1, \ldots, u_k) = (\hat{u}_1, \ldots, \hat{u}_k)$ as follows.

3. For each $s_i$, run through all $S_i$ stored pairs $\{(\hat{\mathbf{x}}(i), \hat{z}(i))\}$, calculate the number of times we have

$$\sum_{j=1}^{k} \hat{u}_j \hat{x}_j = \hat{z}(i),$$

   and denote this by $num$. Update

$$dist \leftarrow dist + (S_i - 2 \cdot num)^2.$$

4. If $dist$ is the highest received value so far, store $(\hat{u}_1, \ldots, \hat{u}_k)$. Set $dist \leftarrow 0$.

Out: Output $(\hat{u}_1, \ldots, \hat{u}_k)$ having the highest value of $dist$.

**Fig. 2.** A description of the basic algorithm

Now, the main observation is that the relation between $U(\hat{\mathbf{x}}(i))$ and $\hat{z}(i)$ can, from our previous arguments, be written in the form

$$U(\hat{\mathbf{x}}(i)) = \hat{z}(i) + e, \tag{11}$$

where $e$ represents the noise having a noise level of $P(e = 0) = 1/2 + 2^{t-1}\epsilon^t$. Now (11) is equivalently expressed as

$$\sum_{j=1}^{k} u_j \hat{x}_j + \sum_{j=k+1}^{l} u_j s_j = \hat{z}(i) + e, \tag{12}$$

and this can be rewritten as

$$\sum_{j=1}^{k} (u_j + \hat{u}_j)\hat{x}_j + \sum_{j=k+1}^{l} u_j s_j + e = \sum_{j=1}^{k} \hat{u}_j \hat{x}_j + \hat{z}(i). \tag{13}$$

Now recall that $W = \sum_{j=k+1}^{l} u_j s_j$ in (13) is a fixed binary random variable for all linear combinations of the special form that we required, i.e., we will have either $W = 0$ for all our $\hat{\mathbf{x}}(i)$'s, or $W = 1$.

Consider a correct hypothesized value and assume that we have all the $S_i$ equations. Then *num* simply counts the number of times the right hand side in (13) is zero. Since $\sum_{j=1}^{k}(u_j + \hat{u}_j)\hat{x}_j = 0$, the probability for the left hand side to be zero is then $P(W + e = 0)$. This probability is either $1/2 - 2^{t-1}\epsilon^t$ or $1/2 + 2^{t-1}\epsilon^t t$ for all equations, depending on whether $W = 0$ or $W = 1$. Thus *num* has a binomial distribution $Bin(S_i, p)$, with $p$ being one of the two probabilities above.

However, if the hypothesized valued was wrong, then $\sum_{j=1}^{k}(u_j + \hat{u}_j)\hat{x}_j \neq 0$, and hence, it will result in *num* being binomial distributed, $Bin(S_i, p)$, with $p = 1/2$.

In order to separate the two hypothesis we measure the difference between the number of times $\sum_{j=1}^{k} \hat{u}_j\hat{x}_j = \hat{z}(i)$ holds and the number of times it does not hold. Then a squared distance $((S_i - 2\cdot num)^2)$ is used. If we have enough points, i.e., we can create enough different $\hat{x}$ as linear combinations of at most $t$ $\mathbf{x}_i$'s, we will be able to separate the two hypotheses. However, the number of linear combinations for a particular $\mathbf{s}_i$ value is limited. Hence, we also run through a lot of different $\mathbf{s}_i$ values. Each gives a squared distance, and we sum them all up to become our overall distance $(dist)$. In Section 7 we show that a squared distance leads to a statistically optimal test, i.e., we pick the candidate having the highest probability.

# 5   A Sequential Reconstruction Algorithm

In this section we want to elaborate around the idea of using $i$-prefixes from Section 3 and modify the proposed algorithm into a sequential algorithm.

Instead of simply selecting the candidate $(\hat{u}_1, \ldots, \hat{u}_k)$ having the highest value of $dist$, we would now want to have a set of surviving candidates. These are then extended by incrementing, in our case, $k$ by one. This extension doubles the number of candidates, since each surviving candidate can be extended in the $(k + 1)$th position by either 0 or 1. But before the next extension, we run a screening procedure that removes a substantial part of the candidates.

This is a straightforward usage of the idea of $i$-prefixes. From our perspective, it does introduce some small practical problems. The major problem is that we now must store a large set of possible candidates. The performance of our algorithms is highly connected with the computational complexity. If a large memory must be used in our algorithm, some degradation in complexity is likely to appear in practice. This is the reason for presenting a slightly different approach. Essentially we use, instead of an $l$ round algorithm, a tree structure for all candidates that are still "alive". The advantage is that, essentially, the memory requirements are removed. Figure 3 shows how a version of such an algorithm may look like.

Note that the set $\Omega$ is only introduced to simplify the presentation. We do not need to store it. It is a lexicographically ordered set, and when we put new values in $\Omega$, we actually do not need to store anything.

In: $\mathbf{z} = (z_1, \ldots, z_N)$, $[\mathbf{x}_1, \mathbf{x}_2, \ldots, \mathbf{x}_N]$, and constants $t$, $\hat{k}$, $n$ and $threshold(k)$. Let $\Omega$ be a list of all $k$-tuples in lexicographical order.

1. (Precomputation) For each value of $k$, $\hat{k} \le k \le l$, set up a screening procedure as given in 2.

2. (Precomputation) Select $n$ different $(l-k)$-tuples $\mathbf{s}_1, \mathbf{s}_2, \ldots, \mathbf{s}_n$. For each $\mathbf{s}_i$, find all linear combinations of the form $\hat{\mathbf{x}}(i) = \sum_{j=1}^{t} \mathbf{x}_{a_j}$ which are of the special form

$$\hat{\mathbf{x}}(i) = (\hat{x}_1, \ldots, \hat{x}_k, \mathbf{s}_i),$$

for arbitrary values of $\hat{x}_1, \ldots, \hat{x}_k$. Store $(\hat{\mathbf{x}}(i), \hat{z}(i) = \sum_{j=1}^{t} z_{a_j})$. Assume that $S_i$ such pairs have been stored.

3. Take the first value in $\Omega$, denoted $(\hat{u}_1, \ldots, \hat{u}_k)$.

4. For each $\mathbf{s}_i$, run through all $S_i$ stored pairs for $\mathbf{s}_i$, calculate the number of times we have

$$\sum_{j=1}^{k} \hat{u}_j \hat{x}_j = \hat{z}(i),$$

and denote this by $num$. Update

$$dist \leftarrow dist + (S_i - 2 \cdot num)^2.$$

5. If $dist > threshold(k)$, put both $(\hat{u}_1, \ldots, \hat{u}_k, 0)$ and $(\hat{u}_1, \ldots, \hat{u}_k, 1)$ in $\Omega$. Set $dist \leftarrow 0$. If $|\Omega| > 1$ go to 3.

Out: Output all values in $\Omega$ that has reached length $l$.

**Fig. 3.** A description of the sequential algorithm.

*Example 1.* Assume that the sequential algorithm is applied with $\hat{k} = 5$. We examine first the value $(u_1, \ldots, u_5) = (0, 0, 0, 0, 0)$. Assume that the received $dist$ is higher than $threshold(5)$. We then extend this vector with the two possible values for $u_6$, giving $(0, 0, 0, 0, 0, 0)$ and $(0, 0, 0, 0, 0, 1)$. We continue to examine the first of these candidates. Assume that $dist < threshold(6)$. We continue with the second of these candidates. Assume that in this case $dist > threshold(6)$. We extend this vector and get $(0, 0, 0, 0, 0, 1, 0)$ and $(0, 0, 0, 0, 0, 1, 1)$ as two new vectors. We continue in this fashion. The tree structure of this procedure is presented in Figure 4.

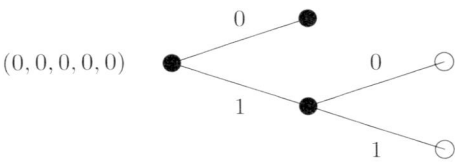

**Fig. 4.** The tree in Example 1.

Whether a candidate will survive the test at level $k$ or not is determined by a threshold value $(threshold(k))$. Increasing the threshold value will throw away more wrong candidates, but will also increase the probability of throwing away the correct candidate. A discussion on how to choose the threshold values is given in Section 7.

Comparing with the algorithm of the previous section, this algorithm will have a better performance (fewer tests on average) if we implement it in an efficient way. One important observation in this direction is the fact that all $\hat{\mathbf{x}}(i)$ vectors for a certain $k$ will appear again as valid $\hat{\mathbf{x}}(i)$ vectors for higher $k$ (assuming that we use the same $\mathbf{s}_i$ vectors). This means that we should not recalculate $num$ on $\hat{\mathbf{x}}(i)$ vectors that have already been used, but rather, we store the value of $num$ for all $\mathbf{s}_i$ values and incorporate this in the calculation for higher $k$ values.

## 6    Performance of the Proposed Algorithm

In this section we present some simulation results for the basic algorithm described in Section 4 and the sequential version in Section 5. Simulations are presented for $t = 2$ and $t = 3$. In general, increasing $t$ will increase the performance at the cost of an increased precomputation time and increased memory requirement in precomputation.

The first simulations are for the same feedback polynomial and the same length of the observed keystream as in [9,10,2]. Table 1 shows the maximum error probability $p = 1/2 - \epsilon$ for the basic algorithm when the received sequence is of length $N = 400000$. The parameter $k$ is varying in the range $13 - 16$ and $n$ is in the set $n \in \{1, 2, 4, 8, \dots, 512\}$. As a particular example, when $k = 16$, $n = 256$

**Table 1.** Maximum $p = 1/2 - \epsilon$ for the basic algorithm with $t = 2$, $k = 13, \dots, 16$, varying $n$, and $N = 400000$.

$$N = 400000$$

| $n$ | $k = 13$ | $k = 14$ | $k = 15$ | $k = 16$ |
|-----|----------|----------|----------|----------|
| 1   | 0.30     | 0.32     | 0.34     | 0.36     |
| 2   | 0.32     | 0.34     | 0.36     | 0.38     |
| 4   | 0.34     | 0.36     | 0.38     | 0.40     |
| 8   | 0.36     | 0.38     | 0.40     | 0.41     |
| 16  | 0.38     | 0.39     | 0.41     | 0.42     |
| 32  | 0.39     | 0.40     | 0.42     | 0.43     |
| 64  | 0.40     | 0.41     | 0.42     | 0.44     |
| 128 | 0.41     | 0.42     | 0.43     | 0.44     |
| 256 | 0.42     | 0.43     | 0.43     | 0.45     |
| 512 | 0.42     | 0.44     | 0.44     | 0.45     |

we reach $p = 0.45$ having 400000 known keystream symbols. The running time is less than 3 minutes, and the precomputation time negligible.

It is important to observe that for a fixed running time, the performance increases with increasing $n$ (up to a certain point). The table entries $\{k = 16, n = 1\}$, $\{k = 15, n = 4\}$, $\{k = 14, n = 16\}$, $\{k = 13, n = 64\}$ all have roughly the same computational complexity, but an increasing performance with $n$ can be observed.

More interesting is perhaps to show simulation results for longer LFSRs, as was done in [3]. We present results for the basic algorithm when $l = 60$ using a feedback polynomial of weight 13. In Table 2 we show the required

$l = 60, t = 2$

| $N$ | $k$ | $n$ | time | $p$ |
|---|---|---|---|---|
| $40 \cdot 10^6$ | 23 | 1 | 96 sec | 0.35 |
| $40 \cdot 10^6$ | 22 | 2 | 48 sec | 0.36 |
| $40 \cdot 10^6$ | 21 | 4 | 25 sec | 0.36 |
| $40 \cdot 10^6$ | 25 | 1 | 26 min | 0.40 |
| $40 \cdot 10^6$ | 24 | 2 | 13 min | 0.40 |
| $40 \cdot 10^6$ | 23 | 4 | 6.5 min | 0.40 |
| $40 \cdot 10^6$ | 22 | 8 | 3.3 min | 0.41 |
| $40 \cdot 10^6$ | 25 | 4 | 106 min | 0.43 |

$l = 60, t = 3$

| $N$ | $k$ | $n$ | time | $p$ |
|---|---|---|---|---|
| $1.5 \cdot 10^5$ | 24 | 1 | 4.5 min | 0.3 |
| $1.5 \cdot 10^5$ | 23 | 2 | 2.3 min | 0.3 |
| $1.5 \cdot 10^5$ | 22 | 4 | 69 sec | 0.3 |
| $1.5 \cdot 10^5$ | 25 | 1 | 18 min | 0.32 |
| $1.5 \cdot 10^5$ | 24 | 2 | 9.2 min | 0.32 |
| $1.5 \cdot 10^5$ | 23 | 4 | 4.6 min | 0.32 |

**Table 2.** Performance of the basic algorithm with $l = 60$ when $t = 2$ and $t = 3$, respectively.

computational complexity and the achieved correlation probability for different algorithm parameters. The implementations were written in C and the running times were measured on a Sun Ultra-80 running under Solaris.

We can compare with other suggested methods. Actually, in the special case of $n = 1$, our proposed algorithm will coincide with the method in [3]. This enables us to see the improvement in Table 2, by observing the decrease of decoding time when $n$ increases (for a fixed $p$). Furthermore, [3] is the only previous work reporting simulation results for $l \geq 60$.

An important advantage for the proposed methods is the storage complexity. The attacks based on convolutional and turbo codes [9,10] uses a trellis with $2^B$ states. Hence, the size of $B$ is limited to $20 - 30$ in practise, due to the fact that it must be kept in memory during decoding. On the other hand, the memory requirements for the algorithms presented in this paper remain constant when $k$ increases. Also, the proposed algorithms are trivially parallelizable, and hence the only limiting factor is the total computational complexity.

Finally, simulation results for the sequential algorithm should be considered. Some initial simulations for the case $N = 400000$, $p = 0.40$, $t = 2$, $n = 64$ indicated a speedup factor of approximately 5. An extensive set of simulations for the sequential algorithm is under progress.

## 7   Theoretical Analysis of the Algorithms

In this section we sketch some results for a theoretical analysis of the proposed algorithms. A complete analysis will appear in the full paper. Here we prove, among other things, that using the squared distance is statistically optimal.

First, we derive an expression for the expected number of linear combinations of the form (9) and (10), i.e., the expected value of the parameter $S_i$ in the algorithm.

**Lemma 1.** *Let $E[S]$ be the expected number of linear combinations of the form (9) and (10) that can be created from $t$ out of $N$ random vectors $\mathbf{x}_1, \ldots, \mathbf{x}_N$. Then $E[S]$ is given as*

$$E[S] = \frac{\binom{N}{t}}{2^{l-k}}.$$

**Proof:** The number of ways we can create a linear combinations of of the form (9) is $\binom{N}{t}$ The probability of getting a particular value of $\mathbf{s}$ is $1/2^{l-k}$. Thus, we have in average $\binom{N}{t}/2^{l-k}$ linear combinations ending with a particular value $\mathbf{s}$. ∎

Next we show that using *dist* as defined in Section 4 gives optimal performance. We start by considering the case when we have one fixed value of $\mathbf{s}$. Then we generalize to the case with several different $\mathbf{s}$ vectors.

Assume that for the given $\mathbf{s}$ we have created $S$ noisy observations of the polynomial $U(\mathbf{x})$. The expected value of $S$ is then given by Lemma 1. Assume further that we are considering a particular candidate $(\hat{u}_1, \ldots, \hat{u}_k)$, and that we have found *num* observations such that $\sum_{j=1}^{k} \hat{u}_j \hat{x}_j = \hat{z}(i)$. Consider two hypothesis $H_0$, and $H_1$. Let $H_1$ be the hypothesis that the candidate is correct, and $H_0$ that the candidate is wrong.

Introduce the random variable $W = \sum_{j=k+1}^{l} u_j s_j$. Define $p_0$ as $p_0 = P(e = 0) = 1/2 + 2^{t-1}\epsilon^t$. Furthermore, $P(W = 0) = 1/2$. We showed in Section 4 the following distribution for *num*:

$$num|H_0 \in Bin(S, 1/2),$$
$$num|H_1, W = 0 \in Bin(S, p_0),$$
$$num|H_1, W = 1 \in Bin(S, (1 - p_0)).$$

Next, we approximate the binomial distribution for *num* with a normal distribution. As long as, $Spq \gg 10$ the approximation will be good. Furthermore, we define the random variable $Y$ as $Y = |2 \cdot num - S|$. If $P((2 \cdot num - S) < 0|H_1, W = 0) = P((2 \cdot num - S) > 0|H_1, W = 1)$ is small then we get the following distribution of $Y$:

$$f_{Y|H_0}(y) = \frac{2}{\sqrt{\pi S}} e^{-y^2/S}$$

$$f_{Y|H_1}(y) = \frac{1}{\sqrt{4\pi S p_0 (1 - p_0)}} e^{-\frac{(y - S p_0)^2}{4 S p_0 (1 - p_0)}}$$

The estimate of $(u_1, \ldots, u_k)$ is taken as the $(\hat{u}_1, \ldots, \hat{u}_k)$ for which $P(H_1|Y)$ is maximal. However, it is not possible to calculate $P(H_1|Y)$ directly. Instead, we can equivalently choose the estimate as the $(\hat{u}_1, \ldots, \hat{u}_k)$ for which the likelihood ratio

$$\Lambda = \frac{P(H_1|Y)}{1 - P(H_1|Y)} = \frac{P(H_1|Y)}{P(H_0|Y)} = \frac{P(Y|H_1)P(H_1)}{P(Y|H_0)P(H_0)},$$

is maximal.

In our case it is more convenient to use the loglikelihood ratio $\lambda = \ln(\Lambda)$. Thus we can formulate the problem of finding the most likely candidate as:

$$\arg\max_{(\hat{u}_1, \ldots, \hat{u}_k)} \left[ \ln P(Y|H_1) + \ln P(H_1) - \ln P(Y|H_0) - \ln P(H_0) \right].$$

It now follows that maximizing $\lambda$ is equivalent to taking the candidate for which $y^2$ is maximum.

This derivation holds when we have one value of $S$. Now we assume that we have $n$ different $S_i$, $(S_1, S_2, \ldots, S_n)$ and corresponding $Y = (Y_1, Y_2, \ldots, Y_n)$. By observing that

$$P(Y|H_0) = P(Y_1|H_0)P(Y_2|H_0) \cdots P(Y_n|H_0)$$

we see that optimality is reached for $dist = dist_1 + dist_2 + \ldots + dist_n$, where $dist_i = y_i^2$. In conclusion, we have showed that the chosen distance is statistically optimal.

To analyze the performance of the algorithm when we use the quadratic distance measure we use the following approach. Assume that we have the correct candidate $(\hat{u}_1, \ldots, \hat{u}_k)$. Then we have

$$|2 \cdot num_i - S_i| \in N(S_i(2p_0 - 1), 2S_i p_0(1 - p_0)).$$

If we instead assume that the candidate $(\hat{u}_1, \ldots, \hat{u}_k)$ is wrong we get

$$(2 \cdot num_i - S_i) \in N(0, S/2).$$

The value of $dist$ is calculated as

$$dist = \sum_{i=1}^{n} (2 \cdot num_i - S_i)^2 = \sum_{i=1}^{n} |2 \cdot num_i - S_i|^2.$$

One sees that $dist$ is calculated by squaring and adding $n$ normal random variables. Hence, $dist$ is a noncentral chi-square distributed random variable with $n$ degrees of freedom.

By using the central limit theorem, we get that for large values of $n$ we will be successful when $E(dist|H_1) > E(dist|H_0)$. Since this inequality always holds, the conclusion is that if $n \to \infty$ we can have $\epsilon \to 0$.

Finally we consider the sequential algorithm of Section 5. When analyzing this algorithm we are now interested in two properties. The first is the probability that we accept a candidate as correct when it actually is wrong, denoted by $P_F$, (false alarm). The second is the probability that we do not accept a correct candidate. Denote this by $P_M$, (miss).

Since we know the distribution of $dist$ under the hypothesis $H_0$ and $H_1$ we can calculate $P_M$ and $P_F$ as follows. Consider a fixed threshold $T$. The probability of miss $P_M$ is then given as

$$P_M = P(dist < T | H_1),$$

and in the same way we get

$$P_F = P(dist > T | H_0).$$

## 8   Conclusions

In this work we have shown how learning theory can be used as a basis for correlation attacks on stream ciphers. Techniques for reconstructing polynomials have been modified and combined with some general techniques from correlation attacks. The performance has been demonstrated through a sketch of a theoretical analysis as well as through simulations. The simulations show a very good performance.

The problem that arises in a standard correlation attack is equivalent to the problem of learning parity with noise, a well known problem in computational learning theory, commonly believed to be a hard problem. This might indicate that it is hard to find further significant improvements on the problem. One interesting idea would be to examine whether recent results on polynomial reconstruction as a decoding tool for certain error correcting codes [20] can be used. Some results in this direction can be found in [8].

### Acknowledgement

The authors are supported by the Foundation for Strategic Research - PCC under Grant 9706-09.

## References

1. A. Blum, M. Furst, M. Kearns, R. Lipton, "Cryptographic primitives based on hard learning problems", *Advances in Cryptology–CRYPTO'93*, Lecture Notes in Computer Science, vol. 773, Springer-Verlag, 1993, pp. 278–291.
2. A. Canteaut, M. Trabbia, "Improved fast correlation attacks using parity-check equations of weight 4 and 5", *Advances in Cryptology–EUROCRYPT'2000*, Lecture Notes in Computer Science, vol. 1807, Springer-Verlag, 2000, pp. 573–588.
3. V. Chepyzhov, T. Johansson, and B. Smeets, "A simple algorithm for fast correlation attacks on stream ciphers", *Fast Software Encryption, FSE'2000*, to appear in Lecture Notes in Computer Science, Springer-Verlag, 2000.
4. V. Chepyzhov, and B. Smeets, "On a fast correlation attack on certain stream ciphers", In *Advances in Cryptology–EUROCRYPT'91*, Lecture Notes in Computer Science, vol. 547, Springer-Verlag, 1991, pp. 176–185.

5. A. Clark, J. Golic, E. Dawson, "A comparison of fast correlation attacks", *Fast Software Encryption, FSE'96*, Lecture Notes in Computer Science, Springer-Verlag, vol. 1039, 1996, pp. 145–158.
6. O. Goldreich and L.A. Levin, "A hard-core predicate for all one-way functions", *Proceedings of the Twenty-First Annual ACM Symposium on Theory of Computing*, Seattle, Washington, 15-17 May 1989, pp. 25–32.
7. O. Goldreich, R. Rubinfeld, M. Sudan, "Learning polynomials with queries: The highly noisy case", *36th Annual Symposium on Foundation of Computer Science*, Milwaukee, Wisconsin, 23-25 October 1995, pp. 294–303.
8. T. Jakobsen, "Higher-Order Cryptanalysis of Block ciphers", Ph.D Thesis, Technical University of Denmark, 1999.
9. T. Johansson, F. Jönsson, "Improved fast correlation attacks on stream ciphers via convolutional codes", *Advances in Cryptology–EUROCRYPT'99*, Lecture Notes in Computer Science, vol. 1592, Springer-Verlag, 1999, pp. 347–362.
10. T. Johansson, F. Jönsson, "Fast correlation attacks based on turbo code techniques", *Advances in Cryptology–CRYPTO'99*, Lecture Notes in Computer Science, vol. 1666, Springer-Verlag, 1999, pp. 181–197.
11. M. Kearns, "Efficient noise-tolerant learning from statistical queries", *Proceedings of the Twenty-Fifth Annual ACM Symposium on Theory of Computing*, San Diego, California, 16-18 May 1993, pp. 392–401.
12. W. Meier, and O. Staffelbach, "Fast correlation attacks on stream ciphers", *Advances in Cryptology–EUROCRYPT'88*, Lecture Notes in Computer Science, vol. 330, Springer-Verlag, 1988, pp. 301–314.
13. W. Meier, and O. Staffelbach, "Fast correlation attacks on certain stream ciphers", *Journal of Cryptology*, vol. 1, 1989, pp. 159–176.
14. A. Menezes, P. van Oorschot, S. Vanstone, *Handbook of Applied Cryptography*, CRC Press, 1997.
15. M. Mihaljevic, M. Fossorier, and H. Imai, "A low-complexity and high-performance algorithm for the fast correlation attack", *Fast Software Encryption, FSE'2000*, to appear in Lecture Notes in Computer Science, Springer-Verlag, 2000.
16. M. Mihaljevic, and J. Golic, "A fast iterative algorithm for a shift register initial state reconstruction given the noisy output sequence", *Advances in Cryptology–AUSCRYPT'90*, Lecture Notes in Computer Science, vol. 453, Springer-Verlag, 1990, pp. 165-175.
17. W. Penzhorn, "Correlation attacks on stream ciphers: Computing low weight parity checks based on error correcting codes", *Fast Software Encryption, FSE'96*, Lecture Notes in Computer Science, vol. 1039, Springer-Verlag, 1996, pp. 159–172.
18. T. Siegenthaler, "Correlation-immunity of nonlinear combining functions for cryptographic applications", *IEEE Trans. on Information Theory*, vol. IT–30, 1984, pp. 776–780.
19. T. Siegenthaler, "Decrypting a class of stream ciphers using ciphertext only", *IEEE Trans. on Computers*, vol. C–34, 1985, pp. 81–85.
20. M. Sudan, "Decoding of Reed Solomon codes beyond the error-correction bound", *Journal of Complexity*, vol. 13(1), March 1997, pp. 180–193.

# Sequential Traitor Tracing

Reihaneh Safavi-Naini and Yejing Wang

School of IT and CS, University of Wollongong,
Wollongong 2522, Australia
[rei/yw17]@uow.edu.au

**Abstract.** Traceability schemes allow detection of at least one traitor when a group of colluders attempt to construct a pirate decoder and gain illegal access to digital content. Fiat and Tassa proposed dynamic traitor tracing schemes that can detect *all* traitors if they attempt to re-broadcast the content after it is decrypted. In their scheme the content is broken into segments and marked so that a re-broadcasted segment can be linked to a particular subgroup of users. Mark allocation for a segment is determined when the re-broadcast from the previous segment is observed. They showed that by careful design of the mark allocation scheme it is possible to detect all traitors.

We consider the same scenario as Fiat and Tassa and propose a new type of traceability scheme, called sequential traitor tracing, that can efficiently detect *all* traitors and does not require any real-time computation. That is, the marking allocation is pre-determined and is independent of the re-broadcasted segment. This is very attractive as it allows segments to be shortened and hence the overall convergence time reduced. We analyse the scheme and give two general constructions one based on a special type of function family, and the other on error correcting codes. We obtain the convergence time of these schemes and show that the scheme based on error correcting codes has a convergence time which is the same as the best known result for dynamic schemes.

## 1 Introduction

In recent years a number of closely related models and schemes with the aim of securing electronic distribution of digital content have been proposed. Services that rely on this kind of distribution, such as pay-TV where the provider needs assurance that only paid customers will receive the service, can only become viable if security of the distribution can be guaranteed.

*Broadcast encryption* systems [6] allow targeting of an encrypted message to a privileged group of receivers. Each receiver has a decoder with his unique key information that allows him to decrypt encrypted messages when he belongs to the target group. The system ensures that the collusion of up to $t$ receivers not belonging to the target group cannot learn anything about the message.

M. Bellare (Ed.): CRYPTO 2000, LNCS 1880, pp. 316–332, 2000.
© Springer-Verlag Berlin Heidelberg 2000

Now assume a group of colluders construct a *pirate decoder* that can decrypt the broadcasted message. A *traitor tracing scheme* [4] allows at least one of the colluders to be identified. Broadcast encryption and traceability systems can be combined [8,10] to produce systems that provide protection against both kinds of attacks.

In [7] a different scenario is considered. This time the traitors' aim is to bypass the security mechanism of the system not by constructing a pirate decoder but by *re-broadcasting the content* after it is decoded. That is the colluders use their decoders to decrypt the content and then once it is in plain-text form, re-broadcast the plain-text to another group of users. In this case the only way to trace traitors is to use different *versions* of the content for different users. This allows a re-broadcasted message to be linked to the subgroup of users who were given that version. The two main characteristics of the new setting are (i) the plaintext content is marked and, (ii) there is a *feedback* from the channel which allows the traitor to become localized. An important feature of this system is that it allows *all* traitors, even without knowing their number beforehand, be traced.

A trivial solution to tracing traitors in the above scenario is to give individual copies to each users. This means that the same content must be sent once per each user and so bandwidth usage is extremely poor. Fiat and Tassa's (FT for short) work showed that by introducing a new dimension to the problem, that is *time*, it is possible to use a small number of versions and hence resulting in a more efficient usage of the communication channel, while still detecting all the traitors. In FT system the content is divided into *consecutive segments* and each segment is *watermarked* to produce $\ell$ versions. For each segment users are partitioned into $\ell$ subgroups, and members of a subgroup receive the same marked version. The system is *dynamic* because the version received by a user, and hence partitioning of the user set, depends on the feedback from previous segment. The number of traitors is not known beforehand and the system adjusts itself such that as long as the total number of traitors is less than $p$, all of them will be traced. The algorithms proposed in [7] allow trade-off between the two efficiency parameters of these systems, that is, bandwidth usage and convergence time.

*Drawbacks of FT model:* There are two major drawbacks in FT model. Firstly, mark allocation in a time slot depends on the *real-time feedback* signal from the previous time slot. This makes the system vulnerable to *delayed rebroadcast attack*. That is, when the attackers do not rebroadcast immediately, but decide to record the content and rebroadcast it at a later time. In this case FT model becomes totally ineffective as the mark allocation in time slots will remain constant.

The second drawback is the high real-time computation required for allocation of marks which means the length of a time-slot cannot be very short. We note that the number of time slots for the convergence of the best proposed algorithm is at least of the order of $\log N$ ($N$ is the number of users) and hence grows with the number of users. For large group sizes, it is desirable to have shorter time slots to obtain reasonable convergence time. However the computation grows with the size of the group which means the length of the time slot cannot be shortened. The conflicting requirements of shorter time slot and higher computation results in systems that not practical for large groups.

## Our Contribution

We consider the same problem as Fiat and Tassa: detecting *all* traitors when traitors re-broadcast the content. However we propose a different solution which does not use the feedback signal for mark allocation and so (i) will not be vulnerable to delayed rebroadcast attack, and (ii) does not require real-time computation for mark allocation and so allow very short time slots.

Similar to Fiat and Tassa, we mark consecutive segments of the content and detect the feedback from the channel. However unlike their scheme the allocation of marks to users in each segment is pre-determined and follows a fixed table. We call this system a *sequential traitor tracing* scheme to emphasise the fact that more than one step is required, and at the same time differentiate it from dynamic schemes. In a dynamic scheme allocation of the marks and delivery of marked versions to users is *after* receiving the feedback signal. That is only after observing the re-broadcasted segment from previous round, re-partitioning of the user set, and calculation and delivery the required keys followed by the marked segment for the next segment can be performed. In our approach because the mark allocation for each segment does not require the feedback of the previous segment, the system will be much more efficient. This means that because no real-time computation is required the length of a segment can be chosen very short. The cost paid for the added security and real-time efficiency could be higher bandwidth or convergence rate (number of steps). However because of shorter time slots the total convergence time could remain comparable or even reduced.

We give a formal definition of the new scheme and derive a bound that relates the number of versions (communication efficiency) and convergence time of the algorithm. We give two general constructions that can be used with any robust watermarking system. In particular we give a construction that allows tracing of all up to $p$ traitors in a group of $f(q)$ users in at most $(p(p+1))^t$ steps and requires only $q$ versions, where $f(q)$ is a polynomial of $q$ with degree $2^t$, and a second construction using error correcting codes that allows all up to $p$ traitors in a group of $N$ users be traced. This construction requires $2p$ versions and has the convergence time equal to $8p \log N + p$ steps which is the same as the best dynamic scheme. We will show that both of these constructions are general and can be used with $p$-frameproof codes and $p$-traceability schemes to construct systems for large groups.

The paper is organised as follows. In section 2 we give the required definitions and review the known results. In section 3 we introduce our model and derive bounds relating efficiency parameters of the system. The two constructions follow in section 4. In section 5 we evaluate our results and their extensions.

## 2   Preliminaries

In this section we briefly review relevant definitions and results.

### Broadcast Encryption

In a *broadcast encryption system* a centre generates a set of *base keys*, and assigns a subset of these keys to every user as his *personal key* such that at a later time it can broadcast an encrypted message that is only accessible to a privileged subgroup and users who are not in the subgroup cannot decrypt the message. The privileged subgroup is not fixed and may be one of a set of possible authorised subsets. *Resilience* of a broadcast encryption system is measured by a parameter $k$ which is the the size of largest colluding group, disjoint from the privileged set, who cannot learn the message.

### Marking Digital Content

*Marking a digital object* has been initially studied in the context of frame-proof codes [2]. Consider a set of $n$ users $\{1, 2, \cdots, n\}$. Let $\Sigma$ be an alphabet. An $(\ell, n)$-code is a set $\Gamma = \{c_1, c_2, \cdots, c_n\} \subseteq \Sigma^\ell$. Let $T$ be a coalition of users, and assume $i \in \{1, 2, \cdots, \ell\}$ is a position. We say that position $i$ is *undetectable* for $T$ if the words assigned to $T$ match in their $i^{th}$ position. Denote by $R$ the set of undetectable positions for $T$. Define the *feasible set* of $T$ as consisting of all $w \in \Sigma^\ell$ such that $w$ and the words assigned to $T$ are matched in $R$. Denote by $F(T)$ the feasible set of $T$.

A code $\Gamma$ is called *p-frameproof* if every set $W \subset \Gamma$, of size at most $p$, satisfies $F(W) \cap \Gamma = W$. Frameproof codes are useful if a software, or a binary file needs to be marked. Detection of a mark requires the mark embedded in the content to be their exact stored values. An extended definition of frameproof code [3], [9] allows the marks to be also deleted however this is again for a mark in the codeword.

Protection of video and audio signal content is through *watermarking systems* that cannot be strictly modelled by a frameproof code. We define a *watermarking code* for audio and video content as a collection of distinct codewords (or marks), $C = \{c_1, c_2, \cdots, c_r\}$, and two algorithms $I$ for *watermark insertion*, and $D$ for *watermark detection* [11]. The insertion algorithm takes a codeword $c_i$ and a content $m$ and produces a *marked version* $m_i$. The detection algorithm takes a content $m'$ and a codeword $c_i$ and produces a *true/false* value depending on success or failure of the detection. In watermark codes a codeword $c_i$ is a distinct *whole* mark and collusion of users may either convert the whole mark into a different mark, or completely remove it. In practice the former type of attack has negligible success probability and it is enough to consider the latter one. A watermarking code is *robust* if no combination of marked objects $\mu = \{m_{i_1}, \cdots, m_{i_\ell}\}$ can produce another marked object $m_i \notin \mu$, or delete the mark. Robust watermarking codes model robust watermarking schemes such as Cox et al [5] and their properties match properties of watermarking systems in practice. They are more general than frameproof codes because there is no restriction on an attacker's operation (feasible set in frameproof codes). This is a realistic model as in practice watermarks are inserted in many different ways and can be subjected to a wide range of attacks.

## Traitor Tracing

Similar to a broadcast encryption system the content provider generates a base set of keys, and assigns subsets of it to each user. The subset of keys received by a user $i$ forms his *personal key* or his *decoder key*, and is denoted by $U_i$. By holding $U_i$ user $i$ will be able to view the content. A colluding set of users can construct a *pirate decoder* which contains a subset of their keys. When a pirate decoder $F$ is found, $|F \cap U_i|$ for all $i$ is calculated and if $|F \cap U_i| \geq |F \cap U_j|$ for all $j$, then user $i$ is called an *exposed user*. Following Definition 1.2 in [9], a scheme is called a *p-traceability scheme* if whenever a pirate decoder $F$ is produced by $T$ and $|T| \leq p$, the exposed user is a member of the coalition $T$.

Suppose there are $\ell$ base keys $\{k_1, k_2, \cdots, k_\ell\}$ and $n$ users. Then a $p$-traceability scheme can be defined by an 0-1 matrix of size $n \times \ell$, such that its $(i, j)^{th}$ element is 1 if and only if user $i$ has key $k_j$ in his decoder.

In a *static traceability scheme* keys are allocated once and remain unchanged through the operation of the system.

## FT Scheme

In FT scheme *content* consists of a number of *segments*, for example one minute of a video or a movie. For each segment a number of *variants* using a *robust watermarking code*, such as *spread spectrum technique* of Cox et al [5], is constructed. A fundamental assumption of the system is that because of the robustness of the watermarking system the re-broadcasted version is one of the versions owned by the members of the colluder group.

For each segment the user set is partitioned into $r$ subsets, each subset receiving the same version. Each user has some key information that allows the content provider to securely give him a session key for his version, or use a broadcast encryption system to securely deliver his version. They proved that for tracing $p$ traitors in any traceability system at least $p + 1$ versions must be used, and gave algorithms that used $p + 1$ and $2p + 1$ versions and required $O(3^p p \log n)$ and $O(p \log n)$ steps to converge, respectively. Their algorithms were improved by Berkman et al [1] who showed an algorithm with $O(p \log n + p^2/c)$ step for convergence and using $p + c + 1$ versions, and a second one with $O(p \log n)$ and $pc + 1$ versions. Again the main emphasis of their work was to find schemes that allow best convergence when close to minimum possible number of versions is used.

## Sequential Traitor Tracing

One of the main drawbacks of FT model is the high real-time computation required for allocation of marks for each segment. Because this computation depends on the feedback from the previous segment it must be performed in real-time and there is no possibility for precomputation which implies that the length of a step must be chosen long enough for the required computation. Because the overall *convergence time* is a product of the number of steps and the *length of the step*, and because more complex algorithms with small number of steps require more computation and so longer segment, it is important to optimise the number of steps versus the required real-time computation.

We propose a sequential tracing scheme in which mark allocation is predetermined and so real-time processing in each segment is not required. Although the mark allocation changes in each step but because it is according to a known table, all the required computation for most of the allocation can be performed as pre-computation. The feedback signal is *only* used for detection of traitors. Although the system has a more limited use of the feedback compared to the full dynamic model of FT and so can be expected to require more steps for convergence, but because of much smaller computation in real-time a much shorter length for segments is possible and so the overall convergence can be expected to be lower.

Before presenting our model, we note that the following attack is outside both FT and our model.

– *Framing a user by re-broadcasting his version:* If a broadcast encryption is used for secure delivery of segments, then a colluder subgroup may construct a pirate decoder and obtain the version $v_p$ of an innocent user. If such a version is re-broadcasted all detection algorithms fail. This would be a feasible attack if the broadcast encryption system used for sending a version to the target subgroup does not provide traceability. Same effect can be obtained if colluders can break the underlying watermarking system and construct the version of another user. We noted that this in general is a very unlikely event.

This means that if a broadcast encryption is used it must be able to trace traitors.

## 3    The Model

Let $\mathcal{U} = \{1, 2, \cdots, N\}$ denote the set of users. A user $i$ has some secret key information, $U_i$, that allows the content provider to identify him and send him a particular version. $U_i$ could be a set of keys in a broadcast encryption scheme or a secret used to encrypt the session key of the user. There is a *mark allocation table* $M$ with $N$ rows and $d$ columns where $M(i, j)$ is the mark allocated to user $i$ in segment $j$. In each segment the content provider sends the $j^{th}$ segment to users according to column $j$ of $M$ and observes the feedback. Traitors can be detected by examining the sequence of feedback signals and after $d$ feedbacks it is possible to trace all the traitors: that is the tracing algorithm *converges*. When a traitor is found, he is *disconnected*. This is by excluding the user from the broadcast encryption system in all future segments. That is, if $i$ is detected as a traitor in segment $j$, then from segment $j + 1$, his reception of segment $M(i, k), k \geq j + 1$, will be blocked.

Assume there is a probability distribution on $U_1 \times U_2 \times \cdots \times U_N$. For $X = \{i_1, i_2, \cdots, i_j\} \in 2^{\mathcal{U}}$ denote by $U_X = U_{i_1} \times U_{i_2} \times \cdots \times U_{i_j}$, $i_1 < i_2 < \cdots < i_j$, the set of secret information given to $X$.

There is a watermarking code, $C = \{c_1, \cdots, c_r\}$, used to mark *segments* of the *protected content*. A segment with a valid mark is called a *variant*. There is a *feedback sequence* $F = ()$ which is initialised to the empty sequence. Let $T \subset \mathcal{U}$ be a set of *traitors* and $\mathcal{P}^{(j)}$ denote the set of all privileged users in the $j^{th}$ segment. The set $\mathcal{P}^{(j)}$ is partitioned into $\mathcal{P}^{(j)} = P_1^{(j)} \cup P_2^{(j)} \cup \cdots \cup P_r^{(j)}$, and each subset $P_i^{(j)}$ is allocated a version marked by $c_i$. Let $\mathcal{V}^{(j)}$ denote the set of all possible versions in segment $j$. In a segment $j$ the content provider uses a vector of $r$ versions, $\mathcal{V}^{(j)} = (V_1^{(j)}, V_2^{(j)}, \cdots, V_r^{(j)}) \in \mathcal{V}^{(j)r}$ where $V_i^{(j)} \neq V_l^{(j)}$, $i, l \in \{1, \cdots, r\}, i \neq l$. There is a probability distribution on $\mathcal{V}^{(j)}$. In each segment $j$ there is a feedback signal $f_j = c_i$ for some $i$ such that $P_i^{(j)} \cap T \neq \emptyset$, which is appended to $F_{j-1}$ to construct $F_j = (f_1, \cdots, f_j)$. This sequence is used to trace traitors. A feedback sequence $F_d$ is *p-consistent* if it can be generated by a colluder set of size at most $p$.

**Definition 1.** *A sequential $(p, d)$-traceability scheme is a family of partitions* $\mathcal{P}^{(j)} = P_1^{(j)} \cup P_2^{(j)} \cup \cdots \cup P_r^{(j)}$, $j = 1, 2, \cdots, d$, *with the following properties*

1. *In each segment $j$, each user receives a version. Formally, for each $U \in P_i^{(j)}$,* $H(\mathcal{V}_i^{(j)} | U) = 0$.
2. *In each segment $j$, a group of users which is disjoint from $P_i^{(j)}$ cannot have any information on versions of members of $P_i^{(j)}$ even if they use all previous feedbacks. Formally, for each $X \in 2^{\mathcal{U}}$, with $X \cap P_i^{(j)} = \emptyset$, $H(\mathcal{V}_i^{(j)} | U_X, F_{j-1})$ $= H(\mathcal{V}_i^{(j)})$.*
3. *After $d$ rounds all up to $p$ traitors can be detected. Formally, any $p$-consistent feedback sequence $F_d$ determines a unique colluder set of at most $p$, that is $H(\mathcal{U}_T | F_d) = 0$.*

The following proposition shows that in a segment $j$, a user does not have any information about the version assigned to another user belonging to a group different from his.

**Proposition 1.** *Let $j, b$ be integers, $b \leq r$, $X_i \subseteq P_i^{(j)}$, $1 \leq i \leq b$ such that $X_i \cap X_k = \emptyset$ for every pair $i \neq k, 1 \leq i, k \leq b$. Then*

$$H(\mathcal{V}_i^{(j)} | \mathcal{V}_1^{(j)}, \cdots, \mathcal{V}_{i-1}^{(j)}, \mathcal{V}_{i+1}^{(j)}, \cdots, \mathcal{V}_b^{(j)}) = H(\mathcal{V}_i^{(j)}).$$

The following theorem gives a lower bound on the number of rounds required to detect all traitors.

**Theorem 1.** *Suppose in a $(p, d)$-traceability scheme there are $N$ users, at most $p$ traitors and $r$ versions. Then*

$$d \geq p \log_r N.$$

# 4   Constructions

In this section we give two constructions for sequential traceability schemes from watermarking codes, one using a special class of functions and the second one using error correcting codes.

## 4.1   A Construction Using a Function Family

This construction uses a robust watermarking code with $n$ codewords and results in a sequential scheme that identifies at least one of the traitors (at most $p$ traitors) in $p^2 + 1$ steps, and *all* the traitors in at most $p^2 + p$ steps. The scheme converges in $p(p + 1)$ steps and so convergence time is independent of the size of the group. This is at the expense of higher number of versions and so less communication efficiency. We will show (section 5) that the scheme can be repeatedly used to increase the number of users with the same number of versions (communication efficiency) while increasing the number of rounds.

Suppose we have a robust watermarking code with $n$ codewords $c_1, c_2, \cdots, c_n$, and let $M$ denote the $n \times 1$ matrix with $c_i$ as its $i^{th}$ row.

Consider a collection of mappings, $\Phi = \{\phi_{ij} : 1 \leq i \leq b, 1 \leq j \leq m\}$,

$$\phi_{ij} : \{1, 2, \cdots, n\} \to \{1, 2, \cdots, n\},$$

that satisfy the following two properties:

(P1) For each $j$ and each pair of the first index $(i_1, i_2)$ with $i_1 \neq i_2$, we have $\phi_{i_1 j}(x) \neq \phi_{i_2 j}(x)$ for all $x \in \{1, 2, \cdots, n\}$.
(P2) For each pair of the first index $(i_1, i_2)$ and each pair of the second index $(j_1, j_2)$ with $j_1 \neq j_2$, we have $\phi_{i_1 j_2}(x) \neq \phi_{i_2 j_2}(y)$ provided that $\phi_{i_1 j_1}(x) = \phi_{i_2 j_1}(y)$.

Given $\Phi$ and a watermarking code with $n$ codewords define a matrix $\widetilde{M}$ as follows.

$$\widetilde{M} = \begin{pmatrix} M & M_1 & \phi_{11}(M) & \phi_{12}(M) & \cdots & \phi_{1m}(M) \\ M & M_2 & \phi_{21}(M) & \phi_{22}(M) & \cdots & \phi_{2m}(M) \\ \vdots & \vdots & \vdots & \vdots & \vdots & \vdots \\ M & M_b & \phi_{b1}(M) & \phi_{b2}(M) & \cdots & \phi_{bm}(M) \end{pmatrix}, \tag{1}$$

where

$$M_i = \begin{pmatrix} c_i \\ c_i \\ \vdots \\ c_i \end{pmatrix} \quad \text{and} \quad \phi_{ij}(M) = \begin{pmatrix} c_{\phi_{ij}(1)} \\ c_{\phi_{ij}(2)} \\ \vdots \\ c_{\phi_{ij}(n)} \end{pmatrix}$$

are $n \times 1$ matrices.

$\widetilde{M}$ has $b$ *block rows*, each block row contains $n$ rows, each row assigned to a user in the sequential scheme. Denote by $(r, k)$ the user holding the $k^{th}$ row of the $r^{th}$ block row.

$\widetilde{M}$ has $m + 2$ *block columns* where the $j^{th}$ block column contains the marks that will be allocated to users in segment $j$.

The scheme works as follows. In each segment (except for the second one, if $b < n$) there are $n$ versions that are marked with codewords of the watermarking code. For each segment the set of users $\mathcal{U}$ is divided into $b$ equal size subgroups and members of each subgroup (of size $n$) will receive a version marked by the same codeword of the watermarking code. In segment $j$, the centre observes the feedback and detects a codeword $f_j \in \{c_1, \cdots, c_n\}$. It updates the feedback sequence and obtains $F_j$. $F_j$ is compared with the rows of $\widetilde{M}$. If a row with $p+1$ 'match' is found, a traitor is detected.

Suppose the center has a feedback sequence $F_d = (f_1, f_2, \cdots, f_d)$, where $f_j$ corresponds to the $j^{th}$ segment. We say $F_d$ *matches* a row $R = (X_1, X_2, \cdots, X_{m+2})$ of $\widetilde{M}$ in $t$ positions, $t < d$, if there exist indices $j_1 < j_2 < \cdots < j_t$ such that $f_{j_1} = X_{j_1}, \cdots, f_{j_t} = X_{j_t}$.

The algorithm can be described as follows. Let $\phi_{i0}$ denote a constant mapping from $\{1, 2, \cdots, n\}$ to $\{1, 2, \cdots, n\}$, that is $\phi_{i0}(x) = i$ for all $x \in \{1, 2, \cdots, n\}$, and $i = 1, 2, \cdots, b$. In the beginning suppose each user has a version according to the first block column in (1). Then the algorithm is as follows.

## The Algorithm

1. Set $h = 0$, and $F$ to an empty list, $F = ()$.
2. Repeat while $h \leq m$:

   (a) For $r = 1, \cdots, b, k = 1, \cdots, n$, send a variant of segment $h$ marked with $c_{\phi_{rh}(k)}$ to the user corresponding to the row $(r-1)n + k$.
   (b) Receive the feedback and extract the mark $f_h$.
   (c) Append the feedback $f_h$ to the list $F$.
   (d) compare $F$ with the first $h + 1$ block columns of $\widetilde{M}$. If a row matches $F$ in $p + 1$ block columns disconnect the corresponding user.
   (e) Increment $h$.

**Theorem 2.** *The scheme described above can correctly detect all traitors.*

The proof of the theorem is based on the following Lemma.

**Lemma 1.** *Suppose there are at most $p$ traitors. Let*

$$F = (f_{j_1}, f_{j_2}, \cdots, f_{j_d}), \quad d \geq p + 1, \tag{2}$$

*be a $p$-consistent feedback sequence. If (2) matches the $(r-1)n + k^{th}$ row (row $(r, k)$ for short) of (1) in $p + 1$ positions, then no collusion excluding $(r, k)$, of at most $p$ users, can produce (2).*

*Proof.* Otherwise suppose collusion $T$ produces (2), $|T| \leq p$, $(r, k) \notin T$. By assumption there are $p+1$ positions where $(r, k)$ matches (2). Among these $p+1$ positions there exist two of them, say $j_{t_1}, j_{t_2}$, such that some $(r', k') \in T$ matches (2), and hence matches $(r, k)$, at $j_{t_1}, j_{t_2}$. By (1) we know that

- either $c_k = c_{k'} = f_{j_{t_1}}$, $c_{\phi_{r,j_{t_2}}(k)} = c_{\phi_{r',j_{t_2}}(k')} = f_{j_{t_2}}$, or
- $c_{\phi_{r,j_{t_1}}(k)} = c_{\phi_{r',j_{t_1}}(k')} = f_{j_{t_1}}$, $c_{\phi_{r,j_{t_2}}(k)} = c_{\phi_{r',j_{t_2}}(k')} = f_{j_{t_2}}$.

The first case implies that $k = k'$ and $\phi_{r,j_{t_2}}(k) = \phi_{r',j_{t_2}}(k')$, and hence $\phi_{r,j_{t_2}}(k) = \phi_{r',j_{t_2}}(k)$, which contradicts (P1) since we know $r \neq r'$ because $k = k'$, and $(r, k) \neq (r', k')$. The second case implies that $\phi_{r,j_{t_1}}(k) = \phi_{r',j_{t_1}}(k')$ and $\phi_{r,j_{t_2}}(k) = \phi_{r',j_{t_2}}(k')$, which contradicts (P2) as $t_1 \neq t_2$.

**Proof of Theorem 2:** (*sketch*) From lemma 1 we know that when $F$ matches a row of $\widetilde{M}$ in $p+1$ positions then a traitor can be identified. The traitor is *disconnected* so that he cannot decrypt future segments and the system continues as before. This means that $p$ traitors can be captured in at most $p(p+1)$ steps.

## Existence of $\Phi$

The construction in section 4.1 relies on the existence of a function family $\Phi$ that satisfies property (P1) and (P2). The number of users in the resulting sequential scheme is $bn$ and so is proportional to the size of the function family. In the following we give a construction for $\Phi$ satisfying properties (P1) and (P2).

**Theorem 3.** *Let $q$ be a prime number. There exists a function family $\Phi = \{\phi_{ij} : 1 \leq i, j \leq (q-1)/2\}$ that satisfies properties (P1) and (P2).*

*Proof.* Let $\mathbf{F}_q$ be a field of $q$ elements, $\mathbf{F}_q^*$ be the set of non-zero elements of $\mathbf{F}_q$. For $i, j \in \{1, 2, \cdots, (q-1)/2\}$ define

$$\phi_{ij} : \mathbf{F}_q^* \to \mathbf{F}_q^*$$

such that $\phi_{ij}(x) = (i+j)x$. Obviously $\phi_{i_1j}(x) \neq \phi_{i_2j}(x)$ for all $x \in \mathbf{F}_q^*$ provided $i_1 \neq i_2$ and so (P1) is satisfied. Now assume $\phi_{i_1,j_1}(x) = \phi_{i_2,j_1}(y)$. Then using the definition of $\phi_{ij}$, we have $i_1x - i_2y = j_1(y - x)$. So for every $j_2 \neq j_1$ we have $i_1x - i_2y \neq j_2(y - x)$, which implies that $\phi_{i_1,j_2}(x) \neq \phi_{i_2,j_2}(y)$. Hence (P2) is satisfied.

An example of this construction is given in Appendix 2.

## Discussion

Combining the above $\Phi$ and a watermarking code with $q-1$ codewords we obtain a sequential scheme in which (i) $q - 1$ variants are used, (ii) $(q - 1)^2/2$ users are accommodated, and (iii) at most $p^2 + p$ rounds are needed for detection of all $p$ traitors.

In choosing $q$, the number of variants, we must consider $N$, the total number of users in the final system and $p$, the maximum number of traitors. We must have $(q - 1)^2/2 \geq N$ and also $p^2 + p \leq 2 + (q - 1)/2$ and so $q \geq \max(1 + \sqrt{2N}, 2p^2 + 2p - 3)$.

## 4.2   A Construction Using Error-Correcting Codes

We can construct a sequential scheme by combining watermarking codes and error-correcting codes. The method is similar to the one given in [3] for constructing frameproof codes.

Let $C_1$ be an $(L, N, D)_n$-ECC error-correcting code over an alphabet of size $n$ with $N$ codewords each of length $L$, and minimum distance $D$. Let $C_2 = \{c_1, c_2, \cdots, c_n\}$ be a watermarking code. Define the *composition* of $C_1$ and $C_2$, denoted by $\Gamma(C_1, C_2)$, as a collection of strings over $C_2$ obtained as

$$C_v = c_{a_1} \parallel c_{a_2} \parallel \cdots \parallel c_{a_L}, \tag{3}$$

for all codewords $v$, $v = a_1 a_2 \cdots a_L \in C_1$. Here $\parallel$ means concatenation of strings.

**Theorem 4.** *Suppose we have $C_1 = \{c_1, c_2, \cdots, c_n\}$ a watermarking code, and $C_2$, a $(L, N, D)_n$-ECC. If*

$$D > (1 - \frac{1}{p})L,$$

*then $\Gamma(C_1, C_2)$ defines a sequential $(p, d)$-traceability scheme in which all, up to $p$, traitors can be traced in at most $p(L - D + 1)$ steps.*

*Proof. (sketch)* Let $C_1 = \{c_1, c_2, \cdots, c_n\}$, then

$$\Gamma(C_1, C_2) = \{(c_{a_1}, c_{a_2}, \cdots, c_{a_L}) \mid (a_1, a_2, \cdots, a_L) \in C_2\}.$$

Assign each string in $\Gamma(C_1, C_2)$ to a user: that is let the marks in the string be the user's $L$ successive marks. Suppose at most $p$ traitors contribute marks to the feedback sequence,

$$F_d = (f_1, f_2, \cdots, f_d). \tag{4}$$

When $d \geq p(L - D + 1)$, there is a traitor whose mark sequence coincides with (4) in at least $L - D + 1$ places. This user can be detected at this stage. To disconnect all up to $p$ traitors, at most $p(L - D + 1)$ steps are required. Here $D \geq (1 - \frac{1}{p})L + 1$ guarantees that $L \geq p(L - D + 1)$.

The following theorem shows that error correcting codes with suitable parameters, as required in Theorem 4, exist.

**Theorem 5.** *(Lemma III.3 of [3]) For any positive integers $p, N$, let $L = 8p \log N$. Then there exists a $(L, N, D)_{2p}$-ECC where $D > (1 - \frac{1}{p})L$.*

Now suppose we have a watermarking code $C_1$ with $2p$ codewords. Let $C_2$ be an $(L, N, D)_{2p}$-ECC and $D > (1 - \frac{1}{p})L$. From Theorem 4, the composition of the two codes is a $(p, d)$-traceability scheme. In this scheme $2p$ versions are used, and the number of rounds to detect all traitors is no more than $p(L - D + 1)$. So we have

$$d \leq p(L - D + 1) < pL - p(1 - \frac{1}{p})L + p$$

$$= L + p = 8p \log N + p \text{ (from Theorem 5)}$$

This is the same order as $O(p \log N + p^2/c)$ which is the best known result [1] for $p + c + 1$ versions.

# 5  Comparison and Discussion

It is not difficult to show that (proof is omitted) the mark allocation of the schemes in section 4 defines a frameproof code.

A very interesting aspect of the constructions given in section 4 is that if watermarking code in these constructions is replaced by a frameproof code or a static traceability scheme, the resulting code will be a frameproof code or traceability scheme, respectively.

The following theorems summarises these results.

**Theorem 6.** *Let $\Phi$ be a family of functions satisfying properties (P1) and (P2), $m > p - 2$.*

1. *If $\mathcal{C} = \{c_1, c_2, \cdots, c_n\}$ is a p-frameproof code, then the code with incidence matrix given by $\widetilde{M}$ as defined in (1) is a p-frameproof code.*
2. *If $\mathcal{C} = \{c_1, c_2, \cdots, c_n\}$ is a p-traceability scheme, then the code with incidence matrix given by $\widetilde{M}$ as defined in (1) is a p-traceability scheme.*

Repeated use of the above theorem results in a $p$-frameproof code ($p$-traceability scheme) with the following parameters: (i) length of the codeword is $(p+1)^t \ell$, and (ii) the number of the codewords is $f(n)$, a polynomial of degree $2^t$, assuming that the original code has $n$ codewords of length $\ell$.

Another interesting observation is that starting from a watermarking code and through the application of constructions in section 4 we will obtain a frameproof code and then using the above theorem we can construct a frameproof code for a much larger group. However because of the underlying watermarking code the resulting matrix can be used for a sequential traceability system. This proves the following corollary.

**Corollary 1.** *There is a sequential $(p, d)$-traceability scheme for $f(q)$ users using $q$ marks and with $d = (p(p+1))^t$, where $f(q)$ is a polynomial in $q$ with degree $2^t$.*

The results will hold even if we relax the restriction on the watermarking code and allow collusion of users to remove the watermark. As noted earlier this is the main type of attack in watermarking systems.

**Theorem 7.** *Let $\mathcal{C}$ be an $(L, N, D)_n$-ECC with $D > (1 - \frac{1}{p})L$.*

1. *If $\mathcal{C}_1 = \{c_1, c_2, \cdots, c_n\}$ is a p-frameproof code, then the composition of $\mathcal{C}$ and $\mathcal{C}_1$ is a p-frameproof code.*
2. *If $\mathcal{C}_2 = \{c_1, c_2, \cdots, c_n\}$ is a p-traceability scheme, then the composition of $\mathcal{C}$ and $\mathcal{C}_2$ is a p-traceability scheme.*

We note that 1 in Theorem 7 was first proved in Lemma III.2 in [3] and is included here for the sake completeness.

## 5.1  Computational Efficiency

The most attractive feature of the above model is the reduced real-time computation. Because allocation of marks is static most of the required keys can be pre-computed and distributed to the users in one initial block. Key updates are required when a traitor is found and needs to be disconnected. An important result of this efficiency is that it is possible to reduce the length of a segment and hence reduce the overall convergence time.

The detection algorithm is also very efficient. This is because we only require detection of partial match. For this it is only required to keep a counter for each user (a row of the mark allocation matrix) that counts the number of matches between the feedback sequence and that row. When a feedback $f_j$ is received the counter for all the rows that have $f_j$ in their $j^{th}$ position are incremented.

## 5.2  Time/Bandwidth Trade-Off

Two important parameters of dynamic traitor tracing schemes are (i) the number of marks, $r$, which determines the communication efficiency of the system, and (ii) the number of steps, $d$, required for convergence, that is finding all the traitors. Fiat and Tassa, and later Berkman et al [1] concentrated on the communication efficiency and so finding the efficient algorithms when $r$ is close to its to theoretical minimum $p + 1$. Berkman et al showed that if $r = pc + 1$ versions are used it is possible to find the traitors in $O(p \log_c N)$ rounds, and if $p + c + 1$ versions are used it is possible to find the traitors in $O(p^2/c + p \log N)$ rounds. Our first scheme guarantees convergence in $(p(p + 1))^t$ steps while the number of users can be about $p^{2^{t+1}}$. This scheme uses $q(\geq p^2 + p - 2)$ versions.

For the second scheme the convergence time is at most $8p \log N + p$ which is of the same order as the best result of [1]. It is important to note that by reducing the segment length the overall convergence time of our scheme would be expected to be lower.

## 5.3  Conclusions

Sequential traceability schemes can be seen as a step between static and dynamic schemes. The attack model in sequential schemes and dynamic schemes are the same and is different from a static traceability scheme. Also the goal of the former two types of systems are the same (tracing all traitors) and is different from static schemes. Sequential schemes do not have the flexibility of dynamic schemes and so in general could require higher bandwidth and/or higher number of convergence steps. However they provide security against delayed rebroadcast attack and are also practically attractive because they do not require real-time computation. We showed a construction that is as good as the best known dynamic construction and so in terms of efficiency measures competes well with dynamic schemes. We also showed that sequential traceability schemes are closely related to frameproof codes and so constructions from frameproof codes can be used for sequential schemes.

Although we gave a bound on the number of steps for convergence, but deriving tight bounds and developing schemes that achieve the bound need further research.

# References

1. O. Berkman, M. Parnas, and J. Sgall. Efficient dynamic traitor tracing. to appear at SODA 2000, 2000.
2. D. Boneh and J. Shaw. Collusion-secure fingerprinting for digital data. In *Advances in Cryptology - CRYPTO'95, Lecture Notes in Computer Science*, volume 963, pages 453–465. Springer-Verlag, Berlin, Heidelberg, New York, 1995.
3. D. Boneh and J. Shaw. Collusion-secure fingerprinting for digital data. *IEEE Transactions on Information Theory*, Vol.44 No.5:1897–1905, 1998.
4. B. Chor, A. Fiat, and M. Naor. Tracing traitors. In *Advances in Cryptology - CRYPTO'94, Lecture Notes in Computer Science*, volume 839, pages 257–270. Springer-Verlag, Berlin, Heidelberg, New York, 1994.
5. I. Cox, J. Killian, T. Leighton, and T. Shamoon. Secure spread spectrum watermarking for multimedia. *IEEE Transaction on Image Processing*, Vol. 6 no. 12:1673–1687, 1997.
6. A. Fiat and M. Naor. Broadcast encryption. In *Advances in Cryptology-CRYPTO'93, Lecture Notes in Computer Science*, volume 773, pages 480–491. Springer-Verlag, Berlin, Heidelberg, New York, 1994.
7. A. Fiat and T. Tassa. Dynamic traitor tracing. In *Advances in Cryptology-CRYPTO'99, Lecture Notes in Computer Science*, volume 1666, pages 354–371. Springer-Verlag, Berlin, Heidelberg, New York, 1999.
8. E. Gafni, J. Staddon, and Y.L. Yin. Efficient methods for intergrating traceability and broadcast encryption. In *Advances in Cryptology-CRYPTO'99, Lecture Notes in Computer Science*, volume 1666, pages 372–387. Springer-Verlag, Berlin, Heidelberg, New York, 1999.
9. D. Stinson and R. Wei. Combinatorial properties and constructions of traceability schemes and framproof codes. *SIAM Journal on Discrete Mathematics*, 11:41–53, 1998.
10. D. R. Stinson and R.Wei. Key preassigned traceability schemes for broadcast encryption. In *Proceedings of SAC'98, Lecture Notes in Computer Science*, volume 1556, pages 144–156. Springer-Verlag, Berlin, Heidelberg, New York, 1999.
11. M. Swanson, M Kobayashi, and A. Tewfik. Multimedia data-embedding and watermarking technologies. *Proceedings of IEEE*, Vol.86 no.6:1064–1087, 1998.

# Appendix 1

**Proof of Proposition 1** Take a subset $Y$ with $Y \cap X_i = \emptyset$ and $Y \cap X_k \neq \emptyset$ for $k \neq i$. Suppose $F_{j-1}$ is a feedback sequence. Then we have

$$
\begin{aligned}
& I(\mathcal{V}_1^{(j)}, \cdots, \mathcal{V}_{i-1}^{(j)}, \mathcal{V}_{i+1}^{(j)}, \cdots, \mathcal{V}_b^{(j)}; \mathcal{V}_i^{(j)} \mid U_Y, F_{j-1}) \\
& = H(\mathcal{V}_1^{(j)}, \cdots, \mathcal{V}_{i-1}^{(j)}, \mathcal{V}_{i+1}^{(j)}, \cdots, \mathcal{V}_b^{(j)} \mid U_Y, F_{j-1}) \\
& \quad - H(\mathcal{V}_1^{(j)}, \cdots, \mathcal{V}_{i-1}^{(j)}, \mathcal{V}_{i+1}^{(j)}, \cdots, \mathcal{V}_b^{(j)} \mid \mathcal{V}_i^{(j)}, U_Y, F_{j-1})
\end{aligned}
$$

$$\leq H(\mathcal{V}_1^{(j)}, \cdots, \mathcal{V}_{i-1}^{(j)}, \mathcal{V}_{i+1}^{(j)}, \cdots, \mathcal{V}_b^{(j)} \mid U_Y, F_{j-1})$$

$$\leq \sum_{k \neq i} H(\mathcal{V}_k^{(j)} \mid U_Y, F_{j-1})$$

$$= 0 \quad (\text{because of } Y \cap X_k \neq \emptyset)$$

So $I(\mathcal{V}_1^{(j)}, \cdots, \mathcal{V}_{i-1}^{(j)}, \mathcal{V}_{i+1}^{(j)}, \cdots, \mathcal{V}_b^{(j)}; \mathcal{V}_i^{(j)} \mid U_Y, F_{j-1}) = 0$. Note that

$$0 = I(\mathcal{V}_1^{(j)}, \cdots, \mathcal{V}_{i-1}^{(j)}, \mathcal{V}_{i+1}^{(j)}, \cdots, \mathcal{V}_b^{(j)}; \mathcal{V}_i^{(j)} \mid U_Y, F_{j-1})$$

$$= I(\mathcal{V}_i^{(j)}; \mathcal{V}_1^{(j)}, \cdots, \mathcal{V}_{i-1}^{(j)}, \mathcal{V}_{i+1}^{(j)}, \cdots, \mathcal{V}_b^{(j)} \mid U_Y, F_{j-1})$$

$$= H(\mathcal{V}_i^{(j)} \mid U_Y, F_{j-1}) - H(\mathcal{V}_i^{(j)} \mid \mathcal{V}_1^{(j)}, \cdots, \mathcal{V}_{i-1}^{(j)}, \mathcal{V}_{i+1}^{(j)}, \cdots, \mathcal{V}_b^{(j)}, U_Y, F_{j-1}).$$

Then we get

$$H(\mathcal{V}_i^{(j)}) \geq H(\mathcal{V}_i^{(j)} \mid \mathcal{V}_1^{(j)}, \cdots, \mathcal{V}_{i-1}^{(j)}, \mathcal{V}_{i+1}^{(j)}, \cdots, \mathcal{V}_b^{(j)})$$

$$\geq H(\mathcal{V}_i^{(j)} \mid \mathcal{V}_1^{(j)}, \cdots, \mathcal{V}_{i-1}^{(j)}, \mathcal{V}_{i+1}^{(j)}, \cdots, \mathcal{V}_b^{(j)}, U_Y, F_{j-1})$$

$$= H(\mathcal{V}_i^{(j)} \mid U_Y, F_{j-1})$$

$$= H(\mathcal{V}_i^{(j)}) \quad (\text{because of } Y \cap X_i = \emptyset).$$

So $H(\mathcal{V}_i^{(j)} \mid \mathcal{V}_1^{(j)}, \cdots, \mathcal{V}_{i-1}^{(j)}, \mathcal{V}_{i+1}^{(j)}, \cdots, \mathcal{V}_b^{(j)}) = H(\mathcal{V}_i^{(j)})$. The proposition is proved.

**Proof of Theorem 1** Suppose $T$ is a set of $p$ traitors. Let $F_d = (f_1, f_2, \cdots, f_d)$ be the corresponding feedback sequence. Then we have

$$H(\mathcal{U}_T) = H(\mathcal{U}_T, F_d) - H(F_d \mid \mathcal{U}_T)$$

$$= H(F_d) + H(\mathcal{U}_T \mid F_d) - H(F_d \mid \mathcal{U}_T)$$

$$= H(F_d) - H(F_d \mid \mathcal{U}_T), \quad (\text{by 1 of definition 1})$$

$$\leq H(F_d) = H(f_1, f_2, \cdots, f_d)$$

$$\leq H(f_1) + H(f_2) + \cdots + H(f_d)$$

$$\leq d \log r$$

Note that $H(\mathcal{U}_T) = pH(\mathcal{U})$. So $pH(\mathcal{U}) \leq d \log r$. It implies that $p \log N \leq d \log r$, here $N, d, r$ is the number of total users, the number of rounds to detect traitors and the number of versions, respectively. The theorem is proved.

**Proof of Theorem 6** Let $T = \{t_1, t_2, \cdots, t_c\}, c \leq p$, be a set of traitors. Suppose they collude to frame a user $U_i \notin T$. Since $m > p - 2$, using Pigeonhole Principle $U_i$ must match one $t_j \in T$ in more than one place. This contradicts (P2). So $T$ can not frame other users. The second result can be proved in a similar way.

# Appendix 2

### An Example

Suppose we want to provide protection for up to 50 users against collusion of up to 2 colluders ($p = 2$). We need $q = 11$ in theorem 3.

For simplicity we use $i$ instead of $c_i$. That is we only list the indices of codewords in watermarking code. Then $M$ consists of the following blocks.

| | | |
|---|---|---|
| $(1,1):$ 1 1 2 3 4 5 | $(2,1):$ 1 2 3 4 5 6 | $(3,1):$ 1 3 4 5 6 7 |
| $(1,2):$ 2 1 4 6 8 10 | $(2,2):$ 2 2 6 8 10 1 | $(3,2):$ 2 3 8 10 1 3 |
| $(1,3):$ 3 1 6 9 1 4 | $(2,3):$ 3 2 9 1 4 7 | $(3,3):$ 3 3 1 4 7 10 |
| $(1,4):$ 4 1 8 1 5 9 | $(2,4):$ 4 2 1 5 9 2 | $(3,4):$ 4 3 5 9 2 6 |
| $(1,5):$ 5 1 10 4 9 3 | $(2,5):$ 5 2 4 9 3 8 | $(3,5):$ 5 3 9 3 8 2 |
| $(1,6):$ 6 1 1 7 2 8 | $(2,6):$ 6 2 7 2 8 3 | $(3,6):$ 6 3 2 8 3 9 |
| $(1,7):$ 7 1 3 10 6 2 | $(2,7):$ 7 2 10 6 2 9 | $(3,7):$ 7 3 6 2 9 5 |
| $(1,8):$ 8 1 5 2 10 7 | $(2,8):$ 8 2 2 10 7 4 | $(3,8):$ 8 3 10 7 4 1 |
| $(1,9):$ 9 1 7 5 3 1 | $(2,9):$ 9 2 5 3 1 10 | $(3,9):$ 9 3 3 1 10 8 |
| $(1,10):$ 10 1 9 8 7 6 | $(2,10):$ 10 2 8 7 6 5 | $(3,10):$ 10 3 7 6 5 4 |

| | |
|---|---|
| $(4,1):$ 1 4 5 6 7 8 | $(5,1):$ 1 5 6 7 8 9 |
| $(4,2):$ 2 4 10 1 3 5 | $(5,2):$ 2 5 1 3 5 7 |
| $(4,3):$ 3 4 4 7 10 2 | $(5,3):$ 3 5 7 10 2 5 |
| $(4,4):$ 4 4 9 2 6 10 | $(5,4):$ 4 5 2 6 10 3 |
| $(4,5):$ 5 4 3 8 2 7 | $(5,5):$ 5 5 8 2 7 1 |
| $(4,6):$ 6 4 8 3 9 4 | $(5,6):$ 6 5 3 9 4 10 |
| $(4,7):$ 7 4 2 9 5 1 | $(5,7):$ 7 5 9 5 1 8 |
| $(4,8):$ 8 4 7 4 1 9 | $(5,8):$ 8 5 4 1 9 6 |
| $(4,9):$ 9 4 1 10 8 6 | $(5,9):$ 9 5 10 8 6 4 |
| $(4,10):$ 10 4 6 5 4 3 | $(5,10):$ 10 5 5 4 3 2 |

Suppose users (1,10) and (4,2) are the two colluders, and assume the feedback sequence is $F = (10, 4, 9, 1, 7, 5)$. We expect to identify the first colluder after observing $p^2 + 1 = 5$ elements, and the second colluder after observing the next element (feedback sequence of length $p^2 + p = p(p + 1) = 6$.) That is after observing 5 elements, there is exactly one user who matches the feedback sequence in $p + 1 = 3$ positions and all other possible traitors match it in at most $p = 2$ positions. By observing the $6^{th}$ element we can find a second colluder that matches 3 times. The following table lists (column) all possible traitors for each element of the feedback sequence. After observing 5 elements of the feedback sequence one of the colluders, (1,10) in this case, will be detected and disconnected. At this stage, the second colluder, (4,2) cannot be identified as he has appeared only twice which is the same number as the innocent user, (4,1). However by observing the $6^{th}$ element of the feedback sequence this colluder can also be identified.

| 10 | 4 | 9 | 1 | 7 | 5 |
|---|---|---|---|---|---|
| ↓ | ↓ | ↓ | ↓ | ↓ | ↓ |
| $(1, 10)$ | $(4, 1)$ | $(1, 10)$ | $(1, 4)$ | $(1, 10)$ | $(1, 1)$ |
| $(2, 10)$ | $(4, 2)$ | $(2, 3)$ | $(2, 3)$ | $(2, 8)$ | $(2, 10)$ |
| $(3, 10)$ | $(4, 3)$ | $(3, 5)$ | $(3, 9)$ | $(3, 3)$ | $(3, 7)$ |
| $(4, 10)$ | $(4, 4)$ | $(4, 4)$ | $(4, 2)$ | $(4, 1)$ | $(4, 2)$ |
| $(5, 10)$ | $(4, 5)$ | $(5, 7)$ | $(5, 8)$ | $(5, 5)$ | $(5, 3)$ |
| | $(4, 6)$ | | | | |
| | $(4, 7)$ | | | | |
| | $(4, 8)$ | | | | |
| | $(4, 9)$ | | | | |
| | $(4, 10)$ | | | | |

# Long-Lived Broadcast Encryption

Juan A. Garay[1], Jessica Staddon[2], and Avishai Wool[1]

[1] Bell Labs, 600 Mountain Ave., Murray Hill, NJ 07974, USA.
{garay,yash}@research.bell-labs.com
[2] Bell Labs Research Silicon Valley, 3180 Porter Drive, Palo Alto, CA 94304, USA.
staddon@research.bell-labs.com

**Abstract.** In a broadcast encryption scheme, digital content is encrypted to ensure that only privileged users can recover the content from the encrypted broadcast. Key material is usually held in a "tamper-resistant," replaceable, smartcard. A coalition of users may attack such a system by breaking their smartcards open, extracting the keys, and building "pirate decoders" based on the decryption keys they extract.
In this paper we suggest the notion of *long-lived broadcast encryption* as a way of adapting broadcast encryption to the presence of pirate decoders and maintaining the security of broadcasts to privileged users while rendering all pirate decoders useless. When a pirate decoder is detected in a long-lived encryption scheme, the keys it contains are viewed as compromised and are no longer used for encrypting content. We provide both empirical and theoretical evidence indicating that there is a long-lived broadcast encryption scheme that achieves a steady state in which only a small fraction of cards need to be replaced in each epoch. That is, for any fraction $\beta$, the parameter values may be chosen in such a way to ensure that eventually, at most $\beta$ of the cards must be replaced in each epoch.
Long-lived broadcast encryption schemes are a more comprehensive solution to piracy than traitor-tracing schemes, because the latter only seek to identify the makers of pirate decoders and don't deal with how to maintain secure broadcasts once keys have been compromised. In addition, long-lived schemes are a more efficient long-term solution than revocation schemes, because their primary goal is to minimize the amount of recarding that must be done in the long term.

## 1 Introduction

Broadcast encryption (BE) schemes define methods for encrypting content so that only privileged users are able to recover the content from the broadcast. Keys are allocated in such a way that users may be prevented on a short-term basis from recovering the message from the encrypted content. This short-term exclusion of users occurs, for example, when a proper subset of users request to view a movie. The long-term exclusion (or, revocation) of a user is necessary when a user leaves the system entirely.

M. Bellare (Ed.): CRYPTO 2000, LNCS 1880, pp. 333–352, 2000.
© Springer-Verlag Berlin Heidelberg 2000

In practice most BE systems are smartcard-based. It has been well documented (see, for example, [19]) that pirate smartcards (also called pirate "decoders") are commonly built to allow non-paying customers to recover the content. Broadcast encryption schemes can be coupled with traceability schemes to offer some protection against piracy. If a scheme has $x$-traceability, then it is possible to identify at least one of the smartcards used to construct a given pirate card provided at most $x$ cards are used in total. When a pirate card is discovered, the keys it contains are necessarily compromised and this must be taken into account when encrypting content. Earlier work in traceability does not deal with this; instead, the analysis stops with the tracing of smartcards (or, traitor users).

In this paper, we introduce the notion of *long-lived broadcast encryption* schemes, whose purpose is to adapt to the presence of compromised keys and continue to broadcast securely to privileged sets of users.

Our basic approach is as follows. Initially, every user has a smartcard with several decryption keys on it, and keys are shared by users according to a predefined scheme. When a pirate decoder is discovered, it is analyzed and the keys it contains are identified. Such keys are called "compromised," and are not used henceforth. Similarly, when a user's contract runs out and she is to be excluded, the keys on her smartcard are considered compromised. Over time, we may arrive at a state in which the number of compromised keys on some legitimate user's smartcard rises above the threshold at which secure communication is possible using the broadcast encryption scheme.[1] In order to restore the ability to securely broadcast to such a user, the service provider *replaces* the user's old smartcard with a new one containing a fresh set of keys.

The events driving the service provider's actions are the card compromises: either due to pirate decoders or the expiration of users' contracts. We use these events to divide time into administrative *epochs* of $d$ compromises each. At the end of an epoch, the service provider computes which legitimate users need their cards replaced, and replaces those cards. Therefore, the primary cost in a long-lived BE scheme is the amount of recarding that needs to be done in each epoch.

We assume in this paper that the schemes use perfect encryption, i.e., a pirate can access the content only by obtaining the key. Hence, we are requiring a strong form of security; information theoretic security. However, we argue that this model is very realistic for encrypted pay TV applications. In fact only a handful of the successful attacks described in [19] can be classified as cryptanalytic attacks, and only against analog equivalents of simple substitution ciphers. All the rest exploit breaches in the smartcard, using attacks such as those described in [2]. Furthermore, since smartcard-based systems are widely used in practice [19] it is natural to study this security model.

---

[1]  It costs little to assume that in addition to the shared broadcast keys, each user's smartcard contains a key unique to him. Thus the service provider can always revert to unicast communication to any user if all the user's broadcast keys are compromised.

In this model, and towards the goal of designing an efficient long-lived BE scheme, we consider the performance of three different schemes that have been suggested for broadcast encryption: two randomized and one deterministic. We start with a short-term analysis[2], which focuses on the first epoch. The parameters we analyze are the total number of keys the service provider needs, and the expected number of compromised cards the scheme can tolerate before replacement cards need to be issued. The analysis shows that the costs of the three schemes are quite similar. Hence, we use as the basis for our long-lived construction the simplest of the three, which is a randomized scheme. We provide both empirical and theoretical evidence using an expected-case analysis, that a steady state is achieved in which only a bounded number of users need to be recarded in any epoch of this long-lived scheme.

*Related work.* Our methods are based on efficient BE schemes. The study of broadcast encryption is initiated in [4,13,15] and the efficiency of BE schemes is studied in [5,6,18]. The model of BE that we consider here is a formalization of the deterministic (i.e. resilient with probability 1) model of [13] and is consistent with [1,18].

We are particularly interested in BE schemes that are based on cover-free families (see Section 2.1). Cover-free families are studied in [12] and BE schemes involving such families are studied in [17,14]. Our long-lived system may be based on (short-term) BE schemes that are tight with the proven lower bounds on the total number of keys in such schemes [12,14].

The recent papers of [20,3] propose novel revocation schemes.[3] In these schemes, in order to maintain the ability to revoke $t$ users, the center must make a private communication to each of the remaining users when a single user is revoked. Our goals are fundamentally different from this in that we seek to minimize the amount of communication (e.g. recarding) that's necessary, and so we adapt to the presence of compromised cards (or equivalently, revoked users) by simply removing the keys on these cards from the encryption process. When this approach is no longer possible due to a large number of compromised keys, we recard the affected users only. We show that through an appropriate choice of the parameters, the affected number of users can be a small fraction of the users in the steady state. Hence, our solutions seem useful in either the smartcard scenario or in a network-based system as studied in [20]. Whereas, it may be difficult to apply the techniques in [20] in a smartcard scenario as the cost of reprogramming or replacing a large number of cards, may be prohibitive.

Broadcast encryption schemes and multicast encryption schemes (see for example, [8,9,10]) are designed with many common goals in mind. BE schemes and multicast schemes are similar in that a pirate smartcard in the former is essentially treated the same as the card of a revoked user in a multicast scheme.

---

[2] Throughout this paper we focus on an expected case analysis. Some justification for this approach comes from the law of large numbers [21].

[3] The term "revocation" is typically used to indicate the permanent exclusion of a user from the system, rather than the prevention of a user from recovering a particular message.

In either case, care must be taken in future broadcasts to ensure that the keys contained in the card are useless for recovering future messages. The main difference is that in the multicast scenario, the primary goal is to maintain a secure group key at all times, whereas, in broadcast encryption schemes, the group itself varies over time as a subset of the universe of users, and so any group key is only established when the privileged subset is identified. Moreover, widespread rekeying of users (typically via a series of encrypted messages) in the multicast group may be required as part of the process of establishing the new group key. For example, the tree-based scheme in [26] specifies rekeying of all users when a single user leaves the system. This is to be contrasted with the approach taken in long-lived BE, in which we view rekeying of users as the most significant cost of the system, and hence, we rekey as infrequently as possible. Independently of our work, a recent paper [22] confirms that rekeying upon each change in the multicast group membership is prohibitively expensive, and proposes rekeying the entire group at fixed intervals at a cost of some latency in membership adjustments.

Our work is related to the goals of traitor-tracing schemes [11,24,7] in that such schemes are also concerned with coalitions of users who conspire to build pirate smartcards (which we refer to as compromised cards). However, we emphasize that there are substantial differences between a traitor-tracing scheme and the schemes we present here. The most important difference is that traceability schemes do not describe how to broadcast securely to privileged sets of users after pirate decoders have been located. The purpose of an $x$-traceability scheme is to make the practice of building pirate smartcards risky. This is accomplished by allocating keys to users in such a way that once a pirate smartcard is confiscated, at least one of the cards that was used to construct it, can be identified. Clearly, the keys in a pirate smartcard must be viewed as compromised. Hence, after a pirate smartcard is confiscated, the set of keys used to encrypt content must be modified to avoid allowing a user with compromised keys to recover the content. Traitor tracing does not deal with this.

Another difference is that the security achieved in traceability schemes is limited by the necessity of having a bound on the number of users in a coalition. Our approach can handle a pirate decoder built by a coalition of any size. Furthermore, in the secret key model that we consider here, there is a sizable gap between the proven lower bound on the number of keys in a $x$-traceability scheme and the number of keys in the best known construction. Since our methods rely on efficient BE schemes we are able to keep the total number of keys tight with the aforementioned lower bounds, while retaining an ability to broadcast securely in the presence of pirate smartcards, and consequently, compromised keys.

Finally, we note that if traceability is desired in the long-lived system, this may be achieved by basing the system on a BE scheme with some traceability (as, for example, in [25,14]).

*Our results.* The contributions of this paper can be summarized as follows:

- We introduce the notion of *long-lived* broadcast encryption, whose purpose is to continue to broadcast securely to privileged sets of users as cards are compromised over time;
- an analysis and comparison of three BE schemes based on cover-free families with respect to total number of keys and expected number of cards that the scheme can tolerate before recarding; and, based on this analysis,
- an efficient long-lived BE scheme. We provide empirical and theoretical evidence that for any fraction $\beta$, there is a scheme that recards at most a $\beta$ fraction of the users in the steady state.

*Organization of the paper.* Definitions and notation, as well as the formalization of the long-lived approach to broadcast encryption, are presented in Section 2. The short-term analysis and comparison of the three BE schemes is given in Section 3. The long-lived BE scheme, analysis and experimental evaluation are in Section 4. We conclude with some final remarks and directions for future work in Section 5.

## 2    Preliminaries

### 2.1    Definitions and Notation

We consider broadcast encryption schemes in the secret key scenario. For our purposes, a broadcast encryption scheme consists of a collection of subsets of keys (one for each user) and a broadcasting protocol, which indicates how to securely distribute content to privileged sets of users. Let $\{u_1, ..., u_n\}$ denote the set of all users, and let the $i$th user's set of keys be denoted by $U_i$. Typically, a user's keys are contained in a card, and consequently, we will often refer to $U_i$ as $u_i$'s *card*.

We denote the universe of keys by $\mathcal{K} = \{k_1, ..., k_K\}$ ($K$ keys in total), and each user has $r$ keys in $\mathcal{K}$ ($\forall i, |U_i| = r$). In this paper, the privileged sets of users are of fixed size $n - m$. A privileged set of users will be denoted by $P$, and the corresponding excluded set of $m$ users will be denoted by $X$.

The *broadcasting protocol* specifies which subsets of keys in $\mathcal{K}$ suffice to recover the content from the encrypted broadcast. In this paper we are interested in *s-threshold* protocols [17] in which a user needs to use $s$ keys out of $r$ in order to decode the content. When $s = 1$, this broadcasting protocol is sometimes called an *OR* protocol and has been studied in [1,14,15,18].

**Definition 1.** *An* $(s, |S_P|)$**-threshold protocol** *is used to broadcast a message,* $M$, *to users* $P = \{u_1, ..., u_{n-m}\}$, *in the following manner.* $K$ *shares of* $M$, $M_{k_1}$, $M_{k_2}, ..., M_{k_K}$, *are created in such a way that any* $s$ *of the shares suffices to recover* $M$. *The shares corresponding to keys held by users in* $X = \{u_{n-m+1}, ..., u_n\}$ *are discarded, and each remaining share is encrypted with its corresponding key and these encrypted messages are broadcast to the universe of users.*

We focus on threshold protocols because they are simple and yield broadcast encryption schemes with maximal *resilience*. A scheme is said to be $m-resilient$ if $m$ excluded (i.e., not privileged) users cannot recover the content even by pooling their keys. A broadcast encryption scheme with $(s, |S_P|)$-threshold protocols for every privileged set $P$, is $m$-resilient. In addition, for some values of $s$, the techniques of [17] can be used to reduce the broadcast transmission length.

When using threshold protocols for broadcasting we must ensure that a user has sufficiently many keys left after the keys of $m$ other users are excluded to recover the content from the broadcast. Traditionally, this has been guaranteed by allocating keys to users in such a way that the set system is a *cover-free family*.

**Definition 2.** *Let $\mathcal{K}$ be a collection of elements. A set of subsets of $\mathcal{K}$, $\{U_1, \ldots \ldots, U_n\}$, is an $(m, \alpha)$-**cover-free family**, if for all $i = 1, \ldots, n$, and for all sets of $m$ indices, $\{j_1, \ldots, j_m\}$ not containing $i$, $|U_i \cap (\cup_{s=1}^m U_{j_s})| \leq (1 - \alpha)|U_i|$.*

Note that $\frac{1}{r} \leq \alpha \leq 1$. In the original construction of [12] $\alpha = \frac{1}{r}$; i.e., no $m$ users cover *all* of another user's keys.

In most of our work, we adhere to the cover-free requirement to allow comparisons with earlier work. Note that this is a very strong requirement: It guarantees that it is impossible for *any* coalition of $m$ cards to cover an $\alpha$-fraction of another card's keys. As a result, the constructions need very large key sets, roughly, $K$ is $\Omega(n^{m/r})$, when $r \geq m$, and $\Omega(n)$, otherwise [14]. These bounds may well be prohibitive for large user populations.

However, in a long-lived system, the cover-free requirement seems less relevant, simply because a cover-free scheme gives no guarantee on the system's behavior after $m + 1$ cards are compromised. In addition, in the randomized attack model that we are considering, it can easily be shown (see Lemma 8) that even in a system with significantly fewer keys than an $m$-cover-free system, a set of $m$ compromised cards will cover another card only with negligible probability. Thus, in the long-term analysis, and in the experimental results, we do not define values of $m$ and adhere to the cover-free requirement for those values. Instead we de-couple the number of users $n$ from the total number of keys $K$ number, and observe the behavior of the resulting schemes in terms of how many cards need to be issued per epoch.

We are interested in how the broadcast encryption scheme is affected by pirate smartcards, which we assume to be cards containing $r$ keys.[4] A card may be *compromised* either because of piracy or simply because a user ceases to be an active subscriber and leaves the system. In either case, the keys on the card become permanently unavailable for use as encryption keys. A compromised card may be a clone of some user's card or may contain a set of $r$ keys that does not exactly match any of the $n$ users in the system.

---

[4] We believe it is reasonable to assume that a pirate decoder contains at least as many keys as legitimate cards. It is sometimes the case that the pirate cards even use *better* technology than the legitimate ones [16], i.e., they can store more keys.

When the keys on a card are all unavailable because it is a compromised card or belongs to an excluded user, we say that the keys it contains and the card itself are *dead*. A key that is not on a compromised card and does not belong to an excluded user is said to be *active*. A card is said to be *clean* if it contains only active keys. We use $d$ as a counter for the number of dead cards (i.e., either due to piracy or exclusion).

We note that the reason behind the unavailability of a key has an effect on our behavior. When a key is dead because it appears on a compromised card, the key is permanently unavailable; whereas if it simply appears on an excluded user's card, its unavailability may be short-term as the excluded user may be a privileged user at a later time. For more on this issue see Section 2.2.

## 2.2   The Long-Lived Approach

In this section we describe our basic method for securely broadcasting to privileged users as cards get compromised. The method is based on knowledge of compromised cards and consists of two basic components:

1.    Adjusting the set $S_P$ of keys that are used to encrypt the broadcast; and

2.    recarding of users.

The method is *reactive* in the sense that actions are taken responding to the number of compromised cards (which, for example, might hamper the continuity of service for privileged users, or bring transmission costs to unacceptable levels). This divides the life of the system into *epochs*. At the end of epoch $i$, $i = 1, 2, \cdots$, a decision is made about which cards need to be replaced, and new cards are issued. We now describe the structure of our long-lived reactive recarding scheme more formally.

A long-lived broadcast encryption scheme consists of:

– Underlying structure: An efficient (short-term) broadcast encryption scheme consisting of an $(m,\alpha)$-cover-free family and a unicast key between each user and the center. An $(\alpha r, |S_P|)$-threshold protocol is used to broadcast to privileged set $P$. If some user is unreachable under the $(\alpha r, |S_P|)$-threshold

**Table 1.** Summary of notation

– $\{u_1, \ldots, u_n\}$ is the set of all users.
– $U_i$ is the set of keys held by $u_i$.
– $\mathcal{K} = \{k_1, ..., k_K\}$ is the set of all keys.
– $S_P$ is the set of keys used to broadcast to privileged set $P$.
– $n$ is the total number of users.
– $K$ is the total number keys.
– $r$ the number of keys per user.
– $m$ is the number of users who are excluded.
– $d$ is the number of unavailable (dead) cards at a certain point in time.
– $C_j^i$ is the set of cards in epoch $i$ that were created in epoch $j$.

protocol (i.e., too many of their keys appear on dead cards) then the unicast key will be used to reach that user.

- A distribution on the compromised cards: In this paper we assume the cards are drawn independently at random (with replacement) from the key space.
- Reaction to a newly compromised card: To render the compromised card useless as a decoder we exclude all its keys from $S_P$, creating a new set $S_P^1$. Broadcasting to privileged set $P$ is with an $(\alpha r, |S_P^1|)$-threshold protocol, relying on unicast keys if necessary.
- Recarding policy: A recarding session is entered whenever $d$ cards become unavailable. During a recarding session, any user with less than $\alpha r$ active keys receives a new card.

The parameter $d$ in the fixed schedule will be based on the number of compromised users and the desired transmission length. In a recarding session, new values are chosen randomly for all dead keys.

Recall that a key is dead either because it belongs to an excluded user or is on a compromised card. In the former case, the key is unavailable on what may be a short-term basis, as an excluded user may well be a privileged user at another time. Hence, we note that our long-term analysis (Section 4) is best applied to a stable privileged set $P$, or to the whole set of users when the number of excluded users, $m$, is small. Given this, it is very likely that users will only be recarded when more than $(1-\alpha)r$ of their keys are *permanently* unavailable (i.e., contained in compromised cards) rather than simply temporarily unavailable, due to the current set of excluded users. As stated in Section 1, the primary motivation for recarding users should be the presence of compromised cards.

A summary of terms and notation is given in Table 1.

## 3    A Short-Term Analysis of Three BE Schemes

In this section we describe three schemes, each based on a cover-free family. The first BE scheme is a randomized bucket-based construction from [17] and the second is a deterministic construction based on polynomials [14,17]. Both constructions yield $(m, \alpha)$-cover-free families. The third scheme is a very simple randomized method for producing $(m, \alpha)$-cover-free families. We present a short-term analysis of all three schemes, which indicates that they are remarkably similar in terms of efficiency. Specifically, the three schemes only differ by a constant fraction in the number of dead cards (i.e., compromised or belonging to excluded users) they can tolerate before recarding is needed. Hence, given the simplicity of the randomized scheme, we choose to focus our long term analysis on it (see Section 4).

### 3.1    A Bucket-Based BE Scheme

In this section we consider a reactive recarding scheme based on the randomized cover-free family construction of [17].[5] In [17], the construction is presented for

---

[5]  See the randomized construction of an "inner code" in Section 4 of [17].

an $(m, 1/2)$-cover-free family. We present the construction of an $(m, \alpha)$-cover-free family for completeness. The set of all keys $\{k_1, ..., k_K\}$ is partitioned into sets of size $K/r$. Each card contains a randomly selected key from each set. To broadcast to a set $P$ of privileged users, we use an $(\alpha r, |S_P|)$-threshold protocol as described in Section 2.1.

The following lemma gives a lower bound on the total number of keys $K$, for which this construction yields a $(m, \alpha)$-cover-free family with high probability. Note that in this scheme $m$ depends on $r$ and $K$.

**Lemma 1.** *Let $\epsilon$ be a positive fraction. If $r = \frac{K \ln(1/\alpha)}{4m}$ and $K$ is $\Omega(m(m \ln n + \ln(1/\epsilon)))$, then the bucket-based construction is an $(m, \alpha)$-cover-free family with probability $1 - \epsilon$.*

**Proof:** Consider $m + 1$ users, $u, u_1, ..., u_m$. We'll calculate the probability that $u_1, .., u_m$ cover enough of user $u$'s keys to violate the cover-free condition. The partitions are each of size $K/r = \frac{4m}{\ln(1/\alpha)}$. The probability that a key $k$ is not in $\cup_{i=1}^{m} U_i$ is $(1 - \frac{\ln(1/\alpha)}{4m})^m$. Therefore,

$$E(|U \cap (\cup_{i=1}^{m} U_i)|) = r(1 - (1 - \frac{\ln(1/\alpha)}{4m})^m)$$

When $m$ is sufficiently large relative to $1/\alpha$, $(1 - \frac{\ln(1/\alpha)}{4m})^{\frac{4m}{\ln(1/\alpha)}} \geq \frac{1}{e^2}$. Hence, the above expected value is at most $r(1 - \sqrt{\alpha})$. Hence,

$$Pr[|U \cap (\cup_{i=1}^{m} U_i)| > (1-\alpha)r] \leq Pr[|U \cap (\cup_{i=1}^{m} U_i)| > (1 + \sqrt{\alpha})\mu \mid \mu = r(1 - \sqrt{\alpha})]$$

Using Chernoff bounds, that probability is at most, $e^{\frac{-\alpha r(1-\sqrt{\alpha})}{3}}$. When $K \geq \frac{4(m+1)m\ln n + 4m\ln(1/\epsilon)}{\alpha(1-\sqrt{\alpha})\ln(1/\alpha)}$, it follows that, $\binom{n}{m+1}e^{\frac{-\alpha r(1-\sqrt{\alpha})}{3}} < \epsilon$.     □

Now we consider the short-term behavior of this scheme. In particular, we are interested in how many compromised cards this scheme can tolerate before recarding is necessary. Since we are interested in an expected-case analysis, we calculate how many dead cards, chosen randomly with replacement, cause a user to need to be recarded. The lower bound proven in Lemma 2 is very close to the bounds proven for the other two reactive recarding schemes (see Lemma 3 and Lemma 5), however, we note that since $r$ is likely to be quite large in this scheme, the tolerable number of dead cards may be fairly small.

**Lemma 2.** *Consider a user, $u$. In the bucket-based construction, the expected number of dead cards that can be tolerated before it is necessary to recard $u$ is greater than $\ln(1/\alpha)(\frac{K}{r} - 1)$.*

**Proof:** First we show that it suffices to only consider dead cards that are clones (i.e., cards that contain exactly one key from each bucket). To see this, note that the probability a key is in a randomly chosen set of $r$ keys (i.e., a cloned card or otherwise) is $\frac{r}{K}$. From Lemma 1, we know that the probability a randomly chosen key is on a cloned card is $\frac{\ln(1/\alpha)}{4m} = \frac{r}{K}$. Hence in our expected case analysis it suffices to assume the dead cards are clones.

Consider $d$ dead cards, $U_1,...,U_d$. As calculated in the previous lemma, for a random user $u$,

$$E(|U \cap (\cup_{i=1}^d U_i|) = r(1 - (1 - \frac{\ln(1/\alpha)}{4m})^d)$$

This quantity is greater than $(1 - \alpha)r$ when $\alpha > (1 - \frac{\ln(1/\alpha)}{4m})^d$. Solving for $d$, we get,

$$d > \frac{\ln(1/\alpha)}{\ln(\frac{4m}{4m-\ln(1/\alpha)})} = \ln(1/\alpha)(\frac{K}{r} - 1)$$

Since $\ln(\frac{4m}{4m-\ln(1/\alpha)}) \leq \frac{\ln(1/\alpha)}{4m-\ln(1/\alpha)}$, it follows that $d > 4m - \ln(1/\alpha)$.  □

## 3.2  A Deterministic BE Scheme

In this section we consider the polynomial-based broadcast encryption scheme of [14]. This scheme differs from the polynomial-based scheme in [17] in that $r$ is an independent parameter, and not a function of the other variables. This scheme uses polynomials to construct a deterministic $(m, \alpha)$-cover-free family. An $(\alpha r, S_P)$-threshold protocol is used to broadcast to a privileged set, $P$.

Let $p$ be a prime larger than $r$, and let $A$ be a subset of the finite field $F_p$ of size $r$. Consider the set of all polynomials over $F_p$ of degree at most $\frac{r(1-\alpha)}{m}$. (For simplicity, we assume that $m|r(1 - \alpha)$.) There are $p^{\frac{r(1-\alpha)}{m}+1}$ such polynomials. We associate each of the $n$ users with a different polynomial. Therefore, $p$ needs to satisfy the condition that $p^{\frac{r(1-\alpha)}{m}+1} \geq n$. For each pair $(x, y)$, where $x \in A$ and $y \in F_p$, we create a unique key $k_{(x,y)}$. Hence, the total number of keys, $K$, is $rp \geq rn^{\frac{m}{r(1-\alpha)+m}}$. If a user $u$ is associated with a given polynomial $f$, $u$'s smartcard contains the keys in the set $\{k_{(x,f(x))}|x \in A\}$. Since any two of the polynomials intersect in at most $\frac{r(1-\alpha)}{m}$ points, it follows that any two users share at most $\frac{r(1-\alpha)}{m}$ keys. This ensures that if all the keys belonging to the $m$ excluded users are removed, each privileged user will still have at least $\alpha r$ keys. Hence, the center can broadcast to users a privileged set $P$, with an $(\alpha r, S_P)$-threshold protocol. A user needs to be recarded when the number of active keys on their card falls below $\alpha r$.

**Lemma 3.** *Consider a user, $u$. For $K$ sufficiently large, the expected number of dead cards that can be tolerated in the deterministic scheme before it is necessary to recard $u$ is at least $\frac{K}{r} \left( \frac{\ln(1/\alpha)}{2^{r(1-\alpha)+2}} \right)$.*

The proof of this lemma is very similar to the proof of Lemma 2.

### 3.3    A Simple Randomized BE Scheme

In this scheme, each user is allocated a randomly selected set of $r$ keys out of a universe of $K$ keys total, where $K$ is chosen large enough to ensure that we have an $(m, \alpha)$-cover-free family with high probability (see Lemma 4). Broadcasting to a set $P$ of privileged users is accomplished as discussed in the previous two sections. Hence, when a user has less than $\alpha r$ active keys, the user's card needs to be replaced. In the recarding procedure, new keys are generated for all dead keys, and active keys are unchanged. The total number of active keys (i.e., keys that need to be stored by the broadcasting center) is unaffected by the recarding procedure. We first prove a lower bound on the total number of keys.

**Lemma 4.** *Given any positive fraction $\epsilon$, if the total number of keys, $K$ is $\Omega(\frac{n^{m+1}}{\epsilon}^{\frac{1}{r(1-\alpha)+1}})$ then the randomized reactive recarding scheme is an $(m, \alpha)$-cover-free family with probability at least $1 - \epsilon$.*

**Proof:** Consider $m + 1$ users, $u$, $u_1$, $u_2$,...,$u_m$. First we bound the probability that $u_1$,...,$u_m$ cover more than $(1 - \alpha)r$ of $u$'s keys. Since $|\cup_{i=1}^m U_i| \leq mr$, we have the following bound:

$$Pr(|U \cap (\cup_{i=1}^m U_i)| > (1 - \alpha)r)$$

$$\leq \frac{\binom{mr}{(1-\alpha)r+1}\binom{K-(1-\alpha)r-1}{\alpha r-1} + \binom{mr}{(1-\alpha)r+2}\binom{K-(1-\alpha)r-2}{\alpha r-2} + ... + \binom{mr}{r}}{\binom{K}{r}}$$

Using binomial bounds and simplifying, we have:
$$Pr(|U \cap (\cup_{i=1}^m U_i)| > (1 - \alpha)r) \leq \frac{(rem)^r e^{\alpha r-1} \alpha r}{K^{r(1-\alpha)+1}}$$

Hence, the probability that $|U \cap (\cup_{i=1}^m U_i)| \leq \alpha r$ is at least $1 - \frac{(rem)^r e^{\alpha r-1} \alpha r}{K^{r(1-\alpha)+1}}$. There are $n - m$ privileged users, therefore the probability that there is at least one privileged user who shares more than $r(1 - \alpha)$ keys with $u_1, ..., u_m$, is at most $1 - [1 - \frac{rem^r e^{\alpha r-1} \alpha r}{K^{r(1-\alpha)+1}}]^{n-m}$. To account for all possible excluded sets of $m$ users, it suffices to multiply by $\binom{n}{m}$:

$$\binom{n}{m}(1 - [1 - \frac{(rem)^r e^{\alpha r-1} \alpha r}{K^{r(1-\alpha)+1}}]^{n-m})$$

Substituting a binomial approximation,

$$1 - (\frac{m}{ne})^m \epsilon \leq (1 - \frac{(rem)^r e^{\alpha r-1} \alpha r}{K^{r(1-\alpha)+1}})^{n-m}$$

If $K > r^{r+1}m^r e^{2r-1}$ (this is reasonable since we expect $r$ to be small), then we can use the fact that $(1 - x)^{n-m} \geq 1 - (n - m)x$ when $x \leq 1$, to simplify this expression. With this substitution, it suffices to show that $(n-m)\frac{(rem)^r e^{\alpha r-1} \alpha r}{K^{r(1-\alpha)+1}} \leq (\frac{m}{ne})^m \epsilon$. Solving for $K$ yields the statement of the lemma.   $\square$

*Remark 1.* Note that the factor of $n^{m+1/(r(1-\alpha)+1)}$ in the bound on $K$ is due to the cover-free requirement, that with very high probability it is *impossible* for $m$

cards to cover another. However, the construction itself remains viable for any value of $K$. In fact, in the randomized attack model, in which the pirates pry open randomly selected cards, much smaller values of $K$ suffice to guarantee a low probability of $m$ cards covering another (see Lemma 8).

**Lemma 5.** *Consider a user $u$. The expected number of dead cards that can be tolerated in the randomized recarding scheme before it is necessary to recard $u$ is at least $\left(\frac{K-r}{r}\right)\ln(1/\alpha)$.*

The proof of this lemma is very similar to the proof of Lemma 2.

### 3.4   Comparison of the BE Schemes

As the previous lemmas show, the three BE schemes have very similar costs. Each yield $(m, \alpha)$-cover-free families with high probability when the total number of keys is close to the optimal bound[6] of $n^{m/r}$. In the deterministic and randomized schemes, this is clear. To see that the bucket-based scheme is close to this bound, note that the proven bound is roughly $\Omega(m^2 \ln n)$ and $K \geq n^{m/r}$ is equivalent to $K \ln K \geq \frac{4m^2 \ln n}{\ln(1/\alpha)}$, when $r$ has the value stated in the lemma.

In addition to the above similarity, all three schemes can tolerate approximately $\frac{K}{r}$ dead cards before recarding a particular user is necessary. We note that this means we expect to need to recard a user only after $\frac{K}{r} - m$ cards are compromised (due to piracy or contract expiration). As the schemes are so close in terms of efficiency and cost, we use the third scheme as the basis of our long-lived system. It is the most simple, as it is entirely random, and it has the advantage over the bucket-based scheme that $r$ and $m$ are independent parameters.

## 4   Long-Lived Broadcast Encryption

In this section we extend the randomized BE scheme from Section 3.3 to a long-lived scheme. The extension is reactive as defined in Section 2.2—recarding is performed once every $d$ dead cards—and for simplicity we consider the $(m, 1/r)$-cover-free family version of the scheme (*OR* protocols). We emphasize that this analysis is best applied to a stable set of privileged users, or to the entire set of users when $m$ is small. In either case, we expect to only have to recard a user when too many of their keys appear on compromised (i.e., permanently unavailable) cards.

The main cost associated with long-lived schemes is the number of cards that must be replaced. We present a scheme in which given a positive fraction $\beta$, the parameters may be chosen so that eventually at most $\beta n$ of the cards need to be replaced during any recarding session. This property is demonstrated both empirically and theoretically.

---

[6] This bound is for $\alpha = \frac{1}{r}$.

We now turn to the description of the long-lived extension and its analysis. Assume that $d$ cards are compromised. The process for generating the new cards is as follows. Let $Z$ be the set of keys these cards contain and let $z = |Z|$; note that $z \leq dr$. The scheme

1.    discards all the keys in $Z$, and

2.    generates a set $Z'$ of new keys, $|Z'| = z$. The new set of all keys becomes $\mathcal{K}' = (\mathcal{K} \backslash Z) \cup Z'$.

The resulting number of keys is again $K$ in total. Every user that needs to be recarded receives the fresh values of the same keys.[7]

As keys become compromised and users are recarded, the users can be partitioned into sets of users with cards with fresh keys, and users with cards containing keys some of which are dead. This process is depicted in Figure 1. We let $C_j^i$ denote the set of cards in epoch $i$ that were created in epoch $j$. Initially (epoch 1), $C_1^1 = \{U_1, ..., U_n\}$. Selecting (randomly) $d$ dead cards from $C_1^1$ yields $C_2^2$, the set of users that need to be recarded, as well as $C_1^2 = C_1^1 \backslash C_2^2$; in epoch 2, selecting $d$ random cards from $C_1^2$ and $C_2^2$ yields $C_3^3$ as well as $C_2^3$ and $C_2^3$; and so on.

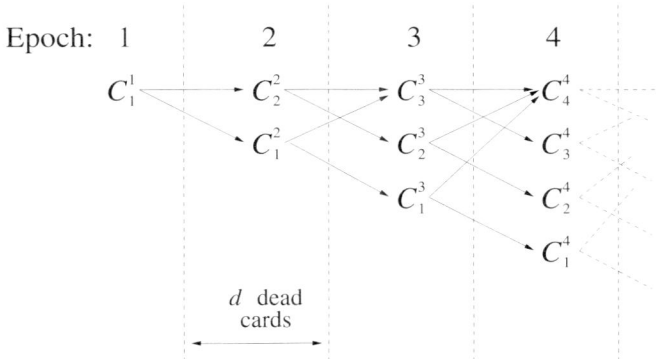

**Fig. 1.** Randomized long-lived BE scheme. $d$ dead cards determine the epochs; $C_j^i$ is the set of cards in epoch $i$ that were created in epoch $j$.

Towards bounding the necessary number of recards per epoch, namely, the (expected) size of set $C_j^j$ in epoch $j$, we first prove recurrence relations relating the expected number of cards in epoch $j$ that were created in epoch $i \leq j$, $E(|C_i^j|)$.

---

[7]    This is to preserve the cover-free property of the scheme used as basis. In the randomized scheme, the same will hold if for every user that needs to be recarded, $r$ keys are again picked at random from the updated set $\mathcal{K}'$.

**Lemma 6.** *In the randomized long-lived BE scheme with $\alpha = 1/r$ and a fixed recarding schedule of once every $d$ dead cards, the following hold for all $i \geq 1$:*

1. $E(|C_{i+1}^{i+1}|) \leq \sum_{j=1}^{i} E(|C_j^i|)[1 - (1 - \frac{r}{K})^{(i+1-j)d}]^r$;
2. $\forall\, j, 1 \leq j \leq i,\ E(|C_j^{i+1}|) = E(|C_j^i|)(1 - [1 - (1 - \frac{r}{K})^{(i+1-j)d}]^r)$.

**Proof:** To see the first inequality, note that if a user is recarded (or created) in epoch $j$, then during the time interval from the beginning of epoch $j$ to the end of epoch $i$, $d(i + 1 - j)$ randomly chosen cards become unavailable. If these cards cover the user's card, then the user must be recarded. Due to the random nature of the scheme, a user is covered with probability, $[1 - (1 - \frac{r}{K})^{(i+1-j)d}]^r$. We have a weak inequality rather than equality, because a user may be covered by fewer than $d(i - j + 1)$ cards.

The second equation is obtained by noting that all users who were recarded (or created) in epoch $j$, and who are not covered by the end of epoch $i$, become the set of users $C_j^{i+1}$. ◻

We now use the first part of Lemma 6 to demonstrate that an upper bound on the number of recards per epoch holds in the limit, and that this upper bound can be made small through appropriate choices of $K$, $r$ and $d$. We emphasize that this is an approximate analysis, and is provided largely to give some intuition for the experimental results in Section 4.1. A more rigorous analysis will appear in the full-length version of this paper.

The analysis contains three components. First, given fixed values of the parameters, we show that there exists an integer $\ell_1$, such that the probability that a card is covered (and hence, needs to be refreshed) within $\ell_1$ epochs, is negligible. The intuition for this result is that if a card has been refreshed recently, then it is unlikely that it will be covered again within a small number of epochs. This result indicates that the contribution to $E(|C_{i+1}^{i+1}|)$ from the first $\ell_1$ terms of inequality 1 in Lemma 6, is fairly small. In addition, the later terms in inequality 1 in Lemma 6 may also not contribute much to the upper bound on $E(|C_{i+1}^{i+1}|)$. In particular, there exists an integer $\ell_2$ (greater than $\ell_1$), such that it is unlikely that a card will *not* be covered within $\ell_2$ epochs. Note that this implies that when $i - j \geq \ell_2$, $E(|C_j^i|)$ is fairly small, and hence, won't contribute much to the upper bounds on $E(|C_{i+1}^{i+1}|)$. Finally, we show that both $\ell_1$ and $\ell_2$ are on the order of $K/rd$, hence the dominating terms are those for which $i - j$ is $\Theta(\frac{K}{rd})$, and this leads to an approximation for the upper bound of the steady state recard rate, $\beta$. The following lemma makes these ideas more precise.

**Lemma 7.** *Assume $n$, $K$, $r$, $d$ and $\epsilon > 0$ are given. The following are true:*

1. *If $\ell_1 \in O(\frac{\epsilon^{1/r}K}{rd})$, then the probability that a card is covered within $\ell_1$ epochs is less than $\epsilon$.*
2. *If $\ell_2 \in \Omega(\frac{(1-\epsilon)^{1/r}K}{rd})$, then the probability that a card survives for more than $\ell_2$ epochs before it is covered, is less than $\epsilon$.*
3. *If $i - j \in \Theta(\frac{K}{rd})$ then the coefficient of $E(|C_j^i|)$ in inequality 1 of Lemma 6, is approximately, $(1 - (\frac{1 - \frac{r}{K}}{e^c})^d)^r$, where $c$ is a constant.*

**Proof:**

1. The probability that a card is covered by $d\ell_1$ randomly chosen cards is $(1 - (1 - \frac{r}{K})^{d\ell_1})^r$. Setting this quantity less than $\epsilon$ and solving for $\ell_1$ yields, $\ell_1 < \frac{\ln(1 - \epsilon^{1/r})}{d\ln(1 - r/K)}$.

2. The probability that a card is not covered within $\ell_2$ epochs is, $1 - (1 - (1 - \frac{r}{K})^{d\ell_2})^r$. Setting this less than $\epsilon$ and solving for $\ell_2$ yields, $\ell_2 > (\frac{K}{rd}(1 - \epsilon)^{1/r})$.

3. Assuming that $i - j = \frac{cK}{rd}$, for some constant $c$, we'll bound the contribution of $E(|C_j^i|)$ to the inequality in Lemma 6 (i.e., we'll bound the coefficient of $E(|C_j^i|)$) — given the earlier results, this bound is an approximate upper bound to $\beta$, the long-term steady state).

    When $i - j = \frac{cK}{rd}$, the coefficient of $E(|C_j^i|)))$, is $(1 - ((1 - r/K)^d)^{cK/rd+1})^r$.

    When $K$ is sufficiently large, this is of the order of $(1 - \frac{(1-r/K)^d}{e^c})^r$.    □

When combined with parts 1 and 2, part 3 of the lemma indicates that the steady state recard rate $\beta$ should decrease with $K$, which agrees with the experimental results that follow. The quantity also increases with $d$, which agrees with the basic intuition that the longer we wait to recard, the more recarding we will have to do.

### 4.1   Numerical Experimentation

In order to get a better understanding of the card replacement dynamics, we present some numerical experiments. In these experiments we evaluate Equation 1 of Lemma 6 (assuming an equality rather than an inequality) for a variety of parameter settings, and track the number of cards that were issued in every epoch.

We focus on the random attack model, and assume that, in each epoch, the dead cards are selected uniformly at random from the set of user cards. We do not require that the system be cover-free, so $K$ is not constrained by the bound of Lemma 4. Instead, we let $K$ be a free parameter which we vary.

To justify this decoupling of $K$ from $n$, we present the following simple lemma that provides a lower bound on $K$ such that with high probability, none of the $n$ user cards are covered by $d$ randomly chosen compromised cards (i.e., some cover-freeness is achieved with high probability). As mentioned in Section 2.1, this lower bound may be much smaller than the size of $K$ in a $d$-cover-free family.

**Lemma 8.** *Let $\epsilon > 0$, and $n$, $r$ and $d$ be given. If $K > \frac{r}{c}$, where $c$ is a constant that depends on $\epsilon$, $n$, $r$, and $d$, then the probability that any user's card is covered by $d$ randomly chosen cards is less than $\epsilon$.*

**Proof:** The probability that $n$ (randomly chosen) user's cards aren't covered by $d$ randomly chosen cards is $[1 - (1 - (1 - \frac{r}{K})^d)^r]^n$. Hence, we solve the following inequality for $K$, $[1 - (1 - (1 - \frac{r}{K})^d)^r]^n > 1 - \epsilon$, which yields, $K > \frac{r}{1-(1-[1-(1-\epsilon)^{1/n}]^{1/r})^{1/d}}$.    □

As an example, for the values of $n$, $r$ and $d$ used in Figure 2 and $\epsilon = .1$, this lemma gives a lower bound on $K$ of approximately 69, far less than the lower bound of approximately $10^{10}$ for a $d$-cover-free family.

In all of our experiments we use a user population of size $n = 100,000$, which we view as being on the low end of real population sizes. The card capacity $r$ ranges between 10 and 50, which is realistic for current smartcards with 8KB of memory and keys requiring, say, 64 bytes each including overhead. We let the epoch length be $10 \leq d \leq 50$ dead cards. We use $1000 \leq K \leq 5000$, hence $K$ may be much smaller than the number of keys required by Lemma 4, which calls for $K \geq n^{m/r}$ keys (note the dependency on $m$, the number of users the underlying BE is able to exclude).

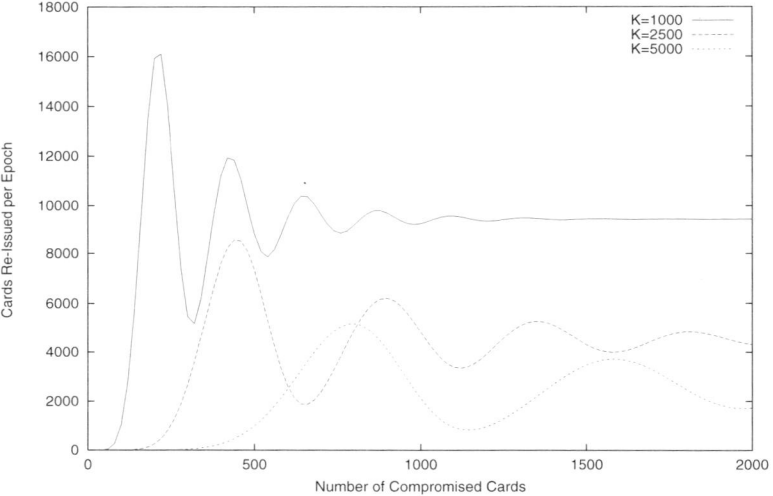

**Fig. 2.** Number of cards re-issued per epoch, with $n = 100,000$, $r = 10$, $d = 20$, for different values of the total number of keys $K$.

Figure 2 shows the dynamics of the card re-issue strategy, and the effect of the total number of keys $K$. We see that the curves begin with oscillations. In the first epochs no cards are re-issued since the first dead cards do not cover any user. But after a certain number $d_c$ of dead cards are discovered, enough keys are compromised and there is a rapid increase in re-issued cards. This in turn "cleans" the card population, and the re-issue rate drops. We see that the oscillations are dampened and a steady state appears fairly quickly.

The parameter $K$ affects several aspects of the dynamics: the first card re-issue point $d_c$ is later for larger $K$ ($d_c \approx 40$ for $K = 1000$ but $d_c \approx 200$ for $K = 5000$); the oscillations are gentler, have a smaller amplitude, and lower peak rate, for larger $K$; and most importantly, the steady state rate of re-issue is lower for larger $K$ ($\approx 9400$ cards per epoch for $K = 1000$ but $\approx 2000$ cards per epoch for $K = 5000$). Overall, we see that increasing $K$ improves all the aspects of the re-issue strategy. Thus we conclude that it is better to use the largest possible $K$ that is within the technological requirements.

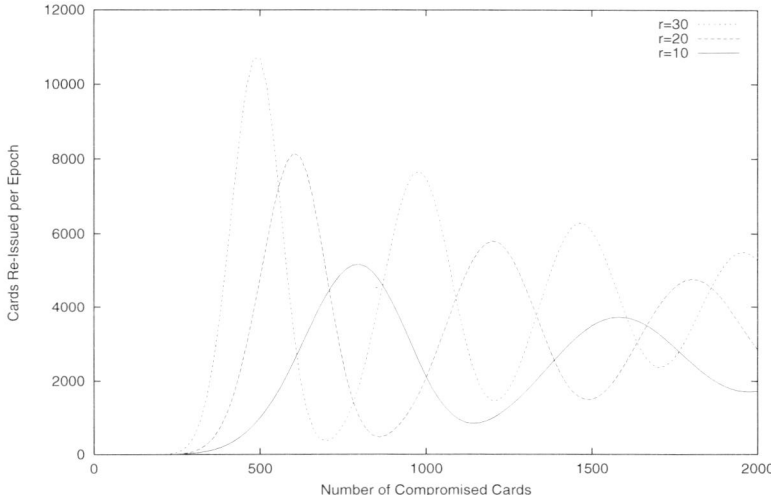

**Fig. 3.** Number of cards re-issued per epoch, with $n = 100,000$, $K = 5000$, $d = 20$, for different values of the number of keys per card, $r$.

Figure 3 shows the effect of increasing the card capacity $r$. The diagram indicates that larger values of $r$ cause greater re-issue costs: larger $r$'s have a higher steady state re-issue rate, and higher peak re-issue rates. This agrees with the fact that as $r$ increases, we expect each key to be contained in more cards, so the effect of a compromised key is more widespread. Also, as indicated by Lemma 5, we expect to have to recard users sooner when $r$ is large (and $K$ is fixed). However, with a smaller $r$ the expected transmission length is longer; at the extreme, setting $r = 1$ gives optimal re-issue rates (no cards need to be re-issued), with very long transmissions.

Figure 4 shows the effect of increasing the epoch length $d$. From the figure, it is clear that a longer epoch results in a smaller total number of re-issued cards. However, a long epoch also means that many keys are compromised during the epoch, and consequently, it may be impossible to broadcast securely to some users during the epoch without unicasts. Hence, recarding costs and transmission costs may influence the choice of $d$.

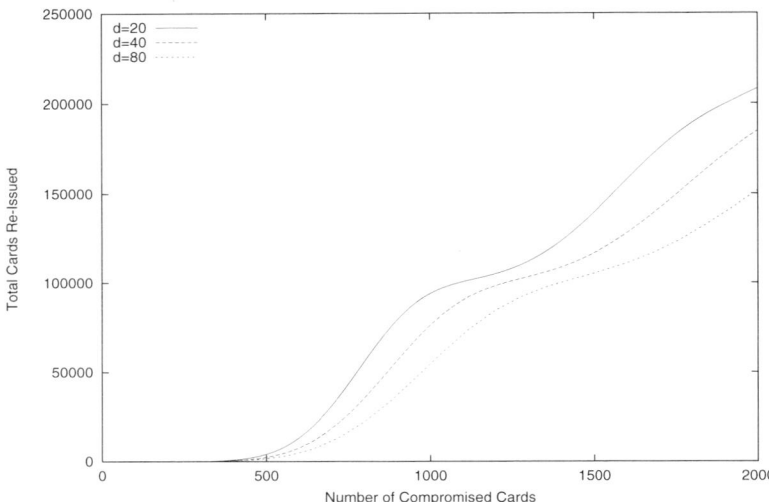

**Fig. 4.** The accumulated total number of cards re-issued, with $n = 100,000$, $K = 5000$, $r = 10$, for different values of the epoch length $d$.

## 5    Summary and Directions for Future Work

In this paper, we consider making broadcast encryption schemes resistant to piracy by introducing a policy of permanently revoking compromised keys. This is to be distinguished from the short-term revocation of keys that is typically done in a BE scheme in order to prevent users from recovering a particular message (e.g., a movie) and is instead more analogous to the revocation of users in a multicast group.

There are many open questions with respect to the analysis of the simple model we've proposed. For example, it would be interesting to look at different distributions on the compromised cards (i.e., other than independently at random) and to determine how transmission length is affected by parameters such as $d$.

It would also be interesting to consider modifications to the overall approach taken here. The long-lived scheme presented in Section 4 is reactive in the sense that actions are taken responding to the number of compromised cards. Are methods that do not count on pirate smartcard intelligence—oblivious, "proactive"—viable?

### Acknowledgements

The authors thank Adi Shamir and the anonymous referees for their many valuable comments.

# References

1. M. Abdalla, Y. Shavitt, and A. Wool. Towards making broadcast encryption practical. In M. Franklin, editor, *Proc. Financial Cryptography'99*, Lecture Notes in Computer Science **1648** (1999), pp. 140–157. To appear in IEEE/ACM Trans. on Networking.

2. R. Anderson and M. Kuhn. Low cost attacks on tamper resistant devices. In *5th Security Protocols Workshop*, Lecture Notes in Computer Science **1361** (1997), pp. 125–136.

3. J. Anzai, N. Matsuzaki and T. Matsumoto. *A Quick Group Key Distribution Scheme with "Entity Revocation"* In Advances in Cryptology - Asiacrypt '99, Lecture Notes in Computer Science (1999), pp. 333-347.

4. S. Berkovits. *How to Broadcast a Secret*. In Advances in Cryptology - Eurocrypt '91, Lecture Notes in Computer Science **547** (1992), pp. 536-541.

5. C. Blundo and A. Cresti. *Space Requirements for Broadcast Encryption*. In Advances in Cryptology - Eurocrypt '94, Lecture Notes in Computer Science **950** (1994), pp. 287-298.

6. C. Blundo, L. A. Frota Mattos and D. Stinson. *Trade-offs Between Communication and Storage in Unconditionally Secure Systems for Broadcast Encryption and Interactive Key Distribution*. In Advances in Cryptology - Crypto '96, Lecture Notes in Computer Science **1109** (1996), pp. 387-400.

7. D. Boneh and M. Franklin. *An Efficient Public Key Traitor Tracing Scheme*. In Advances in Cryptology - Crypto '99, Lecture Notes in Computer Science **1666** (1999), pp. 338-353.

8. R. Canetti, J. Garay, G. Itkis, D. Micciancio, M. Naor and B. Pinkas. *Multicast Security: A Taxonomy and Efficient Constructions*. In Proc. INFOCOM 1999, Vol. 2, pp. 708-716, New York, NY, March 1999..

9. R. Canetti, T. Malkin and K. Nissim. *Efficient Communication-Storage Tradeoffs for Multicast Encryption*. In Advances in Cryptology - Eurocrypt '99, Lecture Notes in Computer Science.

10. R. Canetti and B. Pinkas. *A Taxonomy of Multicast Security Issues*. Internet draft. Available at: `ftp://ftp.ietf.org/internet-drafts/draft-canetti-secure-multicast-taxonomy-00.txt`

11. B. Chor, A. Fiat, M. Naor and B. Pinkas. *Tracing Traitors*. Full version to appear in IEEE Transactions on Information Theory. Preliminary version in Advances in Cryptology - Crypto '94, Lecture Notes in Computer Science **839** (1994), pp. 257-270.

12. P. Erdös, P. Frankl and Z. Füredi. *Families of Finite Sets in which No Set is Covered by the Union of r Others*. Israel Journal of Mathematics **51** (1985), pp.75-89.

13. A. Fiat and M. Naor. *Broadcast Encryption*. In Advances in Cryptology - Crypto '93, Lecture Notes in Computer Science **773** (1994), pp. 480-491.

14. E. Gafni, J. Staddon and Y. Yin. *Efficient Methods for Integrating Braodcast Encryption and Traceability*. In Advances in Cryptology - Crypto '99, Lecture Notes in Computer Science **1666** (1999), pp. 372-387.

15. M. Just, E. Kranakis, D. Krizanc and P. van Oorschot. *On Key Distribution via True Broadcasting*. In Proceedings of 2nd ACM Conference on Computer and Communications Security, November 1994, pp. 81-88.

16. M. Kuhn. Personal communication, 1999.

17. R. Kumar, S. Rajagopalan and A. Sahai. Coding Constructions for Blacklisting Problems without Computational Assumptions. In Advances in Cryptology - Crypto '99, Lecture Notes in Computer Science **1666** (1999), pp. 609-623.
18. M. Luby and J. Staddon. *Combinatorial Bounds for Broadcast Encryption*. In Advances in Cryptology - Eurocrypt '98, Lecture Notes in Computer Science, **1403**(1998), pp. 512-526.
19. J. McCormac. *European Scrambling Systems 5*. Waterford University Press, 1996.
20. M. Naor and B. Pinkas. *Efficient Trace and Revoke Schemes*. In *Proc. Financial Cryptography 2000*, Anguila, February 2000.
21. J. Pitman. *Probability*. Springer-Verlag, 1993.
22. S. Setia, S. Koussih, S. Jajodia and E. Harder. *Kronos: A Scalable Group Re-Keying Approach for Secure Multicast*. In 2000 IEEE Symposium on Security and Privacy, pp. 215-228.
23. D. Stinson. *Cryptography: Theory and Practice*. CRC Press, 1995.
24. D. Stinson and R. Wei. *Combinatorial Properties and Constructions of Traceability Schemes and Frameproof Codes*. SIAM J. Discrete Math, **11** (1998), pp. 41-53.
25. D. Stinson and R. Wei. *Key Preassigned Traceability Schemes for Broadcast Encryption*. In Proc. SAC '98, Lecture Notes in Computer Science **1556** (1999), pp. 144-156.
26. D. Wallner, E. Harder and R. Agee. *Key Management for Multicast: Issues and Architectures*. Internet Request for Comments, **2627** (June 1999). Available at: ftp.ietf.org/rfc/rfc2627.txt.

# Taming the Adversary

Martín Abadi

Bell Labs Research, Lucent Technologies
abadi@lucent.com
www.pa.bell-labs.com/~abadi

**Abstract.** While there is a great deal of sophistication in modern cryptology, simple (and simplistic) explanations of cryptography remain useful and perhaps necessary. Many of the explanations are informal; others are embodied in formal methods, particularly in formal methods for the analysis of security protocols. This note (intended to accompany a talk at the Crypto 2000 conference) describes some of those explanations. It focuses on simple models of attacks, pointing to partial justifications of these models.

## 1  Polite Adversaries

Some of the simplest explanations of cryptography rely on analogies with physical objects, such as safes, locks, and sealed envelopes. These explanations are certainly simplistic. Nevertheless, and in spite of the sophistication of modern cryptology, these and other simplifications can be helpful when used appropriately. The simplifications range from informal metaphors to rigorous abstract models, and include frequent omissions of detail and conceptual conflations (e.g., [28]). They commonly appear in descriptions of systems that employ cryptography, in characterizations of attackers, and correspondingly in statements of security properties. They are sometimes deceptive and dangerous. However, certain simplifications can be justified:

- on pragmatic grounds, when the simplifications enable reasoning (even automated reasoning) that leads to better understanding of systems, yielding increased confidence in some cases and the discovery of weaknesses in others;
- on theoretical grounds, when the simplifications do not hide security flaws (for example, when it can be proved that a simple attacker is as powerful as an arbitrary one).

In particular, in the design and study of security protocols, it is typical to adopt models that entail sensible but substantial simplifying restrictions on attackers. (See for example [13] and most of the references below.) In these models, an adversary may perform the same operations as other principals. For example, all principals, including the adversary, may be allowed to send and receive messages. If the protocol relies explicitly on cryptography, all principals may be allowed to perform cryptographic operations; thus, the adversary may generate keys, encrypt, decrypt, sign, verify signatures, and hash. In addition, the

M. Bellare (Ed.): CRYPTO 2000, LNCS 1880, pp. 353–358, 2000.
© Springer-Verlag Berlin Heidelberg 2000

adversary may have some capabilities not shared by other principals, for example intercepting messages. On the other hand, the adversary may be subject to limitations that tend to make it like other principals, for example:

- it sends messages of only certain forms (e.g., [6, 18]),
- its actions are in a certain order (e.g., [10, 17]),
- it behaves "semi-honestly", that is, it follows the protocol properly but keeps records of all its intermediate computations (e.g., [14]).

Although an actual attacker need not obey them, such conditions are sometimes sound and often convenient.

An even stronger simplification is implicit in these models, namely that the adversary politely respects the constraints and abstractions built into the models. In effect, the adversary does not operate at a lower-level of abstraction than other principals.

- If other principals treat encrypted messages as indivisible units, and only process them by applying decryption operations, the adversary may proceed in the same way. Since those messages are actually bitstrings, many other transformations are possible on them. The adversary will not apply those transformations.
- Similarly, the adversary may generate keys and nonces in stylized ways, like other principals, but may not use arbitrary calculations on bitstrings for this purpose. We may even reason as though all principals obtained keys and nonces from a central entity that can guarantee their distinctness (like an object allocator in a programming language run-time system).
- More radically, the adversary may create and use secure communication channels only through a high-level interface, without touching the cryptographic implementation of these channels.
- Even detailed, concrete models often miss features that could conceivably be helpful for an actual attacker (real-time delays, power consumptions, temperature variations, perhaps others).

In support of such restrictions, we may sometimes be able to argue that operating at a lower-level of abstraction does not permit more successful attacks. In other words, security against polite adversaries that use a limited, high-level suite of operations should imply security against adversaries that use a broader, lower-level vocabulary (e.g., [1]). For instance, the adversary should gain nothing by using the lower-level vocabulary in trying to guess the keys of other principals, since those keys should be strong. (The next section says a little more about these arguments.) A complementary justification is that the restrictions make it easier to reason about protocols, and that in practice this reasoning remains fruitful because the restrictions do not hide many important subtleties and attacks.

## 2    From Politeness to Formality

With varying degrees of explicitness, many of these simplifications have been embodied in symbolic algorithms, proof systems, and other formal methods for the analysis of security protocols (e.g., [4, 9, 11–13, 16, 19, 21, 23, 25, 27, 29, 31]). The power of the formal methods is largely due to these simplifications.

In these methods, keys, nonces, and other fresh quantities are typically not defined as ordinary bitstrings. They may even be given a separate type. While all bitstrings can be enumerated by starting from 0 and adding 1 successively, such an enumeration need not cover keys and nonces. Moreover, an adversary that may non-deterministically choose any bitstring may be unable to pick particular keys or nonces. Keys and nonces are introduced by other means, for example by quantification ("for every key $K$ such that the adversary does not have or invent $K$ . . .") or by a construct for generating new data ("let $K$ be a new key in . . ."). The former approach is common in logical methods (e.g., [29]); the latter, in those based on process calculi such as the pi calculus [26] (e.g., [4, 11]). They both support the separation of keys and nonces from ordinary data.

This separation is an extremely convenient stroke of simplism. Without this separation, it is hard to guarantee that the adversary does not guess keys. At best, we may expect such a guarantee against an adversary of reasonable computational power and only probabilistically. With the separation, we may prevent the adversary from guessing keys without imposing restrictions on the adversary's power to compute on bitstrings, and without mention of computational complexities or probabilities (cf. [7, 15, 32]).

Accordingly, keys and cryptographic operations are manipulated through symbolic rules. These rules do not expose the details of the definitions of cryptographic operations. They reflect only essential properties, for example that decryption can undo encryption. This property is easy to express through an equation, such as $d(e(K, x), K) = x$. The treatment of other properties is sometimes more delicate. For example, suppose that the symbol $f$ represent a one-way function. The one-wayness of $f$ may be modeled, implicitly, by the absence of any sequence of operations that an adversary can use to recover the expression $x$ from the expression $f(x)$.

The set of rules is extensible. For example, it is often possible to incorporate special properties of particular cryptographic functions, such as the commutation of two exponentiations with the same modulus. The awareness of such extensions is quite old—it appears already in Merritt's dissertation [24, page 60]. Nevertheless, we still seem to lack a method for deciding whether a given set of rules captures "enough" properties of an underlying cryptosystem.

More broadly, even if these formal methods are consistent and useful, their account of cryptography is (deliberately) partial; it may not even be always sound. The simplifications undoubtedly imply inaccuracies, perhaps mistakes. Thus, we may wonder whether the separation of keys and nonces from ordinary data does not have any unintended consequences. (In the *Argumentum Ornithologicum* [8], the difference of a fresh quantity from particular integers leads to the conclusion that God exists.)

Some recent research efforts provide limited but rigorous justifications for abstract treatments of cryptography [2, 3, 5, 22, 30] (see also [20]). They establish relations between:

- secure channels, secure message transmission, and other high-level notions,
- formal accounts of cryptography, of the kind discussed in this section, and
- lower-level accounts of cryptography, based on standard concepts of computation on bitstrings (rather than ad hoc concepts of computation on symbolic expressions).

In particular, a formal treatment of encryption is sound with respect to a lower-level computational model [5]. The formal treatment is small but fairly typical: simplistic and symbolic. In the computational model, on the other hand, all keys and other cryptographic data are bitstrings, and adversaries have access to the rich, low-level vocabulary of algorithms on bitstrings. Despite these additional capabilities of the adversaries, the assertions that can be proved formally are also valid in the computational model, not absolutely but with high probability and against adversaries of reasonable computational power (under moderate, meaningful hypotheses). Thus, at least in this case, we obtain a computational foundation for the tame, convenient adversaries of the formal world.

# References

1. Martín Abadi. Protection in programming-language translations. In *Proceedings of the 25th International Colloquium on Automata, Languages and Programming*, volume 1443 of *Lecture Notes in Computer Science*, pages 868–883. Springer-Verlag, July 1998. Also Digital Equipment Corporation Systems Research Center report No. 154, April 1998.
2. Martín Abadi, Cédric Fournet, and Georges Gonthier. Secure implementation of channel abstractions. In *Proceedings of the Thirteenth Annual IEEE Symposium on Logic in Computer Science*, pages 105–116, June 1998.
3. Martín Abadi, Cédric Fournet, and Georges Gonthier. Authentication primitives and their compilation. In *Proceedings of the 27th ACM Symposium on Principles of Programming Languages*, pages 302–315, January 2000.
4. Martín Abadi and Andrew D. Gordon. A calculus for cryptographic protocols: The spi calculus. *Information and Computation*, 148(1):1–70, January 1999. An extended version appeared as Digital Equipment Corporation Systems Research Center report No. 149, January 1998.
5. Martín Abadi and Phillip Rogaway. Reconciling two views of cryptography (The computational soundness of formal encryption). In *Proceedings of the First IFIP International Conference on Theoretical Computer Science*, Lecture Notes in Computer Science. Springer-Verlag, August 2000. To appear.
6. Roberto M. Amadio and Denis Lugiez. On the reachability problem in cryptographic protocols. Technical Report 3915, INRIA, March 2000. Extended abstract to appear in the Proceedings of CONCUR 2000.
7. Manuel Blum and Silvio Micali. How to generate cryptographically strong sequences of pseudo random bits. In *Proceedings of the 23rd Annual Symposium on Foundations of Computer Science (FOCS 82)*, pages 112–117, 1982.
8. Jorge Luis Borges. Argumentum Ornithologicum. In *Obras completas 1923–1972*, page 787. Emecé Editores, Buenos Aires, 1974.
9. Michael Burrows, Martín Abadi, and Roger Needham. A logic of authentication. *Proceedings of the Royal Society of London A*, 426:233–271, 1989. A preliminary

version appeared as Digital Equipment Corporation Systems Research Center report No. 39, February 1989.

10. Edmund Clarke, Somesh Jha, and Will Marrero. Partial order reductions for security protocol verification. In *Tools and Algorithms for the Construction and Analysis of Systems*, volume 1785 of *Lecture Notes in Computer Science*, pages 503–518. Springer-Verlag, March/April 2000.

11. Mads Dam. Proving trust in systems of second-order processes. In *Proceedings of the 31th Hawaii International Conference on System Sciences*, volume VII, pages 255–264, 1998.

12. Richard A. DeMillo, Nancy A. Lynch, and Michael Merritt. Cryptographic protocols. In *Proceedings of the Fourteenth Annual ACM Symposium on Theory of Computing*, pages 383–400, 1982.

13. Danny Dolev and Andrew C. Yao. On the security of public key protocols. *IEEE Transactions on Information Theory*, IT-29(12):198–208, March 1983.

14. Oded Goldreich. Secure multi-party computation (working draft). On the Web at http://theory.lcs.mit.edu/~oded/frag.html, 1998.

15. Shafi Goldwasser and Silvio Micali. Probabilistic encryption. *Journal of Computer and System Sciences*, 28:270–299, April 1984.

16. James W. Gray, III and John McLean. Using temporal logic to specify and verify cryptographic protocols (progress report). In *Proceedings of the 8th IEEE Computer Security Foundations Workshop*, pages 108–116, 1995.

17. Joshua D. Guttman and F. Javier Thayer Fábrega. Authentication tests. In *Proceedings 2000 IEEE Symposium on Security and Privacy*, pages 96–109, May 2000.

18. Antti Huima. Efficient infinite-state analysis of security protocols. Presented at the 1999 Workshop on Formal Methods and Security Protocols, 1999.

19. Richard A. Kemmerer. Analyzing encryption protocols using formal verification techniques. *IEEE Journal on Selected Areas in Communications*, 7(4):448–457, May 1989.

20. P. Lincoln, J. Mitchell, M. Mitchell, and A. Scedrov. A probabilistic poly-time framework for protocol analysis. In *Proceedings of the Fifth ACM Conference on Computer and Communications Security*, pages 112–121, 1998.

21. Gavin Lowe. Breaking and fixing the Needham-Schroeder public-key protocol using FDR. In *Tools and Algorithms for the Construction and Analysis of Systems*, volume 1055 of *Lecture Notes in Computer Science*, pages 147–166. Springer-Verlag, 1996.

22. Nancy Lynch. I/O automaton models and proofs for shared-key communication systems. In *Proceedings of the 12th IEEE Computer Security Foundations Workshop*, pages 14–29, 1999.

23. Catherine Meadows. A system for the specification and analysis of key management protocols. In *Proceedings of the 1991 IEEE Symposium on Research in Security and Privacy*, pages 182–195, 1991.

24. Michael J. Merritt. *Cryptographic Protocols*. PhD thesis, Georgia Institute of Technology, February 1983.

25. Jonathan K. Millen, Sidney C. Clark, and Sheryl B. Freedman. The Interrogator: Protocol security analysis. *IEEE Transactions on Software Engineering*, SE-13(2):274–288, February 1987.

26. Robin Milner, Joachim Parrow, and David Walker. A calculus of mobile processes, parts I and II. *Information and Computation*, 100:1–40 and 41–77, September 1992.

27. John C. Mitchell, Mark Mitchell, and Ulrich Stern. Automated analysis of cryptographic protocols using Murφ. In *Proceedings of the 1997 IEEE Symposium on Security and Privacy*, pages 141–151, 1997.
28. R. M. Needham. Logic and over-simplification. In *Proceedings of the Thirteenth Annual IEEE Symposium on Logic in Computer Science*, pages 2–3, June 1998.
29. Lawrence C. Paulson. The inductive approach to verifying cryptographic protocols. *Journal of Computer Security*, 6(1–2):85–128, 1998.
30. Birgit Pfitzmann, Matthias Schunter, and Michael Waidner. Cryptographic security of reactive systems (extended abstract). *Electronic Notes in Theoretical Computer Science*, 32, April 2000.
31. F. Javier Thayer Fábrega, Jonathan C. Herzog, and Joshua D. Guttman. Strand spaces: Why is a security protocol correct? In *Proceedings 1998 IEEE Symposium on Security and Privacy*, pages 160–171, May 1998.
32. Andrew C. Yao. Theory and applications of trapdoor functions. In *Proceedings of the 23rd Annual Symposium on Foundations of Computer Science (FOCS 82)*, pages 80–91, 1982.

# The Security of All-or-Nothing Encryption: Protecting against Exhaustive Key Search

Anand Desai

Department of Computer Science & Engineering,
University of California at San Diego,
9500 Gilman Drive, La Jolla, California 92093, USA.
adesai@cs.ucsd.edu

**Abstract.** We investigate the all-or-nothing encryption paradigm which was introduced by Rivest as a new mode of operation for block ciphers. The paradigm involves composing an all-or-nothing transform (AONT) with an ordinary encryption mode. The goal is to have secure encryption modes with the additional property that exhaustive key-search attacks on them are slowed down by a factor equal to the number of blocks in the ciphertext. We give a new notion concerned with the privacy of keys that provably captures this key-search resistance property. We suggest a new characterization of AONTs and establish that the resulting all-or-nothing encryption paradigm yields secure encryption modes that also meet this notion of key privacy. A consequence of our new characterization is that we get more efficient ways of instantiating the all-or-nothing encryption paradigm. We describe a simple block-cipher-based AONT and prove it secure in the Shannon Model of a block cipher. We also give attacks against alternate paradigms that were believed to have the above key-search resistance property.

## 1  Introduction

In this paper, we study all-or-nothing transforms in the context of the original application for which they were introduced by Rivest [20]. The goal is to increase the difficulty of an exhaustive key search on symmetric encryption schemes, while keeping the key size the same and not overly burdening the legitimate users.

BACKGROUND AND MOTIVATION. Block ciphers, such as DES, can be vulnerable to exhaustive key-search attacks due to their relatively small key-sizes. The attacks on block ciphers also carry over to symmetric encryption schemes based on the block ciphers (hereafter called, *encryption modes*). One way to get better resistance to key-search attacks, is to use a longer key (either with the existing block cipher or with a next-generation block cipher such as AES). This, however, can be an expensive proposition, since it would necessitate changing the existing cryptographic hardware and software implementing these encryption modes. In some cases, the preferred approach might be to squeeze a little more security out of the existing encryption modes using some efficient pre-processing techniques.

M. Bellare (Ed.): CRYPTO 2000, LNCS 1880, pp. 359–375, 2000.
© Springer-Verlag Berlin Heidelberg 2000

Rivest observed that with most of the popular encryption modes, it is possible to obtain one block of the message by decrypting *just one* block of the ciphertext. With the cipher-block-chaining mode (CBC) [18], for example, given any two consecutive blocks of ciphertext, it is possible to decrypt a single value and obtain one block of the message. Thus the time to check a candidate key is that of just one block cipher operation. Such modes are said to be *separable*. Rivest suggests designing *strongly non-separable* encryption modes. As defined in [20], strongly non-separable encryption means that it should be infeasible to determine even one message block (or any property of a particular message block) without decrypting *all* the ciphertext blocks.

The all-or-nothing encryption paradigm was suggested as a means to achieve strongly non-separable encryption modes. It involves using an *all-or-nothing transform* (AONT) as a pre-processing step to an ordinary encryption mode. As defined in [20], an AONT is an efficiently computable transformation, mapping sequences of blocks to sequences of blocks, with the following properties:

- Given the output sequence of blocks, one can easily obtain the original sequence of blocks.
- Given all but one of the output sequence of blocks, it is computationally infeasible to determine any function of any input block.

It is necessary that an AONT be randomized so that a chosen input does not yield a known output. Note that in spite of the privacy parallel in their definitions, an AONT is distinct from an encryption scheme. In particular, there is no secret-key information associated with an AONT. However, it is suggested that if the output of an AONT is encrypted, with say the codebook mode (ie. a secret-keyed block cipher applied block by block), then the resulting scheme will not only be secure as an encryption scheme but also be strongly non-separable.

We are interested in encryption modes wherein an exhaustive key-search is somehow dependent on the size of the ciphertext. This is the primary motivation for using strongly non-separable modes. The intuition is that brute-force searches on such encryption modes would be slowed down by a factor equal to the number of blocks in the ciphertext. But does strong non-separability really capture this property? A reason to believe otherwise is that the property we want is concerned more with the privacy of the underlying *key* than that of the data. Consider the (admittedly, contrived) example of an encryption mode that, in addition to the encrypted message blocks, always outputs a block that is the result of the underlying block cipher on the string of all 0s. Such a mode could turn out to be strongly non-separable although it clearly does not possess the property we desire: a key-search adversary can test any candidate key by decrypting just the block enciphering the 0 string. One could think of more subtle ways for some other "invariable information" about the key being leaked that would illustrate this point more forcefully. Strong non-separability does capture some strong (data-privacy) property, but that is not the one we are interested in. What we need here instead is a suitable notion of key-privacy. We want encryption modes that have this property, as well as the usual data-privacy ones.

OUR NOTIONS AND MODEL. We give a notion, called non-separability of keys, that formalizes the inability of an adversary to gain *any* information about the underlying key, without "decrypting" *every* block of the ciphertext. The notion can be informally described through the following interactive protocol: an adversary $A$ is first given two randomly selected keys $a_0$ and $a_1$. $A$ then outputs a message $x$ and gets back, based on a hidden bit $b$, the encryption $y$ of $x$ under $a_b$. We ask that it be infeasible for a hereafter "restricted" $A$ to guess $b$ correctly with probability significantly more than 0.5. The restriction we put on $A$ is in limiting how it can use its knowledge of $a_0$ and $a_1$ in trying to guess $b$.

In order to make the above restriction meaningful, we describe our notion in the Shannon Model of a block cipher [21]. This model has been used in similar settings before [17,1]. Roughly speaking, the model instantiates an independent random permutation with every different key. We discuss the limitations of the model and their implications to our results in Section 7.

We show that our notion captures our desired key-search resistance property. That is, we prove that exhaustive key-search attacks on encryption modes secure in the non-separability of keys sense are slowed down by a factor equal to the number of blocks in the ciphertext. Our notion is orthogonal to the standard notions of data-privacy. In particular, the notion by itself does not imply security as an encryption scheme. It can, however, be used in conjunction with any notion of data-privacy to define a new encryption goal.

We want to justify the intuition that all-or-nothing encryption modes are secure encryption modes that also have the key-search resistance property. Recall that an all-or-nothing encryption mode is formed by composing an AONT with an ordinary encryption mode. The definition of an AONT from [20], however, is more of an intuitive nature than of sufficient rigor to establish any claims with it. One problem with the definition, as pointed out by Boyko [10], is that it speaks of information leaked about a *particular* message block. In our context, information leaked about the message as a whole, say the XOR of all the blocks, can be just as damaging. A formal characterization of AONTs was later given by Boyko [10]. He makes a case for defining an AONT with respect to any (and variable amount of) missing information, as opposed to a missing block. While this is certainly more general and probably necessary in some settings, we believe that in the context of designing efficient encryption modes with the key-search resistance property, a formalization with respect to a missing block is preferable. It turns out that even this weaker formalization is enough to realize our goal through the all-or-nothing encryption paradigm. An advantage of a weaker characterization of AONTs, as we will see later, is that we can build more efficient constructions that meet it. Our characterization of AONTs is tailored to their use in designing encryption modes that have the desired key-search resistance property.

OUR SECURITY RESULTS. We establish that all-or-nothing encryption modes (using our definition of an AONT) are secure in the non-separability of keys sense as well as being secure against chosen-plaintext attack. Our analysis relates the security of the all-or-nothing encryption paradigm to the security of the underlying AONT in a precise and quantitative way.

We give an efficient block-cipher-based construction of an AONT. Our construction is a simplified version of Rivest's "package transform". The package transform may well have some stronger security properties than ours, but it turns out that even our simplified version is secure under our definition of an AONT. The proof of this is also in the Shannon Model of a block cipher. With this, we can now get all-or-nothing encryption modes that cost only two times the cost of normal CBC encryption, while with the package transform, the resulting modes had cost about three times the cost of CBC.

In addition, we give attacks against alternate paradigms believed to have the key-search resistance property. We show that a paradigm claimed to capture this property [7] does not actually do so. There seem to be several misconceptions about what it takes to capture this property. One of these is that symmetric encryption schemes secure against chosen-ciphertext attack or some even stronger (data-privacy) notion may already do so. We show otherwise by giving an attack on a scheme secure in the strongest data-privacy sense yet known.

RELATED WORK. Rivest's all-or-nothing encryption is not the only way known to get more security out of a fixed number of key bits. Alternate approaches include DESX (an idea due to Rivest that was analyzed by Kilian and Rogaway [17]) and those favoring a long key set-up time, such as the method of Quisquater et al. [19]. These approaches do not incur the fixed penalty for every encrypted block that all-or-nothing encryption does, but unlike all-or-nothing encryption, they cannot work with existing encryption devices and software without changing the underlying encryption algorithm. In either case, as Rivest points out, the different approaches are complementary and can be easily combined.

Several approaches to the design of AONTs have been discussed by Rivest [20]. Our construction, like the package transform, happens to be based on a block cipher. The hash function based OAEP transform was proven secure in the Random Oracle Model by Boyko [10]. An information-theoretic treatment of a weaker form of AONTs has been given by Stinson [22]. Constructions based solely on the assumption of one-way functions have been given by Canetti et al. [12]. However, these are somewhat inefficient for practice. Applications of AONTs go beyond just the one considered in this work. They can be used to make fixed block-size encryption schemes more efficient [15], reduce communication requirements [20,14], and protect against partial key exposure [12].

## 2    Preliminaries

We use a standard notation for expressing probabilistic experiments and algorithms. Namely, if $A(\cdot, \cdot, \ldots)$ is a probabilistic algorithm then $a \leftarrow A(x_1, x_2, \ldots)$ denotes the experiment of running $A$ on inputs $x_1, x_2, \ldots$ and letting $a$ be the outcome, the probability being over the coins of $A$. Similarly, if $A$ is a set then $a \leftarrow A$ denotes the experiment of selecting a point uniformly from $A$ and assigning $a$ this value.

BLOCK CIPHERS. For any integer $l \geq 1$ let $P_l$ denote the space of all $(2^l)!$ permutations on $l$ bits. A block cipher is a map $F : \{0,1\}^k \times \{0,1\}^l \mapsto \{0,1\}^l$. For every $a \in \{0,1\}^k$, $F(a, \cdot) \in P_l$. We define $F_a$ by $F_a(x) = F(a,x)$. Let $\mathrm{BC}(k,l)$ denote the space of all block ciphers with parameters $k$ and $l$ as above.

We model $F$ as an *ideal* block cipher in the sense of Shannon, in that $F$ is drawn at random from $\mathrm{BC}(k,l)$. Given $F \in \mathrm{BC}(k,l)$, we define $F^{-1} \in \mathrm{BC}(k,l)$ by $F^{-1}(a,y) = F_a^{-1}(y)$ for $a \in \{0,1\}^k$. Note that in the experiments to follow there is no "fixed" cipher; we will refer to an ideal block cipher $F$, access to which will be via oracles for $F(\cdot, \cdot)$ and $F^{-1}(\cdot, \cdot)$.

ENCRYPTION MODES. Formally, an encryption mode based on a block cipher $F$ is given by a triple of algorithms, $\Pi = (\mathcal{K}, \mathcal{E}, \mathcal{D})$, where

- $\mathcal{K}$, the *key generation algorithm*, is a probabilistic algorithm that takes a security parameter $k \in \mathsf{N}$ (provided in unary) and returns a key $a$ specifying permutations $F_a$ and $F_a^{-1}$.

- $\mathcal{E}$, the *encryption algorithm*, is a probabilistic or stateful algorithm that takes permutations $F_a$ and $F_a^{-1}$ (as oracles) and a message $x \in \{0,1\}^*$ to produce a ciphertext $y$.

- $\mathcal{D}$, the *decryption algorithm*, is a deterministic algorithm which takes permutations $F_a$ and $F_a^{-1}$ (as oracles) and ciphertext $y$ to produce either a message $x \in \{0,1\}^*$ or a special symbol $\perp$ to indicate that the ciphertext was invalid.

We require that for all $a$ which can be output by $\mathcal{K}(1^k)$, for all $x \in \{0,1\}^*$, and for all $y$ that can be output by $\mathcal{E}^{F_a, F_a^{-1}}(x)$, we have that $\mathcal{D}^{F_a, F_a^{-1}}(y) = x$. We also require that $\mathcal{K}$, $\mathcal{E}$ and $\mathcal{D}$ can be computed in polynomial time. As the notation indicates, the encryption and decryption algorithms are assumed to have oracle access to the permutations specified by the key $a$ but do not receive the key $a$ itself. This is the distinguishing feature of encryption modes over other types of symmetric encryption schemes.

NOTION OF SECURITY. We recall a notion of security against chosen-plaintext attack for symmetric encryption schemes, due to Bellare et al. [2], suitably modified for encryption modes in the Shannon Model of a block cipher. This itself is an adaptation to the private-key setting of the definition of "polynomial security" for public-key encryption given by Goldwasser and Micali [13].

**Definition 1. [Indistinguishability of Encryptions]**     *Let* $\Pi = (\mathcal{K}, \mathcal{E}, \mathcal{D})$ *be an encryption mode. For an adversary $A$ and $b = 0,1$ define the experiment*

Experiment $\mathrm{Exp}_{\Pi}^{\mathrm{ind}}(A, b)$

$\qquad F \leftarrow \mathrm{BC}(k,l); \ \ a \leftarrow \mathcal{K}(1^k); \ \ (x_0, x_1, s) \leftarrow A^{F, F^{-1}, \mathcal{E}^{F_a, F_a^{-1}}} \ (\mathsf{find});$

$\qquad y \leftarrow \mathcal{E}^{F_a, F_a^{-1}}(x_b); \ \ d \leftarrow A^{F, F^{-1}, \mathcal{E}^{F_a, F_a^{-1}}} \ (\mathsf{guess}, y, s); \ \ \mathsf{return} \ d.$

*Define the advantage of $A$ and the advantage function of $\Pi$ respectfully, as follows:*

$$\mathsf{Adv}_{\Pi}^{\mathrm{ind}}(A) = \Pr\left[\mathrm{Exp}_{\Pi}^{\mathrm{ind}}(A, 1) = 1\right] - \Pr\left[\mathrm{Exp}_{\Pi}^{\mathrm{ind}}(A, 0) = 1\right]$$

$$\mathsf{Adv}_{\Pi}^{\mathrm{ind}}(t, m, p, q, \mu) = \max_{A} \{\mathsf{Adv}_{\Pi}^{\mathrm{ind}}(A)\}$$

*where the maximum is over all $A$ with "time-complexity" $t$, making at most $p$ queries to $F/F^{-1}$, choosing $|x_0| = |x_1|$ such that $|y| = ml$ and making at most $q$ queries to $\mathcal{E}^{F_a, F_a^{-1}}$, these totaling at most $\mu$ bits.* ∎

Here the "time-complexity" is the worst case total execution time of experiment $\mathrm{Exp}_{\Pi}^{\mathrm{ind}}(A, b)$ plus the size of the code of $A$, in some fixed RAM model of computation. This convention is used for other definitions in this paper, as well. The notation $A^{F, F^{-1}, \mathcal{E}^{F_a, F_a^{-1}}}$ indicates an adversary $A$ with access to an encryption oracle $\mathcal{E}^{F_a, F_a^{-1}}$ and oracles for $F$ and $F^{-1}$. The encryption oracle is provided so as to model chosen plaintext attacks, while the $F/F^{-1}$ oracles appear since we are working in the Shannon Model of a block cipher.

## 3   Non-separability of Keys

We give a notion of key-privacy to capture the requirement that *every* block of the ciphertext must be "decrypted" before *any* information about underlying key (including that the key may not be the "right" one) is known. This notion is formally captured through a game in which an adversary $A$ is imagined to run in two stages. In the find stage, $A$ is given two randomly selected keys $a_0$ and $a_1$, and is allowed to choose a message $x$ along with some state information $s$. In the guess stage, it is given a random ciphertext $y$ of the plaintext $x$, under one of the selected keys, along with the state information $s$. Let $m = \frac{|y|}{l}$ be the number of blocks in the challenge $y$. The adversary is given access to oracles for $F$ and $F^{-1}$ in both stages. In the guess stage, we impose a restriction that the adversary may make at most $(m-1)$ queries to $F_{a_0}/F_{a_0}^{-1}$ and at most $(m-1)$ queries to $F_{a_1}/F_{a_1}^{-1}$. The adversary "wins" if it correctly identifies which of the two selected keys was used to encrypt $x$ in the challenge.

**Definition 2. [Non-Separability of Keys]**   *Let $\Pi = (\mathcal{K}, \mathcal{E}, \mathcal{D})$ be an encryption mode. For an adversary $A$ and $b = 0, 1$ define the experiment*

Experiment $\mathrm{Exp}_{\Pi}^{\mathrm{nsk}}(A, b)$
  $F \leftarrow \mathrm{BC}(k, l);\ (a_0, a_1) \leftarrow \mathcal{K}(1^k);\ (x, s) \leftarrow A^{F, F^{-1}}(\mathsf{find}, a_0, a_1);$
  $y \leftarrow \mathcal{E}^{F_{a_b}, F_{a_b}^{-1}}(x);\ d \leftarrow A^{F, F^{-1}}(\mathsf{guess}, y, s);\ \ \mathsf{return}\ d.$

*Define the advantage of $A$ and the advantage function of $\Pi$ respectfully, as follows:*

$$\mathsf{Adv}_{\Pi}^{\mathrm{nsk}}(A) = \Pr\left[\mathrm{Exp}_{\Pi}^{\mathrm{nsk}}(A, 1) = 1\right] - \Pr\left[\mathrm{Exp}_{\Pi}^{\mathrm{nsk}}(A, 0) = 1\right]$$

$$\mathsf{Adv}_{\Pi}^{\mathrm{nsk}}(t, m, p) = \max_{A} \{\mathsf{Adv}_{\Pi}^{\mathrm{nsk}}(A)\}$$

*where the maximum is over all $A$ with time complexity $t$, making at most $p$ queries to $F/F^{-1}$ such that, for $m = \frac{|y|}{l}$, at most $(m-1)$ of these are to $F_{a_0}/F_{a_0}^{-1}$ and at most $(m-1)$ of these are to $F_{a_1}/F_{a_1}^{-1}$ in the guess stage.* ∎

Note that this definition only captures a notion concerned with the privacy of the underlying key. It does not imply security as an encryption scheme. The notion can be used in conjunction with the data-privacy notions of encryption schemes. Indeed, it also makes sense to talk about the key-privacy of encryption modes that are secure under data-privacy notions that are stronger than the one captured by Definition 1.

NON-SEPARABILITY OF KEYS VERSUS KEY-SEARCH. We show that security in the non-separability of keys sense implies that "key-search" attacks are slowed down by a factor proportional to the number of blocks in the ciphertext. Indeed this is the primary motivation of using encryption modes secure in the non-separability of keys sense. Thus this implication may be taken as evidence of having a "correct" definition in Definition 2.

In the key-search notion, we measure the success of an adversary $A$ in guessing the underlying key $a$ given a ciphertext $y$ (of a plaintext $x$ of its choice). The insecurity of an encryption mode in the key-search sense is given by the maximum success over all adversaries using similar resources.

**Definition 3. [Key-Search]**   *Let* $\Pi = (\mathcal{K}, \mathcal{E}, \mathcal{D})$ *be an encryption mode. For an adversary $A$ define the experiment*

Experiment $\mathrm{Exp}_{\Pi}^{\mathrm{ks}}(A)$
   $F \leftarrow \mathrm{BC}(k, l); \quad a \leftarrow \mathcal{K}(1^k); \quad (x, s) \leftarrow A^{F, F^{-1}}(\mathsf{select}); \quad y \leftarrow \mathcal{E}^{F_a, F_a^{-1}}(x);$
   $a' \leftarrow A^{F, F^{-1}}(\mathsf{predict}, y, s); \quad \text{If } a' = a \text{ then } d \leftarrow 1 \text{ else } d \leftarrow 0; \quad \text{return } d.$

*Define the success of $A$ and the success function of $\Pi$ respectfully, as follows:*

$$\mathsf{Succ}_{\Pi}^{\mathrm{ks}}(A) = \Pr\left[\mathrm{Exp}_{\Pi}^{\mathrm{ks}}(A) = 1\right]$$

$$\mathsf{Succ}_{\Pi}^{\mathrm{ks}}(t, m, p) = \max_{A}\{\mathsf{Succ}_{\Pi}^{\mathrm{ks}}(A)\}$$

*where the maximum is over all $A$ with time complexity $t$, making at most $p$ queries to $F/F^{-1}$ and choosing $|x|$ such that $|y| = ml$.* ∎

Note that there are no restrictions (for any key) on how many of $A$'s $p$ queries to $F/F^{-1}$ are in the predict stage.

Our first theorem establishes our claim about the implication. We emphasize that this result, and every other result (on encryption) in this work, are on encryption modes. In particular, we assume that the encryption and decryption algorithms can be described given just oracle access to permutations and do not need the (block-cipher) key specifying these permutations.

**Theorem 1. [Non-Separability of Keys Slows Down Key-Search]** *Suppose $\Pi$ is an encryption mode using an ideal cipher of key length $k$ and block length $l$. Then*

$$\mathsf{Succ}_{\Pi}^{\mathrm{ks}}(t, m, p) \leq \mathsf{Adv}_{\Pi}^{\mathrm{nsk}}(t', m, p) + \left(2 \cdot \left\lfloor \frac{p}{m} \right\rfloor + 4\right) \cdot \frac{1}{2^k - 1}$$

*where $t' = t + \mathcal{O}(k + ml + pl)$.* ∎

The proof of Theorem 1 appears in the full version of this paper [11]. We sketch only the basic idea here. The proof uses a fairly standard contradiction argument. Assume $B$ is an adversary in the key-search sense. We construct a non-separability of keys adversary $A$, that uses $B$ and has the claimed complexity. $A$ will run $B$ using its oracles to answer $B$'s queries and then make its guess based on how $B$ behaves. A complication arises due to the fact that there is a restriction on the number of queries $A$ can make with the two keys it is given in the find stage and that $B$ is not subject to this restriction. We get around this by having $A$ keep track of how many queries $B$ makes using these keys. If $B$ ever exceeds the amount $A$ is restricted to for one of these keys, then $A$ guesses that its challenge was encrypted under this key. We then show that the probability of a false positive is small.

We next give an interpretation of Theorem 1. Say $\Pi$ is secure in the sense of Definition 2. Then we know that for reasonable values of $t', m, p$ the value of the $\mathsf{Adv}_\Pi^{\mathrm{nsk}}(t', m, p)$ is negligible. The theorem says that for a reasonable value of $t$ we could expect $\mathsf{Succ}_\Pi^{\mathrm{ks}}(t, m, p)$ to be not much more than $\left(2 \cdot \left\lfloor \frac{p}{m} \right\rfloor + 4\right) \cdot \frac{1}{2^k - 1}$. This means that after $p$ queries to $F/F^{-1}$ there is roughly only a $\left(\frac{p}{m} \cdot 2^{-k}\right)$ chance of finding the key. Contrast this with an encryption mode where each query to $F/F^{-1}$ could potentially rule out a candidate key. Then we would expect an $(p \cdot 2^{-k})$ chance of finding the underlying key. Thus we have succeeded in reducing the success of a key-search attack by a factor of $m$, as promised. (The factor of 2 in the theorem comes about due to the scaling factors implicit in the advantage function of Definition 2.)

## 4     Separable Attacks

It is easy to check that none of the commonly used encryption modes, such as the cipher-block-chaining (CBC) mode and the counter mode (CTR) (see [2] for a description of these modes) have the key-search resistance property we desire. There seems to be a belief that some of the existing notions and schemes may already capture this property. We show that this is unlikely by giving attacks on some paradigms that cover a large number of "promising" candidates.

ENCODE-THEN-ENCIPHER ENCRYPTION. The variable-input-length (VIL) enciphering paradigm has been suggested in [7] as a practical solution to the problem of "encrypting" messages of variable and arbitrary lengths to a ciphertext of the same length. (Since enciphering is deterministic, it cannot be considered to be secure encryption. However, as pointed out in [8], simply encoding messages with some randomness, prior to enciphering, is enough to guarantee security as an encryption scheme.) It is claimed in [7] that the VIL paradigm also provides a way to provably achieve the goal of exhaustive key-search being slowed down proportional to the length of the ciphertext. However, we show that this is *not* the case by describing a simple but effective attack on their VIL mode of operation. The attack is effective even when the messages are encoded before enciphering. We point out here (deferring details to the full version of this paper [11]) that even "super" VIL modes [9] would be susceptible to this attack.

We describe a simplified version of an example of a VIL mode given in [7]. The construction first computes a pseudorandom value by applying a CBC-MAC on the plaintext. In the second step, the counter mode is used to "encrypt" the plaintext using the pseudorandom value from the first step as the "counter". We now describe a simple attack on this example. Our attack exploits the fact that for messages longer than a few blocks, most of the blocks in the VIL mode are being encrypted in the CTR mode. The main ideas of the VIL mode are on how to pick the "counter" for the CTR mode and on how to format the last few blocks so as to enable message recoverability while still maintaining the length requirement. We observe that the attack is effective given any two blocks of a challenge ciphertext, and moreover, is independent of the "counter" value. Given, say, just $y_i = x_i \oplus F_a(r + i)$ and $y_j = x_j \oplus F_a(r + j)$, for some plaintext $x = x_1 \cdots x_n$, counter $r$ and indices $1 \leq i < j \leq n$, there is a test for any candidate key $a'$ that requires just two queries to $F^{-1}(\cdot, \cdot)$. The test is that the following relationship hold: $F^{-1}(a', y_j \oplus x_j) - F^{-1}(a', y_i \oplus x_i) = j - i$. This test can be carried out effectively in the VIL mode example and serves to show that this paradigm in general does not capture the goal of slowing down exhaustive key-search.

AUTHENTICATED ENCRYPTION. The most common misconception seems to be that some of the stronger notions of data-privacy or data-integrity for symmetric encryption capture the key-search resistance property that we do in Definition 2. We claim that all of these notions, however strong they may be, are orthogonal to our notion of key-privacy. We argue this for the case of authenticated encryption, which is one of the strongest notions of security considered in symmetric encryption. In particular, this notion implies other strong notions, including security against chosen-ciphertext attack. Informally described, authenticated encryption requires that it be infeasible for an adversary to get the receiver to accept as authentic a string $C$ where the adversary has not already witnessed $C$. Formal definitions appear in [4,8,16] along with methods to construct such schemes. One of the generic methods shown to be secure in the authenticated encryption sense is the "encrypt-then-MAC" paradigm [4]. In this paradigm, a ciphertext is formed by encrypting the plaintext to a string $C$ using a generic symmetric encryption scheme secure in the indistinguishability of encryptions sense and then appending to $C$ the output of a MAC on $C$. Clearly, if the underlying generic encryption scheme used does not have the property captured by our key-privacy notion, then neither would the resulting authenticated encryption scheme.

## 5    All-or-Nothing Transforms

The notion of an all-or-nothing transform (AONT) was suggested by Rivest [20] to enable a paradigm for realizing encryption modes with the key-search resistance property. The paradigm consists of pre-processing a message with an AONT and encrypting the result by an "ordinary" encryption mode. We give a systematic treatment of AONTs in this section. The paradigm itself will be discussed in Section 6.

SYNTAX. Formally, the syntax of an un-keyed AONT is given by a pair of algorithms, $\Pi = (\mathcal{E}, \mathcal{D})$, where

- $\mathcal{E}$, the *encoding algorithm*, is a probabilistic algorithm that takes a message $x \in \{0, 1\}^*$ to produce a pseudo-ciphertext $y$.
- $\mathcal{D}$, the *decoding algorithm*, is a deterministic algorithm which takes a pseudo-ciphertext $y$ to produce either a message $x \in \{0, 1\}^*$ or a special symbol $\perp$ to indicate that the pseudo-ciphertext was invalid.

We require that for all $x \in \{0, 1\}^*$, and for all $y$ that can be output by $\mathcal{E}(x)$, we have that $\mathcal{D}(y) = x$. We also require that $\mathcal{E}$ and $\mathcal{D}$ be polynomial-time computable.

NOTION OF SECURITY. We give a new definition of security for AONTs. A block-length $l$ will be associated with an AONT. During the adversary's find stage it comes up with a message and some state information. The challenge is either a pseudo-ciphertext $y_0$ corresponding to the chosen plaintext $x$ or a random string $y_1$ of the same length as $y_0$. In the guess stage, it is allowed to adaptively see all but one of the challenge blocks and guess whether the part of challenge it received corresponds to $y_0$ or $y_1$.

**Definition 4. [All-Or-Nothing Transforms]**   *Let $\Pi = (\mathcal{E}, \mathcal{D})$ be an AONT of block length $l$. For an adversary $A$ and $b = 0, 1$ define the experiment*

Experiment $\mathrm{Exp}_{\Pi}^{\mathrm{aon}}(A, b)$
   $(x, s) \leftarrow A(\mathsf{find})$;
   $y_0 \leftarrow \mathcal{E}(x)$;
   $y_1 \leftarrow \{0, 1\}^{|y_0|}$;   // $(y_b = y_b[1] \cdots y_b[m]$ where $|y_b[i]| = l$ for $i \in \{1, \cdots, m\})$
   $d \leftarrow A^{\mathcal{Y}}(\mathsf{guess}, s)$;   // $(\mathcal{Y}$ takes an index $j \in \{1, \ldots, m\}$ and returns $y_b[j])$
   return $d$.

*Define the advantage of $A$ and the advantage function of $\Pi$ respectfully, as follows:*

$$\mathsf{Adv}_{\Pi}^{\mathrm{aon}}(A) = \Pr\left[\mathrm{Exp}_{\Pi}^{\mathrm{aon}}(A, 1) = 1\right] - \Pr\left[\mathrm{Exp}_{\Pi}^{\mathrm{aon}}(A, 0) = 1\right]$$

$$\mathsf{Adv}_{\Pi}^{\mathrm{aon}}(t, m) = \max_{A}\{\mathsf{Adv}_{\Pi}^{\mathrm{aon}}(A)\}$$

*where the maximum is over all $A$ with time complexity $t$, choosing $|x|$ such that $|y_0| = ml$ and making at most $(m - 1)$ queries to $\mathcal{Y}$.* ∎

Our formalization differs from that given by Boyko [10] in some significant ways. We require that the missing information be a block as opposed to some variable number of bits anywhere in the output. This captures a weaker notion, but as argued earlier, this is not necessarily a drawback. A consequence of this is that we are able to design more efficient AONTs. Notice that, in our missing-block formalization, we ask for the indistinguishability of the AONT output (with a missing block) from a *random string* of the same length. This is in contrast to the typical "indistinguishability of outputs on two inputs" required by [10] or any of the indistinguishability-based notions of encryption. We give some intuition for the need for this strengthening here. Consider a transform that added some known redundancy to every block (say, the first bit of every output block was always a 0). This alone would not make a transform insecure if we had used the "indistinguishability of outputs on two inputs" formulation for capturing all-or-nothingness, since the outputs on every input would have this same redundancy. However under our formulation we will find such a transform to be insecure since the random string would not necessarily have this redundancy. It turns out that if the all-or-nothing encryption paradigm is to have the key-search resistance property then such transforms cannot be considered to be secure as AONTs. Recall that the paradigm is to use an "ordinary" encryption mode on the output of an AONT. It is easy to see that it is essential that a key-search adversary "decrypting" one block of ciphertext should not be able to figure out if the decrypted value was the output of an AONT or not.

We have so far assumed the standard model, but we will often want to consider AONTs in some stronger model like the Random Oracle Model or the Shannon Model. Definition 4 can be suitably modified to accommodate these. For example, with the Shannon Model of an ideal block cipher $F$, we will assume that all parties concerned have access to $F$ and $F^{-1}$ oracles. The queries made to $F/F^{-1}$ become a part of the definition. We will define $\mathsf{Adv}_\Pi^{\mathrm{aon}}(t, m, p)$ rather than $\mathsf{Adv}_\Pi^{\mathrm{aon}}(t, m)$, where in addition to the usual parameters, $p$ is the maximum number of queries allowed to $F/F^{-1}$.

CONSTRUCTION. We give a construction based on the CTR mode of encryption. We describe the transform $\mathrm{CTRT} = (\mathcal{E}\text{-}\mathrm{CTRT}, \mathcal{D}\text{-}\mathrm{CTRT})$ of block length $l$, using an ideal cipher $F$ with key length $k$ and block length $l$, where $k \leq l$. (This condition can be easily removed and is made here only for the sake of exposition.) The message $x$ to be transformed is regarded as a sequence of $l$-bit blocks, $x = x[1] \ldots x[n]$ (padding is done first, if necessary). We define $\mathcal{E}\text{-}\mathrm{CTRT}^{F, F^{-1}}(x)$ and $\mathcal{D}\text{-}\mathrm{CTRT}^{F, F^{-1}}(x')$, as follows:

| Algorithm $\mathcal{E}\text{-}\mathrm{CTRT}^{F, F^{-1}}(x[1] \ldots x[n])$ | Algorithm $\mathcal{D}\text{-}\mathrm{CTRT}^{F, F^{-1}}(x')$ |
|---|---|
| $K' \leftarrow \{0, 1\}^l$ | Parse $x'$ as $x'[1] \ldots x'[n+1]$ |
| $K = K' \bmod 2^k$  // $(\lvert K \rvert = k)$ | $K' = x'[1] \oplus \cdots \oplus x'[n+1]$ |
| for $i = 1, \ldots, n$ do | $K = K' \bmod 2^k$  // $(\lvert K \rvert = k)$ |
| $\quad x'[i] = x[i] \oplus F_K(i)$ | for $i = 1, \ldots, n$ do |
| $x'[n+1] = K' \oplus x'[1] \oplus \ldots \oplus x'[n]$ | $\quad x[i] = x'[i] \oplus F_K(i)$ |
| return $x'[1] \ldots x'[n+1]$ | return $x[1] \ldots x[n]$ |

CTRT is a variant of Rivest's package transform [20] where one "pass" has been skipped altogether. Yet we find it to be secure in the sense of Definition 4.

**Theorem 2. [Security of CTRT]** *Suppose transform CTRT of block length $l$ uses an ideal cipher of key length $k$ and block length $l$ (where $k \leq l$). Then for any $t, m, p$ such that $m + p \leq 2^{k-1}$,*

$$\mathsf{Adv}_{\mathrm{CTRT}}^{\mathrm{aon}}(t, m, p) \leq \frac{m^2 + 8 \cdot p}{2^k}$$

The proof of this theorem appears in the full version of this paper [11]. We mention here only some of the key aspects of the proof. As long as at least one block of the output is missing, the key $K$ used to "encrypt" the message blocks is information theoretically hidden. The main step in the analysis is to bound the probability of an adversary calling its oracles with key $K$. In doing this, we need to be particularly careful with the fact that we allow adversaries to be adaptive. Another issue that complicates matters is the injectivity of ideal ciphers. For example, we cannot conclude that if an adversary has never queried its oracles with key $K$, then its "challenges" must be indistinguishable to it. The injectivity makes certain conditions impossible with the "AONT-derived" challenge that are possible with the "random" challenge.

It is conceivable that the package transform of Rivest [20] is actually secure in the strong sense captured by Boyko [10]. CTRT, on the other hand, is clearly *insecure* in that strong sense. (Note that CTRT would also have been secure in the sense given by Rivest [20].) However, as we will see next, it turns out that CTRT is strong enough that when used to realize the all-or-nothing encryption paradigm, the resulting mode will have the properties we desire.

# 6     All-or-Nothing Encryption

The all-or-nothing encryption paradigm consists of composing an AONT with an "ordinary" encryption mode. We study the particular case when the ordinary encryption mode is the codebook (ie. ECB) mode. It is easy to see that the codebook mode by itself is not secure encryption. However it does have many advantages over some of the other modes. In particular, it is simple, efficient, length-preserving, and admits an efficient parallel implementation. Following [20], we will refer to an AONT followed by the codebook mode as the "all-or-nothing codebook mode". We will establish the security of the all-or-nothing codebook mode in the theorems to follow. Similar results can be derived when some other reasonable mode is used in place of the codebook mode.

The all-or-nothing codebook mode is first and foremost a secure encryption scheme. We establish this in the following theorem.

**Theorem 3. [Security in the** indistinguishability of encryptions **sense]** *Suppose $\Pi = (\mathcal{K}, \mathcal{E}, \mathcal{D})$ is an all-or-nothing codebook mode using an ideal cipher of key length $k$ and block length $l$ and an all-or-nothing transform $\Pi' = (\mathcal{E}', \mathcal{D}')$ of block length $l$. Let $T$ be the time to decode a $ml$ bit string using $\mathcal{D}'$ and $nl$ be the length of a decoded $ml$ bit string. Then for any $n \geq 2$ and any $p, q, \mu$,*

$$\mathsf{Adv}_{\Pi}^{\mathrm{ind}}(t, m, p, q, \mu) \leq 2m \cdot \mathsf{Adv}_{\Pi'}^{\mathrm{aon}}(t', m) + \frac{2mp}{2^k} + \frac{2m}{2^l}$$

*where $t' = t + (\frac{\mu}{l} + m - 1) \cdot T + \mathcal{O}(ml + pl + \mu)$.* ∎

The proof of this theorem appears in the full version of this paper [11]. The intuition behind the result is as follows. An AONT has the property that the chances of a collision amongst the blocks of its output (even across multiple queries made by an adaptive adversary) is small. Thus when an AONT is composed with the codebook mode, we have that with high probability each block of the ciphertext is a result of having enciphered on a new point. Note that although the codebook mode is deterministic, the fact that the AONT itself is probabilistic makes the resulting all-or-nothing codebook mode probabilistic. The main part of the proof is in formalizing and establishing this property of AONTs. We show that if this property did not hold for some transform, then that transform could not be secure as an AONT.

We next show that the all-or-nothing codebook mode also has the desired key-search resistance property.

**Theorem 4. [Security in the non-separability of keys sense]** *Suppose* $\Pi = (\mathcal{K}, \mathcal{E}, \mathcal{D})$ *is an all-or-nothing codebook mode using an ideal cipher of key length* $k$ *and block length* $l$ *and an all-or-nothing transform* $\Pi' = (\mathcal{E}', \mathcal{D}')$ *of block length* $l$. *Then*

$$\mathsf{Adv}_{\Pi}^{\mathrm{nsk}}(t, m, p) \leq 2m \cdot \mathsf{Adv}_{\Pi'}^{\mathrm{aon}}(t', m)$$

*where* $t' = t + \mathcal{O}(ml + pl)$.

*Proof.* We use a contradiction argument to prove this. Let $B$ be an adversary in the **non-separability of keys** sense. We construct an **all-or-nothing** adversary $A$, that uses $B$ and has the claimed complexity.

Since we are assuming an **all-or-nothing** adversary $A$ that does not receive $F/F^{-1}$ oracles, these must be simulated when running $B$. The cost of this simulation will appear in the time complexity of $A$.

We use the notation $(+, K, z)$ (respectively, $(-, K, z)$) to indicate a query to $F$ (respectively, $F^{-1}$) with key $K$ and a $l$-bit string $z$; the expected response being $F_K(z)$ (respectively, $F_K^{-1}(z)$).

For an integer $m$ let $[m] = \{1, \cdots, m\}$. For an $ml$ bit string $z$ let $z = z[1] \cdots z[m]$ such that $|z[i]| = l$ for $i \in [m]$.

The adversary $A$ using adversary $B$ is given in Figure 1. The idea is the following: $A$ first picks two keys $a_0, a_1$ to simulate the experiment underlying **non-separability of keys** for $B$. It runs $B$, answering all of its queries by simulating the oracles, until $B$ ends its find stage by returning some string $x$. $A$ returns $x$ as the output of its own find stage. In its guess stage, $A$ will pick a random bit $d$ and an index $j$. $A$ will then ask its oracle $\mathcal{Y}$ for all but the $j$-th challenge block. It then uses the key $a_d$ to encipher these blocks to get all blocks of a string $z$ other than $z[j]$. $A$ assigns a random value to $z[j]$ and then runs $B$'s guess stage with the challenge being $z$. It will simulate $B$'s oracles as before, but halt if $B$ ever asks the query $(-, a_d, z[j])$ or $(+, a_d, F_{a_d}^{-1}(z[j]))$. If $B$ was halted, then $A$ outputs a random bit as its guess. Otherwise, it checks to see if $B$ was correct. $A$ guesses that its challenge must have been "real" if $B$ is correct and outputs a random bit otherwise.

| Algorithm $A(\mathsf{find})$ | Algorithm $A^{\mathcal{Y}}(\mathsf{guess},(s,x,a_0,a_1))$ |
|---|---|
| $a_0, a_1 \leftarrow \mathcal{K}(1^k)$ <br> run $B(\mathsf{find}, a_0, a_1)$ using SimBOr <br> let $(x,s)$ be the output of $B$ <br> $s' \leftarrow (s, x, a_0, a_1)$ <br> return $(x, s')$ <br><br> Subroutine SimBOr <br>   if $B$ makes a query $(+, K, u)$ <br>     then answer $B$ with $F_K(u)$ <br>   if $B$ makes a query $(-, K, u)$ <br>     then answer $B$ with $F_K^{-1}(u)$ | $d \leftarrow \{0,1\}$ <br> $j \leftarrow [m]$ <br> for $(i \in [m]) \wedge (i \neq j)$ do <br>   $y[i] \leftarrow \mathcal{Y}(i)$ <br>   $z[i] \leftarrow F_{a_d}(y[i])$ <br> $z[j] \leftarrow \{0,1\}^l$ <br> run $B(\mathsf{guess}, z, s)$ using SimBOr until <br>   $B$ makes a query $(-, a_d, z[j])$ or <br>   $B$ makes a query $(+, a_d, F_{a_d}^{-1}(z[j]))$ or <br>   $B$ halts <br> if $B$ halts then let $d'$ be its output <br>   else $b' \leftarrow \{0,1\}$ <br> if $d' = d$ then $b' \leftarrow 0$ else $b' \leftarrow \{0,1\}$ <br> return $b'$ |

**Fig. 1.** An all-or-nothing adversary using a non-separability of keys adversary

From the description, we have that the time complexity $t' = t + \mathcal{O}(ml + pl)$. Next we compute the advantage function. For $b \in \{0,1\}$ let Probability Space $b$ be that of the following underlying experiment:

$$(x, s') \leftarrow A(\mathsf{find}); \ y_0 \leftarrow \mathcal{E}'(x); \ y_1 \leftarrow \{0,1\}^{ml}; \ \mathcal{Y}(i) = y_b[i] \text{ for } i \in [m]:$$

For $b \in \{0,1\}$ let $\Pr_b[\cdot]$ denote the probability under Probability Space $b$.

$$\mathsf{Adv}_{\Pi'}^{\mathrm{aon}}(A) \stackrel{\mathrm{def}}{=} \Pr_0[\, A^{\mathcal{Y}}(\mathsf{guess}, s') = 0\,] - \Pr_1[\, A^{\mathcal{Y}}(\mathsf{guess}, s') = 0\,]$$

Hereafter, we suppress the superscripts and parenthesized parts for clarity.

Let Fail be the event that $B$ makes a query $(-, a_d, z[j])$ or $(+, a_d, F_{a_d}^{-1}(z[j]))$ where $j$ is the index of the block in $y$ that $A$ does not receive, $z[j]$ is the random block picked by $A$ and $d \in \{0,1\}$ is the bit that selects the key $a_0$ or $a_1$. We have

$$\mathsf{Adv}_{\Pi'}^{\mathrm{aon}}(A) \stackrel{\mathrm{def}}{=} \Pr_0[\, A = 0\,] - \Pr_1[\, A = 0\,]$$

$$= \Pr_0[\, A = 0 \mid \mathsf{Fail}\,] \cdot \Pr_0[\mathsf{Fail}] + \Pr_0[\, A = 0 \mid \overline{\mathsf{Fail}}\,] \cdot \Pr_0[\overline{\mathsf{Fail}}] -$$

$$\Pr_1[\, A = 0 \mid \mathsf{Fail}\,] \cdot \Pr_1[\mathsf{Fail}] - \Pr_1[\, A = 0 \mid \overline{\mathsf{Fail}}\,] \cdot \Pr_1[\overline{\mathsf{Fail}}]$$

Now from the description of $A$ we have:

$$\Pr_0[\mathsf{Fail}] = \Pr_1[\mathsf{Fail}] = \frac{m-1}{m}$$

$$\Pr_0[\, A = 0 \mid \mathsf{Fail}\,] = \Pr_1[\, A = 0 \mid \mathsf{Fail}\,] = 0.5$$

$$\Pr_0[\, A = 0 \mid \overline{\mathsf{Fail}}\,] = \Pr_0[B \text{ guesses correctly}] = 0.5 + 0.5 \cdot \mathsf{Adv}_{\Pi}^{\mathrm{nsk}}(B)$$

$$\Pr_1[\, A = 0 \mid \overline{\mathsf{Fail}}\,] = \Pr_1[B \text{ guesses correctly}] = 0.5$$

The derivation of these equalities is quite straightforward. The only one requiring explanation is the last one. To determine $\Pr_1[\,B$ guesses correctly$\,]$ we recall that in **Probability Space** 1 the challenge $z$ that $B$ receives is the codebook output with key $a_d$ on a random string $y_1$. The probability we want is that of $B$ guessing $d$ correctly. Given that $B$ does not see $y_1$, but just the output $z$ under the codebook mode with an ideal cipher, it is easy to see that the probability is exactly as claimed. Continuing with the advantage, we have

$$\mathsf{Adv}_{\Pi'}^{\mathrm{aon}}(A) = \frac{1}{m} \cdot \left( \frac{1}{2} + \frac{1}{2} \cdot \mathsf{Adv}_{\Pi}^{\mathrm{nsk}}(B) - \frac{1}{2} \right) = \frac{1}{2m} \cdot \mathsf{Adv}_{\Pi}^{\mathrm{nsk}}(B)$$

From this, we get the claimed relationship in the advantage functions. ∎

We have assumed in our treatment of the all-or-nothing encryption paradigm that the block size associated with the AONT is the same as that of the encryption mode following it. This assumption could be easily removed by making a few changes to our framework. However, in this form, it has certain implications when a block-cipher based construction, like CTRT, is used as the underlying AONT. We would require that the key space for the AONT block cipher be sufficiently large that brute force searches are infeasible. The block cipher used in the AONT does not have to be the same as that used to encrypt its output, though it certainly could be.

# 7   Comments and Open Problems

Our notions and proofs are in the Shannon Model of a block cipher. The model makes some strong assumptions and may be unsuitable for some of today's block ciphers with their delicate key schedules. This raises the concern that our results may not be telling us much about the "real-world". However, we claim that the results proven in this model are still meaningful since they permit "generic" attacks (ie. attacks that assume the underlying primitives to be "ideal"). In practice, most attacks disregard the cryptanalytic specifics of the block cipher anyway and instead treat it as a black-box transformation.

Our use of the Shannon Model for capturing non-separability of keys was driven by the need to correctly and usefully formalize the notion. It is hard to see how this could have been done in the standard model. In establishing our claims about all-or-nothing encryption modes, we used a definition of an AONT from the standard model. (Things would change very little if we had instead started with a definition from one of the other models.) However, to prove that our CTRT transform was secure as an AONT, we needed to work in the Shannon Model. It may be possible to prove such constructions in the standard model or perhaps some other weaker model, but this is something that is currently unknown. Note that we do know that there are AONTs that can be proven secure in the standard model [12]. However, for efficiency reasons, we do not consider these to be viable options in practice.

In the case of CTRT being used as the AONT, the cost of the resulting all-or-nothing codebook mode would be a factor of only two greater than CBC. From our results, we get that the resulting all-or-nothing codebook mode would be secure in the non-separability of keys sense, as well as being secure against chosen-plaintext attack. It would be interesting to see if all-or-nothing encryption modes (with some modifications, if required) could be shown to be secure against chosen-ciphertext attack.

## Acknowledgements

I am indebted to Mihir Bellare for providing invaluable support and direction with this work. Many of the ideas found here are due to him. I would also like to thank Jee Hea Lee, Sara Miner and the CRYPTO 2000 program committee and reviewers for their very helpful comments.

The author was supported in part by Mihir Bellare's 1996 Packard Foundation Fellowship in Science and Engineering and NSF CAREER Award CCR-9624439.

## References

1. W. AIELLO, M. BELLARE, G. DI CRESCENZO AND R. VENKATESAN, "Security amplification by composition: The case of doubly-iterated, ideal ciphers," *Advances in Cryptology - Crypto '98*, Lecture Notes in Computer Science Vol. 1462, H. Krawczyk ed., Springer-Verlag, 1998.
2. M. BELLARE, A. DESAI, E. JOKIPII AND P. ROGAWAY, "A concrete security treatment of symmetric encryption," *Proceedings of the 38th Symposium on Foundations of Computer Science*, IEEE, 1997.
3. M. BELLARE, J. KILIAN AND P. ROGAWAY, "The security of cipher block chaining," *Advances in Cryptology - Crypto '94*, Lecture Notes in Computer Science Vol. 839, Y. Desmedt ed., Springer-Verlag, 1994.
4. M. BELLARE AND C. NAMPREMPRE, "Authenticated encryption: Relations among notions and analysis of the generic composition paradigm," Report 2000/025, *Cryptology ePrint Archive*, http://eprint.iacr.org/, May 2000.
5. M. BELLARE AND P. ROGAWAY, "Random oracles are practical: A paradigm for designing efficient protocols," *Proceedings of the 1st Annual Conference on Computer and Communications Security*, ACM, 1993.
6. M. BELLARE AND P. ROGAWAY, "Optimal asymmetric encryption," *Advances in Cryptology - Eurocrypt '94*, Lecture Notes in Computer Science Vol. 950, A. De Santis ed., Springer-Verlag, 1994
7. M. BELLARE AND P. ROGAWAY, "On the construction of variable-input-length ciphers," *Fast Software Encryption '99*, Lecture Notes in Computer Science Vol. 1636, L. Knudsen ed., Springer-Verlag, 1999.
8. M. BELLARE AND P. ROGAWAY, "Encode-then-encipher encryption: How to exploit nonces or redundancy in plaintexts for efficient cryptography," Manuscript, December 1998, available from authors.
9. D. BLEICHENBACHER AND A. DESAI, "A construction of super-pseudorandom cipher," Manuscript, May 1999, available from authors.

10. V. Boyko, "On the security properties of OAEP as an all-or-nothing trans-
    form," *Advances in Cryptology - Crypto '99*, Lecture Notes in Computer Science
    Vol. 1666, M. Wiener ed., Springer-Verlag, 1999.
11. A. Desai, "The security of all-or-nothing encryption," Full version of this paper,
    available via: http://www-cse.ucsd.edu/users/adesai/.
12. R. Canetti, Y. Dodis, S. Halevi, E. Kushilevitz and A. Sahai, "Exposure-
    Resilient Cryptography: Constructions for the All-Or-Nothing Transform without
    Random Oracles," *Advances in Cryptology - Eurocrypt '00*, Lecture Notes in
    Computer Science Vol. 1807, B. Preneel ed., Springer-Verlag, 2000.
13. S. Goldwasser and S. Micali, "Probabilistic encryption," *J. of Computer and
    System Sciences*, Vol. 28, April 1984, pp. 270–299.
14. M. Jakobsson, J. Stern and M. Yung, "Scramble All, Encrypt Small," *Fast
    Software Encryption '99*, Lecture Notes in Computer Science Vol. 1636, L. Knud-
    sen ed., Springer-Verlag, 1999.
15. D. Johnson, S. Matyas, and M. Peyravian, "Encryption of long blocks using
    a short-block encryption procedure," Submission to IEEE P1363a, available via:
    http://grouper.ieee.org/groups/1363/contributions/peyrav.ps, Nov. 1996.
16. J. Katz and M. Yung, "Unforgeable Encryption and Adaptively Secure Modes
    of Operation," *Fast Software Encryption '00*, Lecture Notes in Computer Science
    Vol. ??, B. Schneier ed., Springer-Verlag, 2000.
17. J. Kilian and P. Rogaway, "How to protect DES against exhaustive key
    search," *Advances in Cryptology - Crypto '96*, Lecture Notes in Computer Science
    Vol. 1109, N. Koblitz ed., Springer-Verlag, 1996.
18. National Bureau of Standards, NBS FIPS PUB 81, "DES modes of operation,"
    U.S Department of Commerce, 1980.
19. J.-J. Quisquater, Y. Desmedt and M. Davio, "The importance of "good"
    key scheduling schemes (how to make a secure DES scheme with $\leq$ 48 bit keys),"
    *Advances in Cryptology - Crypto '85*, Lecture Notes in Computer Science Vol. 218,
    H. Williams ed., Springer-Verlag, 1985.
20. R. Rivest, "All-or-nothing encryption and the package transform," *Fast Software
    Encryption '97*, Lecture Notes in Computer Science Vol. 1267, E. Biham ed.,
    Springer-Verlag, 1997.
21. C. Shannon, "Communication theory of secrecy systems," *Bell Systems Technical
    Journal*, Vol. 28, No. 4, 1949, pp. 656-715.
22. D. Stinson, "Something about all-or-nothing (transforms)," Manuscript. Avail-
    able from: http://www.cacr.math.uwaterloo.ca/~dstinson/, June 1999.

# On the Round Security
# of Symmetric-Key Cryptographic Primitives

Zulfikar Ramzan and Leonid Reyzin

Laboratory for Computer Science
Massachusetts Institute of Technology
Cambridge, MA 02139
{zulfikar, reyzin}@theory.lcs.mit.edu
http://theory.lcs.mit.edu/{~zulfikar, ~reyzin}

**Abstract.** We put forward a new model for understanding the security of symmetric-key primitives, such as block ciphers. The model captures the fact that many such primitives often consist of iterating simpler constructs for a number of rounds, and may provide insight into the security of such designs.

We completely characterize the security of four-round Luby-Rackoff ciphers in our model, and show that the ciphers remain secure *even if the adversary is given black-box access to the middle two round functions*. A similar result can be obtained for message authentication codes based on universal hash functions.

## 1   Introduction

### 1.1   Block Ciphers

A *block cipher* is a family of permutations on a message space indexed by a secret key. Each permutation in the family deterministically maps *plaintext* blocks of some fixed length to *ciphertext* blocks of the same length; both the permutation and its inverse are efficiently computable given the key.

Motivated originally by the study of security of the block cipher DES [16], Luby and Rackoff provided a formal model for the security of block ciphers in their seminal paper [14]. They consider a block cipher to be secure ("super pseudorandom," or secure under both "chosen plaintext" and "chosen ciphertext" attacks) if, without knowing the key, a polynomial-time adversary with oracle access to both directions of the permutation is unable to distinguish it from a truly random permutation on the same message space. This definition is an extension of the definition of a pseudorandom function generator from [12], where the adversary has oracle access only to the forward direction of the function.[1]

---

[1]  The paper [14] also considers block ciphers that are just *pseudorandom*, or secure against chosen plaintext attack only, where the adversary has access only to the forward direction of the permutation.

M. Bellare (Ed.): CRYPTO 2000, LNCS 1880, pp. 376–393, 2000.
© Springer-Verlag Berlin Heidelberg 2000

## 1.2   The Natural Round Structure of Symmetric-Key Primitives

In addition to defining security of block ciphers, Luby and Rackoff also provided a construction of a secure block cipher based on a pseudorandom function generator. Their block cipher consists of four rounds of *Feistel [11] permutations*, each of which consists of an application of a pseudorandom function and an exclusive-or operation. Each round's output is used for the next round's input, except for the last round, whose output is the output of the block cipher.

Much of the theoretical research that followed the work of [14] focused on efficiency improvements to this construction (e.g., see [15], [18] and references therein). All of these variations can also be naturally broken up into rounds.

This theme of an inherent round structure in block ciphers is also seen extensively in practice. For example, a number of ciphers, including DES [16] and many of the AES submissions [17] have an inherent round structure (though not necessarily involving Feistel permutations), where the output of one round is used as input to the next.

In addition to block ciphers, constructions of other cryptographic primitives often also proceed in rounds. For example, universal-hash-function-based message authentication codes (UHF MACs) [6], [22], [9] can be viewed as consisting of two rounds. Moreover, cryptographic hash functions (e.g., MD-5 [19]), and the various message authentication schemes that are built on top of them (e.g., HMAC [1]), have an induced round structure as well.

Consequently, it should come as little surprise that cryptanalysts have often considered looking at individual rounds in order to better understand the security properties of a given design; for example, a large number of papers have been written analyzing reduced-round variants of block ciphers and hash functions (see [5], [21], and the references therein).

It thus seems that a theoretical framework incorporating the notion of rounds would be desirable. This paper proposes such a framework. Although our model is a simple extension of the classical models of security for symmetric primitives ([14], [12], [2]), it allows one to obtain a number of interesting results not captured by the traditional models. In particular, we analyze the security of the original Luby-Rackoff construction, some of its variants, and UHF MACs within our framework.

## 1.3   Our Contributions

**A New Model.** The definition of a secure block cipher from [14], or of a secure MAC from [3], allows the adversary only black-box access to the primitive. We develop the notion *round security*, which considers what happens when the adversary has additional access to some of the internal rounds of the computation of the primitive. We focus on block ciphers, but our techniques can be extended to other primitives such as MACs.

For example, in the case of block ciphers, we study what happens when the adversary is allowed, in addition to its chosen-plaintext and chosen-ciphertext queries, to input a value directly to some round $i$ of the block cipher and view the output after some round $j$, with restrictions on $i$ and $j$. The adversary's job is still the same: to distinguish whether the chosen-ciphertext and chosen-plaintext queries are being answered by the block cipher or by a random permutation. The queries to internal rounds are always answered by the block cipher.

As discussed below, this model allows us gain a better understanding of what makes symmetric constructions secure, and enables us to make statements about security that are not captured by the traditional model.

**Round Security of Luby-Rackoff Ciphers.** We completely characterize the round security of the Luby-Rackoff construction and its more efficient variants from [15] and [18]. That is, we precisely specify the sets of rounds that the adversary can access for the cipher to remain secure, and show that access to other sets of rounds will make the cipher insecure.

The cipher proposed by Luby and Rackoff [14] operates on a $2n$-bit string $(L, R)$ and can be described simply as follows:

$$S = L \oplus h_1(R)$$
$$T = R \oplus f_1(S)$$
$$V = S \oplus f_2(T)$$
$$W = T \oplus h_2(V),$$

where $h_1, h_2, f_1, f_2$ are pseudorandom functions, $\oplus$ represents the exclusive-or, and the output is $(V, W)$.

Naor and Reingold [15] demonstrated that pseudorandom functions $h_1$ and $h_2$ can be replaced by XOR-universal hash functions, thus suggesting that strong randomness is important only in the middle two rounds. We extend their observation by showing that, in fact, secrecy is important in the first and last rounds, while randomness (but no secrecy) is needed in the middle two rounds. Specifically, we show that:

– The cipher *remains secure* even if the adversary has oracle access to both $f_1$ and $f_2$.
– The cipher becomes *insecure* if the adversary is allowed access to any other round oracles.

Moreover, we demonstrate that instantiating $h_1$ and $h_2$ as hash functions instead of as pseudorandom functions does not significantly lower the round security of the block cipher, thus supporting the observation that strong randomness is not needed in the first and last rounds of the Luby-Rackoff construction.

**Round Security of Universal Hash Function MACs.** Using techniques in our paper, one can also characterize the round security of a class of Universal-Hash Function-based Message Authentication Codes (UHF MACs). In the first round, these UHF MACs apply a universal hash function $h$ to a relatively large message, to get a shorter intermediary string. Then, in the second round, they use a pseudorandom function $f$ on the shorter string to get a final tag. It turns out that:

- A UHF MAC remains *secure* if the adversary has oracle access to $f$.
- A UHF MAC is, in general, *insecure* if the adversary has oracle to $h$.

**Implications for the Random Oracle Model.** Our work has interesting implications for Luby-Rackoff ciphers and UHF MACs in the random oracle model. One can easily define security of block ciphers and MACs in this model given the work of [4]: one simply allows all parties (including the adversary) access to the same oracle, and the adversary has to succeed for a random choice of the oracle.

Our results imply that the Luby-Rackoff cipher remains secure in the random oracle model if one replaces the functions $f_1$ and $f_2$ with random oracles. That is, in the random oracle model, keying material will only be necessary for $h_1$ and $h_2$, which, as shown in [15] and [18], can be just (variants of) universal hash functions.

Similarly, the UHF MAC remains secure if the pseudorandom function, used in the second round, is replaced with a random oracle. Thus, again, in the random oracle model, keying material is needed only for the hash function.

Block ciphers have been analyzed in the random-oracle model before. For example, Even and Mansour [10] construct a cipher using a public random *permutation* oracle $P$ (essentially, the construction is $y = P(k_1 \oplus x) \oplus k_2$, where $k_1$ and $k_2$ constitute the key, $x$ is the plaintext, and $y$ is the resulting ciphertext). They show their construction is hard to invert and to existentially forge. We can recast their construction in our model, as a three-round cipher, where the adversary has access to the second round. Using the techniques in our paper, we can, in fact, obtain a stronger result; namely, that their cipher is super pseudorandom.

Of course, whether a scheme in the random oracle model can be instantiated securely in the real world (that is, with polynomial-time computable functions in place of random oracles) is uncertain, particularly in light of the results of Canetti, Goldreich and Halevi [7]. However, our results open up an interesting direction: is it possible to replace pseudorandom functions with unkeyed functions in any of the constructions we discuss?

## 2   Prior Definitions and Constructions

Below we describe the relevant definitions and prior constructions. Our presentation is in the "concrete" (or "exact") security model as opposed to the asymptotic model (though our results can be made to hold for either). Our treatment follows that of Bellare, Kilian, and Rogaway [3], and Bellare, Canetti, Krawczyk [2].

### 2.1   Definitions

**Notation.** For a bit string $x$, we let $|x|$ denote its length. If $x$ has even length, then $x^L$ and $x^R$ denote the left and right halves of the bits respectively; we sometimes write $x = (x^L, x^R)$. If $x$ and $y$ are two bit strings of the same length,

$x \oplus y$ denotes their bitwise exclusive OR. If $S$ is a probability space, then $x \xleftarrow{R} S$ denotes the process of picking an element from $S$ according to the underlying probability distribution. Unless otherwise specified, the underlying distribution is assumed to be uniform. We let $I_n$ denote the set of bit strings of length $n$: $\{0,1\}^n$.

By a finite function (or permutation) family $\mathcal{F}$, we denote a set of functions with common domain and common range. Let $\mathsf{Rand}^{k \to l}$ be the set of all functions going from $I_k$ to $I_l$, and let $\mathsf{Perm}^m$ be the set of all permutations on $I_m$. We call a finite function (or permutation) family *keyed* if every function in it can be specified (not necessarily uniquely) by a key $a$. We denote the function given by $a$ as $f_a$. We assume that given $a$, it is possible to efficiently evaluate $f_a$ at any point (as well as $f_a^{-1}$ in case of a keyed permutation family). For a given keyed function family, a key can be any string from $I_s$, where $s$ is known as "key length." (Sometimes it is convenient to have keys from a set other than $I_s$; we do not consider such function families simply for clarity of exposition—our results do not change in such a case.) For functions $f$ and $g$, $g \circ f$ denotes the function $x \mapsto g(f(x))$.

**Model of Computation.** The adversary $\mathcal{A}$ is modeled as a program for a Random Access Machine (RAM) that has black-box access to some number $k$ of oracles, each of which computes some specified function. If $(f_1, \ldots, f_k)$ is a $k$-tuple of functions, then $\mathcal{A}^{f_1, \ldots, f_k}$ denotes a $k$-oracle adversary who is given black-box oracle access to each of the functions $f_1, \ldots, f_k$. We define $\mathcal{A}$'s "running time" to be the number of time steps it takes plus the length of its description (to prevent one from embedding arbitrarily large lookup tables in $\mathcal{A}$'s description).

**Pseudorandom Functions and Block Ciphers.** The pseudorandomness of a keyed function family $\mathcal{F}$ with domain $I_k$ and range $I_l$ captures its computational indistinguishability from $\mathsf{Rand}^{k \to l}$. This definition is a slightly modified version of the one given by Goldreich, Goldwasser and Micali [12].

**Definition 1.** *A pseudorandom function family $\mathcal{F}$ is a keyed function family with domain $I_k$, range $I_l$, and key length $s$. Let $\mathcal{A}$ be a 1-oracle adversary. Then we define $\mathcal{A}$'s advantage as*

$$\mathsf{Adv}_{\mathcal{F}}^{\mathsf{prf}}(\mathcal{A}) = \left| \Pr[a \xleftarrow{R} I_s : \mathcal{A}^{f_a} = 1] - \Pr[f \xleftarrow{R} \mathsf{Rand}^{k \to l} : \mathcal{A}^f = 1] \right|.$$

*For any integers $q, t \geq 0$, we define an insecurity function $\mathsf{Adv}_{\mathcal{F}}^{\mathsf{prf}}(q,t)$:*

$$\mathsf{Adv}_{\mathcal{F}}^{\mathsf{prf}}(q,t) = \max_{\mathcal{A}} \{ \mathsf{Adv}_{\mathcal{F}}^{\mathsf{prf}}(\mathcal{A}) \}.$$

*The above maximum is taken over choices of adversary $\mathcal{A}$ such that:*

- *$\mathcal{A}$ makes at most $q$ oracle queries, and*

– the running time of $\mathcal{A}$, plus the time necessary to select a $\xleftarrow{R} I_s$ and answer $\mathcal{A}$'s queries, is at most $t$.

We are now ready to define a secure block cipher, or what Luby and Rackoff [14] call a *super pseudorandom* permutation. The notion captures the pseudorandomness of a permutation family on $I_l$ in terms of its indistinguishability from $\mathsf{Perm}^l$, where the adversary is given access to both directions of the permutation. In other words, it measures security of a block cipher against chosen plaintext and ciphertext attacks.

**Definition 2.** *A block cipher $\mathcal{F}$ is a keyed permutation family with domain and range $I_l$ and key length $s$. Let $\mathcal{A}$ be a 2-oracle adversary. Then we define $\mathcal{A}$'s advantage as*

$$\mathsf{Adv}_{\mathcal{F}}^{\mathsf{sprp}}(\mathcal{A}) = \left| \Pr[a \xleftarrow{R} I_s : \mathcal{A}^{f_a, f_a^{-1}} = 1] - \Pr[f \xleftarrow{R} \mathsf{Perm}^l : \mathcal{A}^{f, f^{-1}} = 1] \right|.$$

*For any integers $q, t \geq 0$, we define an insecurity function $\mathsf{Adv}_{\mathcal{F}}^{\mathsf{sprp}}(q, t)$ similarly to Definition 1.*

**Hash Functions.** Our definitions of hash functions follow those given in [8], [18], [22], [13], [20].

**Definition 3.** *Let $H$ be a keyed function family with domain $I_k$, range $I_l$, and key length $s$. Let $\epsilon_1, \epsilon_2, \epsilon_3, \epsilon_4 \geq 2^{-l}$. $H$ is an $\epsilon_1$-uniform family of hash functions if for all $x \in I_k, z \in I_l$, $\Pr[a \xleftarrow{R} I_s : h_a(x) = z] \leq \epsilon_1$. $H$ is $\epsilon_2$-XOR-universal if for all $x \neq y \in I_k, z \in I_l$, $\Pr[a \xleftarrow{R} I_s : h_a(x) \oplus h_a(y) = z] \leq \epsilon_2$. It is $\epsilon_3$-bisymmetric if for all $x, y \in I_k$ (here we allow $x = y$), $z \in I_l$, $\Pr[a_1 \xleftarrow{R} I_s, a_2 \xleftarrow{R} I_s : h_{a_1}(x) \oplus h_{a_2}(y) = z] \leq \epsilon_3$. It is $\epsilon_4$-universal if for all $x \neq y \in I_k$, $\Pr[a \xleftarrow{R} I_s : h_a(x) = h_a(y)] \leq \epsilon_4$.*

We note that in some of the past literature, hash functions are assumed to be uniform by default. We prefer to separate uniformity from other properties.

An example of a family that has all four properties for $\epsilon_1 = \epsilon_2 = \epsilon_3 = \epsilon_4 = 2^{-l}$ is a family keyed by a random $l \times k$ matrix $A$ over $GF(2)$ and a random $l$-bit vector $v$, with $h_{A,v}(x) = Ax + v$ [8].

*Remark 1.* We will use the phrase "$h$ is a uniform (XOR-universal, bisymmetric, universal) hash function" to mean "$h$ is drawn from a uniform (XOR-universal, bisymmetric, universal) family of hash functions."

## 2.2   Constructions of Luby-Rackoff Ciphers

We now define Feistel structures, which are the main tool for constructing pseudorandom permutations on $2n$ bits from functions on $n$ bits.

**Definition 4 (Basic Feistel Permutation).** *Let $f$ be a mapping from $I_n$ to $I_n$. Let $x = (x^L, x^R)$ with $x^L, x^R \in I_n$. We denote by $\overline{f}$ the permutation on $I_{2n}$ defined as $\overline{f}(x) = (x^R, x^L \oplus f(x^R))$. Note that it is a permutation because $\overline{f}^{-1}(y) = (y^R \oplus f(y^L), y^L)$.*

**Definition 5 (Feistel Network).** *If $f_1, \ldots, f_s$ are mappings with domain and range $I_n$, then we denote by $\Psi(f_1, \ldots, f_s)$ the permutation on $I_{2n}$ defined as $\Psi(f_1, \ldots, f_s) = \overline{f_s} \circ \ldots \circ \overline{f_1}$*

Luby and Rackoff [14] were the first to construct pseudorandom permutations. They did so using four independently-keyed pseudorandom functions. The main theorem in their paper is:

**Theorem 1 (Luby-Rackoff).** *Let $h_1, f_1, f_2, h_2$ be independently-keyed functions from a keyed function family $\mathcal{F}$ with domain and range $I_n$ and key space $I_s$. Let $\mathcal{P}$ be the family of permutations on $I_{2n}$ with key space $I_{4s}$ defined by $\mathcal{P} = \Psi(h_1, f_1, f_2, h_2)$ (the key for an element of $\mathcal{P}$ is simply the concatenation of keys for $h_1, f_1, f_2, h_2$). Then*

$$\mathsf{Adv}_{\mathcal{P}}^{\mathsf{sprp}}(q, t) \leq \mathsf{Adv}_{\mathcal{F}}^{\mathsf{prf}}(q, t) + \binom{q}{2}\left(2^{-n+1} + 2^{-2n+1}\right).$$

Naor and Reingold [15] optimized the above construction by enabling the use of $XOR$-universal hash functions in the first and last rounds.

**Theorem 2 (Naor-Reingold).** *Let $f_1$ and $f_2$ be independently-keyed functions from a keyed function family $\mathcal{F}$ with domain and range $I_n$ and key space $I_{s_1}$. Let $h_1, h_2$ be $\epsilon$-XOR-universal hash functions, keyed independently of each other and of $f_1, f_2$, from a keyed function family $H$ with domain and range $I_n$ and key space $I_{s_2}$. Let $\mathcal{P}$ be the family of permutations on $I_{2n}$ with key space $I_{2s_1+2s_2}$ defined by $p = \Psi(h_1, f_1, f_2, h_2)$. Then*

$$\mathsf{Adv}_{\mathcal{P}}^{\mathsf{sprp}}(q, t) \leq \mathsf{Adv}_{\mathcal{F}}^{\mathsf{prf}}(q, t) + \binom{q}{2}\left(2\epsilon + 2^{-2n+1}\right).$$

Patel, Ramzan, and Sundaram [18], following a suggestion in [15], optimized the construction further by allowing the same pseudorandom function to be used in the middle rounds, thus reducing the key size. This required an additional condition on the hash function.

**Theorem 3 (Patel-Ramzan-Sundaram).** *Let $f$ be a function from a keyed function family $\mathcal{F}$ with domain and range $I_n$ and key space $I_{s_1}$. Let $h_1, h_2$ be $\epsilon_1$-bisymmetric $\epsilon_2$-XOR-universal hash functions, keyed independently of each other and of $f$, from a keyed function family $H$ with domain and range $I_n$ and key space $I_{s_2}$. Let $\mathcal{P}$ be the family of permutations on $I_{2n}$ with key space $I_{s_1+2s_2}$ defined by $\mathcal{P} = \Psi(h_1, f, f, h_2)$. Then*

$$\mathsf{Adv}_{\mathcal{P}}^{\mathsf{sprp}}(q, t) \leq \mathsf{Adv}_{\mathcal{F}}^{\mathsf{prf}}(2q, t) + q^2\epsilon_1 + \binom{q}{2}\left(2\epsilon_2 + 2^{-2n+1}\right)$$

# 3    New Model: Round Security

Having presented the classical definitions and constructions of block ciphers, we are now ready to define the new model of round security. The definitions can be easily extended to other symmetric primitives, such as MACs.

Let $\mathcal{P}, \mathcal{F}^1, \mathcal{F}^2, \ldots, \mathcal{F}^r$ be keyed permutation families, each with domain and range $I_l$ and key length $s$, such that for any key $a \in I_s$, $p_a = f_a^r \circ \ldots \circ f_a^1$. Then $\mathcal{F}^1, \ldots, \mathcal{F}^r$ is called an *r-round decomposition* for $\mathcal{P}$. For $i \leq j$, denote by $(i \to j)_a$ the permutation $f_a^j \circ \ldots \circ f_a^i$, and by $(i \leftarrow j)_a$ the permutation $\left( f_a^j \circ \ldots \circ f_a^i \right)^{-1}$. Denote by $i \to j$ and $i \leftarrow j$ the corresponding keyed function families.

Note that having oracle access to a member of $i \to j$ means being able to give inputs to round $i$ of the forward direction of a block cipher and view outputs after round $j$. Likewise, having oracle access to $i \leftarrow j$ corresponds to being able to give inputs to round $j$ of the *reverse* direction of the block cipher and view outputs after round $i$. Thus, the oracle for $1 \to r = \mathcal{P}$ corresponds to the oracle for chosen plaintext attack, and the oracle for $1 \leftarrow r$ corresponds to the oracle for chosen ciphertext attack.

We are now ready to define security in this round-based model. This definition closely mimics Definition 2. The difference is that the adversary is allowed oracle access to some subset $K$ of the set $\{i \to j, i \leftarrow j : 1 \leq i \leq j \leq r\}$, and the insecurity function additionally depends on $K$.

**Definition 6.** *Let $\mathcal{P}$ be a block cipher with domain and range $I_l$, key length $s$ and some r-round decomposition $\mathcal{F}^1, \ldots, \mathcal{F}^r$. Fix some subset $K = \{\pi^1, \ldots, \pi^k\}$ of the set $\{i \to j, i \leftarrow j : 1 \leq i \leq j \leq r\}$, and let $\mathcal{A}$ be a $k + 2$-oracle adversary. Then we define $\mathcal{A}$'s advantage as*

$$\mathsf{Adv}^{\mathsf{sprp}}_{\mathcal{P},\mathcal{F}^1,\ldots,\mathcal{F}^r,K}(\mathcal{A}) =$$

$$\left| \Pr[a \xleftarrow{R} I_s : \mathcal{A}^{p_a, p_a^{-1}, \pi_a^1, \ldots, \pi_a^k} = 1] - \Pr[p \xleftarrow{R} \mathsf{Perm}^l, a \xleftarrow{R} I_s : \mathcal{A}^{p, p^{-1}, \pi_a^1, \ldots, \pi_a^k} = 1] \right|$$

*For any integers $q, t \geq 0$ and set $K$, we define an insecurity function*

$$\mathsf{Adv}^{\mathsf{sprp}}_{\mathcal{P},\mathcal{F}^1,\ldots,\mathcal{F}^r}(q, t, K)$$

*similarly to Definition 2.*

# 4    Round Security of Luby-Rackoff Ciphers

Having developed a round security framework for block ciphers, we examine the specific case of a four-round cipher described in Section 2.2. Our goal is to characterize the insecurity function defined above depending on the set $K$ of oracles.

We are able to do so completely, in the following sense. We place every set $K$ in one of two categories: either the insecurity function is unacceptably high, or it is almost as low as in the standard model. That is, we completely characterize the acceptable sets of oracles for the construction to remain secure in our model.

Moreover, we do so for all three ciphers presented in Section 2.2 (although we need to add an $\epsilon$-uniformity condition on the hash functions in the second and third constructions in order for them to remain secure; this is a mild condition, often already achieved by a hash function family). As it turns out, the round security of the three constructions is the same. Specifically, all three ciphers remain secure if the adversary is given access to the second and third rounds. These results suggest, in some sense, that the so-called "whitening" steps, performed in the first and last rounds, require secrecy but only weak randomness, whereas the middle rounds require strong randomness but no secrecy.

We present our results in two parts. First, in Section 4.1, we examine what combinations of oracles make the cipher insecure. Then, in Section 4.2, we show that any other combination leaves it secure.

## 4.1   Negative Results

In this section we demonstrate which oracles make the cipher insecure. Our negative results are strong, in the sense that they hold regardless of what internal functions $h_1, h_2, f_1, f_2$ are used. That is, the cipher can be distinguished from a random permutation even if each of these functions is chosen truly at random. Thus, our results hold for all three ciphers presented in Section 2.2.

**Theorem 4.** *Regardless of how the functions $h_1, f_1, f_2, h_2$ are chosen from the set of all functions with domain and range $I_n$, let $P = \Psi(h_1, f_1, f_2, h_2)$. Let $t$ be the time required to compute 17 n-bit XOR operations, a comparison of two n-bit strings, and 9 oracle queries.[2] Then*

$$\mathsf{Adv}^{\mathsf{sprp}}_{\mathcal{P}, \overline{h_1}, \overline{f_1}, \overline{f_2}, \overline{h_2}}(9, t, K) \geq 1 - 2^{-n},$$

*as long as $K$ is not a subset of $\{2 \to 2, 2 \leftarrow 2, 3 \to 3, 3 \leftarrow 3, 2 \to 3, 2 \leftarrow 3\}$. That is, $P$ is insecure as long as the adversary has access to an oracle that includes the first or fourth rounds.*

We will prove the theorem by eliminating oracles that allow the adversary to distinguish the cipher from a random permutation. This involves using the attack against a three-round cipher from [14]. The complete proof is given in Appendix A.

## 4.2   Positive Results

In this section, we prove what is essentially the converse of the results of the previous section. Namely, we show that if $K$ is the set given in Theorem 4, then the cipher is secure. Of course, if $K$ is a subset of it, then the cipher is also secure.

---

[2] The values 17 and 9 can be reduced by more careful counting; it is unclear, however, if there is any reason to expend effort finding the minimal numbers that work.

**Theorem 5.** *Suppose* $K \subseteq \{2 \to 2, 2 \leftarrow 2, 3 \to 3, 3 \leftarrow 3, 2 \to 3, 2 \leftarrow 3\}$. *Let* $h_1, f_1, f_2, h_2$ *and* $P$ *be as in Theorem 1. Then*

$$\mathsf{Adv}^{\mathsf{sprp}}_{P,\overline{h_1},\overline{f_1},\overline{f_2},\overline{h_2}}(q, t, K) \le \mathsf{Adv}^{\mathsf{prf}}_{\mathcal{F}}(q, t) + \binom{q}{2}\left(2^{-n+1} + 2^{-2n+1}\right) + q^2\left(2^{-n-1}\right).$$

*If* $h_1, f_1, f_2, h_2$ *and* $P$ *are as in Theorem 2, with the additional condition that* $h_1$ *and* $h_2$ *be* $\epsilon_3$*-uniform, then*

$$\mathsf{Adv}^{\mathsf{sprp}}_{P,\overline{h_1},\overline{f_1},\overline{f_2},\overline{h_2}}(q, t, K) \le \mathsf{Adv}^{\mathsf{prf}}_{\mathcal{F}}(q, t) + \binom{q}{2}\left(2\epsilon + 2^{-2n+1}\right) + q^2\epsilon_3/2.$$

*Finally, if* $h_1, f, h_2$ *and* $P$ *are as in Theorem 3, with the additional condition that* $h_1$ *and* $h_2$ *be* $\epsilon_3$*-uniform, then*

$$\mathsf{Adv}^{\mathsf{sprp}}_{P,\overline{h_1},\overline{f},\overline{f},\overline{h_2}}(q, t) \le \mathsf{Adv}^{\mathsf{prf}}_{\mathcal{F}}(2q, t) + q^2(\epsilon_1 + \epsilon_3) + \binom{q}{2}\left(2\epsilon_2 + 2^{-2n+1}\right).$$

We focus our proof on the last part of the theorem. The proofs of other cases are very similar. Our proof technique is a generalization of the techniques of Naor and Reingold [15] designed to deal with the extra queries. Moreover, we analyze concrete, rather than asymptotic, security.

First, in the following simple claim, we reduce the statement to the case when $f$ is a truly random function.

*Claim.* Suppose

$$\mathsf{Adv}^{\mathsf{sprp}}_{P,\overline{h_1},\overline{f},\overline{f},\overline{h_2}}(q, t) \le \delta$$

when $f$ is picked from $\mathsf{Rand}^{n \to n}$, rather than from a pseudorandom family. Then

$$\mathsf{Adv}^{\mathsf{sprp}}_{P,\overline{h_1},\overline{f},\overline{f},\overline{h_2}}(q, t) \le \delta + \mathsf{Adv}^{\mathsf{prf}}_{\mathcal{F}}(2q, t)$$

when is $f$ picked from $\mathcal{F}$.

*Proof.* Indeed, suppose $\mathcal{A}$ is an adversary for the block cipher $P$, with advantage $\gamma$. Build an adversary $\mathcal{A}'$ for pseudorandom function family $\mathcal{F}$ as follows: $\mathcal{A}'$ selects at random $h_1$ and $h_2$ from a suitable family, and runs $\mathcal{A}$ on the cipher $\Psi(h_1, f, f, h_2)$. In order to answer the queries of $\mathcal{A}$, $\mathcal{A}'$ simply queries $f$ where appropriate and computes the answer according to the Feistel structure. $\mathcal{A}'$ then outputs the same result as $\mathcal{A}$.

Note that $\mathcal{A}$ has advantage at least $\gamma$ if $f$ is from $\mathcal{F}$, and at most $\delta$ for a truly random $f$. By a standard application of the triangle inequality, $\mathsf{Adv}^{\mathsf{prf}}_{\mathcal{F}}(\mathcal{A}') \ge \gamma - \delta$. $\qquad\square$

We note that access to the oracles of $K$ is equivalent to access to the oracle for $f$ (although one query to $2 \to 3$ or $3 \to 2$ can be simulated by two queries to $f$). Thus, it suffices to prove the following theorem.

**Theorem 6.** *Let $f$ be a random function, and let $h_1, h_2$ be $\epsilon_1$-bisymmetric $\epsilon_2$-XOR-universal $\epsilon_3$-uniform hash functions with domain and range $I_n$, $\Psi = \Psi(h_1, f, f, h_2)$, and $R$ be a random permutation on $I_{2n}$. Then, for any 3-oracle adversary $\mathcal{A}$ (we do not restrict the running time of $\mathcal{A}$) that makes at most $q_c$ queries to its first two oracles and at most $q_o$ queries to its third oracle,*

$$\left| \Pr[\mathcal{A}^{\Psi(h_1, f, f, h_2), \Psi^{-1}(h_1, f, f, h_2), f} = 1] - \Pr[\mathcal{A}^{R, R^{-1}, f} = 1] \right|$$

$$\leq q_c^2 \epsilon_1 + 2 q_o q_c \epsilon_3 + \binom{q_c}{2} \left( 2 \epsilon_2 + 2^{-2n+1} \right).$$

The remainder of this section gives the proof of this theorem. To summarize, the first part of the proof focuses on the transcript (a.k.a. the "view") of the adversary, and shows that each possible transcript is about as likely to occur when $\mathcal{A}$ is given $\Psi$ as when $\mathcal{A}$ is given $R$. The second part uses a probability argument to show that this implies that $\mathcal{A}$ will have a small advantage in distinguishing $\Psi$ from $R$.

**Proof of Theorem 6.** To start with, let $P$ denote the permutation oracle (either $\Psi(h_1, f, f, h_2)$ or $R$) that $\mathcal{A}$ accesses. Let $\mathcal{O}^f$ denote the oracle that computes the function $f$ (note that when $\mathcal{A}$ gets $\Psi$ as its permutation oracle, $f$ is actually used as the round function in the computation of the oracle $P = \Psi$; when $\mathcal{A}$ gets $R$ as its permutation oracle, $f$ is completely independent of $P = R$). The machine $\mathcal{A}$ has two possibilities for queries to the oracle $P$: $(+, x)$ which asks to obtain the value of $P(x)$, or $(-, y)$ which asks to obtain the value of $P^{-1}(y)$ – where both $x$ and $y$ are in $I_{2n}$. We call these cipher queries. We define the query-answer pair for the $i^{th}$ cipher query as $\langle x_i, y_i \rangle \in I_{2n} \times I_{2n}$ if $\mathcal{A}$'s query was $(+, x)$ and $y$ is the answer it received from $P$ or its query was $(-, y)$ and $x$ is the answer it received. We assume that $\mathcal{A}$ makes exactly $q_c$ queries and we call the sequence $\{\langle x_1, y_1 \rangle, \ldots, \langle x_{q_c}, y_{q_c} \rangle\}_P$ the cipher-transcript of $\mathcal{A}$.

In addition, $\mathcal{A}$ can make queries to $\mathcal{O}^f$. We call these oracle queries. We denote these queries as: $(\mathcal{O}^f, x')$ which asks to obtain $f(x')$. We define the query-answer pair for the $i^{th}$ oracle query as $\langle x_i', y_i' \rangle \in I_n \times I_n$ if $\mathcal{A}$'s query was $(\mathcal{O}^f, x')$ and the answer it received was $y'$. We assume that $\mathcal{A}$ makes $q_o$ queries to this oracle. We call the sequence $\{\langle x_1', y_1' \rangle, \ldots, \langle x_{q_o}', y_{q_o}' \rangle\}_{\mathcal{O}^f}$ the oracle-transcript of $\mathcal{A}$.

Note that since $\mathcal{A}$ is computationally unbounded, we can make the standard assumption that $\mathcal{A}$ is a deterministic machine. Under this assumption, the exact next query made by $\mathcal{A}$ can be determined by the previous queries and the answers received. We formalize this as follows:

**Definition 7.** *Let $C_{\mathcal{A}}[\{\langle x_1, y_1 \rangle, \ldots, \langle x_i, y_i \rangle\}_P, \{\langle x_1', y_1' \rangle, \ldots, \langle x_j', y_j' \rangle\}_{\mathcal{O}^f}]$, where either $i < q_c$ or $j < q_o$, denote the $i + j + 1^{st}$ query $\mathcal{A}$ makes as a function of the first $i + j$ query-answer pairs in $\mathcal{A}$'s cipher and oracle transcripts. Let $C_{\mathcal{A}}[\{\langle x_1, y_1 \rangle, \ldots, \langle x_{q_c}, y_{q_c} \rangle\}_P, \{\langle x_1', y_1' \rangle, \ldots, \langle x_{q_o}', y_{q_o}' \rangle\}_{\mathcal{O}^f}]$ denote the output $\mathcal{A}$ gives as a function of its cipher and oracle transcripts.*

**Definition 8.** *Let $\sigma$ be the pair of sequences*

$$(\{\langle x_1, y_1\rangle, \ldots, \langle x_{q_c}, y_{q_c}\rangle\}_P, \{\langle x_1', y_1'\rangle, \ldots, \langle x_{q_o}', y_{q_o}'\rangle\}_{\mathcal{O}^f}),$$

*where for $1 \leq i \leq q_c$ we have that $\langle x_1, y_1\rangle \in I_{2n} \times I_{2n}$, and for $1 \leq j \leq q_o$, we have that $\langle x', y'\rangle \in I_n$. Then, $\sigma$ is a consistent $\mathcal{A}$-transcript if for every $1 \leq i \leq q_c$ :*

$$C_{\mathcal{A}}[\{\langle x_1, y_1\rangle, \ldots, \langle x_i, y_i\rangle\}_P, \{\langle x_1', y_1'\rangle, \ldots, \langle x_j', y_j'\rangle\}_{\mathcal{O}^f}] \in$$
$$\{(+, x_{i+1}), (-, y_{i+1}), (\mathcal{O}^f, x_{j+1}')\}.$$

We now consider another process for answering $\mathcal{A}$'s cipher queries that will be useful to us.

**Definition 9.** *The random process $\tilde{R}$ answers the $i^{th}$ cipher query of $\mathcal{A}$ as follows:*

1. *If $\mathcal{A}$'s query is $(+, x_i)$ and for some $1 \leq j < i$ the $j^{th}$ query-answer pair is $\langle x_i, y_i\rangle$, then $\tilde{R}$ answers with $y_i$.*
2. *If $\mathcal{A}$'s query is $(-, y_i)$ and for some $1 \leq j < i$ the $j^{th}$ query-answer pair is $\langle x_i, y_i\rangle$, then $\tilde{R}$ answers with $x_i$.*
3. *If neither of the above happens, then $\tilde{R}$ answers with a uniformly chosen element in $I_{2n}$.*

Note that $\tilde{R}$'s answers may not be consistent with any function, let alone any permutation. We formalize this concept.

**Definition 10.** *Let $\sigma = \{\langle x_1, y_1\rangle, \ldots, \langle x_{q_c}, y_{q_c}\rangle\}_P$ be any possible $\mathcal{A}$-cipher transcript. We say that $\sigma$ is inconsistent if for some $1 \leq j < i \leq q_c$ the corresponding query-answer pairs satisfy $x_i = x_j$ but $y_i \neq y_j$, or $x_i \neq x_j$ but $y_i = y_j$.*

*Note 1.* If $\sigma = (\{\langle x_1, y_1\rangle, \ldots, \langle x_{q_c}, y_{q_c}\rangle\}_P, \{\langle x_1', y_1'\rangle, \ldots, \langle x_{q_o}', y_{q_o}'\rangle\}_{\mathcal{O}^f})$ is a possible $\mathcal{A}$-transcript, we assume from now on that if $\sigma$ is consistent and if $i \neq j$ then $x_i \neq x_j$, $y_i \neq y_j$, and $x_i' \neq x_j'$. This formalizes the concept that $\mathcal{A}$ never repeats a query if it can determine the answer from a previous query-answer pair.

Fortunately, we can show that the process $\tilde{R}$ often "behaves" exactly like a permutation. It turns out that if $\mathcal{A}$ is given oracle access to either $\tilde{R}$ or $R$ to answer its cipher queries, it will have a negligible advantage in distinguishing between the two. We prove this more formally in proposition 1. Before doing so, we first consider the distributions on the various transcripts seen by $\mathcal{A}$ as a function of the different distributions on answers it can get.

**Definition 11.** *The random variables $T_\Psi, T_R, T_{\tilde{R}}$ denote the cipher-transcript / oracle transcript pair seen by $\mathcal{A}$ when its cipher queries are answered by $\Psi$, $R$, $\tilde{R}$ respectively, and its oracle queries are all answered by $\mathcal{O}^f$.*

*Remark 2.* Observe that according to our definitions and assumptions, $\mathcal{A}^{\Psi,\Psi^{-1},f}$ and $C_{\mathcal{A}}(T_{\Psi})$ denote the same random variable. The same is true for $\mathcal{A}^{R,R^{-1},f}$ and $C_{\mathcal{A}}(T_R)$.

**Proposition 1.** $|\Pr_{\tilde{R}}[C_{\mathcal{A}}(T_{\tilde{R}}) = 1] - \Pr_R[C_{\mathcal{A}}(T_R) = 1]| \leq \binom{q_c}{2} \cdot 2^{-2n}$

*Proof.* For any possible and consistent $\mathcal{A}$-transcript $\sigma$ we have that:

$$\Pr_R[T_R = \sigma] = \frac{(2^{2n} - q_c)!}{2^{2n}!} \cdot 2^{-q_o n} = \Pr_{\tilde{R}}[T_{\tilde{R}} = \sigma \mid T_{\tilde{R}} \text{ is consistent}].$$

Thus $T_R$ and $T_{\tilde{R}}$ have the same distribution conditioned on $T_{\tilde{R}}$ being consistent. We now bound the probability that $T_{\tilde{R}}$ is inconsistent. Recall that $T_{\tilde{R}}$ is inconsistent if there exists an $i$ and $j$ with $1 \leq j < i \leq q_c$ for which $x_i = x_j$ but $y_i \neq y_j$, or $x_i \neq x_j$ but $y_i = y_j$. For a particular $i$ and $j$ this event happens with probability $2^{-2n}$. So,

$$\Pr_{\tilde{R}}[T_{\tilde{R}} \text{ is inconsistent}] \leq \binom{q_c}{2} \cdot 2^{-2n}.$$

We complete the proof via a standard argument:

$$\left| \Pr_{\tilde{R}}[C_M(T_{\tilde{R}}) = 1] - \Pr_R[C_M(T_R) = 1] \right|$$

$$\leq \left| \Pr_{\tilde{R}}[T_{\tilde{R}} = \sigma \mid T_{\tilde{R}} \text{ is consistent}] - \Pr_R[C_M(T_R) = 1] \right| \cdot \Pr_{\tilde{R}}[T_{\tilde{R}} \text{ is consistent}]$$

$$+ \left| \Pr_{\tilde{R}}[T_{\tilde{R}} = \sigma \mid T_{\tilde{R}} \text{ is inconsistent}] - \Pr_R[C_M(T_R) = 1] \right| \cdot \Pr_{\tilde{R}}[T_{\tilde{R}} \text{ is inconsistent}]$$

$$\leq \Pr_{\tilde{R}}[T_{\tilde{R}} \text{ is inconsistent}] \leq \binom{q_c}{2} \cdot 2^{-2n}.$$

This completes the proof of the proposition. $\qquad\square$

We now proceed to obtain a bound on the advantage that $\mathcal{A}$ will have in distinguishing between $T_{\Psi}$ and $T_{\tilde{R}}$. It turns out that $T_{\Psi}$ and $T_{\tilde{R}}$ are identically distributed unless the same value is input to $f$ on two different occasions (we show this in Lemma 1). This depends *only* on the choice of $h_1$ and $h_2$. We call this event "BAD" (in the next definition) and obtain a bound on the probability that it actually occurs (in Proposition 2).

**Definition 12.** *For every specific pair of functions $h_1, h_2$ define $BAD(h_1, h_2)$ to be the set of all possible and consistent transcripts*

$$\sigma = (\{\langle x_1, y_1 \rangle, \dots, \langle x_{q_c}, y_{q_c} \rangle\}_P, \{\langle x_1', y_1' \rangle, \dots, \langle x_{q_o}', y_{q_o}' \rangle\}_{\mathcal{O}f})$$

*satisfying at least one of the following events:*
- **B1:** *there exists $1 \leq i < j \leq q_c$ such that $h_1(x_i^R) \oplus x_i^L = h_1(x_j^R) \oplus x_j^L$, or*

- **B2:** *there exists* $1 \leq i < j \leq q_c$ *such that* $y_i^R \oplus h_2(y_i^L) = y_j^R \oplus h_2(y_j^L)$, *or*
- **B3:** *there exists* $1 \leq i, j \leq q_c$ *such that* $h_1(x_i^R) \oplus x_i^L = y_j^R \oplus h_2(y_j^L)$, *or*
- **B4:** *there exists* $1 \leq i \leq q_c$, $1 \leq j \leq q_o$ *such that* $h_1(x_i^R) \oplus x_i^L = x_j'$, *or*
- **B5:** *there exists* $1 \leq i \leq q_c$, $1 \leq j \leq q_o$ *such that* $y_i^R \oplus h_2(y_i^L) = x_j'$.

**Proposition 2.** *Let* $h_1, h_2$ *be* $\epsilon_1$-*bisymmetric* $\epsilon_2$-*XOR-universal* $\epsilon_3$-*uniform hash functions. Then, for any possible and consistent* $\mathcal{A} - transcript$ $\sigma$, *we have that*

$$\Pr_{h_1, h_2}\left[\sigma \in BAD(h_1, h_2)\right] \leq q_c^2 \epsilon_1 + 2 q_o q_c \epsilon_3 + \binom{q_c}{2} \cdot 2\epsilon_2$$

*Proof.* Recall that a transcript $\sigma \in BAD(h_1, h_2)$ if one of the events $B_i$ occur. It is straightforward to determine the individual probabilities of each of these events separately by using the properties of $h$, and apply the union bound to add up the probabilities for each event. □

**Lemma 1.** *Let* $\sigma = (\{\langle x_1, y_1 \rangle, \ldots, \langle x_{q_c}, y_{q_c} \rangle\}_P, \{\langle x_1', y_1' \rangle, \ldots, \langle x_{q_o}', y_{q_o}' \rangle\}_{\mathcal{O}f})$ *be any possible and consistent* $M - transcript$, *then*

$$\Pr_\Psi[T_\Psi = \sigma | \sigma \notin BAD(h_1, h_2)] = \Pr_{\tilde{R}}[T_{\tilde{R}} = \sigma].$$

*Proof.* It is not hard to see that $\Pr_{\tilde{R}}[T_{\tilde{R}} = \sigma] = 2^{-(2q_c + q_o)n}$ (see [15] for more details).

Now, fix $h_1, h_2$ to be such that $\sigma \notin BAD(h_1, h_2)$. We will now compute $\Pr_f[T_\Psi = \sigma]$ (note that the probability is now only over the choice of $f$). Since $\sigma$ is a possible $\mathcal{A}$-transcript, it follows that $T_{\Psi(h_1, f, f, h_2)} = \sigma$ iff $y_i = \Psi(h_1, f, f, h_2)(x_i)$ for all $1 \leq i \leq q_c$ and $y_j' = f(x_j')$ for all $1 \leq j \leq q_o$. If we define

$$S_i = x_i^L \oplus h_1(x_i^R)$$
$$T_i = y_i^R \oplus h_2(y_i^L),$$

then

$$(y_i^L, y_i^R) = \Psi(x_i^L, x_i^R) \Leftrightarrow f(S_i) = T_i \oplus x_i^R \text{ and } f(T_i) = y_i^L \oplus S_i.$$

Now observe that for all $1 \leq i < j \leq q_c$, $S_i \neq S_j$ and $T_i \neq T_j$ (otherwise $\sigma \in BAD(h_1, h_2)$). Similarly, for all $1 < i, j < q_c$, $S_i \neq T_j$. In addition, it follows again from the fact that $\sigma \notin BAD(h_1, h_2)$ that for all $1 \leq i \leq q_c$ and $1 \leq j \leq q_o$, $x_j' \neq S_j$ and $x_j' \neq T_j$. So, if $\sigma \notin BAD(h_1, h_2)$ all the inputs to $f$ are distinct. Since $f$ is a random function, $\Pr_f[T_\Psi = \sigma] = 2^{-(2q_c + q_o)n}$ (The cipher transcript contributes $2^{-2nq_c}$ and the oracle transcript contributes $2^{-q_o n}$ to the probability).

Thus, for every choice of $h_1, h_2$ such that $\sigma \notin BAD(h_1, h_2)$, the probability that $T_\Psi = \sigma$ is exactly the same: $2^{-(2q_c + q_o)n}$. Therefore:

$$\Pr_\Psi[T_\Psi = \sigma | \sigma \notin BAD(h_1, h_2)] = 2^{-(2q_c + q_o)n}.$$

which completes the proof of the lemma. □

The rest of the proof consists of using the above lemma and Propositions 1 and 2 in a probability argument.

Let $\Gamma$ be the set of all possible and consistent transcripts $\sigma$ such that $C_{\mathcal{A}}(\sigma) = 1$. Then

$$\left| \Pr_{\Psi}[\mathcal{A}^{\Psi,\Psi^{-1},f} = 1] - \Pr_{R}[\mathcal{A}^{R,R^{-1},f} = 1] \right|$$

$$= \left| \Pr_{\Psi}[C_{\mathcal{A}}(T_{\Psi}) = 1] - \Pr_{R}[C_{\mathcal{A}}(T_R) = 1] \right|$$

$$\leq \left| \Pr_{\Psi}[C_{\mathcal{A}}(T_{\Psi}) = 1] - \Pr_{\tilde{R}}[C_{\mathcal{A}}(T_{\tilde{R}}) = 1] \right| + \binom{q_c}{2} \cdot 2^{-2n}$$

The last inequality follows from the previous by proposition 1. Now, let $\mathcal{T}$ denote the set of all possible transcripts (whether or not they are consistent), and let $\Delta$ denote the set of all possible inconsistent transcripts $\sigma$ such that $C_{\mathcal{A}}(\sigma) = 1$. Notice that $\Gamma \cup \Delta$ contains all the possible transcripts such that $C_{\mathcal{A}}(\sigma) = 1$, and $\mathcal{T} - (\Gamma \cup \Delta)$ contains all the possible transcripts such that $C_{\mathcal{A}}(\sigma) = 0$. Then:

$$\left| \Pr_{\Psi}[C_{\mathcal{A}}(T_{\Psi}) = 1] - \Pr_{\tilde{R}}[C_{\mathcal{A}}(T_{\tilde{R}}) = 1] \right|$$

$$= \left| \sum_{\sigma \in \mathcal{T}} \Pr_{\Psi}[C_{\mathcal{A}}(\sigma) = 1] \cdot \Pr_{\Psi}[T_{\Psi} = \sigma] - \sum_{\sigma \in \mathcal{T}} \Pr_{\tilde{R}}[C_{\mathcal{A}}(\sigma) = 1] \cdot \Pr_{\tilde{R}}[T_{\tilde{R}} = \sigma] \right|$$

$$\leq \left| \sum_{\sigma \in \Gamma} (\Pr_{\Psi}[T_{\Psi} = \sigma] - \Pr_{\tilde{R}}[T_{\tilde{R}} = \sigma]) \right| + \left| \sum_{\sigma \in \Delta} (\Pr_{\Psi}[T_{\Psi} = \sigma] - \Pr_{\tilde{R}}[T_{\tilde{R}} = \sigma]) \right|$$

$$\leq \left| \sum_{\sigma \in \Gamma} (\Pr_{\Psi}[T_{\Psi} = \sigma] - \Pr_{\tilde{R}}[T_{\tilde{R}} = \sigma]) \right| + \Pr_{\tilde{R}}[T_{\tilde{R}} \text{ is inconsistent}].$$

Recall (from the proof of Proposition 1) that $\Pr_{\tilde{R}}[T_{\tilde{R}} \text{ is inconsistent}] \leq \binom{q_c}{2} \cdot 2^{-2n}$. We now want to bound the first term of the above expression.

$$\left| \sum_{\sigma \in \Gamma} (\Pr_{\Psi}[T_{\Psi} = \sigma] - \Pr_{\tilde{R}}[T_{\tilde{R}} = \sigma]) \right|$$

$$\leq \left| \sum_{\sigma \in \Gamma} (\Pr_{\Psi}[T_{\Psi} = \sigma | \sigma \in BAD(h_1, h_2)] - Pr_{\tilde{R}}[T_{\tilde{R}} = \sigma]) \cdot \Pr_{\Psi}[\sigma \in BAD(h_1, h_2)] \right|$$

$$+ \left| \sum_{\sigma \in \Gamma} (\Pr_{\Psi}[T_{\Psi} = \sigma | \sigma \notin BAD(h_1, h_2)] - \Pr_{\tilde{R}}[T_{\tilde{R}} = \sigma]) \cdot \Pr_{\Psi}[\sigma \notin BAD(h_1, h_2)] \right|$$

Now, we can apply Lemma 1 to get that the last term of the above expression is equal to 0. All that remains is to find a bound for the first term:

$$\left| \sum_{\sigma \in \Gamma} (\Pr_{\Psi}[T_{\Psi} = \sigma | \sigma \in BAD(h_1, h_2)] - \Pr_{\tilde{R}}[T_{\tilde{R}} = \sigma]) \cdot \Pr_{\Psi}[\sigma \in BAD(h_1, h_2)] \right|$$

$$\leq \max_{\sigma} \Pr_{\Psi}[\sigma \in BAD(h_1, h_2)] \times$$

$$\max\left\{\sum_{\sigma \in \Gamma}(\Pr_{\Psi}[T_{\Psi} = \sigma | \sigma \in BAD(h_1, h_2)], \sum_{\sigma \in \Gamma}\Pr_{\tilde{R}}[T_{\tilde{R}} = \sigma])\right\}.$$

Note that the last two sums of probabilities are both between 0 and 1, so the above expression is bounded by $\max_{\sigma}\Pr_{\Psi}[\sigma \in BAD(h_1, h_2)]$, which is, by Proposition 2, bounded by $q_c^2\epsilon_1 + 2q_o q_c \epsilon_3 + \binom{q_c}{2} \cdot 2\epsilon_2$.

Finally, combining the above computations, we get:

$$\left|\Pr_{\Psi}[\mathcal{A}^{\Psi, \Psi^{-1}, f} = 1] - \Pr_{R}[\mathcal{A}^{R, R^{-1}, f} = 1]\right| \leq q_c^2 \epsilon_1 + 2q_o q_c \epsilon_3 + \binom{q_c}{2}(2\epsilon_2 + 2^{-2n+1}),$$

which completes the proof of Theorem 6. $\qquad\qquad\qquad\qquad\qquad\square$

## Acknowledgments

The authors would like to thank Ron Rivest and Salil Vadhan for helpful discussions, and the anonymous referees for many detailed suggestions. Both authors were supported by National Science Foundation Graduate Research Fellowships. The first author was supported by a grant from Merrill-Lynch, and the second author was supported by a grant from the NTT corporation.

# References

1. M. Bellare, R. Canetti, and H. Krawczyk. Keying hash functions for message authentication. In *Advances in Cryptology—CRYPTO '96*. Springer-Verlag.
2. Mihir Bellare, Ran Canetti, and Hugo Krawczyk. Pseudorandom functions revisited: The cascade construction and its concrete security. In *37th Annual Symposium on Foundations of Computer Science*, pages 514–523. IEEE, 1996.
3. Mihir Bellare, Joe Kilian, and Phillip Rogaway. The security of cipher block chaining. In Yvo G. Desmedt, editor, *Advances in Cryptology—CRYPTO '94*, volume 839 of *Lecture Notes in Computer Science*, pages 341–358. Springer-Verlag, 21–25 August 1994.
4. Mihir Bellare and Phillip Rogaway. Random oracles are practical: A paradigm for designing efficient protocols. In *First ACM Conference on Computer and Communications Security*, pages 62–73, Fairfax, 1993.
5. E. Biham and A. Shamir. *Differential Cryptanalysis of the Data Encryption Standard*. Springer Verlag, 1993. ISBN: 0-387-97930-1, 3-540-97930.
6. J. Black, S. Halevi, H. Krawczyk, T. Krovetz, and P. Rogaway. UMAC: fast and secure message authentication. In M. Wiener, editor, *Proc. CRYPTO 99*, volume 1666 of *Springer-Verlag*, pages 216–233, August 1999. Full version can be found at: http://www.cs.ucdavis.edu/~rogaway/umac.
7. R. Canetti, O. Goldreich, and S. Halevi. The random oracle methodology, revisited. In *Proc. 30th ACM Symp. on Theory of Computing*, 1998.
8. J. L. Carter and M. N. Wegman. Universal classes of hash functions. *JCSS*, 18(2):143–154, April 1979.

9. M. Etzel, S. Patel, and Z. Ramzan. Square hash: Fast message authentication via optimized universal hash functions. In *Proc. CRYPTO 99*, Lecture Notes in Computer Science. Springer-Verlag, 1999.

10. Shimon Even and Yishay Mansour. A construction of a cipher from a single pseudorandom permutation. *Journal of Cryptology*, 10(3):151–162, Summer 1997.

11. H. Feistel. Cryptography and computer privacy. *Scientific American*, 228(5):15–23, May 1973.

12. O. Goldreich, S. Goldwasser, and S. Micali. How to construct random functions. *Journal of the ACM*, 33(4):792–807, October 1984.

13. H. Krawczyk. LFSR-based hashing and authentication. In *Proc. CRYPTO 94*, Lecture Notes in Computer Science. Springer-Verlag, 1994.

14. M. Luby and C. Rackoff. How to construct pseudorandom permutations and pseudorandom functions. *SIAM J. Computing*, 17(2):373–386, April 1988.

15. M. Naor and O. Reingold. On the construction of pseudo-random permutations: Luby-Rackoff revisited. *J. of Cryptology*, 12:29–66, 1999. Preliminary version in: *Proc. STOC 97*.

16. National Bureau of Standards. FIPS publication 46: Data encryption standard, 1977. Federal Information Processing Standards Publication 46.

17. National Institute of Standards and Technology. Advanced encryption standard home page. http://www.nist.gov/aes.

18. S. Patel, Z. Ramzan, and G. Sundaram. Towards making Luby-Rackoff ciphers optimal and practical. In *Proc. Fast Software Encryption 99*, Lecture Notes in Computer Science. Springer-Verlag, 1999.

19. R. Rivest. The MD5 message digest algorithm. IETF RFC-1321, 1992.

20. P. Rogaway. Bucket hashing and its application to fast message authentication. In *Proc. CRYPTO 95*, Lecture Notes in Computer Science. Springer-Verlag, 1995.

21. B. Schneier. A self-study course in block cipher cryptanalysis. Available from: http://www.counterpane.com/self-study.html, 1998.

22. Mark N. Wegman and J. Lawrence Carter. New hash functions and their use in authentication and set equality. *JCSS*, 22(3):265–279, June 1981.

# A    Proof of Theorem 4

First, we note the following fact.

**Lemma 2.** *If we give the adversary $\mathcal{A}$ a way to compute the values of $h_1$ on arbitrary inputs, then there exists $\mathcal{A}$ that asks three queries to $h_1$, two queries to the chosen-plaintext oracle $p$, and one query to the chosen-ciphertext oracle $p^{-1}$, performs 8 XOR operations, and has an advantage of $1 - 2^{-n}$.*

*Proof.* This is so because access to $h_1$ allows the adversary to "peel off" the first round of the cipher, and then use the attack of [14] against a three-round cipher.

Consider an adversary who performs the following steps:

1. pick three arbitrary $n$-bit strings $L_1, R_1, R_2$;
2. query the plaintext oracle on $(L_1, R_1)$ to get $(V_1, W_1)$
3. query the plaintext oracle on $(L_1 \oplus h_1(R_1) \oplus h_1(R_2), R_2)$ to get $(V_2, W_2)$
4. query the ciphertext oracle on $(V_2, W_2 \oplus R_1 \oplus R_2)$
5. output 1 if $h_1(R_3) \oplus L_3 = V_1 \oplus V_2 \oplus L_1 \oplus h_1(R_1)$

Recall the the goal of the adversary is to output 1 when given the plaintext and ciphertext oracles for a random permutation with noticeably different probability than when given oracles for the block cipher.

Clearly, if the plaintext and ciphertext oracles are truly random, then the adversary will output 1 with probability $2^{-n}$, because $V_1$ and $L_3$ are then random and independent of the rest of the terms. However, if the plaintext and ciphertext oracles are for the block cipher, then the adversary would output 1 with probability 1. Here is why.

Let $S_i, T_i$ $(1 \leq i \leq 3)$ be the intermediate values computed in rounds 1 and 2 of the block cipher for the three queries. Let $L_2 = L_1 \oplus h_1(R_1) \oplus h_1(R_2)$, $V_3 = V_2$ and $W_3 = W_2 \oplus R_1 \oplus R_2$. Note that $S_1 = L_1 \oplus h_1(R_1) = L_2 \oplus h_1(R_2) = S_2$. Then $T_3 = W_3 \oplus h_2(V_3) = W_2 \oplus R_1 \oplus R_2 \oplus h_2(V_2) = T_2 \oplus R_1 \oplus R_2 = f_2(S_2) \oplus R_2 \oplus R_1 \oplus R_2 = f_2(S_1) \oplus R_1 = T_1$. Finally, $h_1(R_3) \oplus L_3 = S_3 = V_3 \oplus f_3(T_3) = V_2 \oplus f_3(T_1) = V_2 \oplus V_1 \oplus S_1 = V_2 \oplus V_1 \oplus L_1 \oplus h_1(R_1)$. □

Note that this fact can be similarly shown for $h_2$. The lemma above allows us to easily prove the following result.

**Lemma 3.** *If $K$ contains at least one of the following oracles: $1 \rightarrow 4$, $1 \leftarrow 4$, $2 \rightarrow 4$, $2 \leftarrow 4$, $1 \rightarrow 3$, $1 \leftarrow 3$, $1 \rightarrow 1$, $1 \rightarrow 2$, $1 \leftarrow 1$, $1 \leftarrow 2$, $4 \leftarrow 4$, $3 \leftarrow 4$, $4 \rightarrow 4$ or $3 \leftarrow 4$, then there exists $\mathcal{A}$ making no more than 9 queries to the oracles and performing no more than 17 XOR operations whose advantage is $1 - 2^{-n}$.*

*Proof.* If $K$ contains $1 \rightarrow 4$ or $1 \rightarrow 3$, then $\mathcal{A}$ can input an arbitrary pair $(L, R)$ to either of these and receive $(V, W)$ or $(T, V)$. $\mathcal{A}$ then inputs $(L, R)$ to the chosen plaintext oracle $p$ to receive $(V', W')$, and checks if $V = V'$.

Similarly for $1 \leftarrow 4$ or $2 \leftarrow 4$.

If $K$ contains $2 \rightarrow 4$, then $\mathcal{A}$ can input an arbitrary pair $(R, S)$ to it to receive $(V, W)$. $\mathcal{A}$ then inputs $(V, W)$ to the chosen ciphertext oracle $p^{-1}$ to receive $(L, R')$ and checks if $R = R'$. Similarly for $1 \leftarrow 3$.

If $K$ contains $1 \rightarrow 1$ or $1 \rightarrow 2$, then $\mathcal{A}$ can input $(L, R)$ and receive, in particular, $S = L \oplus h_1(R)$. $\mathcal{A}$ can then compute $h_1(R) = S \oplus L$, and use the procedure of Lemma 2.

Access to $1 \leftarrow 1$ allows $\mathcal{A}$ to input $(R, S)$ and receive $(L = S \oplus h_1(R), R)$. $\mathcal{A}$ can then compute $h_1(R) = L \oplus S$.

Access to $1 \leftarrow 2$ allows $\mathcal{A}$ to compute $h_1(R)$ as follows:

1. query the $1 \leftarrow 2$ oracle on an arbitrary pair $(S_1, T_1)$ to get $(L_1, R_1)$;
2. let $T_2 = T_1 \oplus R_1 \oplus R$ and $S_2 = S_1$;
3. query the $1 \leftarrow 2$ oracle on $(S_2, T_2)$ to get $(L_2, R_2)$; then $R_2 = T_2 \oplus f_1(S_2) = (T_1 \oplus R_1 \oplus R) \oplus (R_1 \oplus T_1) = R$;
4. compute $h_1(R) = L_2 \oplus S_2$.

Thus, any of the oracles $1 \rightarrow 1, 1 \rightarrow 2, 1 \leftarrow 1, 1 \leftarrow 2$ gives $\mathcal{A}$ access to $h_1$ and thus makes the cipher insecure.

Similarly for $4 \leftarrow 4, 3 \leftarrow 4, 4 \rightarrow 4$ and $3 \rightarrow 4$. □

Finally, to prove Theorem 4, note that there are 20 possible oracles. Of those, 14 are ruled out by the above lemma, leaving only 6 possible oracles to choose from.

# New Paradigms for Constructing
# Symmetric Encryption Schemes Secure
# against Chosen-Ciphertext Attack

Anand Desai

Department of Computer Science & Engineering,
University of California at San Diego,
9500 Gilman Drive, La Jolla, California 92093, USA.
adesai@cs.ucsd.edu

**Abstract.** The paradigms currently used to realize symmetric encryption schemes secure against adaptive chosen ciphertext attack (CCA) try to make it infeasible for an attacker to forge "valid" ciphertexts. This is achieved by either encoding the plaintext with some redundancy before encrypting or by appending a MAC to the ciphertext. We suggest schemes which are provably secure against CCA, and yet every string is a "valid" ciphertext. Consequently, our schemes have a smaller ciphertext expansion than any other scheme known to be secure against CCA. Our most efficient scheme is based on a novel use of "variable-length" pseudorandom functions and can be efficiently implemented using block ciphers. We relate the difficulty of breaking our schemes to that of breaking the underlying primitives in a precise and quantitative way.

## 1 Introduction

Our goal in this paper is to design efficient symmetric (ie. private-key) encryption schemes that are secure against adaptive chosen-ciphertext attack (CCA). Rather than directly applying the paradigm used in designing public-key encryption schemes secure against CCA, we develop new ones which take advantage of the peculiarities of the symmetric setting. As a result we manage to do what may not have been known to be possible: constructing encryption schemes secure against CCA wherein every string of appropriate length is a "valid" ciphertext and has a corresponding plaintext. The practical significance of this is that our schemes have a smaller ciphertext expansion than that of any other scheme known to be secure against CCA.

### 1.1 Privacy under Chosen-Ciphertext Attack

The most basic goal of encryption is to ensure that an adversary does not learn any useful information from the ciphertexts. The first rigorous formalizations of this goal were described for the public-key setting by Goldwasser and Micali [15]. Their goal of indistinguishability for public-key encryption has been considered

M. Bellare (Ed.): CRYPTO 2000, LNCS 1880, pp. 394–412, 2000.
© Springer-Verlag Berlin Heidelberg 2000

under attacks of increasing severity: chosen-plaintext attack, and two kinds of chosen-ciphertext attacks [20,23]. The strongest of these attacks, due to Rack-off and Simon, is known as the adaptive chosen-ciphertext attack (referred to as CCA in this work). Under this attack, the adversary is given the ability to obtain plaintexts of ciphertexts of its choice (with the restriction that it not ask for the decryption of the "challenge" ciphertext itself). The combination of the goal of indistinguishability and CCA gives rise to a very strong notion of privacy, known as IND-CCA. A second goal, called non-malleability, introduced by Dolev, Dwork and Naor [13], can also be considered in this framework. This goal formalizes the inability of an adversary given a challenge ciphertext to modify it into another, in such a way that the underlying plaintexts are somehow "meaningfully related". The notion of indistinguishability under chosen-plaintext attack was adapted to the symmetric setting by Bellare, Desai, Jokipii and Rogaway [2]. Their paradigm of giving the adversary "encryption oracles" can be used to "lift" any of the notions for the public-key setting to the symmetric setting. Studies on relations among the various possible notions have established that IND-CCA implies all these other notions in the public-key setting [3,13], as well as, in the symmetric setting [16].

Symmetric encryption schemes are widely used in practice and form the basis of many security protocols used on the Internet. The use of schemes secure in the IND-CCA sense is often mandated by the way they are to be used in these protocols. Consequently, there has been an increasing focus on designing encryption schemes that are secure in this strong sense. A commonly used privacy mechanism is to use a public-key encryption scheme to send session keys and then use these keys and a symmetric encryption scheme to actually encrypt the data. This method is attractive since symmetric encryption is significantly more efficient than its public-key counterpart. The security of such "hybrid" encryption schemes is as weak as its weakest link. In particular, if we want a hybrid encryption scheme secure in the IND-CCA sense, then we must use a symmetric encryption scheme that is also secure in the IND-CCA sense.

Barring a few exceptions, most of the recent work on encryption has concentrated on the public-key setting alone. The prevailing intuition seems to be that the ideas from the public-key setting "extend" to the symmetric setting. Indeed there are many cases where this is true, and there are often paradigms in one setting that have a counterpart in the other. However, this viewpoint ignores the important differences in the settings and we usually pay for this in terms of efficiency. We take a direct approach to the problem of designing symmetric encryption schemes secure in the IND-CCA sense. For practical reasons, we are particularly interested in block-cipher-based schemes (ie. encryption modes).

## 1.2   Our Paradigms

We describe two new paradigms for realizing symmetric encryption schemes secure against CCA: the Unbalanced Feistel paradigm and the Encode-then-Encipher paradigm.

UNBALANCED FEISTEL. Our first paradigm is described in terms of "variable-length" pseudorandom functions. These extend the notion of "fixed-length" pseudorandom functions (PRFs) introduced by Bellare, Kilian and Rogaway [4] so as to model block ciphers. A variable-length input pseudorandom function (VI-PRF) is a function that takes inputs of any pre-specified length or of variable length and produces an output of some fixed length. A variable-length output pseudorandom function (VO-PRF), on the other hand, is a function whose *output* can be of some pre-specified length or of variable length. The input consists of a fixed-length part and a part specifying the length of the required output.

Our paradigm is illustrated in Figure 2. It is interesting that there is a similarity between our scheme and the "simple probabilistic encoding scheme" used by Bellare and Rogaway in their OAEP scheme [7]. Their encoding scheme is defined as: $M \oplus G(r) \| r \oplus H(M \oplus G(r))$, where $M$ is the message to be encrypted, $r$ is a randomly chosen quantity, $G$ is a "generator" random oracle and $H$ is a "hash function" random oracle. They show that applying a trapdoor permutation, such as RSA, to such an encoded string constitutes asymmetric encryption secure against chosen-plaintext attack. One can view our scheme as the above "encoding scheme" with $G$ replaced by a VO-PRF and $H$ replaced by a VI-PRF. We show that this alone constitutes symmetric encryption secure against CCA.

Constructions for VI-PRFs and VO-PRFs could be based on one-way or trapdoor functions. For practical reasons, we are more interested in constructions that can be based on more efficient cryptographic primitives. Some efficient constructions of VI-PRFs based on PRFs are the CBC-MAC variant analyzed by Petrank and Rackoff [22] and the "three-key" variants of Black and Rogaway [10]. We give a simple and efficient construction of a VO-PRF from a PRF. See Figure 1. There could be many other ways of instantiating VO-PRFs using ideas from the constructions of VI-PRFs and "key-derivation" functions.

We give a quantitative analysis of our scheme to establish its security against CCA. Our analysis relates the difficulty of breaking the scheme to that of breaking the underlying VI-PRF and VO-PRF. We also give a quantitative security analysis of our VO-PRF example. The security of the VI-PRF examples have already been established by similar analyses, as discussed earlier.

We give a concrete example instantiating the Unbalanced Feistel paradigm using a block cipher. The encryption is done in two steps. In the first step, we encrypt the plaintext $M$ to a string $C \| r$ using a modified form of the counter mode of encryption. Here the "counter" $r$ is picked to be random and we have $|C| = |M|$. In the second step, we mask $r$ by XORing it with a modified form of a CBC-MAC on $C$ to get a string $\sigma$. The ciphertext output is $C \| \sigma$.

ENCODE-THEN-ENCIPHER. This is a rather well-known (but not particularly well-understood) method of encrypting. Recent work by Bellare and Rogaway [8] has tried to remedy this by giving a precise treatment of this idea. Encryption, in this paradigm, is a process in which the plaintext is first "encoded" and then sent through a secret-keyed length-preserving permutation in a process known as "enciphering". The privacy of the resulting encryption schemes for different security interpretations of "encoding" and "enciphering" are given in [8]. We concentrate in this paper on one particular combination of the encoding and enciphering interpretations that was not considered in [8]. We consider "encoding" of a message to be simply the message with some randomness appended to it. We take "enciphering" to mean the application of a variable-length input super-pseudorandom permutation (VI-SPRP). We show that with these meanings, the Encode-then-Encipher paradigm yields symmetric encryption schemes that are secure against CCA. Note that a super-pseudorandom permutation (SPRP) [18] alone will not do since we need a permutation that can work with variable and arbitrary length inputs. Also, the very efficient constructions of Naor and Reingold [19] cannot be used here since they are not "full-fledged" VI-SPRPs. The problem of constructing VI-SPRPs has been explored by Bleichenbacher and Desai [11] and Patel et al. [21]. See Section 4 for more details. The encryption schemes resulting from this paradigm are quite practical, but given the current state-of-art, this approach does not match the Unbalanced Feistel paradigm for efficiency.

## 1.3   Related Work and Discussion

The idea behind the paradigms currently used in practice for designing encryption schemes secure in the IND-CCA sense is to make it infeasible to create a "valid" ciphertext (unless the ciphertext was created by encrypting some known plaintext). The intuition is that doing this makes the decryption access ability all but useless. There are a couple of different methods used in symmetric encryption based on this idea.

ALTERNATE PARADIGMS. The most commonly used approach of getting security in the IND-CCA sense is to authenticate ciphertexts using a message authentication code (MAC). Bellare and Namprempre have shown that of the various possible ways of composing a generic MAC and a generic symmetric encryption scheme, the one consisting of first encrypting the plaintext and then appending to the result a MAC of the result, is the only one that is secure in the IND-CCA sense [5]. Another approach is to add some known redundancy to the plaintext before encrypting. The idea is that most strings of the length of the ciphertext will be "invalid" and that they will be recognized as such, since their "decryption" will not have the expected redundancy. A recently suggested encryption mode, the RPC mode of Katz and Yung [17], uses this idea. Yet another approach that uses this idea is to apply a VI-SPRP to plaintexts that are encoded with randomness *and* redundancy [8].

COMPARISONS. An unavoidable consequence of the paradigms used by the methods above is that the ciphertexts generated are longer than those possible using schemes that are only secure against chosen-plaintext attack. In particular, they are longer by the size of the output of the MAC or by the amount of redundancy used. To begin with, we have that any secure encryption scheme will be length-increasing. For short plaintexts these increases in the length of the ciphertext can be a significant overhead. Avoiding any overhead other than that absolutely necessary would also be useful in any environment where bandwidth is at a premium. In our approach, we avoid the part of the overhead due to the MAC or redundancy. The ciphertext expansion due to the randomness used is unavoidable in the model we consider (ie. where the sender and receiver do not share any state other than the key).

We point out that the methods above achieve something more than privacy against CCA. They achieve privacy as well as *integrity*. There are many levels of integrity that one can consider (see [5,17]). The strongest one exactly coincides with the idea used by the methods above. Namely, that it be infeasible to create a "valid" (new) ciphertext. This is clearly not achievable by our method or by any other where every string of appropriate length corresponds to some plaintext. A slightly weaker form of integrity requires that it be infeasible to create a (new) ciphertext such that something may be known about the underlying plaintext. Our methods can be shown to have this integrity property. Should the strongest integrity property be required, we could encode the plaintexts with some redundancy and then apply our paradigm. This would mean losing some of its advantages but it would still be a competitive alternative. These claims are substantiated in the full version of this paper [12].

## 2    Preliminaries

We adopt a standard notation with respect to probabilistic algorithms and sets. If $A(\cdot, \cdot, \ldots)$ is a probabilistic algorithm then $x \leftarrow A(x_1, x_2, \ldots)$ denotes the experiment of running $A$ on inputs $x_1, x_2, \ldots$ and letting $x$ be the outcome. Similarly, if $A$ is a set then $x \leftarrow A$ denotes the experiment of selecting a point uniformly from $A$ and assigning $x$ this value.

SYMMETRIC ENCRYPTION. A *symmetric encryption scheme*, $\Pi = (\mathcal{E}, \mathcal{D}, \mathcal{K})$, is a three-tuple of algorithms where:

- $\mathcal{K}$ is a randomized *key generation* algorithm. It returns a key $a$; we write $a \leftarrow \mathcal{K}$.
- $\mathcal{E}$ is a randomized or stateful *encryption* algorithm. It takes the key $a$ and a *plaintext* $x$ and returns a *ciphertext* $y$; we write $y \leftarrow \mathcal{E}_a(x)$.
- $\mathcal{D}$ is a deterministic *decryption* algorithm. It takes a key $a$ and string $y$ and returns either the corresponding plaintext $x$ or the symbol $\perp$; we write $x \leftarrow \mathcal{D}_a(y)$ where $x \in \{0,1\}^* \cup \perp$.

We require that $\mathcal{D}_a(\mathcal{E}_a(x)) = x$ for all $x \in \{0,1\}^*$.

SECURITY AGAINST CHOSEN-CIPHERTEXT ATTACK. The formalization we give is an adaptation of the "find-then-guess" definition of Bellare et al. [2] so as to model adaptive chosen-ciphertext attack in the sense of Rackoff and Simon [23]. In the indistinguishability of encryptions under chosen-ciphertext attack the adversary $A$ is imagined to run in two phases. In the find phase, given adaptive access to an encryption and decryption oracle, $A$ comes up with a pair of messages $x_0, x_1$ along with some state information $s$ to help in the second phase. In the guess phase, given the encryption $y$ of one of the messages and $s$, it must identify which of the two messages goes with $y$. $A$ may not use its decryption oracle on $y$ in the **guess** phase.

**Definition 1.** [IND-CCA] *Let* $\Pi = (\mathcal{K}, \mathcal{E}, \mathcal{D})$ *be a symmetric encryption scheme. For an adversary $A$ and $b = 0, 1$ define the experiment*

$Experiment$ $\mathsf{Exp}_{\Pi}^{\text{ind-cca}}(A, b)$
    $a \leftarrow \mathcal{K}; \ (x_0, x_1, s) \leftarrow A^{\mathcal{E}_a, \mathcal{D}_a}(\mathsf{find}); \ y \leftarrow \mathcal{E}_a(x_b); \ d \leftarrow A^{\mathcal{E}_a, \mathcal{D}_a}(\mathsf{guess}, y, s);$
    Return $d$.

*It is mandated that $|x_0| = |x_1|$ above and that $A$ does not query $\mathcal{D}_a(\cdot)$ on ciphertext $y$ in the* **guess** *phase. Define the advantage of $A$ and the advantage function of $\Pi$ respectfully, as follows:*

$$\mathsf{Adv}_{\Pi}^{\text{ind-cca}}(A) = \Pr[\, \mathsf{Exp}_{\Pi}^{\text{ind-cca}}(A, 0) = 0 \,] - \Pr[\, \mathsf{Exp}_{\Pi}^{\text{ind-cca}}(A, 1) = 0 \,]$$

$$\mathsf{Adv}_{\Pi}^{\text{ind-cca}}(t, q_e, q_d, \mu, \nu) = \max_A \{ \mathsf{Adv}_{\Pi}^{\text{ind-cca}}(A) \}$$

*where the maximum is over all $A$ with "time-complexity" $t$, making at most $q_e$ encryption oracle queries and at most $q_d$ decryption oracle queries, these together totalling at most $\mu$ bits and choosing $|x_0| = |x_1| = \nu$ bits.* ∎

Here the "time-complexity" is the worst case total execution time of experiment $\mathsf{Exp}_{\Pi}^{\text{ind-cca}}(A, b)$ plus the size of the code of $A$, in some fixed RAM model of computation. This convention is used for other definitions in this paper, as well.

# 3   Unbalanced Feistel Encryption

We begin with a block-cipher-based instantiation of the Unbalanced Feistel paradigm. A reader interested in seeing the paradigm first may skip this example and go to Section 3.2, without any loss of understanding. A security analysis for this paradigm is given in Section 3.3.

## 3.1   A Concrete Example

Our starting point is a block cipher $F : \{0, 1\}^k \times \{0, 1\}^l \mapsto \{0, 1\}^l$. The scheme $\Pi[F] = (\mathcal{K}, \mathcal{E}, \mathcal{D})$ has a key generation algorithm $\mathcal{K}$ that specifies a key $K = (K1 \| K2 \| K3 \| K4) \leftarrow \{0, 1\}^{4k}$, partitioned into 4 equal pieces. We have:

Algorithm $\mathcal{E}_{K1 \| K2 \| K3 \| K4}(M)$
(1)    Let $r \leftarrow \{0, 1\}^l$ be a random initial vector.
(2)    Let $s = F_{K1}(r)$.
(3)    Let $P$ be the first $|M|$ bits of $F_{K2}(s + 1) \| F_{K2}(s + 2) \| F_{K2}(s + 3) \| \cdots$.
(4)    Let $C = P \oplus M$.
(5)    Let $\mathsf{pad} = 10^m$ such that $m$ is the smallest integer making $|C| + |\mathsf{pad}|$ divisible by $l$.

(6)    Parse $C\|\mathsf{pad}$ as $C_1 \ldots C_n$ such that $|C_i| = l$ for all $1 \leq i \leq n$.

(7)    Let $C'_0 = 0^l$, and let $C'_i = F_{K3}(C'_{i-1} \oplus C_i)$ for all $1 \leq i \leq n-1$.

(8)    Let $\sigma = r \oplus F_{K4}(C'_{n-1} \oplus C_n)$

(9)    Return ciphertext $C\|\sigma$.

Algorithm $\mathcal{D}_{K1\|K2\|K3\|K4}(C'')$

(1)    Parse $C''$ as $C\|\sigma$ such that $|\sigma| = l$.

(2)    Let $\mathsf{pad} = 10^m$ such that $m$ is the smallest integer making $|C| + |\mathsf{pad}|$ divisible by $l$.

(3)    Parse $C\|\mathsf{pad}$ as $C_1 \ldots C_n$ such that $|C_i| = l$ for all $1 \leq i \leq n$.

(4)    Let $C'_0 = 0^l$, and let $C'_i = F_{K3}(C'_{i-1} \oplus C_i)$ for all $1 \leq i \leq n-1$.

(5)    Let $r = \sigma \oplus F_{K4}(C'_{n-1} \oplus C_n)$

(6)    Let $s = F_{K1}(r)$.

(7)    Let $P$ be the first $|C|$ bits of $F_{K2}(s+1)\|F_{K2}(s+2)\|F_{K2}(s+3)\| \cdots$.

(8)    Let $M = P \oplus C$.

(9)    Return plaintext $M$.

This example can be seen as having two stages. In the first stage we encrypt the plaintext $M$ to a string $C\|r$ using a modified form of the counter mode. Here $r$ is the randomness used to encrypt and $|C| = |M|$, even though $|M|$ may not be an integral multiple of the block length $l$. In the second stage we run $C$ through a modified form of the CBC-MAC and XOR this value with $r$ to get $\sigma$. The ciphertext is defined to be $C\|\sigma$. Although we do not make $r$ a part of the ciphertext, it is still possible to retrieve it if the secret key is known. Thus the scheme is invertible. While the other counter mode variants can easily be shown to be insecure against CCA, the claim is that "masking" $r$ in this manner, makes our mode secure against CCA.

The difference between the standard (randomized) counter mode, analyzed by Bellare et al. [2], and the variant we use here is that instead of using $r$ directly, we use a block-cipher "encrypted" value of $r$. This has the effect of neutralizing simple attacks where there may be some "control" over $r$. We cannot use the standard "one-key" CBC-MAC in our construction, given that it is known to be secure only on fixed-length messages (with this length being an integral multiple of the block length) [4]. Instead we use a "two-key" CBC-MAC, wherein the last block of the message is processed by the block cipher with an independent key. This is a variant of a construction analyzed by Petrank and Rackoff [22] which first computes a regular CBC-MAC on the entire message and then applies a block-cipher with an independent key to the output of the CBC-MAC. We also use a standard padding method with our MAC, since $|M|$ (and hence $|C|$) may not be an integral multiple of the block length. Note that we use the padding method in a way that does not cause an increase in the length of the ciphertext.

## 3.2   The General Approach

We begin with a description of the primitives used to realize the general scheme and some definitions to understand the security claims.

FIXED-LENGTH PSEUDORANDOM FUNCTIONS. A fixed-length (finite) function family is a keyed multi-set $F$ of functions where all the functions have the same domain and range. To pick a function $f$ from family $F$ means to pick a key $a$, uniformly from the key space of $F$, and let $f = F_a$. A family $F$ has input length $l$ and output length $L$ if each $f \in F$ maps $\{0,1\}^l$ to $\{0,1\}^L$. We let Func($l$) denote a reference family consisting of all functions with input length $l$ and output length $l$. A function $f \leftarrow$ Func($l$) is defined as follows: for each $M \in \{0,1\}^l$, let $f(M)$ be a random string in $\{0,1\}^l$.

A finite function family $F$ is pseudorandom if the input-output behavior of $F_a$ is indistinguishable from the behavior of a random function of the same domain and range. This is formalized via the notion of distinguishers [14]. Our concrete security formalization is that of [4].

**Definition 2.** [PRF] *Let* $F : \mathcal{K} \times \{0,1\}^l \mapsto \{0,1\}^l$ *be a function. For a distinguisher $A$ and $b = 0,1$ define the experiment*

*Experiment* $\mathsf{Exp}_F^{\mathrm{prf}}(A,b)$
$\quad a \leftarrow \mathcal{K}; \ \mathcal{O}_0 \leftarrow F_a; \ \mathcal{O}_1 \leftarrow \mathrm{Func}(l); \ d \leftarrow A^{\mathcal{O}_b}; \ $ Return $d$.

*Define the advantage of $A$ and the advantage function of $F$ respectfully, as follows:*

$$\mathsf{Adv}_F^{\mathrm{prf}}(A) = \Pr[\,\mathsf{Exp}_F^{\mathrm{prf}}(A,0) = 0\,] - \Pr[\,\mathsf{Exp}_F^{\mathrm{prf}}(A,1) = 0\,]$$

$$\mathsf{Adv}_F^{\mathrm{prf}}(t,q) = \max_A \{\mathsf{Adv}_F^{\mathrm{prf}}(A)\}$$

*where the maximum is over all $A$ with time complexity $t$ and making at most $q$ oracle queries.* ∎

VARIABLE-LENGTH INPUT PSEUDORANDOM FUNCTIONS. These functions take an input of variable and arbitrary length and produce a fixed-length output. We define a reference family VI-Func($l$). A random variable-length input function $h \leftarrow$ VI-Func($l$) is defined as follows: for each $M \in \{0,1\}^*$, let $h(M)$ be a random string in $\{0,1\}^l$.

**Definition 3.** [VI-PRF] *Let* $H : \mathcal{K} \times \{0,1\}^* \mapsto \{0,1\}^l$ *be a function. For a distinguisher $A$ and $b = 0,1$ define the experiment*

*Experiment* $\mathsf{Exp}_H^{\mathrm{vi\text{-}prf}}(A,b)$
$\quad a \leftarrow \mathcal{K}; \ \mathcal{O}_0 \leftarrow H_a; \ \mathcal{O}_1 \leftarrow \mathrm{VI\text{-}Func}(l); \ d \leftarrow A^{\mathcal{O}_b}; \ $ Return $d$.

*Define the advantage of $A$ and the advantage function of $H$ respectfully, as follows:*

$$\mathsf{Adv}_H^{\mathrm{vi\text{-}prf}}(A) = \Pr[\,\mathsf{Exp}_H^{\mathrm{vi\text{-}prf}}(A,0) = 0\,] - \Pr[\,\mathsf{Exp}_H^{\mathrm{vi\text{-}prf}}(A,1) = 0\,]$$

$$\mathsf{Adv}_H^{\mathrm{vi\text{-}prf}}(t,q,\mu) = \max_A \{\mathsf{Adv}_H^{\mathrm{vi\text{-}prf}}(A)\}$$

*where the maximum is over all $A$ with time complexity $t$ and making at most $q$ oracle queries, these totalling at most $\mu$ bits.* ∎

Many of the variable-length input MAC constructions are VI-PRFs. Our "two-key" CBC-MAC, discussed earlier, is such an example. The security of this construction follows from that of the CBC-MAC variant analyzed by Petrank and Rackoff [22]. Black and Rogaway have suggested several constructions of VI-PRFs that are computationally more efficient than this one [10]. There are efficient variable-length input MAC constructions, such as the protected counter sum construction of Bernstein [9] and the cascade construction of Bellare et al. [1] that are not strictly VI-PRF due to their probabilistic nature, but which could be used in their place in our paradigm.

VARIABLE-LENGTH OUTPUT PSEUDORANDOM FUNCTIONS. These are functions

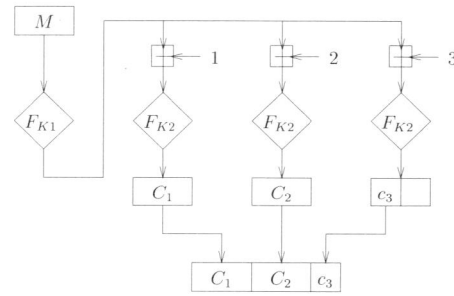

**Fig. 1.** *The* XORG *function.*

that can generate an output of arbitrary and variable length. We think of a function from a VO-PRF family as taking two inputs: a fixed-length binary string and a unary string, and producing an output of a size specified by the unary input. We define a reference family VO-Func($l$). A random variable-length output function $g \leftarrow$ VO-Func($l$) is defined as follows. For each $M \in \{0,1\}^l$ let $R_l(M)$ be a random string in $\{0,1\}^\infty$. Then for each $M \in \{0,1\}^l$ and $L \in 1^*$, let $g(M\|L)$ be the first $|L|$ bits of $R_l(M)$. One can think of a "random variable-length output function" as a process that answers a query $M\|L$ as follows: if $M$ is "new" then return a random element $C \in \{0,1\}^{|L|}$; if $M$ has already appeared in a past query $M\|L'$ (to which the response was $C$) and $|L'| \leq |L|$ then return the first $|L'|$ bits of $C$; and if $M$ has already appeared in a past query $M\|L'$ (to which the response was $C$) and $|L'| > |L|$ then return $C\|C'$ where $C' \leftarrow \{0,1\}^{|L'|-|L|}$.

**Definition 4.** [VO-PRF] *Let $G : \mathcal{K} \times \{0,1\}^l \times 1^* \mapsto \{0,1\}^*$ be a function. For a distinguisher $A$ and $b = 0,1$ define the experiment*

*Experiment* $\mathsf{Exp}_G^{\text{vo-prf}}(A,b)$
    $a \leftarrow \mathcal{K}$; $\mathcal{O}_0 \leftarrow G_a$; $\mathcal{O}_1 \leftarrow$ VO-Func; $d \leftarrow A^{\mathcal{O}_b}$; Return $d$.

*Define the advantage of $A$ and the advantage function of $G$ respectfully, as follows:*

$$\mathsf{Adv}_G^{\text{vo-prf}}(A) = \Pr[\,\mathsf{Exp}_G^{\text{vo-prf}}(A,0) = 0\,] - \Pr[\,\mathsf{Exp}_G^{\text{vo-prf}}(A,1) = 0\,]$$

$$\mathsf{Adv}_G^{\text{vo-prf}}(t,q,\mu) = \max_A \{\mathsf{Adv}_G^{\text{vo-prf}}(A)\}$$

*where the maximum is over all $A$ with time complexity $t$ and making at most $q$ queries, these totalling at most $\mu$ bits.* ∎

A somewhat similar (but weaker) primitive is implicit in the counter mode of operation. Hence this is a good starting point for constructing full-fledged VO-PRFs. The result is a construction we call XORG. See Figure 1 for a picture. Let $F$ be a block cipher and $a = (K1\|K2)$ be a key specifying permutations $F_{K1}$ and $F_{K2}$. Then for any $L \in 1^*$ and $M \in \{0,1\}^l$, the output of XORG is defined as the first $|L|$ bits of $F_{K2}(M'+1)\|F_{K2}(M'+2)\|F_{K2}(M'+3)\|\cdots$, where $M' = F_{K1}(M)$. A security analysis of XORG is given in Section 3.3.

THE UFE SCHEME. We now describe our general scheme $\text{UFE}[G,H] = (\mathcal{K}, \mathcal{E}, \mathcal{D})$

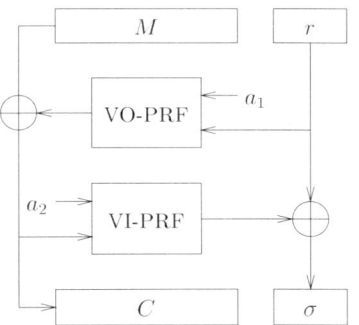

**Fig. 2.** *The UFE scheme.*

where $G : \mathcal{K}_{\text{vo-prf}} \times \{0,1\}^l \times 1^* \mapsto \{0,1\}^*$ is a VO-PRF and $H : \mathcal{K}_{\text{vi-prf}} \times \{0,1\}^* \mapsto \{0,1\}^l$ is a VI-PRF. The key generation algorithm $\mathcal{K} = \mathcal{K}_{\text{vo-prf}} \times \mathcal{K}_{\text{vi-prf}}$ specifies a key $a = a_1\|a_2$ where $a_1 \leftarrow \mathcal{K}_{\text{vo-prf}}$; $a_2 \leftarrow \mathcal{K}_{\text{vi-prf}}$. The encryption and decryption algorithms are defined as:

| Algorithm $\mathcal{E}_{a_1\|a_2}(M)$ | Algorithm $\mathcal{D}_{a_1\|a_2}(C')$ |
|---|---|
| $r \leftarrow \{0,1\}^l$ | parse $C'$ as $C\|\sigma$ where $|\sigma| = l$ |
| $C \leftarrow M \oplus G_{a_1}(r)$ | $r \leftarrow \sigma \oplus H_{a_2}(C)$ |
| $\sigma \leftarrow r \oplus H_{a_2}(C)$ | $M \leftarrow C \oplus G_{a_1}(r)$ |
| return $C\|\sigma$ | return $M$ |

A picture for the UFE scheme is given in Figure 2. We analyze the security of this scheme in Section 3.3.

## 3.3    Analysis

We begin with an analysis of our VO-PRF example. See Figure 1. The theorem says that XORG is secure as a VO-PRF as long as the underlying PRF is secure.

**Theorem 1. [Security of** XORG**]** *Let* $G = \text{XORG}[F]$ *where* $F = \text{PRF}(l)$. *Then,*

$$\text{Adv}_G^{\text{vo-prf}}(t, q, \mu) \leq 2 \cdot \text{Adv}_F^{\text{prf}}(t', q') + \frac{2(q-1)(q + \frac{\mu}{l})}{2^l}$$

*where* $t' = t + \mathcal{O}(q + \mu + l)$ *and* $q' = \lceil \frac{\mu}{l} \rceil + q$.

*Proof.* Let $A$ be an adversary attacking $G$ in the VO-PRF sense, and let $t, q, \mu$ be the resources associated with $\text{Exp}_G^{\text{vo-prf}}(A, b)$. Let $\mathcal{K}_{\text{prf}}$ be the key generation algorithm of $F$.

We assume without loss of generality that $A$ does not repeat queries. (A query consists of a string $M \in \{0,1\}^l$ and a string $L \in 1^*$. Our assumption is that $A$ picks a different $M$ for each query). We consider various probabilities related to running $A$ under different experiments:

$$p_1 = \Pr[\, K1, K2 \leftarrow \mathcal{K}_{\text{prf}} : \ A^{G_{K1\|K2}} = 1 \,]$$

$$p_2 = \Pr[\, f \leftarrow \text{Func}(l); \ K2 \leftarrow \mathcal{K}_{\text{prf}} : \ A^{G_{K2}^f} = 1 \,]$$

$$p_3 = \Pr[\, f, h \leftarrow \text{Func}(l) : \ A^{G^{f,h}} = 1 \,]$$

$$p_4 = \Pr[\, g \leftarrow \text{VO-Func}(l) : \ A^g = 1 \,]$$

The notation above is as follows: In the experiment defining $p_2$, $A$'s oracle, on query $M$ and $L \in 1^*$ responds by returning the first $|L|$ bits of $F_{K2}(M' + 1)\|F_{K2}(M'+2)\|F_{K2}(M'+3)\| \cdots$, where $M' = f(M)$. In the experiment defining $p_3$, $A$'s oracle, on query $M$ and $L \in 1^*$ responds by returning the first $|L|$ bits of $h(M' + 1)\|h(M' + 2)\|h(M' + 3)\| \cdots$, where $M' = f(M)$.

We want to upper bound $\text{Adv}_G^{\text{vo-prf}}(A) = p_1 - p_4$. We do this in steps.

Our first claim is that $p_1 - p_2 \leq \text{Adv}_F^{\text{prf}}(t', q)$.

Consider the following distinguisher $D$ for $F$. It has an oracle $\mathcal{O} : \{0,1\}^l \mapsto \{0,1\}^l$. It picks $K2 \leftarrow \mathcal{K}_{\text{prf}}$ and runs $A$. When $A$ makes a query $M$ and $L \in 1^*$, it returns $G_{K2}^{\mathcal{O}}(M)$ as the answer. $D$ outputs whatever $A$ outputs at the end. It is clear that $\text{Adv}_F^{\text{prf}}(D) = p_1 - p_2$. The claim follows.

Next we show that $p_2 - p_3 \leq \text{Adv}_F^{\text{prf}}(t', q')$.

Consider the following distinguisher $D$ for $F$. It has an oracle $\mathcal{O} : \{0,1\}^l \mapsto \{0,1\}^l$. It simulates $f \leftarrow \text{Func}$ and runs $A$. When $A$ makes a query $M$ and $L \in 1^*$, it returns $G^{f,\mathcal{O}}(M)$ as the answer. For any query $M_i\|L_i$ of $A$, $D$ must make $\lceil \frac{L_i}{l} \rceil$ queries to $\mathcal{O}$. $D$ outputs whatever $A$ outputs at the end. It follows that $\text{Adv}_F^{\text{prf}}(D) = p_2 - p_3$.

Finally, we show that $p_3 - p_4 \leq \frac{2(q-1)\cdot(q+\frac{\mu}{l})}{2^l}$.

We introduce some more notation to justify this. For any integer $t$ let $[t] = \{1, \cdots, t\}$. Let $(M_1, C_1), \ldots, (M_q, C_q)$ be the transcript of $A$'s interaction with its oracle, where for $i \in [q]$, $(M_i, C_i)$ represents an oracle query $M_i$ (such that $|M_i| = l$) and $L_i \in 1^*$ and the response $C_i$ (such that $|C_i| = |L_i|$). Let $n_i = \lceil \frac{L_i}{l} \rceil$ for $i \in [q]$. Let AC (Adversary is Correct) be the event that $A$ correctly guesses whether the oracle is $G^{f,h}$ or $g$, where these are as defined in the experiments underlying $p_3$ and $p_4$. In answering the $i$-th query $M_i \| L_i$, the oracle computes $r_i \leftarrow f(M_i)$ and applies a random function $h$ to the $n_i$ strings $r_i + 1, \ldots, r_i + n_i \in \{0, 1\}^l$. We call these strings the $i$-th $sequence$, and $r_i + k$ is the $k$-th point in this sequence, where $k \in [n_i]$.

Let Bad be event that $(r_i + k = r_j + k')$ for some $(i, k) \neq (j, k')$, and $(i, j \in [q]) \wedge (k \in [n_i]) \wedge (k' \in [n_j])$. That is Bad is the event that there are overlapping sequences. We have

$$\Pr[\,\mathsf{AC}\,] = \Pr[\,\mathsf{AC}\,|\,\overline{\mathsf{Bad}}\,] \cdot \Pr[\,\overline{\mathsf{Bad}}\,] + \Pr[\,\mathsf{AC}\,|\,\mathsf{Bad}\,] \cdot \Pr[\,\mathsf{Bad}\,]$$
$$\leq \Pr[\,\mathsf{AC}\,|\,\overline{\mathsf{Bad}}\,] + \Pr[\,\mathsf{Bad}\,]$$

Given the event $\overline{\mathsf{Bad}}$, we have that, in replying to $M_i \| L_i$, the output is randomly and uniformly distributed over $\{0, 1\}^{|L_i|}$. It follows that $\Pr[\,\mathsf{AC}\,|\,\overline{\mathsf{Bad}}\,] = \frac{1}{2}$.

Next, we bound $\Pr[\,\mathsf{Bad}\,]$. For $i \in [q]$, let $\mathsf{Bad}_i$ be the event that $r_i$ causes event Bad. We have

$\Pr[\,\mathsf{Bad}\,] \leq \Pr[\,\mathsf{Bad}_1\,] + \sum_{i=2}^{q} \Pr\left[\,\mathsf{Bad}_i \mid \overline{\mathsf{Bad}}_{i-1}\,\right]$. By definition, $\Pr[\,\mathsf{Bad}_1\,] = 0$. Since we are assuming that $M_i \neq M_j$ for any $(i \neq j) \wedge (i, j \in [q])$, we have that $r_i$ is randomly and uniformly distributed in $\{0, 1\}^l$. We observe that the chance of overlapping sequences is maximized if all the $i - 1$ previous queries resulted in $i - 1$ sequences that were no less than $n_i - 1$ blocks apart. We have a collision if the $i$-th sequence begins in a block that is $n_i - 1$ blocks before any other previous query $j$ or in a block occupied by that sequence $j$. We have that for $i > 1$,

$$\Pr\left[\,\mathsf{Bad}_i \mid \overline{\mathsf{Bad}}_{i-1}\,\right] \leq \frac{\sum_{j=1}^{i-1}(n_j + n_i - 1)}{2^l} = \frac{(i-1)(n_i - 1) + \sum_{j=1}^{i-1} n_j}{2^l}$$

Continuing,

$$\Pr[\,\mathsf{Bad}\,] \leq \sum_{i=2}^{q} \Pr\left[\,\mathsf{Bad}_i \mid \overline{\mathsf{Bad}}_{i-1}\,\right]$$
$$\leq \sum_{i=2}^{q} \frac{(i-1)(n_i - 1) + \sum_{j=1}^{i-1} n_j}{2^l} \leq \frac{(q-1)(q + \frac{\mu}{l})}{2^l}$$
$$p_3 - p_4 \leq 2 \cdot \Pr[\,\mathsf{AC}\,] - 1 \leq 2 \cdot \Pr[\,\mathsf{Bad}\,] \leq \frac{2(q-1)(q + \frac{\mu}{l})}{2^l}$$

Using the above bounds and that $\mathsf{Adv}_G^{\text{vo-prf}}(A) = p_1 - p_4 = (p_1 - p_2) + (p_2 - p_3) + (p_3 - p_4)$, we get the claimed result. ∎

We next turn to the security of our general scheme. We first establish the security of UFE assuming the underlying primitives are ideal.

**Lemma 1. [Upper bound on insecurity of UFE using random functions]**
*Let $\Pi = (\mathcal{K}, \mathcal{E}, \mathcal{D})$ be the scheme UFE$[G, H]$ where $G = $ VO-Func$(l)$ and $H = $ VI-Func$(l)$. Then for any $t, \mu, \nu$,*

$$\mathsf{Adv}_{\Pi}^{\text{ind-cca}}(t, q_e, q_d, \mu, \nu) \leq \delta_{\Pi} \overset{\text{def}}{=} \frac{(q_e + q_d)(q_e + q_d + 1)}{2^l}$$

*Proof.* Let $A$ be an adversary attacking $\Pi$ in the IND-CCA sense, and let $t, q_e, q_d, \mu, \nu$ be the resources associated with $\mathsf{Exp}_{\Pi}^{\text{ind-cca}}(A, b)$. We show that,

$$\mathsf{Adv}_{\Pi}^{\text{ind-cca}}(A) \overset{\text{def}}{=} 2 \cdot \Pr[\mathsf{Exp}_{\Pi}^{\text{ind-cca}}(A, b) = b] - 1 \leq \frac{(q_e + q_d)(q_e + q_d + 1)}{2^l}$$

We refer to the event $\mathsf{Exp}_{\Pi}^{\text{ind-cca}}(A, b) = b$ as event AC (Adversary is Correct). In the rest of the proof we will freely refer to random variables from $\mathsf{Exp}_{\Pi}^{\text{ind-cca}}(A, b)$.

Let $g \leftarrow G$ and $h \leftarrow H$ be the variable-length output function and variable-length input function, respectively, specified by the key $a$ in the experiment.

We assume without loss of generality that $A$ does not make "redundant" decryption oracle queries. That is, we are assuming that $A$ does not ask a decryption query $v$ if it had already made the query $v$ to its decryption oracle, or if it had obtained $v$ in response to some earlier encryption oracle query. Note that since encryption is probabilistic, $A$ may want to repeat encryption oracle queries.

Let $q$ be the total number of distinct plaintext-ciphertext pairs resulting from $A$'s interaction with its oracles. The following inequality holds: $q \leq q_e + q_d$. For simplicity, we assume that this is an equality. That is each query results in a unique plaintext-ciphertext pair. We are interested in an upper bound for $\Pr[\mathsf{AC}]$, and this assumption only increases this probability.

For any integer $t$ let $[t] = \{1, \cdots, t\}$.

Let $(M_1, C_1 \| \sigma_1), \ldots, (M_k, C_k \| \sigma_k), \ldots, (M_{q+1}, C_{q+1} \| \sigma_{q+1})$ be plaintext and ciphertext pairs, such that for $(i \in [q+1]) \wedge (i \neq k)$, $(M_i, C_i \| \sigma_i)$ represents an oracle query and the corresponding response. We have that $A$ picks plaintexts $x_0, x_1$ such that $|x_0| = |x_1| = \nu$ at the end of the find stage and receives $y \leftarrow \mathcal{E}_a(x_b)$ for some $b \in \{0, 1\}$. Let $M_k = x_b$ and $C_k \| \sigma_k = y$ where $|\sigma_k| = l$.

Let $r_i$ be the $l$-bit IV associated to $(M_i, C_i \| \sigma_i)$, for $i \in [q+1]$.

Let Bad be event that $r_i = r_j$ for some $(i, j \in [q+1]) \wedge (i \neq j)$. We have

$$\Pr[\mathsf{AC}] = \Pr[\mathsf{AC} \mid \overline{\mathsf{Bad}}] \cdot \Pr[\overline{\mathsf{Bad}}] + \Pr[\mathsf{AC} \mid \mathsf{Bad}] \cdot \Pr[\mathsf{Bad}]$$
$$\leq \Pr[\mathsf{AC} \mid \overline{\mathsf{Bad}}] + \Pr[\mathsf{Bad}]$$

Given the event $\overline{\mathsf{Bad}}$, we have that, in computing $y$, the output of $G$ is randomly and uniformly distributed over $\{0, 1\}^\nu$. Since this value is XORed with $x_b$, it follows that $\Pr[\mathsf{AC} \mid \overline{\mathsf{Bad}}] = \frac{1}{2}$. Next, we turn to a bound for $\Pr[\mathsf{Bad}]$.

For $i \in [q+1]$, let $\mathsf{Bad}_i$ be the event that $r_i$ causes event $\mathsf{Bad}$. We have $\Pr[\mathsf{Bad}] \leq \Pr[\mathsf{Bad}_1] + \sum_{i=2}^{q+1} \Pr[\mathsf{Bad}_i \mid \overline{\mathsf{Bad}}_{i-1}]$. By definition, $\Pr[\mathsf{Bad}_1] = 0$. We will consider $A$'s view just before it makes its $i$-th query, for $i \in [q+1]$. (If $i = k$, then we take the "query" to be an encryption query $x_b$.) Let us assume that this includes the knowledge that the event $\overline{\mathsf{Bad}}_{i-1}$ holds. Now depending on the nature of $A$'s $i$-th query, there are two cases we can consider: either $(M_i, C_i \| \sigma_i)$ results from an encryption query $M_i$ or from a decryption query $C_i \| \sigma_i$.

First we consider the case of the $i$-th query being an encryption query. The IV $r_i$, in this case, will be randomly and uniformly distributed in $\{0,1\}^l$. We have that the chance of a collision is at most $\frac{i-1}{2^l}$. Next we consider the case of the $i$-th query being a decryption query.

For $(i \in [q+1]) \wedge (i > 1)$, consider $A$'s view just before it makes its $i$-th query. We know that this includes $(M_1, C_1 \| \sigma_1), \ldots, (M_{i-1}, C_{i-1} \| \sigma_{i-1})$. However, given $\overline{\mathsf{Bad}}_{i-1}$, we claim that $h(C_j)$, for any $1 \leq j < i$, is information theoretically hidden in $A$'s view. With $\overline{\mathsf{Bad}}_{i-1}$, the only potential ways $A$ can learn something about $h(C_j)$ is through $\sigma_j$ or $r_j$. However we have that $r_j$ never becomes a part of $A$'s view (the IV is not returned in a decryption query). And we have $\sigma_j = r_j \oplus h(C_j)$. Since $r_j$ is unknown, we have that $\sigma_j$ does not leak any information about $h(C_j)$.

There are two sub-cases we can consider (when the $i$-th query is a decryption query). The first sub-case is that $C_i \neq C_j$, for all $1 \leq j < i$. Since $h$ is being invoked on a "new" string, the value $h(C_i)$ will be randomly and uniformly distributed in $\{0,1\}^l$. Consequently, $r_i$ will also be randomly and uniformly distributed in $\{0,1\}^l$. As in the previous case, we have for $i > 1$, the chance of a collision to be at most $\frac{i-1}{2^l}$.

The other sub-case is that $C_i = C_j$, for some $1 \leq j < i$. We want to bound the probability of $A$ picking a $\sigma_i$ that causes $\mathsf{Bad}_i$. We know that $A$ cannot pick $\sigma_i = \sigma_j$, since we are assuming that it does not make redundant queries. Moreover, we know that in $A$'s view, the value of $h(C_k)$ is information theoretically hidden, for any $1 \leq k < i$. Hence $A$'s only strategy in picking a $\sigma_i$ that causes a collision (other than choosing the value $\sigma_j$) can be to guess a value. It follows that $A$'s chances of causing a collision are smaller in this sub-case than if it had picked a new $C_i$. So here too, we have for $i > 1$, the chance of a collision to be at most $\frac{i-1}{2^l}$.

Continuing,

$$\Pr[\mathsf{Bad}] \leq \sum_{i=2}^{q+1} \Pr[\mathsf{Bad}_i \mid \overline{\mathsf{Bad}}_{i-1}] \leq \sum_{i=2}^{q+1} \frac{i-1}{2^l} \leq \frac{q(q+1)}{2 \cdot 2^l}$$

Using this in the bound for $\Pr[\mathsf{AC}]$ and doing a little arithmetic we get the claimed result. $\blacksquare$

The actual security of UFE is easily derived given Lemma 1.

**Theorem 2. [Security of** UFE**]** *Let* $\Pi = (\mathcal{K}, \mathcal{E}, \mathcal{D})$ *be the encryption scheme* UFE$[G, H]$ *where* $G = $ VO-PRF$(l)$ *and* $H = $ VI-PRF$(l)$. *Then,*

$$\mathsf{Adv}_{\Pi}^{\mathrm{ind\text{-}cca}}(t, q_e, q_d, \mu, \nu) \leq \mathsf{Adv}_{G}^{\mathrm{vo\text{-}prf}}(t', q', \mu') + \mathsf{Adv}_{H}^{\mathrm{vi\text{-}prf}}(t', q', \mu') + \delta_{\Pi}$$

*where* $t' = t + \mathcal{O}(\mu + \nu + lq_e + lq_d)$ *and* $q' = q_e + q_d$ *and* $\mu' = \mu + \nu$ *and* $\delta_{\Pi} \stackrel{\mathrm{def}}{=} \frac{(q_e + q_d)(q_e + q_d + 1)}{2^l}$.

*Proof.* Lemma 1 says that $\Pi[$VO-Func, VI-Func$]$ is secure. The intuition is that this implies that $\Pi[G, H]$ is secure, since otherwise it would mean that $G$ is not secure as a VO-PRF or that $H$ is not secure as a VI-PRF. Formally, we prove it using a contradiction argument.

Let $A$ be an adversary attacking $\Pi[G, H]$ in the IND-CCA sense. Let $t, q_e, q_d, \mu, \nu$ be the resources associated with $\mathsf{Exp}_{\Pi}^{\mathrm{ind\text{-}cca}}(A, b)$.

We will run $A$ under different experiments. We will refer to these experiments as $\mathsf{Exp}_{\Pi_1}^{\mathrm{ind\text{-}cca}}(A, b)$ and $\mathsf{Exp}_{\Pi_2}^{\mathrm{ind\text{-}cca}}(A, b)$ and $\mathsf{Exp}_{\Pi_3}^{\mathrm{ind\text{-}cca}}(A, b)$ where $\Pi_1 = \Pi[G, H]$ and $\Pi_2 = \Pi[$VO-Func$, H]$ and $\Pi_3 = \Pi[$VO-Func, VI-Func$]$. We will also refer to the corresponding advantage functions, which will follow the natural notation and interpretation, given the above. We are interested in an upper bound for $\mathsf{Adv}_{\Pi_1}^{\mathrm{ind\text{-}cca}}(t, q_e, q_d, \mu, \nu)$. We do this in steps.

Our first claim is

$$\mathsf{Adv}_{\Pi_1}^{\mathrm{ind\text{-}cca}}(t, q_e, q_d, \mu, \nu) \leq \mathsf{Adv}_{\Pi_2}^{\mathrm{ind\text{-}cca}}(t, q_e, q_d, \mu, \nu) + \mathsf{Adv}_{G}^{\mathrm{vo\text{-}prf}}(t', q', \mu')$$

Consider the following distinguisher $D$ for $G$. It has an oracle $\mathcal{O} : \{0, 1\}^l \times 1^* \mapsto \{0, 1\}^*$. It first picks a key for $H$ that specifies a function $h$. It then runs $A$ answering $A$'s oracle queries as follows. If $A$ makes an encryption oracle query $M$, then it picks a random $r \in \{0, 1\}^l$ and makes a query $r \| 1^{|M|}$ to $\mathcal{O}$. It then takes the response $P$ and computes $C = M \oplus P$. It returns to $A$ as its response the string $C \| (r \oplus h(C))$. Similarly to a decryption query $C \| \sigma$, it returns the string $C \oplus \mathcal{O}(\sigma \oplus h(C))$. Note that it is important that $D$ is able to correctly do the encryption and decryption using its oracle. It simulates the experiment defining the advantage of $A$. If $A$ is successful in the end, then it guesses that $\mathcal{O}$ was from VO-Func, otherwise it guesses that it was from VO-PRF.

We get $\mathsf{Adv}_{G}^{\mathrm{vo\text{-}prf}}(D) = \mathsf{Adv}_{\Pi_1}^{\mathrm{ind\text{-}cca}}(A) - \mathsf{Adv}_{\Pi_2}^{\mathrm{ind\text{-}cca}}(A)$. One can check that the number of queries $q'$ made by $D$ is at most $q_e + q_d$. Also the length $\mu'$ of all of $D$'s queries is at most the sum of the length $\mu$ of all the queries of $A$ and the length $\nu$ of the challenge that $D$ has to prepare for $A$. This proves the claim.

The next claim is

$$\mathsf{Adv}_{\Pi_2}^{\mathrm{ind\text{-}cca}}(t, q_e, q_d, \mu, \nu) \leq \mathsf{Adv}_{\Pi_3}^{\mathrm{ind\text{-}cca}}(t, q_e, q_d, \mu, \nu) + \mathsf{Adv}_{H}^{\mathrm{vi\text{-}prf}}(t', q', \mu')$$

We can construct a distinguisher $D$ for $H$ along similar lines as above. The main difference is that $D$ must simulate a random function from VO-Func in its simulation for $A$. We omit the details to prove this claim.

Combining our claims and substituting the bound for $\mathsf{Adv}_{\Pi_3}^{\mathrm{ind\text{-}cca}}(t, q_e, q_d, \mu, \nu)$ from Lemma 1, we get the claimed result. ∎

# 4    Encode-Then-Encipher Encryption

Bellare and Rogaway show that if messages are encoded with randomness and redundancy and then enciphered with a VI-SPRP then the resulting scheme is secure in a sense that implies security in the IND-CCA sense [8]. We show in this section that to achieve security in just the IND-CCA sense, the redundancy is unnecessary.

VARIABLE-LENGTH INPUT SUPER-PSEUDORANDOM PERMUTATIONS. These are permutations that take an input of variable and arbitrary length and produce an output of the same length. We define a reference family VI-Perm. A random variable-length input permutation $(f, f^{-1}) \leftarrow$ VI-Perm is defined as follows: for each number $i$, let $f_i$ be a random permutation on $\{0, 1\}^i$, and for each $M \in \{0, 1\}^*$, let $f(M) = f_i(M)$, where $i = |M|$. Let $f^{-1}$ be the inverse of $f$.

**Definition 5.** [VI-SPRP] *Let* $S : \mathcal{K} \times \{0, 1\}^* \mapsto \{0, 1\}^*$ *be a permutation. For a distinguisher* $A$ *and* $b = 0, 1$ *define the experiment*

*Experiment* $\mathsf{Exp}_S^{\mathrm{vi\text{-}sprp}}(A, b)$

$\qquad a \leftarrow \mathcal{K}; \quad \mathcal{O}_0, \mathcal{O}_0^{-1} \leftarrow S_a, S_a^{-1}; \quad \mathcal{O}_1, \mathcal{O}_1^{-1} \leftarrow \text{VI-Perm}; \quad d \leftarrow A^{\mathcal{O}_b, \mathcal{O}_b^{-1}}; \quad$ Return $d$.

*Define the advantage of* $A$ *and the advantage function of* $S$ *respectfully, as follows:*

$$\mathsf{Adv}_S^{\mathrm{vi\text{-}sprp}}(A) = \Pr[\, \mathsf{Exp}_S^{\mathrm{vi\text{-}sprp}}(A, 0) = 0 \,] - \Pr[\, \mathsf{Exp}_S^{\mathrm{vi\text{-}sprp}}(A, 1) = 0 \,]$$

$$\mathsf{Adv}_S^{\mathrm{vi\text{-}sprp}}(t, q_e, q_d, \mu) = \max_A \{ \mathsf{Adv}_S^{\mathrm{vi\text{-}sprp}}(A) \}$$

*where the maximum is over all* $A$ *with time complexity* $t$ *and making at most* $q_e$ *queries to* $S_a$ *and* $q_d$ *queries to* $S_a^{-1}$, *these together totalling at most* $\mu$ *bits.* ∎

Constructions of these "full-fledged" pseudorandom permutations are relatively rare. Naor and Reingold show how to efficiently construct an SPRP that can work with any large input-length given an SPRP (or a PRF) of some fixed smaller input-length [19]. However, their constructions cannot work with inputs of arbitrary and variable length, and it is unclear how they can be extended to do so. Bleichenbacher and Desai suggest a construction for a VI-SPRP using a block cipher (modeled as an SPRP) that has a cost of about three applications of the block cipher per message block [11]. Patel, Ramzan and Sundaram have a construction that is computationally less expensive than this but wherein the key-length varies with the message-length [21].

THE EEE SCHEME. We now describe the scheme EEE$[S] = (\mathcal{K}, \mathcal{E}, \mathcal{D})$ where $S : \mathcal{K}_{\mathrm{vi\text{-}sprp}} \times \{0, 1\}^* \mapsto \{0, 1\}^*$ is a VI-SPRP. The key generation algorithm $\mathcal{K} = \mathcal{K}_{\mathrm{vi\text{-}sprp}}$ specifies a key $a$. For any positive integer $l$, the encryption and decryption algorithms are defined as:

| Algorithm $\mathcal{E}_a(M)$ | Algorithm $\mathcal{D}_a(C)$ |
|---|---|
| $\quad r \leftarrow \{0, 1\}^l$ | if $|C| \leq l$ then $M \leftarrow \perp$ else |
| $\quad C \leftarrow S_a(M \| r)$ | $\quad (M \| r) \leftarrow S_a^{-1}(C)$ where $|r| = l$ |
| $\quad$ return $C$ | $\quad$ return $M$ |

We give the security of this scheme next.

**Theorem 3. [Security of EEE]** *Let $\Pi = (\mathcal{K}, \mathcal{E}, \mathcal{D})$ be the encryption scheme* EEE[$S$] *where $S$ = VI-SPRP. Then,*

$$\mathsf{Adv}_{\Pi}^{\text{ind-cca}}(t, q_e, q_d, \mu, \nu) \leq 2 \cdot \mathsf{Adv}_{S}^{\text{vi-sprp}}(t', q_e', q_d', \mu') + \frac{2(q_e + q_d)}{2^l}$$

*where $t' = t + \mathcal{O}(\mu + \nu + lq_e + lq_d)$ and $q_e' = q_e + 1$ and $q_d' = q_d$ and $\mu' = \mu + \nu$.*

*Proof.* Let $A$ be an adversary attacking $\Pi$ in the IND-CCA sense, and let $t, q_e, q_d, \mu, \nu$ be the resources associated with $\mathsf{Exp}_{\Pi}^{\text{ind-cca}}(A, b)$.

We assume without loss of generality that $A$ does not make "redundant" queries to its decryption oracle. That is, we are assuming that $A$ does not ask a decryption oracle query $v$ if it had already made the query $v$ to its decryption oracle, or if it had obtained $v$ in response to some earlier encryption oracle query.

Our goal is to bound $\mathsf{Adv}_{\Pi}^{\text{ind-cca}}(A, b)$. To this end we introduce an algorithm $D$.

Algorithm $D$ is a distinguisher for $S$. It is given oracles for permutations $f, f^{-1}$ that are either from a VI-SPRP family or from the random family VI-Perm. It runs $A$, answering $A$'s queries as follows: If $A$ makes an encryption oracle query $M$ then $D$ picks $r \leftarrow \{0, 1\}^l$ and computes $C \leftarrow f(M\|r)$. It returns $C$ as the response to the query. If $A$ makes a decryption oracle query $C$ then $D$ first checks if $|C| \leq l$. If it is then $D$ returns "invalid" as its response. Otherwise it computes $(M\|r) \leftarrow f^{-1}(C)$ (where $|r| = l$) and returns $M$ as the response to the query. $A$ eventually stops (at the end its find stage) and outputs $(x_0, x_1, s)$. $D$ then chooses $d \leftarrow \{0, 1\}$ and $r_0 \leftarrow \{0, 1\}^l$ and computes $y \leftarrow f(x_d\|r_0)$. (If $D$ has already queried on this point before or ever received this in response to some previous decryption oracle query, then it does not have to use its oracle to compute $y$). $D$ then runs $A$ with the parameters (guess, $y, s$), answering $A$'s oracle queries as before. When $A$ terminates, $D$ checks to see if it was correct. If it was, then $D$ guesses that its oracles were "real", otherwise it guesses that they were "random".

We develop some notation to simplify the exposition of the analysis. Let $\Pr_1[\cdot]$ denote a probability in the probability space where the oracles given to $D$ are "real". Similarly, let $\Pr_0[\cdot]$ denote a probability in the probability space where the oracles given to $D$ are "random". We will suppress showing explicit access to the oracles since they will be obvious from context. We have

$$\mathsf{Adv}_{S}^{\text{vi-sprp}}(D) \overset{\text{def}}{=} \Pr_1[D = 1] - \Pr_0[D = 1]$$

From the description of $D$, we see that $\Pr_1[D = 1]$ is exactly the probability of $A$ being correct in an experiment defining the advantage in the IND-CCA sense. Thus we get,

$$\Pr_1[D = 1] = \frac{1}{2} + \frac{1}{2} \cdot \mathsf{Adv}_{\Pi}^{\text{ind-cca}}(A)$$

Next we upper bound $\Pr_0[D = 1]$. Let Coll be the event that there is a collision of one of the nonces resulting from $A$'s queries with the one in the challenge.

More precisely, $\mathsf{Coll}$ is the event that $\exists i \in [q_e + q_d] : r_i = r_0$.

$$\mathrm{Pr}_0[\,D = 1\,] = \mathrm{Pr}_0\,[\,D = 1 \mid \mathsf{Coll}\,] \cdot \mathrm{Pr}_0[\,\mathsf{Coll}\,] + \mathrm{Pr}_0\,[\,D = 1 \mid \overline{\mathsf{Coll}}\,] \cdot \mathrm{Pr}_0[\,\overline{\mathsf{Coll}}\,]$$

$$\leq \mathrm{Pr}_0[\,\mathsf{Coll}\,] + \mathrm{Pr}_0\,[\,D = 1 \mid \overline{\mathsf{Coll}}\,]$$

Since the permutations underlying $\mathrm{Pr}_0[\cdot]$ are "random", we get

$$\mathrm{Pr}_0[\,\mathsf{Coll}\,] \leq \frac{q_e + q_d}{2^l}; \quad \mathrm{Pr}_0\,[\,D = 1 \mid \overline{\mathsf{Coll}}\,] \; = \; \frac{1}{2}$$

Using the above to lower bound the advantage of $D$ and completing the argument in the standard way, we get the claimed result. ∎

## Acknowledgements

This paper benefited a great deal from help and advice received from Mihir Bellare. Many of the ideas and motivation for the problem considered here came out of collaboration with Daniel Bleichenbacher. I would also like to thank Sara Miner, Chanathip Namprempre, Bogdan Warinschi and the CRYPTO 2000 program committee for their very helpful comments.

The author was supported in part by Mihir Bellare's 1996 Packard Foundation Fellowship in Science and Engineering and NSF CAREER Award CCR-9624439.

## References

1. M. BELLARE, R. CANETTI AND H. KRAWCZYK, "Pseudorandom functions revisited: The cascade construction and its concrete security," *Proceedings of the 37th Symposium on Foundations of Computer Science*, IEEE, 1996.

2. M. BELLARE, A. DESAI, E. JOKIPII AND P. ROGAWAY, "A concrete security treatment of symmetric encryption," *Proceedings of the 38th Symposium on Foundations of Computer Science*, IEEE, 1997.

3. M. BELLARE, A. DESAI, D. POINTCHEVAL AND P. ROGAWAY, "Relations among notions of security for public-key encryption schemes," *Advances in Cryptology - Crypto '98*, LNCS Vol. 1462, H. Krawczyk ed., Springer-Verlag, 1998.

4. M. BELLARE, J. KILIAN AND P. ROGAWAY, "The security of the cipher block chaining message authentication code," *Advances in Cryptology - Crypto '94*, LNCS Vol. 839, Y. Desmedt ed., Springer-Verlag, 1994.

5. M. BELLARE AND C. NAMPREMPRE, "Authenticated encryption: Relations among notions and analysis of the generic composition paradigm," Report 2000/025, *Cryptology ePrint Archive*, http://eprint.iacr.org/, May 2000.

6. M. BELLARE AND P. ROGAWAY, "Entity authentication and key distribution," *Advances in Cryptology - Crypto '93*, LNCS Vol. 773, D. Stinson ed., Springer-Verlag, 1993.

7. M. BELLARE AND P. ROGAWAY, "Optimal asymmetric encryption: How to encrypt with RSA," *Advances in Cryptology - Eurocrypt '95*, LNCS Vol. 921, L. Guillou and J. Quisquater ed., Springer-Verlag, 1995.

8. M. BELLARE AND P. ROGAWAY, "Encode-then-encipher encryption: How to exploit nonces or redundancy in plaintexts for efficient cryptography," Manuscript, December 1998, available from authors.

9. D. BERNSTEIN, "How to stretch random functions: The security of protected counter sums," *J. of Cryptology*, Vol. 12, No. 3, 1999.

10. J. BLACK AND P. ROGAWAY, "CBC MACs for Arbitrary Length Messages: The Three-Key Constructions," *Advances in Cryptology - Crypto '00*, LNCS Vol. ??, M. Bellare ed., Springer-Verlag, 2000.

11. D. BLEICHENBACHER AND A. DESAI, "A construction of a super-pseudorandom cipher," Manuscript, May 1999, available from authors.

12. A. DESAI, "New paradigms for constructing symmetric encryption schemes secure against chosen-ciphertext attack," Full version of this paper, available via: http://www-cse.ucsd.edu/users/adesai/.

13. D. DOLEV, C. DWORK AND M. NAOR, "Non-malleable cryptography," *SIAM J. of Computing*, to appear. Preliminary version in *Proceedings of the 23rd Annual Symposium on the Theory of Computing*, ACM, 1991.

14. O. GOLDREICH, S. GOLDWASSER AND S. MICALI, How to construct random functions. *Journal of the ACM*, Vol. 33, N0. 4, 1986, pp. 210-217.

15. S. GOLDWASSER AND S. MICALI, "Probabilistic encryption," *Journal of Computer and System Science*, Vol. 28, 1984, pp. 270–299.

16. J. KATZ AND M. YUNG, "Complete characterization of security notions for probabilistic private-key encryption," *Proceedings of the 32nd Annual Symposium on the Theory of Computing*, ACM, 2000.

17. J. KATZ AND M. YUNG, "Unforgeable Encryption and Adaptively Secure Modes of Operation," *Fast Software Encryption '00*, LNCS Vol. ??, B. Schneier ed., Springer-Verlag, 2000.

18. M. LUBY AND C. RACKOFF, "How to construct pseudorandom permutations from pseudorandom functions," *SIAM J. Computing*, Vol. 17, No. 2, April 1988.

19. M. NAOR AND O. REINGOLD, "On the construction of pseudorandom permutations: Luby-Rackoff revisited," *J. of Cryptology*, Vol. 12, No. 1, 1999.

20. M. NAOR AND M. YUNG, "Public-key cryptosystems provably secure against chosen-ciphertext attackss," *Proceedings of the 22nd Annual Symposium on the Theory of Computing*, ACM, 1990.

21. S. PATEL, Z. RAMZAN, AND G. SUNDARAM, "Efficient Variable-Input-Length Cryptographic Primitives," Manuscript, 2000.

22. E. PETRANK AND C. RACKOFF, "CBC MAC for Real-Time Data Sources," *Dimacs Technical Report*, 97-26, 1997.

23. C. RACKOFF AND D. SIMON, "Non-interactive zero-knowledge proof of knowledge and chosen-ciphertext attack," *Advances in Cryptology - Crypto '91*, LNCS Vol. 576, J. Feigenbaum ed., Springer-Verlag, 1991.

# Efficient Non-malleable Commitment Schemes

Marc Fischlin and Roger Fischlin

Fachbereich Mathematik (AG 7.2)
Johann Wolfgang Goethe-Universität Frankfurt am Main
Postfach 111932
D–60054 Frankfurt/Main, Germany

{marc,fischlin} @ mi.informatik.uni-frankfurt.de
http://www.mi.informatik.uni-frankfurt.de/

**Abstract.** We present efficient non-malleable commitment schemes based on standard assumptions such as RSA and Discrete-Log, and under the condition that the network provides publicly available RSA or Discrete-Log parameters generated by a trusted party. Our protocols require only three rounds and a few modular exponentiations. We also discuss the difference between the notion of non-malleable commitment schemes used by Dolev, Dwork and Naor [DDN00] and the one given by Di Crescenzo, Ishai and Ostrovsky [DIO98].

## 1 Introduction

Loosely speaking, a commitment scheme is non-malleable if one cannot transform the commitment of another person's secret into one of a related secret. Such non-malleable schemes are for example important for auctions over the Internet: it is necessary that one cannot generate a valid commitment of a bid $b + 1$ after seeing the commitment of an unknown bid $b$ of another participant. Unfortunately, this property is not achieved by commitment schemes in general, because ordinary schemes are only designated to hide the secret. Even worse, most known commitment schemes are in fact provably malleable.

The concept of non-malleability has been introduced by Dolev et al. [DDN00]. They present a non-malleable public-key encryption scheme (based on any trapdoor permutation) and a non-malleable commitment scheme with logarithmically many rounds based on any one-way function. Yet, their solutions involve cumbersome non-interactive and interactive zero-knowledge proofs, respectively. While efficient non-malleable encryption schemes under various assumptions have appeared since then [BR93,BR94,CS98], as far as we know more efficient non-malleable commitment protocols are still missing. Di Crescenzo et al. [DIO98] present a non-interactive and non-malleable commitment scheme based on any one-way function in the common random string model. Though being non-interactive, their system is rather theoretical as it excessively applies an ordinary commitment scheme to non-malleably commit to a single bit.

Here, we present efficient statistically-secret non-malleable schemes based on standard assumptions, such as the RSA assumption and the hardness of

M. Bellare (Ed.): CRYPTO 2000, LNCS 1880, pp. 413–431, 2000.
© Springer-Verlag Berlin Heidelberg 2000

computing discrete logarithms. Our schemes are designed in the public parameter model (a.k.a. auxilary string model). That is, public parameters like a random prime $p$ and generators of some subgroup of $\mathbb{Z}_p^*$ are generated and published by a trusted party. We stress that, in contrast to public-key infrastructure, this model does not require the participants to put any trapdoor information into the parameters. The public parameter model relies on a slightly stronger assumption than the common random string model. Yet, the difference is minor as modern networks are likely to provide public parameters for standard crypto systems. Moreover, as for the example of the discrete logarithm, the public parameter model can be formally reduced to the common random string model if we let the participants map the random string via standard procedures to a prime and appropriate generators.

In our schemes the sender basically commits to his message using an ordinary, possibly malleable DLog- or RSA-based commitment scheme and performs a three-round witness-independent proof of knowledge, both times using the public parameters. While the straightforward solution of a standard proof of knowledge fails (because the adversary may in addition to the commitment also transform the proof of knowledge), we force the adversary to give his "own" proof of knowledge without being able to adapt the one of the original sender. Similar ideas have also been used in [DDN00,DIO98]. In our case, the proof of knowledge guarantees that the adversary already knows the message he has committed to. This means that he is aware of some information about the related message of the original sender, contradicting the secrecy property of the ordinary commitment scheme.

We also address definitional issues. We show that the notion of non-malleability used by Di Crescenzo et al. [DIO98] is weaker than the one presented in [DDN00]. According to the definition of [DIO98], a scheme is non-malleable if the adversary cannot construct a commitment from a given one, such that after having seen the opening of the original commitment, the adversary is able to correctly open his commitment with a related message. But the definition of Dolev et al. [DDN00] demands more: if there is a one-to-one correspondence between the commitment and the message (say, if the commitment binds unconditionally), then they define that such a scheme is non-malleable if one cannot even generate a commitment of a related message. We call schemes having the latter property *non-malleable with respect to commitment*. For these schemes to contradict non-malleability it suffices to come up with a commitment such that there exists a related opening. Schemes satisfying the former definition are called *non-malleable with respect to decommitment* or, for sake of distinctiveness, *with respect to opening*. In this case, the adversary must also be able to open the modified commitment correctly given the decommitment of the original commitment. The scheme in [DDN00] achieves the stronger notion, whereas we do not know if the scheme in [DIO98] is also non-malleable with respect to commitment.

Clearly, a commitment scheme which is non-malleable in the strong sense is non-malleable with respect to opening, too. We stress that the other direction does not hold in general. That is, given a statistically-secret commitment scheme which is secure with respect to opening, we can devise a commitment scheme satisfying the weak notion, but not the strong definition. Since our statistically-secret schemes based on standard assumptions like RSA or Discrete-Log achieve non-malleability with respect to opening, both notions are *not* equivalent under these assumptions. The proof of this claim is deferred from this abstract.

We believe that non-malleability with respect to opening is the appropriate notion for statistically-secret schemes like ours. The reason is that for such schemes virtually any commitment can be opened with any message. Hence, finding a commitment of a related message to a given commitment is easy: any valid commitment works with very high probability. Recently, Yehuda Lindell informed us about an application of non-malleable commitment schemes to authenticated key-exchange where non-malleability with respect to commitment is necessary [L00]. Yet, non-malleability with respect to opening still seems to be adequate for most applications. For instance, recall the example of Internet auctions. The commitments of the bids are collected and then, after a deadline has passed, are requested to be opened. Any secret which is not correctly revealed is banned. Therefore, security with respect to opening suffices in this setting.

Our schemes as well as the one by [DDN00] use proof-of-knowledge techniques. But since we are merely interested in non-malleability with respect to opening, we do not need proofs of knowledge to the full extent. Namely, it suffices that the proof of knowledge is verifiable by the receiver after the sender has de-committed. Since the adversary must be able to open his commitment correctly, we can presume in the commitment phase that the proof of knowledge is indeed valid. This enables us to speed up our proofs of knowledge, i.e., we introduce new techniques for such *a-posteriori verifiable* proofs of knowledge based on the Chinese Remainder Theorem. As a side effect, this proof of knowledge allows to hash longer messages before committing and the resulting scheme still achieves non-malleability. In contrast to this, non-malleable schemes based on well-known proofs of knowledge do not seem to support the hash-and-commit paradigm in general.

The paper is organized as follows. In Section 2 we introduce basic notations and definitions of commitment schemes as well as the notions of non-malleability. In Section 3 we present efficient schemes in the public parameter model based on the discrete-log assumption, and, finally, in Section 4 we show how to speed up the proof of knowledge.

## 2   Preliminaries

Unless stated otherwise all parties and algorithms are probabilistic polynomial-time. Throughout this paper, we use the notion of uniform algorithms; all results transfer to the non-uniform model of computation. A function $\delta(n)$ is said to be *negligible* if $\delta(n) < 1/p(n)$ for every polynomial $p(n)$ and sufficiently large $n$. A function $\delta(n)$ is called *overwhelming* if $1 - \delta(n)$ is negligible. A function is *noticeable* if it is not negligible.

Two sequences $(X_n)_{n \in \mathbb{N}}$ and $(Y_n)_{n \in \mathbb{N}}$ of random variables are called *computationally indistinguishable* if for any probabilistic polynomial-time algorithm $\mathcal{D}$ the advantage

$$|\text{Prob}\,[\mathcal{D}(1^n, X_n) = 1] - \text{Prob}\,[\mathcal{D}(1^n, Y_n) = 1]|$$

of $\mathcal{D}$ is negligible, where the probabilities are taken over the coin tosses of $\mathcal{D}$ and the random choice of $X_n$ and $Y_n$, respectively. The sequences are called *statistically close* or *statistically indistinguishable* if

$$\tfrac{1}{2} \cdot \sum_{s \in S_n} |\text{Prob}\,[X_n = s] - \text{Prob}\,[Y_n = s]|$$

is negligible, where $S_n$ is the union of the supports of $X_n$ and $Y_n$.

## 2.1   Commitment Schemes

We give a rather informal definition of commitment schemes. For a formalization we refer the reader to [G98]. A commitment scheme is a two-phase interactive protocol between two parties, the sender $\mathcal{S}$ holding a message $m$ and a random string $r$, and the receiver $\mathcal{R}$.

In the first phase, called the commitment phase, $\mathcal{S}$ gives some information derived from $m, r$ to $\mathcal{R}$ such that, on one hand, $\mathcal{R}$ does not gain any information about $m$, and on the other hand, $\mathcal{S}$ cannot later change his mind about $m$. We call the whole communication in this phase the commitment of $\mathcal{S}$. Of course, both parties should check (if possible) that the values of the other party satisfy structural properties, e.g., that a value belongs to a subgroup of $\mathbb{Z}_p^*$, and should reject immediately if not. In the following, we do not mention such verification steps explicitly. We say that a commitment, i.e., the communication, is *valid* if the honest receiver does not reject during the commitment phase.

In the decommitment stage, the sender communicates the message $m$ and the randomness $r$ to the receiver, who verifies that $m, r$ match the communication of the first phase. If the sender obeys the protocol description, then the commitment is valid and $\mathcal{R}$ always accepts the decommitment.

There are two fundamental kinds of commitment schemes:

- A scheme is *statistically-binding (and computationally-secret)* if any arbitrary powerful malicious $\mathcal{S}^*$ cannot open a valid commitment ambiguously except with negligible probability (over the coin tosses of $\mathcal{R}$), and two commitments are computationally indistinguishable for every probabilistic polynomial-time (possibly malicious) $\mathcal{R}^*$. If the binding property holds unconditionally and not only with high probability, then we call the scheme unconditionally-binding.

- A scheme is *(computationally-binding and) statistically-secret* if it satisfies the "dual" properties, that is, if the distribution of the commitments are statistically close for any arbitrary powerful $\mathcal{R}^*$, and yet opening a valid commitment ambiguously contradicts the hardness of some cryptographic assumption. If the distribution of the commitments of any messages are identical, then a statistically-secret schemes is called perfectly-secret.

## 2.2   Non-malleability

As mentioned in the introduction, different notions of non-malleability have been used implicitly in the literature. To highlight the difference we give a formal definition of non-malleable commitment schemes, following the approach of [DDN00]. For non-interactive commitment schemes, all the adversary can do is modify a given commitment. In the interactive case, though, the adversary might gain advantage from the interaction. We adopt this worst-case scenario and assume that the adversary interacts with the original sender, while at the same time he is trying to commit to a related message to the original receiver.

A pictorial description of a so-called *person-in-the-middle attack* (PIM attack) on an interactive protocol is given in Figure 1. The adversary $\mathcal{A}$ intercepts the messages of the sender $\mathcal{S}$. Then $\mathcal{A}$ may modify the messages before passing them to the receiver $\mathcal{R}$ and proceeds accordingly with the answers. In particular, $\mathcal{A}$ decides to whom he sends the next message, i.e., to the sender or to the receiver. This is the setting where $\mathcal{A}$ has full control over the parties $\mathcal{R}_1$ and $\mathcal{S}_2$ in two supposedly independent executions $\langle \mathcal{S}_1, \mathcal{R}_1 \rangle(m)$, $\langle \mathcal{S}_2, \mathcal{R}_2 \rangle(m^*)$ of the same interactive protocol. Here and in the rest of this paper, we usually mark values sent by the adversary with an asterisk.

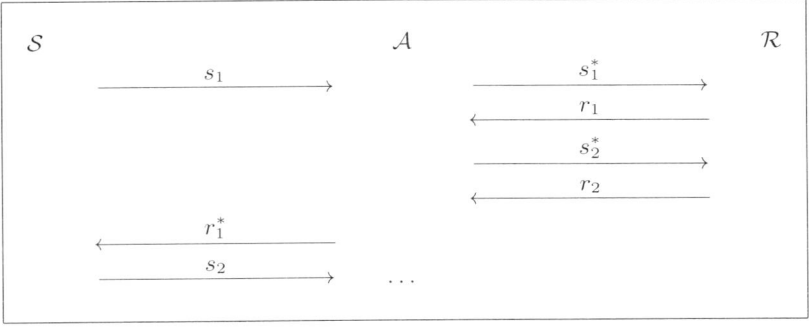

**Fig. 1.** Person-In-The-Middle Attack on Interactive Protocols

Apparently, the adversary can always commit to the same message by forwarding the communication. In many applications, this can be prevented by letting the sender append his identity to the committed message. The messages of the sender and the adversary are taken from a space $\mathbb{M}$ chosen by the adversary. Abusing notations, we view $\mathbb{M}$ also as a distribution, and write $m \in_R \mathbb{M}$ for a randomly drawn message according to $\mathbb{M}$. The adversary is deemed to be successful if he commits to a related message, where related messages are identified by so-called interesting relations: a relation $R \subseteq \mathbb{M} \times \mathbb{M}$ is called *interesting* if it is non-reflexive (to exclude copying) and efficiently computable. Let $\mathrm{hist}(\cdot)$ be a polynomial-time computable function, representing the a-priori information that the adversary has about the sender's message. In the sequel, we view $\mathrm{hist}(\cdot)$ as a part of the adversary's description, and usually omit mentioning it explicitly.

We describe the attack in detail. First, the adversary $\mathcal{A}$ generates a description of $\mathbb{M}$. Then the public parameters are generated by a trusted party according to a publicly known distribution (if a protocol does not need public information then this step is skipped).[1] The sender $\mathcal{S}$ is initialized with $m \in_R \mathbb{M}$. Now $\mathcal{A}$, given $\mathrm{hist}(m)$, mounts a PIM attack with $\mathcal{S}(m)$ and $\mathcal{R}$. Let $\pi_{\mathsf{com}}(\mathcal{A}, R)$ denote the probability that, at the end of the commitment phase, the protocol execution between $\mathcal{A}$ and $\mathcal{R}$ constitutes a valid commitment for a message $m^*$ satisfying $(m, m^*) \in R$. Let $\pi_{\mathsf{open}}(\mathcal{A}, R)$ denote the probability that $\mathcal{A}$ is also able to successfully open the commitment after $\mathcal{S}$ has decommitted.

In a second experiment, a simulator $\mathcal{A}'$ tries to commit to a related message without the help of the sender. That is, $\mathcal{A}'$ first generates a description of $\mathbb{M}'$ and the public parameters and then, given $\mathrm{hist}(m)$ for some $m \in_R \mathbb{M}'$, it outputs a commitment communication without interacting with $\mathcal{S}(m)$. Let $\pi'_{\mathsf{com}}(\mathcal{A}', R)$ denote the probability that this communication is a valid commitment of a related message $m'$. By $\pi'_{\mathsf{open}}(\mathcal{A}', R)$ we denote the probability that $\mathcal{A}'$ additionally reveals a correct decommitment.

Note that all probabilities are implicit functions of a security parameter. For the definition we assume that messages contain a description of the distributions $\mathbb{M}$ and $\mathbb{M}'$, respectively, as prefix. This prevents $\mathcal{A}'$ from taking a trivial set $\mathbb{M}'$, since these sets can be ruled out by $R$. If the simulator $\mathcal{A}'$ sets $\mathbb{M}' = \mathbb{M}$ —which is the case in all the schemes we know of— we can omit the description portion.

**Definition 1.** *A commitment scheme is called*

*b) non-malleable with respect to commitment if for every adversary $\mathcal{A}$ there exists a simulator $\mathcal{A}'$ such that for all interesting relations $R$ the difference $|\pi_{\mathsf{com}}(\mathcal{A}, R) - \pi'_{\mathsf{com}}(\mathcal{A}', R)|$ is negligible.*

*a) non-malleable with respect to opening if for every adversary $\mathcal{A}$ there exists a simulator $\mathcal{A}'$ such that for all interesting relations $R$ the difference $|\pi_{\mathsf{open}}(\mathcal{A}, R) - \pi'_{\mathsf{open}}(\mathcal{A}', R)|$ is negligible.*

Slightly relaxing the definition, we admit an *expected* polynomial-time simulator $\mathcal{A}'$. In fact, we are only able to prove our schemes non-malleable with this deviation. The reason for this is that we apply proofs of knowledge, so in order to make the success probability of $\mathcal{A}'$ negligibly close to the adversary's success probability, we run a knowledge extractor taking expected polynomial-time.[2] Following the terminology in [DDN00], we call such schemes *liberal non-malleable* with respect to commitment and opening, respectively.

---

[1] In a stronger requirement the order of these steps is swapped, i.e., the adversary chooses the message space in dependence of the public parameters. Although our scheme achieves this stronger notion, we defer this from this abstract.

[2] The same problem occurs in [DDN00]. Alternatively, the authors of [DDN00] also propose a definition of $\epsilon$-malleability, which basically says that for given $\epsilon$ there is a strict polynomial-time simulator (polynomial in the security parameter $n$ and $\epsilon^{-1}(n)$) whose success probability is only $\epsilon$-far from the adversary's probability.

Consider a computationally-binding and perfectly-secret commitment scheme. There, every valid commitment is correctly openable with every message (it is, however, infeasible to find different messages that work). Thus, we believe that non-malleability with respect to opening is the interesting property in this case. On the other hand, non-malleability with respect to commitment is also a concern for statistically-binding commitment schemes: with overwhelming probability there do not exist distinct messages that allow to decommit correctly. This holds for *any dishonest* sender and, in particular, for the person-in-the-middle adversary. We can therefore admit this negligible error and still demand non-malleability with respect to commitment.

# 3     Efficient Non-malleable Commitment Schemes

In this section we introduce our commitment schemes which are non-malleable with respect to opening. For lack of space, we only present the discrete-log scheme; the RSA-based protocol is omitted. In Section 3.1 we start with an instructive attempt to achieve non-malleability by standard proof-of-knowledge techniques. We show that this approach yields a scheme which is only non-malleable with respect to opening against static adversaries, i.e., adversaries that try to find a commitment after passively observing a commitment between the original sender and receiver and such that the adversary can later correctly open the commitment after learning the decommitment of the sender. In Section 3.2 we develop out of this our scheme which is non-malleable against the stronger PIM adversaries.

## 3.1     Non-malleability with Respect to Static Adversaries

Consider Pedersen's well-known discrete-log-based perfectly-secret scheme [P91]. Let $G_q \subseteq \mathbb{Z}_p^*$ be a group of prime order $q$ and $g_0, h_0$ two random generators of $G_q$. Assume that computing the discrete logarithm $\log_{g_0} h_0$ is intractable. To commit to a message $m \in \mathbb{Z}_q$, choose $r \in_R \mathbb{Z}_q$ and set $M := g_0^m h_0^r$. To open this commitment, reveal $m$ and $r$. Obviously, the scheme is perfectly-secret as $M$ is uniformly distributed in $G_q$, independently of the message. It is computationally-binding because opening a commitment with distinct messages requires computing $\log_{g_0} h_0$.

Unfortunately, Pedersen's scheme is malleable: given a commitment $M$ of some message $m$ an adversary obtains a commitment for $m + 1 \bmod q$ by multiplying $M$ with $g$. Later, the adversary reveals $m + 1 \bmod q$ and $r$ after learning the original decommitment $m, r$. This holds even for static adversaries. Such adversaries do not try to inject messages in executions, but rather learn a protocol execution of $\mathcal{S}$ and $\mathcal{R}$ —which they cannot influence— and afterwards try to commit to a related message to $\mathcal{R}$. As for non-malleability with respect to opening, the adversary must also be able to open the commitment after the sender has decommitted.

A possible fix that might come to one's mind are proofs of knowledge showing that the sender actually knows the message encapsulated in the commitment.

For the discrete-log case such a proof of knowledge consists of the following steps [O92]: the sender transmits a commitment $S := g_0^s h_0^t$ of a random value $s \in_R \mathbb{Z}_q$, the receiver replies with a random challenge $c \in_R \mathbb{Z}_q$ and the sender answers with $y := s + cm \bmod q$ and $z := t + cr \bmod q$. The receiver finally checks that $SM^c = g_0^y h_0^z$.

If we add a proof of knowledge to Pedersen's scheme we obtain a protocol which is non-malleable with respect to opening against static adversaries. This follows from the fact that any static adversary merely sees a commitment of an unknown message before trying to find an appropriate commitment of a related message. Since the proof of knowledge between $S$ and $R$ is already finished at this time, the static adversary cannot rely on the help of $S$ and transfer the proof of knowledge. We leave further details to the reader and focus instead on the non-malleable protocol against PIM adversaries in the next section.

### 3.2   Non-malleability with Respect to PIM Adversaries

The technique of assimilating a proof of knowledge as in the previous section does not thwart PIM attacks. Consider again the PIM adversary committing to $m + 1 \bmod q$ by multiplying $M$ with $g$. First, this adversary forwards the sender's commitment $S$ for the proof of knowledge to the receiver and hands the challenge $c$ of the receiver to the sender. Conclusively, he modifies the answer $y, z$ of the sender to $y^* := y + c \bmod q$ and $z^* := z$. See Figure 2. Clearly, this is a valid proof of knowledge for $m + 1 \bmod q$ and this PIM adversary successfully commits and later decommits to a related message.

Coin-flipping comes to rescue. In a coin flipping protocol one party commits to a random value $a$, then the other party publishes a random value $b$, and finally the first party decommits to $a$. The result of this coin flipping protocol is set to $c := a \oplus b$ or, in our case, to $c := a + b \bmod q$ for $a, b \in \mathbb{Z}_q$. If at least one party is honest, then the outcome $c$ is uniformly distributed (if the commitment scheme is "secure").

The idea is now to let the challenge in our proof of knowledge be determined by such a coin-flipping protocol. But if we too use Pedersen's commitment scheme with the public generators $g_0, h_0$ to commit to value $a$ in this coin-flipping protocol, we do not achieve any progress: the adversary might be able to commit to a related $a^*$ and thus bias the outcome of the coin-flipping to a suitable challenge $c^*$. The solution is to apply Pedersen's scheme in this sub protocol with the commitment $M$ as one of the generators, together with an independent generator $h_1$ instead of $g_0, h_0$; for technical reasons we rather use $(g_1 M)$ and $h_1$ for another generator $g_1$. As we will show, since the coin-flipping in the proof of knowledge between $A$ and $R$ is based on generators $g_1 M^*$ and $h_1$ instead of $g_1 M, h_1$ as in the sender's proof of knowledge, this prevents the adversary from adopting the sender's and receiver's values and therefore to transfer the proof of knowledge. Details follow.

We describe the protocol given in Figure 3 which combines the aforementioned ideas. The public parameters are primes $p, q$ with $q|(p-1)$ together with four random generators $g_0, g_1, h_0, h_1$ of a subgroup $G_q \subseteq \mathbb{Z}_p^*$ of prime order

| sender $S$ | adversary $A$ | receiver $R$ |
|---|---|---|

message $m \in \mathbb{Z}_q$      public: $p, q, g_0, h_0$

*a) commitment phase:*

choose $r, s, t \in_R \mathbb{Z}_q$
set $M := g_0^m h_0^r$
set $S := g_0^s h_0^t$

$\xrightarrow{\quad M, S \quad}$

$S^* := S$
$M^* := gM$

$\xrightarrow{\quad M^*, S^* \quad}$ choose $c \in_R \mathbb{Z}_q$

$\xleftarrow{\qquad c \qquad}$

$c^* := c$

$\xleftarrow{\quad c^* \quad}$

$y := s + c^* m \ (q)$
$z := t + c^* r \ (q)$

$\xrightarrow{\quad y, z \quad}$

$z^* := z$
$y^* := y + c \ (q)$

$\xrightarrow{\quad y^*, z^* \quad}$ verify that
$S^* (M^*)^c \overset{!}{=} g_0^{y^*} h_0^{z^*}$

*b) decommitment phase:*

$\xrightarrow{\quad m, r \quad}$

$r^* := r$
$m^* := m + 1 \ (q)$

$\xrightarrow{\quad m^*, r^* \quad}$ verify that
$M^* \overset{!}{=} g_0^{m^*} h_0^{r^*}$

**Fig. 2.** PIM Attack on Pedersen's Commitment Scheme with Proof of Knowledge

$q$. Our protocol also works for other cyclic groups of prime order $q$ like elliptic curves, but we explain for the case $G_q \subseteq \mathbb{Z}_p^*$ only. Basically, the sender $S$ commits to his message $m \in \mathbb{Z}_q^*$ with Pedersen's scheme[3] by computing $M = g_0^m h_0^r$ and proves by a proof of knowledge (values $S, c, y, z$ in Figure 3) that he is aware of a valid opening of the commitment. The challenge $c$ in this proof of knowledge is determined by a coin-flipping protocol with values $A, a, u, b$.

It is clear that our protocol is computationally-binding under the discrete-log assumption, and perfectly-secret as the additional proof of knowledge for $m$ is *witness-independent* (a.k.a. perfectly witness-indistinguishable) [FS90], i.e., for any challenge $c$ the transmitted values $S, y, z$ are distributed independently of the actual message [O92].

**Proposition 1.** *The commitment scheme in Figure 3 is perfectly-secret and, under the discrete-log assumption, computationally-binding.*

It remains to show that our scheme is non-malleable. We present the proof from a bird's eye view and fill in more details in Appendix A, yet remain sketchily

---

[3] Note that as opposed to Pedersen's scheme we require that $m \neq 0$; the technical reason is that in the security proof we need to invert the message modulo $q$.

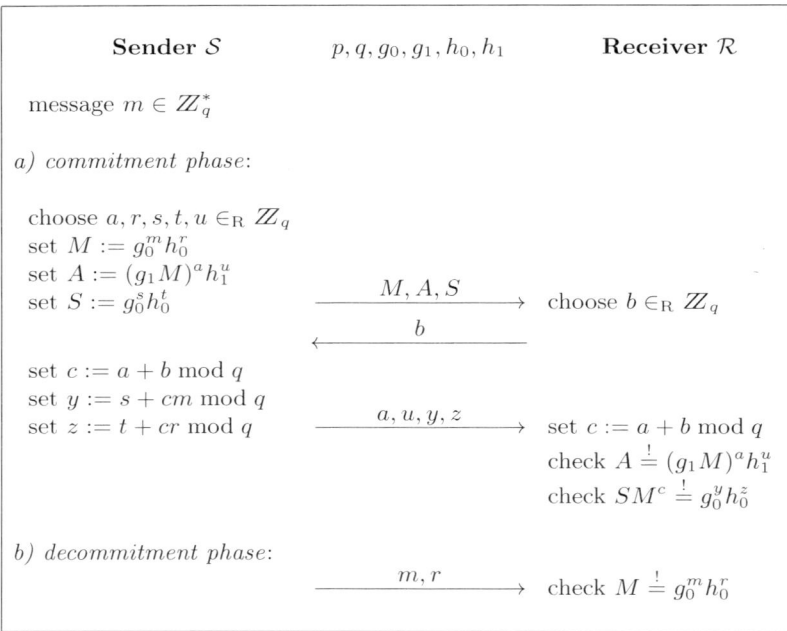

**Fig. 3.** DLog-Based Non-Malleable Commitment Scheme

in this version. By now, we already remark that the non-malleability property of our scheme also relies on the hardness of computing discrete logarithms. This dependency is not surprising: after all, any adversary being able to compute discrete logarithms with noticeable probability also refutes the binding property of Pedersen's scheme and can thus decommit for any related message with this probability.

The idea of the proof is as follows. Given a commitment $M$ of some unknown message $m$ (together with a witness-independent proof of knowledge described by $S, c, y, z$) with respect to parameters $p, q, g_0, h_0$ we show how to employ the PIM adversary $\mathcal{A}$ to derive some information about $m$. Namely, if we are able to learn the related message $m^*$ of the adversary, then we know that $m$ satisfies $(m, m^*) \in R$ for the relation $R$. This, of course, contradicts the perfect secrecy of the commitment scheme.

In this discussion here, we make two simplifications concerning the adversary: first, we assume that the PIM adversary always catches up concerning the order of the transmissions, i.e., sends his first message after learning the first message of $\mathcal{S}$ and answers to $\mathcal{S}$ after having seen $\mathcal{R}$'s response etc. Second, let the adversary *always* successfully commit and decommit to a related message, rather than with small probability. Both restrictions can be removed.

The fact that we learn the adversary's message $m^*$ follows from the proof of knowledge. Intuitively, a proof of knowledge guarantees that the prover knows the message, i.e., one can extract the message by running experiments with the prover. Specifically, we inject values $p, q, g_0, h_0, M, S, c, y, z$ into a simulated PIM attack with $\mathcal{A}$ and impersonate $\mathcal{S}$ and $\mathcal{R}$. Additionally, we choose $g_1$ at random and set $h_1 := (g_1 M)^w$ for a random $w \in_R \mathbb{Z}_q$. We also compute random $a_0, u_0 \in_R \mathbb{Z}_q$ and insert $g_1, h_1$ and $A := (g_1 M)^{a_0} h_1^{u_0}$ into the experiment with $\mathcal{A}$. We start with the extraction procedure by committing to $m, s, a_0$ via $M, S, A$ on behalf of the sender. Then, by the predetermination about the order of the transmissions, the adversary sends $M^*, S^*, A^*$ (possibly by changing $M, S, A$ and without knowing explicitly the corresponding values $m^*, r^*$ etc.). See Figure 5 on page 430 for a pictorial description.

We play the rest of the commitment phase twice by rewinding it to the step where the receiver chooses $b$ and sends it to the adversary $\mathcal{A}$. To distinguish the values in both repetitions we add the number of the loop as subscript and write $a_1, a_1^*, a_2, a_2^*$ etc. The first time, the adversary, upon receiving $b_1$, passes some $b_1^*$ to the (simulated) sender $\mathcal{S}$, and expects $\mathcal{S}$ to open the commitment for $a$ and supplement the proof of knowledge for $M$ with respect to the challenge $a_1 + b_1^* \bmod q$. By the trapdoor property of Pedersen's commitment scheme [BCC88] we are able to open $A$ with any value for $a_1$ since we know $\log_{(g_1 M)} h_1$. That is, to decommit $A$ with some $a_1$ reveal $a_1$ and $u_1 = u_0 + (a_0 - a_1)/\log_{(g_1 M)} h_1 \bmod q$; it is easy to verify that indeed $A = (g_1 M)^{a_1} h_1^{u_1}$. In particular, we choose $a_1$ such that $a_1 + b_1^* \bmod q$ equals the given value $c$. Hence, $y$ and $z$ are proper values to complement the proof of knowledge for $M$. Finally, the adversary answers with the decommitment $a_1^*, u_1^*$ for $A^*$ and the rest of the proof of knowledge for $M^*$ with respect to challenge $a_1^* + b_1 \bmod q$. Now we rewind the execution and select another random challenge $b_2$. The adversary then decides upon his value $b_2^*$ (possibly different from his previous choice $b_1^*$) and hands it to $\mathcal{S}$. Again, we open $A$ with $a_2$ such that $c = a_2 + b_2^* \bmod q$. The adversary finishes his commitment with $a_2^*, u_2^*$ as opening for $A^*$ and the missing values for the proof of knowledge.

The fundamental proof-of-knowledge paradigm [FFS88] says that we can extract the message $m^*$ if we learn two valid executions between $\mathcal{A}$ and $\mathcal{R}$ with the same commitment $M^*, S^*, A^*$ but different challenges. Hence, if the adversary's decommitments satisfy $a_1^* = a_2^*$ and we have $b_1 \neq b_2$ (which happens with probability $1 - 1/q$), then this yields different challenges $a_1^* + b_1, a_2^* + b_2$ in the executions between $\mathcal{A}$ and $\mathcal{R}$ and we get to know the message $m^*$. We are therefore interested in the event that the adversary is able to "cheat" by presenting different openings $a_1^* \neq a_2^*$. In Appendix A we prove that if the adversary finds different openings for commitment $A^*$ with noticeable probability, then we derive a contradiction to the intractability of the discrete-log problem. Hence, under the discrete-log assumption the probability that this event occurs is negligible and we extract $m^*$ with overwhelming probability.

Note that that in the repetitions we force the coin-flipping protocol between $\mathcal{S}$ and $\mathcal{A}$ to result in the same challenge both times. The latter is necessary because if we were able to answer a different challenge than $c$ then we could extract the unknown message $m$ and would thus know $m$ (which is of course not the case).

In conclusion, if the adversary is able to commit and decommit to a related message $m^*$, then we can extract $m^*$ and learn something about $m$. For details and further discussion we refer to Appendix A. Altogether,

**Theorem 1.** *Under the discrete-logarithm assumption, the scheme in Figure 3 is a perfectly-secret commitment scheme which is liberal non-malleable with respect to opening.*

It is worthwhile to point out that we cannot hash longer messages to $\mathbb{Z}_q^*$ before applying our non-malleable commitment scheme. Because then we extract the hash value and not the message $m^*$ itself. But this could be insufficient, since it might be impossible to deduce anything about $m$ via $R(m, m^*)$ given solely the hash value of $m^*$. We stress that the schemes in Section 4 with the faster a-posteriori verifiable proofs of knowledge do not suffer from this problem. There, one can first hash the message as the proof of knowledge operates on the original message instead of the hash value.

# 4    Speeding Up the Proof of Knowledge

The DLog-based scheme in the previous section (as well as the RSA-based one) uses Okamoto's witness-independent proof of knowledge. But since we are interested in non-malleability with respect to opening, the proof of knowledge need not be verifiable immediately in the commitment phase. It suffices that the sender convinces the receiver of the proof's validitiy in the decommitment stage. To refute non-malleability, the adversary must open his commitment correctly, and particularly, the proof must be shown to be right then. Therefore, the simulator can already in the commitment phase assume that the proof is indeed valid. We call such a proof of knowledge *a-posteriori verifiable*.

Using the Chinese Remainder Theorem, we present a very fast a-posteriori verifiable proof of knowledge and thus a faster variant of the non-malleable commitment scheme given in Section 3. Assume that we hash messages of polynomial length to $\mathbb{Z}_q^*$ with a collision-intractable hash function $\mathcal{H}$ (whose description is part of the public parameters). Given the DLog commitment $M = g_0^{\mathcal{H}(m)} h_0^r$ of the hash value $\mathcal{H}(m)$, the proof of knowledge consists of two steps: the receiver selects a small random prime $P$ as challenge and the sender answers with $(m, r)$ —viewed as a natural number— reduced $\bmod P$. If we repeat this for several primes, then we reconstruct the number $(m, r)$ using the Chinese Remainder Theorem. This corresponds to the case of RSA or DLog proofs of knowledge, where distinct challenges yield a representation. Yet, our proof reveals some information about $(m, r)$. To prevent this, we add a sufficiently large random prefix $s \gg P$ to the message $m$ and use $\mathcal{H}(s, m)$ instead of $\mathcal{H}(m)$ in $M$. Then $(s, m, r) \bmod P$ hides $m, r$ statistically. Note that the receiver cannot check immediately that the answer to the challenge is right. But a simulator, repeating the proof of knowledge with different primes, is able to verify that it has reconstructed the correct value $(s, m, r)$ by comparing $g_0^{\mathcal{H}(s', m')} h_0^{r'}$ for the extracted values $s', m', r'$ to $M$.

We introduce some notations. Let $\boldsymbol{x} \cdot \boldsymbol{y}$ denote the concatenation of two binary strings $\boldsymbol{x}, \boldsymbol{y} \in \{0,1\}^*$ and $\mathrm{val}(\boldsymbol{x}) := \sum_{i \geq 1} x_i 2^{i-1}$ the value of string $\boldsymbol{x} = x_1 x_2 \ldots x_n$. Conversely, denote by $\mathrm{bin}(r)$ the standard binary representation of number $r$. Suppose that we are given an efficiently computable function $\mathbb{P}_{K,k/2}$ mapping $K$-bit strings to $k/2$-bit primes such that for $\boldsymbol{x} \in_R \{0,1\}^K$ the prime $\mathbb{P}_{K,k/2}(\boldsymbol{x})$ is uniformly distributed in a superpolynomial subset of all $k/2$-bit primes. See [CS99,M95] for fast algorithms to generate primes from random strings. We will later present an alternative proof technique based on polynomials instead of primes.

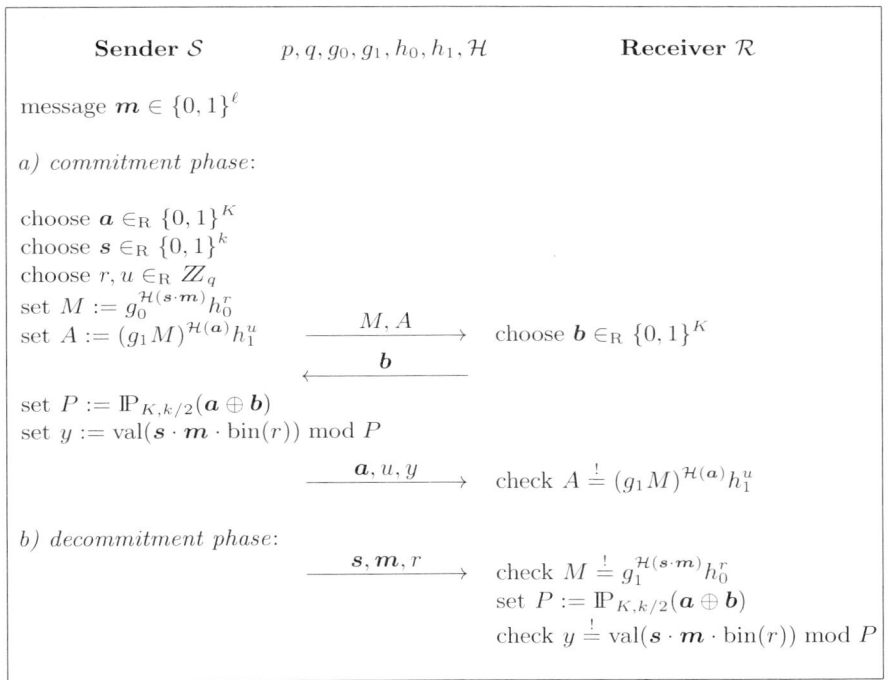

**Fig. 4.** Non-Malleable Commitment Scheme with fast a-posteriori verifiable POK

The modified commitment scheme is given in Figure 4. We replace the commitment $M = g_0^m h_0^r$ by $M = g_0^{\mathcal{H}(\boldsymbol{s} \cdot \boldsymbol{m})} h_0^r$. Additionally, the proof of the DLog-based scheme shows that we can also hash the value $\boldsymbol{a}$ for the coin flips before computing the commitment $A$. This enables us to generate longer random strings from a single commitment. Moreover, we now use the Chinese-Remainder-based proof of knowledge instead of the one by Okamoto. To commit to $\boldsymbol{m} \in \{0,1\}^\ell$ the sender $\mathcal{S}$ prepends $k$ random bits $\boldsymbol{s} \in_R \{0,1\}^k$ to the message. To prove knowledge of $\mathrm{val}(\boldsymbol{s} \cdot \boldsymbol{m} \cdot \mathrm{bin}(r))$ the sender reveals the residue

$$\mathrm{val}(\boldsymbol{s} \cdot \boldsymbol{m} \cdot \mathrm{bin}(r)) \equiv 2^k \, \mathrm{val}(\boldsymbol{m} \cdot \mathrm{bin}(r)) + \mathrm{val}(\boldsymbol{s}) \pmod{P}$$

modulo a $k/2$-bit prime $P$. The value $\mathrm{val}(\boldsymbol{s}) \in [0, 2^k)$ acts as a one-time pad since its distribution modulo $P$ is statistically close to the uniform distribution on $[0, P)$.

We omit the formal proof that the scheme is liberal non-malleable with respect to opening under the discrete-logarithm assumption and given that $\mathcal{H}$ is collision-intractable (or, more precisely, given that $\mathcal{H}$ is a family of collision-intractable hash functions). As for the proof of knowledge, the extractor rewinds the adversary $\mathcal{A}$ which mounts a PIM attack, and forces $\mathcal{A}$ to answer different primes while the original sender always has to answer to the same prime. Suppose that $k = \ell = \log q$. Thus, if $\mathcal{A}$ correctly reveals the residues of $\mathrm{val}(\boldsymbol{s} \cdot \boldsymbol{m} \cdot \mathrm{bin}(r))$ modulo seven different primes $P_1, \ldots, P_7$, then using the Chinese Remainder Theorem we retrieve $\mathrm{val}(\boldsymbol{s} \cdot \boldsymbol{m} \cdot \mathrm{bin}(r))$, because $\mathrm{val}(\boldsymbol{s} \cdot \boldsymbol{m} \cdot \mathrm{bin}(r)) \leq 2^{3k} < P_1 \cdots P_7$. If $\mathcal{A}$ has only a noticeable success probability, then the extractor creates a list of residues modulo a polynomial number of primes and applies the Chinese Remainder Theorem to all subsets of seven residues finding the right values with overwhelming probability. Using the more sophisticated algorithm by Goldreich, Ron and Sudan [GRS99] or the improved one by Boneh [B00], we derive:

**Theorem 2.** *Under the discrete-logarithm assumption and if $\mathcal{H}$ is a (family of) collision-intractable hash function(s), then for all polynomials $K, k, \ell$ with $K, \ell = poly(k)$ and $k \geq (\log q)^{1-\epsilon}$ for a constant $\epsilon > 0$ the scheme in Figure 4 is a statistically-secret commitment scheme which is liberal non-malleable with respect to opening.*

The bottleneck of this scheme is the generation of primes. An even faster approach is based on polynomials over the finite Field $\mathbb{F}$ with $2^{\sqrt{k}}$ elements. Let $K = k$ and $\Pi : \{0,1\}^* \to \mathbb{F}[\tau]$ denote the mapping which maps a bit string of length $d\sqrt{k}$ to a monic polynomial of degree $d$ over $\mathbb{F}$ by taking every block of $\sqrt{k}$ bits as a coefficient of the polynomial. In Figure 4, we replace the prime $P$ by the polynomial $P(\tau) := \Pi(\boldsymbol{a} \oplus \boldsymbol{b})$ of degree $\sqrt{k}$ and set $y(\tau) := \Pi(\boldsymbol{s} \cdot \boldsymbol{m} \cdot \mathrm{bin}(r)) \bmod P(\tau)$. To retrieve $\boldsymbol{s}, \boldsymbol{m}, r$ from the proof of knowledge, we apply the Chinese Remainder Theorem for polynomials [K98, 4.6.2, Ex. 3]. Two randomly chosen monic polynomials of same degree over $\mathbb{F}$ are co-prime with probability $1 - 2^{-\sqrt{k}}$ [K98, 4.6.5, Ex. 5]. For instance, if $\log q = \ell = k = K$, then given $\Pi(\boldsymbol{s} \cdot \boldsymbol{m} \cdot \mathrm{bin}(r))$ modulo four randomly chosen monic polynomials $P_1, P_2, P_3, P_4 \in \mathbb{F}[\tau]$ of degree $\sqrt{k}$, we retrieve with overwhelming probability the polynomial $\Pi(\boldsymbol{s} \cdot \boldsymbol{m} \cdot \mathrm{bin}(r))$ modulo $(P_1 P_2 P_3 P_4)$. This yields $\boldsymbol{s}, \boldsymbol{m}$ and $r$, because $\deg \Pi(\boldsymbol{s} \cdot \boldsymbol{m} \cdot \mathrm{bin}(r)) \leq 3\sqrt{k} < \deg(P_1 P_2 P_3 P_4)$.

## Acknowledgments

We are indebted to Cynthia Dwork for discussions about non-malleability. We also thank the participants of the Luminy 1999 crypto workshop for stimulating discussions, as well as the Crypto 2000 reviewers and program committee, especially Shai Halevi. We are also grateful to Yehuda Lindell.

# References

BG92.       M. BELLARE and O. GOLDREICH: *On Defining Proofs of Knowledge*, Advances in Cryptology — Proceedings Crypto '92, Lecture Notes in Computer Science, vol. 740, pp. 390–420, Springer Verlag, 1993.

BR94.       M. BELLARE and P. ROGAWAY: *Optimal Asymmetric Encryption*, Advances in Cryptology — Proceedings Eurocrypt '94, Lecture Notes in Computer Science, vol. 950, pp. 92–111, Springer Verlag, 1993.

BR93.       M. BELLARE and P. ROGAWAY: *Random Oracles are Practical: a Paradigm for Designing Efficient Protocols*, First ACM Conference on Computer and Communication Security, ACM Press, pp. 62–73, 1993.

B00.        D. BONEH: *Finding Smooth Integers Using CRT Decoding*, to appear in Proceedings of the 32$^{nd}$ Annual ACM Symposium on Theory of Computing (STOC), ACM Press, 2000.

BCC88.      G. BRASSARD, D. CHAUM and C. CRÉPEAU: *Minimum Disclosure Proofs of Knowledge*, Journal of Computer and Systems Science, vol. 37(2), pp. 156–189, 1988.

CS98.       R. CRAMER and V. SHOUP: *A Practical Public Key Cryptosystem Provable Secure Against Adaptive Chosen Ciphertext Attack*, Advances in Cryptology — Proceedings Crypto '98, Lecture Notes in Computer Science, vol. 1492, pp. 13–25, Springer Verlag, 1998.

CS99.       R. CRAMER and V. SHOUP: *Signature Schemes Based on the Strong RSA Assumption*, ACM Conference on Computer and Communication Security, ACM Press, 1999.

DIO98.      G. DI CRESCENZO, Y. ISHAI and R. OSTROVSKY: *Non-interactive and Non-Malleable Commitment*, Proceedings of the 30$^{th}$ Annual ACM Symposium on Theory of Computing (STOC), pp. 141–150, ACM Press, 1998.

DDN00.      D. DOLEV, C. DWORK and M. NAOR: *Non-Malleable Cryptography*, manuscript, to appear in SIAM Jornal on Computing, January 2000. Preliminary version in Proceedings of the 21$^{st}$ Annual ACM Symposium on Theory of Computing (STOC), pp. 542–552, ACM Press, 1991.

FFS88.      U. FEIGE, A. FIAT and A. SHAMIR: *Zero-Knowledge Proofs of Identity*, Journal of Cryptology, vol. 1(2), pp. 77–94, Springer-Verlag, 1988.

FS90.       A. FIAT and A. SHAMIR: *Witness Indistinguishable and Witness Hiding Protocols* Proceedings of the 22$^{nd}$ Annual ACM Symposium on the Theory of Computing (STOC), pp. 416–426, ACM Press, 1990.

G98.        O. GOLDREICH: **Foundations of Cryptography**, Fragments of a Book, Version 2.03, 1998.

GRS99.      O. GOLDREICH, D. RON and M. SUDAN: *Chinese Remainder With Errors*, Proceedings of the 31$^{st}$ Annual ACM Symposium on Theory of Computing (STOC), pp. 225–234, ACM Press, 1999.

K98.        D.E. KNUTH: **Seminumerical Algorithms**, The Art of Computer Programming, vol. 2, 3$^{rd}$ edition, Addison Wesley, 1998.

L00.        Y. LINDELL: Personal communication, based on work on authenticated key-exchange with Oded Goldreich. May 2000.

M95.        U. MAURER: *Fast Generation of Prime Numbers and Secure Public-Key Cryptographic Parameters*, Journal of Cryptology, vol. 8, pp. 123–155, Springer-Verlag, 1995.

O92.          T. OKAMOTO: *Provable Secure and Practical Identification Schemes and Corresponding Signature Schemes*, Advances in Cryptology — Proceedings Crypto '92, Lecture Notes in Computer Science, vol. 740, pp. 31–53, Springer Verlag, 1993.

P91.          T.P. PEDERSEN: *Non-Interactive and Information-Theoretical Secure Verifiable Secret Sharing*, Crypto '91, Lecture Notes in Computer Science, Vol. 576, Springer-Verlag, pp. 129–140, 1991.

RSA78.        R. RIVEST, A. SHAMIR and L. ADLEMAN: *A Method for Obtaining Digital Signatures and Public-Key Cryptosystems*, Communication of the ACM, vol. 21(2), pp. 120–126, 1978.

# A    Sketch of Proof of Theorem 1

We address a more formal proof that our protocol constitutes a non-malleable commitment scheme. Our aim is to extract the adversary's message within a negligibly close bound to the adversary's success probability $\pi_{\mathsf{open}}(\mathcal{A}, R)$ to derive some information about the unknown message for a given commitment. We omit a proof that this implies non-malleability of the commitment scheme because this proof already appears in [DDN00]. Instead, we show how to extract the adversary's message in our scheme. To this end, we repeat some basic facts about proofs of knowledge and knowledge extractors [FFS88,BG92]; we discuss them for the example of Okamoto's discrete-log-based proof of knowledge [O92] for a given $M = g_0^m h_0^r$. The knowledge extractor interacting with the prover works in two phases. Namely, it first generates a random conversation $S, c, y, z$ by running the prover to obtain $S$, by selecting $c$ and by letting the prover answer with $y, z$ to $S, c$. If this communication is invalid, then the extractor aborts. Else the extractor also stops with probability $1/q$. Otherwise it extracts at all costs. That is, the extractor fixes this communication up to the challenge, and then loops (till success) to seek another accepting conversation with the same communication prefix $S$ and different $c$. This is done by rewinding the execution to the choice of the challenge and re-selecting other random challenges. The extractor runs in expected polynomial time and outputs a representation of $M$ with respect to $g_0, h_0$ with probability $\pi - 1/q$. Here, $\pi$ denotes the probability that the prover makes the verifier accept, and $1/q$ is called the error of the protocol.

Next, we transfer the proof-of-knowledge technique to our setting. As in Section 3.2 we, too, adopt the convention that the adversary $\mathcal{A}$ does not "mix" the order of messages but rather catches up. We show how to remove this restriction in the final version of the paper.

Assume that we communicate with some party $\mathcal{C}$ which is going to commit to an unknown message $m \in_R \mathbb{M}$. We choose random primes $p, q$ and two generators $g_0, h_0$ and send them to $\mathcal{C}$. Party $\mathcal{C}$ selects $r, s, t \in_R \mathbb{Z}_q$ and sends $M := g_0^m h_0^r$, $S := g_0^s h_0^t$. We answer with a random challenge $c \in_R \mathbb{Z}_q$ and $\mathcal{C}$ returns $y := s + cm, z := t + cr \bmod q$. Finally, we check the correctness. Put differently, we perform all the steps of the sender in our protocol except for the coin flipping.

We want to use the PIM adversary to learn some information about $\mathcal{C}$'s message $m$. To this end, we incorporate the values of $\mathcal{C}$'s commitment in a knowledge extraction procedure for $M$. The extractor chooses additional generators $g_1, h_1$ by setting $g_1 := g_0^v$ and $h_1 := (g_1 M)^w$ for random $v, w \in_R \mathbb{Z}_q^*$, and computes $A := (g_1 M)^{a_0} h_1^{u_0}$ according to the protocol description for random $a_0, u_0 \in_R \mathbb{Z}_q$. Then it starts to emulate the PIM attack by pretending to be $\mathcal{S}$ and $\mathcal{R}$ and with values $p, q, g_0, g_1, h_0, h_1, M, S, A$. Because of the assumption about the order of messages, the adversary commits then to $M^*, S^*, A^*$. Next, we use the same stop-or-extract technique as in [O92]. In our case, the rewind point (if we do rewind) is the step where the receiver sends $b$. In each repetition, we send a random value $b_i \in_R \mathbb{Z}_q$ —the subscript denotes the number $i = 1, 2, \ldots$ of the loop— on behalf of the receiver and the adversary hands some value $b_i^*$ to the simulated sender. Knowing the trapdoor $w = \log_{(g_1 M)} h_1$ we open $A$ with $a_i, u_i = u_0 + (a_0 - a_i)/w \bmod q$ such that $a_i + b_i^*$ equals the given value $c$, and send the valid answer $y, z$ to the challenge $c$ in the proof of knowledge for $M$. The adversary replies with $a_i^*, u_i^*, y_i^*, z_i^*$ to the receiver. A description is shown in Figure 5.

An important modification of the knowledge extractor in comparison to the one in [FFS88,O92] is that, once having entered the loop phase, not only does our extractor stop in case of success; it also aborts with no output if in some repetitions $i, j$ the adversary both times successfully finishes the commitment phase —which includes a correct decommitment of $A^*$— but opens $A^*$ with distinct values $a_i^* \neq a_j^*$. We say that $\mathcal{A}$ *wins* if this happens. In this case, the extractor fails to extract a message.

Our first observation is that our knowledge extractor stops (either with success or aborting prematurely) in expected polynomial-time. This follows as in [FFS88,O92]. Let us analyze the success probability of our extractor. We assert that the extractor succeeds in outputting a message with probability at least $\pi_{\text{open}}(\mathcal{A}, R) - 1/q - \delta(n)$, where $\delta(n)$ denotes the probability that $\mathcal{A}$ wins (for security parameter $n$). The reason for this is that, given $\mathcal{A}$ does not win, the adversary's openings $a_{i_1}^* = a_{i_2}^* = \ldots$ in the valid commitment conversations are all equal. But then the values $b_{i_j} + a_{i_j}^* \bmod q$ for $j = 1, 2, \ldots$ of challenges in the proof of knowledge between $\mathcal{A}$ and $\mathcal{R}$ are uniformly and independently distributed. Analogously to [FFS88,O92] it follows that the extractor finds a message with probability $\pi_{\text{open}}(\mathcal{A}, R) - 1/q$ in this case.

It remains to bound the probability $\delta(n)$ that $\mathcal{A}$ wins. We will prove that $\delta(n)$ is negligible under the discrete-log assumption. For this, we first remark that we are only interested in the case that $\mathcal{A}$ sends distinct openings of $A^*$ in *accepting* executions, because the extractor only relies on these executions. In order to derive a contradiction to the intractability of the discrete-log problem we observe that the notion of non-malleability with respect to opening requires that $\mathcal{A}$ also reveals a valid decommitment. Hence, we view the decommitment phase as an additional step of the proof of knowledge. In other words, a correct decommitment is part of an accepting conversation of the proof of knowledge. Yet, this step has an extra property: the adversary must finish his commitment before he is allowed to ask $\mathcal{S}$ to open the original commitment. This corresponds

| simulation of $\mathcal{S}$ | adversary $\mathcal{A}$ | simulation of $\mathcal{R}$ |
|---|---|---|

*given parameters:*

$p, q, g_0, h_0$
$M, S, c, y, z$

*additional parameters:*

choose $a_0, u_0, v, w \in_{\mathrm{R}} \mathbb{Z}_q$
set $g_1 := g_0^v$
set $h_1 := (g_1 M)^w$
set $A := (g_1 M)^{a_0} h_1^{u_0}$

*frozen simulation:*                        $p, q, g_0, g_1, h_0, h_1$

$\xrightarrow{\quad M, A, S \quad}$

$\qquad\qquad\qquad\xrightarrow{\quad M^*, A^*, S^* \quad}$

*rewind point (loop $i = 1, 2, \dots$):* .................................................

choose $b_i \in_{\mathrm{R}} \mathbb{Z}_q$

$\xleftarrow{\qquad b_i \qquad}$

$\xleftarrow{\qquad b_i^* \qquad}$

set $a_i := c - b_i^* \bmod q$
set $u_i := u_0 + (a_0 - a_i)/w \bmod q$    $\xrightarrow{\quad a_i, u_i, y, z \quad}$

$\qquad\qquad\qquad\xrightarrow{\quad a_i^*, u_i^*, y_i^*, z_i^* \quad}$

**Fig. 5.** Knowledge Extraction

to the fact that the decommitment phase of both parties $\mathcal{S}$ and $\mathcal{A}$ is delayed until both commitment phases are complete.

**Lemma 1.** *The probability that $\mathcal{A}$ wins is negligible.*

*Proof.* We show that if the claim of Lemma 1 does not hold this contradicts the intractability of the discrete-log problem. We are given randomly generated primes $p, q$, a generator $g$, and a value $X \in G_q$ for which we are supposed to compute $\log_g X$. We show how to use $\mathcal{A}$ to do so.

The first observation is that if $\mathcal{A}$ wins with noticeable probability within the (expected) polynomial number of loops, then by standard techniques it follows that $\mathcal{A}$ does so with noticeable probability in the first $p(n)$ repetitions for an appropriate polynomial $p(n)$. Thus, we simply truncate any loop beyond this, and the running time of our derived discrete-log algorithm is strictly polynomial.

Instead of using the commitment $M$ of the third party $\mathcal{C}$, this time we run the knowledge extraction procedure incorporating the given values $p, q, g, X$, but generating the same distribution as the extractor. That is, select a message $m \in_R \mathbb{M}$, as well $v, w \in_R \mathbb{Z}_q^*$, set

$$g_0 := g^{-1/m} X, \qquad g_1 := g, \qquad h_0 := X^v, \qquad h_1 := X^w,$$

and compute $M, A, S, c, y, z$ according to the protocol description. Wlog. assume that $X \neq 1$ and $X^m \neq g$, else we already know the discrete log of $X$. Then $g_0$, $g_1$, $h_0$ and $h_1$ are random generators of the subgroup $G_q$. Furthermore, $g_1 M = gg_0^m h_0^r = X^{m+rv}$ and thus $\log_{(g_1 M)} h_1 = (m + rv)/w \bmod q$. Next we emulate $\mathcal{A}$ on values $p, q, g_0, g_1, h_0, h_1$ and $M, A, S$ by running the extraction procedure above. Note that this, too, means that we may abort before even starting to loop. Once we have entered the rewind phase, whenever the extractor is supposed to open $A$ to determine the challenge $c$ in the loop, we also open the commitment such that the coin flipping protocol always yields the same value $c$. This is possible as we know $\log_{(g_1 M)} h_1$ and are therefore able to open $A$ ambiguously. Observe that the communication here is identically distributed to the one in the extraction procedure. Hence, given that $\mathcal{A}$ wins with noticeable probability in the extraction procedure, $\mathcal{A}$ finds some $a_i^* \neq a_j^*$ for two accepting executions $i, j$ with the same probability in this experiment here. Let $u_i^*, u_j^*$ denote the corresponding portions of the decommitment for $A^*$ in these loops. Recall that we take the decommitment stage as an additional step of the proof of knowledge. Therefore, after having revealed $m, r$ in place of the sender in loop no. $j$, we also obtain some $m^*, r^*$ satisfying the verification equation $M^* = g_0^{m^*} h_0^{r^*}$ from the adversary. Particularly, we have:

$$h_1^{(u_i^* - u_j^*)/(a_j^* - a_i^*)} = g_1 M^* = g_1 g_0^{m^*} h_0^{r^*} = g^{1-m^*/m} X^{m^*+r^*v}$$

Since $h_1 = X^w$ we can transform this into

$$g^{1-m^*/m} = X^x \qquad \text{for } x = w(u_i^* - u_j^*)/(a_j^* - a_i^*) - (m^* + r^*v) \bmod q$$

Observe that $x$ is computable from the data that we have gathered so far. From $m^* \neq m$ we conclude that $1 - m^*/m \neq 0 \bmod q$ and therefore $x \neq 0 \bmod q$ has an inverse modulo $q$. Thus the discrete logarithm of $X$ to base $g$ equals $(1 - m^*/m)/x \bmod q$. $\qquad\square$

Summerizing, with probability $\pi_{\mathsf{open}}(\mathcal{A}, R) - 1/q - \delta(n)$ (which is negligibly close to the adversary's success probability) we extract some message $m'$. The final step in the proof is to show that indeed $m'$ equals the adversary's decommitment $m^*$ except with negligible probability; this follows by standard techniques and is omitted.

# Improved Non-committing Encryption Schemes Based on a General Complexity Assumption

Ivan Damgård and Jesper Buus Nielsen

**BRICS**⋆ Department of Computer Science
University of Aarhus
Ny Munkegade
DK-8000 Arhus C, Denmark
{ivan,buus}@brics.dk

**Abstract.** Non-committing encryption enables the construction of multiparty computation protocols secure against an *adaptive* adversary in the computational setting where private channels between players are not assumed. While any non-committing encryption scheme must be secure in the ordinary semantic sense, the converse is not necessarily true. We propose a construction of non-committing encryption that can be based on any public-key system which is secure in the ordinary sense and which has an extra property we call *simulatability*. This generalises an earlier scheme proposed by Beaver based on the Diffie-Hellman problem, and we propose another implementation based on RSA. In a more general setting, our construction can be based on any collection of trapdoor permutations with a certain simulatability property. This offers a considerable efficiency improvement over the first non-committing encryption scheme proposed by Canetti et al. Finally, at some loss of efficiency, our scheme can be based on general collections of trapdoor permutations without the simulatability assumption, and without the common-domain assumption of Canetti et al. In showing this last result, we identify and correct a bug in a key generation protocol from Canetti et al.

## 1 Introduction

The problem of multiparty computation dates back to the papers by Yao [20] and Goldreich et al. [15]. What was proved there was basically that a collection of $n$ players can efficiently compute the value of an $n$-input function, such that everyone learns the correct result, but no other new information. More precisely, these protocols can be proved secure against a polynomial time bounded adversary who can *corrupt* a set of less than $n/2$ players initially, and then make them behave as he likes. Even so, the adversary should not be able to prevent the correct result from being computed and should learn nothing more than the result and the inputs of corrupted players. Because the set of corrupted players is fixed from the start, such an adversary is called *static* or non-adaptive.

---

⋆ Basic Research in Computer Science,
  Centre of the Danish National Research Foundation.

M. Bellare (Ed.): CRYPTO 2000, LNCS 1880, pp. 432–450, 2000.
© Springer-Verlag Berlin Heidelberg 2000

There are several different proposals on how to define formally the security of such protocols [19,3,8], but common to them all is the idea that security means that the adversary's view can be *simulated* efficiently by a machine that has access to only those data that the adversary is *entitled* to know. Proving correctness of a simulation in the case of [15] requires a complexity assumption, such as existence of trapdoor permutations. Later, *unconditionally* secure MPC protocols were proposed by Ben-Or et al. and Chaum et al.[6,10], in the model where *private* channels are assumed between every pair of players. These protocols are in fact secure, even if the adversary is *adaptive*, i.e. can choose dynamically throughout the protocol who to corrupt, as long as the total number of corruptions is not too large. It is widely accepted that adaptive adversaries model realistic attacks much better than static ones. Thus it is natural to ask whether adaptive security can also be obtained in the computational setting?

If one is willing to trust that honest players can erase sensitive information such that the adversary can find no trace of it, should he break in, then such adaptive security can be obtained quite efficiently [5]. Such secure erasure can be too much to hope for in realistic scenarios, and one would like to be able to do without them. But without erasure, protocols such as the one from [15] is not known to be adaptively secure. The original simulation based security proof for [15] fails completely against an adaptive adversary.

However, in [9], Canetti et al. introduce a new concept called *non-committing encryption* and observe that if one replaces messages on the secure channels used in [6,10] by non-committing encryptions sent on an open network, one obtains adaptively secure MPC in the computational setting. They also showed how to implement non-committing encryption based on so called common-domain trapdoor permutations. The special property of non-committing encryption (which ordinary public-key encryption lacks) is the following: although a normal ciphertext determines a plaintext uniquely, encrypted communication can nevertheless be simulated with an indistinguishable distribution such that the simulator can later "open" a ciphertext to reveal any plaintext it desires. In an MPC setting, this is what allows to simulate the adversary's view before *and after* a player is corrupted. The scheme from [9] has expansion factor at least $k^2$, i.e., it needs to send $\Omega(k^2)$ bits for each plaintext bit communicated.

Subsequently, Beaver [4] proposed a much simpler scheme based on the Decisional Diffie-Hellman assumption (DDH) with expansion factor $O(k)$. Recently, Jarecki and Lysyanskaya [17] have proposed an even more efficient scheme also based on DDH with constant expansion factor, which however is only non-committing if the receiver of a message is later corrupted. This is sufficient for their particular application to threshold cryptography, but not for constructing adaptively secure protocols in general.

## 2   Our Results

In this paper, we first present a definition of *simulatable* public-key systems. This captures some essential properties allowing for construction of non-committing encryption schemes based on ordinary semantically secure public-key encryption. Roughly speaking, a public-key scheme is simulatable if, in addition to the normal key generation procedure, there is an algorithm to generate a public key, without getting to know the corresponding secret key. Moreover, it must be possible to sample efficiently a random ciphertext without getting to know the corresponding plaintext (we give precise definitions later in the paper)

We then describe a general way to build non-committing encryption from simulatable and semantically secure public-key encryptions schemes. Our method offers a major improvement over [9] in the number of bits we need to send. It may be seen as a generalisation of Beaver's plug-and-play approach from [4]. Beaver pointed out that it should be possible to generalise his approach to more general assumptions than DDH. But no such generalisation appears to have been published before.

The idea that it could be useful to generate a public key without knowing the secret key is not new. It seems to date back to De Santis et al.[12] where it was used in another context. The idea also appears in [9], but was only used there to improve the key generation procedure is some special cases (namely based on discrete logarithms and factoring). Here, we show the following

- From any semantically secure and simulatable public-key system, one can construct a non-committing encryption scheme.
- The scheme requires 3 messages to communicate $k$ encrypted bits, where $k$ is the security parameter. The total amount of communication is $O(k)$ public keys, $O(k)$ encryptions of a $k$-bit plaintext (in the original scheme), and $k$ bits.
- Only the final $k$ bits of communication depend on the actual message to be sent, and hence nearly all the work needed can be done in a preprocessing phase.

As mentioned, the DDH assumption is sufficient to support this construction. We propose an alternative implementation based on the RSA assumption, which is somewhat slower than the DDH solution[1].

We then look at general families of trapdoor permutations. We call such a family simulatable if one can efficiently generate a permutation in the family without getting to know the trapdoor, and if the domain can be sampled in an invertible manner. Invertible sampling is a technical condition which we discuss in more detail later. All known examples of trapdoor permutations have invertible sampling. Although this condition seems to be necessary for all applications of the type discussed in [9,4] and here, it does not seem to have been identified explicitly before.

We show that such a simulatable family implies immediately a simulatable public-key system with no further assumptions. The non-committing encryption scheme we obtain from this requires per encrypted bit communicated that we send $O(1)$ descriptions of a permutation in the family and $O(k)$ bits (where the

---

[1] A proposal with a similar idea for the key generation but with a less efficient encryption operation was made in [9].

hidden constant only has to be larger than 2, and where all bits except one can be sent in a preprocessing phase). With the same assumption, the scheme from [9] requires $\Omega(1)$ permutation descriptions and $\Omega(k^2)$ bits. Moreover, the $\Omega(k^2)$ bits depend on the message communicated and so cannot be pushed into a preprocessing phase. On the other hand it should be noted that the scheme from [9] needs only 2 messages (rather than 3 as in our scheme). It is not known if the same improvement in bandwidth can be obtained with only 2 messages.

Our final main result is an implementation of non-committing encryption based on any family of trapdoor permutations, assuming only invertible sampling, i.e., without assuming full simulatability or the common-domain assumption of [9].

At first sight, this seems to follow quite easily from the results we already mentioned, if we use as subroutine a key generation protocol from [9]. This protocol is based on oblivious transfer and can easily be modified to work based on any family of trapdoor permutations, assuming only invertible sampling. The protocol was intended to establish a situation where a player knows the trapdoor for one out of two public trapdoor permutations, without revealing which one he knows. It turns out that our scheme can start from this situation and work with no extra assumptions.

However, as we explain later, we have identified a bug in the key generation protocol of [9] causing it to be insecure. Basically, there is a certain way to deviate from the protocol which will enable the adversary to find out which of the two involved trapdoors is known to an honest player. We suggest a modification that solves this problem. While the modification is very simple and just consists of having players prove correctness of their actions by standard zero-knowledge protocols, it is perhaps somewhat surprising that it works. Standard rewindable zero-knowledge often cannot be used against an adaptive adversary: the simulator can get stuck when rewinding if the adversary changes its mind about who to corrupt. However, in our case, we show that the simulator will never need to rewind.

We note that the key generation of [9] needs invertible sampling in any case, and thus our assumption of existence of trapdoor permutations with invertible sampling is the weakest known assumption sufficient for non-committing encryption.

## 3   A Quick and Dirty Explanation

Before going into formal details, we give a completely informal description of some main ideas. Let $R, S$ be the two players who want to communicate in a non-committing way.

First $S$ chooses a bit $c$ and generates a pair of public keys $(P_0, P_1)$ such that he only knows the secret $S_c$ key corresponding to $P_c$. He sends $(P_0, P_1)$ to $R$. Then $R$ chooses a bit $d$ at random and sends to $S$ two pairs of ciphertext/plaintext $(C_0, M_0), (C_1, M_1)$. This is done such that only one pair is valid, i.e., $C_d$ is an encryption of $M_d$ under $P_d$, whereas $C_{1-d}$ is a random ciphertext for which $R$ does not know the plaintext, and $M_{1-d}$ is a random plaintext chosen independently.

Now $S$ can decrypt $C_c$ and test whether the result equals $M_c$. This with almost certainty determines $d$. Finally, $S$ sends a bit $s = c \oplus d$ to $R$ telling him whether $c = d$. If $c = d$, the parties will use this secret bit to communicate message bit $m$ securely as $f = m \oplus c$. If $c \neq d$, we say that this attempt to communicate a bit has failed, and none of the bits $c, d$ are used later.

The intuition now is that successful attempts can be faked by a simulator that chooses to know all secret keys sand plaintexts involved, generates only valid ciphertext/plaintext pairs but leaves $c$ undefined for the moment. Then, if for instance $S$ is later corrupted, the simulator can choose to reveal the secret key corresponding to either $P_0$ or $P_1$ depending on the value it wants for $c$ at that point. It will then be clear that the pair $(C_c, M_c)$ is valid — the other pair is valid too, but the adversary cannot see this if the simulator can convincingly claim that $P_{1-c}$ was chosen without learning the corresponding secret key.

A main part of the following is devoted to showing that if we define appropriately what it means to generate public keys and ciphertexts with no knowledge of the corresponding secrets, then this intuition is good.

## 4    Simulatable Public-Key Systems

Throughout the paper we will use the following notation. For a probabilistic algorithm $\mathcal{A}$ we will by $\mathcal{R}_{\mathcal{A}}$ denote a sufficiently large set $\{0,1\}^l$ from which the random bits for $\mathcal{A}$ are drawn. We let $r \leftarrow \mathcal{R}_{\mathcal{A}}$ denote a $r$ drawn uniformly random from $\mathcal{R}_{\mathcal{A}}$, let $a \leftarrow \mathcal{A}(x, r)$ denote the result $a$ of evaluating $\mathcal{A}$ on input $x$ using random bits $r$, and denote by $a \leftarrow \mathcal{A}(x)$ a value $a$ drawn from the random variable $\mathcal{A}(x)$ describing $\mathcal{A}(x, r)$ when $r$ is uniform over $\mathcal{R}_{\mathcal{A}}$.

We now want to define a public-key encryption scheme where one can generate a public key without getting to know the matching secret key. So in addition to the normal key generation algorithm $\mathcal{K}$ that outputs a public and secret key $(P, S)$, we assume that there is another algorithm which we call the oblivious public-key-generator $\tilde{\mathcal{K}}$ which outputs only a public key $P$ with a distribution similar to public keys produced by $\mathcal{K}$. However, this condition is not sufficient to capture what we want. $\tilde{\mathcal{K}}$ could satisfy it by just running the same algorithm as $\mathcal{K}$ but output only $P$. We therefore also ask that based only on a public key $P$, there is an efficient algorithm $\tilde{\mathcal{K}}^{-1}$ that comes up with a set of random choices $r'$ for $\tilde{\mathcal{K}}$ such that $P = \tilde{\mathcal{K}}(r')$ and that $P, r'$ cannot be distinguished from a normal set of random choices and resulting output from $\tilde{\mathcal{K}}$. This ensures that whatever side information you get from producing $P$ using $\tilde{\mathcal{K}}$, you could also compute efficiently from only $P$ itself. In a similar way we can define what it means to produce a random ciphertext with no knowledge of the plaintext.

The property of being able to reconstruct the random bits used by an algorithm, we call invertible sampling. We define this notion first.

**Definition 1 (Invertible sampling).** *Let $A : X \times \{0,1\}^* \to Y$ be a PPT algorithm. We say that $A$ has* invertible sampling *and that $A$ is a* PPTIS *algorithm, if* *there exists a PPT* random-bits-faking-algorithm $A^{-1} : Y \times X \rightarrow \{0,1\}^*$ *such that for all input* $x \in X$, *uniformly random bits* $r \leftarrow \mathcal{R}_A$, *output value* $y \leftarrow A(x,r)$, *and* fake random bits $r' \leftarrow A^{-1}(y,x)$ *the random variables* $(x,y,r')$ *and* $(x,y,r)$ *are computationally indistinguishable.*

Invertible sampling seems to be closely connected to non-committing encryption and adaptively secure computation in the non-erasure model.

As will be discussed further in chapter 6.2, the security of the non-committing encryption scheme in [9] relies on a invertible sampling property of the domains of the permutation. Also, the non-committing encryption scheme in [4], although not treated explicitly there, relies on the fact that you can invertible sample a quadratic residue in a specific group.

Invertible sampling is closely connected to adaptive security in models, where security is defined by requiring that an adversary's view of a real-life execution of a protocol can be simulated given just the data the adversary is entitled to, and where erasures are not allowed. Consider the protocol, where a party $P_1$ receives input $x$, computes $y \leftarrow f(x,r)$, where $r$ is some uniformly random string, and outputs $x$. Assume that all other parties do nothing. After a corruption of $P_1$ a real-life adversary sees the input $x$, the output $y$, and the random bits $r$. By definition of security there exists an ideal-evaluation adversary $\mathcal{S}$, that given just $x$ and $y$ will output the same view, i.e. compute $r'$ such that $(x,y,r')$ are computationally indistinguishable from $(x,y,r)$. Since $\mathcal{S}$ is a PPT algorithms it then follows that the protocol is secure iff $f$ has invertible sampling. Why it is indeed meaningful to deem such a protocol insecure if $f$ does not have invertible sampling, even though the protocol only has local computations, will not be discussed here.

**Definition 2 (Simulatable public-key system).** *Let* $(\mathcal{K}, \mathcal{E}, \mathcal{D}, \mathcal{M})$ *be a public-key system with key-generator* $\mathcal{K}$, *encryption algorithm* $\mathcal{E}$, *decryption algorithm* $\mathcal{D}$, *message-generator* $\mathcal{M}$, *and security parameter* $k$ $(1^k$ *is implicitly given as input to all algorithms in the following.) We say that* $(\mathcal{K}, \mathcal{E}, \mathcal{D}, \mathcal{M})$ *is a* simulatable public-key system *if besides fulfilling the usual requirements there exists PPTIS algorithms* $\tilde{\mathcal{K}}$ *and* $\mathcal{C}$, *called the* oblivious public-key-generator *resp. the* oblivious ciphertext-generator *such that the following holds.*

**Oblivious public-key generation** *For* $(P,S) \leftarrow \mathcal{K}$ *and* $\tilde{P} \leftarrow \tilde{\mathcal{K}}$ *the random variables* $P$ *and* $\tilde{P}$ *are computationally indistinguishable.*

**Oblivious ciphertext generation** *For* $(P,S) \leftarrow \mathcal{K}$, $C_1 \leftarrow \mathcal{C}_P$ *and* $M \leftarrow \mathcal{M}_P$, $C_2 \leftarrow \mathcal{E}_P(M)$, *the random variables* $(P,C_1)$ *and* $(P,C_2)$ *are computationally indistinguishable.*

**Semantic security** *For* $(P,S) \leftarrow \mathcal{K}$, *and for* $i = 0,1$: $M_i \leftarrow \mathcal{M}_P$, $C_i \leftarrow \mathcal{E}_P(M_i)$, *the random variables* $(P, M_0, M_1, C_0)$ *and* $(P, M_0, M_1, C_1)$ *are computationally indistinguishable.*

## 5   Non-committing Encryption

Non-committing encryption was defined in [9] as the problem of communicating a bitstring from a party $S$ to party $R$ in a $n$-party network with insecure authenticated channels between all parties. It is required that the protocol for doing this is secure against an adaptive adversary, who can corrupt up to $n - 1$ parties.

We introduce and provide examples of protocols adhering to a stronger notion of non-committing encryption which is resilient against a corruption of all parties and which involves only the communicating parties.

**Definition 3 (Strong non-committing encryption).** *Let $f(m, \epsilon) = (\epsilon, m)$ be the two-party function for communicating a message $m \in \{0,1\}^*$. Let $\pi$ be a two-party protocol. We say that $\pi$ is a strong non-committing encryption scheme if it 2-adaptively securely computes $f$.*

A reason for preferring a protocol meeting definition 3 is first of all that a protocol meeting the strong notion allows two parties to communicate independently of the other parties. We can think of such a protocol as a channel between two parties maintained by just these parties. This provides more flexibility for use in sparse network topologies and with arbitrary adversary structures.

Many proposals for the definition of secure multiparty computation has appeared in the literature presently culminating in the proposal of [8] which as the first definition allows for general security preserving modular composition of protocols in the computational setting. We will use this model of secure multiparty computation.

In general the model defines an ideal-evaluation of a function $f$ and requires that whatever a PPT real-life adversary $\mathcal{A}$ might obtain from attacking a real-life execution of the protocol $\pi$ a corresponding ideal-evaluation adversary $\mathcal{S}$ could obtain from attacking only the ideal-evaluation.

In our case the ideal-evaluation is functionally as follows. There are three active parties, all PPT algorithms. The sender $S$, the receiver $R$, and the adversary $\mathcal{S}$. The sender and receiver shares a *secure* channel and $S$ simply sends the message $m$ to $R$. The adversary sees no communication, but can corrupt the parties adaptively. If so he learns $m$ (either as the senders input or the receivers output) and can control the corrupted party for the remaining evaluation. I.e. if $S$ is corrupted before sending $m$ the adversary might send a different message.

In the real-life execution the adversary $\mathcal{A}$ sees all communication, and if he corrupts a party he receives that parties input and output (here $m$) *and* that parties random bits. All communication, inputs values, and all random bits are enough that the adversary can reconstruct the entire execution history of the corrupted party. This is what captures the non-erasure property of the model.

We then define security by requiring that for any real-life adversary $\mathcal{A}$ there exists an ideal-evaluation adversary $\mathcal{S}$, such that the collective output of all uncorrupted parties and $\mathcal{S}$ after attacking an ideal-evaluation of sending $m$ is distributed computationally indistinguishable from the collective output of all uncorrupted parties and $\mathcal{A}$ after attacking a real-life execution of the protocol with input $m$.

A complete definition and a summary of previous definitional work appears in [8]. A sketch of the part of the model used in this paper appears in our technical report [11].

## 5.1   The Main Idea

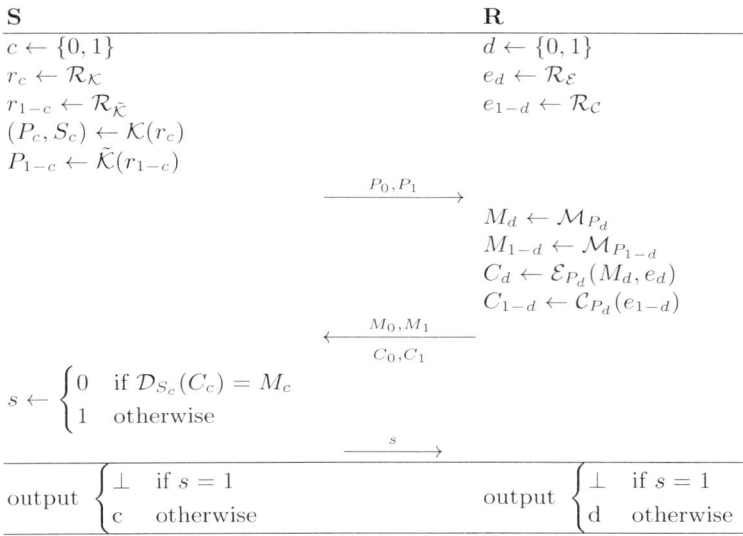

**Fig. 1.** One attempt to establish a shared random bit.

The main idea in the protocol is — like in all previous proposals — that we have our parties learn less information than is actually possible. This opens the possibility that a simulator can choose to learn full information and exploit this to its advantage. The main building block of the protocol, which we call an **attempt**, is sketched in Fig. 1.

Let $r_S$ and $r_R$ be the random inputs of $S$ resp. $R$. We write the values obtained by an attempt as

$$\text{Attempt}(r_S, r_R) = (r_c, P_c, M_c, e_c, C_c), S_c, (r_{1-c}, P_{1-c}, M_{1-c}, e_{1-c}, C_{1-c}), (c, d, s)$$

Let Attempt denote the random variable describing $\text{Attempt}(r_S, r_R)$ when $r_S$ and $r_R$ are chosen uniformly random. Let $\text{Attempt}_i$ for $i = 0, 1$ denote the distribution of Attempt under the condition that $s = i$. An attempt where $s = 0$ is called a **successful attempt** and an attempt where $s = 1$ is called a **failed attempt**.

For later use in the simulator and for illustration of the main idea we now show how we can produce a distribution computationally indistinguishable from that of $\text{Attempt}_0$, but where the common value $b = c = d$ of the shared secret bit can later be changed. We say that the simulation is non-committing to $b$.

Let SimSuccess be the values produced as follows: $s \leftarrow 0, i = 0, 1: r_i \leftarrow \mathcal{R}_\mathcal{K}$, $(P_i, S_i) \leftarrow \mathcal{K}(r_i), M_i \leftarrow \mathcal{M}, e_i \leftarrow \mathcal{R}_\mathcal{E}, C_i \leftarrow \mathcal{E}_{P_i}(M_i, e_i)$. The only difference compared to $\text{Attempt}_0$ is that in SimSuccess, we choose to learn the corresponding private key of $P_{1-c}$, choose $C_{1-d}$ as an encryption of $M_{1-d}$, and do not fix $c$ and $d$ yet.

For patching successful attempts we define the function $\text{Patch}(A, b)$, which for an element $A$ drawn from SimSuccess and a bit $b \in \{0, 1\}$ produces values similar to those in $\text{Attempt}_0$ by computing $c$ and $d$ as $c \leftarrow b, d \leftarrow b$, and patching $r_{1-c}$ and $e_{1-d}$ by $r'_{1-c} \leftarrow \tilde{\mathcal{K}}^{-1}(P_{1-c}), e'_{1-d} \leftarrow \mathcal{C}^{-1}(C_{1-d}, P_{1-d})$.

Let $\text{Patch} = (r_c, P_c, M_c, e_c, C_c), S_c, (r'_{1-c}, P_{1-c}, M_{1-c}, e'_{1-c}, C_{1-c}), (c, d, s)$ denote the random variable describing $\text{Patch}(A, b)$ when $A$ is drawn randomly from SimSuccess and $b$ is chosen uniformly random from $\{0, 1\}$.

**Lemma 1.** *The distribution of* $\text{Patch}$ *is computationally indistinguishable from the distribution of* $\text{Attempt}_0$.

**Proof:** Let $b$ denote the common value of $c$ and $d$ and observe that $\Pr[b = 0]$ and $\Pr[b = 1]$ is negligible close to $\frac{1}{2}$ in both Patch and $\text{Attempt}_0$. It is therefore enough to show that the conditional distributions under $b = 0$ and $b = 1$ are computationally indistinguishable.

For fixed $b$ the variables $c$, $d$, and $s$ are constants and has the same values in the two distributions, so we can exclude them from the analysis. Furthermore $(r_c, P_c, M_c, e_c, C_c), S_c$ can be seen to have the same distribution in the two distributions and is independent of $(r_{1-c}, P_{1-c}, M_{1-c}, e_{1-c}, C_{1-c})$, so all that remains is to show that these $(1 - c)$-values are distributed computationally indistinguishable in $\text{Attempt}_0$ and Patch. In $\text{Attempt}_0$ these values are distributed as

$$(\tilde{r} \leftarrow \mathcal{R}_{\tilde{\mathcal{K}}}, \tilde{P} \leftarrow \tilde{\mathcal{K}}(\tilde{r}), M \leftarrow \mathcal{M}_{\tilde{P}}, e \leftarrow \mathcal{R}_\mathcal{C}, C \leftarrow \mathcal{C}_{\tilde{P}}(e)) \tag{1}$$

and in Patch they are distributed as

$$(r', P, M \leftarrow \mathcal{M}_P, e', C \leftarrow \mathcal{E}_P(M, e)) \tag{2}$$

where $r \leftarrow \mathcal{R}_\mathcal{K}, (P, S) \leftarrow \mathcal{K}(r), r' \leftarrow \tilde{\mathcal{K}}^{-1}(P)$ and $e \leftarrow \mathcal{R}_\mathcal{E}, e' \leftarrow \mathcal{C}^{-1}(C, P)$. That these distributions are computationally indistinguishable follows by a hybrids argument, going from (1) to (2) using (in this order) the oblivious public-key generation including the invertible sampling of $\tilde{\mathcal{K}}$, the oblivious ciphertext generation including the invertible sampling of $\mathcal{C}$, and finally the semantic security. For more details see the technical report [11]. ∎

## Why Failed Attempts Cannot Be Simulated without Committing.
Consider the situation where $c \neq d$. The secret key $S_c$ is always known by

$S$. If this key becomes known to the adversary by corrupting $S$, he can check whether $\mathcal{D}_{S_c}(C_{1-d}) \neq M_{1-d}$, as it should be with high probability. The simulator can therefore pick at most one message/encryption pair such that $C$ is an encryption of $M$. On the other hand the adversary when corrupting $R$ expects to see a value $d$ and random bits $e_d$ such that $C_d \leftarrow \mathcal{E}_{P_d}(M_d, e_d)$. Thus at least one message/encryption pair should be correct. All in all exactly one pair is correct, which commits the simulator to $d$ (and thus $c$).

## 5.2   The Full Protocol

We will here analyse the protocol in Fig. 2, where we execute attempts in sequence until we have a successful one and then use the shared secret value $b$ of $c$ and $d$ to communicate a message bit $m$ as $f = m \oplus b$.

Each attempt has probability $\frac{1}{2}$ of being successful, so the expected number of attempts is two.

To prove security of the protocol we construct a simulator.

| **S** | | **R** |
|---|---|---|
| input $m \in \{0,1\}$ | | input $\epsilon$ |
| $b \leftarrow$ Attempt | | $b \leftarrow$ Attempt |
| if $b = \perp$ then retry | | if $b = \perp$ then retry |
| $f \leftarrow b \oplus m$ | | |
| | $\xrightarrow{\quad f \quad}$ | |
| output $\epsilon$ | | output $f \oplus b$ |

**Fig. 2.** Sequential 1-bit protocol.

## 5.3   The Simulator

Let $\mathcal{A}$ be any real-life adversary. We construct a corresponding ideal-evaluation adversary $I(\mathcal{A})$ as follows. The ideal-evaluation adversary $I(\mathcal{A})$ initialises $\mathcal{A}$ with a sufficiently large random string $r_{\mathcal{A}}$. The real-life adversary $\mathcal{A}$ will now start attacking. It expects to attack a real-life execution, but $I(\mathcal{A})$ is attacking an ideal-evaluation. We describe how to handle this.

**The Basic Simulation.** As long as $\mathcal{A}$ does not corrupt any parties $I(\mathcal{A})$ proceeds as follows. Before simulating each attempt decide whether the attempt should be a success or should fail by drawing $s$ uniformly random from $\{0, 1\}$.

If $s = 1$ then start by preprocessing values to be revealed to $\mathcal{A}$. Simply execute the protocol for a failed attempt. I.e. draw $c$ uniformly random from $\{0, 1\}$, set $d = 1 - c$, and then execute the attempt protocol in Fig. 1. This provides the values $(r_c, P_c, M_c, e_c, C_c), S_c, (r_{c-1}, P_{c-1}, M_{c-1}, e_{c-1}, C_{c-1}), (c, d = 1-c, s = 1)$. The adversary expects to see all communication in a real-life execution, so show him $(P_0, P_1)$, $(M_0, M_1, C_0, C_1)$, and $s$, in that order — we will later deal with the issue of how $I(\mathcal{A})$ should act on corruption requests from $\mathcal{A}$.

If $s = 0$, then $I(\mathcal{A})$ simulates a successful attempt. Again values for the communication is preprocessed. This time by running the algorithm $A \leftarrow$ SimSuccess. This provides values for all communication except $f$. We pick $f$ uniformly random from $\{0, 1\}$ and reveal the communication to $\mathcal{A}$ as above.

When a successful attempt has been simulated the actual simulation is over, but $I(\mathcal{A})$ keeps interacting with $\mathcal{A}$, which might still corrupt more parties — we do return to this right ahead. At some point $\mathcal{A}$ terminates with some output value, which we take to be the output value of $I(\mathcal{A})$.

**Dealing with Corruption Requests.** Below we describe how to handle the first corruption request. We look at two points, where the first corruption might take place. During the failed attempts and during the successful attempt. We prove that in either case $I(\mathcal{A})$ can patch the internal simulated state of $S$ and $R$ and simulate the corruption such that the entire state of $S$, $R$, and $\mathcal{A}$ is computationally indistinguishable from the state that would have been produced by running a real-life execution on the same input with adversary $\mathcal{A}$.

From this it follows that the simulator can then complete the simulation by just running the remaining honest party according to the real-life protocol with the simulated state as a starting point. Since the starting point is computationally indistinguishable from that of a real-life execution at the same point of execution and all participating algorithms are PPT it follows directly from the definition of computational indistinguishability that the final output of $\mathcal{A}$, and thereby $I(\mathcal{A})$, will be computationally indistinguishable from the output of $\mathcal{A}$ produce by an execution of the real-life protocol [2].

If $\mathcal{A}$ corrupts a party during the simulation of a failed attempt $I(\mathcal{A})$ corrupts the corresponding party in the ideal-evaluation and learns $m$. Observe that the simulated communication values *and* all preprocessed internal values in failed attempts are distributed *identically* to the values produced by a real-life execution. Therefore after obtaining $m$ the simulator $I(\mathcal{A})$ can just pass this along to $\mathcal{A}$ and the obtained global state is distributed *identically* to that of a real-life execution.

If $\mathcal{A}$ corrupts a party during the simulation of a successful attempt, we again have a number of cases as there is three rounds of communication. We first look at the case where the corruption occurs after $s$ and $f$ have been communicated. Here $I(\mathcal{A})$ again corrupts the same party in the ideal-evaluation, obtains $m$, and passes it on to $\mathcal{A}$. Now $I(\mathcal{A})$ must decide on values for $c$ and $d$. We pick the common value $b \leftarrow m \oplus f$. This value is consistent with all other values since with $c = b$ and $d = b$ we have that $f = m \oplus c$ and $m = f \oplus d$ as required. The simulator now patches the preprocessed values using $\text{Patch}(A, m \oplus f)$ and hands out the patched values thus produced to $\mathcal{A}$. Observe that $m \oplus f$ is uniformly distributed over $\{0, 1\}$ as we picked $f$ uniformly random. It then follows directly from lemma 1 and the fact that $c$ and $d$ is chosen consistent with $f$ and $m$ that the global state of $S$, $R$, and $\mathcal{A}$ is computationally indistinguishable from that

---

[2] Note that what allows this simple argument is that in contrast to simulators for more involved protocols not only have we obtained that the values revealed to $\mathcal{A}$ up to the first corruption is computationally indistinguishable from those of a real-life execution. We have managed to produce a complete global state of communication and the state of corrupted *and* uncorrupted parties that is computationally indistinguishable from that of a real-life execution.

which would have been produced from an execution of the real-life protocol on the same inputs.

If the corruption occurs before $f$ is revealed to $\mathcal{A}$, just obtain $m$ as before and patch the preprocessed values with a uniformly random common value $b$ for $c$ and $d$. The value of $f$ will then be given by the real-life protocol when $I(\mathcal{A})$ starts the execution of the remaining honest party. Earlier corruptions are handled similarly.

**Theorem 1.** *If simulatable public-key systems exist, then the protocol in Fig. 2 is a strong non-committing encryption scheme for communication one bit.*

**Proof:** We have to prove that for all real-life adversaries $\mathcal{A}$ there exists an ideal-evaluation adversary $\mathcal{S}$ such that the output of $\mathcal{A}$ after attacking the real-life protocol is computationally indistinguishable from the output of $\mathcal{S}$ after attacking an ideal-evaluation of the same inputs.

Given $\mathcal{A}$ we simply set $\mathcal{S} = I(\mathcal{A})$ an the claim follows from the above analysis of $I(\mathcal{A})$. ∎

**Theorem 2.** *If simulatable public-key systems exist, then strong non-committing encryption schemes exist. The scheme requires 3 messages to communicate $k$ encrypted bits, where $k$ is the security parameter. The total amount of communication is $O(k)$ public keys, $O(k)$ encryptions of a $k$-bit plaintext (in the original scheme), and $k$ bits.*

**Proof:** It follows directly from the Markov inequality that $a = 4k$ parallel attempts will give $k$ successful ones for communication except with probability $\exp(-\frac{k}{2})$, which is certainly negligible in $k$. This protocol uses three rounds for the attempts and we can communicate $f$ in round three together with $s$. We thereby obtain the claimed round complexity. The claimed communication complexity is trivially correct.

Since the simulator for attempts does not rewind the real-life adversary $\mathcal{A}$, we can obtain a simulator for the parallel protocol by simply running $a$ 'copies' of the simulator for the attempts in parallel. See the technical report [11] for more details. ∎

## 6   Implementations

The following theorem provides the first example of a simulatable public-key system.

**Theorem 3.** *The ElGamal public-key system is simulatable assuming that it it semanticly secure.*

**Proof:** Recall that a public key is a triple $(p, g, h)$, where $p$ is a prime such that the discrete log problem in $\mathbf{Z}_p^*$ is intractable, $\langle g \rangle = \mathbf{Z}_p^*$, and $h = g^x$ for some random $x$, which is the private key. Now simply let the oblivious public-key-generator pick $h$ directly in $\mathbf{Z}_p^*$ without learning its discrete log base $g$ and let

$\tilde{\mathcal{K}}^{-1}(p,g,h) = (r_p, r_g, r_h)$, where $r_p$, $r_g$, and $r_h$ are random bits similar to those used to pick $p$, $g$, resp. $h$.

How to reconstruct $r_p$ depends of course on the algorithm used to pick $p$. For simplicity, say that we pick $p$ by drawing random numbers in some interval $I$ until we get a number that tests to primality by some probabilistic test. We will then have to reconstruct, from $p$, a distribution similar to the prefix of numbers that were not primes. This can trivially be done by drawing random numbers in $I$ until a prime is found and use the prefix of non-primes and the random bits used to test them non-prime. The value $r_p$ is set to be these bits, $p$, and bits used to test $p$ prime using the primality test. This value $r_p$ is trivially distributed computationally indistinguishable from the bits originally used to pick $p$. The oblivious public-key generation is then trivially fulfilled. The values $r_g$ and $r_h$ are trivial to reconstruct if $g$ and $h$ is chosen in a natural way.

A message $x \in \mathbf{Z}_p^*$ is encrypted as $(g^k, xh^k)$, where $k$ is chosen uniformly random in $\mathbf{Z}_{p-1}$. It is obvious, that a ciphertext can be generated obliviously as $(y_1, y_2)$, where $y_1$ and $y_2$ are picked uniformly random and independent in $\mathbf{Z}_p^*$. Invertible sampling is trivial. ∎

## 6.1  Trapdoor Permutations

Before presenting the next example of a simulatable public-key system, we define the concept of a simulatable collection of trapdoor permutations and prove that the existence of such a collection implies the existence of simulatable public-key systems.

We first recall the standard definition of collections of trapdoor permutations:

**Definition 4 (Collection of trapdoor permutations).** *We call* $(I, F, \mathcal{G}, \mathcal{X})$ *a collection of trapdoor permutations with security parameter $k$, if $I$ is an infinite index set, $F = \{f_i : D_i \to D_i\}_{i \in I}$ is a set of permutations, the index/trapdoor-generator $\mathcal{G}$ and the domain-generator $\mathcal{X}$ are PPT algorithms, and the following hold:*

**Easy to generate and compute** *$\mathcal{G}$ generates pairs of indices and trapdoors, $(i, t_i) \leftarrow \mathcal{G}(1^k)$, where $i \in I \cap \{0,1\}^{p(k)}$ for some fixed polynomial $p(k)$. Furthermore, there is a polynomial time algorithm which on input $i, x \in D_i$ computes $f_i(x)$.*

**Easy to sample domain** *$\mathcal{X}$ samples elements in the domains of the permutations, i.e. $x \leftarrow \mathcal{X}(i)$, where $x$ is uniformly random in $D_i$.*

**Hard to invert** *For $(i, t_i) \leftarrow \mathcal{G}(1^k)$, $x \leftarrow \mathcal{X}(i)$, and for any PPT algorithm $A$ the probability that $A(i, f_i(x)) = x$ is negligible in $k$.*

**But easy with trapdoor** *There is a polynomial time algorithm which on input $i, t_i, f_i(x)$ computes $x$, for all $x \in D_i$.*

The next definition, of simulatable collections, is built along the lines of the definition of simulatable public-key systems. It basically defines a collection of trapdoor permutations where in addition it is easy to generate a permutation $f$ in the collection without getting to know the trapdoor. Further more we need that the domain of the trapdoors has invertible sampling. This is to allow oblivious ciphertext generation.

Invertible sampling is trivial if the domain of $f$ is, for instance, the set of $k$-bit strings and sampling is done in the natural way. But it may in general be an extra requirement which, however, seems to be necessary for any application of the kind we consider here. It is easy to construct artificial domains without invertible sampling, but all collections of trapdoor permutations we know of have domains with invertible sampling.

**Definition 5 (Simulatable collection of trapdoor permutations).** *Let* $(I, F, \mathcal{G}, \mathcal{X})$ *be a collection of trapdoor permutations with security parameter* $k$. *We say that* $(I, F, \mathcal{G}, \mathcal{X})$ *is a* simulatable collection of trapdoor permutations *with* oblivious index-generator $\tilde{\mathcal{G}}$, *if* $\tilde{\mathcal{G}}$ *and* $\mathcal{X}$ *are PPTIS algorithm and the random variables* $i$ *and* $\tilde{i}$ *are computationally indistinguishable, where* $(i, t_i) \leftarrow \mathcal{G}$ *and* $\tilde{i} \leftarrow \tilde{\mathcal{G}}$.

Given $F$ a collection of trapdoor permutations one can construct a semantically secure public-key system using the construction in [7]. We review the construction here and observe that it preserves simulatability.

Let $B$ be a hard-core predicate of the collection of trapdoor permutations. If no such $B$ is known one can construct a new simulatable collection of trapdoor permutations following the construction in [7]. The key-generator is set to be $\mathcal{K} = \mathcal{G}$, i.e. for $(i, t_i) \leftarrow \mathcal{G}$ we set $(P, S) = (i, t_i)$. The message space can be set to $\mathcal{M} = \{0, 1\}^{p(k)}$ for any polynomial $p(k)$ and the ciphertext space for $P = i$ is $\mathcal{M} \times D_i$, where $D_i$ is the domain of $f_i$.

Let $X(i, x, n) = B(x)B(f_i(x))B(f_i^2(x)) \ldots B(f_i^{n-1}(x))$ be the usual pseudorandom string generated from $x$. Then the encryption of $m \in \mathcal{M}$ under $i$ is $(m \oplus X(i, x, |m|), f_i^{|m|}(x))$ for random $x \leftarrow \mathcal{X}(i)$. The decryption is trivial given $t_i$. To pick such a ciphertext obliviously for a given key $P$ generate $m \leftarrow \mathcal{M}$ and $x \leftarrow \mathcal{X}(i)$ and let $C = (m, x)$. This will be distributed exactly as $C \leftarrow \mathcal{E}_P(m')$ for $m' \leftarrow \mathcal{M}$. Invertible sampling is given by $\mathcal{C}^{-1}(i, m, x) = (m, \mathcal{X}^{-1}(i, x))$.

The oblivious public-key generation is given by setting $\tilde{\mathcal{K}} = \tilde{\mathcal{G}}$ and $\tilde{\mathcal{K}}^{-1} = \tilde{\mathcal{G}}^{-1}$ and the semantic security is proven in [7].

**Theorem 4.** *Let* $\mathcal{F} = (I, F, \mathcal{G}, \mathcal{X})$ *be a simulatable collection of trapdoor permutations and let* $E_{\mathcal{F}} = (\mathcal{K}, \mathcal{E}, \mathcal{D}, \mathcal{M})$ *be the public-key system described above. Then* $E_{\mathcal{F}}$ *is simulatable.*

We proceed to construct a simulatable collection of trapdoor permutations based on RSA. We cannot use the standard collection of RSA-trapdoors as it has not been proven to have oblivious public-key generation. If the oblivious public-key generation learns the factorisation of $n$, the random-bits-faking-algorithm would have to factor $n$, which is hopefully hard. If the oblivious public-key generation does not learn the factorisation of $n$ it would have to test in PPT whether $n$ is a factor of two large primes, which we do not know how to do. We therefore need a modification.

**Assumption 1** *Let* $I = \{(n, e) | n = pqr, p, q \text{ are primes and } |p|, |q| \geq k, |n| = k \log k, \text{ and } n < e < 2n \text{ is a prime}\}$. *Here,* $k$ *is as usual the security parameter. For* $(n, e) \in I$ *let* $t_{(n,e)} = d$ *where* $ed = 1 \mod \phi(n)$. *Let* $f_{(n,e)} : \boldsymbol{Z}_n^* \to \boldsymbol{Z}_n^*, x \mapsto x^e \mod n$. *Then* $F = \{f_i\}_{i \in I}$ *is a collection of trapdoor permutations.*

Observe, that there is a non-negligible chance that a random integer $n$ contains two large primefactors. I.e. if we pick $n$ at random and $e$ as a prime larger than $n$, then $x \mapsto x^e \mod n$ is a weak trapdoor permutation over $\boldsymbol{Z}_n^*$ (relative to assumption 1.) The same observation was used in [9], where they refer to general amplification results[21,13] to obtain a collection of strong trapdoor permutations from this collection of weak ones. Here we apply an explicit amplification procedure, which is slightly more efficient, and prove that it gives us a simulatable collection of trapdoor permutations.

Let $l$ be an amplification parameter, which we fix later.

An index with corresponding trapdoor is given by $i = (e, n_1, \ldots, n_l)$ and $t_i = (d_1, \ldots, d_l)$, where $e$ is a $(k \log k)$-bit random prime and for $j = 1, \ldots, l$ the number $n_j$ is a uniformly random $(k \log k)$-bit number and $d_j = e^{-1} \mod \phi(n_j)$. To compute $d_j$ the key-generator $\mathcal{G}$ must generate uniformly (or indistinguishably close to uniformly) random $n_j$ in such a way that $\phi(n_j)$ is known. In [2] it was shown how to do this.

An oblivious index $(e, n_1, \ldots, n_l) \leftarrow \tilde{\mathcal{G}}$ is simply generated by picking $e$ as before and picking the $n_j$ uniformly random. The only problem for $\tilde{\mathcal{G}}^{-1}$ in faking bits for the index $(e, n_1, \ldots, n_l)$ is the prime $e$. On how to do this see the proof of theorem 3.

The domain for the index $i = (e, n_1, \ldots, n_l)$ will be $D_i = \prod_{j=1}^{l} \boldsymbol{Z}_{n_j}^*$ and the corresponding permutation will be $f_{(e,n_1,\ldots,n_l)}(x_1, \ldots, x_l) = (x_1^e \mod n_1, \ldots, x_l^e \mod n_l)$. Since $e$ is relatively prime to all $n_j$ our functions are indeed permutations and are invertible in PPT using the trapdoor information $t_i = (d_1, \ldots, d_l)$.

We pick a uniformly random element $x$ from $D_i$ by picking a uniformly random element $x_j$ from each $\boldsymbol{Z}_{n_j}^*$. These elements should be chosen in a way that allows $\mathcal{X}^{-1}$ to reconstruct the random bits used. One way is to pick uniformly random elements from $\boldsymbol{Z}_{n_j}$ until an element from $\boldsymbol{Z}_{n_j}^*$ is found. This gives us $\mathcal{X}^{-1}$ by following the construction for primes — see the proof of theorem 3.

What remains is to prove the one-wayness of our collection. In [18] the probability that the $i$'th largest primefactor of a random number $n$ is larger than $n^c$ for a given constant $c$ is investigated. It is shown to approach a constant as $n$ approaches infinity. In particular, the probability that the second largest primefactor is smaller than $n^c$ is approximately linear for small $c$, in fact it is about $2c$ for $c \leq 0.4$. It follows that the probability that a number of length $k \log k$ bits has its second largest prime factor shorter than $k$ bits is $O(1/\log k)$. If we set $l$ to $\log k$, we then have that the probability that there does not exist $j \in \{1, \ldots, l\}$ such that $(n_j, e) \in I$, where $I$ is the index set of assumption 1, is $O((\frac{1}{\log k})^{\log k})$ and so is negligible. So we have:

**Theorem 5.** *Under assumption 1, the set* $SRSA = \{f_i : D_i \to D_i\}$ *is a simulatable collection of trapdoor permutations.*

We note that $(\frac{1}{\log k})^{\log k}$ is only slightly below what is needed to be negligible. To obtain a security preserving[13] amplification we could use $k$ $k$-bit moduli. Another approach would be to remove the need for amplification by finding an invertible way to produce integers with two large primefactors and use just one such modulus for encryption.

## 6.2   Doing without Oblivious Index Generation

We now proceed to prove that one can do without the oblivious index generation.

We basicly remove the oblivious index generation assumption by using the key generation protocol from [9] applying a fix and a twist. The fix is necessary as we have found an attack against the protocol used in [9]. The twist is applied to remove the common-domain assumption which is needed by the construction in [9].

**The Key Generation Protocol.** In [9] a non-committing encryption scheme was built consisting of two phases. The first phase is a key generation protocol which is intended to create a situation, where players $S$ and $R$ share two trapdoor permutations from what is called a common-domain trapdoor system. Moreover, $S$ knows exactly one of the corresponding trapdoors, and if $S$ remaines honest in this phase, a simulator is able to make a simulated computation, where both trapdoors are learned and which can later (in case $S$ is corrupted) be convincingly patched to look as if either of the trapdoors were known to $S$. One immediate consequence is that the adversary must not know which of the two trapdoors is known to $S$, before corrupting $S$.

The key generation requires participation of all $n$ parties of the protocol and proceeds as follows: Each player $P_i$ chooses at random two permutations $(g_0^i, g_1^i)$ and send these to $S$. Next $S$ chooses $c = 0$ or $1$ at random, and execute the oblivious transfer (OT) protocol of [14] with $P_i$ as sender using the trapdoors of $(g_0^i, g_1^i)$ as input and $S$ as receiver using $c$ as input, and such that $S$ receives the trapdoor of $g_c^i$. The OT protocol of [14] has a non-binding property that allows a simulator to learn both trapdoors when it is playing $S$'s part and later to claim that either trapdoor was received.

In the above, there is no guarantee that $P_i$ really uses the trapdoors of $(g_0^i, g_1^i)$ as input to the OT, but, as pointed out in [9] one may assume that the trapdoor of a permutation consists of all inputs required to generate it so that $S$ can verify what he receives. Finally, $S$ publishes the subset $A$ of players from whom he got correct trapdoors, and we define $f_0$ to be the composition of the permutations $\{g_0^i\}_{i \in A}$ in some canonical order, and similarly for $f_1$.

**The Attack and a Fix.** We describe an attack against the above key generation protocol. If $S$ is still honest, but $P_i$ is corrupt, the adversary may choose to let $P_i$ use as inputs to the OT a correct trapdoor for $g_a^i$ but garbage for $g_b^i$. When the adversary sees the set $A$ he can then determine the value of $c$. If $i \notin A$ the sender must have chosen $c = b$ and detected $P_i$s fraud. If $i \in A$ then the sender must have chosen $c = a$. In any case the adversary learns $c$ and in $\frac{1}{2}$ of the cases even without being detected. But this is a piece of information that the adversary should not be able to get. The simulator's freedom to set $c$ after corruption is exactly what makes the simulation go through.

We solve this by requiring that a sender in an OT always proves in zero-knowledge to the receiver he inputs correct information to the OT. I.e. prove that there exists $r$ such that $(g, t_g) = \mathcal{G}(r)$ and that there exists a bitstring $r'$ such that $(g, t_g, r')$ is the inputs to the OT ($r'$ being the random bits used in the OT.) Since this is an NP statement and $R$ knows both witnesses $r$ and $r'$, this is always possible to prove[14]. This will imply that except with negligible probability, $P_i$ will have to supply correct trapdoors to both permutations or be disqualified. Normally, the use of such ZK proofs leads to problems against adaptive adversaries because of the rewinding needed to simulate the proofs. However, in this protocol, it happens to be the case that the simulator never needs to "prove" a statement for which it doesn't know a witness, and so rewinding is not needed.

**The Twist.** Having executed all the OT's, $S$ would in the protocol from [9] compose the permutations from honest parties to obtained two permutations, one with a known trapdoor and one with an unknown trapdoor. This requires and produces common-domain trapdoors. Instead of composing we simply concatenate.

Let $g_c^1, \ldots, g_c^l$ be the permutations that was correctly received and for which the corresponding trapdoor was received. From these permutations $S$ defines a new permutation $f_c$, where $f_c(x^1, \ldots, x^l) = (g_c^1(x^1), \ldots, g_c^l(x^l))$. Let $B$ be a hard-core predicate for the collection of trapdoor permutations used. Then $B(x^1, \ldots, x^l) = \bigoplus_{i=1}^l B(x^i)$ is a hard-core predicate for $f_c$. Let $x^1, \ldots, x^l$ be random and let $X(g^i, x^i, n) = B(x^i)B(g^i(x^i))B((g^i)^2(x^i)) \ldots B((g^i)^{n-1}(x^i))$ be the usual pseudo-random string generated from $x^i$. Then the encryption of $m \in \{0,1\}^*$ under $f_c$ using the above hard-core predicate is seen to be $((g^1)^{|m|}(x^1), \ldots, (g^l)^{|m|}(x^l)), m \oplus X(g^1, x^1, |m|) \oplus \ldots \oplus X(g^l, x^l, |m|)$. Similar for $f_{1-c}$.

In the following we call the $(g_1, \ldots, g_l)$ tuples public keys and the $(t_{g_1}, \ldots, t_{g_l})$ tuples private keys to distinguish from the individual permutations and trapdoors.

**Using the Key Generation in Our Protocol.** After an execution of the key generation protocol an honest $S$ has two public keys where he knows the private key to exactly one of them, but where the adversary cannot tell which one he knows. This is exactly the scenario that the first round of our protocol described earlier creates. Thus, one attempt to exchange a secret key can be done by running the key generation protocol followed by our communication phase. Our technical report [11] contains an analysis of the security. We sketch it here.

As before all failed attempts are simulated by simply following the real-life protocol. In the simulation of the successful attempt the simulator makes sure that it knows both private keys by learning all involved trapdoors: for each OT, if the sender is honest it chooses the trapdoors itself, if not, it chooses to learn both trapdoors during the OT (this succeeds except with negligible probability by the ZK proofs we introduced.) On corruption the simulator can patch the view of $S$ to be consistent with either private key being learned.

To show this simulation is indistinguishable from a real execution we observe that the only event in which there is a difference between the distributions is when $R$ or $S$ are corrupted after the message is sent. Here, the adversary will see a message/ciphertext pair which is valid w.r.t. to a given public key in the simulation but is invalid in a real execution. Since the adversary cannot corrupt all players, there is at least one involved trapdoor he does not know, so he should not be able to tell the difference. To prove this, we can take a permutation $f$ with unknown trapdoor, and choose a random player $P_i$. We then run the simulation pretending that $P_i$ chose $f$ as one of its inputs to the OT and hoping that the adversary will not corrupt $P_i$ but will corrupt $S$ or $R$ later. If simulation and execution can be distinguished at all, this must happen with non-negligible probability. It now follows that a successful distinguisher must be able to break encryption using $f$ as public key.

**Theorem 6.** *If there exist collections of trapdoor permutations for which the domains have invertible sampling, then non-committing encryption schemes exist.*

We note that the OT protocol which we use as an essential tool is itself based on trapdoor permutations. Moreover, in order for the OT to be non-committing, the domain of permutations must have invertible sampling. This property is also necessary in the original key generation protocol from [9], where also a common-domain property was needed, so assuming only invertible sampling is a weaker assumption. Further more, the discussion in chapter 4 of invertible sampling might indicate, that a protocol using ideas similar to those presented in this paper will need the invertible sampling property of crucial domains sampled *or* use a $n$-party protocol for sampling the domains, as we did for the key space in this chapter.

# References

1. *Proceedings of the Twentieth Annual ACM Symposium on Theory of Computing,* Chicago, Illinois, 2–4 May 1988.
2. Eric Bach. How to generate factored random numbers. *SIAM Journal on Computing,* 17(2):179–193, April 1988.
3. D. Beaver. Foundations of secure interactive computing. In Joan Feigenbaum, editor, *Advances in Cryptology - Crypto '91,* pages 377–391, Berlin, 1991. Springer-Verlag. Lecture Notes in Computer Science Volume 576.
4. D. Beaver. Plug and play encryption. In Burt Kaliski, editor, *Advances in Cryptology - Crypto '97,* pages 75–89, Berlin, 1997. Springer-Verlag. Lecture Notes in Computer Science Volume 1294.
5. D. Beaver and S. Haber. Cryptographic protocols provably secure against dynamic adversaries. In Rainer A. Rueppel, editor, *Advances in Cryptology - EuroCrypt '92,* pages 307–323, Berlin, 1992. Springer-Verlag. Lecture Notes in Computer Science Volume 658.

6. Michael Ben-Or, Shafi Goldwasser, and Avi Wigderson. Completeness theorems for non-cryptographic fault-tolerant distributed computation (extended abstract). In ACM [1], pages 1–10.
7. M. Blum and S. Goldwasser. An efficient probabilistic public key encryption scheme which hides all partial information. In G. R. Blakley and David Chaum, editors, *Advances in Cryptology: Proceedings of Crypto '84*, pages 289–302, Berlin, 1985. Springer-Verlag. Lecture Notes in Computer Science Volume 196.
8. Ran Canetti. Security and composition of multi-party cryptographic protocols. Obtainable from the Theory of Cryptography Library, august 1999.
9. Ran Canetti, Uri Feige, Oded Goldreich, and Moni Naor. Adaptively secure multi-party computation. In *Proceedings of the Twenty-Eighth Annual ACM Symposium on the Theory of Computing*, pages 639–648, Philadelphia, Pennsylvania, 22–24 May 1996.
10. David Chaum, Claude Crépeau, and Ivan Damgård. Multiparty unconditionally secure protocols (extended abstract). In ACM [1], pages 11–19.
11. Ivan B. Damgård and Jesper Buus Nielsen. Improved non-committing encryption schemes based on a general complexity assumption. Research Series RS-00-6, BRICS, Department of Computer Science, University of Aarhus, March 2000.
12. Alfredo De Santis and Giuseppe Persiano. Zero-knowledge proofs of knowledge without interaction (extended abstract). In *33rd Annual Symposium on Foundations of Computer Science*, pages 427–436, Pittsburgh, Pennsylvania, 24–27 October 1992. IEEE.
13. Oded Goldreich, Russell Impagliazzo, Leonid Levin, Ramarathnam Venkatesan, and David Zuckerman. Security preserving amplification of hardness. In *31st Annual Symposium on Foundations of Computer Science*, volume I, pages 318–326, St. Louis, Missouri, 22–24 October 1990. IEEE.
14. Oded Goldreich, Silvio Micali, and Avi Wigderson. Proofs that yield nothing but their validity and a methodology of cryptographic protocol design (extended abstract). In *27th Annual Symposium on Foundations of Computer Science*, pages 174–187, Toronto, Ontario, Canada, 27–29 October 1986. IEEE.
15. Oded Goldreich, Silvio Micali, and Avi Wigderson. How to play any mental game or a completeness theorem for protocols with honest majority. In *Proceedings of the Nineteenth Annual ACM Symposium on Theory of Computing*, pages 218–229, New York City, 25–27 May 1987.
16. IEEE. *23rd Annual Symposium on Foundations of Computer Science*, Chicago, Illinois, 3–5 November 1982.
17. Stanislaw Jarecki and Anna Lysyanskaya. Adaptively secure threshold cryptography: introducing concurrency, removing erasures. In Bart Preneel, editor, *Advances in Cryptology - EuroCrypt 2000*, pages 221–242, Berlin, 2000. Springer-Verlag. Lecture Notes in Computer Science Volume 1807.
18. D. E. Knuth and L. Trabb Pardo. Analysis of a simple factorization algorithm. *Theoretical Computer Science*, 3(3):321–348, 1976.
19. S. Micali and P. Rogaway. Secure computation. In Joan Feigenbaum, editor, *Advances in Cryptology - Crypto '91*, pages 392–404, Berlin, 1991. Springer-Verlag. Lecture Notes in Computer Science Volume 576.
20. Andrew C. Yao. Protocols for secure computations (extended abstract). In *23rd Annual Symposium on Foundations of Computer Science* [16], pages 160–164.
21. Andrew C. Yao. Theory and applications of trapdoor functions (extended abstract). In *23rd Annual Symposium on Foundations of Computer Science* [16], pages 80–91.

# A Note on the Round-Complexity of Concurrent Zero-Knowledge

Alon Rosen

Department of Computer Science
Weizmann Institute of Science
Rehovot 76100, Israel
alon@wisdom.weizmann.ac.il

**Abstract.** We present a lower bound on the number of rounds required by Concurrent Zero-Knowledge proofs for languages in $\mathcal{NP}$. It is shown that in the context of Concurrent Zero-Knowledge, at least eight rounds of interaction are essential for black-box simulation of non-trivial proof systems (i.e., systems for languages that are not in $\mathcal{BPP}$). This improves previously known lower bounds, and rules out several candidates for constant-round Concurrent Zero-Knowledge. In particular, we investigate the Richardson-Kilian protocol [20] (which is the only protocol known to be Concurrent Zero-Knowledge in the vanilla model), and show that for an apparently natural choice of its main parameter (which yields a 9-round protocol), the protocol is not likely to be Concurrent Zero-Knowledge.

## 1 Introduction

Zero-knowledge proof systems, introduced by Goldwasser, Micali and Rackoff [14] are efficient interactive proofs which have the remarkable property of yielding nothing beyond the validity of the assertion being proved. The generality of zero-knowledge proofs has been demonstrated by Goldreich, Micali and Wigderson [12], who showed that every NP-statement can be proved in zero-knowledge provided that one-way functions exist [16,19]. Since then, zero-knowledge protocols have turned out to be an extremely useful tool in the design of various cryptographic tasks.

The original setting in which zero-knowledge proofs were investigated consisted of a single prover and verifier which execute only one instance of the protocol at a time. A more realistic setting, especially in the time of the internet, is one which allows the concurrent execution of zero-knowledge protocols. In the concurrent setting (first considered by Dwork, Naor and Sahai [6]), many protocols (sessions) are executed at the same time, involving many verifiers which may be talking with the same (or many) provers simultaneously (the so-called parallel composition considered in [11] is a special case). This presents the new risk of an overall adversary which controls the verifiers, interleaving the executions and choosing verifiers queries based on other partial executions of the protocol. Since it seems unrealistic for the honest provers to coordinate their action so that

M. Bellare (Ed.): CRYPTO 2000, LNCS 1880, pp. 451–468, 2000.
© Springer-Verlag Berlin Heidelberg 2000

zero-knowledge is preserved, we must assume that in each prover-verifier pair the prover acts independently. A zero-knowledge proof is said to be *concurrent zero-knowledge* if it remains zero-knowledge even when executed in the concurrent setting. Recall that in order to prove that a certain protocol is zero-knowledge it is required to demonstrate that every probabilistic polynomial-time adversary interacting with the prover can be simulated by a probabilistic polynomial-time machine (a.k.a. the *simulator*) which is in solitude. In the concurrent setting, the simulation task becomes even more complicated, as the adversary may have control over multiple sessions at the same time, and is thus able to determine their scheduling (i.e., the order in which the interleaved execution of these sessions should be conducted).

## 1.1   Previous Work

Coming up with an efficient concurrent zero-knowledge protocol for all languages in $\mathcal{NP}$ seems to be a challenging task. Indications on the difficulty of this problem were already given in [6], where it was argued that for a specific recursive scheduling of $n$ sessions a particular (natural) simulation of a particular 4-round protocol may require time which is exponential in $n$. Further evidence on the difficulty was given by Kilian, Petrank and Rackoff [18]. Using the same recursive scheduling as in [6], they were able to prove that for every language outside $\mathcal{BPP}$ there is no 4-round protocol whose concurrent execution is simulatable in polynomial-time (by a *black-box simulator*).

Recent works have (successfully) attempted to overcome the above difficulties by augmenting the communication model with the so-called *timing assumption* [6,7] or, alternatively, by using various set-up assumptions (such as the *public-key model* [4,5]).[1] For a while it was not clear whether it is even possible to come up with a concurrent zero-knowledge protocol (not to mention an efficient one) without making any kind of timing or set-up assumptions. It was therefore a remarkable achievement when Richardson and Kilian [20] proposed a concurrent zero-knowledge protocol for all languages in $\mathcal{NP}$ (in the vanilla model).[2]

Unfortunately, the simulator shown by Richardson-Kilian is polynomial-time only when a non-constant round version of their protocol is considered. This leaves a considerable gap between the currently known upper and lower bounds on the number of rounds required by concurrent zero-knowledge [20,18]. We note that narrowing the above gap is not only of theoretical interest but has also practical consequences. Since the number of rounds is an important resource for protocols, establishing whether constant-round concurrent zero-knowledge exists is a well motivated problem.

---

[1]  The lower bound of [18] (as well as our own work) applies only in a cleaner model, in which no timing or set-up assumptions are allowed (the so-called *vanilla* model).

[2]  In fact, their solution is a family of protocols where the number of rounds is determined by a special parameter.

## 1.2    A Closer Look at the Richardson-Kilian Protocol

Being the only protocol known to be concurrent zero-knowledge in the vanilla model, versions of the Richardson-Kilian protocol are natural candidates for constant-round concurrent zero-knowledge. That is, it is still conceivable that there exists a (different) polynomial-time simulator for one of the protocol's constant-round versions.

**The Protocol:**    The Richardson-Kilian (RK for short) protocol [20] consists of two stages. In the *first stage*, which is independent of the actual common input, the verifier commits to $k$ random bit sequences, $v_1, ..., v_k \in \{0, 1\}^n$, where $n$ is the "security" parameter of the protocol and $k$ is a special parameter which determines the number of rounds. This is followed by $k$ iterations so that in each iteration the prover commits to a random bit sequence, $p_i$, and the verifier decommits to the corresponding $v_i$. The *result* of the $i^{\text{th}}$ iteration is defined as $v_i \oplus p_i$ and is known only to the prover. In the *second stage*, the prover provides a witness indistinguishable (WI) proof [8] that either the common input is in the language or that the result of one of the $k$ iterations is the all-zero string (i.e., $v_i = p_i$ for some $i$). Intuitively, since the latter case is unlikely to happen in an actual execution of the protocol, the protocol constitutes a proof system for the language. However, the latter case is the key to the simulation of the protocol in the concurrent zero-knowledge model: Whenever the simulator may cause $v_i = p_i$ to happen for some $i$ (this is done by the means of *rewinding* the verifier after the value $v_i$ has been revealed), it can simulate the rest of the protocol (and specifically Stage 2) by merely running the WI proof system with $v_i$ (and the prover's coins) as a witness.

**The Simulator:**    The RK protocol was designed to overcome the main difficulty encountered whenever many sessions are to be simulated in the concurrent setting. As observed by Dwork, Naor and Sahai [6], rewinding a specific session in the concurrent setting may result in loss of work done for other sessions, and cause the simulator to do the same amount of work again. In particular, all simulation work done for sessions starting after the point to which we rewind may be lost. Considering a specific session of the RK protocol (out of $m = \text{poly}(n)$ concurrent sessions), there must be an iteration (i.e., an $i \in \{1, ..., k\}$) so that at most $(m-1)/k$ sessions start in the interval corresponding to the $i^{\text{th}}$ iteration (of this specific session). So if we try to rewind on the correct $i$, we will invest (and so waste) only work proportional to $(m-1)/k$ sessions. The idea is to abort the rewinding attempt on the $i^{\text{th}}$ iteration if more than $(m-1)/k$ sessions are initiated in the corresponding interval (this will rule out the incorrect $i$'s). The same reasoning applies *recursively* (i.e., to the rewinding in these $(m-1)/k$ sessions). Denoting by $W(m)$ the amount of work invested in $m$ sessions, we obtain the recursion $W(m) = \text{poly}(m) \cdot W(\frac{m-1}{k})$, which solves to $W(m) = m^{\Theta(\log_k m)}$. Thus, whenever $k = n$, we get $W(m) = m^{O(1)}$, whereas taking $k$ to be a constant will cause $W(m)$ to be quasi-polynomial.

## 1.3   Our First Result

Given the above state of affairs, one may be tempted to think that a better simulation method would improve the recursion into something of the form $W(m) = O(W(\frac{m-1}{k}))$. In such a case, taking $k$ to be a constant (greater than 1) would imply a constant-round protocol whose simulation in the concurrent setting requires polynomial-time (i.e., $W(m) = O(W(\frac{m-1}{k}))$) solves to $W(m) = m^{O(1)}$). This should hold in particular for $k = 2$ (which gives a 9-round version of the RK protocol). However, as we show in the sequel, this is not likely to be the case.

**Theorem 1 (informal) :** *If $L$ is a language such that concurrent executions of the 9-round version of the RK protocol (i.e., for $k = 2$) can be black-box simulated in polynomial-time, then $L \in \mathcal{BPP}$.*

Thus, in general, the RK protocol is unlikely to be simulatable by a recursive procedure (as above) that satisfies the work recursion $W(m) = O(W(\frac{m-1}{k}))$.

## 1.4   Our Second Result

The proof of Theorem 1 is obtained by extending the proof of the following general result.

**Theorem 2 :** *Suppose that $(P, V)$ is a 7-round proof system for a language $L$ (i.e., on input $x$, the number of messages exchanged is at most 7), and that concurrent executions of $P$ can be simulated in polynomial-time using black-box simulation. Then $L \in \mathcal{BPP}$. This holds even if the proof system is only computationally-sound (with negligible soundness error) and the simulation is only computationally-indistinguishable (from the actual executions).*

In addition to shedding more light on the reasons that make the problem of constant-round concurrent zero-knowledge so difficult to solve, Theorem 2 rules out several constant-round protocols which may have been previously considered as candidates. These include 5-round zero-knowledge proofs for $\mathcal{NP}$ [10], as well as 6-round perfect zero-knowledge *arguments* for $\mathcal{NP}$ [3].

## 1.5   Techniques

The proof of Theorem 2 builds on the works of Goldreich and Krawczyk [11] and Kilian, Petrank and Rackoff [18]. It utilizes a fixed scheduling of the concurrent executions. This scheduling is defined recursively and is more sophisticated than the one proposed by [6] and used by [18]. It also exploits a special property of the first message sent by the verifier.

Note that since the scheduling considered here is fixed (rather than dynamic), both Theorems 1 and 2 are actually stronger than stated. Furthermore, our argument refers to verifier strategies that never refuse to answer the prover's queries. Simulating a concurrent interaction in which the verifier may occasionally refuse to answer (depending on its coin tosses and on the history of the current and/or other conversations) seems even more challenging than the simulation task which is treated in this work. Thus, it is conceivable that one may use the extra power of the adversary verifier to prove stronger lower bounds.

## 1.6   Organization of the Paper

Theorem 2 is proved in Section 2. We then demonstrate (in Section 3) how to modify the proof so it will work for the 9-round version of the Richardson-Kilian protocol (i.e., with $k = 2$). We conclude with Section 4 by discussing additional issues and recent work.

## 2   Proof of Theorem 2

In this section we prove that in the context of concurrent zero-knowledge, at least eight rounds of interaction are essential for black-box simulation of non-trivial proof systems (i.e., systems for languages that are not in $\mathcal{BPP}$). We note that in all known protocols, the zero-knowledge feature is demonstrated via a black-box simulator, and that it is hard to conceive of an alternative (for demonstrating zero-knowledge).[3]

*Definitions:* We use the standard definitions of interactive proofs [14] and arguments (a.k.a computationally-sound proofs) [2], black-box simulation (allowing non-uniform, deterministic verifier strategies, cf. [11,18]) and concurrent zero-knowledge (cf. [20,18]). Furthermore, since we consider a fixed scheduling of sessions, there is no need to use formalism for specifying to which session the next message of the verifier belongs. Finally, by (computationally-sound) interactive proof systems we mean systems in which the soundness error is negligible.[4]

*Preliminary conventions:* We consider protocols in which 8 messages are exchanged subject to the following conventions. The first message is an initiation message by the prover[5], denoted $p_1$, which is answered by the verifier's first message denoted $v_1$. The following prover and verifier messages are denoted $p_2, v_2, ..., p_4, v_4$, where the last message (i.e., $v_4$) is a single bit indicating whether the verifier has accepted the input (and will not be counted as an actual message). Clearly, any 7-round protocol can be modified to fit this form. Next, we consider black-box simulators which are restricted in several ways (but claim

---

[3] The interesting work of Hada and Tanaka [15] is supposedly an exception; but not really: They show that such a non-black-box simulation can be conducted if one makes an assumption of a similar nature (i.e., that for every machine which does X there exists a machine which does X along with Y). In contrast, starting from a more standard assumption (such as "it is infeasible to do X"), it is hard to conceive how one may use non-black-box simulators in places where black-box ones fail.

[4] We do not know whether this condition can be relaxed. Whereas we may consider polynomially-many parallel interactions of a proof system in order to decrease soundness error in interactive proofs (as such may occur anyhow in the concurrent model), this is not necessarily sufficient in order to decrease the soundness error in the case of arguments (cf. [1]).

[5] Being in control of the schedule, it would be more natural to let the verifier initiate each session. However, since the schedule is fixed, we choose to simplify the exposition by letting the prover send the first message of the session.

that each of these restrictions can be easily satisfied): Firstly, we allow only simulators running in *strict* polynomial-time, but allow them to produce output that deviates from the actual execution by at most a gap of $1/6$ (rather than requiring the deviation to be negligible).[6] (The latter relaxation enables a simple transformation of any expected polynomial-time simulator into a simulator running in strict polynomial-time.) Secondly, we assume, without loss of generality that the simulator never repeats the same query. As usual (cf. [11]), the queries of the simulator are prefixes of possible execution transcripts (in the concurrent setting[7]). Such a prefix is a sequence of alternating prover and verifier messages (which may belong to different sessions as determined by the fixed schedule). Thirdly, we assume that before making a query $\overline{q} = (a_1, b_1, ..., a_t, b_t, a_{t+1})$, where the $a$'s are prover messages, the simulator makes queries to all relevant prefixes (i.e., $(a_1, b_1, ..., a_{i-1}, b_{i-1}, a_i)$, for every $i \leq t$), and indeed has obtained the $b_i$'s as answers. Lastly, we assume that before producing output $(a_1, b_1, ..., a_T, b_T)$, the simulator makes the query $(a_1, b_1, ..., a_T)$.

## 2.1   The Schedule, Aversary Verifiers, and Decision Procedure

**The Fixed Schedule:** For each $x \in \{0,1\}^n$, we consider the following concurrent scheduling of $n$ sessions all run on common input $x$. The scheduling is defined recursively, where the scheduling of $m$ sessions (denoted $\mathcal{R}_m$) proceeds in 3 phases:

**First phase:** Each of the first $m/\log m$ sessions exchanges three messages (i.e., $\mathsf{p}_1, \mathsf{v}_1, \mathsf{p}_2$), this is followed by a recursive application of the scheduling on the next $m/\log m$ sessions.

**Second phase:** Each of the first $m/\log m$ sessions exchanges two additional messages (i.e., $\mathsf{v}_2, \mathsf{p}_3$), this is followed by a recursive application of the scheduling on the last $m - 2 \cdot \frac{m}{\log m}$ sessions.

**Third phase:** Each of the first $m/\log m$ sessions exchanges the remaining messages (i.e., $\mathsf{v}_3, \mathsf{p}_4, \mathsf{v}_4$).

The schedule is depicted in Figure 1. We stress that the verifier typically postpones its answer (i.e., $\mathsf{v}_j^{(i)}$) to the last prover's message (i.e., $\mathsf{p}_j^{(i)}$) till after a recursive sub-schedule is executed, and that it is crucial that in the first phase each session will finish exchanging its messages before the next sessions begins (whereas the order in which the messages are exchanged in the second and third phases is immaterial).

---

[6] We refer to the deviation gap, as viewed by any polynomial-time distinguisher. Such a distinguisher is required to decide whether its input consists of a conversation corresponding to real ececutions of the protocol, or rather to a transcript that was produced by the simulator. The computational deviation consists of the fraction of inputs which are accpted by the distinguisher in one case but rejected in the other.

[7] Indeed, for sake of clarity, we adopt a redundant representation. Alternatively, one may consider the subsequence of all prover's messages appearing in such transcripts.

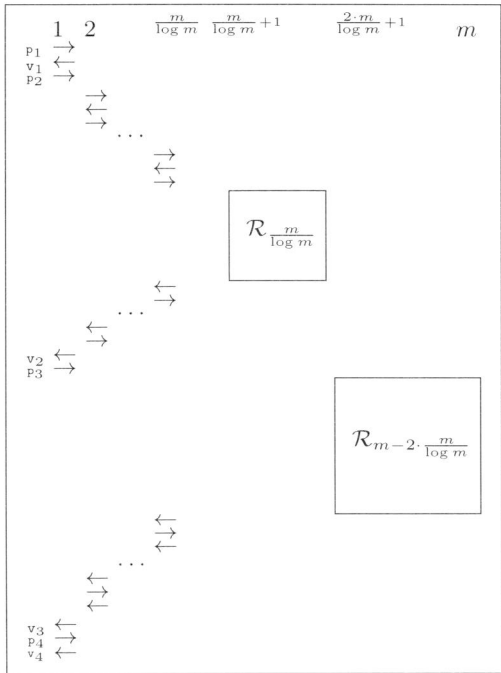

**Fig. 1.** The fixed schedule – recursive structure for $m$ sessions.

**Definition 3** (identifiers of next message): *The fixed schedule defines a mapping from partial execution transcripts ending with a prover message to the* identifiers of the next verifier message; *that is, the session and round number to which the next verifier message belongs.* (Recall that such partial execution transcripts correspond to queries of a black-box simulator and so the mapping defines the identifier of the answer:) *For such a query* $\bar{q} = (a_1, b_1, ..., a_t, b_t, a_{t+1})$, *we let* $\pi_{\mathrm{sn}}(\bar{q}) \in \{1, ..., n\}$ *denote the session to which the next verifier message belongs, and by* $\pi_{\mathrm{msg}}(\bar{q}) \in \{1, ..., 4\}$ *its index within the verifier's messages in this session.*

**Definition 4** (initiation-prefix): *The* initiation-prefix $\overline{ip}$ *of a query* $\bar{q}$ *is the prefix of* $\bar{q}$ *ending with the prover's initiation message of session* $\pi_{\mathrm{sn}}(\bar{q})$. *More formally,* $\overline{ip} = a_1, b_1, ..., a_\ell, b_\ell, a_{\ell+1}$, *is the initiation-prefix of* $\bar{q} = (a_1, b_1, ..., a_t, b_t, a_{t+1})$ *if* $a_{\ell+1}$ *is of the form* $\mathrm{p}_1^{(i)}$ *for* $i = \pi_{\mathrm{sn}}(\bar{q})$. *(Note that* $\pi_{\mathrm{msg}}(\bar{q})$ *may be any index in* $\{1, ..., 4\}$, *and that* $a_{t+1}$ *need not belong to session* $i$.)

**Definition 5** (prover-sequence): *The* prover-sequence *of a query* $\bar{q}$ *is the sequence of all prover's messages in session* $\pi_{\mathrm{sn}}(\bar{q})$ *that appear in the query* $\bar{q}$. *The length of such a sequence is* $\pi_{\mathrm{msg}}(\bar{q}) \in \{1, \ldots, 4\}$. *In case the length of the prover-sequence equals 4, both query* $\bar{q}$ *and its prover-sequence are said to be* terminating (*otherwise, they are called* non-terminating). *The prover-sequence is said to correspond to the initiation-prefix $\overline{ip}$ of the query $\overline{q}$.* (Note that all queries having the same initiation-prefix agree on the first element of their prover-sequence, since this message is part of the initiation-prefix.)

We consider what happens when a black-box simulator (for the above schedule) is given oracle access to a verifier strategy $V_h$ defined as follows (depending on a hash function $h$ and the input $x$).

**The Verifier Strategy $V_h$:** On query $\overline{q} = (a_1, b_1, ..., a_t, b_t, a_{t+1})$, where the $a$'s are prover messages (and $x$ is implicit in $V_h$), the verifier answers as follows:

1. First, $V_h$ checks if the execution transcript given by the query is legal (i.e., consistent with $V_h$'s prior answers), and answers with an error message if the query is not legal. (In fact this is not necessary since by our convention the simulator only makes legal queries. From this point on we ignore this case.)
2. More importantly, $V_h$ checks whether the query contains the transcript of a session in which the last verifier message indicates rejecting the input. In case such a session exists, $V_h$ refuses to answer (i.e., answers with some special "refuse" symbol).
3. Next, $V_h$ determines the initiation-prefix, denoted $a_1, b_1, ..., a_\ell, b_\ell, a_{\ell+1}$, of query $\overline{q}$. It also determines $i = \pi_{\mathrm{sn}}(\overline{q})$, $j = \pi_{\mathrm{msg}}(\overline{q})$, and the prover-sequence of query $\overline{q}$, denoted $\mathrm{p}_1^{(i)}, ..., \mathrm{p}_j^{(i)}$.
4. Finally, $V_h$ determines $r_i = h(a_1, b_1, ..., a_\ell, b_\ell, a_{\ell+1})$ (as coins to be used by $V$), and answers with the message $V(x, r_i; \mathrm{p}_1^{(i)}, ..., \mathrm{p}_j^{(i)})$ that would have been sent by the honest verifier on common input $x$, random-pad $r_i$, and prover's messages $\mathrm{p}_1^{(i)}, ..., \mathrm{p}_j^{(i)}$.

Assuming towards the contradiction that a black-box simulator, denoted $S$, contradicting Theorem 2 exists, we now descibe a probabilistic polynomial-time decision procedure for $L$, based on $S$. Recall that we may assume that $S$ runs in strict polynomial time: we denote such time bound by $t_S(\cdot)$. On input $x \in L \cap \{0,1\}^n$ and oracle access to any (probabilistic polynomial-time) $V^*$, the simulator $S$ must output transcipts with distribution having computational deviation of at most $1/6$ from the distribution of transcripts in the actual concurrent executions of $V^*$ with $P$.

*A slight modification of the simulator:* Before presenting the procedure, we slightly modify the simulator so that it never makes a query that is refused by a verifier $V_h$. Note that this condition can be easily checked by the simulator, and that the modification does not effect the simulator's output. From this point on, when we talk of the simulator (which we continue to denote by $S$) we mean the modified one.

**Decision Procedure for $L$:** On input $x \in \{0,1\}^n$, proceed as follows:

1. Uniformly select a function $h$ out of a small family of $t_S(n)$-wise independent hash functions mapping poly$(n)$-bit long sequences to $\rho_V(n)$-bit sequences, where $\rho_V(n)$ is the number of random bits used by $V$ on an input $x \in \{0,1\}^n$.
2. Invoke $S$ on input $x$ providing it black-box access to $V_h$ (as defined above). That is, the procedure emulates the execution of the oracle machine $S$ on input $x$ along with emulating the answers of $V_h$.
3. Accept if and only if all sessions in the transcript output by $S$ are accepting.

By our hypothesis, the above procedure runs in probabilistic polynomial-time. We next analyze its performance.

**Lemma 6** (performance on YES-instances): *For all but finitely many $x \in L$, the above procedure acccepts $x$ with probability at least $2/3$.*

**Proof Sketch:** The key observation is that for uniformly selected $h$, the behavior of $V_h$ in actual (concurrent) interactions with $P$ is identical to the behavior of $V$ in such interactions. The reason is that, in such actual interactions, a randomly selected $h$ determines uniformly and independently distributed random-pads for all $n$ sessions. Since with high probability (say at least $5/6$), $V$ accepts in all $n$ concurrent sessions, the same must be true for $V_h$, when $h$ is uniformly selected. Since the simulation deviation of $S$ is at most $1/6$, it follows that for every $h$ the probability that $S^{V_h}(x)$ is a transcript in which all sessions accept is lower bounded by $p_h - 1/6$, where $p_h$ denotes the probability that $V_h$ accepts $x$ (in all sessions) when interacting with $P$. Taking expectation over all possible $h$'s, the lemma follows. ∎

**Lemma 7** (performance on NO-instances): *For all but finitely many $x \notin L$, the above procedure rejects $x$ with probability at least $2/3$.*

We can actually prove that for every polynomial $p$ and all but finitely many $x \notin L$, the above procedure accepts $x$ with probability at most $1/p(|x|)$. Assuming towards the contradiction that this is not the case, we will construct a (probabilistic polynomial-time) strategy for a cheating prover that fools the honest verifier $V$ with success probability at least $1/\text{poly}(n)$ (in contradiction to the computational-soundness of the proof system). Loosely speaking, the argument capitalizes on the fact that rewinding of a session requires the simulator to work on a new simulation sub-problem (one level down in the recursive construction). New work is required since each different message for the rewinded session forms an unrelated instance of the simulation sub-problem (by virtue of definition of $V_h$). The schedule causes work involved in such rewinding to accumulate to too much, and so it must be the case that the simulator does not rewind some (full instance of some) session. In this case the cheating prover may use such a session in order to fool the verifier.

## 2.2   Proof of Lemma 7 (Performance on NO-Instances)

Let us fix an $x \in \{0,1\}^n \setminus L$ as above.[8] Define by $\mathsf{AC} = \mathsf{AC}_x$ the set of pairs $(\sigma, h)$ so that on input $x$, coins $\sigma$ and oracle access to $V_h$, the simulator outputs a transcript, denoted $S_\sigma^{V_h}(x)$, in which all $n$ sessions accept. Recall that our contradiction assumption is that $\Pr_{\sigma, h}[(\sigma, h) \in \mathsf{AC}] > 1/p(n)$, for some fixed polynomial $p(\cdot)$.

**The Cheating Prover:** The cheating prover starts by uniformly selecting a pair $(\sigma, h)$ and hoping that $(\sigma, h)$ is in $\mathsf{AC}$. It next selects uniformly two elements $\xi$ and $\zeta$ in $\{1, ..., q_S(n)\}$, where $q_S(n) < t_S(n)$ is a bound on the number of queries made by $S$ on input $x \in \{0,1\}^n$. The prover next emulates an execution of $S_\sigma^{V_{h'}}(x)$ (where $h'$, which is essentially equivalent to $h$, will be defined below), while interacting with the honest verifier $V$. The prover handles the simulator's queries as well as the communication with the verifier as follows: Suppose that the simulator makes query $\bar{q} = (a_1, b_1, ..., a_t, b_t, a_{t+1})$, where the $a$'s are prover messages.

1. Operating as $V_h$, the cheating prover first determines the initiation-prefix, $\overline{ip} = a_1, b_1, ..., a_\ell, b_\ell, a_{\ell+1}$, corresponding to the current query $\bar{q}$. (Note that by our convention and the modification of the simulator there is no need to perform Steps 1 and 2 of $V_h$.)
2. If $\overline{ip}$ is the $\xi^{\text{th}}$ distinct initiation-prefix resulting from the simulator's queries so far then the cheating prover operates as follows:
   (a) The cheating prover determines $i = \pi_{\text{sn}}(\bar{q})$, $j = \pi_{\text{msg}}(\bar{q})$, and the prover-sequence of $\bar{q}$, denoted $\mathsf{p}_1^{(i)}, ..., \mathsf{p}_j^{(i)}$ (as done by $V_h$ in Step 3).
   (b) If the query $\bar{q}$ is non-terminating (i.e., $j \leq 3$), and the cheating prover has only sent $j-1$ messages to the actual verifier then it forwards $\mathsf{p}_j^{(i)}$ to the verifier, and feeds the simulator with the verifier's response (i.e., which is of the form $\mathsf{v}_j^{(i)}$).[9]
   (c) If the query $\bar{q}$ is non-terminating (i.e., $j \leq 3$), and the cheating prover has already sent $j$ messages to the actual verifier, the prover retrieves the $j^{\text{th}}$ message it has received and feeds it to the simulator.[10]

---

[8] In a formal proof we need to consider infinitely many such $x$'s.

[9] We comment that by our conventions regarding the simulator, it cannot be the case that the cheating prover has sent less than $j-1$ messages to the actual verifier: The prefixes of the current query dictate $j-1$ such messages.

[10] We comment that the cheating prover may fail to conduct Step 2c. This will happen whenever the simulator makes two queries with the same initiation-prefix and the same number of prover messages in the corresponding session, but with a different sequence of such messages. Whereas this will never happen when $j = 1$ (as once the initiation-prefix is fixed then so is the value of $\mathsf{p}_1^{(i)}$), it may very well be the case that for $j \in \{2, 3\}$ a previous query regarding initiation-prefix $\overline{ip}$ had a different $\mathsf{p}_j^{(i)}$ message. In such a case the cheating prover will indeed fail. The punchline of the analysis is that with noticeable probability this will not happen.

(d) Whenever the query $\bar{q}$ is terminating (i.e., $j = 4$), the cheating prover operates as follows:

    i. As long as the $\zeta^{\text{th}}$ terminating query corresponding to the above initiation-prefix has not been made, the cheating prover feeds the simulator with $v_4^{(i)} = 0$ (i.e., session rejected).

    ii. Otherwise, the cheating prover operates as in Step 2b (i.e., it forwards $p_4^{(i)}$ to the verifier, and feeds the simulator with the verifier's response – some $v_4^{(i)}$ message).[11]

3. If $\overline{ip}$ is NOT the $\xi^{\text{th}}$ distinct initiation-prefix resulting from the queries so far then the prover emulates $V_h$ in the obvious manner (i.e., as in Step 4 of $V_h$): It first determines $r_i = h(a_1, b_1, ..., a_\ell, b_\ell, a_{\ell+1})$, and then answers with $V(x, r_i; p_1^{(i)}, ..., p_j^{(i)})$, where all notations are as above.

*Defining $h'$ (mentioned above):* Let $(\sigma, h)$ and $\xi$ be the initial choices made by the cheating prover, and suppose that the honest verifier uses coins $r$. Then, the function $h'$ is defined to be uniformly distributed among the functions $h''$ which satisfy the following conditions: The value of $h''$ on the $\xi^{\text{th}}$ initiation-prefix equals $r$, whereas for every $\xi' \neq \xi$, the value of $h''$ on the $\xi'^{\text{th}}$ initiation-prefix equals the value of $h$ on this prefix. (Here we use the hypothesis that the functions are selected in a family of $t_S(n)$-wise independent hash functions. We note that replacing $h$ by $h'$ does not effect Step 3 of the cheating prover, and that the prover does not know $h'$.)

    The probability that the cheating prover makes the honest verifier accept is lower bounded by the probability that both $(\sigma, h') \in \text{AC}$ and the messages forwarded by the cheating prover in Step 2 are consistent with an accepting conversation with $V_{h'}$. For the latter event to occur, it is necessary that the $\xi^{\text{th}}$ distinct initiation-prefix will be useful (in the sense hinted above and defined now). It is also necessary that $\zeta$ was "successfully" chosen (i.e., the $\zeta^{\text{th}}$ terminating query which corresponds to the $\xi^{\text{th}}$ initiation-prefix is accepted by $V_{h'}$).

**Definition 8** (accepting query): *A terminating query $\bar{q} = (a_1, b_1, ..., a_t, b_t, a_{t+1})$ (i.e., for which $\pi_{\text{msg}}(\bar{q}) = 4$) is said to be* accepting *if $V_{h'}(a_1, b_1, ..., a_t, b_t, a_{t+1})$ equals 1 (i.e., session $\pi_{\text{sn}}(\bar{q})$ is accepted by $V_{h'}$).*

**Definition 9** (useful initiation-prefix): *A specific initiation-prefix $\overline{ip}$ in an execution of $S_\sigma^{V_{h'}}(x)$ is called* useful *if the following conditions hold:*

1. *During its execution, $S_\sigma^{V_{h'}}(x)$ made at least one accepting query which corresponds to the initiation-prefix $\overline{ip}$.*

---

[11] We note that once the cheating prover arrives to this point, then it either succeds in the cheating task or completely fails (depending on the verifier's response). As a consequence, it is not essential to define the cheating prover's actions from this point on (as in both cases the algorithm will be terminated).

2. *As long as no accepting query corresponding to the initiation-prefix $\overline{ip}$ was made during the execution of $S_\sigma^{V_{h'}}(x)$, the number of (non-terminating) different prover-sequences that correspond to $\overline{ip}$ is at most 3, and these prover-sequences are prefixes of one another.*[12]

*Otherwise, the prefix is called* unuseful.

**The Success Probability:** Define a Boolean indicator $\chi(\sigma, h', \xi)$ to be true if and only if the $\xi^{\text{th}}$ distinct initiation-prefix in an execution of $S_\sigma^{V_{h'}}(x)$ is useful. Define an additional Boolean indicator $\psi(\sigma, h', \xi, \zeta)$ to be true if and only if the $\zeta^{\text{th}}$ terminating query among all terminating queries that correspond to the $\xi^{\text{th}}$ distinct initiation-prefix (in an execution of $S_\sigma^{V_{h'}}(x)$) is the first one to be accepting. It follows that if the cheating prover happens to select $(\sigma, h, \xi, \zeta)$ so that both $\chi(\sigma, h', \xi)$ and $\psi(\sigma, h', \xi, \zeta)$ hold then it convinces $V(x, r)$; the first reason being that the $\zeta^{th}$ such query is answered by an accept message[13], and the second reason being that the emulation does not get into trouble (in Steps 2c and 2d). To see this, notice that all first $(\zeta - 1)$ queries having the $\xi^{\text{th}}$ distinct initiation-prefix satisfy exactly one of the following conditions:

1. They have non-terminating prover-sequences that are prefixes of one another (which implies that the cheating prover never has to forward such queries to the verifier twice).
2. They have terminating prover-sequences which should be rejected (recall that as long as the $\zeta^{\text{th}}$ terminating query has not been asked by $S_\sigma^{V_{h'}}(x)$, the cheating prover automatically rejects any terminating query).

Thus, the probability that when selecting $(\sigma, h, \xi, \zeta)$ the cheating prover convinces $V(x, r)$ is at least

$$\Pr\left[\psi(\sigma, h', \xi, \zeta) \;\&\; \chi(\sigma, h', \xi)\right]$$
$$= \Pr\left[\psi(\sigma, h', \xi, \zeta) \mid \chi(\sigma, h', \xi)\right] \cdot \Pr\left[\chi(\sigma, h', \xi)\right]$$
$$\geq \Pr\left[\psi(\sigma, h', \xi, \zeta) \mid \chi(\sigma, h', \xi)\right] \cdot \Pr\left[(\sigma, h') \in \mathsf{AC} \;\&\; \chi(\sigma, h', \xi)\right] \qquad (1)$$

Note that if the $\xi^{\text{th}}$ distinct initiation-prefix is useful, and $\zeta$ is uniformly (and independently) selected in $\{1, ..., q_S(n)\}$, the probability that the $\zeta^{\text{th}}$ query corresponding to the $\xi^{\text{th}}$ distinct initiation–prefix is the first to be accepting is at least $1/q_S(n)$. Thus, Eq. (1) is lower bounded by

$$\frac{\Pr\left[(\sigma, h') \in \mathsf{AC} \;\&\; \chi(\sigma, h', \xi)\right]}{q_S(n)} \qquad (2)$$

---

[12] In other words, we allow for many different terminating queries to occur (as long as they are not accepting). On the other hand, for $j \in \{1, 2, 3\}$ only a single query that has a prover sequence of length $j$ is allowed. This requirement will enable us to avoid situations in which the cheating prover will fail (as described in Footnote 10).

[13] We use the fact that $V(x, r)$ behaves exactly as $V_{h'}(x)$ behaves on queries for the $\xi^{\text{th}}$ distinct initiation-prefix.

Using the fact that, for every value of $\xi$ and $\sigma$, when $h$ and $r$ are uniformly selected the function $h'$ is uniformly distributed, we infer that $\xi$ is distributed independently of $(\sigma, h')$. Thus, Eq. (2) is lower bounded by

$$\Pr[(\sigma, h') \in \mathsf{AC}] \cdot \frac{\Pr[\exists i \text{ s.t. } \chi(\sigma, h', i) \mid (\sigma, h') \in \mathsf{AC}]}{q_S(n)^2} \tag{3}$$

Thus, Eq. (3) is noticeable (i.e., at least $1/\mathrm{poly}(n)$) provided that so is the value of $\Pr[\exists i \text{ s.t. } \chi(\sigma, h', i) \mid (\sigma, h') \in \mathsf{AC}]$. The rest of the proof is devoted to establishing the last hypothesis. In fact we prove a much stronger statement:

**Lemma 10** *For every* $(\sigma, h') \in \mathsf{AC}$, *the execution of* $S_\sigma^{V_{h'}}(x)$ *contains a useful initiation-prefix (that is, there exists an* $i$ *s.t.* $\chi(\sigma, h', i)$ *holds).*

### 2.3   Proof of Lemma 10 (Existence of Useful Initiation Prefixes)

The proof of Lemma 10 is by contradiction. We assume the existence of a pair $(\sigma, h') \in \mathsf{AC}$ so that all initiation-prefixes in the execution of $S_\sigma^{V_{h'}}(x)$ are unuseful and show that this implies that $S_\sigma^{V_{h'}}(x)$ made at least $n^{\Omega\left(\frac{\log n}{\log \log n}\right)} \gg \mathrm{poly}(n)$ queries which contradicts the assumption that it runs in polynomial-time.

**The Query–and–Answer Tree:** Throughout the rest of the proof, we fix an arbitrary $(\sigma, h') \in \mathsf{AC}$ so that all initiation-prefixes in the execution of $S_\sigma^{V_{h'}}(x)$ are unuseful, and study this execution. A key vehicle in this study is the notion of a query–and–answer tree introduced in [18]. This is a rooted tree in which vertices are labeled with verifier messages and edges are labeled by prover's messages. The root is labeled by the empty string, and it has outgoing edges corresponding to the possible prover's messages initializing the first session. In general, paths down the tree (i.e., from the root to some vertices) correspond to queries. The query associated with such a path is obtained by concatenating the labeling of the vertices and edges in the order traversed. We stress that each vertex in the tree corresponds to a query actually made by the simulator.

*Satisfied sub-path:* A sub-path from one node in the tree to some of its descendants is said to **satisfy session** $i$ if the sub-path contains edges (resp., vertices) for each of the messages sent by the prover (resp., verifier) in session $i$, and if the last such message (i.e., $\mathbf{v}_4^{(i)}$) indicates that the verifier accepts session $i$. A sub-path is called **satisfied** if it satisfies all sessions for which the first prover's message appears on the sub-path.

*Forking sub-tree:* For any $i$ and $j \in \{2, 3, 4\}$, we say that a sub-tree $(i, j)$-**forks** if it contains two sub-paths, $\bar{p}$ and $\bar{r}$, having the same initiation-prefix, so that

1. Sub-paths $\bar{p}$ and $\bar{r}$ differ in the edge representing the $j^{\mathrm{th}}$ prover message for session $i$ (i.e., a $\mathbf{p}_j^{(i)}$ message).
2. Each of the sub-paths $\bar{p}$ and $\bar{r}$ reaches a vertex representing the $j^{\mathrm{th}}$ verifier message (i.e., some $\mathbf{v}_j^{(i)}$).

In such a case, we may also say that the sub-tree $(i, j)$-**forks on** $\bar{p}$ (or on $\bar{r}$).

*Good sub-tree:* Consider an arbitrary sub-tree rooted at a vertex corresponding to the first message in some session so that this session is the first at some level of the recursive construction of the schedule. The full tree is indeed such a tree, but we will need to consider sub-trees which correspond to $m$ sessions in the recursive schedule construction. We call such a sub-tree $m$-**good** if it contains a sub-path satisfying all $m$ sessions for which the prover's first message appears in the sub-tree (all these first messages are in particular contained in the sub-path). Since $(\sigma, h') \in \text{AC}$ it follows that the full tree contains a path from the root to a leaf representing an accepting transcript. The path from the root to this leaf thus satisfies all sessions (i.e., 1 through $n$) which implies that the full tree is $n$-good. The crux of the entire proof is given in the following lemma.

**Lemma 11** *Let $T$ be an $m$-good sub-tree, then at least one of the following holds:*

1. *$T$ contains at least two different $\left(m - 2 \cdot \frac{m}{\log m}\right)$-good sub-trees.*

2. *$T$ contains at least $\frac{m}{\log m}$ different $\left(\frac{m}{\log m}\right)$-good sub-trees.*

Denote by $W(m)$ the size of an $m$-good sub-tree (where $W(m)$ stands for the work actually performed by the simulator on $m$ concurrent sessions in our fixed scheduling). It follows (from Lemma 11) that any $m$-good sub-tree must satisfy

$$W(m) \geq \min \left\{ \frac{m}{\log m} \cdot W\left(\frac{m}{\log m}\right), 2 \cdot W\left(m - 2 \cdot \frac{m}{\log m}\right)\right\} \quad (4)$$

Since Eq. (4) solves to $n^{\Omega\left(\frac{\log n}{\log \log n}\right)}$ (proof omitted), and since every vertex in the query–and–answer tree corresponds to a query actually made by the simulator, then the assumption that the simulator runs in $\text{poly}(n)$-time (and hence the tree is of $\text{poly}(n)$ size) is contradicted. Thus, Lemma 10 follows from Lemma 11.

## 2.4  Proof of Lemma 11 (The Structure of Good Sub-trees)

Considering the $m$ sessions corresponding to an $m$-good sub-tree, we focus on the $m/\log m$ sessions dealt explicitly at this level of the recursive construction (i.e., the first $m/\log m$ sessions, which we denote by $\mathcal{F} \overset{\text{def}}{=} \{1, ..., m/\log m\}$).

**Claim 12** *Let $T$ be an $m$-good sub-tree. Then for any session $i \in \mathcal{F}$, there exists $j \in \{2, 3\}$ such that the sub-tree $(i, j)$-forks.*

**Proof:**  Consider some $i \in \mathcal{F}$, and let $\bar{p}_i$ be the first sub-path reached during the execution of $S_\sigma^{V_{h'}}(x)$ which satisfies session $i$ (since the sub-tree is $m$-good such a sub-path must exist, and since $i \in \mathcal{F}$ every such sub-path must be contained in the sub-tree). Recall that by the contradiction assumption for the proof of Lemma 10, all initiation-prefixes in the execution of $S_\sigma^{V_{h'}}(x)$ are unuseful. In particular, the initiation-prefix corresponding to sub-path $\bar{p}_i$ is unuseful. Still, path $\bar{p}_i$ contains vertices for each prover message in session $i$ and contains an accepting message by the verifier. So the only thing which may prevent the above initiation-prefix from being useful is having two (non-terminating) queries with the very same initiation-prefix (non-terminating) prover-sequences of the same length. Say that these sequences first differ at their $j^{\text{th}}$ element, and note that $j \in \{2, 3\}$ (as the prover-sequences are non-terminating and the first prover message, $\mathsf{p}_1^{(i)}$, is constant once the initiation-prefix is fixed). Also note that the two (non-terminating) queries were answered by the verifier (rather than refused), since the (modified) simulator avoids queries which will be refused. By associating a sub-path to each one of the above queries we obtain two different sub-paths (having the same initiation-prefix), that differ in some $\mathsf{p}_j^{(i)}$ edge and eventually reach a $\mathsf{v}_j^{(i)}$ vertex (for $j \in \{2, 3\}$). The required $(i, j)$-forking follows.   ■

**Claim 13** *If there exists a session $i \in \mathcal{F}$ such that the sub-tree $(i, 3)$-forks, then the sub-tree contains two different $(m - 2 \cdot \frac{m}{\log m})$-good sub-trees.*

**Proof:**   Let $i \in \mathcal{F}$ such that the sub-tree $(i, 3)$-forks. That is, there exist two sub-paths, $\bar{p}_i$ and $\bar{r}_i$, that differ in the edge representing a $\mathsf{p}_3^{(i)}$ message, and that eventually reach some $\mathsf{v}_3^{(i)}$ vertex. In particular, paths $\bar{p}_i$ and $\bar{r}_i$ split from each other before the edge which corresponds to the $\mathsf{p}_3^{(i)}$ message occurs along these paths (as otherwise the $\mathsf{p}_3^{(i)}$ edge would have been identical in both paths). By nature of the fixed scheduling, the vertex in which the above splitting occurs precedes the first message of all (nested) sessions in the second recursive construction (that is, sessions $2 \cdot \frac{m}{\log m} + 1, ..., m$). It follows that both $\bar{p}_i$ and $\bar{r}_i$ contain the first and last messages of each of these (nested) sessions (as they both reach a $\mathsf{v}_3^{(i)}$ vertex). Therefore, by definition of $V_h$, all these sessions must be satisfied by both these paths (or else $V_h$ would have not answered with a $\mathsf{v}_3^{(i)}$ message but rather with a "refuse" symbol). Consider now the corresponding sub-paths of $\bar{p}_i$ and $\bar{r}_i$ which begin at edge $\mathsf{p}_1^{(k)}$ where $k = 2 \cdot \frac{m}{\log m} + 1$ (i.e., $\mathsf{p}_1^{(k)}$ is the edge which represents the first message of the first session in the second recursive construction). Each of these new sub-paths is contained in a disjoint sub-tree corresponding to the recursive construction, and satisfies all of its $(m - 2 \cdot \frac{m}{\log m})$ sessions. It follows that the (original) sub-tree contains two different $(m - 2 \cdot \frac{m}{\log m})$-good sub-trees and the claim follows.   ■

**Claim 14** *If for every session $i \in \mathcal{F}$ the sub-tree $(i, 2)$-forks, then the sub-tree contains at least $|\mathcal{F}| = \frac{m}{\log m}$ different $(\frac{m}{\log m})$-good sub-trees.*

In the proof of Claim 14 we use a special property of $(i, 2)$-forking: The only location in which the splitting of path $\bar{r}_i$ from path $\bar{p}_i$ may occur, is a vertex which represents a $\mathsf{v}_1^{(i)}$ message. Any splitting which has occured at a vertex which precedes the $\mathsf{v}_1^{(i)}$ vertex would have caused the initiation-prefixes of (session $i$ along) paths $\bar{p}_i$ and $\bar{r}_i$ to be different (by virtue of the definition of $V_h$, and since all vertices preceding $\mathsf{v}_1^{(i)}$ are part of the initiation-prefix of session $i$).

**Proof:**    Since for all sessions $i \in \mathcal{F}$ the sub-tree $(i, 2)$-forks, then for every such $i$ there exist two sub-paths, $\bar{p}_i$ and $\bar{r}_i$, that split from each other in a $\mathrm{v}_1^{(i)}$ vertex and that eventually reach some $\mathrm{v}_2^{(i)}$ vertex. Similarly to the proof of Claim 13, we can claim that each one of the above paths contains a "special" sub-path (denoted $\bar{\bar{p}}_i$ and $\bar{\bar{r}}_i$ respectively), that starts at a $\mathrm{v}_1^{(i)}$ vertex, ends at a $\mathrm{v}_2^{(i)}$ vertex, and satisfies all $\frac{m}{\log m}$ sessions in the first recursive construction (that is, sessions $\frac{m}{\log m} + 1, \dots, 2 \cdot \frac{m}{\log m}$). Note that paths $\bar{\bar{p}}_i$ and $\bar{\bar{r}}_i$ are completely disjoint. Let $i_1, i_2$ be two different sesions in $\mathcal{F}$ (without loss of generality $i_1 < i_2$), and let $\bar{\bar{p}}_{i_1}, \bar{\bar{r}}_{i_1}, \bar{\bar{p}}_{i_2}, \bar{\bar{r}}_{i_2}$ be their corresponding "special" sub-paths. The key point is that for every $i_1, i_2$ as above, it cannot be the case that both "special" sub-paths corresponding to session $i_2$ are contained in the sub-paths corresponding to session $i_1$ (to justify this, we use the fact that $\bar{\bar{p}}_{i_2}$ and $\bar{\bar{r}}_{i_2}$ split from each other in a $\mathrm{v}_1^{(i_2)}$ vertex and that for every $i \in \{i_1, i_2\}$, paths $\bar{\bar{p}}_i$ and $\bar{\bar{r}}_i$ are disjoint).

This enables us to associate a distinct $(\frac{m}{\log m})$-good sub-tree to every $i \in \mathcal{F}$ (i.e., which either corresponds to path $\bar{\bar{p}}_i$, or to path $\bar{\bar{r}}_i$). Which in particular means that the tree contains at least $|\mathcal{F}|$ different $(\frac{m}{\log m})$-good sub-trees. ∎

We are finally ready to analyze the structure of the sub-tree $T$. Since for every $i \in \mathcal{F}$ there must exist $j \in \{2, 3\}$ such that the sub-tree $(i, j)$-forks (Claim 12), then it must be the case that either $T$ contains two distinct $(m - 2 \cdot \frac{m}{\log m})$-good sub-trees (Claim 13), or $T$ contains at least $\frac{m}{\log m}$ distinct $(\frac{m}{\log m})$-good sub-trees (Claim 14). This completes the proof of Lemma 11 which in turn implies Lemmata 10 and 7. The proof of Theorem 2 is complete.

# 3    Extending the Proof for the Richardson-Kilian Protocol

Recall that the Richardson-Kilian protocol [20] consists of two stages. We will treat the first stage of the RK protocol (which consists of 6 rounds) as if it were the first 6 rounds of any 7-round protocol, and the second stage (which consists of a 3-round WI proof) as if it were the remaining $7^{\mathrm{th}}$ message. An important property which is satisfied by the RK protocol is that the coin tosses used by the verifier in the second stage are independent of the coins used by the verifier in the first stage. We can therefore define and take advantage of two (different) types of initiation-prefixes. A first-stage initiation prefix and a second-stage initiation prefix (which is well defined only given the first one). These initiation-prefixes will determine the coin tosses to be used by $V_h$ in each corresponding stage of the protocol (analogously to the proof of Theorem 2).

The cheating prover will pick a random index for each of the above types of initiation-prefixes (corresponding to $\xi$ and $\zeta$ in the proof of Theorem 2). The first index (i.e., $\xi$) is treated exactly as in the proof of Theorem 2, whereas the second index (i.e., $\zeta$) will determine which of the WI session corresponding to the second-phase initiation-prefix (and which also correspond to the very same $\xi^{\mathrm{th}}$ first-phase initiation-prefix) will be actually executed between the cheating prover and the verifier. As long as the $\zeta^{\mathrm{th}}$ second-stage initiation prefix will not be encountered, the cheating prover will be able to impersonate $V_h$ while always deciding correctly whether to reject or to accept the corresponding "dummy" WI session (as the second-stage initiation-prefix completely determines the coins to be used by $V_h$ in the second stage of the protocol). As in the proof of Theorem 2, the probability that the $\zeta^{\text{th}}$ second-stage initiation prefix (that correponds to the $\xi^{\text{th}}$ first-phase initiation-prefix) will make the verifier accept is non-negligible. The existence of a useful pair of initiation-prefixes (i.e., $\xi$ and $\zeta$) is proved essentially in the same way as in the proof of Theorem 2.

## 4   Concluding Remarks

*Summary:* In this work we have pointed out the impossibility of black-box simulation of non-trivial 7-round protocols in the concurrent setting. The result which is proved is actually stronger than stated. Not only because we consider a fixed scheduling in which the adversarial verifier never refuses to answer (and thus should have been easier to simulate, as argued in Section 1.5), but also because we are considering simulators which may have as much as a constant deviation from actual executions of the protocol (rather than negligible deviation, as typically required in the definition of Zero-Knowledge).

*On the applicability of the RK protocol:* We note that the above discussion does not imply that the $k = 2$ version of the RK protocol is completely useless. As noted by Richardson and Kilian, for security parameter $n$, the simulation of $m = \text{poly}(n)$ concurrent sessions may be performed in quasi-polynomial time (recall that the simulation work required for $m$ sessions is $m^{\Theta(\log_k m)}$). Thus, the advantage that a polynomial-time adversary verifier may gain from executing $m$ concurrent sessions is not significant. As a matter of fact, if one is willing to settle for less than polynomially-many (in $n$) concurrent sessions, then the RK protocol may be secure in an even stronger sense. Specifically, as long as the number of sessions is $m = 2^{O(\sqrt{\log_2 n})}$, then simulation of the RK protocol can be performed in polynomial-time even if $k = 2$. This is considerably larger than the logarithmic number of concurrent sessions enabled by straightforward simulation of previously known constant-round protocols.

*Improved simulation of the RK protocol:* Recently, Kilian and Petrank [17] have proved that the Richardson-Kilian protocol will remain concurrent zero-knowledge even if $k = O(g(n) \cdot \log^2 n)$, where $g(\cdot)$ is any non-constant function (e.g., $g(n) = \log n$). Thus, the huge gap between the known upper and lower bounds on the number of rounds required by concurrent zero-knowledge has been considerably narrowed.

### Acknowledgements

I would like to thank Oded Goldreich for introducing me to the subject, and for the generous assistance in writing this manuscript. I would also like to thank Yoav Rodeh for significant discussions, and Yehuda Lindell (as well as the anonymous referees) for commenting on the earlier version of this paper.

# References

1. M. Bellare, R. Impagliazzo and M. Naor. Does Parallel Repetition Lower the Error in Computationally Sound Protocols? In *38th FOCS*, pages 374–383, 1997.
2. G. Brassard, D. Chaum and C. Crépeau. Minimum Disclosure Proofs of Knowledge. *JCSS*, Vol. 37, No. 2, pages 156–189, 1988.
3. G. Brassard, C. Crépeau and M. Yung. Constant-Round Perfect Zero-Knowledge Computationally Convincing Protocols. *Theoret. Comput. Sci.* , Vol. 84, pp. 23-52, 1991.
4. R. Canetti, O. Goldreich, S. Goldwasser, and S. Micali. Resettable Zero-Knowledge. In *32nd STOC*, 2000.
5. I. Damgard. Efficient Concurrent Zero-Knowledge in the Auxiliary String Model. In *EuroCrypt2000*.
6. C. Dwork, M. Naor, and A. Sahai. Concurrent Zero-Knowledge. In *30th STOC*, pages 409–418, 1998.
7. C. Dwork, and A. Sahai. Concurrent Zero-Knowledge: Reducing the Need for Timing Constraints. In *Crypto98*, Springer LNCS 1462 , pages 442–457, 1998.
8. U. Feige and A. Shamir. Witness Indistinguishability and Witness Hiding Protocols. In *22nd STOC*, pages 416–426, 1990.
9. O. Goldreich. Foundations of Cryptography – Fragments of a Book. Available from http://theory.lcs.mit.edu/~oded/frag.html.
10. O. Goldreich and A. Kahan. How to Construct Constant-Round Zero-Knowledge Proof Systems for NP. *Jour. of Cryptology*, Vol. 9, No. 2, pages 167–189, 1996.
11. O. Goldreich and H. Krawczyk. On the Composition of Zero-Knowledge Proof Systems. *SIAM J. Computing*, Vol. 25, No. 1, pages 169–192, 1996.
12. O. Goldreich, S. Micali and A. Wigderson. Proofs that Yield Nothing But Their Validity or All Languages in NP Have Zero-Knowledge Proof Systems. *JACM*, Vol. 38, No. 1, pp. 691–729, 1991.
13. O. Goldreich and Y. Oren. Definitions and Properties of Zero-Knowledge Proof Systems. *Jour. of Cryptology*, Vol. 7, No. 1, pages 1–32, 1994.
14. S. Goldwasser, S. Micali and C. Rackoff. The Knowledge Complexity of Interactive Proof Systems. *SIAM J. Comput.*, Vol. 18, No. 1, pp. 186–208, 1989.
15. S. Hada and T. Tanaka. On the Existence of 3-Round Zero-Knowledge Protocols. In *Crypto98*, Springer LNCS 1462, pages 408–423, 1998.
16. J. Hastad, R. Impagliazzo, L.A. Levin and M. Luby. Construction of Pseudorandom Generator from any One-Way Function. *SIAM Jour. on Computing*, Vol. 28 (4), pages 1364–1396, 1999.
17. J. Kilian and E. Petrank. Concurrent Zero-Knowledge in Poly-logarithmic Rounds. In Cryptology ePrint Archive: Report 2000/013. Available from http://eprint.iacr.org/2000/013
18. J. Kilian, E. Petrank, and C. Rackoff. Lower Bounds for Zero-Knowledge on the Internet. In *39th FOCS*, pages 484–492, 1998.
19. M. Naor. Bit Commitment using Pseudorandomness. *Jour. of Cryptology*, Vol. 4, pages 151–158, 1991.
20. R. Richardson and J. Kilian. On the Concurrent Composition of Zero-Knowledge Proofs. In *EuroCrypt99*, Springer LNCS 1592, pages 415–431, 1999.

# An Improved Pseudo-random Generator
# Based on Discrete Log

Rosario Gennaro

IBM T.J.Watson Research Center, P.O. Box 704, Yorktown Heights, NY 10598,
rosario@watson.ibm.com

**Abstract.** Under the assumption that solving the discrete logarithm problem modulo an $n$-bit prime $p$ is hard even when the exponent is a small $c$-bit number, we construct a new and improved pseudo-random bit generator. This new generator outputs $n - c - 1$ bits per exponentiation with a $c$-bit exponent.

Using typical parameters, $n = 1024$ and $c = 160$, this yields roughly 860 pseudo-random bits per small exponentiations. Using an implementation with quite small precomputation tables, this yields a rate of more than 20 bits per modular multiplication, thus much faster than the the squaring (BBS) generator with similar parameters.

## 1 Introduction

Many (if not all) cryptographic algorithms rely on the availability of truly random bits. However perfect randomness is a scarce resource. Fortunately for almost all cryptographic applications, it is sufficient to use pseudo-random bits, i.e. sources of randomness that "look" sufficiently random to the adversary.

This notion can be made more formal. The concept of cryptographically strong pseudo-random bit generators (PRBG) was introduced in papers by Blum and Micali [4] and Yao [20]. Informally a PRBG is cryptographically strong if it passes all polynomial-time statistical tests or, in other words, if the distribution of sequences output by the generator cannot be distinguished from truly random sequences by any polynomial-time judge.

Blum and Micali [4] presented the first cryptographically strong PRBG under the assumption that modular exponentiation modulo a prime $p$ is a one-way function. This breakthrough result was followed by a series of papers that culminated in [8] where it was shown that secure PRBGs exists if any one-way function does.

To extract a single pseudo-random bit, the Blum-Micali generator requires a full modular exponentiation in $Z_p^*$. This was improved by Long and Wigderson [14] and Peralta [17], who showed that up to $O(\log \log p)$ bits could be extracted by a single iteration (i.e. a modular exponentiation) of the Blum-Micali generator. Håstad *et al.* [10] show that if one considers discrete-log modulo a composite then almost $n/2$ pseudo-random bits can be extracted per modular exponentiation.

M. Bellare (Ed.): CRYPTO 2000, LNCS 1880, pp. 469–481, 2000.
© Springer-Verlag Berlin Heidelberg 2000

Better efficiency can be gained by looking at the quadratic residuosity problem in $Z_N^*$ where $N$ is a Blum integer (i.e. product of two primes of identical size and both $\equiv 3 \bmod 4$.) Under this assumption, Blum *et al.* [3] construct a secure PRBG for which each iteration consists of a single squaring in $Z_N^*$ and outputs a pseudo-random bit. Alexi *et al.* [2] showed that one can improve this to $O(\log \log N)$ bits and rely only the intractability of factoring as the underlying assumption. Up to this date, this is the most efficient provably secure PRBG.

In [16] Patel and Sundaram propose a very interesting variation on the Blum-Micali generator. They showed that if solving the discrete log problem modulo an $n$-bit prime $p$ is hard even when the exponent is small (say only $c$ bits long with $c < n$) then it is possible to extract up to $n - c - 1$ bits from one iteration of the Blum-Micali generator. However the iterated function of the generator itself remains the same, which means that one gets $n - c - 1$ bits per full modular exponentiations. Patel and Sundaram left open the question if it was possible to modify their generator so that each iteration consisted of an exponentiation with a small $c$-bit exponent. We answer their question in the affirmative.

OUR CONTRIBUTION. In this paper we show that it is possible to construct a high-rate discrete-log based secure PRBG. Under the same assumption introduced in [16] we present a generator that outputs $n - c - 1$ bits per iteration, which consists of a single exponentiation with a $c$-bit exponent.

The basic idea of the new scheme is to show that if the function $f : \{0,1\}^c \longrightarrow Z_p^*$ defined as $f(x) = g^x \bmod p$ is is a one-way function then it also has also strong pseudo-randomness properties over $Z_p^*$. In particular it is possible to think of it as pseudo-random generator itself. By iterating the above function and outputting the appropriate bits, we obtain an efficient pseudo-random bit generator.

Another attractive feature of this generator (which is shared by the Blum-Micali and Patel-Sundaram generators as well) is that all the exponentiations are computed over a fixed basis, and thus precomputation tables can be used to speed them up.

Using typical parameters $n = 1024$ and $c = 160$ we obtain roughly 860 pseudo-random bits per 160-bit exponent exponentiations. Using the precomputation scheme proposed in [13] one can show that such exponentiation will cost on average roughly 40 multiplications, using a table of only 12 Kbytes. Thus we obtain a rate of more than 21 pseudo-random bits per modular multiplication. Different tradeoffs between memory and efficiency can be obtained.

## 2   Preliminaries

In this section we summarize notations, definitions and prior work which is relevant to our result. In the following we denote with $\{0,1\}^n$ the set of $n$-bit strings. If $x \in \{0,1\}^n$ then we write $x = x_n x_{n-1} \ldots x_1$ where each $x_i \in \{0,1\}$. If we think of $x$ as an integer then we have $x = \sum_i x_i 2^{i-1}$ (that is $x_n$ is the most significant bit). With $R_n$ we denote the uniform distribution over $\{0,1\}^n$.

## 2.1   Pseudo-random Number Generators

Let $X_n, Y_n$ be two arbitrary probability ensembles over $\{0,1\}^n$. In the following we denote with $x \leftarrow X_n$ the selection of an element $x$ in $\{0,1\}^n$ according to the distribution $X_n$.

We say that $X_n$ and $Y_n$ have *statistical distance* bounded by $\Delta(n)$ if the following holds:

$$\sum_{x \in \{0,1\}^n} |Prob_{X_n}[x] - Prob_{Y_n}[x]| \leq \Delta(n)$$

We say that $X_n$ and $Y_n$ are *statistically indistinguishable* if for every polynomial $P(\cdot)$ and for sufficiently large $n$ we have that

$$\Delta(n) \leq \frac{1}{P(n)}$$

We say that $X_n$ and $Y_n$ are *computationally indistinguishable* (a concept introduced in [7]) if any polynomial time machine cannot distinguish between samples drawn according to $X_n$ or according to $Y_n$. More formally:

**Definition 1.** *Let $X_n, Y_n$ be two families of probability distributions over $\{0,1\}^n$. Given a Turing machine $\mathcal{D}$ consider the following quantities*

$$\delta_{\mathcal{D}, X_n} = Prob[x \leftarrow X_n \; ; \; \mathcal{D}(x) = 1]$$

$$\delta_{\mathcal{D}, Y_n} = Prob[y \leftarrow Y_n \; ; \; \mathcal{D}(y) = 1]$$

*We say that $X_n$ and $Y_n$ are* computationally indistinguishable *if for every probabilistic polynomial time $\mathcal{D}$, for every polynomial $P(\cdot)$, and for sufficiently large $n$ we have that*

$$|\delta_{\mathcal{D}, X_n} - \delta_{\mathcal{D}, Y_n}| \leq \frac{1}{P(n)}$$

We now move to define pseudo-random number generators [4,20]. There are several equivalent definitions, but the following one is sufficient for our purposes. Consider a family of functions

$$\mathsf{G}_n : \{0,1\}^{k_n} \longrightarrow \{0,1\}^n$$

where $k_n < n$. $\mathsf{G}_n$ induces a family of probability distributions (which we denote with $G_n$) over $\{0,1\}^n$ as follows

$$Prob_{G_n}[y] = Prob[y = \mathsf{G}_n(s) \; ; \; s \leftarrow R_{k_n}]$$

**Definition 2.** *We say that $\mathsf{G}_n$ is a* cryptographically strong pseudo-random bit generator *if the function $\mathsf{G}_n$ can be computed in polynomial time and the two families of probability distributions $R_n$ and $G_n$ are computationally indistinguishable.*

The input of a pseudo-random generator is usually called the *seed*.

## 2.2  Pseudo-randomness over Arbitrary Sets

Let $A_n$ be a family of sets such that for each $n$ we have $2^{n-1} \le |A_n| < 2^n$ (i.e. we need $n$ bits to describe elements of $A_n$). We denote with $U_n$ the uniform distribution over $A_n$ . Also let $k_n$ be a sequence of numbers such that for each $n$, $k_n < n$. Consider a family of functions

$$\mathsf{AG}_n : \{0,1\}^{k_n} \longrightarrow A_n$$

$\mathsf{AG}_n$ induces a family of probability distributions (which we denote with $AG_n$) over $A_n$ as follows

$$Prob_{AG_n}[y] = Prob[y = \mathsf{AG}_n(s) \; ; \; s \leftarrow R_{k_n}]$$

**Definition 3.** *We say that* $\mathsf{AG}_n$ *is a* cryptographically strong pseudo-random generator *over* $A_n$ *if the function* $\mathsf{AG}_n$ *can be computed in polynomial time and the two families of probability distributions* $U_n$ *and* $AG_n$ *are computationally indistinguishable.*

A secure pseudo-random generator over $A_n$ is already useful for applications in which one needs pseudo-random elements of that domain. Indeed no adversary will be able to distinguish if $y \in A_n$ was truly sampled at random or if it was computed as $\mathsf{AG}_n(s)$ starting from a much shorter seed $s$. An example of this is to consider $A_n$ to be $Z_p^*$ for an $n$-bit prime number $p$. If our application requires pseudo-random elements of $Z_p^*$ then such a generator would be sufficient.

However as *bit* generators they may not be perfect, since if we look at the bits of an encoding of the elements of $A_n$, then their distribution may be biased. This however is not going to be a problem for us since we will use pseudo-random generators over arbitrary sets as a tool in the proof of our main pseudo-random *bit* generator.

## 2.3  The Discrete Logarithm Problem

Let $p$ be a prime. We denote with $n$ the binary length of $p$. It is well known that $Z_p^* = \{x : 1 \le x \le p-1\}$ is a cyclic group under multiplication mod$p$. Let $g$ be a generator of $Z_p^*$. Thus the function

$$f : Z_{p-1} \longrightarrow Z_p^*$$

$$f(x) = g^x \bmod p$$

is a permutation. The inverse of $f$ (called the *discrete logarithm* function) is conjectured to be a function hard to compute (the cryptographic relevance of this conjecture first appears in the seminal paper by Diffie and Hellman [5] on public-key cryptography). The best known algorithm to compute discrete logarithms is the so-called *index calculus* method [1] which however runs in time sub-exponential in $n$.

In some applications (like the one we are going to describe in this paper) it is important to speed up the computation of the function $f(x) = g^x$. One possible way to do this is to restrict its input to small values of $x$. Let $c$ be a integer which we can think as depending on $n$ ($c = c(n)$). Assume now that we are given $y = g^x \bmod p$ with $x \leq 2^c$. It appears to be reasonable to assume that computing the discrete logarithm of $y$ is still hard even if we know that $x \leq 2^c$. Indeed the running time of the index-calculus method depends only on the size $n$ of the whole group. Depending on the size of $c$, different methods may actually be more efficient. Indeed the so-called *baby-step giant-step* algorithm by Shanks [12] or the *rho* algorithm by Pollard [18] can compute the discrete log of $y$ in $O(2^{c/2})$ time.

Thus if we set $c = \omega(\log n)$, there are no known polynomial time algorithms that can compute the discrete log of $y = g^x \bmod p$ when $x \leq 2^c$. In [16] it is explicitly assumed that *no* such efficient algorithm can exist. This is called the *Discrete Logarithm with Short c-Bit Exponents (c-DLSE)* Assumption and we will adopt it as the basis of our results as well.

**Assumption 1 (c-DLSE [16])** *Let $PRIMES(n)$ be the set of n-bit primes and let $c$ be a quantity that grows faster than $\log n$ (i.e. $c = \omega(\log n)$). For every probabilistic polynomial time Turing machine $\mathcal{I}$, for every polynomial $P(\cdot)$ and for sufficiently large $n$ we have that*

$$\Pr \begin{bmatrix} p \leftarrow PRIMES(n); \\ x \leftarrow R_c; \\ \mathcal{I}(p, g, g^x, c) = x \end{bmatrix} \leq \frac{1}{P(n)}$$

This assumption is somewhat supported by a result by Schnorr [19] who proves that no *generic* algorithm can compute $c$-bits discrete logarithms in less than $2^{c/2}$ generic steps. A generic algorithm is restricted to only perform group operations and cannot take advantage of specific properties of the encoding of group elements.

In practice, given today's computing power and discrete-log computing algorithms, it seems to be sufficient to set $n = 1024$ and $c = 160$. This implies a "security level" of $2^{80}$ (intended as work needed in order to "break" 160-DLSE).

## 2.4   Hard Bits for Discrete Logarithm

The function $f(x) = g^x \bmod p$ is widely considered to be one-way (i.e. a function easy to compute but not to invert). It is well known that even if $f$ is a one-way function, it does not hide all information about its preimages. For the specific case of the discrete logarithm, it is well known that given $y = g^x \bmod p$ it is easy to guess the least significant bit of $x \in Z_{p-1}$ by testing to see if $y$ is a quadratic residue or not in $Z_p^*$ (there is a polynomial-time test to determine that).

A Boolean predicate $\Pi$ is said to be *hard* for a one-way function $f$ if any algorithm $\mathcal{A}$ that given $y = f(x)$ guesses $\Pi(x)$ with probability substantially better than $1/2$, can be used to build another algorithm $\mathcal{A}'$ that on input $y$ computes $x$ with non-negligible probability.

Blum and Micali in [4] prove that the predicate

$$\Pi : Z_{p-1} \longrightarrow \{0,1\}$$

$$\Pi(x) = (x \leq \frac{p-1}{2})$$

is hard for the discrete logarithm function. Recently Håstad and Näslund [9] proved that every bit of the binary representation of $x$ (except the least significant one) is hard for the discrete log function.

In terms of *simultaneous* security of several bits, Long and Wigderson [14] and Peralta [17] showed that there are $O(\log \log p)$ predicates which are *simultaneously hard* for discrete log. Simultaneously hard means that the whole collection of bits looks "random" even when given $y = g^x$. A way to formalize this (following [20]) is to say that it is not possible to guess the value of the $j^{th}$ predicate even after seeing $g^x$ and the value of the previous $j-1$ predicates over $x$. Formally: there exists $O(\log \log p)$ Boolean predicates

$$\Pi_i : Z_{p-1} \longrightarrow \{0,1\} \quad \text{for } i = 1, \ldots, O(\log \log p)$$

such that for every $1 \leq j \leq O(\log \log p)$, if there exists a probabilistic polynomial-time algorithm $\mathcal{A}$ and a polynomial $P(\cdot)$ such that

$$Prob[x \leftarrow Z_{p-1} \; ; \; \mathcal{A}(g^x, \Pi_1(x), \ldots, \Pi_{j-1}(x)) = \Pi_j(x)] \geq \frac{1}{2} + \frac{1}{P(n)}$$

then there exists a probabilistic polynomial time algorithm $\mathcal{A}'$ which on input $g^x$ computes $x$ with non-negligible probability.

## 2.5   The Patel-Sundaram Generator

Let $p$ be a $n$-bit prime such that $p \equiv 3 \bmod 4$ and $g$ a generator of $Z_p^*$. Denote with $c$ a quantity that grows faster than $\log n$, i.e. $c = \omega(\log n)$.

In [16] Patel and Sundaram prove that under the $c$-DLSE Assumption the bits $x_2, x_3, \ldots, x_{n-c}$ are simultaneously hard for the function $f(x) = g^x \bmod p$. More formally[1]:

**Theorem 1 ([16]).** *For sufficiently large $n$, if $p$ is a $n$-bit prime such that $p \equiv 3 \bmod 4$ and if the c-DLSE Assumption holds, then for every $j$, $2 \leq j \leq n-c$, for every polynomial time Turing machine $\mathcal{A}$, for every polynomial $P(\cdot)$ and for sufficiently large $n$ we have that*

$$|Prob[x \leftarrow Z_{p-1} \; ; \; \mathcal{A}(g^x, x_2, \ldots, x_{j-1}) = x_j] - \frac{1}{2}| \leq \frac{1}{P(n)}$$

---

[1]   We point out that [16] requires that $p$ be a *safe* prime, i.e. such that $(p-1)/2$ is also a prime; but a close look at their proof reveals that $p \equiv 3 \bmod 4$ suffices.

We refer the reader to [16] for a proof of this Theorem.

Theorem 1 immediately yields a secure PRBG. Start with $x^{(0)} \in_R Z_{p-1}$. Set $x^{(i)} = g^{x^{(i-1)}} \mod p$. Set also $r^{(i)} = x_2^{(i)}, x_3^{(i)}, \ldots, x_{n-c}^{(i)}$. The output of the generator will be $r^{(0)}, r^{(1)}, \ldots, r^{(k)}$ where $k$ is the number of iterations.

Notice that this generator outputs $n - c - 1$ pseudo-random bits at the cost of a modular exponentiation with a random $n$-bit exponent.

## 3   Our New Generator

We now show that under the DLSE Assumption it is possible to construct a PRBG which is much faster than the Patel-Sundaram one. In order to do this we first revisit the construction of Patel and Sundaram to show how one can obtain a pseudo-random generator over $Z_p^* \times \{0,1\}^{n-c-1}$.

Then we construct a function from $Z_{p-1}$ to $Z_p^*$ which induces a pseudo-random distribution over $Z_p^*$. The proof of this fact is by reduction to the security of the modified Patel-Sundaram generator. This function is not a generator yet, since it does not stretch its input.

We finally show how to obtain a pseudo-random *bit* generator, by iterating the above function and outputting the appropriate bits.

### 3.1   The Patel-Sundaram Generator Revisited

As usual let $p$ be a $n$-bit prime, $p \equiv 3 \mod 4$, and $c = \omega(\log n)$. Consider the following function (which we call PSG for Patel-Sundaramam Generator):

$$\mathsf{PSG}_{n,c} : Z_{p-1} \longrightarrow Z_p^* \times \{0,1\}^{n-c-1}$$

$$\mathsf{PSG}_{n,c}(x) = (g^x \mod p, x_2, \ldots, x_{n-c})$$

That is, on input a random seed $x \in Z_{p-1}$, the generator outputs $g^x$ and $n-c-1$ consecutive bits of $x$, starting from the second least significant.

An immediate consequence of the result in [16], is that under the $c$-DLSE assumption $\mathsf{PSG}_{n,c}$ is a secure pseudo-random generator over the set $Z_p^* \times \{0,1\}^{n-c-1}$. More formally, if $U_n$ is the uniform distribution over $Z_p^*$, then the distribution induced by $\mathsf{PSG}_{n,c}$ over $Z_p^* \times \{0,1\}^{n-c-1}$ is computationally indistinguishable from the distribution $U_n \times R_{n-c-1}$.

In other words, for any probabilistic polynomial time Turing machine $\mathcal{D}$, we can define

$$\delta_{\mathcal{D}, UR_n} = Prob[y \leftarrow Z_p^* ; r \leftarrow R_{n-c-1} ; \mathcal{D}(y, r) = 1]$$

$$\delta_{\mathcal{D}, PSG_{n,c}} = Prob[x \leftarrow Z_{p-1} ; \mathcal{D}(\mathsf{PSG}_{n,c}(x)) = 1]$$

then for any polynomial $P(\cdot)$ and for sufficiently large $n$, we have that

$$|\delta_{\mathcal{D}, UR_n} - \delta_{\mathcal{D}, PSG_{n,c}}| \leq \frac{1}{P(n)}$$

In the next section we show our new generator and we prove that if it is not secure that we can show the existence of a distinguisher $\mathcal{D}$ that contradicts the above.

## 3.2    A Preliminary Lemma

We also assume that $p$ is a $n$-bit prime, $p \equiv 3 \bmod 4$ and $c = \omega(\log n)$. Let $g$ be a generator of $Z_p^*$ and denote with $\hat{g} = g^{2^{n-c}} \bmod p$. Recall that if $s$ is an integer we denote with $s_i$ the $i^{th}$-bit in its binary representation.

The function we consider is the following.

$$\mathsf{RG}_{n,c} : Z_{p-1} \longrightarrow Z_p^*$$

$$\mathsf{RG}_{n,c}(s) = \hat{g}^{(s \text{ div } 2^{n-c})} g^{s_1} \bmod p$$

That is we consider modular exponentiation in $Z_p^*$ with base $g$, but only after zeroing the bits in positions $2, \ldots, n - c$ of the input $s$ (these bits are basically ignored).

The function $\mathsf{RG}$ induces a distribution over $Z_p^*$ in the usual way. We denote it with $RG_{n,c}$ the following probability distribution over $Z_p^*$

$$Prob_{RG_{n,c}}[y] = Prob[y = \mathsf{RG}_{n,c}(s) \; ; \; s \leftarrow Z_{p-1}]$$

The following Lemma states that the distribution $RG_{n,c}$ is computationally indistinguishable from the uniform distribution over $Z_p^*$ if the $c$-DLSE assumption holds.

**Lemma 1.** *Let $p$ be a $n$-bit prime, with $p \equiv 3 \bmod 4$ and let $U_n$ be the uniform distribution over $Z_p^*$. If the $c$-DLSE Assumption holds, then the two distributions $U_n$ and $RG_{n,c}$ are computationally indistinguishable (see Definition 1).*

The proof of the Lemma goes by contradiction. We show that if $RG_{n,c}$ can be distinguished from $U_n$, then the modified Patel-Sundaram generator $\mathsf{PSG}$ is not secure. We do this by showing that any efficient distinguisher between $RG_{n,c}$ and the uniform distribution over $Z_p^*$ can be transformed into a distinguisher for $\mathsf{PSG}_{n,c}$. This will contradict Theorem 1 and ultimately the $c$-DLSE Assumption.

**Sketch of Proof**    Assume for the sake of contradiction that there exists a distinguisher $\mathcal{D}$ and a polynomial $P(\cdot)$ such that for infinitely many $n$'s we have that

$$\delta_{\mathcal{D},U_n} - \delta_{\mathcal{D},RG_{n,c}} \geq \frac{1}{P(n)}$$

where

$$\delta_{\mathcal{D},U_n} = Prob[x \leftarrow Z_p^* \; ; \; \mathcal{D}(p,g,x,c) = 1]$$

$$\delta_{\mathcal{D},RG_{n,c}} = Prob[s \leftarrow Z_{p-1} \; ; \; \mathcal{D}(p,g,\mathsf{RG}_{n,c}(s),c) = 1]$$

We show how to construct a distinguisher $\hat{\mathcal{D}}$ that "breaks" $\mathsf{PSG}$.

In order to break $\mathsf{PSG}_{n,c}$ we are given as input $(p,g,y,r,c)$ with $y \in Z_p^*$ and $r \in \{0,1\}^{n-c-1}$ and we want to guess if it comes from the distribution $U_n \times R_{n-c-1}$ or from the distribution $PSG_{n,c}$ of outputs of the generator $\mathsf{PSG}_{n,c}$. The distinguisher $\hat{\mathcal{D}}$ will follow this algorithm:

1. Consider the integer $z := r \circ 0$ where $\circ$ means concatenation. Set $w := yg^{-z} \bmod p$;
2. Output $\mathcal{D}(p, g, w, c)$

Why does this work? Assume that $(y, r)$ was drawn according to $\mathsf{PSG}_{n,c}(x)$ for some random $x \in Z_{p-1}$. Then $w = g^u$ where $u = 2^{n-c}(x \operatorname{div} 2^{n-c}) + x_1 \bmod p - 1$. That is, the discrete log of $w$ in base $g$ has the $n - c - 1$ bits in position $2, \ldots, n - c$ equal to 0 (this is because $r$ is identical to those $n - c - 1$ bits of the discrete log of $y$ by the assumption that $(y, r)$ follows the $PSG_{n,c}$ distribution). Thus once we set $\hat{g} = g^{2^{n-c}}$ we get $w = \hat{g}^{x \operatorname{div} 2^{n-c}} g^{x_1} \bmod p$, i.e. $w = \mathsf{RG}_{n,c}(x)$. Thus if $(y, r)$ is drawn according to $PSG_n$ then $w$ follows the same distribution as $RG_n$.

On the other hand if $(y, r)$ was drawn with $y$ randomly chosen in $Z_p^*$ and $r$ randomly chosen in $\{0, 1\}^{n-c-1}$, then all we know is that $w$ is a random element of $Z_p^*$.

Thus $\hat{\mathcal{D}}$ will guess the correct distribution with the same advantage as $\mathcal{D}$ does. Which contradicts the security of the PSG generator.     $\square$

## 3.3 The New Generator

It is now straightforward to construct the new generator. The algorithm receives as a seed a random element $s$ in $Z_{p-1}$ and then it iterates the function RG on it. The pseudo-random bits outputted by the generator are the bits ignored by the function RG. The output of the function RG will serve as the new input for the next iteration.

More in detail, the algorithm $\mathsf{IRG}_{n,c}$ (for Iterated-RG generator) works as follows. Start with $x^{(0)} \in_R Z_{p-1}$. Set $x^{(i)} = \mathsf{RG}_{n,c}(x^{(i-1)})$. Set also $r^{(i)} = x_2^{(i)}, x_3^{(i)}, \ldots, x_{n-c}^{(i)}$. The output of the generator will be $r^{(0)}, r^{(1)}, \ldots, r^{(k)}$ where $k$ is the number of iterations (chosen such that $k = poly(n)$ and $k(n - c - 1) > n$).

Notice that this generator outputs $n - c - 1$ pseudo-random bits at the cost of a modular exponentiation with a random $c$-bit exponent (i.e. the cost of the computation of the function RG).

**Theorem 2.** *Under the c-DLSE Assumption, $\mathsf{IRG}_{n,c}$ is a secure pseudo-random bit generator (see Definition 2).*

**Sketch of Proof**     We first notice that, for sufficiently large $n$, $r^{(0)}$ is an almost uniformly distributed $(n - c - 1)$-bit string. This is because $r^{(0)}$ is composed of the bits in position $2, 3, \ldots, n - c$ of a random element of $Z_{p-1}$ and thus their bias is bounded by $2^{-c}$ (i.e. the statistical distance between the distribution of $r^{(0)}$ and the uniform distribution over $\{0, 1\}^{n-c-1}$ is bounded by $2^{-c}$).

Now by virtue of Lemma 1 we know that all the values $x^{(i)}$ follow a distribution which is computationally indistinguishable from the uniform one on $Z_p^*$. By the same argument as above it follows that all the $r^{(i)}$ must follow a distribution which is computationally indistinguishable from $R_{n-c-1}$.

More formally, the proof follows a hybrid argument. If there is a distinguisher $\mathcal{D}$ between the distribution induced by $\mathsf{IRG}_{n,c}$ and the distribution $R_{k(n-c-1)}$, then for a specific index $i$ we must have a distinguisher $\mathcal{D}_1$ between the distribution followed by $r^{(i)}$ and the uniform distribution $R_{n-c-1}$. Now that implies that it is possible to distinguish the distribution followed by $x^{(i)}$ and the uniform distribution over $Z_p^*$ (just take the bits in position $2, 3, \ldots, n - c$ of the input and pass them to $\mathcal{D}_2$). This contradicts Lemma 1 and ultimately the $c$-DLSE Assumption. $\qquad\square$

## 4   Efficiency Analysis

Our new generator is very efficient. It outputs $n - c - 1$ pseudo-random bits at the cost of a modular exponentiation with a random $c$-bit exponent, or roughly $1.5c$ modular multiplications in $Z_p^*$. Compare this with the Patel-Sundaram generator where the same number of pseudo-random bits would cost $1.5n$ modular multiplications. Moreover the security of our scheme is tightly related to the security of the Patel-Sundaram one, since the reduction from our scheme to theirs is quite immediate.

So far we have discussed security in asymptotic terms. If we want to instantiate practical parameters we need to analyze more closely the *concrete* security of the proposed scheme.

A close look at the proof of security in [16] shows the following. If we assume that Theorem 1 fails, i.e. that for some $j$, $2 \le j \le n-c$, there exists an algorithm $\mathcal{A}$ which runs in time $T(n)$, and a polynomial $P(\cdot)$ such that w.l.o.g.

$$Prob[x \leftarrow Z_{p-1} \; ; \; \mathcal{A}(g^x, x_2, \ldots, x_{j-1}) = x_j] > \frac{1}{2} + \frac{1}{P(n)}$$

then we have an algorithm $\mathcal{I}^{\mathcal{A}}$ to break $c$-DLSE which runs in time $O((n - c)cP^2(n)T(n))$ if $2 \le j < n - c - \log P(n)$ and in time $O((n - c)cP^3(n)T(n))$ if $n - c - \log P(n) \le j \le n - c$ (the hidden constant is very small). This is a very crude analysis of the efficiency of the reduction in [16] and it is quite possible to improve on it (details in the final paper).

In order to be able to say that the PRBG is secure we need to make sure that the complexity of this reduction is smaller than the time to break $c$-DLSE with the best known algorithm (which we know today is $2^{c/2}$).

COMPARISON WITH THE BBS GENERATOR. The BBS generator was introduced by Blum *et al.* in [3] under the assumption that deciding quadratic residuosity modulo a composite is hard. The generator works by repeatedly squaring mod $N$ a random seed in $Z_N^*$ where $N$ is a Blum integer ($N = PQ$ with $P, Q$ both primes of identical size and $\equiv 3 \bmod 4$.) At each iteration it outputs the least significant bit of the current value. The rate of this generator is thus of 1 bit/squaring. In [2], Alexi *et al.* showed that one can output up to $k = O(\log \log N)$ bits per iteration of the squaring generator (and this while also relaxing the underlying assumption to the hardness of factoring). The actual number $k$ of bits that can be outputted depends on the concrete parameters adopted.

The [2] reduction is not very tight and it was recently improved by Fischlin and Schnorr in [6]. The complexity of the reduction quoted there is

$$O(n \log n P^2(n) T(n) + n^2 P^4(n) \log n)$$

(here $P(n), T(n)$ refers to a machine which guesses the next bit in one iteration of the BBS generator in time $T(n)$ and with advantage $1/P(n)$).

If we want to output $k$ bits per iteration, the complexity grows by a factor of $2^{2k}$ and the reduction quickly becomes more expensive than known factoring algorithms. Notice instead that the reduction in [16] (and thus in ours) depends only linearly on the number of bits outputted.

CONCRETE PARAMETERS. Let's fix $n = 1024$ and $c = 160$. With these parameters we can safely assume that the complexity of the best known algorithms to break $c$-DLSE [1,12,17] is beyond the reach of today's computing capabilities.

For moduli of size $n = 1024$, the results in [6] seem to indicate that in practice, for $n = 1024$ we can output around 4 bits per iteration of the BBS generator, if we want to rule out adversaries which run in time $T = 1$ MIPS-year (roughly $10^{13}$ instructions) and predict the next bit with advantage $1/100$ (which is quite high). This yields a rate of 4 bits per modular squaring.

Using the same indicative parameters suggested in [6], we can see that we can safely output all $n - c - 1 \approx 860$ bits in one iteration of the Patel-Sundaram generator. Since the security of our scheme is basically the same as their generator we can also output all 860 bits in our scheme as well.

Thus we obtain 860 bits at the cost of roughly 240 multiplications, which yields a rate of about 3.5 bits per modular multiplication. Thus the basic implementation of our scheme has efficiency comparable to the BBS generator. In the next section we show how to improve on this, by using precomputation tables.

## 4.1   Using Precomputed Tables

The most expensive part of the computation of our generator is to compute $\hat{g}^s \bmod p$ where $s$ is a $c$-bit value.

We can take advantage of the fact that in our generator[2] the modular exponentiations are all computed over the same basis $\hat{g}$. This feature allows us to precompute powers of $\hat{g}$ and store them in a table, and then use this values to compute fastly $\hat{g}^s$ for any $s$.

The simplest approach is to precompute a table $T$

$$T = \{\hat{g}^{2^i} \bmod p \; ; \; i = 0, \dots, c\}$$

Now, one exponentiation with base $\hat{g}$ and a random $c$-bit exponent can be computed using only $.5c$ multiplications on average. The cost is an increase to $O(cn)$

---

[2] As well as in the Patel-Sundaram one, or in the Blum-Micali one

bits of required memory. With this simple improvement one iteration of our generator will require roughly 80 multiplications, which yields a rate of more that 10 pseudo-random bits per multiplication. The size of the table is about 20 Kbytes.

Lim and Lee [13] present more flexible trade-offs between memory and computation time to compute exponentiations over a fixed basis. Their approach is applicable to our scheme as well. In short, the [13] precomputation scheme is governed by two parameters $h, v$. The storage requirement is $(2^h - 1)v$ elements of the field. The number of multiplications required to exponentiate to a $c$-bit exponent is $\lceil \frac{c}{h} \rceil + \lceil \frac{c}{hv} \rceil - 2$ in the worst case.

Using the choice of parameters for 160-bit exponents suggested in [13] we can get roughly 40 multiplications with a table of only 12 Kbytes. This yields a rate of more than 21 pseudo-random bits per multiplication. A large memory implementation (300 Kbytes) will yield a rate of roughly 43 pseudo-random bits per multiplication.

In the final version of this paper we will present a more complete concrete analysis of our new scheme.

## 5    Conclusions

In this paper we presented a secure pseudo-random bit generator whose efficiency is comparable to the squaring (BBS) generator. The security of our scheme is based on the assumption that solving discrete logarithms remains hard even when the exponent is small. This assumption was first used by Patel and Sundaram in [16]. Our construction however is much faster than theirs since it only uses exponentiations with small inputs.

An alternative way to look at our construction is the following. Under the $c$-DLSE assumption the function $f : \{0,1\}^c \longrightarrow Z_p^*$ defined as $f(x) = g^x$ is a one-way function. Our results indicate that $f$ has also strong pseudo-randomness properties over $Z_p^*$. In particular it is possible to think of it as pseudo-random generator itself. We are aware of only one other example in the literature of a one-way function with this properties, in [11] based on the hardness of subset-sum problems.

The DLSE Assumption is not as widely studied as the regular discrete log assumption so it needs to be handled with care. However it seems a reasonable assumption to make.

It would be nice to see if there are other cryptographic primitives that could benefit in efficiency from the adoption of stronger (but not unreasonable) number-theoretic assumptions. Examples of this are already present in the literature (e.g. the efficient construction of pseudo-random functions based on the Decisional Diffie-Hellman problem in [15].) It would be particularly interesting to see a pseudo-random bit generator that beats the rate of the squaring generator, even if at the cost of a stronger assumption on factoring or RSA.

## Acknowledgments

This paper owes much to the suggestions and advices of Shai Halevi. Thanks also to the other members of the CRYPTO committee for their suggestions and to Dario Catalano for reading early drafts of the paper.

## References

1. L. Adleman. *A Subexponential Algorithm for the Discrete Logarithm Problem with Applications to Cryptography*. IEEE FOCS, pp.55-60, 1979.
2. W. Alexi, B. Chor, O. Goldreich and C. Schnorr. *RSA and Rabin Functions: Certain Parts are as Hard as the Whole*. SIAM J. Computing, 17(2):194–209, April 1988.
3. L. Blum, M. Blum and M. Shub. *A Simple Únpredictable Pseudo-Random Number Generator*. SIAM J.Computing, 15(2):364–383, May 1986.
4. M. Blum and S. Micali. *How to Generate Cryptographically Strong Sequences of Pseudo-Random Bits*. SIAM J.Computing, 13(4):850–864, November 1984.
5. W. Diffie and M. Hellman. *New Directions in Cryptography*. IEEE Trans. Inf. Theory, IT-22:644–654, November 1976.
6. R. Fischlin and C. Schnorr. *Stronger Security Proofs for RSA and Rabin Bits*. J.Crypt., 13(2):221–244, Spring 2000.
7. S. Goldwasser and S. Micali. *Probabilistic Encryption*. JCSS, 28:270–299, 1988.
8. J. Håstad, R. Impagliazzo, L. Levin and M. Luby. *A Pseudo-Random Generator from any One-Way Function*. SIAM J.Computing, 28(4):1364-1396, 1999.
9. J. Håstad and M. Näslund. The Security of Individual RSA Bits. IEEE FOCS, pp.510–519, 1998.
10. J. Håstad, A. Schrift and A. Shamir. *The Discrete Logarithm Modulo a Composite Hides $O(n)$ Bits*. JCSS, 47:376-404, 1993.
11. R. Impagliazzo and M. Naor. *Efficient Cryptographic Schemes Provably as Secure as Subset Sum*. J.Crypt., 9(4):199–216, 1996.
12. D. Knuth. *The Art of Computer Programming (vol.3): Sorting and Searching*. Addison-Wesley, 1973.
13. C.H. Lim and P.J. Lee. *More Flexible Exponentiation with Precomputation*. CRYPTO'94, LNCS 839, pp.95–107.
14. D. Long and A. Wigderson. *The Discrete Log Hides $O(\log n)$ Bits*. SIAM J.Computing, 17:363–372, 1988.
15. M. Naor and O. Reingold. *Number-Theoretic Constructions of Efficient Pseudo-Random Functions*. IEEE FOCS, pp.458–467, 1997.
16. S. Patel and G. Sundaram. *An Efficient Discrete Log Pseudo Random Generator*. CRYPTO'98, LNCS 1462, pp.304–317, 1998.
17. R. Peralta. *Simultaneous Security of Bits in the Discrete Log*. EUROCRYPT'85, LNCS 219, pp.62–72, 1986.
18. J. Pollard. *Monte-Carlo Methods for Index Computation (mod p)*. Mathematics of Computation, 32(143):918–924, 1978.
19. C. Schnorr *Security of Allmost ALL Discrete Log Bits*. Electronic Colloquium on Computational Complexity. Report TR98-033. Available at `http://www.eccc.uni-trier.de/eccc/`.
20. A. Yao. *Theory and Applications of Trapdoor Functions*. IEEE FOCS, 1982.

# Linking Classical and Quantum Key Agreement: Is There "Bound Information"?

Nicolas Gisin[1] and Stefan Wolf[2]

[1] Group of Applied Physics, University of Geneva, CH-1211 Geneva, Switzerland.
`Nicolas.Gisin@physics.unige.ch`
[2] Department of Computer Science, ETH Zürich, CH-8092 Zürich, Switzerland.
`wolf@inf.ethz.ch`

**Abstract.** After carrying out a protocol for quantum key agreement over a noisy quantum channel, the parties Alice and Bob must process the raw key in order to end up with identical keys about which the adversary has virtually no information. In principle, both classical and quantum protocols can be used for this processing. It is a natural question which type of protocols is more powerful. We show that the limits of tolerable noise are identical for classical and quantum protocols in many cases. More specifically, we prove that a quantum state between two parties is entangled if and only if the classical random variables resulting from optimal measurements provide some mutual classical information between the parties. In addition, we present evidence which strongly suggests that the potentials of classical and of quantum protocols are equal in every situation. An important consequence, in the purely classical regime, of such a correspondence would be the existence of a classical counterpart of so-called bound entanglement, namely "bound information" that cannot be used for generating a secret key by any protocol. This stands in sharp contrast to what was previously believed.

**Keywords.** Secret-key agreement, intrinsic information, secret-key rate, quantum privacy amplification, purification, entanglement.

## 1 Introduction

In modern cryptography there are mainly two security paradigms, namely computational and information-theoretic security. The latter is sometimes also called unconditional security. Computational security is based on the assumed hardness of certain computational problems (e.g., the integer-factoring or discrete-logarithm problems). However, since a computationally sufficiently powerful adversary can solve any computational problem, hence break any such system, and because no useful general lower bounds are known in complexity theory, computational security is always conditional and, in addition to this, in danger by progress in the theory of efficient algorithms as well as in hardware engineering (e.g., quantum computing). Information-theoretic security on the other hand is based on probability theory and on the fact that an adversary's information is

M. Bellare (Ed.): CRYPTO 2000, LNCS 1880, pp. 482–500, 2000.
© Springer-Verlag Berlin Heidelberg 2000

limited. Such a limitation can for instance come from noise in communication channels or from the laws of quantum mechanics.

Many different settings based on noisy channels have been described and analyzed. Examples are Wyner's wire-tap channel [30], Csiszár and Körner's broadcast channel [7], or Maurer's model of key agreement from joint randomness [20], [22].

Quantum cryptography on the other hand lies in the intersection of two of the major scientific achievements of the 20th century, namely quantum physics and information theory. Various protocols for so-called quantum key agreement have been proposed (e.g., [3], [10]), and the possibility and impossibility of purification in different settings has been studied by many authors.

The goal of this paper is to derive parallels between classical and quantum key agreement and thus to show that the two paradigms are more closely related than previously recognized. These connections allow for investigating questions and solving open problems of purely classical information theory with quantum-mechanic methods. One of the possible consequences is that, in contrast to what was previously believed, there exists a classical counterpart to so-called *bound entanglement* (i.e., entanglement that cannot be purified by any quantum protocol), namely mutual information between Alice and Bob which they cannot use for generating a secret key by any classical protocol.

The outline of this paper is as follows. In Section 2 we introduce the classical (Section 2.1) and quantum (Section 2.2) models of information-theoretic key agreement and the crucial concepts and quantities, such as secret-key rate and intrinsic information on one side, and measurements, entanglement, and quantum privacy amplification on the other. In Section 3 we show the mentioned links between these two models, more precisely, between entanglement and intrinsic information (Section 3.1) as well as between quantum purification and the secret-key rate (Section 3.4). We illustrate the statements and their consequences with a number of examples (Sections 3.2 and 3.5). In Section 3.6 we define and characterize the classical counterpart of bound entanglement, called bound intrinsic information. Finally we show that not only problems in classical information theory can be addressed by quantum-mechanical methods, but that the inverse is also true: In Section 3.3 we propose a new measure for entanglement based on the intrinsic information measure.

# 2    Models of Information-Theoretically Secure Key Agreement

## 2.1    Key Agreement from Classical Information: Intrinsic Information and Secret-Key Rate

In this section we describe Maurer's general model of classical key agreement by public discussion from common information [20]. Here, two parties Alice and Bob who are willing to generate a secret key have access to repeated independent realizations of (classical) random variables $X$ and $Y$, respectively, whereas an adversary Eve learns the outcomes of a random variable $Z$. Let $P_{XYZ}$ be the joint distribution of the three random variables. In addition, Alice and Bob are connected by a noiseless and authentic but otherwise completely insecure channel. In this situation, the secret-key rate $S(X;Y\|Z)$ has been defined as the maximal rate at which Alice and Bob can generate a secret key that is equal for Alice and Bob with overwhelming probability and about which Eve has only a negligible amount of (Shannon) information. For a detailed discussion of the general scenario and the secret-key rate as well as for various bounds on $S(X;Y\|Z)$, see [20], [21], [22].

Bound (1) implies that if Bob's random variable $Y$ provides more information about Alice's $X$ than Eve's $Z$ does (or vice versa), then this advantage can be exploited for generating a secret key:

$$S(X;Y\|Z) \geq \max\left\{I(X;Y) - I(X;Z),\, I(Y;X) - I(Y;Z)\right\}. \tag{1}$$

This is a consequence of a result by Csiszár and Körner [7]. It is somewhat surprising that this bound is not tight, in particular, that secret-key agreement can even be possible when the right-hand side of (1) vanishes or is negative. However, the positivity of the expression on the right-hand side of (1) is a necessary and sufficient condition for the possibility of secret-key agreement by *one-way communication*: Whenever Alice and Bob start in a disadvantageous situation with respect to Eve, *feedback* is necessary. The corresponding initial phase of the key-agreement protocol is then often called *advantage distillation* [20], [29].

The following upper bound on $S(X;Y\|Z)$ is a generalization of Shannon's well-known impracticality theorem [28] and quantifies the intuitive fact that no information-theoretically secure key agreement is possible when Bob's information is independent from Alice's random variable, given Eve's information: $S(X;Y\|Z) \leq I(X;Y|Z)$. However, this bound is not tight. Because it is a possible strategy of the adversary Eve to process $Z$, i.e., to send $Z$ over some channel characterized by $P_{\overline{Z}|Z}$, we have for such a new random variable $\overline{Z}$ that $S(X;Y\|Z) \leq I(X;Y|\overline{Z})$, and hence

$$S(X;Y\|Z) \leq \min\nolimits_{P_{\overline{Z}|Z}}\{I(X;Y|\overline{Z})\} =: I(X;Y\downarrow Z) \tag{2}$$

holds. The quantity $I(X;Y\downarrow Z)$ has been called the *intrinsic conditional information between $X$ and $Y$ given $Z$* [22]. It was conjectured, and evidence supporting this belief was given, that $S(X;Y\|Z) > 0$ holds if $I(X;Y\downarrow Z) > 0$ does [22]. Some of the results below strongly suggest that this is true if one of the random variables $X$ and $Y$ is binary and the other one at most ternary, but false in general.

## 2.2   Quantum Key Agreement: Measurements, Entanglement, Purification

We assume that the reader is familiar with the basic quantum-theoretic concepts and notations. For an introduction, see for example [24].

In the context of quantum key agreement, the classical scenario $P_{XYZ}$ is replaced by a quantum state vector[1] $\Psi \in \mathcal{H}_A \otimes \mathcal{H}_B \otimes \mathcal{H}_E$, where $\mathcal{H}_A$, $\mathcal{H}_B$, and $\mathcal{H}_E$ are Hilbert spaces describing the systems in Alice's, Bob's, and Eve's hands, respectively. Then, measuring this quantum state by the three parties leads to a classical probability distribution. In the following, we assume that Eve is free to carry out so-called *generalized measurements* (POVMs) [24]. In other words, the set $\{|z\rangle\}$ will not be assumed to be an orthonormal basis, but any set generating the Hilbert space $\mathcal{H}_E$ and satisfying the condition $\sum_z |z\rangle\langle z| = \mathbb{1}_{\mathcal{H}_E}$. Then, if the three parties carry out measurements in certain (orthonormal) bases $\{|x\rangle\}$ and $\{|y\rangle\}$, and in the set $\{|z\rangle\}$, respectively, they end up with the classical scenario $P_{XYZ} = |\langle x,y,z|\Psi\rangle|^2$. Since this distribution depends on the chosen bases and set, a given quantum state $\Psi$ does *not uniquely* determine a classical scenario: some measurements may lead to scenarios useful for Alice and Bob, whereas for Eve, some others may.

The analog of Alice and Bob's marginal distribution $P_{XY}$ is the partial state $\rho_{AB}$, obtained by tracing over Eve's Hilbert space $\mathcal{H}_E$. More precisely, let $\Psi = \sum_{xyz} c_{xyz}|x,y,z\rangle$, where $|x,y,z\rangle$ is short for $|x\rangle \otimes |y\rangle \otimes |z\rangle$. We can write $\Psi = \sum_z \sqrt{P_Z(z)}\,\psi_z \otimes |z\rangle$, where $P_Z$ denotes Eve's marginal distribution of $P_{XYZ}$. Then $\rho_{AB} = \mathrm{Tr}_{\mathcal{H}_E}(P_\Psi) := \sum_z P_Z(z)P_{\psi_z}$, where $P_{\psi_z}$ is the projector to the state vector $\psi_z$.

An important property is that $\rho_{AB}$ is pure (i.e., $\rho_{AB}^2 = \rho_{AB}$) if and only if the global state $\Psi$ factorizes, i.e., $\Psi = \psi_{AB} \otimes \psi_E$, where $\psi_{AB} \in \mathcal{H}_A \otimes \mathcal{H}_B$ and $\psi_E \in \mathcal{H}_E$. In this case Alice and Bob are independent of Eve: Eve cannot obtain any information on Alice's and Bob's states by measuring her system.

After a measurement, Alice and Bob obtain a classical distribution $P_{XY}$. In accordance with Landauer's principle that all information is ultimately physical, the classical scenario arises from a physical process, namely the measurements performed. Thus the quantum state $\Psi$, and not the distribution $P_{XYZ}$, is the true primitive. Note that only if also Eve performs a measurement, $P_{XYZ}$ is at all defined. It is clear however that it might be advantageous (if technologically possible) for the adversary not to do any measurements before the public discussion. Because of this, staying in the quantum regime can simplify the analysis.

When Alice and Bob share many independent systems[2] $\rho_{AB}$, there are basically two possibilities for generating a secret key. Either they first measure their systems and then run a classical protocol (process classical information) secure against all measurements Eve could possibly perform (i.e., against all possible distributions $P_{XYZ}$ that can result after Eve's measurement). Or they first run a quantum protocol (i.e., process the information in the quantum domain) and then perform their measurements. The idea of quantum protocols is to process the systems in state $\rho_{AB}$ and to produce fewer systems in a pure state (i.e., to

---

[1] We consider pure states, since it is natural to assume that Eve controls all the environment outside Alice and Bob's systems.

[2] Here we do not consider the possibility that Eve coherently processes several of her systems. This corresponds to the assumption in the classical scenario that repeated realizations of $X$, $Y$, and $Z$ are independent of each other.

*purify* $\rho_{AB}$), thus to eliminate Eve from the scenario. Moreover, the pure state Alice and Bob end up with should be maximally entangled (i.e., even for some different and incompatible measurements, Alice's and Bob's results are perfectly correlated). Finally, Alice and Bob measure their maximally entangled systems and establish a secret key. This way of obtaining a key directly from a quantum state $\Psi$, without any error correction nor classical privacy amplification, is called *quantum privacy amplification*[3] (QPA for short) [8], [2]. Note that the procedure described in [8] and [2] guarantees that Eve's *relative* information (relative to the key length) is arbitrarily small, but not that her *absolute* information is negligible. The analog of this problem in the classical case is discussed in [21].

The precise conditions under which a general state $\rho_{AB}$ can be purified are not known. However, the two following conditions are necessary. First, the state must be *entangled* or, equivalently, *not separable*. A state $\rho_{AB}$ is separable if and only if it can be written as a mixture of product states, i.e., $\rho_{AB} = \sum_j p_j \rho_{Aj} \otimes \rho_{Bj}$. Separable states can be generated by purely classical communication, hence it follows from bound (2) that entanglement is a necessary condition. The second condition is more subtle: The matrix $\rho^t_{AB}$ obtained from $\rho_{AB}$ by *partial transposition* must have at least one negative eigenvalue [17], [16]. The partial transposition of the density matrix $\rho_{AB}$ is defined as $(\rho^t_{AB})_{i,j;\mu,\nu} := (\rho_{AB})_{i,\nu;\mu,j}$, where the indices $i$ and $\mu$ [$j$ and $\nu$] run through a basis of $\mathcal{H}_A$ [$\mathcal{H}_B$]. Note that this definition is base-dependent. However, the *eigenvalues* of $\rho^t_{AB}$ are not [25]. The second of these conditions implies the first one: Negative partial transposition (i.e., at least one eigenvalue is negative) implies entanglement.

In the binary case ($\mathcal{H}_A$ and $\mathcal{H}_B$ both have dimension two), the above two conditions are equivalent and sufficient for the possibility of quantum key agreement: all entangled binary states can be purified. The same even holds if one Hilbert space is of dimension 2 and the other one of dimension 3. However, for larger dimensions there are examples showing that these conditions are not equivalent: There are entangled states whose partial transpose has no negative eigenvalue, hence cannot be purified [17]. Such states are called *bound entangled*, in contrast to *free entangled* states, which can be purified. Moreover, it is believed that there even exist entangled states which cannot be purified although they have negative partial transposition [9].

# 3   Linking Classical and Quantum Key Agreement

In this section we derive a close connection between the possibilities offered by classical and quantum protocols for key agreement. The intuition is as follows. As described in Section 2.2, there is a very natural connection between quantum states $\Psi$ and classical distributions $P_{XYZ}$ which can be thought of as arising

---

[3]   The term "quantum privacy amplification" is somewhat unfortunate since it does not correspond to classical privacy amplification, but includes advantage distillation and error correction.

from $\Psi$ by measuring in a certain basis, e.g., the standard basis[4]. (Note however that the connection is not unique even for fixed bases: For a given distribution $P_{XYZ}$, there are many states $\Psi$ leading to $P_{XYZ}$ by carrying out measurements.) When given a state $\Psi$ between three parties Alice, Bob, and Eve, and if $\rho_{AB}$ denotes the resulting mixed state after Eve is traced out, then the corresponding classical distribution $P_{XYZ}$ has positive intrinsic information if and only if $\rho_{AB}$ is entangled. However, this correspondence clearly depends on the measurement bases used by Alice, Bob, and Eve. If for instance $\rho_{AB}$ is entangled, but Alice and Bob do very unclever measurements, then the intrinsic information may vanish. If on the other hand $\rho_{AB}$ is separable, Eve may do such bad measurements that the intrinsic information becomes positive, despite the fact that $\rho_{AB}$ could have been established by public discussion without any prior correlation (see Example 4). Consequently, the correspondence between intrinsic information and entanglement must involve some optimization over all possible measurements on all sides.

A similar correspondence on the protocol level is supported by many examples, but not rigorously proven: The distribution $P_{XYZ}$ allows for classical key agreement if and only if quantum key agreement is possible starting from the state $\rho_{AB}$.

We show how these parallels allow for addressing problems of purely classical information-theoretic nature with the methods of quantum information theory, and vice versa.

### 3.1  Entanglement and Intrinsic Information

Let us first establish the connection between intrinsic information and entanglement. Theorem 1 states that if $\rho_{AB}$ is separable, then Eve can "force" the information between Alice's and Bob's classical random variables (given Eve's classical random variable) to be zero (whatever strategy Alice and Bob use[5]). In particular, Eve can prevent classical key agreement.

**Theorem 1** *Let $\Psi \in \mathcal{H}_A \otimes \mathcal{H}_B \otimes \mathcal{H}_E$ and $\rho_{AB} = \mathrm{Tr}_{\mathcal{H}_E}(P_\Psi)$. If $\rho_{AB}$ is separable, then there exists a generating set $\{|z\rangle\}$ of $\mathcal{H}_{\mathcal{E}}$ such that for all bases $\{|x\rangle\}$ and $\{|y\rangle\}$ of $\mathcal{H}_{\mathcal{A}}$ and $\mathcal{H}_{\mathcal{B}}$, respectively, $I(X;Y|Z) = 0$ holds for $P_{XYZ}(x,y,z) := |\langle x,y,z|\Psi\rangle|^2$.*

*Proof.* If $\rho_{AB}$ is separable, then there exist vectors $|\alpha_z\rangle$ and $|\beta_z\rangle$ such that $\rho_{AB} = \sum_{z=1}^{n_z} p_z P_{\alpha_z} \otimes P_{\beta_z}$, where $P_{\alpha_z}$ denotes the one-dimensional projector onto the subspace spanned by $|\alpha_z\rangle$.

---

[4] A priori, there is no privileged basis. However, physicists often write states like $\rho_{AB}$ in a basis which seems to be more natural than others. We refer to this as the standard basis. Somewhat surprisingly, this basis is generally easy to identify, though not precisely defined. One could characterize the standard basis as the basis for which as many coefficients as possible of $\Psi$ are real and positive. We usually represent quantum states with respect to the standard basis.

[5] The statement of Theorem 1 also holds when Alice and Bob are allowed to do generalized measurements.

Let us first assume that $n_z \leq \dim \mathcal{H}_E$. Then there exists a basis $\{|z\rangle\}$ of $\mathcal{H}_E$ such that $\Psi = \sum_z \sqrt{p_z} \, |\alpha_z, \beta_z, z\rangle$ holds [23], [12], [19].

If $n_z > \dim \mathcal{H}_E$, then Eve can add an auxiliary system $\mathcal{H}_{aux}$ to hers (usually called an *ancilla*) and we have $\Psi \otimes |\gamma_0\rangle = \sum_z \sqrt{p_z} \, |\alpha_z, \beta_z, \gamma_z\rangle$, where $|\gamma_0\rangle \in \mathcal{H}_{aux}$ is the state of Eve's auxiliary system, and $\{|\gamma_z\rangle\}$ is a basis of $\mathcal{H}_E \otimes \mathcal{H}_{aux}$. We define the (not necessarily orthonormalized) vectors $|z\rangle$ by $|z, \gamma_0\rangle = \mathbf{1}_{\mathcal{H}_E} \otimes P_{\gamma_0} |\gamma_z\rangle$. These vectors determine a generalized measurement with positive operators $O_z = |z\rangle\langle z|$. Since $\sum_z O_z \otimes P_{\gamma_0} = \sum_z |z, \gamma_0\rangle\langle z, \gamma_0| = \sum_z \mathbf{1}_{\mathcal{H}_E} \otimes P_{\gamma_0} |\gamma_z\rangle\langle\gamma_z| \mathbf{1}_{\mathcal{H}_E} \otimes P_{\gamma_0} = \mathbf{1}_{\mathcal{H}_E} \otimes P_{\gamma_0}$, the $O_z$ satisfy $\sum_z O_z = \mathbf{1}_{\mathcal{H}_E}$, as they should in order to define a generalized measurement [24]. Note that the first case ($n_z \leq \dim \mathcal{H}_E$) is a special case of the second one, with $|\gamma_z\rangle = |z, \gamma_0\rangle$. If Eve now performs the measurement, then we have $P_{XYZ}(x, y, z) = |\langle x, y, z|\Psi\rangle|^2 = |\langle x, y, \gamma_z|\Psi, \gamma_0\rangle|^2$, and

$$P_{XY|Z}(x, y, z) = |\langle x, y|\alpha_z, \beta_z\rangle|^2 = |\langle x|\alpha_z\rangle|^2 \, |\langle y|\beta_z\rangle|^2 = P_{X|Z}(x, z) P_{Y|Z}(y, z)$$

holds for all $|z\rangle$ and for all $|x, y\rangle \in \mathcal{H}_A \otimes \mathcal{H}_B$. Consequently, $I(X; Y|Z) = 0$.    □

Theorem 2 states that if $\rho_{AB}$ is entangled, then Eve *cannot* force the intrinsic information to be zero: Whatever she does (i.e., whatever generalized measurements she carries out), there is something Alice and Bob can do such that the intrinsic information is positive. Note that this does *not*, a priori, imply that secret-key agreement is possible in every case. Indeed, we will provide evidence for the fact that this implication does generally *not* hold.

**Theorem 2** *Let $\Psi \in \mathcal{H}_A \otimes \mathcal{H}_B \otimes \mathcal{H}_E$ and $\rho_{AB} = \mathrm{Tr}_{\mathcal{H}_E}(P_\Psi)$. If $\rho_{AB}$ is entangled, then for all generating sets $\{|z\rangle\}$ of $\mathcal{H}_\mathcal{E}$, there are bases $\{|x\rangle\}$ and $\{|y\rangle\}$ of $\mathcal{H}_\mathcal{A}$ and $\mathcal{H}_\mathcal{B}$, respectively, such that $I(X; Y \downarrow Z) > 0$ holds for $P_{XYZ}(x, y, z) := |\langle x, y, z|\Psi\rangle|^2$.*

*Proof.* We prove this by contradiction. Assume that there exists a generating set $\{|z\rangle\}$ of $\mathcal{H}_E$ such that for all bases $\{|x\rangle\}$ of $\mathcal{H}_A$ and $\{|y\rangle\}$ of $\mathcal{H}_B$, we have $I(X; Y \downarrow Z) = 0$ for the resulting distribution. For such a distribution, there exists a channel, characterized by $P_{\overline{Z}|Z}$, such that $I(X; Y|\overline{Z}) = 0$ holds, i.e.,

$$P_{XY|\overline{Z}}(x, y, \overline{z}) = P_{X|\overline{Z}}(x, \overline{z}) P_{Y|\overline{Z}}(y, \overline{z}) . \tag{3}$$

Let $\rho_{\overline{z}} := (1/p_{\overline{z}}) \sum_z p_z P_{\overline{Z}|Z}(\overline{z}, z) P_{\psi_z}$, $p_z = P_Z(z)$, and $p_{\overline{z}} = \sum_z P_{\overline{Z}|Z}(\overline{z}, z) p_z$, where $\psi_z$ is the state of Alice's and Bob's system conditioned on Eve's result $z$: $\Psi \otimes |\gamma_0\rangle = \sum_z \psi_z \otimes |\gamma_z\rangle$ (see the proof of Theorem 1).

From (3) we can conclude $\mathrm{Tr}(P_x \otimes P_y \rho_{\overline{z}}) = \mathrm{Tr}(P_x \otimes \mathbf{1}\rho_{\overline{z}}) \, \mathrm{Tr}(\mathbf{1} \otimes P_y \rho_{\overline{z}})$ for all one-dimensional projectors $P_x$ and $P_y$ acting in $\mathcal{H}_A$ and $\mathcal{H}_B$, respectively. Consequently, the states $\rho_{\overline{z}}$ are products, i.e., $\rho_{\overline{z}} = \rho_{\alpha_{\overline{z}}} \otimes \rho_{\beta_{\overline{z}}}$, and $\rho_{AB} = \sum_{\overline{z}} p_{\overline{z}} \rho_{\overline{z}}$ is separable.    □

Theorem 2 can be formulated in a more positive way. Let us first introduce the concept of a set of bases $(\{|x\rangle\}_j, \{|y\rangle\}_j)$, where the $j$ label the different bases, as they are used in the 4-state (2 bases) and the 6-state (3 bases) protocols [3], [4], [1]. Then if $\rho_{AB}$ is entangled there exists a set $(\{|x\rangle\}_j, \{|y\rangle\}_j)_{j=1,\dots,N}$ of $N$ bases such that for all generalized measurements $\{|z\rangle\}$, $I(X;Y \downarrow [Z,j]) > 0$ holds. The idea is that Alice and Bob randomly choose a basis and, after the transmission, publicly restrict to the (possibly few) cases where they happen to have chosen the same basis. Hence Eve knows $j$, and one has

$$I(X;Y \downarrow [Z,j]) = \frac{1}{N} \sum_{j=1}^{N} I(X^j; Y^j \downarrow Z) \ .$$

If the set of bases is large enough, then for all $\{|z\rangle\}$ there is a basis with positive intrinsic information, hence the mean is also positive. Clearly, this result is stronger if the set of bases is small. Nothing is proven about the achievable size of such sets of bases, but it is conceivable that $\max\{\dim \mathcal{H}_A, \dim \mathcal{H}_B\}$ bases are always sufficient.

**Corollary 3** *Let $\Psi \in \mathcal{H}_A \otimes \mathcal{H}_B \otimes \mathcal{H}_E$ and $\rho_{AB} = \mathrm{Tr}_{\mathcal{H}_E}(P_\Psi)$. Then the following statements are equivalent:*

*(i) $\rho_{AB}$ is entangled,*

*(ii) for all generating sets $\{|z\rangle\}$ of $\mathcal{H}_E$, there exist bases $\{|x\rangle\}$ of $\mathcal{H}_A$ and $\{|y\rangle\}$ of $\mathcal{H}_B$ such that the distribution $P_{XYZ}(x,y,z) := |\langle x,y,z|\Psi\rangle|^2$ satisfies $I(X;Y\downarrow Z) > 0$,*

*(iii) for all generating sets $\{|z\rangle\}$ of $\mathcal{H}_E$, there exist bases $\{|x\rangle\}$ of $\mathcal{H}_A$ and $\{|y\rangle\}$ of $\mathcal{H}_B$ such that the distribution $P_{XYZ}(x,y,z) := |\langle x,y,z|\Psi\rangle|^2$ satisfies $I(X;Y|Z) > 0$.*

A first consequence of the fact that Corollary 3 often holds with respect to the standard bases (see below) is that it yields, at least in the binary case, a criterion for $I(X;Y\downarrow Z) > 0$ that is efficiently verifiable since it is based on the positivity of the eigenvalues of a $4\times4$ matrix. Previously, the quantity $I(X;Y\downarrow Z)$ has been considered hard to handle.

## 3.2    Examples I

The following examples illustrate the correspondence established in Section 3.1. They show in particular that very often (Examples 1, 2, and 3), but not always (Example 4), the direct connection between entanglement and positive intrinsic information holds with respect to the standard bases (i.e., the bases physicists use by commodity and intuition). Example 1 was already analyzed in [15]. The examples of this section will be discussed further in Section 3.5 under the aspect of the existence of key-agreement protocols in the classical and quantum regimes.

*Example 1.* Let us consider the so-called 4-state protocol of [3]. The analysis of the 6-state protocol [1] is analogous and leads to similar results. We compare the possibility of quantum and classical key agreement given the quantum state and the corresponding classical distribution, respectively, arising from this protocol. The conclusion is, under the assumption of incoherent eavesdropping, that key agreement in one setting is possible if and only if this is true also for the other.

After carrying out the 4-state protocol, and under the assumption of optimal eavesdropping (in terms of Shannon information), the resulting quantum state is [11]

$$\Psi = \sqrt{F/2}\,|0,0\rangle \otimes \xi_{00} + \sqrt{D/2}\,|0,1\rangle \otimes \xi_{01} + \sqrt{D/2}\,|1,0\rangle \otimes \xi_{10} + \sqrt{F/2}\,|1,1\rangle \otimes \xi_{11}\ ,$$

where $D$ (the *disturbance*) is the probability that $X \neq Y$ holds if $X$ and $Y$ are the classical random variables of Alice and Bob, respectively, where $F = 1 - D$ (the *fidelity*), and where the $\xi_{ij}$ satisfy $\langle \xi_{00}|\xi_{11}\rangle = \langle \xi_{01}|\xi_{10}\rangle = 1 - 2D$ and $\langle \xi_{ii}|\xi_{ij}\rangle = 0$ for all $i \neq j$. Then the state $\rho_{AB}$ is (in the basis $\{|00\rangle,\ |01\rangle,\ |10\rangle,\ |11\rangle\}$)

$$\rho_{AB} = \frac{1}{2}\begin{pmatrix} D & 0 & 0 & -D(1-2D) \\ 0 & 1-D & -(1-D)(1-2D) & 0 \\ 0 & -(1-D)(1-2D) & 1-D & 0 \\ -D(1-2D) & 0 & 0 & D \end{pmatrix},$$

and its partial transpose

$$\rho_{AB}^{t} = \frac{1}{2}\begin{pmatrix} D & 0 & 0 & -(1-D)(1-2D) \\ 0 & 1-D & -D(1-2D) & 0 \\ 0 & -D(1-2D) & 1-D & 0 \\ -(1-D)(1-2D) & 0 & 0 & D \end{pmatrix}$$

has the eigenvalues $(1/2)(D \pm (1-D)(1-2D))$ and $(1/2)((1-D) \pm D(1-2D))$, which are all non-negative (i.e., $\rho_{AB}$ is separable) if

$$D \geq 1 - \frac{1}{\sqrt{2}}\ . \tag{4}$$

From the classical viewpoint, the corresponding distributions (arising from measuring the above quantum system in the standard bases) are as follows. First, $X$ and $Y$ are both symmetric bits with $\mathrm{Prob}\,[X \neq Y] = D$. Eve's random variable $Z = [Z_1, Z_2]$ is composed of 2 bits $Z_1$ and $Z_2$, where $Z_1 = X \oplus Y$, i.e., $Z_1$ tells Eve whether Bob received the qubit disturbed ($Z_1 = 1$) or not ($Z_1 = 0$) (this is a consequence of the fact that the $\xi_{ii}$ and $\xi_{ij}$ ($i \neq j$) states generate orthogonal subspaces), and where the probability that Eve's second bit indicates the correct value of Bob's bit is $\mathrm{Prob}[Z_2 = Y] = \delta = (1 + \sqrt{1 - \langle \xi_{00}|\xi_{11}\rangle^2})/2 = 1/2 + \sqrt{D(1-D)}$. We now prove that for this distribution, the intrinsic information is zero if and only if

$$\frac{D}{1-D} \geq 2\sqrt{(1-\delta)\delta} = 1 - 2D \tag{5}$$

holds. We show that if the condition (5) is satisfied, then $I(X; Y \downarrow Z) = 0$ holds. The inverse implication follows from the existence of a key-agreement protocol in all other cases (see Example 1 (cont'd) in Section 3.5). If (5) holds, we can construct a random variable $\overline{Z}$, that is generated by sending $Z$ over a channel characterized by $P_{\overline{Z}|Z}$, for which $I(X;Y|\overline{Z}) = 0$ holds. We can restrict ourselves to the case of equality in (5) because Eve can always increase $\delta$ by adding noise.

Consider now the channel characterized by the following conditional distribution $P_{\overline{Z}|Z}$ (where $\overline{Z} = \{u, v\}$):

$$P_{\overline{Z}|Z}(u, [0,0]) = P_{\overline{Z}|Z}(v, [0,1]) = 1 \ ,$$
$$P_{\overline{Z}|Z}(l, [1,0]) = P_{\overline{Z}|Z}(l, [1,1]) = 1/2$$

for $l \in \{u, v\}$. We show $I(X;Y|\overline{Z}) = E_{\overline{Z}}[I(X;Y|\overline{Z} = \overline{z})] = 0$, i.e., that $I(X;Y|\overline{Z} = u) = 0$ and $I(X;Y|\overline{Z} = v) = 0$ hold. By symmetry it is sufficient to show the first equality. For $a_{ij} := P_{XY\overline{Z}}(i, j, u)$, we get

$$a_{00} = (1-D)(1-\delta)/2\,, \ a_{11} = (1-D)\delta/2\,, \ a_{01} = a_{10} = (D(1-\delta)/2 + D\delta/2)/2 = D/4\,.$$

From equality in (5) we conclude $a_{00}a_{11} = a_{01}a_{10}$, which is equivalent to the fact that $X$ and $Y$ are independent, given $\overline{Z} = u$.

Finally, note that the conditions (4) and (5) are equivalent for $D \in [0, 1/2]$. This shows that the bounds of tolerable noise are indeed the same for the quantum and classical scenarios. $\diamond$

*Example 2.* We consider the bound entangled state presented in [17]. This example received quite a lot of attention by the quantum-information community because it was the first known example of bound entanglement (i.e., entanglement without the possibility of quantum key agreement). We show that its classical counterpart seems to have similarly surprising properties. Let $0 < a < 1$ and

$$\Psi = \sqrt{\frac{3a}{8a+1}}\ \psi \otimes |0\rangle + \sqrt{\frac{1}{8a+1}}\ \phi_a \otimes |1\rangle + \sqrt{\frac{a}{8a+1}}\ (|122\rangle + |133\rangle + |214\rangle + |235\rangle + |326\rangle)\,,$$

where $\psi = (|11\rangle + |22\rangle + |33\rangle)/\sqrt{3}$ and $\phi_a = \sqrt{(1+a)/2}\,|31\rangle + \sqrt{(1-a)/2}\,|33\rangle$. It has been shown in [17] that the resulting state $\rho_{AB}$ is entangled.

The corresponding classical distribution is as follows. The ranges are $\mathcal{X} = \mathcal{Y} = \{1, 2, 3\}$ and $\mathcal{Z} = \{0, 1, 2, 3, 4, 5, 6\}$. We write $(ijk) = P_{XYZ}(i, j, k)$. Then we have $(110) = (220) = (330) = (122) = (133) = (214) = (235) = (326) = 2a/(16a+2)$, $(311) = (1+a)/(16a+2)$, and $(331) = (1-a)/(16a+2)$. We study the special case $a = 1/2$. Consider the following representation of the resulting distribution (to be normalized). For instance, the entry "(0) 1 , (1) 1/2" for $X = Y = 3$ means $P_{XYZ}(3,3,0) = 1/10$ (normalized), $P_{XYZ}(3,3,1) = 1/20$, and $P_{XYZ}(3,3,z) = 0$ for all $z \notin \{0, 1\}$.

| X <br> Y (Z) | 1 | 2 | 3 |
|---|---|---|---|
| 1 | (0) 1 | (4) 1 | (1) 3/2 |
| 2 | (2) 1 | (0) 1 | (6) 1 |
| 3 | (3) 1 | (5) 1 | (0) 1 <br> (1) 1/2 |

As we would expect, the intrinsic information is positive in this scenario. This can be seen by contradiction as follows. Assume $I(X;Y{\downarrow}Z) = 0$. Hence there exists a discrete channel, characterized by the conditional distribution $P_{\overline{Z}|Z}$, such that $I(X;Y|\overline{Z}) = 0$ holds. Let $\overline{\mathcal{Z}} \subseteq \mathbf{N}$ be the range of $\overline{Z}$, and let $P_{\overline{Z}|Z}(i,0) =: a_i$, $P_{\overline{Z}|Z}(i,1) =: x_i$, $P_{\overline{Z}|Z}(i,6) =: s_i$. Then we must have $a_i, x_i, s_i \in [0,1]$ and $\sum_i a_i = \sum_i x_i = \sum_i s_i = 1$. Using $I(X;Y|\overline{Z}) = 0$, we obtain the following distributions $P_{XY|\overline{Z}=i}$ (to be normalized):

| X \ Y | 1 | 2 | 3 |
|---|---|---|---|
| 1 | $a_i$ | $\frac{3a_i x_i}{2s_i}$ | $\frac{3x_i}{2}$ |
| 2 | $\frac{2a_i s_i}{3x_i}$ | $a_i$ | $s_i$ |
| 3 | $\frac{2a_i(a_i+x_i/2)}{3x_i}$ | $\frac{a_i(a_i+x_i/2)}{s_i}$ | $a_i + \frac{x_i}{2}$ |

By comparing the $(2,3)$-entries of the two tables above, we obtain

$$1 \geq \sum_i \frac{a_i(a_i + x_i/2)}{s_i} \ . \tag{6}$$

We prove that (6) implies $s_i \equiv a_i$ (i.e., $s_i = a_i$ for all $i$) and $x_i \equiv 0$. Clearly, this does not lead to a solution and is hence a contradiction. For instance, $P_{XY|\overline{Z}=i}(1,2) = 2a_i s_i/3x_i$ is not even defined in this case if $a_i > 0$.

It remains to show that (6) implies $a_i \equiv s_i$ and $x_i \equiv 0$. We show that whenever $\sum_i a_i = \sum_i s_i = 1$ and $a_i \not\equiv s_i$, then $\sum_i a_i^2/s_i > 1$ . First, note that $\sum_i a_i^2/s_i = \sum_i a_i = 1$ for $a_i \equiv s_i$. Let now $s_{i_1} \leq a_{i_1}$ and $s_{i_2} \geq a_{i_2}$. We show that $a_{i_1}^2/s_{i_1} + a_{i_2}^2/s_{i_2} < a_{i_1}^2/(s_{i_1} - \varepsilon) + a_{i_2}^2/(s_{i_2} + \varepsilon)$ holds for every $\varepsilon > 0$, which obviously implies the above statement. It is straightforward to see that this is equivalent to $a_{i_1}^2 s_{i_2}(s_{i_2} + \varepsilon) > a_{i_2}^2 s_{i_1}(s_{i_1} - \varepsilon)$, and holds because of $a_{i_1}^2 s_{i_2}(s_{i_2} + \varepsilon) > a_{i_1}^2 a_{i_2}^2$ and $a_{i_2}^2 s_{i_1}(s_{i_1} - \varepsilon) < a_{i_1}^2 a_{i_2}^2$. This concludes the proof of $I(X;Y{\downarrow}Z) > 0$.    ◇

As mentioned, the interesting point about Example 2 is that the quantum state is bound entangled, and that also classical key agreement seems impossible despite the fact that $I(X;Y{\downarrow}Z) > 0$ holds. This is a contradiction to a conjecture stated in [22]. The classical translation of the bound entangled state leads to a classical distribution with very strange properties as well! (See Example 2 (cont'd) in Section 3.5).

In Example 3, another bound entangled state (first proposed in [18]) is discussed. The example is particularly nice because, depending on the choice of a parameter $\alpha$, the quantum state can be made separable, bound entangled, and free entangled.

*Example 3.* We consider the following distribution (to be normalized). Let $2 \leq \alpha \leq 5$.

| X<br>Y (Z) | 1 | 2 | 3 |
|---|---|---|---|
| 1 | (0) 2 | (4) $5-\alpha$ | (3) $\alpha$ |
| 2 | (1) $\alpha$ | (0) 2 | (5) $5-\alpha$ |
| 3 | (6) $5-\alpha$ | (2) $\alpha$ | (0) 2 |

This distribution arises when measuring the following quantum state. Let $\psi :=$ $(1/\sqrt{3})\,(|11\rangle + |22\rangle + |33\rangle)$. Then

$$\Psi = \sqrt{\frac{2}{7}}\,\psi \otimes |0\rangle + \sqrt{\frac{a}{21}}\,(|12\rangle \otimes |1\rangle + |23\rangle \otimes |2\rangle + |31\rangle \otimes |3\rangle)$$

$$+\sqrt{\frac{5-a}{21}}\,(|21\rangle \otimes |4\rangle + |32\rangle \otimes |5\rangle + |13\rangle \otimes |6\rangle), \qquad \text{and}$$

$$\rho_{AB} = \frac{2}{7}\,P_{\psi} + \frac{a}{21}\,(P_{12} + P_{23} + P_{31}) + \frac{5-a}{21}\,(P_{21} + P_{32} + P_{13})$$

is separable if and only if $\alpha \in [2,3]$, bound entangled for $\alpha \in (3,4]$, and free entangled if $\alpha \in (4,5]$ [18] (see Figure 1).

Let us consider the quantity $I(X;Y{\downarrow}Z)$. First of all, it is clear that $I(X;Y{\downarrow}Z) = 0$ holds for $\alpha \in [2,3]$. The reason is that $\alpha \geq 2$ and $5-\alpha \geq 2$ together imply that Eve can "mix" her symbol $Z = 0$ with the remaining symbols in such a way that when given that $\overline{Z}$ takes the "mixed value," then $XY$ is uniformly distributed; in particular, $X$ and $Y$ are independent. Moreover, it can be shown in analogy to Example 2 that $I(X;Y{\downarrow}Z) > 0$ holds for $\alpha > 3$. $\diamondsuit$

Examples 1, 2, and 3 suggest that the correspondence between separability and entanglement on one side and vanishing and non-vanishing intrinsic information on the other always holds with respect to the standard bases or even arbitrary bases. This is however not true in general: Alice and Bob as well as Eve can perform bad measurements and give away an initial advantage. The following is a simple example where measuring in the standard basis is a bad choice for Eve.

*Example 4.* Let us consider the quantum states

$$\Psi = \frac{1}{\sqrt{5}}\,((|00+01+10\rangle) \otimes |0\rangle + |00+11\rangle \otimes |1\rangle)\,, \quad \rho_{AB} = \frac{3}{5}\,P_{|00+01+10\rangle} + \frac{2}{5}\,P_{|00+11\rangle}\,.$$

If Alice, Bob, and Eve measure in the standard bases, we get the classical distribution (to be normalized)

| X<br>Y (Z) | 0 | 1 |
|---|---|---|
| 0 | (0) 1<br>(1) 1 | (0) 1<br>(1) 0 |
| 1 | (0) 1<br>(1) 0 | (0) 0<br>(1) 1 |

For this distribution, $I(X;Y{\downarrow}Z) > 0$ holds. Indeed, even $S(X;Y\|Z) > 0$ holds. This is not surprising since both $X$ and $Y$ are binary, and since the described parallels suggest that in this case, positive intrinsic information implies that a secret-key agreement protocol exists.

The proof of $S(X;Y\|Z) > 0$ in this situation is analogous to the proof of this fact in Example 3. The protocol consists of Alice and Bob independently making their bits symmetric. Then the repeat-code protocol can be applied.

However, the partial-transpose condition shows that $\rho_{AB}$ is separable. This means that measuring in the standard basis is bad for Eve. Indeed, let us rewrite $\Psi$ and $\rho_{AB}$ as

$$\Psi = \sqrt{\Lambda}\,|m,m\rangle \otimes |\tilde{0}\rangle + \sqrt{1-\Lambda}\,|-m,-m\rangle \otimes |\tilde{1}\rangle \ ,$$

$$\rho_{AB} = \frac{5+\sqrt{5}}{10}\,P_{|m,m\rangle} + \frac{5-\sqrt{5}}{10}\,P_{|-m,-m\rangle}\ ,$$

where $\Lambda = (5+\sqrt{5})/10$, $|m,m\rangle = |m\rangle \otimes |m\rangle$, $|\pm m\rangle = \sqrt{(1\pm\eta)/2}\,|0\rangle \pm \sqrt{(1\mp\eta)/2}\,|1\rangle$, and $\eta = 1/\sqrt{5}$.

In this representation, $\rho_{AB}$ is obviously separable. It also means that Eve's optimal measurement basis is

$$|\tilde{0}\rangle = \sqrt{\Lambda}\,|0\rangle - \frac{1}{\sqrt{5\Lambda}}\,|1\rangle\ , \quad |\tilde{1}\rangle = -\sqrt{1-\Lambda}\,|0\rangle - \frac{1}{\sqrt{5(1-\Lambda)}}\,|1\rangle\ .$$

Then, $I(X;Y{\downarrow}Z) = 0$ holds for the resulting classical distribution. $\diamondsuit$

## 3.3  A Classical Measure for Quantum Entanglement

It is a challenging problem of theoretical quantum physics to find good measures for entanglement [26]. Corollary 3 above suggests the following measure, which is based on classical information theory.

**Definition 1** Let for a quantum state $\rho_{AB}$

$$\mu(\rho_{AB}) := \min_{\{|z\rangle\}} \left( \max_{\{|x\rangle\},\{|y\rangle\}} (I(X;Y{\downarrow}Z)) \right)\ ,$$

where the minimum is taken over all $\Psi = \sum_z \sqrt{p_z}\psi_z \otimes |z\rangle$ such that $\rho_{AB} = \mathrm{Tr}_{\mathcal{H}_E}(P_\Psi)$ holds and over all generating sets $\{|z\rangle\}$ of $\mathcal{H}_E$, the maximum is over all bases $\{|x\rangle\}$ of $\mathcal{H}_A$ and $\{|y\rangle\}$ of $\mathcal{H}_B$, and where $P_{XYZ}(x,y,z) := |\langle x,y,z|\Psi\rangle|^2$. $\bigcirc$

The function $\mu$ has all the properties required from such a measure. If $\rho_{AB}$ is pure, i.e., $\rho_{AB} = |\psi_{AB}\rangle\langle\psi_{AB}|$, then we have in the Schmidt basis (see for example [24]) $\psi_{AB} = \sum_j c_j|x_j,y_j\rangle$, and $\mu(\rho_{AB}) = -\mathrm{Tr}(\rho_A \log \rho_A)$ (where $\rho_A = \mathrm{Tr}_B(\rho_{AB})$) as it should [26]. It is obvious that $\mu$ is convex, i.e., $\mu(\lambda\rho_1 + (1-\lambda)\rho_2) \leq \lambda\mu(\rho_1) + (1-\lambda)\mu(\rho_2)$.

*Example 5.* This example is based on Werner's state. Let $\Psi = \sqrt{\lambda}\,\psi^{(-)} \otimes |0\rangle +$ $\sqrt{(1-\lambda)/4}\,|001 + 012 + 103 + 114\rangle$, where $\psi^{(-)} = |10 - 01\rangle/\sqrt{2}$, and $\rho_{AB} = \lambda P_{\psi^{(-)}} + ((1-\lambda)/4)\mathbb{1}$. It is well-known that $\rho_{AB}$ is separable if and only if $\lambda \le 1/3$. Then the classical distribution is $P(010) = P(100) = \lambda/2$ and $P(001) = P(012) = P(103) = P(114) = (1-\lambda)/4$.

If $\lambda \le 1/3$, then consider the channel $P_{\overline{Z}|Z}(0,0) = P_{\overline{Z}|Z}(2,2) = P_{\overline{Z}|Z}(3,3) = 1$, $P_{\overline{Z}|Z}(0,1) = P_{\overline{Z}|Z}(0,4) = \xi$, $P_{\overline{Z}|Z}(1,1) = P_{\overline{Z}|Z}(4,4) = 1 - \xi$, where $\xi = 2\lambda/(1-\lambda) \le 1$. Then $\mu(\rho_{AB}) = I(X;Y\!\downarrow\!Z) = I(X;Y|\overline{Z}) = 0$ holds, as it should.

If $\lambda > 1/3$, then consider the (obviously optimal) channel $P_{\overline{Z}|Z}(0,0) = P_{\overline{Z}|Z}(2,2) = P_{\overline{Z}|Z}(3,3) = P_{\overline{Z}|Z}(0,1) = P_{\overline{Z}|Z}(0,4) = 1$. Then

$$\mu(\rho_{AB}) = I(X;Y\!\downarrow\!Z) = I(X;Y|\overline{Z}) = P_{\overline{Z}}(0) \cdot I(X;Y|\overline{Z} = 0)$$
$$= \frac{1+\lambda}{2} \cdot (1 - q\log_2 q - (1-q)\log_2(1-q)) ,$$

where $q = 2\lambda/(1+\lambda)$.     $\diamondsuit$

## 3.4  Classical Protocols and Quantum Privacy Amplification

It is a natural question whether the analogy between entanglement and intrinsic information (see Section 3.1) carries over to the protocol level. The examples given in Section 3.5 support this belief. A quite interesting and surprising consequence would be that there exists a classical counterpart to bound entanglement, namely intrinsic information that cannot be distilled into a secret key by any classical protocol, if $|\mathcal{X}| + |\mathcal{Y}| > 5$, where $\mathcal{X}$ and $\mathcal{Y}$ are the ranges of $X$ and $Y$, respectively. In other words, the conjecture in [22] that such information can always be distilled would be *proved* for $|\mathcal{X}| + |\mathcal{Y}| \le 5$, but *disproved* otherwise.

**Conjecture 1** *Let $\Psi \in \mathcal{H}_A \otimes \mathcal{H}_B \otimes \mathcal{H}_E$ and $\rho_{AB} = \mathrm{Tr}_{\mathcal{H}_E}(P_\Psi)$. Assume that for all generating sets $\{|z\rangle\}$ of $\mathcal{H}_E$ there are bases $\{|x\rangle\}$ and $\{|y\rangle\}$ of $\mathcal{H}_A$ and $\mathcal{H}_B$, respectively, such that $S(X;Y||Z) > 0$ holds for the distribution $P_{XYZ}(x,y,z) := |\langle x, y, z|\Psi\rangle|^2$. Then quantum privacy amplification is possible with the state $\rho_{AB}$, i.e., $\rho_{AB}$ is free entangled.*

**Conjecture 2** *Let $\Psi \in \mathcal{H}_A \otimes \mathcal{H}_B \otimes \mathcal{H}_E$ and $\rho_{AB} = \mathrm{Tr}_{\mathcal{H}_E}(P_\Psi)$. Assume that there exists a generating set $\{|z\rangle\}$ of $\mathcal{H}_E$ such that for all bases $\{|x\rangle\}$ and $\{|y\rangle\}$ of $\mathcal{H}_A$ and $\mathcal{H}_B$, respectively, $S(X;Y||Z) = 0$ holds for the distribution $P_{XYZ}(x,y,z) := |\langle x, y, z|\Psi\rangle|^2$. Then quantum privacy amplification is impossible with the state $\rho_{AB}$, i.e., $\rho_{AB}$ is bound entangled or separable.*

## 3.5  Examples II

The following examples support Conjectures 1 and 2 and illustrate their consequences. We consider mainly the same distributions as in Section 3.2, but this time under the aspect of the existence of classical and quantum key-agreement protocols.

*Example 1 (cont'd).* We have shown in Section 3.2 that the resulting quantum state is entangled if and only if the intrinsic information of the corresponding classical situation (with respect to the standard bases) is non-zero. Such a correspondence also holds on the protocol level. First of all, it is clear for the quantum state that QPA is possible whenever the state is entangled because both $\mathcal{H}_A$ and $\mathcal{H}_B$ have dimension two. On the other hand, the same is also true for the corresponding classical situation, i.e., secret-key agreement is possible whenever $D/(1 - D) < 2\sqrt{(1 - \delta)\delta}$ holds, i.e., if the intrinsic information is positive. The necessary protocol includes an interactive phase, called *advantage distillation*, based on a repeat code or on parity checks (see [20] or [29]).     ◇

*Example 2 (cont'd).* The quantum state $\rho_{AB}$ in this example is bound entangled, meaning that the entanglement cannot be used for QPA. Interestingly, but not surprisingly given the discussion above, the corresponding classical distribution has the property that $I(X;Y{\downarrow}Z) > 0$, but nevertheless, all the known classical advantage-distillation protocols [20], [22] fail for this distribution! It seems that $S(X;Y||Z) = 0$ holds (although it is not clear how this fact could be rigorously proven).     ◇

*Example 3 (cont'd).* We have seen already that for $2 \leq \alpha \leq 3$, the quantum state is separable and the corresponding classical distribution (with respect to the standard bases) has vanishing intrinsic information. Moreover, it has been shown that for the quantum situation, $3 < \alpha \leq 4$ corresponds to bound entanglement, whereas for $\alpha > 4$, QPA is possible and allows for generating a secret key [18]. We describe a classical protocol here which suggests that the situation for the classical translation of the scenario is totally analogous: The protocol allows classical key agreement exactly for $\alpha > 4$. However, this does not imply (although it appears very plausible) that no classical protocol exists at all for the case $\alpha \leq 4$.

Let $\alpha > 4$. We consider the following protocol for classical key agreement. First of all, Alice and Bob both restrict their ranges to $\{1, 2\}$ (i.e., publicly reject a realization unless $X \in \{1, 2\}$ and $Y \in \{1, 2\}$). The resulting distribution is as follows (to be normalized):

| X<br>Y (Z) | 1 | 2 |
|:---:|:---:|:---:|
| 1 | (0) 2 | (4) $5 - \alpha$ |
| 2 | (2) $\alpha$ | (0) 2 |

Then, Alice and Bob both send their bits locally over channels $P_{\overline{X}|X}$ and $P_{\overline{Y}|Y}$, respectively, such that the resulting bits $\overline{X}$ and $\overline{Y}$ are symmetric. The channel $P_{\overline{X}|X}$ [$P_{\overline{Y}|Y}$] sends $X = 0$ [$Y = 1$] to $\overline{X} = 1$ [$\overline{Y} = 0$] with probability $(2\alpha - 5)/(2\alpha + 4)$, and leaves $X$ [$Y$] unchanged otherwise. The distribution $P_{\overline{X}\overline{Y}Z}$ is then

| $\overline{X}$ <br> $\overline{Y}\ (Z)$ | 1 | 2 |
|---|---|---|
| 1 | (0) $2 \cdot \frac{9}{2\alpha+4}$ <br> (2) $\alpha \cdot \frac{9}{2\alpha+4} \cdot \frac{2\alpha-5}{2\alpha+4}$ | (1) $5-\alpha$ <br> (2) $\alpha \left(\frac{2\alpha-5}{2\alpha+4}\right)^2$ <br> (0) $2 \cdot 2 \cdot \frac{2\alpha-5}{2\alpha+4}$ |
| 2 | (2) $\alpha \left(\frac{9}{2\alpha+4}\right)^2$ | (0) $2 \cdot \frac{9}{2\alpha+4}$ <br> (2) $\alpha \cdot \frac{9}{2\alpha+4} \cdot \frac{2\alpha-5}{2\alpha+4}$ |

It is not difficult to see that for $\alpha > 4$, we have $\mathrm{Prob}\left[\overline{X} = \overline{Y}\right] > 1/2$ and that, given that $\overline{X} = \overline{Y}$ holds, Eve has no information at all about what this bit is. This means that the repeat-code protocol mentioned in Example 1 allows for classical key agreement in this situation [20], [29]. For $\alpha \leq 4$, classical key agreement, like quantum key agreement, seems impossible however. The results of Example 3 are illustrated in Figure 1. $\diamondsuit$

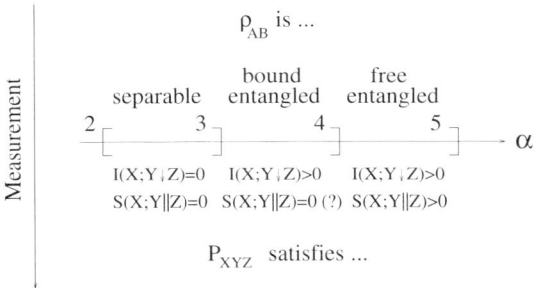

Fig. 1. The Results of Example 3

### 3.6 Bound Intrinsic Information

Examples 2 and 3 suggest that, in analogy to bound entanglement of a quantum state, *bound classical information* exists, i.e., conditional intrinsic information which cannot be used to generate a secret key in the classical scenario. We give a formal definition of bound intrinsic information.

**Definition 2** Let $P_{XYZ}$ be a distribution with $I(X;Y{\downarrow}Z) > 0$. If $S(X;Y||Z) > 0$ holds for this distribution, the intrinsic information between $X$ and $Y$, given $Z$, is called *free*. Otherwise, if $S(X;Y||Z) = 0$, the intrinsic information is called *bound*. $\bigcirc$

Note that the existence of bound intrinsic information could not be proven so far. However, all known examples of bound entanglement, combined with all known advantage-distillation protocols, do not lead to a contradiction to Conjecture 1! Clearly, it would be very interesting to rigorously prove this conjecture because then, all pessimistic results known for the quantum scenario would immediately carry over to the classical setting (where such results appear to be much harder to prove).

Examples 2 and 3 also illustrate nicely what the nature of bound information is. Of course, $I(X; Y \downarrow Z) > 0$ implies both $I(X; Y) > 0$ and $I(X; Y|Z) > 0$. However, if $|\mathcal{X}| + |\mathcal{Y}| > 5$, it is possible that the dependence between $X$ and $Y$ and the dependence between $X$ and $Y$, given $\overline{Z}$, are "orthogonal." By the latter we mean that for all fixed (deterministic or probabilistic) functions $f : \mathcal{X} \to \{0, 1\}$ and $g : \mathcal{Y} \to \{0, 1\}$ for which the correlation of $f(X)$ and $g(Y)$ is positive, i.e.,

$$P_{f(X)g(Y)}(0,0) \cdot P_{f(X)g(Y)}(1,1) > P_{f(X)g(Y)}(0,1) \cdot P_{f(X)g(Y)}(1,0) ,$$

the correlation between the same binary random variables, given $\overline{Z} = \overline{z}$, is negative (or "zero") for all $\overline{z} \in \overline{\mathcal{Z}}$, where $\overline{Z}$ is the random variable generated by sending $Z$ over Eve's optimal channel $P_{\overline{Z}|Z}$.

A complete understanding of bound intrinsic information is of interest also because it automatically leads to a better understanding of bound entanglement in quantum information theory.

## 4    Concluding Remarks

We have considered the model of information-theoretic key agreement by public discussion from correlated information. More precisely, we have compared scenarios where the joint information is given by classical random variables and by quantum states (e.g., after execution of a quantum protocol). We proved a close connection between such classical and quantum information, namely between intrinsic information and entanglement. As an application, the derived parallels lead to an efficiently verifiable criterion for the fact that the intrinsic information vanishes. Previously, this quantity was considered to be quite hard to handle.

Furthermore, we have presented examples providing evidence for the fact that the close connections between classical and quantum information extend to the level of the protocols. A consequence would be that the powerful tools and statements on the existence or rather non-existence of quantum-privacy-amplification protocols immediately carry over to the classical scenario, where it is often unclear how to show that no protocol exists. Many examples (only some of which are presented above due to space limitations) coming from measuring bound entangled states, and for which none of the known classical secret-key agreement protocols is successful, strongly suggest that bound entanglement has a classical counterpart: intrinsic information which cannot be distilled to a secret key. This stands in sharp contrast to what was previously believed about classical key agreement. We state as an open problem to rigorously prove Conjectures 1 and 2.

Finally, we have proposed a measure for entanglement, based on classical information theory, with all the properties required for such a measure.

## Acknowledgments

The authors thank Claude Crépeau, Artur Ekert, Bruno Huttner, Itoshi Inamori, Ueli Maurer, and Sandu Popescu for interesting discussions, and the referees for their helpful comments. This work was partially supported by the Swiss National Science Foundation (SNF).

## References

1. H. Bechmann-Pasquinucci and N. Gisin, Incoherent and coherent eavesdropping in the six-state protocol of quantum cryptography, *Phys. Rev. A*, Vol. 59, No. 6, pp. 4238–4248, 1999.
2. C. H. Bennett, G. Brassard, S. Popescu, B. Schumacher, J. A. Smolin, and W. K. Wooters, Purification of noisy entanglement and faithful teleportation via noisy channels, *Phys. Rev. Lett.*, Vol. 76, pp. 722–725, 1996.
3. C. H. Bennett and G. Brassard, Quantum cryptography: public key distribution and coin tossing, *Proceedings of the IEEE International Conference on Computer, Systems, and Signal Processing*, IEEE, pp. 175–179, 1984.
4. D. Bruss, Optimal eavesdropping in quantum cryptography with six states, *Phys. Rev. Lett.*, Vol. 81, No. 14, pp. 3018–3021, 1998.
5. V. Bužek and M. Hillery, Quantum copying: beyond the no-cloning theorem, *Phys. Rev. A*, Vol. 54, pp. 1844–1852, 1996.
6. J. F. Clauser, M. A. Horne, A. Shimony and R. A. Holt, Proposed experiment to test local hidden-variable theories, *Phys. Rev. Lett.*, Vol. 23, pp. 880–884, 1969.
7. I. Csiszár and J. Körner, Broadcast channels with confidential messages, *IEEE Transactions on Information Theory*, Vol. IT-24, pp. 339–348, 1978.
8. D. Deutsch, A. Ekert, R. Jozsa, C. Macchiavello, S. Popescu, and A. Sanpera, Quantum privacy amplification and the security of quantum cryptography over noisy channels, *Phys. Rev. Lett.*, Vol. 77, pp. 2818–2821, 1996.
9. D. P. DiVincenzo, P. W. Shor, J. A. Smolin, B. M. Terhal, and A. V. Thapliyal, Evidence for bound entangled states with negative partial transpose, quant-ph/9910026, 1999.
10. A. E. Ekert, Quantum cryptography based on Bell's theorem, *Phys. Rev. Lett.*, Vol. 67, pp. 661–663, 1991. See also *Physics World*, March 1998.
11. C. Fuchs, N. Gisin, R. B. Griffiths, C. S. Niu, and A. Peres, Optimal eavesdropping in quantum cryptography – I: information bound and optimal strategy, *Phys. Rev. A*, Vol. 56, pp. 1163–1172, 1997.
12. N. Gisin, Stochastic quantum dynamics and relativity, *Helv. Phys. Acta*, Vol. 62, pp. 363–371, 1989.
13. N. Gisin and B. Huttner, Quantum cloning, eavesdropping, and Bell inequality, *Phys. Lett. A*, Vol. 228, pp. 13–21, 1997.
14. N. Gisin and S. Massar, Optimal quantum cloning machines, *Phys. Rev. Lett.*, Vol. 79, pp. 2153–2156, 1997.
15. N. Gisin and S. Wolf, Quantum cryptography on noisy channels: quantum versus classical key agreement protocols, *Phys. Rev. Lett.*, Vol. 83, pp. 4200–4203, 1999.

16. M. Horodecki, P. Horodecki, and R. Horodecki, Mixed-state entanglement and distillation: is there a "bound" entanglement in nature?, *Phys. Rev. Lett.*, Vol. 80, pp. 5239–5242, 1998.

17. P. Horodecki, Separability criterion and inseparable mixed states with positive partial transposition, *Phys. Lett. A*, Vol. 232, p. 333, 1997.

18. P. Horodecki, M. Horodecki, and R. Horodecki, Bound entanglement can be activated, *Phys. Rev. Lett.*, Vol. 82, pp. 1056–1059, 1999. quant-ph/9806058.

19. L. P. Hughston, R. Jozsa, and W. K. Wootters, A complete classification of quantum ensembles having a given density matrix, *Phys. Lett. A*, Vol. 183, pp. 14–18, 1993.

20. U. Maurer, Secret key agreement by public discussion from common information, *IEEE Transactions on Information Theory*, Vol. 39, No. 3, pp. 733–742, 1993.

21. U. Maurer and S. Wolf, Information-theoretic key agreement: from weak to strong secrecy for free, *Proceedings of EUROCRYPT 2000*, Lecture Notes in Computer Science, Vol. 1807, pp. 352–368, Springer-Verlag, 2000.

22. U. Maurer and S. Wolf, Unconditionally secure key agreement and the intrinsic conditional information, *IEEE Transactions on Information Theory*, Vol. 45, No. 2, pp. 499–514, 1999.

23. N. D. Mermin, The Ithaca interpretation of quantum mechanics, *Pramana*, Vol. 51, pp. 549–565, 1998.

24. A. Peres, *Quantum theory: concepts and methods*, Kluwer Academic Publishers, 1993.

25. A. Peres, Separability criterion for density matrices, *Phys. Rev. Lett.*, Vol. 77, pp. 1413–1415, 1996.

26. S. Popescu and D. Rohrlich, Thermodynamics and the measure of entanglement, quant-ph/9610044, 1996.

27. G. Ribordy, J. D. Gautier, N. Gisin, O. Guinnard, and H. Zbinden, Automated plug and play quantum key distribution, *Electron. Lett.*, Vol. 34, pp. 2116–2117, 1998.

28. C. E. Shannon, Communication theory of secrecy systems, *Bell System Technical Journal*, Vol. 28, pp. 656–715, 1949.

29. S. Wolf, *Information-theoretically and computationally secure key agreement in cryptography*, ETH dissertation No. 13138, Swiss Federal Institute of Technology (ETH Zurich), May 1999.

30. A. D. Wyner, The wire-tap channel, *Bell System Technical Journal*, Vol. 54, No. 8, pp. 1355–1387, 1975.

31. H. Zbinden, H. Bechmann, G. Ribordy, and N. Gisin, Quantum cryptography, *Applied Physics B*, Vol. 67, pp. 743–748, 1998.

# Maximum Correlation Analysis of Nonlinear S-boxes in Stream Ciphers

Muxiang Zhang[1] and Agnes Chan[2]

[1] GTE Laboratories Inc., 40 Sylvan Road LA0MS59, Waltham, MA 02451
mzhang@gte.com
[2] College of Computer Science, Northeastern University, Boston, MA 02115
ahchan@ccs.neu.edu

**Abstract.** This paper investigates the design of S-boxes used for combining linear feedback shift register (LFSR) sequences in combination generators. Such combination generators have higher throughput than those using Boolean functions as the combining functions. However, S-boxes tend to leak more information about the LFSR sequences than Boolean functions. To study the information leakage, the notion of maximum correlation is introduced, which is based on the correlation between linear functions of the input and all the Boolean functions (linear and nonlinear) of the output of an S-box. Using Walsh transform, a spectral characterization of the maximum correlation coefficients, together with their upper and lower bounds, are established. For the perfect nonlinear S-boxes designed for block ciphers, an upper bound on the maximum correlation coefficients is presented.

## 1 Introduction

Stream ciphers have a long history and still play an important role in secure communications. Typically, a stream cipher consists of a keystream generator whose output sequence is added modulo-2 to the plaintext sequence. So far, many kinds of keystream generators have been proposed, among which combination generators [15] and filter generators [14] are two of the most widely used. A combination generator consists of several linear feedback shift registers whose output sequences are combined by a nonlinear Boolean function (also called a nonlinear combining function or combiner). A filter generator consists of a single LFSR and uses a nonlinear Boolean function to filter the content of the shift register. It is clear that a filter generator is a special case of the combination generator, where all the combined sequences are produced by the same LFSR. The security of these keystream generators relies heavily on the nonlinear combining functions. In [17] Siegenthaler has shown that if the nonlinear combining function of a combination generator leaks information about the individual LFSR sequences into the output sequence, the LFSR sequences can be analyzed from a known segment of the keystream sequence. This kind of attacks are referred to as correlation attacks. To prevent correlation attacks, the nonlinear combining function should not leak information about its input. However, it has been shown in [9] that the output of a Boolean function is always correlated to some

M. Bellare (Ed.): CRYPTO 2000, LNCS 1880, pp. 501–514, 2000.
© Springer-Verlag Berlin Heidelberg 2000

linear functions of its input, in fact, the sum of the squares of the correlation coefficients is always 1. Thus, zero correlation to some linear functions of the input necessarily implies higher correlation to other linear functions of the input. The best one can do is to make the correlation between the output and every linear function of the input uniformly small.

In hardware, combination generators and filter generators have fast speed and simple VLSI circuitry. With respect to software implementation, however, there are two major problems for LFSR-base keystream generators. First, the speed of a software implemented LFSR is much slower than that of a hardware implemented one. Keystream generators consisting of several LFSRs make the speed of the software implementation even slower. Second, combination generators and filter generators only output one bit at every clock, which again makes the software implementation inefficient. To increase the throughput, a direct approach is to use nonlinear combining functions that output several bits at a time. Nonlinear functions with multiple-bit input and multiple-bit output are referred to as S-boxes in block ciphers and have been extensively studied [1,3,4,10,16]. In this paper, we investigate the design of S-boxes for stream ciphers. Compared with a combination generator using a Boolean function as the combiner, a combination generator utilizing an S-box as the combiner might be much easier to attack since every output bit of the S-box leaks information about the input. How to control the information leakage is a crucial problem for the design of keystream generators that produce several bits at a time. To mitigate the information leakage, we investigate the maximum correlation between linear functions of the input and all Boolean functions, linear and nonlinear, of the output of an S-box and introduce the notion of maximum correlation coefficient. It is shown that the mutual information between the output of an S-box and linear functions of the input is bounded by the maximum correlation coefficients. In terms of the Walsh transform, a spectral characterization of the maximum correlation coefficients is developed. Based on the spectral characterization bounds on the maximum correlation coefficients are developed, as well as the relationship between maximum correlation and nonlinearity [11] of S-boxes. For the perfect nonlinear S-boxes [10] designed for block ciphers to defend against differential cryptanalysis, an upper bound on the maximum correlation coefficient is presented.

## 2     Maximum Correlation of S-boxes

An S-box of $n$-bit input and $m$-bit output can be described by a function $F$ from $GF(2)^n$ to $GF(2)^m$. Let $x = (x_0, x_1, \ldots, x_{n-1}) \in GF(2)^n$ and $z = (z_0, z_1, \ldots, z_{m-1}) \in GF(2)^m$ denote the input and output of the S-box, i.e., $z = F(x)$. Then $F$ can be represented by a vector, $(f_0, f_1, \ldots, f_{m-1})$, of $m$ Boolean functions, where $z_i = f_i(x)$. Each Boolean function is called a component function of $F$. When $F$ is used to combine $n$ LFSR-sequences, we have a keystream generator that outputs $m$ binary sequences simultaneously. The individual binary sequence is produced by a combination generator in which a component function of $F$ is used as the combiner. Obviously, each binary sequence can be used to perform correlation attacks. As a consequence, the first design rule for the S-box is that every component function of $F$ has small correlation to linear functions. In addition, the $m$ binary sequences can also be combined together to perform correlation attacks. In this case, larger correlation to the LFSR-sequences may be exploited. To defend against this kind of attacks, the second design rule for the S-box is that every combination (linear and nonlinear) of the output has small correlation to linear functions of the input. It is clear that the second design rule is a generalization of the first one. To investigate the correlation properties of S-boxes, let's first review the notion of correlation coefficient of Boolean functions.

**Definition 1.** Let $f, g : GF(2)^n \rightarrow GF(2)$ be Boolean functions and $X$ be a uniformly distributed random variable over $GF(2)^n$. Then $Z = f(X)$ and $Z' = g(X)$ are random variables over $GF(2)$. The correlation coefficient of $f$ and $g$, denoted by $c(f, g)$, is defined as follows:

$$c(f, g) = P(Z = Z') - P(Z \neq Z'). \tag{1}$$

The correlation with linear functions is of special interest in the analysis and design of stream ciphers. A linear function of $n$ variables can be expressed as an inner product, $w \cdot x = w_1 x_1 \oplus w_2 x_2 \oplus \ldots \oplus w_n x_n$. Such a linear function is often denoted by $l_w(x)$. The correlation coefficient $c(f, l_w)$ describes the statistical dependency between $f$ and $l_w$, and is interpreted as the nonlinearity of $f$ with respect to $l_w$.

**Definition 2.** Let $F$ be a function from $GF(2)^n$ to $GF(2)^m$ and let $\mathcal{G}$ denote the set of all Boolean functions defined on $GF(2)^m$. For any $w \in GF(2)^n$, the maximum correlation coefficient between $F$ and the linear function $l_w$ is defined by

$$\mathcal{C}_F(w) = \max_{g \in \mathcal{G}} c(g \circ F, l_w),$$

where $g \circ F$ is the composition of $g$ and $F$, that is, $g \circ F(x) = g(F(x))$. If $g \in \mathcal{G}$ and $c(g \circ F, l_w)$ is maximum, then $g$ is called the maximum correlator of $F$ to $l_w$.

Nyberg [11] has investigated a special case where the set $\mathcal{G}$ contains only linear and affine functions. Based on Hamming distance, Nyberg defined the nonlinearity of S-boxes. The Hamming distance between two Boolean functions $f, g : GF(2)^n \rightarrow GF(2)$ is defined by

$$d(f, g) = |\{x \in GF(2)^n : f(x) \neq g(x)\}|.$$

It is easy to prove [15] that the Hamming distance $d(f, g)$ is related to the correlation coefficient $c(f, g)$ by

$$c(f, g) = 1 - 2^{-n+1} d(f, g). \tag{2}$$

**Definition 3.** Let $F$ be a function from $GF(2)^n$ to $GF(2)^m$. The nonlinearity of $F$ is defined as

$$\mathcal{N}_F = \min_{\substack{v \in GF(2)^m \\ v \neq 0}} \min_{\substack{w \in GF(2)^n \\ a \in GF(2)}} d(l_v \circ F, a \oplus l_w). \tag{3}$$

Assume that $\mathcal{N}_F = d(l_v \circ F, a \oplus l_w)$ for some nonzero $v \in GF(2)^m$ and some affine function $a \oplus l_w$. It is clear that $\mathcal{N}_F$ is also equal to $d(a \oplus l_v \circ F, l_w)$. By (2) and (3), $c(a \oplus l_v \circ F, l_w)$ is the maximum correlation between linear and affine functions of the output and linear functions of the input of $F$. By Definition 2, it is obvious that $c(a \oplus l_v \circ F, l_w) \leq \mathcal{C}_F(w)$, with strict inequality if the maximum correlator of $F$ to $l_w$ is not linear. Hence, the nonlinearity of $F$ does not necessarily imply maximum correlation between the output and linear functions of the input.

In general, it is difficult to figure out the maximum correlation coefficients since there are $2^{2^m}$ functions in $\mathcal{G}$. The following theorem provides a method to compute the maximum correlation coefficients.

**Theorem 1.** Let $F$ be a function from $GF(2)^n$ to $GF(2)^m$ and $X$ be a uniformly distributed random variable over $GF(2)^n$, $Z = F(X)$. For $w \in GF(2)^n$ and $z \in GF(2)^m$, let $e_w(z)$ denote the conditional probability difference between $w \cdot X = 1$ and $w \cdot X = 0$ under the condition $Z = z$, namely,

$$e_w(z) = P(w \cdot X = 1 | Z = z) - P(w \cdot X = 0 | Z = z). \tag{4}$$

Then

$$\mathcal{C}_F(w) = \sum_{z \in GF(2)^m} |e_w(z)| P(Z = z).$$

Moreover, the function $g(z) = sgn(e_w(z))$ is the maximum correlator of $F$ to $l_w$, where

$$sgn(x) = \begin{cases} 1, & x > 0, \\ 0 \text{ or } 1, & x = 0, \\ 0, & x < 0. \end{cases}$$

*Proof.* For any $w \in GF(2)^n$, and $g \in \mathcal{G}$, where $\mathcal{G}$ denotes the set of all Boolean functions on $GF(2)^m$, by (1),

$$c(g \circ F, l_w) = P(w \cdot X = g(Z)) - P(w \cdot X \neq g(Z)).$$

Since $P(w \cdot X = g(Z)) + P(w \cdot X \neq g(Z)) = 1$, $c(g \circ F, l_w)$ can be represented as follows

$$
\begin{aligned}
c(g \circ F, l_w) &= 2P(w \cdot X = g(Z)) - 1 \\
&= \sum_{z \in GF(2)^m} 2P(w \cdot X = g(Z) | Z = z) P(Z = z) - 1 \\
&= \sum_{z \in GF(2)^m} (2P(w \cdot X = g(z) | Z = z) - 1) P(Z = z).
\end{aligned}
$$

Therefore,

$$\max_{g \in \mathcal{G}} c(g \circ F, l_w) = \max_{g \in \mathcal{G}} \sum_{z \in GF(2)^n} (2P(w \cdot X = g(z) |_{Z=z}) - 1) P(Z = z). \tag{5}$$

Note that maximizing the sum in (5) is equivalent to maximizing every term in the sum, i.e.,

$$\max_{g \in \mathcal{G}} c(g \circ F, l_w) = \sum_{z \in GF(2)^n} \max_{g(z) \in GF(2)} (2P(w \cdot X = g(z)|Z = z) - 1)P(Z = z).$$

As $g(z) \in GF(2)$ and $P(w \cdot X = 1|Z = z) + P(w \cdot X = 0|Z = z) = 1$, it can be concluded that

$$\max_{g(z) \in GF(2)} (2P(w \cdot X = g(z)|Z=z)-1) = \begin{cases} 2P(w \cdot X = 1|Z=z) - 1, & \text{if } e_w(z) \geq 0, \\ 2P(w \cdot X = 0|Z=z) - 1, & \text{if } e_w(z) < 0. \end{cases}$$

On the other hand,

$$2P(w \cdot X = 1|Z = z) - 1 = P(w \cdot X = 1|Z = z) - P(w \cdot X = 0|Z = z),$$

while

$$2P(w \cdot X = 0|Z = z) - 1 = P(w \cdot X = 0|Z = z) - P(w \cdot X = 1|Z = z).$$

Hence,

$$\max_{g(z) \in GF(2)} (2P(w \cdot X = g(z)|Z = z) - 1) = |e_w(z)|,$$

the maximum value is reached if $g(z) = sgn(e_w(z))$. Therefore,

$$\mathcal{C}_F(w) = \max_{g \in \mathcal{G}} c(g \circ F, l_w) = \sum_{z \in GF(2)^m} |e_w(z)|P(Z = z),$$

and $g(z) = sgn(e_w(z))$ is the maximum correlator of $F$ to $l_w$.

Based on Theorem 1, the maximum correlation of $F$ to $l_w$ can be computed when the conditional probability difference $e_w(z)$ is known for every $z \in GF(2)^m$. The conditional probability difference can be calculated from the algebraic expression or from the truth table of $F$, with a complexity $2^{m+n}$, which is far below $2^n 2^{2^m}$ as required by exhaustive search. Furthermore, if $Z = F(X)$ is uniformly distributed over $GF(2)^m$, Theorem 1 implies the following bounds for $\mathcal{C}_F(w)$,

$$\min_{z \in GF(2)^m} |e_w(z)| \leq \mathcal{C}_F(w) \leq \max_{z \in GF(2)^m} |e_w(z)|.$$

Theoretically, the correlation between $Z = F(X)$ and $w \cdot X$ is measured by the mutual information $I(w \cdot X; Z)$. If $I(w \cdot X; Z)$ is small, we can not get much information about $w \cdot X$ from $Z$. In contrast, if $I(w \cdot X; Z)$ is large, we should be able to get some information about $w \cdot X$. However, the mutual information $I(w \cdot X; Z)$ does not tell us how to get information about $w \cdot X$ from $Z$. Using maximum correlation, we can approximate the random variable $w \cdot X$ by a Boolean function of $Z$. The successful rate of the approximation is measured by the maximum correlation coefficient. According to the Data Processing Inequality [5] of information theory, we have

$$I(w \cdot X; g(Z)) \leq I(w \cdot X; Z).$$

So we actually lose information about $w \cdot X$ when we perform maximum correlation. In the following, we investigate the relationship between mutual information and maximum correlation.

**Lemma 1.** Let $h(x)$ denote the binary entropy function, i.e.,

$$h(x) = -x \log_2 x - (1 - x) \log_2(1 - x), \quad 0 \le x \le 1. \tag{6}$$

Then, for $-0.5 \le x \le 0.5$, $1 - 2|x| \le h(0.5 + x) \le 1 - 2(\log_2 e)x^2$.

*Proof.* Let $\psi(x) = h(0.5 + x) - (1 - 2|x|)$. Since $h(0.5 + x)$ is a convex function, $\psi(x)$ is convex in both intervals $(-0.5, 0)$ and $(0, 0.5)$. Also, since $\psi(-0.5) = \psi(0) = \psi(0.5) = 0$, it can be concluded that $\psi(x) \ge 0$, for $-0.5 \le x \le 0.5$, i.e., $h(0.5 + x) \ge 1 - 2|x|$.

Next, let $\varphi(x) = 1 - 2(\log_2 e)x^2 - h(0.5 + x)$. Then

$$\varphi'(x) = -4x \log_2 e + (\ln(0.5 + x) - \ln(0.5 - x)) \log_2 e$$

and

$$\varphi''(x) = -4 \log_2 e + \frac{4 \log_2 e}{1 - (2x)^2}.$$

Since $0 \le 1 - (2x)^2 \le 1$, $\varphi''(x) \ge 0$. Hence, $\varphi(x)$ is a convex function. Moreover, $\varphi'(0) = 0$, which implies that $x = 0$ is the stationary point of $\varphi(x)$. Thus, $\varphi(x) \ge \varphi(0) = 0$.

**Definition 4.** Let $F$ be a function from $GF(2)^n$ to $GF(2)^m$ and $X$ be a uniformly distributed random variable over $GF(2)^n$. If $Z = F(X)$ is uniformly distributed over $GF(2)^m$, then $F$ is called a balanced function.

**Theorem 2.** Let $F$ be a balanced function from $GF(2)^n$ to $GF(2)^m$ and $X$ be a uniformly distributed random variable over $GF(2)^n$, $Z = F(X)$. For any nonzero $w \in GF(2)^n$,

$$I(w \cdot X; Z) \le C_F(w) \le \sqrt{2(\ln 2)I(w \cdot X; Z)}.$$

*Proof.* For any nonzero $w$, it is easy to prove that the random variable $w \cdot X$ is uniformly distributed over $GF(2)$. Thus, $H(w \cdot X) = 1$, and

$$I(w \cdot X; Z) = H(w \cdot X) - H(w \cdot X|Z)$$
$$= 1 + \sum_{\substack{y \in GF(2) \\ z \in GF(2)^n}} P(w \cdot X = y|Z = z)P(Z = z) \log_2 P(w \cdot X = y|Z = z).$$

Using the binary entropy function $h()$ defined in (6), the mutual information $I(w \cdot X; Z)$ can be expressed as

$$I(w \cdot X; Z) = 1 - \frac{1}{2^m} \sum_{z \in GF(2)^m} h(P(w \cdot x = 1|Z = z)). \tag{7}$$

By (4), $P(w \cdot X = 1|Z = z)$ can be replaced by $0.5 + e_w(z)/2$. Thus,

$$I(w \cdot X; Z) = 1 - \frac{1}{2^m} \sum_{z \in GF(2)^m} h(0.5 + e_w(z)/2).$$

By Lemma 1,

$$1 - |e_w(z)| \leq h(0.5 + e_w(z)/2) \leq 1 - \frac{1}{2}(e_w(z))^2 \log_2 e.$$

Therefore,

$$\frac{1}{2^m} \sum_{z \in GF(2)^m} \frac{\log_2 e}{2}(e_w(z))^2 \leq I(w \cdot X; Z) \leq \frac{1}{2^m} \sum_{z \in GF(2)^m} |e_w(z)|. \tag{8}$$

By Theorem 1, it is clear that $I(w \cdot X; Z) \leq \mathcal{C}_F(w)$.

Next, by Cauchy's inequality,

$$\sum_{z \in GF(2)^m} (e_w(z))^2 \geq \frac{1}{2^m}(\sum_{z \in GF(2)^m} |e_w(z)|)^2.$$

From (8), it follows that

$$(\frac{1}{2^m} \sum_{z \in GF(2)^m} |e_w(z)|)^2 \leq 2(\ln 2)I(w \cdot X; Z).$$

Again, by Theorem 1, $\mathcal{C}_F(w) \leq \sqrt{2(\ln 2)I(w \cdot X; Z)}$.

Theorem 2 establishes a connection between the mutual information and the maximum correlation. This connection provides us flexibility for the analysis and design of S-boxes. Conceptually, mutual information provides us a tool to analyze the information leakage from the output bits while maximum correlation explicitly describes the correlation properties of S-boxes. For example, to design a balanced function with $I(w \cdot X; Z) \leq \delta$, we only need to design a balanced function $F$ with $\mathcal{C}_F(w) < \delta$.

## 3 A Spectral Characterization of Maximum Correlation Coefficients

In the analysis and design of Boolean functions, Walsh transform [6] has played an important role. The merit of Walsh transform lies in that it provides illustrative description of Boolean functions having certain cryptographic properties [12,18]. The Walsh transform of a real-valued function $f$ on $GF(2)^n$ is defined as follows:

$$S_f(w) = 2^{-n} \sum_{x \in GF(2)^n} f(x)(-1)^{w \cdot x}.$$

The function $f(x)$ can be recovered from the inverse Walsh transform,

$$f(x) = \sum_{w \in GF(2)^n} S_f(w)(-1)^{w \cdot x}.$$

When $f$ is a Boolean function, we often turn it into a function $\hat{f}(x) = (-1)^{f(x)}$ and define the Walsh transform of $f$ as that of $\hat{f}$.

Let $F$ be a function from $GF(2)^n$ to $GF(2)^m$. For any $v \in GF(2)^m$, $l_v \circ F$ defines a Boolean function on $GF(2)^n$. For $x \in GF(2)^n$, $l_v \circ F(x)$ can be expressed by the inner product $v \cdot F(x)$. The Walsh transform of $l_v \circ F$, denoted by $S_F(v, w)$, is called the Walsh transform of $F$ and is given by

$$S_F(v, w) = \frac{1}{2^n} \sum_{x \in GF(2)^n} (-1)^{v \cdot F(x) + w \cdot x}.$$

**Theorem 3.** Let $F$ be a function from $GF(2)^n$ to $GF(2)^m$. For any $w \in GF(2)^n$,

$$\mathcal{C}_F(w) = \frac{1}{2^m} \sum_{z \in GF(2)^m} | \sum_{v \in GF(2)^m} S_F(v, w)(-1)^{v \cdot z}|.$$

*Proof.* Let $X$ be a uniformly distributed random variable over $GF(2)^n$ and $Z = F(X)$. For any $w \in GF(2)^n$ and $v \in GF(2)^m$,

$$P(w \cdot X = v \cdot Z) = P(w \cdot X = 0, v \cdot Z = 0) + P(w \cdot X = 1, v \cdot Z = 1). \quad (9)$$

Since $P(v \cdot Z = 0) = P(w \cdot X = 0, v \cdot Z = 0) + P(w \cdot X = 1, v \cdot Z = 0)$, we have

$$P(v \cdot Z = 0) - P(w \cdot X = v \cdot Z) = P(w \cdot X = 1, v \cdot Z = 0) - P(w \cdot X = 1, v \cdot Z = 1). \quad (10)$$

Note that the right-hand side of (10) is equal to the sum,

$$\sum_{z \in GF(2)^m} P(w \cdot X = 1, Z = z)(-1)^{v \cdot z},$$

which implies that $P(v \cdot Z = 0) - P(w \cdot X = v \cdot Z)$ is the Walsh transform of $2^m P(w \cdot X = 1, Z = z)$. Taking the inverse Walsh transform,

$$P(w \cdot X = 1, Z = z) = \frac{1}{2^m} \sum_{v \in GF(2)^m} (P(v \cdot Z = 0) - P(w \cdot X = v \cdot Z))(-1)^{v \cdot z}. \quad (11)$$

Next, the probability $P(v \cdot Z = 0)$ can also be expressed by the following sum,

$$P(v \cdot Z = 0) = \sum_{z \in GF(2)^m} P(Z = z, v \cdot z = 0)$$

$$= \sum_{z \in GF(2)^m} P(Z = z)(1 + (-1)^{v \cdot z})/2.$$

Thus,

$$2P(v \cdot Z = 0) - 1 = \sum_{z \in GF(2)^m} P(Z = z)(-1)^{v \cdot z},$$

which implies that $2^{-m}(2P(v \cdot Z = 0) - 1)$ is the Walsh transform of $2^m P(Z = z)$. Therefore,

$$P(Z = z) = \frac{1}{2^m} \sum_{v \in GF(2)^m} (2P(v \cdot Z = 0) - 1)(-1)^{v \cdot z}. \quad (12)$$

From (11) and (12), it follows that

$$2P(w \cdot X = 1, Z = z) - P(Z = z) = \frac{1}{2^m} \sum_{v \in GF(2)^m} (1 - 2P(w \cdot X = v \cdot Z))(-1)^{v \cdot z}.$$

(13)

By Theorem 1,

$$\mathcal{C}_F(w) = \sum_{z \in GF(2)^m} |e_w(z)| P(Z = z) = \sum_{z \in GF(2)^m} |2P(w \cdot X = 1, Z = z) - P(Z = z)|.$$

Hence, by (13),

$$\mathcal{C}_F(w) = \frac{1}{2^m} \sum_{z \in GF(2)^m} | \sum_{v \in GF2^m} (1 - 2P(w \cdot X = v \cdot Z))(-1)^{v \cdot z}|. \quad (14)$$

Since $X$ is uniformly distributed over $GF(2)^n$, $P(w \cdot X = v \cdot Z) - P(w \cdot X \neq v \cdot Z)$ equals to

$$2^{-n}(|\{x \in GF(2)^n : w \cdot x = v \cdot F(x)\}| - |\{x \in GF(2)^n : w \cdot x \neq v \cdot F(x)\}|),$$

which can be represented as

$$\frac{1}{2^n} \sum_{x \in GF(2)^n} (-1)^{v \cdot F(x) + w \cdot x}.$$

Therefore,

$$P(w \cdot X = v \cdot Z) - P(w \cdot X \neq v \cdot Z) = S_F(v, w).$$

As a consequence,

$$2P(w \cdot X = v \cdot Z) - 1 = P(w \cdot X = v \cdot Z) - P(w \cdot X \neq v \cdot Z) = S_F(v, w). \quad (15)$$

Substituting (15) into (14),

$$\mathcal{C}_F(w) = \frac{1}{2^m} \sum_{z \in GF(2)^m} | \sum_{v \in GF(2)^m} S_F(v, w)(-1)^{v \cdot z}|,$$

Theorem 3 relates the maximum correlation coefficient to the Walsh transforms of a set of Boolean functions. Since Boolean functions have been extensively studied with respect to various cryptographic properties. Using Theorem 3, we can make use of the known results about Boolean functions to study the correlation properties of S-boxes.

**Theorem 4.** Let $F$ be a function from $GF(2)^n$ to $GF(2)^m$. For any $w \in GF(2)^n$,

$$\mathcal{C}_F(w) \leq 2^{m/2} \max_{v \in GF(2)^m} |S_F(v, w)|.$$

Moreover,

$$1 \leq \sum_{w \in GF(2)^n} \mathcal{C}_F^2(w) \leq 2^m.$$

*Proof.* By definition 2, it is obvious that $\mathcal{C}_F(0) = 1$. Hence,

$$\sum_{w \in GF(2)^n} \mathcal{C}_F^2(w) \geq 1.$$

By Theorem 3, the maximum correlation coefficient $\mathcal{C}_F(w)$ can be expressed as

$$\mathcal{C}_F(w) = \frac{1}{2^m} \sum_{z \in GF(2)^m} |b_z(w)|,$$

where

$$b_z(w) = \sum_{v \in GF(2)^m} S_F(v, w)(-1)^{v \cdot z}.$$

The sum of the squares of $b_z(w)$ over $z \in GF(2)^m$ is described by

$$\sum_{z \in GF(2)^m} b_z^2(w) = \sum_{z \in GF(2)^m} \left( \sum_{u \in GF(2)^m} S_F(u, w)(-1)^{u \cdot z} \right) \left( \sum_{v \in GF(2)^m} S_F(v, w)(-1)^{v \cdot z} \right)$$

$$= \sum_{\substack{u \in GF(2)^m \\ v \in GF(2)^m}} S_F(u, w) S_F(v, w) \sum_{z \in GF(2)^m} (-1)^{(u \oplus v) \cdot z}.$$

According to the orthogonal property of Walsh function [7],

$$\sum_{z \in GF(2)^m} (-1)^{(u \oplus v) \cdot z} = \begin{cases} 2^m, & \text{if } u = v, \\ 0, & \text{otherwise.} \end{cases}$$

Hence,

$$\sum_{z \in GF(2)^m} b_z^2(w) = 2^m \sum_{v \in GF(2)^m} S_F^2(v, w). \qquad (16)$$

By Cauchy inequality,

$$\mathcal{C}_F^2(w) = \frac{1}{2^{2m}} \left( \sum_{z \in GF(2)^m} |b_z(w)| \right)^2 \leq \frac{1}{2^m} \sum_{z \in GF(2)^m} b_z^2(w).$$

By (16), it follows that

$$\mathcal{C}_F^2(w) \leq \sum_{v \in GF(2)^m} S_F^2(v, w).$$

Hence,

$$\mathcal{C}_F(w) \leq 2^{m/2} \max_{v \in GF(2)^m} |S_F(v, w)|,$$

and

$$\sum_{w \in GF(2)^n} \mathcal{C}_F^2(w) \leq \sum_{v \in GF(2)^m} \sum_{w \in GF(2)^n} S_F^2(v, w).$$

By Parsevals's Theorem [7],

$$\sum_{w \in GF(2)^m} S_F^2(v, w) = 1.$$

Therefore,

$$\sum_{w \in GF(2)^n} \mathcal{C}_F^2(w) \leq 2^m,$$

For Booleans functions, it is well known [9] that the sum of the squares of the correlation coefficients is always 1. For S-boxes, however, the sum of the squares of the maximum correlation coefficients might be greater than 1. To defend against correlation attacks, the maximum correlation coefficients should be uniformly small for all nonzero $w$. Theorem 4 indicates that the maximum correlation coefficients can be controlled through the Walsh spectral coefficients. It has been shown [9] that for any nonzero $w$, the maximum value of $|S_F(v, w)|$ is at least $2^{-n/2}$. As a consequence, the tightest upper bound deduced from Theorem 4 is $2^{(m-n)/2}$, which means that the number of output bits $m$ can not be too large compared with the number of input bits $n$. Obviously, increasing the value of $m$ will introduce extra source that leaks information about the input. In the extreme case when $m = n$ and $F$ is a one-to-one function, $\mathcal{C}_F(w) = 1$. Hence, nonlinear permutations are weak combining functions.

**Theorem 5.** Let $F$ be a function from $GF(2)^n$ to $GF(2)^m$. For any $w \in GF(2)^n, w \neq 0$,

$$\mathcal{C}_F(w) \leq 2^{m/2}(1 - 2^{-n+1}\mathcal{N}_F),$$

where $\mathcal{N}_F$ is the nonlinearity of $F$ defined by (3).

*Proof.* As has been shown in [9], the nonlinearity $\mathcal{N}_F$ can be expressed in terms of the Walsh transform of $F$,

$$\mathcal{N}_F = \min_{\substack{v \in GF(2)^m \\ v \neq 0}} \min_{\substack{w \in GF(2)^n \\ a \in GF(2)}} d(l_v \circ F, a \oplus l_w)$$

$$= \min_{0 \neq v \in GF(2)^m} 2^{n-1}(1 - \max_{w \in GF(2)^n} |S_F(v, w)|).$$

Thus,

$$\max_{w \in GF(2)^n} \max_{0 \neq v \in GF(2)^m} |S_F(v, w)| = 1 - 2^{-n+1}\mathcal{N}_F.$$

For any nonzero $w \in GF(2)^n$, $S_F(0, w) = 0$. Hence,

$$\max_{v \in GF(2)^m} |S_F(v, w)| = \max_{0 \neq v \in GF(2)^m} |S_F(v, w)|.$$

By Theorem 4,

$$\mathcal{C}_F(w) \leq 2^{m/2} \max_{0 \neq v \in GF(2)^m} |S_F(v, w)|$$

$$\leq 2^{m/2} \max_{w \in GF(2)^n} \max_{0 \neq v \in GF(2)^m} |S_F(v, w)|$$

$$= 2^{m/2}(1 - 2^{-n+1}\mathcal{N}_F),$$

Theorem 5 demonstrates that the maximum correlation coefficients are small if the nonlinearity is large. Based on Theorem 5, we can control the maximum correlation coefficients by controlling the nonlinearity of S-boxes.

# 4    Maximum Correlation Analysis of Perfect Nonlinear S-boxes

Originally S-boxes were used in the American Data Encryption Standard (DES). The security analysis of DES has resulted in a series of design criteria [1,3] for S-boxes. These design criteria reflect the capability of DES-like block ciphers to defend against the known attacks. So far, the most successful attacks on DES-like block ciphers are differential cryptanalysis developed by Biham and Shamir [2] and linear cryptanalysis developed by Matsui [8]. Differential cryptanalysis makes uses of the property that with certain changes in the input of an S-box the change in the output is known with high probability. To immune against differential cryptanalysis, S-boxes should have evenly distributed output changes corresponding to any input changes. Nyberg [10] defines this type of S-boxes as perfect nonlinear S-boxes.

**Definition 5.** A function $F$ from $GF(2)^n$ to $GF(2)^m$ is called a perfect non-linear S-box if for every fixed $w \in GF(2)^n$, the difference $F(x+w) - F(x)$ takes on each value $z \in GF(2)^n$ for $2^{n-m}$ values of $x$.

When $m = 1$, the perfect nonlinear S-box $F$ is also called a perfect nonlinear Boolean function. In [9] Meier and Staffelbach proved that perfect nonlinear Boolean functions are actually a class of previously known bent functions introduced by Rothaus [13] in combinatorial theory.

**Definition 6.** A Boolean function $f$ defined on $GF(2)^n$ is called a bent function if for every $w \in GF(2)^n$, $|S_f(w)| = 2^{-n/2}$.

Nyberg [10] has shown that an S-box is perfect nonlinear if and only if every nonzero linear function of the output variables is a bent functions.

**Lemma 2.** A function $F$ from $GF(2)^n$ to $GF(2)^m$ is perfect nonlinear if and only if for every nonzero $v \in GF(2)^m$ the function $l_v \circ F$ is a perfect nonlinear Boolean function or bent function.

Based on Lemma 2, two construction methods of perfect nonlinear S-boxes were given by Nyberg. In the following we will study the maximum correlation properties of these S-boxes.

**Theorem 6.** Let $F$ be a perfect nonlinear S-box from $GF(2)^n$ to $GF(2)^m$. Then for any non-zero $w \in GF(2)^n$,

$$C_F(w) \leq 2^{(m-n)/2}.$$

*Proof.* By Lemma 2, $v \cdot F(x)$ is a bent function for every nonzero $v \in GF(2)^m$, thus, for every $w \in GF(2)^n$,

$$|S_F(v, w)| = 2^{-n/2}.$$

When $v = 0$, $S_F(v, w) = 0$ for nonzero $w$. Hence, for every nonzero $w \in GF(2)^n$,

$$\max_{v \in GF(2)^m} |S_F(v, w)| = \max_{0 \neq v \in GF(2)^m} |S_F(v, w)| = 2^{-n/2}.$$

By Theorem 4, it is clear that

$$\mathcal{C}_F(w) \leq 2^{m/2} \max_{v \in GF(2)^m} |S_F(v, w)| = 2^{(m-n)/2}.$$

For a perfect nonlinear S-box of $n$-bit input and $m$-bit output, it is known [9] that the correlation coefficient between each output bit and every nonzero linear function of the input bits is $2^{-n/2}$, which is very small when $n$ is large. Lemma 2 implies that linear functions of the output bits do not help increasing the correlation coefficients. However, Theorem 6 demonstrates that the correlation coefficients may be increased by a factor as large as $2^{m/2}$ if nonlinear functions of the output bits are used. Hence, when used in stream ciphers, we need to consider the number of bits a perfect nonlinear S-box should output. Nyberg [10] has shown that perfect nonlinear S-boxes only exist if $m \leq n/2$. With respect to correlation attacks, we also want to make sure that $2^{(m-n)/2}$ is a very small number. On the other hand, large value of $m$ means large throughput of the data streams generated by keystream generators. Hence, there is a trade-off between the capability to defend against correlation attacks and the throughput of the keystream sequences. In the design of stream ciphers, we choose $n$ and $m$ according to the expected security strength and the throughput requirement.

## 5    Conclusion

This paper is devoted to the design of S-boxes for stream ciphers. When used to combine several LFSR sequences in a combination generator, S-boxes leak more information about the LFSR sequences than Boolean functions. To control the information leakage, the notion of maximum correlation is introduced. It is a generalization of Nyberg's nonlinearity of S-boxes. The merit with maximum correlation is that more information about linear functions of the input may be obtained when all Boolean functions, instead of just linear functions, of the output of the S-box are exploited. In terms of Walsh transform, a spectral characterization of the maximum correlation coefficients is presented, which can be used to investigate upper and lower bounds on the maximum correlation coefficients as well as the relationship between maximum correlation and nonlinearity. For a perfect nonlinear S-box with n-bit input and m-bit output, it is shown that the maximum correlation coefficients are upper bounded by $2^{(m-n)/2}$.

# References

1. C.M. Adams and S.E. Tavares. The structured design of cryptographically good S-boxes. *Journal of Cryptology*, vol. 3, pp. 27-41, 1990.
2. E. Biham and A. Shamir. Differential cryptanalysis of DES-like cryptosystems. *Journal of Cryptology*, vol. 4, no. 1, pp. 3-72, 1991.
3. F. Chabaud and S. Vaudenay. Links between differential and linear cryptanalysis. In *Lectures in Computer Science, Advances in Cryptology-EUROCRYPT'94*, vol. 950, pp. 356-365, Springer-Verlag, 1995.
4. J.H. Cheon, S. Chee, and C. Park. S-boxes with controllable nonlinearity. In *Lecture Notes in Computer Science, Advances in Cryptology-EUROCRYPT'99*, vol. 1592, pp. 286-294, Springer-Verlag, 1999.
5. T.M. Cover and J.A. Thomas. *Elements of Information Theory*. John Wiley & Sons Inc., 1991.
6. S.W. Golomb. *Shift Register Sequences*. Holden-Day, San Francisco, 1976. Reprinted by Aegean Park Press, 1982.
7. M.G. Karpovsky. *Finite Orthogonal Series in the Design of Digital Devices*. New York and Jerudalem: Wiley and TUP, 1976.
8. M. Matsui. Linear cryptanalysis method for DES ciphers. In *Lecture Notes in Computer Science, Advances in Cryptology-EUROCRYPT'93*, vol. 765, pp. 386-397, Springer-Verlag, 1994.
9. W. Meier and O. Staffelbach. Nonlinear criteria for cryptographic functions. In *Lecture Notes of Computer Science, Advances in Cryptology: Proceedings of EURO-CRYPT'89*, vol. 434, pp. 549–562, Springer–Verlag, 1990.
10. K. Nyberg. Perfect nonlinear S-boxes. In *Lecture Notes in Computer Science, Advance in Cryptology-EUROCRYPT'91*, vol. 547, pp. 378-385, Springer-Verlag, 1991.
11. K. Nyberg. On the construction of highly nonlinear permutations. In *Lecture Notes in Computer Science, Advance in Cryptology-EUROCRYPT'92*, vol. 658, pp. 92-98, Springer-Verlag, 1993.
12. B. Preneel, W.V. Leekwijck, L.V. Linden, R. Govaerts, and J. Vandewalle. Propagation characteristics of Boolean functions. In *Lecture Notes in Computer Science, Advance in Cryptology-EUROCRYPT'90*, vol. 473, pp. 161-173, Springer-Verlag, 1991.
13. O.S. Rothaus. On bent functions. *J. Combinatorial Theory*, Series A, vol. 20, pp. 300–305, 1976.
14. R. Rueppel. *Analysis and Design of Stream Ciphers*. Springer-Verlag, Berlin, 1986.
15. R.A. Rueppel. Stream ciphers. In *Contemporary Cryptology: The Science of Information Integrity*, G. Simmons ed., New York: IEEE Press, pp. 65–134, 1991.
16. J. Seberry, X.M. Zhang, and Y. Zheng. Systematic generation of cryptographically robust S-boxes. In *Proceedings of the first ACM Conference on Computer and Communications Security*, pp. 172-182, 1993.
17. T. Siegenthaler. Decrypting a class of stream ciphers using ciphertext only. *IEEE Trans. Computer*, vol. C-34, no. 1, pp. 81–85, 1985.
18. Guozhen Xiao and J.L. Massey. A spectral characterization of correlation–immune combining functions. *IEEE Trans. on Information Theory*, vol. IT–34, no. 3, pp. 564–571, 1988.

# Nonlinearity Bounds and Constructions of Resilient Boolean Functions

Palash Sarkar[1] and Subhamoy Maitra[2]

[1] Applied Statistics Unit, Indian Statistical Institute,
203, B T Road, Calcutta 700 035, INDIA
palash@isical.ac.in
[2] Computer and Statistical Service Center, Indian Statistical Institute,
203, B T Road, Calcutta 700 035, INDIA
subho@isical.ac.in

**Abstract.** In this paper we investigate the relationship between the nonlinearity and the order of resiliency of a Boolean function. We first prove a sharper version of McEliece theorem for Reed-Muller codes as applied to resilient functions, which also generalizes the well known Xiao-Massey characterization. As a consequence, a nontrivial upper bound on the nonlinearity of resilient functions is obtained. This result coupled with Siegenthaler's inequality leads to the notion of best possible trade-off among the parameters: number of variables, order of resiliency, nonlinearity and algebraic degree. We further show that functions achieving the best possible trade-off can be constructed by the Maiorana-McFarland like technique. Also we provide constructions of some previously unknown functions.

**Keywords:** Boolean functions, Balancedness, Algebraic Degree, Nonlinearity, Correlation Immunity, Resiliency, Stream Ciphers, Combinatorial Cryptography.

## 1 Introduction

Stream cipher cryptosystems are extensively used for defence communications worldwide and provide a reliable and efficient method of secure communication. In the standard model of stream cipher the outputs of several independent Linear Feedback Shift Register (LFSR) sequences are combined using a nonlinear Boolean function to produce the keystream. This keystream is bitwise XORed with the message bitstream to produce the cipher. The decryption machinery is identical to the encryption machinery.

Siegenthaler [23] pointed out that if the combining function is not chosen properly then the whole system is susceptible to a divide-and-conquer attack. He also defined the class of functions which can resist such attacks [22]. Moreover, such functions must also provide resistance against other well known attacks [4]. Later work on stream ciphers with memoryless Boolean functions have proceeded along two lines. In one direction, Siegenthaler's attack has been successively

M. Bellare (Ed.): CRYPTO 2000, LNCS 1880, pp. 515–532, 2000.
© Springer-Verlag Berlin Heidelberg 2000

refined and sharpened in a series of papers [14,11,10,15]. On the other hand, in another direction, researchers have tried to design better and better Boolean functions for use in stream cipher systems. Here we concentrate on this second direction of research.

It is now generally accepted that for a Boolean function to be used in stream cipher systems it must satisfy several properties - balancedness, high nonlinearity, high algebraic degree and high order of correlation immunity (see Section 2 for definitions). Also a balanced correlation immune function is called a resilient function. However, we would like to point out that though the necessity of these properties is undisputed, it is by no means clear that these are also sufficient to resist all kinds of attacks. In fact, for practical stream cipher systems, security is judged by the ability of the system to guard against the currently known attacks. In such a situation, it is important to understand the exact degree of protection that a particular system provides. The present effort should be viewed as a contribution to the development of this understanding.

Each of the above mentioned parameters provide protection against a class of attacks. Also it is known that there are certain trade-offs involved among the above parameters. For example, Siegenthaler showed [22] that for an $n$-variable function, of degree $d$ and order of correlation immunity $m$, the following holds: $m + d \leq n$. Further, if the function is balanced then $m + d \leq n - 1$. However, the exact nature of trade-off between order of correlation immunity and nonlinearity has not been previously investigated. A series of papers [1,21,3,5,13,16,19] have approached the construction problem in the following fashion. Fix the number of variables and the order of correlation immunity (and possibly the algebraic degree) and try to design balanced functions with as high nonlinearity as possible. Many interesting ideas have been used and successively better results have been proved.

Thus, the natural question that arises is, what is the maximum nonlinearity achievable with a fixed number of variables and a fixed order of correlation immunity? More generally, the crucial question is when can we say that a balanced Boolean function achieves *the best possible trade-off* among the following parameters: number of variables, order of correlation immunity, nonlinearity and algebraic degree? From a practical point of view, a designer of a stream cipher system will typically try to achieve a balance between the size of the hardware and the security of the system. The size of the hardware is directly proportional to the number of input variables of the combining Boolean function. On the other hand, the parameters nonlinearity, algebraic degree and order of resiliency have to be chosen to be large enough so that the current attacks do not succeed in reasonable time. Thus it is important to have *good* functions on *small* number of variables. A natural choice for good functions are those which achieve the best possible trade-off among the above mentioned parameters. Thus the ability to identify, construct and implement such functions is very important from the designer's point of view.

In a more theoretical direction, one of the main results we prove is that if $f$ is an $n$-variable, $m$-resilient function, then $W_f(\overline{\omega}) \equiv 0 \bmod 2^{m+2}$, for all

$\overline{\omega} \in \{0,1\}^n$, where $W_f()$ is the Walsh transform of $f$. This is a generalization of the famous Xiao-Massey characterization of correlation immune functions. More importantly, the result has a root in coding theory. From Siegenthaler's inequality it is known that any $n$-variable, $m$-resilient function has degree at most $n-m-1$ and hence is in Reed-Muller code $\mathcal{R}(n-m-1, n)$. The famous McEliece theorem [12, Page 447] when applied to Reed-Muller code $\mathcal{R}(n-m-1, n)$ guarantees that $W_f(\overline{\omega}) \equiv 0 \bmod 2^{1+\lfloor \frac{n-1}{n-m-1} \rfloor}$. The above mentioned result that we prove is much sharper. From this result we obtain a nontrivial upper bound on the nonlinearity of $n$-variable, $m$-resilient functions. In a series of papers Hou [8,7,9], has investigated the covering radius problem for Reed-Muller codes. The covering radius of first order Reed-Muller code is equal to the maximum possible nonlinearity of $n$-variable functions. As observed before, an $m$-resilient function is in $\mathcal{R}(n-m-1, n)$, but it does not seem that the current results on the covering radius of higher order Reed-Muller codes can be applied to obtain an upper bound on the maximum possible nonlinearity of $m$-resilient functions.

We show that one of the existing construction methods (the Maiorana-McFarland like construction technique) can provide all but finitely many functions of certain infinite sequences of functions which provide best possible trade-off among the parameters: number of variables, order of resiliency, nonlinearity and algebraic degree. However, the Maiorana-McFarland like construction technique does not work in all cases. In such cases, we introduce a new sharper construction method to obtain such functions. Functions with these parameters were not known earlier. We also discuss important issues on functions with small number of variables in Section 5.

Future work on resilient Boolean functions should proceed along the following lines. It is not clear whether the upper bounds on nonlinearity of resilient functions obtained in Theorem 2 are tight. It will be a major task to show that in certain cases the upper bounds are not tight and to obtain sharper upper bounds. However, in significantly many cases these upper bounds can be shown to be tight (for example see Table 1 in Section 3). Based on these upper bounds, we introduce concepts of Type-I and Type-II resilient functions (see Section 4). Type-II resilient functions with maximum possible algebraic degree are well suited for use in stream ciphers. We have used existing and new techniques to construct such functions. Also it seems that the construction of these functions are difficult in some cases. Either obtaining new construction methods for these functions or showing their non-existence should be the main theme of any further work. On one hand these are combinatorially challenging problems and on the other hand their answers have direct relevance to the task of designing secure stream cipher systems.

## 2   Preliminaries

In this section we introduce a few basic concepts. Note that we denote the addition operator over $GF(2)$ by $\oplus$.

**Definition 1.** *For binary strings $S_1, S_2$ of same length $\lambda$, we denote by $\#(S_1 = S_2)$ (respectively $\#(S_1 \neq S_2)$), the number of places where $S_1$ and $S_2$ are equal (respectively unequal). The Hamming distance between $S_1, S_2$ is denoted by $d(S_1, S_2)$, i.e. $d(S_1, S_2) = \#(S_1 \neq S_2)$. The Walsh Distance $wd(S_1, S_2)$, between $S_1$ and $S_2$, is defined as, $wd(S_1, S_2) = \#(S_1 = S_2) - \#(S_1 \neq S_2)$. Note that, $wd(S_1, S_2) = \lambda - 2\,d(S_1, S_2)$. Also the Hamming weight or simply the weight of a binary string $S$ is the number of ones in $S$. This is denoted by $wt(S)$. An n-variable function $f$ is said to be balanced if its output column in the truth table contains equal number of 0's and 1's (i.e. $wt(f) = 2^{n-1}$).*

**Definition 2.** *An n-variable Boolean function $f(X_n, \ldots, X_1)$ can be considered to be a multivariate polynomial over $GF(2)$. This polynomial can be expressed as a sum of products representation of all distinct k-th order products $(0 \leq k \leq n)$ of the variables. More precisely, $f(X_n, \ldots, X_1)$ can be written as $a_0 \oplus (\bigoplus_{i=1}^{i=n} a_i X_i) \oplus (\bigoplus_{1 \leq i \neq j \leq n} a_{ij} X_i X_j) \oplus \ldots \oplus a_{12\ldots n} X_1 X_2 \ldots X_n$ where the coefficients $a_0, a_{ij}, \ldots, a_{12\ldots n} \in \{0, 1\}$. This representation of $f$ is called the algebraic normal form (ANF) of $f$. The number of variables in the highest order product term with nonzero coefficient is called the algebraic degree, or simply degree of $f$.*

In the stream cipher model, the combining function $f$ must be so chosen that it increases the linear complexity [17] of the resulting key stream. High algebraic degree provides high linear complexity [18,4] and hence it is desirable for $f$ to have high algebraic degree. Another important cryptographic property for a Boolean function is high nonlinearity. A function with low nonlinearity is prone to *Best Affine Approximation* (BAA) [4, Chapter 3] attack.

**Definition 3.** *Functions of degree at most one are called affine functions. An affine function with constant term equal to zero is called a linear function. The set of all n-variable affine (respectively linear) functions is denoted by $A(n)$ (respectively $L(n)$). The nonlinearity of an n variable function $f$ is $nl(f) = min_{g \in A(n)}(d(f, g))$, i.e. the distance from the set of all n-variable affine functions.*

An important tool for the analysis of Boolean function is its Walsh transform, which we define next.

**Definition 4.** *Let $\overline{X} = (X_n, \ldots, X_1)$ and $\overline{\omega} = (\omega_n, \ldots, \omega_1)$ both belong to $\{0, 1\}^n$ and $\overline{X}.\overline{\omega} = X_n \omega_n \oplus \ldots \oplus X_1 \omega_1$. Let $f(\overline{X})$ be a Boolean function on n variables. Then the Walsh transform of $f(\overline{X})$ is a real valued function over $\{0, 1\}^n$ that can be defined as $W_f(\overline{\omega}) = \sum_{\overline{X} \in \{0,1\}^n} (-1)^{f(\overline{X}) \oplus \overline{X}.\overline{\omega}}$. The Walsh transform is sometimes called the spectral distribution or simply the spectrum of a Boolean function.*

Xiao and Massey [6] provided a spectral characterization of correlation immune functions. Here we state this characterization as the definition of correlation immunity.

**Definition 5.** *A function $f(X_n, \ldots, X_1)$ is m-th order correlation immune (CI) iff its Walsh transform $W_f$ satisfies $W_f(\overline{\omega}) = 0$, for $1 \leq wt(\overline{\omega}) \leq m$. Further, if $f$ is balanced then $W_f(\overline{0}) = 0$. Balanced m-th order correlation immune functions are called m-resilient functions. Thus, a function $f(X_n, \ldots, X_1)$ is m-resilient iff its Walsh transform $W_f$ satisfies $W_f(\overline{\omega}) = 0$, for $0 \leq wt(\overline{\omega}) \leq m$.*

The relationship between Walsh transform and Walsh distance is [13] $W_f(\overline{\omega}) = wd(f, \bigoplus_{i=1}^{i=n} \omega_i X_i)$.

We will require the following basic result, which is known [12, Page 8], but we give a proof for the sake of completeness. Let $f \times g$ denote the Boolean function $h$ whose ANF is the product (over $GF(2)$) of the ANFs (which are polynomials over $GF(2)$) of $f$ and $g$, i.e., $h(X_n, \ldots, X_1) = f(X_n, \ldots, X_1) \times g(X_n, \ldots, X_1)$.

**Lemma 1.** *Let $f(X_n, \ldots, X_1)$ and $g(X_n, \ldots, X_1)$ be two n-variable functions. Then $d(f, g) = wt(f) + wt(g) - 2wt(f \times g)$.*

*Proof.* Let $F_2^n = \{0, 1\}^n$. The function $f$ can be completely described by a subset $A$ of $F_2^n$, such that $(b_n, \ldots, b_1) \in F_2^n$ is in $A$ iff $f(b_n, \ldots, b_1) = 1$. This set $A$ is usually called the support of $f$. We can get a similar support $B$ for $g$. The support of $f \oplus g$ is $A \Delta B$ (symmetric difference) and the support of $f \times g$ is $A \cap B$. The result follows from the fact that $d(f, g) = wt(f \oplus g) = | A\Delta B | = | A | + | B | - 2| A \cap B |$.  $\square$

### 2.1  Some Useful Notations

Before proceeding, we would like to introduce a few notations for future convenience. Recall from Definition 3 that $nl(f)$ denotes the nonlinearity of a Boolean function $f$. We use $nlmax(n)$ to denote the maximum possible nonlinearity of an $n$-variable function. The maximum possible nonlinearity of an $n$-variable, $m$-resilient function is denoted by $nlr(n, m)$ and the maximum possible nonlinearity of an $n$-variable function which is CI of order $m$ is denoted by $nlc(n, m)$.

By an $(n, m, d, x)$ function we mean an $n$-variable, $m$-resilient function with degree $d$ and nonlinearity $x$. By $(n, 0, d, x)$ function we mean a balanced $n$-variable function with degree $d$ and nonlinearity $x$. In the above notation the degree component is replaced by a '$-$', i.e., $(n, m, -, x)$, if we do not want to specify the degree.

Further, given an affine function $l \in A(n)$, by $ndg(l)$ we denote the number of variables on which $l$ is nondegenerate.

### 2.2  Maiorana-McFarland Like Construction Technique

There are several construction methods for resilient Boolean functions in the literature. Perhaps the most important of all these is the Maiorana-McFarland like construction technique which has been investigated in a number of previous papers [1,21,3,2,19]. Here we briefly describe the basic method. Let $\pi$ be a map from $\{0, 1\}^r$ to $\{0, 1\}^k$, where for any $\overline{X} \in \{0, 1\}^r$, $wt(\pi(\overline{X})) \geq m + 1$. Let $f : \{0, 1\}^{r+k} \rightarrow \{0, 1\}$ be a Boolean function defined as $f(\overline{X}, \overline{Y}) = \overline{Y}.\pi(\overline{X}) \oplus g(\overline{X})$, where $\overline{X} \in \{0,1\}^r$, $\overline{Y} \in \{0,1\}^k$ and $\overline{Y}.\pi(\overline{X})$ is the inner product of $\overline{Y}$ and $\pi(\overline{X})$. Then $f$ is $m$-resilient. It is possible to interpret $f$ as a concatenation of $2^r$ affine functions $l_0, \ldots, l_{2^r-1}$ from $A(k)$, the set of $k$-variable affine functions, where $ndg(l_i) \geq m+1$. Later we will use this method to construct certain sequences of resilient functions.

## 3  Spectral Weights of CI and Resilient Functions

In this section we prove a crucial result on the divisibility properties of the spectral weights of correlation immune and resilient functions. Such a result has an analogue in the McEliece Theorem [12] for Reed-Muller codes: *the weight of any function in $\mathcal{R}(r,n)$ is divisible by $2^{\lfloor \frac{n-1}{r} \rfloor}$, where $\mathcal{R}(r,n)$ is the set of all $n$-variable Boolean functions of degree at most $r$.* If $f$ is an $n$-variable, $m$-resilient function, using Siegenthaler's inequality we know that the degree of $f$ is at most $n-m-1$. For any linear function $l \in L(n)$, we have $f \oplus l$ is in $\mathcal{R}(n-m-1,n)$ and so $wt(f \oplus l) = d(f,l)$ is divisible by $2^{\lfloor \frac{n-1}{n-m-1} \rfloor}$. However, this result is not sharp enough to prove a nontrivial upper bound on the nonlinearity of resilient functions. In Theorem 1 we prove that for any $n$-variable, $m$-resilient function $f$ and $l \in L(n)$, $d(f,l)$ is divisible by $2^{m+1}$. This is a much stronger result. For example, if $n=7$ and $m=3$, McEliece Theorem guarantees that $d(f,l)$ is divisible by $2^2$. On the other hand Theorem 1 establishes that $d(f,l)$ is divisible by $2^4$.

Theorem 1 also sharpens the Xiao-Massey characterization [6] of correlation immune functions. A Boolean function $f$ is $m$-th order CI iff $wd(f,l) = 0$ for all $l \in L(n)$ with $1 \leq ndg(l) \leq m$. However, this characterization does not state anything about $wd(f,l)$ with $ndg(l) > m$. We show in Theorem 3 that $2^{m+1}$ divides $wd(f,l)$ for all $l$ in $L(n)$ with $ndg(l) > m$. For resilient functions the Xiao-Massey characterization can only be extended to include the condition that Walsh distance between $f$ and the all zero function is 0. However, Theorem 1 shows that $2^{m+2}$ divides $wd(f,l)$ for all $l$ in $L(n)$ with $ndg(l) > m$.

Using Theorem 1 and Theorem 3 we prove nontrivial upper bounds on the nonlinearity of resilient and correlation immune functions. Previous works related to upper bound on nonlinearity of resilient functions were attempted in [3,16]. In [3] an upper bound was obtained for a very small subset of resilient functions. It was shown in [19], that it is possible to construct resilient functions, outside the subset of [3], with nonlinearity more than the upper bound obtained in [3]. In [16], the maximum nonlinearity issue for 6-variable resilient functions has been completely settled by exhaustive computer search technique. Corollary 1 provides a simple proof of the same result.

**Lemma 2.** *Let $f$ be an $n$-variable function of even weight and $l \in L(n)$. Then $d(f,l)$ (respectively $wd(f,l)$) is congruent to $0 \bmod 2$ (respectively $0 \bmod 4$).*

*Proof.* From Lemma 1 we know that $d(f,l) = wt(f) + wt(l) - 2wt(f \times l)$. Since all the terms on the right are even it follows that $d(f,l)$ is also even. □

The next result is a simple consequence of the fact that $\mathcal{R}(m,n)$ is orthogonal to $\mathcal{R}(n-m-1,n)$ [12, Page 375].

**Lemma 3.** *Let $f$ be an $n$-variable ($n \geq 3$), 1-resilient function and $l \in L(n)$. Then $d(f,l)$ (respectively $wd(f,l)$) is congruent to $0 \bmod 4$ (respectively $0 \bmod 8$).*

*Proof.* Since $f$ is 1-resilient, by Siegenthaler's inequality we know that degree of $f$ is at most $n-2$. If $l$ is in $L(n)$, then $f \times l$ is a function of degree at most $n-1$ and hence $wt(f \times l)$ is even. Thus $d(f,l) = wt(f) + wt(l) - 2wt(f \times l) \equiv wt(f) \bmod 4$. As $f$ is balanced, $wt(f) \equiv 0 \bmod 4$, and consequently $d(f,l) \equiv 0 \bmod 4$.    □

**Corollary 1.** *The maximum nonlinearity for a six variable 1-resilient function is 24.*

*Proof.* Using Lemma 3, we know that for any $l \in L(6)$ and any 1-resilient function $f$, $d(f,l) \equiv 0 \bmod 4$. Thus the possible values for $d(f,l)$ are $32 \pm 4k$, for some $k \geq 0$. If for every $l$, $k \leq 1$, then $f$ must be bent and hence cannot be resilient. So there must be some $l$, such that $d(f,l) = 32 \pm 8$. But then the nonlinearity is at most 24.    □

Next we present the major result on the spectral weights of resilient functions.

**Theorem 1.** *Let $f$ be an $n$-variable, $m$-resilient (with $n \geq 3$ and $m \leq n - 3$) function and $l \in L(n)$. Then $d(f,l)$ (respectively $wd(f,l)$) is congruent to $0 \bmod 2^{m+1}$ (respectively $0 \bmod 2^{m+2}$).*

*Proof.* There are three inductions involved : on the number of variables $n$, on the order of resiliency $m$ and on the number of variables in the linear function $l$, which we denote by $k = ndg(l)$.
*Base for induction on $n$:* It is possible to verify the result for $n = 3$. Assume the result is true for all functions on less than $n$ variables (with $n \geq 4$).
*Inductive Step for induction on $n$:* Let $f$ be an $n$-variable function.
   Now we use induction on $m$. The induction on $m$ is carried out separately for odd and even values.
*Base for induction on $m$:* If $m = 0$, then $f$ is a balanced function and Lemma 2 provides the base case.
If $m = 1$, then Lemma 3 provides the base case.
   Next we make the induction hypothesis that if $f$ is $(m-2)$-resilient (with $m - 2 \geq 0$), and $l \in L(n)$, then $d(f,l) \equiv 0 \bmod 2^{m-1}$.
*Inductive Step for induction on $m$:* Let $f$ be $m$-resilient and let $l$ be any function in $L(n)$. We now use induction on the number of variables $k$ in $l$ (i.e., $l \in L(n)$ is nondegenerate on exactly $k$ variables).
*Base for induction on $k$:* $k \leq m$, since $f$ is $m$-resilient $d(f,l) = 2^{n-1} \equiv 0 \bmod 2^{m+1}$.
*Inductive Step for induction on $k$:* Let $k > m$ and using Lemma 2 and Lemma 3 we can assume $k \geq 2$. Without loss of generality assume $X_n$ and $X_{n-1}$ are present

in $l$. Write $l = X_n \oplus X_{n-1} \oplus \lambda$, where $\lambda$ is nondegenerate on at most $k-2$ variables. Also define $\lambda_1 = X_n \oplus \lambda$ and $\lambda_2 = X_{n-1} \oplus \lambda$. Using induction hypothesis on $k$, we know $d(f, \lambda) \equiv d(f, \lambda_1) \equiv d(f, \lambda_2) \equiv 0 \bmod 2^{m+1}$. Let $g_{00}, g_{01}, g_{10}, g_{11}$ be $(n-2)$-variable functions defined by $g_{ij}(X_{n-2}, \ldots, X_1) = f(i, j, X_{n-2}, \ldots, X_1)$. Since $\lambda$ has at most $n-2$ variables, there is a function $\mu \in L(n-2)$ which has the same set of variables as $\lambda$. Denote by $a_{ij}$ the value $d(g_{ij}, \mu)$. Since $\lambda, \lambda_1, \lambda_2$ have less than $k$ variables, using the induction hypothesis on $k$ we have the following equations.

1. $d(f, \lambda) = a_{00} + a_{01} + a_{10} + a_{11} = k_1 2^{m+1}$,
2. $d(f, \lambda_1) = a_{00} + a_{01} - a_{10} - a_{11} = k_2 2^{m+1}$,
3. $d(f, \lambda_2) = a_{00} - a_{01} + a_{10} - a_{11} = k_3 2^{m+1}$, and
4. $d(f, l) = a_{00} - a_{01} - a_{10} + a_{11}$.

From the first three equations, we can express $a_{01}, a_{10}$ and $a_{11}$ in terms of $a_{00}$. This gives us

$a_{01} = (k_1 + k_3)2^m - a_{00}$, $a_{10} = (k_1 + k_2)2^m - a_{00}$ and $a_{11} = -(k_2 + k_3)2^m + a_{00}$.

Now using equation 4, we get $d(f, l) = 4a_{00} - (k_1 + k_2 + k_3)2^{m+1}$. Since $f$ is $m$-resilient and $g$ is obtained from $f$ by setting two variables to constant values, $g$ is an $(n-2)$-variable, $(m-2)$-resilient function. First assume $m$ is even, then $m-2$ is also even. Using the induction hypothesis on $n$ and the induction hypothesis on even $m$ we have $a_{00} = d(g, \mu) \equiv 0 \bmod 2^{m-1}$. The argument is similar for odd $m$. (This is the reason for choosing the base cases separately for $m = 0$ and $m = 1$.) Hence $d(f, l) \equiv 0 \bmod 2^{m+1}$. $\qquad\square$

Using Theorem 1, it is possible to obtain an upper bound on the nonlinearity of an $n$-variable, $m$-resilient function.

**Theorem 2.** *1. If $n$ is even and $m + 1 > \frac{n}{2} - 1$, then $nlr(n, m) \leq 2^{n-1} - 2^{m+1}$.*
*2. If $n$ is even and $m + 1 \leq \frac{n}{2} - 1$, then $nlr(n, m) \leq 2^{n-1} - 2^{\frac{n}{2}-1} - 2^{m+1}$.*
*3. If $n$ is odd and $2^{m+1} > 2^{n-1} - nlmax(n)$, then $nlr(n, m) \leq 2^{n-1} - 2^{m+1}$.*
*4. If $n$ is odd and $2^{m+1} \leq 2^{n-1} - nlmax(n)$, then $nlr(n, m)$ is the highest multiple of $2^{m+1}$ which is less than or equal to $2^{n-1} - nlmax(n)$.*
*Further in cases 1 and 3, the spectrum of any function achieving the stated bound must be three valued, i.e. the values of the Walsh distances must be $0, \pm 2^{m+2}$.*

*Proof.* We prove only cases 1 and 2, the other cases being similar.
1. Using Theorem 1 for any $n$-variable, $m$-resilient function $f$ and $l \in L(n)$, we have $d(f, l) \equiv 0 \bmod 2^{m+1}$. Thus $d(f, l) = 2^{n-1} \pm k2^{m+1}$ for some $k$. Clearly $k$ cannot be 0 for all $l$ and hence the nonlinearity of $f$ is at most $2^{n-1} - 2^{m+1}$.
2. As in 1, we have $d(f, l) = 2^{n-1} \pm k2^{m+1}$ for some $k$. Let $2^{\frac{n}{2}-1} = p2^{m+1}$ (we can write in this way as $m < \frac{n}{2} - 1$). If for all $l$ we have $k \leq p$, then $f$ must necessarily be bent and hence cannot be resilient. Thus there must be some $l$ such that the corresponding $k > p$. This shows that the nonlinearity of $f$ is at most $2^{n-1} - 2^{\frac{n}{2}-1} - 2^{m+1}$.

The proof of the last statement follows from the fact that if the Walsh distances are not three valued $0, \pm 2^{m+2}$, then $\pm 2^{m+i}$ must be a Walsh distance value for $i \geq 3$. The nonlinearity for such a function is clearly less than the stated bound. $\qquad\square$

In Table 1 we provide some examples of the upper bound provided in Theorem 2. The boundary case of Theorem 2 is given in the following corollary (see also [3,16]).

**Corollary 2.** *For $n \geq 4$, $nlr(n, n-3) = 2^{n-2}$.*

*Proof.* From Theorem 2 it is clear that $nlr(n, n-3) \leq 2^{n-1} - 2^{n-2} = 2^{n-2}$. Moreover, it is easy to construct an $(n, n-3, 2, 2^{n-2})$ function by concatenating two distinct linear functions from $L(n-1)$, each of which are nondegenerate on $n-2$ variables.                                                           □

We also need the following corollary which will be used to define the concept of *saturated function* in Section 4.

**Corollary 3.** *Let $m > \lfloor \frac{n}{2} \rfloor - 2$. Then, $nlr(n, m) \leq 2^{n-1} - 2^{m+1} \leq 2^{n-1} - 2^{\lfloor \frac{n-1}{2} \rfloor}$. Further, the spectrum of any $(n, m, -, 2^{n-1} - 2^{m+1})$ function is necessarily three valued.*

**Table 1.** The entries represent the upper bound on $nlr(n, m)$ given by Theorem 2, where $n$ is the number of variables and $m$ is the order of resiliency. Entries with $*$ represent bounds which have not yet been constructed. Entries with $\#$ represent bounds which have been constructed here for the first time.

| $n$ \ $m$ | 1 | 2 | 3 | 4 | 5 | 6 | 7 | 8 |
|---|---|---|---|---|---|---|---|---|
| 5 | 12 | 8 | 0 | | | | | |
| 6 | 24 | 24 | 16 | 0 | | | | |
| 7 | 56 | 56* | 48 | 32 | 0 | | | |
| 8 | 116* | 112 | 112# | 96 | 64 | 0 | | |
| 9 | 244* | 240 | 240* | 224# | 192 | 128 | 0 | |
| 10 | 492* | 480 | 480 | 480* | 448 | 384 | 256 | 0 |

The set of $n$-variable $m$-th order correlation immune functions is a superset of $n$-variable $m$-resilient functions. The following two results are for correlation immune functions and are similar to Theorem 1 and 2.

**Theorem 3.** *Let $f$ be an $n$-variable, $m$-th order correlation immune (with $n \geq 3$ and $m \leq n - 2$) function and $l \in L(n)$. Then $d(f, l)$ (respectively $wd(f, l)$) is congruent to $0 \bmod 2^m$ (respectively $0 \bmod 2^{m+1}$).*

*Proof.* We have to note that if a function $f$ is 1st order correlation immune (CI) then $d(f, l)$ is even ($wd(f, l) \equiv 0 \bmod 4$) for any linear function $l$. Now given a 2nd order CI function, by Siegenthaler's inequality we know that degree of $f$ is at most $n-2$. Thus, similar to the proof of Lemma 3, we get $d(f, l)$ (respectively $wd(f, l)$) is congruent to $0 \bmod 4$ (respectively $0 \bmod 8$). Using these as the base cases, the proof is similar to the proof of Theorem 1.            □

**Theorem 4.** *1. If $n$ is even and $m > \frac{n}{2} - 1$, then $nlc(n, m) \leq 2^{n-1} - 2^m$.*
*2. If $n$ is even and $m \leq \frac{n}{2} - 1$, then $nlc(n, m) \leq 2^{n-1} - 2^{\frac{n}{2}-1} - 2^m$.*
*3. If $n$ is odd and $2^m > 2^{n-1} - nlmax(n)$, then $nlc(n, m) \leq 2^{n-1} - 2^m$.*
*4. If $n$ is odd and $2^m \leq 2^{n-1} - nlmax(n)$, then $nlc(n, m)$ is the highest multiple of $2^m$ which is less than or equal to $2^{n-1} - nlmax(n)$.*
*Further in cases 1 and 3, the spectrum of any function achieving the stated bound must be three valued, i.e. the values of the Walsh distances must be $0, \pm 2^{m+1}$.*

The nonlinearity bounds proved in this section have the following important consequences.
1. These bounds set up a *benchmark* by which one can measure the efficacy of any new construction method for resilient functions. It will also be a major task to show that in certain cases the upper bound of Theorem 2 is not tight.
2. Based on Theorem 2 and Siegenthaler's inequality, we are able to satisfactorily identify the class of Boolean functions achieving the best possible trade-off among the parameters : number of variables, order of resiliency, nonlinearity and algebraic degree.

## 4    Construction of Resilient Functions

Motivated by Theorem 2, we introduce two classes of resilient functions. An $(n, m, d, x)$ function is said to be of Type-I if $x$ is the upper bound on $nlr(n, m)$ provided in Theorem 2. Note that, given an $n$-variable function, there may be more than one possible values of order of resiliency $m$, such that the upper bound on $nlr(n, m)$ is same using Theorem 2. We call an $n$-variable, $m$-resilient function having nonlinearity $x$ to be of Type-II if the function is of Type-I and further for any $p > m$ the upper bound on $nlr(n, p)$ in Theorem 2 is strictly less than $x$. These notions of trade-offs can be further strengthened by requiring the degree to be the maximum possible. For this we require Siegenthaler's inequality for resilient functions: $m + d \leq n - 1$, for any $n$-variable, $m$-resilient, degree $d$ function. Thus $(n, m, n - m - 1, x)$ *Type-II functions achieve the best possible trade-off among the parameters : number of variables, order of resiliency, degree and nonlinearity.*

*Example 1.* An $(8, 2, 5, 112)$ function is of Type-I. Moreover, $(8, 2, -, 112)$ functions are not of Type-II since $nlr(8, 3) \leq 112$. However, an $(8, 3, -, 112)$ function is of Type-II since $nlr(8, 4) \leq 96$. Also an $(8, 3, 4, 112)$ function maximizes the algebraic degree and hence provides best possible trade-off among the parameters we consider here. From Theorem 2, the spectrum of any $(8, 3, -, 112)$ function is necessarily three valued. However, this may not necessarily be true for any Type-II function. For example, an $(8, 1, 6, 116)$ function (if one exists) will be of Type-II, but its spectrum will not be three valued.

The way we have defined Type I and Type II functions, it is not guaranteed that such functions always exist. The tightness of the upper bounds in Theorem 2 is contingent on the existence of such functions. However, we will show for certain sequences of Type-II functions, it is possible to construct all but finitely many functions of any such sequence.

We call a Type-II function to be *saturated* if its spectrum is three valued according to Corollary 3. Thus an $(n, m, n-m-1, x)$-function is called a *saturated maximum degree* function if it is of Type-II and its spectrum is three valued. For such a function we must necessarily have $m > \lfloor \frac{n}{2} \rfloor - 2$. Therefore, the $(8, 3, 4, 112)$ functions are of Type II and are also saturated maximum degree functions. However, the $(8, 1, 6, 116)$ Type-II functions (if they exist) can not have a three valued Walsh spectrum. From Parseval's theorem, if it has a three valued Walsh spectrum, then $24^2 \times z = 2^{16}$, which is not possible for integral $z$. Thus, $(8, 1, 6, 116)$ functions are of Type-II and have maximum degree but are not saturated.

**Lemma 4.** *If an $(n, m, n-m-1, x)$ function $f$ is a saturated function, then so is an $(n+1, m+1, n-m-1, 2x)$ function $g$.*

*Proof.* Since $f$ is saturated, $x = 2^{n-1} - 2^{m+1}$ and so $2x = 2^n - 2^{m+2}$. From Corollary 3, $nlr(n+1, m+1) \le 2^n - 2^{m+2}$ and hence the spectrum of $g$ is three valued. □

This naturally leads to a notion of a sequence of Boolean functions, each of which is a saturated maximum degree function. More precisely, a *saturated function sequence* (an SS for short), is an infinite sequence of Boolean functions $f_0, f_1, \ldots$, where $f_0$ is an $(n_0, m_0, n_0 - m_0 - 1, x_0)$ function which is a Type II, saturated maximum degree function and the upper bound on $nlr(n_0 - 1, m_0 - 1)$ in Theorem 2 is strictly less than $\frac{x_0}{2}$. Also for $j \ge 0$, $f_{j+1}$ is an $(n_j + 1, m_j + 1, n_j - m_j - 1, 2x_j)$ function (and hence is also saturated from Lemma 4). Note that $n_j - m_j - 1 = n_0 - m_0 - 1$ and so the degree of all the functions in an SS are same. Thus an SS is completely defined by specifying the parameters of a function $f_0$. Note that the functions which form an SS is not unique, i.e., there can be more than one distinct $(n_0, m_0, n_0 - m_0 - 1, x_0)$ functions and all of them are possible representatives for $f_0$. Thus a particular SS is characterized by several parameters and any sequence of functions satisfying these parameters is said to form the particular SS.

*Example 2.* The following seqences are SS's.
1. $f_0, f_1, \ldots$, where $f_0$ is a $(3, 0, 2, 2)$ function.
2. $f_0, f_1, \ldots$, where $f_0$ is a $(5, 1, 3, 12)$ function.
3. $f_0, f_1, \ldots$, where $f_0$ is a $(7, 2, 4, 56)$ function.
It is not known whether $(7, 2, 4, 56)$ functions exists. However, we show how to construct an $(8, 3, 4, 112)$ function, which is $f_1$ in this SS.

**Definition 6.** *For $i \ge 0$ we define $SS(i)$ as follows. An $SS(0)$ is a sequence $f_{0,0}, f_{0,1}, \ldots$, where $f_{0,0}$ is a $(3, 0, 2, 2)$ function and $f_{0,j}$ is a $(3 + j, j, 2, 2^{j+1})$ function for $j > 0$. For $i > 0$, an $SS(i)$ is a sequence $f_{i,0}, f_{i,1}, \ldots$, where $f_{i,0}$ is a $(3 + 2i, i, 2 + i, 2^{2+2i} - 2^{1+i})$ function which is a Type II, saturated maximum degree function. Also for $j > 0$, $f_{i,j}$ is a $(3 + 2i + j, i + j, 2 + i, 2^{2+2i+j} - 2^{1+i+j})$ function.*

Note that all functions in an $SS(i)$ have the same degree $2 + i$. Construction of $SS(0)$ and $SS(1)$ are already known. Unfortunately, it is not known whether the initial functions for an $SS(i)$ exist for $i > 1$. In the next subsection we show how to construct all but finitely many initial functions of any $SS(i)$.

Now we will concentrate on the construction problem of saturated sequences. In defining SS we stated that any function in an SS must be a saturated function. However, the converse that given any saturated function, it must occur in some $SS(i)$ is not immediate. The following result proves this and justifies the fact that we can restrict our attention to the construction problem for $SS(i)$ only.

**Lemma 5.** *Any saturated function must occur in some $SS(i)$.*

*Proof.* First note that any function of $SS(i)$ has algebraic degree $2 + i$. Any saturated function $f$ must be an $(n, m, n - m - 1, 2^{n-1} - 2^{m+1})$ function having degree $d = n - m - 1$. Hence $f$ must occur in $SS(d - 2)$, i.e., in $SS(n - m - 3)$. $\square$

## 4.1 Construction of $SS(i)$

Here we show that the Maiorana-McFarland like construction procedure can be used to construct all but finitely many functions of any $SS(i)$. First we state the following result which is easy to prove using Lemma 4.

**Lemma 6.** *Let $f_{i,j}$ be a $j$-th function of $SS(i)$. Then the function $g = Y \oplus f_{i,j}$ (where the variable $Y$ does not occur in $f_{i,j}$) is an $f_{i,j+1}$ function of $SS(i)$. Consequently, if one can construct $f_{i,j}$, then one can construct $f_{i,k}$ for all $k > j$.*

This shows that if one can construct any one of the functions in $SS(i)$, then it is possible to construct any function in the succeeding part of the sequence. Thus it is enough if we can construct the first function of each sequence. This is possible for $SS(0)$ and $SS(1)$ since construction of $(3, 0, 2, 2)$ and $(5, 1, 3, 12)$ functions are known. However, the construction problem for the first function of $SS(i)$ for $i > 1$ is an ongoing research problem. Here we show that the Maiorana-McFarland like construction procedure can be used to construct all but finitely many functions of any $SS(i)$. More precisely, if $SS(i) = f_{i,0}, f_{i,1} \ldots$, then we show how to construct $f_{i,t}$ for all $t \geq t_0$, where $t_0$ is such that $2^{1+i} = 3 + i + t_0$. For $SS(2)$, this gives $t_0 = 3$. Moreover, in Subsection 4.2, we show how to construct $f_{2,1}$ and $f_{2,2}$. This leaves open the problem of constructing $f_{i,t}$, with $t < t_0$ and $i \geq 3$ as a challenging research problem.

**Theorem 5.** *For any $SS(i) = f_{i,0}, f_{i,1}, \ldots$, it is possible to construct $f_{i,t}$ for all $t$ greater than or equal to some $t_0$.*

*Proof.* The first function $f_{i,0}$ is a $(3 + 2i, i, 2 + i, 2^{2+2i} - 2^{1+i})$ function. We show that for some $j$, $f_{i,j}$ is constructible by Maiorana-McFarland like construction techniques. Let $j$ be such that $2^{1+i} = 3 + i + j$. A function $f_{i,j}$ is to be an $(n = 3 + 2i + j, i + j, 2 + i, 2^{2+2i+j} - 2^{1+i+j})$ function. We show how to construct such a function. Consider the set $\Lambda$ of all $k = 2 + i + j$-variable linear functions which are nondegenerate on at least $1 + i + j$ variables. Clearly there are $\binom{2+i+j}{2+i+j} +$

$\binom{2+i+j}{1+i+j} = 3 + i + j$ such linear functions. Consider an $n$-variable function $f$ (a string of length $2^n$) formed by concatenating $2^{n-k}$ functions from $\Lambda$. Since $2^{n-k} = 2^{1+i} = 3+i+j = |\Lambda|$, we use each of the functions in $\Lambda$ exactly once in the formation of $f$. Since each function in $\Lambda$ is nondegenerate on $1+i+j$ variables each of these is $(i+j)$-resilient. Let $V = \{X_{2+i+j}, \ldots, X_1\}$ be the set of variables which are involved in the linear functions in $\Lambda$. Each of the variables in $V$ occur in $2^{1+i} - 1$ of the linear functions in $\Lambda$. Thus each variable occurs an odd number of times and hence the degree of $f$ is $n-k+1 = 2+i$. Moreover, this implies that each of the $n$ input variables of the function occurs in the maximum degree term. Since each linear function is used once, the nonlinearity of $f$ is $2^{n-1} - 2^{k-1} = 2^{2+2i+j} - 2^{1+i+j}$. Thus $f$ is a $(3+2i+j, i+j, 2+i, 2^{2+2i+j} - 2^{1+i+j})$ function and can be taken as $f_{i,j}$. Take $t_0 = j$. Then using Lemma 6 it is possible to construct $f_{i,t}$ for all $t > t_0 = j$.  □

In the proof of the above theorem we use Lemma 6 to construct $f_{i,t}$ for all $t > j$, given the function $f_{i,j}$. Thus $f_{i,t}(Y_{t-j}, \ldots, Y_1, \overline{X}) = Y_{t-j} \oplus \ldots \oplus Y_1 \oplus f_{i,j}(\overline{X})$. This results in the function $f_{i,t}$ depending linearly on the variables $Y_{t-j}, \ldots, Y_1$. This is not recommendable from cryptographic point of view. There are two ways to avoid this situation.

**(I)** The above proof of Theorem 5 can be modified so that Lemma 6 is not required at all. In fact, the linear concatenation technique used to construct $f_{i,j}$ can directly be used to construct $f_{i,t}$. In $f_{i,j}$, a total of $2^{1+i}$ slots were filled up using the $3+i+j$ different linear functions (each exactly once) and this was made possible by the fact that $2^{1+i} = 3+i+j$. In constructing $f_{i,t}$ directly we will still have to fill $2^{1+i}$ slots but the number of linear functions that can be used will increase to $3+i+t$. Hence no linear function need to be used more than once and as a result the nonlinearity obtained will achieve the upper bound of Theorem 2. The ANF of the resulting $f_{i,t}$ will depend nonlinearly on all the variables $Y_{t-j}, \ldots, Y_1$.

**(II)** After obtaining $f_{i,j}$, instead of using Lemma 6 we can use a more powerful construction provided in [13]. The method of [13] shows that if $f$ is an $m$-resilient function, then $g$ defined as $g(Y, \overline{X}) = (1 \oplus Y)f(\overline{X}) \oplus Y(a \oplus f(\overline{X} \oplus \overline{\alpha}))$, is an $(m+1)$-resilient function, where $\overline{\alpha}$ is an all one vector and $a = m \bmod 2$. This also guarantees that $g$ does not depend linearly on $Y$. Hence if we use this technique repeatedly to construct $f_{i,t}$ from $f_{i,j}$, then the ANF of the resulting $f_{i,t}$ will depend nonlinearly on all the variables $Y_{t-j}, \ldots, Y_1$.

## 4.2   A Sharper Construction

For $SS(2) = f_{2,0}, f_{2,1}, f_{2,2}, \ldots$, Theorem 5 can be used to construct $f_{2,t}$ for all $t \geq 3$. Here we show how to construct $f_{2,1}$ (an $(8, 3, 4, 112)$ function). However, the construction of $f_{2,0}$, the $(7, 2, 4, 56)$ Type-II function, is not yet known.

For a Boolean function $f$, we define $NZ(f) = \{\overline{\omega} \mid W_f(\overline{\omega}) \neq 0\}$, where $W_f$ is the Walsh transform of $f$. The following result is the first step in the construction of $(8, 3, 4, 112)$ function.

**Lemma 7.** *Let $f_1, f_2$ be two $(7, 3, -, 48)$ functions such that $NZ(f_1) \cap NZ(f_2) = \emptyset$. Let $f = (1 \oplus X_8)f_1 \oplus X_8 f_2$. Then, $f$ is an $(8, 3, -, 112)$ function.*

First let us construct the function $f_2$ using concatenation of linear functions. We take four 5-variable linear functions with each of them nondegenerate on at least 4 variables : $l_{51} = X_1 \oplus X_2 \oplus X_3 \oplus X_4$, $l_{52} = X_1 \oplus X_2 \oplus X_3 \oplus X_5$, $l_{53} = X_1 \oplus X_2 \oplus X_4 \oplus X_5$ and $l_{54} = X_1 \oplus X_3 \oplus X_4 \oplus X_5$. We consider $f_2 = l_{51}l_{52}l_{53}l_{54}$, concatenation of the four linear functions. It is easy to see that since each $l_{5i}$ is 3-resilient, $f_2$ is also 3-resilient. Note that each of the variables $X_2, X_3, X_4, X_5$ occurs in exactly three linear functions, so algebraic degree of $f_2$ is 3. Moreover, nonlinearity of $f_2$ is $3 \times 16 = 48$.

Now let us analyze the Walsh spectrum of $f_2$. Note that for the linear functions $\lambda$ of the form $a_7 X_7 \oplus a_6 X_6 \oplus l_{5i}$, $a_7, a_6 \in \{0, 1\}, 1 \leq i \leq 4$, $wd(f_2, \lambda)$ is nonzero. There are 16 such functions in $L(7)$. For the rest of the functions $\lambda_1$ in $L(7)$, $wd(f_2, \lambda_1)$ is zero. Also, note that according to the Theorem 2, this is a three valued Walsh spectrum.

Next we need to use the following basic idea. *When $d(f_2, l)$ is minimum, then $d(f_1, l)$ must be 64, i.e., when $wd(f_2, l)$ is maximum, then $wd(f_1, l)$ must be 0.* We now construct another $(7, 3, 3, 48)$ function, having a three valued Walsh spectrum such that $wd(f_1, \lambda)$ is zero for all $\lambda$ of the form $a_7 X_7 \oplus a_6 X_6 \oplus l_{5i}$, $a_7, a_6 \in \{0, 1\}, 1 \leq i \leq 4$.

We start from a $(5, 1, 3, 12)$ function $g$. The Walsh spectrum of the function need to be such that $wd(g, l_{5i}) = 0$ for $1 \leq i \leq 4$. We choose $g$ to be 00000111011111001110010110100010 by running a computer program. Then we construct $f_1 = X_7 \oplus X_6 \oplus g$. Note that $f_1$ is a $(7, 3, 3, 48)$ function and the Walsh spectrum of $f_1$ is such that $wd(f_1, \lambda)$ is zero for all $\lambda$ of the form $a_7 X_7 \oplus a_6 X_6 \oplus l_{5i}$, $a_7, a_6 \in \{0, 1\}, 1 \leq i \leq 4$. Thus, $NZ(f_1) \cap NZ(f_2) = \emptyset$. Also there are degree three terms in $f_1$ (respectively $f_2$) which are not in $f_2$ (respectively $f_1$). Hence, $f = (1 \oplus X_8)f_1 \oplus X_8 f_2$ is an $(8, 3, 4, 112)$ function. The output column of the function is a 256-bit string and is as follows in hexadecimal format.

$$
\begin{array}{cccccccc}
077C & E5A2 & F883 & 1A5D & F883 & 1A5D & 077C & E5A2 \\
6996 & 6996 & 6969 & 9696 & 6699 & 9966 & 5AA5 & A55A
\end{array}
$$

**Theorem 6.** *It is possible to construct $(8, 3, 4, 112)$ and $(9, 4, 4, 224)$ functions.*

*Proof.* Above we discussed how to construct an $(8, 3, 4, 112)$ function $f$. Further a $(9, 4, 4, 224)$ function can be constructed as either $(1 \oplus X_9)f(X_8, \ldots, X_1) \oplus X_9(1 \oplus f(1 \oplus X_8, \ldots, 1 \oplus X_1))$ or $X_9 \oplus f$.     □

## 5   On Construction of Small Functions

First we consider balanced functions. The maximum possible nonlinearities for balanced functions on 7, 8, 9 and 10 variables are 56, 118, 244 and 494 respectively. In [20], construction of nonlinear balanced functions on even number of variables was considered. The values obtained for 8 and 10 variables are respectively 116 and 492. In [19], the degree was considered and construction of

$(7, 0, 6, 56), (8, 0, 7, 116), (9, 0, 8, 240)$ and $(10, 0, 9, 492)$ functions were presented. The existence of $(8, 0, -, 118)$ functions have been open for quite some time. We next present a result which could be an important step in solving this problem.

**Theorem 7.** *Let if possible $f$ be a $(8, 0, -, 118)$ function. Then degree of $f$ must be 7 and it is possible to write $f = (1 \oplus X_8) f_1 \oplus X_8 f_2$, where $f_1$ and $f_2$ are 7-variable functions each having nonlinearity 55 and degree 7.*

*Proof.* First we prove that the degree of $f$ must be 7. If the degree of $f$ is less than 7, then using a result of Hou [7, Lemma 2.1], we can perform an affine transformation on the varibles of $f$ to obtain an 8-variable function $g$, such that $g(X_8, X_7, \ldots, X_1) = (1 \oplus X_8) g_1(X_7, \ldots, X_1) \oplus X_8 g_2(X_7, \ldots, X_1)$ and the degrees of $g_1$ and $g_2$ are each less than or equal to 5. The affine transformation preserves the weight and nonlinearity of $f$ and so $wt(f) = wt(g) = wt(g_1) + wt(g_2)$ and $nl(f) = nl(g)$. Since $f$ is balanced, $wt(g_1) + wt(g_2) = wt(g) = wt(f) = 128 \equiv 0 \bmod 4$. Also $wt(g_1)$ and $wt(g_2)$ are both even since their degrees are less than or equal to 5. Hence $wt(g_1) \equiv wt(g_2) \equiv 0 \bmod 4$ or $wt(g_1) \equiv wt(g_2) \equiv 2 \bmod 4$. Since $g_1, g_2$ are 7-variable functions with degree $\leq 5$, it follows that (see [12]) for any linear function $l \in L(7)$, $d(g_1, l) \equiv wt(g_1) \bmod 4$ and $d(g_2, l) \equiv wt(g_2) \bmod 4$. Hence for any $l \in L(7)$, $d(g_1, l) \equiv d(g_2, l) \bmod 4$ and so $d(g_1, l) + d(g_2, l) \equiv 0 \bmod 4$     (\*\*).
Since the nonlinearity of $g$ is 118, there exists $\lambda \in L(7)$ such that one of the following must hold: (1) $d(g, \lambda\lambda) = 118$, (2) $d(g, \lambda\lambda) = 138$, (3) $d(g, \lambda\lambda^c) = 118$, (4) $d(g, \lambda\lambda^c) = 138$. Here we consider only case (1), other ones being similar. From (1) we have $118 = d(g, \lambda\lambda) = d(g_1, \lambda) + d(g_2, \lambda)$ and so $d(g_1, \lambda) + d(g_2, \lambda) = 118 \equiv 2 \bmod 4$ which is a contradiction to equation (\*\*).

Thus the degree of $f$ is 7. Without loss of generality we consider $X_7 \ldots X_1$ is a degree 7 term in the ANF of $f$. We put $f_1(X_7, \ldots, X_1) = f(X_8 = 0, X_7, \ldots, X_1)$ and $f_2(X_7, \ldots, X_1) = f(X_8 = 1, X_7, \ldots, X_1)$. Thus both $f_1, f_2$ are of degree 7 and hence of odd weight and so $nl(f_1), nl(f_2) \leq 55$. It can be proved that if any of $nl(f_1)$ or $nl(f_2)$ is $\leq 53$, then $nl(f) < 118$.    □

The major implication of Theorem 7 is that if it is not possible to construct $(8, 0, 7, 118)$ function by concatenating two 7-variable, degree 7, nonlinearity 55 functions, then the maximum nonlinearity of balanced 8-variable functions is 116.

Now we turn to resilient functions. We first present a construction of a previously unknown function.

**Theorem 8.** *It is possible to construct $(10, 3, 6, 480)$ functions.*

*Proof.* We construct a function $f$ by concatenating linear functions from $L(5)$ as follows. There are 10 functions $\mu_0, \ldots, \mu_9$ in $L(5)$ which are nondegenerate on exactly 3 variables. Also there are 5 functions $\lambda_0, \ldots, \lambda_4$ in $L(5)$ which are nondegenerate on exactly 4 variables. The function $f$ is the concatenation of the following sequence of functions,
$\lambda_0 \lambda_0 \lambda_0 \lambda_0^c \lambda_1 \lambda_1 \lambda_1 \lambda_1^c \lambda_2 \lambda_2 \lambda_3 \lambda_4 \mu_0 \mu_0^c \mu_1 \mu_1^c \mu_2 \mu_2^c \mu_3 \mu_3^c \mu_4 \mu_4^c \mu_5 \mu_5^c \mu_6 \mu_6^c \mu_7 \mu_7^c \mu_8 \mu_8^c \mu_9 \mu_9^c$.
The functions $\lambda_i$ and $\mu_j \mu_j^c$ are both 3-resilient and hence $f$ is 3-resilient too.

It can be checked that there are variables between $X_5, \ldots, X_1$ which occur odd number of times overall in the above sequence. Hence the degree of $f$ is 6. Also the nonlinearity of $f$ can be shown to be 480.    □

Note that the constructed function is not a saturated function and its Walsh spectrum is five-valued $(0, \pm 32, \pm 64)$.

In Table 2, we list some of the best known functions. Also Table 3 provides some open problems.

**Table 2.** Some best known functions. The $(8, 3, 4, 112)$ and $(9, 4, 4, 224)$ functions are from Theorem 6 and the $(10, 3, 6, 480)$ function is from Theorem 8. All the other constructions were known previously [21,19].

| $n$ | |
|---|---|
| 7 | (7,1,5,56), (7,3,3,48), (7,4,2,32) |
| 8 | (8,1,6,112), (8,2,5,112), (8,3,4,112), (8,4,3,96), (8,5,2,64) |
| 9 | (9,1,7,240), (9,2,5,240), (9,3,5,224), (9,4,4,224), (9,5,3,192), (9,6,2,128) |
| 10 | (10,1,8,484), (10,2,7,480), (10,3,6,480), (10,4,5,448), (10,5,4,448), (10,6,3,384), (10,7,2,256) |

**Table 3.** Existence of functions with these parameters is not known.

| $n$ | |
|---|---|
| 7 | (7,2,−,56) |
| 8 | (8,1,−,116) |
| 9 | (9,1,−,244), (9,2,6,240) |
| 10 | (10,1,−,492), (10,1,−,488), (10,2,−,488), (10,4,−,480) |

**Notes :** In a recent work [24], Tarannikov showed that the maximum possible nonlinearity of an $n$-variable, $m$-resilient function is $2^{n-1} - 2^{m+1}$ for $\frac{2n-7}{3} \le m \le n-2$ and functions achieving this nonlinearity must have maximum possible algebraic degree $n - m - 1$. Also a construction method for such $n$-variable functions with the additional restriction that each variable occurs in a maximum degree term is provided for $m$ in the range $\frac{2n-7}{3} \le m \le n - \log_2 \frac{n-2}{3} - 2$.

**Acknowledgement**

The authors are grateful to the anonymous referees for many comments which helped to improve the presentation of the paper.

# References

1. P. Camion, C. Carlet, P. Charpin, and N. Sendrier. On correlation immune functions. In *Advances in Cryptology - CRYPTO'91*, pages 86–100. Springer-Verlag, 1992.
2. C. Carlet. More correlation immune and resilient functions over Galois fields and Galois rings. In *Advances in Cryptology - EUROCRYPT'97*, pages 422–433. Springer-Verlag, May 1997.
3. S. Chee, S. Lee, D. Lee, and S. H. Sung. On the correlation immune functions and their nonlinearity. In *Advances in Cryptology, Asiacrypt 96*, number 1163 in Lecture Notes in Computer Science, pages 232–243. Springer-Verlag, 1996.
4. C. Ding, G. Xiao, and W. Shan. *The Stability Theory of Stream Ciphers*. Number 561 in Lecture Notes in Computer Science. Springer-Verlag, 1991.
5. E. Filiol and C. Fontaine. Highly nonlinear balanced Boolean functions with a good correlation-immunity. In *Advances in Cryptology - EUROCRYPT'98*. Springer-Verlag, 1998.
6. X. Guo-Zhen and J. Massey. A spectral characterization of correlation immune combining functions. *IEEE Transactions on Information Theory*, 34(3):569–571, May 1988.
7. X. Hou. Covering radius of the Reed-Muller code $R(1, 7)$ - a simpler proof. *Journal of Combinatorial Theory, Series A*, 74(3):337–341, 1996.
8. X. Hou. On the covering radius of $R(1, m)$ in $R(3, m)$. *IEEE Transactions on Information Theory*, 42(3):1035–1037, 1996.
9. X. Hou. On the norm and covering radius of the first order Reed-Muller codes. *IEEE Transactions on Information Theory*, 43(3):1025–1027, 1997.
10. T. Johansson and F. Jonsson. Fast correlation attacks based on turbo code techniques. In *Advances in Cryptology - CRYPTO'99*, number 1666 in Lecture Notes in Computer Science, pages 181–197. Springer-Verlag, August 1999.
11. T. Johansson and F. Jonsson. Improved fast correlation attacks on stream ciphers via convolutional codes. In *Advances in Cryptology - EUROCRYPT'99*, number 1592 in Lecture Notes in Computer Science, pages 347–362. Springer-Verlag, May 1999.
12. F. J. MacWilliams and N. J. A. Sloane. *The Theory of Error Correcting Codes*. North Holland, 1977.
13. S. Maitra and P. Sarkar. Highly nonlinear resilient functions optimizing Siegenthaler's inequality. In *Advances in Cryptology - CRYPTO'99*, number 1666 in Lecture Notes in Computer Science, pages 198–215. Springer Verlag, August 1999.
14. W. Meier and O. Staffelbach. Fast correlation attack on stream ciphers. In *Advances in Cryptology - EUROCRYPT'88*, volume 330, pages 301–314. Springer-Verlag, May 1988.
15. S. Palit and B. K. Roy. Cryptanalysis of LFSR-encrypted codes with unknown combining functions. In *Advances in Cryptology - ASIACRYPT'99*, number 1716 in Lecture Notes in Computer Science, pages 306–320. Springer Verlag, November 1999.
16. E. Pasalic and T. Johansson. Further results on the relation between nonlinearity and resiliency of Boolean functions. In *IMA Conference on Cryptography and Coding*, number 1746 in Lecture Notes in Computer Science, pages 35–45. Springer-Verlag, 1999.
17. R. A. Rueppel. *Analysis and Design of Stream Ciphers*. Springer Verlag, 1986.

18. R. A. Rueppel and O. J. Staffelbach. Products of linear recurring sequences with maximum complexity. *IEEE Transactions on Information Theory*, IT-33:124–131, January 1987.

19. P. Sarkar and S. Maitra. Construction of nonlinear Boolean functions with important cryptographic properties. In *Advances in Cryptology - EUROCRYPT 2000*, number 1807 in Lecture Notes in Computer Science, pages 491–512. Springer Verlag, 2000.

20. J. Seberry, X. M. Zhang, and Y. Zheng. Nonlinearly balanced Boolean functions and their propagation characteristics. In *Advances in Cryptology - CRYPTO'93*, pages 49–60. Springer-Verlag, 1994.

21. J. Seberry, X. M. Zhang, and Y. Zheng. On constructions and nonlinearity of correlation immune Boolean functions. In *Advances in Cryptology - EUROCRYPT'93*, pages 181–199. Springer-Verlag, 1994.

22. T. Siegenthaler. Correlation-immunity of nonlinear combining functions for cryptographic applications. *IEEE Transactions on Information Theory*, IT-30(5):776–780, September 1984.

23. T. Siegenthaler. Decrypting a class of stream ciphers using ciphertext only. *IEEE Transactions on Computers*, C-34(1):81–85, January 1985.

24. Y. V. Tarannikov. On resilient Boolean functions with maximum possible nonlinearity. *Cryptology ePrint Archive, eprint.iacr.org, No. 2000/005*, 2000.

# Almost Independent and Weakly Biased Arrays: Efficient Constructions and Cryptologic Applications

Jürgen Bierbrauer[1] and Holger Schellwat[2]

[1] Department of Mathematical Sciences, Michigan Technological University,
Houghton, Michigan 49931, USA
jbierbra@mtu.edu
[2] Department of Natural Sciences, Örebro University, SE-70182 Örebro, Sweden
holger.schellwat@nat.oru.se

**Abstract.** The best known constructions for arrays with low bias are those from [1] and the exponential sum method based on the Weil-Carlitz-Uchiyama bound. They all yield essentially the same parameters. We present new efficient coding-theoretic constructions, which allow far-reaching generalizations and improvements. The classical constructions can be described as making use of Reed-Solomon codes. Our recursive construction yields greatly improved parameters even when applied to Reed-Solomon codes. Use of algebraic-geometric codes leads to even better results, which are optimal in an asymptotic sense. The applications comprise universal hashing, authentication, resilient functions and pseudorandomness.

**Key Words:** Low bias, almost independent arrays, Reed-Solomon codes, Hermitian codes, Suzuki codes, Fourier transform, Weil-Carlitz-Uchiyama bound, exponential sum method, Zyablov bound, hashing, authentication, resiliency.

## 1 Introduction

The concepts of limited dependence and low bias have manifold applications in cryptography and complexity theory. We mention universal hashing, authentication, resiliency against correlation attacks, pseudorandomness, block ciphers, derandomization, two-point based sampling, zero-knowledge, span programs, testing of combinatorial circuits, intersecting codes, oblivious transfer, interactive proof systems, resiliency (see [19,16,18,17,1,11,25,10,6,9,7,13,16]). A basic notion underlying these concepts are families of $\epsilon-$biased random variables. The Weil-Carlitz-Uchiyama bound and several constructions from the influential papers by Naor and Naor [18] and by Alon, Goldreich, Håstad and Peralta [1] provide families of $\epsilon-$biased random variables. All these classical constructions yield very similar parameters. In this paper we describe methods, which generalize these constructions and yield far-reaching improvements. Essential ingredients are linear codes and the Fourier transform.

M. Bellare (Ed.): CRYPTO 2000, LNCS 1880, pp. 533–543, 2000.
© Springer-Verlag Berlin Heidelberg 2000

## 2   Bias and Dependency

We use neutral notation which is suited to describe all the applications (hashing, authentication, derandomization, pseudorandomness, ...).

**Definition 1.** *Let $p$ be a prime. An $(n,k)_p-$array $\mathcal{A}$ is an array with $n$ rows and $k$ columns, where the entries are taken from a set with $p$ elements.*

**Definition 2.** *Let $p$ be a prime, $v = (v_1, v_2, \ldots, v_n) \in \mathbb{F}_p^n$. For every $i \in \mathbb{F}_p = \mathbb{Z}/p\mathbb{Z}$ let $\nu_i(v)$ be the frequency of $i$ as an entry of $v$. Let $\zeta$ be a primitive complex $p-$th root of unity. The* **bias** *of $v$ is defined as*

$$bias(v) = \frac{1}{n} \left| \sum_{i \in \mathbb{F}_p} \nu_i(v) \zeta^i \right|$$

We have $0 \le bias(v) \le 1$. As $\sum_{i \in \mathbb{F}_p} \zeta^i = 0$ the bias is low if all elements of $\mathbb{F}_p$ occur with approximately the same frequency as entries in $v$.

**Definition 3.** *Let $0 \le \epsilon < 1$. An $(n,k)_p-$array is $\epsilon-$***biased** *if every nontrivial linear combination of its columns has bias $\le \epsilon$.*

The bias of an array is a property of the $\mathbb{F}_p-$linear code generated by the columns. The bias of the array is low if and only if every nonzero word of the code has low bias.

While the bias of a vector depends on the choice of the root of unity, the bias of an array is independent of this choice.

**Definition 4.** *Let $0 \le \epsilon < 1$. An $(n,k)_p-$array is $t$-***wise** $\epsilon-$**biased** *if every nontrivial linear combination of at most $t$ of its columns has bias $\le \epsilon$.*

**Definition 5.** *Let $0 \le \epsilon < 1$. An $(n,k)_p-$array $\mathcal{A}$ is $t$-***wise** $\epsilon$ -**dependent** *if for every set $U$ of $s \le t$ columns and every $a \in \mathbb{F}_p^s$ the frequency $\nu_U(a)$ of rows of $\mathcal{A}$, whose projection onto $U$ equals $a$ satisfies*

$$\left| \frac{\nu_U(a)}{n} - 1/p^s \right| \le \epsilon.$$

The notion of a $t$-wise $\epsilon$ -dependent array generalizes the combinatorial notion of an orthogonal array of strength $t$ (equivalently: $t-$universal family of hash functions in the sense of Carter/Wegman [11]). An array is $t$-wise independent (=0-dependent) if and only if it is an orthogonal array of strength $t$.

The most important of these concepts from the point of view of applications is $t$-wise $\epsilon$ -dependency. It captures the familiar theme of representing a family of random variables (the columns of the array) on a small sample space (the rows of the array, with uniform distribution) such that any $t$ of the random variables are almost statistically independent.

We want to point out in the sequel that the construction problem of $t$-wise $\epsilon$-dependent arrays can be efficiently reduced to the construction of $\epsilon$-biased arrays. This is the basic idea behind [18].

The following construction of $t$-wise $\epsilon$-biased arrays is essentially from [18].

**Theorem 1.** *Let the following be given:*

- *An $(n,k)_p$-array $B$, which is $\epsilon$-biased.*
- *A linear code $[N, N-k, t+1]_p$.*

*Then we can construct an $(n,N)_p$-array, which is $t$-wise $\epsilon$-biased.*

**Theorem 2.** *An array, which is $t$-wise $\epsilon$-biased, is also $t$-wise $\epsilon'$-dependent for some $\epsilon' < \epsilon$.*

The fundamental Theorem 2 is proved in a nontrivial but standard way by using the Fourier transform, see [5]. The following construction from the journal version of [16] is obvious and useful:

**Theorem 3.** *If there is an array $(n,k')_p$, which is $t'$-wise $\epsilon$-dependent, and $t \leq t'/l$, $k \leq k'/l$, then there is an array $(n,k)_{p^l}$, which is $t$-wise $\epsilon$-dependent.*

We see that indeed the central problem is to efficiently construct $\epsilon$-biased arrays. Linear codes are then used to construct $t'$-wise $\epsilon$-biased arrays via Theorem 1. The standard method is to use BCH codes. The resulting $t'$-wise $\epsilon$-biased arrays are also $t'$-wise $\epsilon$-dependent by Theorem 2. Because of Theorem 3 it is possible to concentrate entirely on binary arrays. We turn to the basic problem of constructing weakly biased arrays.

**Definition 6.** *Denote by $f_p(b,e)$ the minimum $a$ such that there is an array $(p^a, p^b)_p$, which is $p^{-e}$-biased.*

Clearly $f_p(b,e)$ is weakly monotonely increasing in both arguments. The construction from [1] shows the following:

**Theorem 4.** *There is an efficient construction showing*

$$f_p(b,e) \leq 2(b+e).$$

## 3    The Weil-Carlitz-Uchiyama Construction

The celebrated Weil-Carlitz-Uchiyama bound [8] may be understood as a limit on the bias of dual BCH-codes. More precisely, let $(a_j)$ be a basis of $\mathbb{F}_{p^f} | \mathbb{F}_p$ and $Tr : \mathbb{F}_{p^f} \longrightarrow \mathbb{F}_p$ the trace. Consider the array $\mathcal{A}$ whose rows are indexed by the elements $\alpha \in \mathbb{F}_{p^f}$ and whose columns are indexed by $a_j X^i$, where $i \leq n$ and $i$ is not a multiple of $p$. The corresponding entry is $Tr(a_j \alpha^i)$. The WCU bound asserts that this $(p^f, f(n - \lfloor n/p \rfloor))_p$-array has bias $\leq (n-1)p^{-f/2}$.

Comparison reveals that the WCU construction (exponential sum method) yields parameters which are very similar to (a little better than) Theorem 4. All constructions based on one of these classical methods will produce about the same parameters.

## 4    The Zyablov Construction

As remarked earlier Theorem 3 makes it possible to base the construction on
**binary** $\epsilon-$biased arrays. This has the advantage that a direct link to coding
theory can be used. An array $(n, k)_2$ is $\epsilon-$biased if and only if the code generated
by the columns has dimension $k$ and the relative weights of all nonzero codewords
are in the interval of length $\epsilon$ centered at $1/2$. This elementary observation yields
an immediate reduction of the construction problem of binary weakly biased
arrays to the construction of linear codes containing the all-1-vector.

**Theorem 5.** *Let $0 \leq \epsilon < 1$. The following are equivalent:*

- *An $(n, k)_2$-array, which is $\epsilon-$biased.*
- *A binary linear code of length $n$ and dimension $k + 1$, which contains $\mathbf{1}$ and
  whose minimal distance $d$ satisfies*

$$\frac{d}{n} \geq \frac{1 - \epsilon}{2}$$

Constructing families of $\epsilon-$biased $(n, k)_2-$arrays which are asymptotically
nontrivial (meaning that $\epsilon$ is fixed and $k/n \geq R > 0$) is equivalent to constructing
asymptotically nontrivial families of binary linear codes containing the all-1-
vector.

The question of determining the asymptotics of binary codes is one of the
most famous and most well-studied problems in coding theory. The question is
how incisive the additional condition is. A famous simple result is the **Gilbert-
Varshamov bound:** for every prime-power $q$ and $\delta < (q - 1)/q$ the rate $R = 1 - H_q(\delta)$ can be asymptotically reached by families of $q-$ary linear codes. It can
be managed that the all-1-word is contained in all these codes. Unfortunately
this bound is not constructive.

The construction given in [15] does not yield linear codes. The **Justesen-
method** [14,21] is constructive, but the all-1-word is not contained in the resulting codes. The Justesen method when applied to families of algebraic-geometric
codes yields precisely the **Zyablov bound.** However, for the same reason as
above this does not yield families of binary $\epsilon-$biased arrays.

More interesting for our problem is the original semi-constructive proof of
the Zyablov bound [27]. In fact, apply concatenation to a Reed-Solomon code
$[q^m, rq^m, (1 - r)q^m]_{q^m}$ as outer code and a code $[n, m, d]_q$ as inner code, where
it is assumed that the inner code asymptotically meets the Gilbert-Varshamov
bound $(d/n = \mu, m/n = 1 - H_q(\mu))$. The concatenated code has parameters
$[q^m n, rq^m m, (1 - r)q^m d]_q$, with relative distance $\delta = (1 - r)\mu$ and rate $R = r(1 - H_q(\mu))$. This construction shows that for every $\mu < (q - 1)/q$ and $\delta < \mu$
we can construct families of $q-$ary linear codes with relative distance $\delta$ and
rate $R \geq (1 - H_q(\mu))(1 - \delta/\mu)$. The only drawback is that this is not really
constructive. However, for short inner codes this may be feasible. Let us explore
the situation in more detail.

We aim at a lower bound for $f_2(b, e)$. Choose $r = 2^{-(e+1)}, \mu = \frac{1}{2} - 2^{-(e+2)}$.
As the relative distance of the concatenated code is $(1 - r)\mu$ we obtain as bias

$\epsilon = 1 - 2(1-r)\mu = 1 - (1 - 2^{-(e+1)})^2 = 2^{-e} - 2^{-(2e+2)}$. It follows that $\epsilon < 2^{-e}$. We have $b = m + \log(m) - e - 1$ and $a = m + \log(n)$. What is the order of magnitude of the rate $S = 1 - H_2(\mu)$ of the inner code guaranteed by Gilbert-Varshamov?

We have $\mu = \frac{1}{2} - 2^{-(e+2)} = (2^{e+1} - 1)/2^{e+2}, 1 - \mu = (2^{e+1} + 1)/2^{e+2}$ and $S = 1 - 2^{-(e+2)}((2^{e+1} - 1)(e+2 - \log(2^{e+1} - 1)) + (2^{e+1} + 1)(e+2 - \log(2^{e+1} + 1))$. Collecting the terms without log yields $S = 2^{-(e+2)}((2^{e+1} - 1)\log(2^{e+1} - 1) + (2^{e+1} + 1)\log(2^{e+1} + 1)) - (e + 1)$. Divide the arguments of the $\log$ −terms by $2^{e+1}$. The term obtained from compensating for that is $e + 1$ and cancels against the last summand. We obtain $S = 2^{-(e+2)}((2^{e+1} - 1)\log(1 - 2^{-(e+1)}) + (2^{e+1} + 1)\log(1 + 2^{-(e+1)}))$. Using the series for $\ln(1 \pm x)$ we obtain

$$S = \frac{2^{-(e+2)}}{\ln(2)}(((2^{e+1} - 1)(-2^{-(e+1)} - 2^{-2e-3} - \ldots) + (2^{e+1} + 1)(2^{-(e+1)} - 2^{-2e-3} +)$$

$$= 2^{-(e+2)}\frac{1}{\ln(2)}(-1 + 2^{-(e+1)} - 2^{-e-2} \ldots + 1 + 2^{-(e+1)} - 2^{-e-2} \ldots),$$

where terms involving $-2e$ in the exponent and higher have been omitted. This yields $S \sim 2^{-(2e+3)}/\ln(2)$.

**Theorem 6.** *The Zyablov method needs the construction of binary $[n, m, d]_2$ codes, where $n = \ln(2)2^{2e+3}m$ and $d/n = \frac{1}{2} - 2^{-(e+2)}$. The output is a weakly biased array showing*

$$f_2(m + \log(m) - e - 1, e) \le m + \log(m) + 2e + 3.$$

Theorem 6 states $f_2(b, e) \le b + 3e + 4$. It improves on the bound from Theorem 4 when $b > e$.

## 5    A Coding-Theoretic Construction of Weakly Biased Arrays

In Section 4 we used an equivalent coding-theoretic interpretation of binary weakly biased arrays to obtain constructions. Observe however that this does not seem to lead to explicit asymptotic constructions. The Zyablov method pre-supposes exhaustive search for codes of moderate length attaining the Gilbert-Varshamov bound.

When $p > 2$ an equivalent reduction to coding theory is not available. Our next theorem provides a general link, which allows the use of linear codes in the construction of $p$−ary weakly biased arrays. As this leads to efficient constructions, it is interesting even in the binary case.

**Theorem 7.** *Let $\mathcal{C}$ be a code $[n, k, d]_q$, where $q = p^m$ and $\mathcal{B}$ an $(n_0, m)_p$-array of bias $\epsilon_0$. We can construct an $(nn_0, km)_p$−array with bias $\epsilon = 1 - \delta + \delta\epsilon_0 < 1 - \delta + \epsilon_0$, where $\delta = d/n$ is the relative distance of code $\mathcal{C}$.*

A proof of Theorem 7 is in [5]. Application of Theorem 7 to Reed-Solomon codes $[p^m, Rp^m, (1 - R)p^m]_{p^m}$ and inner unbiased arrays $(p^m, m)_p$ (consisting of all $m$-tuples) yields the following:

**Theorem 8.** *For every natural number $m$ and every rational number $0 < \epsilon_0 < 1$ with denominator $p^m$ we can construct an array $(p^{2m}, m\epsilon_0 p^m)_p$ with bias $\leq \epsilon_0$.*

In particular Theorem 8 yields yet another proof for the parameters from Theorem 4 and from the WCU construction.

Our Theorem 7 is much more general. In order to obtain essential improvements on Theorem 4 let us consider a recursive application. Apply Theorem 7 with a Reed-Solomon code $[p^m, Rp^m, (1 - R)p^m]_{p^m}$, where $R = \epsilon/2$ and an $\epsilon/2-$biased $(4m^2/\epsilon^2, m)_p-$array. We obtain the following:

**Theorem 9.** *We can construct arrays $(4m^2 p^m/\epsilon^2, m\epsilon p^m/2)_p$, which are $\epsilon-biased$. The choice $m = p^j, \epsilon = p^{-e}$ yields $f_p(p^j + j - e - 1, e) \leq p^j + 2j + 2e + 2$.*

Theorem 9 states in particular $f_p(b, e) \leq b + 3e + j + 3$, where $j \sim \log(b + e)$. In the binary case this is very close to Theorem 6 and it yields an essential improvement over Theorem 4 when $b > e$.

*Example 1.* Apply Theorem 7 to a $p^4-$ary Reed-Solomon code of dimension $p^3$ (relative minimum distance $> 1 - (1/p)$) and an inner array $(p^2, 4)_p$, which is $(1/p)-$biased. Such an array follows from the WCU construction. We can describe it as follows: Its rows are $(x, y, xy, x^2 + cy^2)$, where $x, y \in \mathbb{F}_p$ and $c$ is a non-square. The result is an

$$\frac{2}{p} - \text{ biased } (p^6, 4p^3)_p - \text{ array,}$$

which is better than what results from the WCU construction.

*Example 2.* In the same style apply Theorem 7 to a $p^m-$ary Reed-Solomon code of dimension $p^{m-1}$ and an $(1/p)-$biased array $(m^2 p^2, m)_p$, whose existence is guaranteed by Theorem 4. We obtain an

$$\frac{2}{p} - \text{ biased } (m^2 p^{m+2}, m p^{m-1})_p - \text{ array.}$$

This is much better than a corresponding WCU-array. Theorem 4 with the same bias and the same number of columns would use $m^2 p^{2m}/4$ rows.

So far the only ingredients used in our constructions have been Reed-Solomon codes. Next we want to show that algebraic-geometric codes can be used to great advantage. Let us start by pointing out that many important classes of algebraic-geometric codes can be just as efficiently implemented as Reed-Solomon codes. In the next section this is exemplified in the case of the Hermitian codes.

## 6    Hermitian Codes for the User

We describe how to obtain generator matrices for the Hermitian codes. Consider the field extension $\mathbb{F}_{q^2} \mid \mathbb{F}_q$ and the corresponding trace $tr$ and norm $N$, where $tr(x) = x + x^q, N(x) = x^{q+1}$. Our codes are defined over $\mathbb{F}_{q^2}$ and have length $q^3$ (see [24]).

**The coordinates** are parametrized by the pairs $(\alpha, \beta)$, where $N(\alpha) = tr(\beta)$. So we need to calculate traces and norms of all elements in the field and to list all these pairs in some order. There are $q^3$ such pairs.

**The general build-up:** We construct a $(q^3 - g, q^3)-$ matrix $G$ with entries from $\mathbb{F}_{q^2}$. Here $g = \binom{q}{2}$. The first $k$ rows of $G$ generate the $k-$dimensional Hermitian code. It has parameters

$$[q^3, k, q^3 - k + 1 - g]_{q^2}.$$

**The pole-order test:** For $n = 0, 1, 2, \ldots$ we have to decide if $n$ is a **pole-order** or not. If $n$ is a pole-order we determine its **coordinate vector** $(i, j)$. This is done as follows: Let $r$ be the remainder of $n \bmod q$, where $0 \le r \le q - 1$ and $-s$ the (negative) remainder of $n \bmod q + 1$, where $0 \le s \le q$. Then $n$ is a pole-order if and only if

$$x = \frac{n - r}{q} \ge \frac{n + s}{q + 1} = y.$$

If $n \ge 2g$, then the pole-order test does not need to be performed. Every such number is a pole-order. If $n$ is a pole-order, then $n = (q + 1)i + qj$, where $i = (x - y)q + r, j = s$. The coordinate vector of $n$ is $(i, j)$.

**Constructing the rows of** $G$ : Let $u_1 = 0, u_2 = q, u_3 = q + 1 \ldots$ be the first pole-orders. If $u_k$ has coordinate-vector $(i, j)$, then the entry of row $k$ of $G$ in coordinate $(\alpha, \beta)$ is $\beta^i \alpha^j$.

We conclude that the use of Hermitian codes requires the usual field arithmetic, just as Reed-Solomon codes.

## 7   Using Hermitian and Suzuki Codes

Use Theorem 7 with the Hermitian codes as ingredients, $q = p^m$. The codes have parameters

$$[p^{3m}, k, p^{3m} - (k + p^{2m}/2)]_{p^{2m}}.$$

Use as inner arrays the unbiased arrays $(p^{2m}, 2m)_p$. Choose $e \le m$ and $k \sim p^{3m-e} - p^{2m}/2$. With this choice the resulting array has bias $\epsilon \le p^{-e}$. As we have an array $(p^{5m}, 2km)_p$ and $\log_p(2km) \sim 3m - e + \log_p(m)$ it follows $f_p(3m - e + \log_p(m), e) \le 5i$, where $m \ge e$.

Let now $e$ and $b$ be given, where $b \ge 2e$. Determine $m \ge e$ such that $b + e = 3m$ (provided $b + e$ is a multiple of 3). We have seen that $f_p(b, e) \le 5m = \frac{5}{3}(b + e)$, which clearly represents an improvent on Theorem 4 and on the $WCU-$construction. If $b < 2e$, then $f_p(b, e) \le f_p(2e, e) \le 5e$, still an improvement upon Theorem 4 when $b \ge \frac{3}{2}e$.

The **Suzuki codes** in characteric 2 (see [12]) have parameters

$$[2^{4f+2}, 2^j, 2^{4f+2} - (2^j + 2^{3f+1})]_{2^{2f+1}}.$$

Use Theorem 7 with an unbiased array as inner array. If $f \ge e$ and $j = 4f - e + 1$ we obtain $\epsilon \le 2^{-e}$, and hence $f_2(4f - e + 1, e) \le 6f + 3$. This presupposes $b + e = 4f + 1 > 4e$, hence $b > 3e$.

**Theorem 10.** *The Hermitian codes show*

$$f_p(b, e) \leq \frac{5}{3}(b + e) \ if \ b \geq 2e.$$

*The Suzuki codes show*

$$f_2(b, e) \leq \frac{3}{2}(b + e) + 2 \ if \ b > 3e.$$

The results of Theorem 10 are superior to all the constructions discussed earlier, for the parameter range when Theorem 10 applies. The strength of Theorem 4 is its universality and simplicity. For $b < e$ it seems to be hard to obtain improvements upon the WCU-construction. Another construction principle for weakly biased arrays, first introduced in [18], uses expander graphs and asymptotically nontrivial families of codes as ingredients. However, this construction seems to work best when $k$ is large with respect to $1/\epsilon$ ($b$ large with respect to $e$) and it cannot improve upon the results presented above in that parameter range.

We conclude this section with an application of Theorem 7 to Hermitian codes. The $p^2$–ary Hermitian code of dimension $k \sim p^2/2$ has relative minimum weight $\delta = 1 - 1/p$. The unbiased $(p^2, 2)_p$–array yields an $(1/p)$–biased $(p^5, p^2)_p$–array.

*Example 3.* For every odd prime $p$ we can produce an $(1/p)$–biased $(p^5, p^2)_p$–array by applying Theorem 7 to a Hermitian code and an unbiased array.

Observe that the WCU construction when applied in the case of $p^5$ rows and $\epsilon = 1/p$ produces a number of columns of the order of magnitude $p^{3/2}$.

## 8    Construction of Authentication Schemes

Unconditional authentication was originally introduced by Simmons [22,23]. An $(n, k)_q$–array is $\epsilon$–**almost strongly universal**$_2$ (ASU$_2$) if each column has bias 0 and for any two different columns $c, c'$ and any entries $e, e'$ the conditional probability $Pr(c_i = e \mid c'_i = e')$ is bounded by $\epsilon$, where the probability refers to a choice of a row $i$ according to the uniform distribution of rows. In the application rows are keys, columns are source states and entries are authentication tags. A composition construction based on codes is used in [4,2,3]. In [13] a direct link is established between the WCU construction of weakly biased arrays and ASU$_2$-arrays. We generalize this construction as follows:

**Theorem 11.** *If there is an $\epsilon_0$–biased $(n, k)_p$–array then for every $t \leq k$ there is an $\epsilon$–$ASU_2$ array $(p^t n, p^k)_{p^t}$, where $\epsilon = p^{-t} + \epsilon_0$.*

*Proof.* Let $\mathcal{C}$ be the linear $[n, k]_p$-code generated by the columns of the $\epsilon_0$-biased array. The columns of the ASU$_2$–array $\mathcal{A}$ are indexed by $f \in \mathcal{C}$, the rows are indexed by tuples $(i, \alpha_1, \ldots, \alpha_t)$, where $i$ is a coordinate of $\mathcal{C}$ and $\alpha_r \in \mathbb{F}_p$. It is easy to see that we can find linear mappings $M_r : \mathcal{C} \longrightarrow \mathcal{C}$, $r = 1, 2 \ldots, t$ such that every nontrivial $\mathbb{F}_p$−linear combination of the $M_r$ is non-singular. Define the entry of $\mathcal{A}$ in row $(i, \alpha_1, \ldots, \alpha_t)$ and column $f$ as $(M_1(f)(i) + \alpha_1, \ldots, M_t(f)(i) + \alpha_r)$.

It is obvious that each column of $\mathcal{A}$ is unbiased. Let $f, g$ be different columns and $(\beta_r), (\gamma_r)$ be two entries. Let $\nu$ be the number of rows $i$ of the original array such that $M_r(f - g)(i) = \beta_r - \gamma_r$ for all $r$. We have to show that $\nu/n \leq p^{-t} + \epsilon_0$. This follows from Theorem 2 and the linear independence of the $M_r(f - g)$.

We see that via Theorem 11 essential improvements upon the parameters of weakly biased arrays yield improved authentication$_2$ codes.

*Example 4.* Continuing from Example 3 we obtain $(p^6, p^{p^2})_p$ arrays, which are $(2/p)-\text{ASU}_2$. Not surprisingly this is better than the constructions from [4,13] based on Reed-Solomon codes and it reproduces the parameters of the construction from [2] based on Hermitian codes.

*Example 5.* An application of Theorem 11 to the arrays from Example 2 produces arrays $(m^2 p^{m+3}, p^{m p^{m-1}})_p$, which are $(3/p)-\text{ASU}_2$.

A refinement of the theory of unconditional authentication is introduced in [16]. An $(N, m)_p$−array is $(\delta, t)-$**almost strongly universal** (short $(\delta, t)-\text{ASU}$) if for every set $U = U_0 \cup \{u\}$ of $t$ columns and every $a' \in \mathbb{F}_p^{t-1}, x \in \mathbb{F}_p$ the frequencies $\nu_{U_0}(a')$ and $\nu_U(a', x)$ satisfy

$$| \nu_U(a', x)/\nu_{U_0}(a') | \leq \delta.$$

The idea is to use the same key for $t$ subsequent messages while still bounding the opponent's probability of success. The link between almost independent arrays and $(\delta, t)-\text{ASU}$ codes has been established in [16] (and is almost obvious):

**Theorem 12.** *A $t$-wise $\epsilon-$ dependent array is $(\delta, t)-ASU$, where $\delta = (p^{-t} + \epsilon)/(p^{-(t-1)} - \epsilon)$.*

The following theorem generalizes the method used in [16].

**Theorem 13.** *Let $f_2(b, lt) \leq a$. Then there is an $(2^{-(l-1)}, t)-ASU$ with $2^l$ entries, $2^b/(lt) - \log_2(l)$ source bits and $a$ key bits.*

*Proof.* A BCH-code $[2^j, 2^j - ljt, lt + 1]_2$, where $jlt = 2^b$, yields an $lt$−wise $2^{-e}$−biased array $(2^a, 2^j)$. By Theorem 3 this yields an array $(2^a, \frac{1}{l} 2^j)_{2^l}$, which is $t$−wise $2^{-lt}$−biased. Apply Theorem 12. We obtain $\delta < 2/(2^l - 1) \sim 2^{-(l-1)}$. The number of rows is still $2^a$.

## 9   Resiliency

A number of interesting applications of the WCU construction are in [16]. They can all be generalized to admit the use of arbitrary weakly biased arrays. We consider the case of almost resilient functions. The construction from [16] is an application of Theorem 1 to check matrices of binary BCH codes. A straightforward generalization is as follows:

**Theorem 14.** *Assume the following exist:*
- *A systematic $\epsilon-$biased $(2^t, s)_2-$array, and*
- *a linear code $[m, m - s, k + 1]_2$.*
   *Then there exists a systematic $k-$wise $\epsilon-$dependent $(2^t, m)_2-$array*

   The proof is similar to the proof for the special case used in [16]. The end product of Theorem 14 allows the construction of a function : $\mathbb{F}_2^m \longrightarrow \mathbb{F}_2^{m-t}$ such that whenever $k$ of the input parameters are fixed the output is close to being unbiased (for details see [16]). Note that the study of almost resilient functions can be motivated from an analysis of the wire-tap channel of type II [20]. A discussion of that aspect is in [26], where the close link to the coding-theoretic and geometric notion of generalized Hamming weights is pointed out.

# 10    Conclusion

The concepts of sample spaces which are statistically close to being unbiased or independent is fundamental for large areas of computer science and cryptology. The best known constructions all yield very similar parameters. The various constructions from [1] excel by their simplicity and universality, whereas the Weil-Carlitz-Uchiyama construction yields slightly better parameters. In this paper we used several new coding-theoretic construction procedures to obtain essential improvements for vast parameter ranges. These improvements can already be obtained by restricting the ingredients to Reed-Solomon codes. Algebraic-geometric codes produce further improvements in suitable parameter ranges. We pointed out that Hermitian codes, a particularly useful class of AG codes, are just as efficiently computable as Reed-Solomon codes.

   In the applications we concentrated on universal hashing, unconditional authentication and almost resilient functions. A large number of applications are documented in the literature. It is expected that more applications will be discovered.

# References

1. Alon, N., Goldreich, O., Håstad, J., Peralta, R.: Simple constructions of almost $k$-wise independent random variables, Random Structures and Algorithms **3** (1992), 289-304, preliminary version: Symposium 31st FOCS 1990, 544-553
2. Bierbrauer, J.: Universal hashing and geometric codes, Designs, Codes and Cryptography **11** (1997), 207-221
3. Bierbrauer, J.: Authentication via algebraic-geometric codes, in: Recent Progress in Geometry, Supplemento ai Rendiconti del Circolo Matematico di Palermo **51** (1998), 139-152
4. Bierbrauer, J., Johansson, T., Kabatiansky, G., Smeets, B.: On families of hash functions via geometric codes and concatenation, Proceedings CRYPTO 93, Lecture Notes in Computer Science **773** (1994), 331-342
5. Bierbrauer, J., Schellwat, H.: Weakly biased arrays, almost independent arrays and error-correcting codes, submitted for publication in the Proceedings of AMS-DIMACS.

6. Boyar, J., Brassard, G., Peralta, R.: Subquadratic zero-knowledge, JACM **42** (1995), 1169-1193
7. Brassard, G., Crépeau, C., Santha,M.: Oblivious transfers and intersecting codes, IEEE Transactions on Information Theory **42** (1996), 1769-1780
8. Carlitz, L., Uchiyama, S.: Bounds for exponential sums, Duke Mathematical Journal **24** (1957), 37-41
9. Cohen, G. D., Zémor, G.: Intersecting codes and independent families, IEEE Transactions on Information Theory **40** (1994), 1872-1881
10. Gal, A.: A characterization of span program size and improved lower bounds for monotone span programs, Proceedings $13^{th}$ Symposium of the Theory of Computing (1998), 429-437
11. Carter, J. L., Wegman, M. N.: Universal Classes of Hash Functions, J.Computer and System Sci. **18** (1979), 143-154
12. Hansen, J. P., Stichtenoth, H.: Group codes on certain algebraic curves with many rational points, AAECC **1** (1990), 67-77
13. Helleseth, T., Johansson, T.: Universal hash functions from exponential sums over finite fields and Galois rings, Lecture Notes in Computer Science **1109** (1996), 31-44 (CRYPTO 96)
14. Justesen, J.: A class of asymptotically good algebraic codes, IEEE Transactions on Information Theory **18** (1972), 652-656
15. Katsman, G. L., Tsfasman, M. A., Vladut, S. G.: Modular curves and codes with a polynomial construction, IEEE Transaction on Information Theory **30** (1984), 353-355
16. Kurosawa, K., Johansson, T., Stinson, D.: Almost $k$-wise independent sample spaces and their cryptologic applications, Lecture Notes in Computer Science **1233** (1997), 409-421 (Advances in Cryptology, Eurocrypt 97)
17. Lu, C. J.: Improved pseudorandom generators for combinatorial rectangles, Proceedings of the 25th International Colloquium on Automata, Languages and Programming (1998), 223-234
18. Naor, J., Naor, M.: Small-bias probability spaces: efficient constructions and applications, SIAM Journal on Computing **22** (1993), 838-856, preliminary version: Proceedings STOC 1990, 213-223
19. Naor, M., Reingold, O.: On the construction of pseudo-random permutations: Luby-Rackoff revisited, Proceedings STOC **29** (1997), 189-199
20. Ozarow, L. H., Wyner, A. D.: Wire-Tap Channel II, AT&T Bell Laboratories Technical Journal **63** (1984), 2135-2157
21. Shen, B. Z.: A Justesen construction of binary concatenated codes that asymptotically meet the Zyablov bound for low rate, IEEE Transactions on Information Theory **39** (1993), 239-242
22. Simmons, G. J.: A game theory model of digital message authentication, Congressus Numerantium **34** (1992), 413-424
23. Simmons, G. J.: Authentication theory/coding theory, in: Advances in Cryptology, Proceedings of Crypto 84, Lecture Notes in Computer Science **196** (1985), 411-431
24. Stichtenoth, H.: Algebraic function fields and codes, Springer 1993.
25. Wegman, M. N., Carter, J. L.: New Hash Functions and Their Use in Authentication and Set Equality, J.Computer and System Sci. **22** (1981), 265-279
26. Wei, V. K.: Generalized Hamming weights for linear codes, IEEE Transactions on Information Theory **37** (1991), 1412-1418
27. Zyablov, V. V.: An estimate of the complexity of constructing binary linear cascade codes, Problems in Information transmission **7** (1971), 3-10

# Author Index